ANNALS *of* THE NEW YORK ACADEMY OF SCIENCES

The New York Academy of Sciences
7 World Trade Center
250 Greenwich Street, 40th Floor
New York, NY 10007-2157

annals@nyas.org
www.nyas.org/annals

The Pierfranco and Luisa Mariani Foundation

Since its beginnings in 1985, the Mariani Foundation has established itself as a leading organization in the field of pediatric neurology by organizing a variety of advanced courses, providing research grants, and supporting specialized care. The foundation works in close cooperation with major public healthcare institutions, complementing their scientific and educational programs, as well as providing additional services. In 2009 it became the first private entity in Italy to join the founding members of the Carlo Besta Neurologic Institute in Milan.

In the last decade, the Mariani Foundation has added a new branch to its initiatives: fostering the study of the multiple links between the neurosciences and music, first and foremost through a series of conferences that have become the most comprehensive events in the field, eagerly awaited by established and new researchers. The first three conferences—in Venice (2002), Leipzig (2005), and Montreal (2008)—have led to regular publications in *Annals of the New York Academy of Sciences*. The last congress was held in 2011 in Edinburgh, where the Institute for Music in Human and Social Development is active. By providing the most recent information in this stimulating neurologic field, the Mariani Foundation intends to be a reliable and informative source for specialists in this new area of the neurosciences.

Published by Blackwell Publishing
On behalf of the New York Academy of Sciences

Boston, Massachusetts
2012

ANNALS *of* THE NEW YORK ACADEMY OF SCIENCES

VOLUME 1252

ISSUE

The Neurosciences and Music IV

Learning and Memory

ISSUE EDITORS

Katie Overy,[a] Isabelle Peretz,[b] Robert J. Zatorre,[c] Luisa Lopez,[d] and Maria Majno[e]

[a]University of Edinburgh, [b]University of Montreal, [c]McGill University, [d]University of Rome, and [e]Pierfranco and Luisa Mariani Foundation

TABLE OF CONTENTS

Ann. N.Y. Acad. Sci. ISSN 0077-8923

ANNALS OF THE NEW YORK ACADEMY OF SCIENCES
Issue: *The Neurosciences and Music IV: Learning and Memory*

Introduction to *The Neurosciences and Music IV: Learning and Memory*

E. Altenmüller,[1] S.M. Demorest,[2] T. Fujioka,[3] A.R. Halpern,[4] E.E. Hannon,[5] P. Loui,[6] M. Majno,[7] M.S. Oechslin,[8] N. Osborne,[9] K. Overy,[9] C. Palmer,[10] I. Peretz,[11] P.Q. Pfordresher,[12] T. Särkämö,[13] C.Y. Wan,[14] and R.J. Zatorre[15]

[1]Institute of Music Physiology and Musician's Medicine, Hannover University of Music, Drama and Media, Hannover, Germany. [2]School of Music, University of Washington, Seattle, Washington. [3]Rotman Research Institute, Baycrest, University of Toronto, Canada. [4]Department of Psychology, Bucknell University, Lewisburg, Pennsylvania. [5]Department of Psychology, University of Nevada, Las Vegas, Nevada. [6]Beth Israel Deaconess Medical Center and Harvard Medical School, Boston, Massachusetts. [7]Fondazine Pierfranco e Luisa Mariani, Milan, Italy. [8]Geneva Neuroscience Center, University of Geneva, Geneva, Switzerland. [9]Institute for Music in Human and Social Development, University of Edinburgh, Edinburgh, United Kingdom. [10]Department of Psychology, McGill University, Montreal, Canada. [11]BRAMS, University of Montreal, Canada. [12]Department of Psychology, University at Buffalo State University of New York, Buffalo, New York. [13]Institute of Behavioral Sciences, University of Helsinki, Helsinki, Finland. [14]Beth Israel Deaconess Medical Center and Harvard Medical School, Boston, Massachusetts. [15]BRAMS, McGill University, Montreal, Canada

Address for correspondence: Katie Overy, Institute for Music in Human and Social Development (IMHSD), University of Edinburgh, Alison House, 12 Nicolson Square, Edinburgh EH8 9DF, United Kingdom. k.overy@ed.ac.uk

The conference entitled "The Neurosciences and Music-IV: Learning and Memory" was held at the University of Edinburgh from June 9–12, 2011, jointly hosted by the Mariani Foundation and the Institute for Music in Human and Social Development, and involving nearly 500 international delegates. Two opening workshops, three large and vibrant poster sessions, and nine invited symposia introduced a diverse range of recent research findings and discussed current research directions. Here, the proceedings are introduced by the workshop and symposia leaders on topics including working with children, rhythm perception, language processing, cultural learning, memory, musical imagery, neural plasticity, stroke rehabilitation, autism, and amusia. The rich diversity of the interdisciplinary research presented suggests that the future of music neuroscience looks both exciting and promising, and that important implications for music rehabilitation and therapy are being discovered.

Keywords: music; neuroscience; learning; memory; children

Introduction

Music neuroscience research has expanded beyond recognition in recent years, driven not only by increasingly advanced, available, and affordable brain imaging technology and analysis software, but also by a growing interest in musical behavior within the wider disciplines of neuroscience and psychology. The field is diversifying to such an extent that what used to be considered specialized topics can now be considered entire research areas, from specific aspects of music (such as rhythm or imagery) to focused population groups (such as infants or patients), to state-of-the-art techniques (such as EEG, MEG, TMS, or MRI). Students are entering the field with both undergraduate and postgraduate training in music psychology and music neuroscience, something that could hardly have been imagined ten years ago. There is increasing scientific and public interest in how music neuroscience research can potentially inform, and be informed by, the disciplines of music therapy, music education, and music performance. The future of such research clearly captures the imagination and is not showing any signs of diminishing.

A substantial contributing factor to this successful expansion of the field of music neuroscience is, of course, the long-standing support of the Mariani

doi: 10.1111/j.1749-6632.2012.06474.x

Foundation, a charity dedicated to child neurology. The Mariani Foundation's pioneering conferences have created unique opportunities for researchers in the field, attracting students and professors alike from around the globe. Beginning with The Biological Foundations of Music[1] in New York City in 2000, followed by The Neurosciences and Music[2] in Venice in 2002, The Biological Foundations of Music: From Perception to Performance[3] in Leipzig in 2005, and The Neurosciences and Music III: Disorders and Plasticity[4] in Montreal in 2008, these international conferences have become key events, providing an invaluable opportunity for meeting like-minded scientists, exchanging recent findings, and developing new collaborations. The contribution to the expansion of music neuroscience from some key individuals behind these conferences (Maria Majno, Luisa Lopez, and Giuliano Avanzini), along with the continued support of *Annals of the New York Academy of Sciences* in publishing the proceedings, has been hugely significant.

For the Neurosciences and Music-IV conference, hosted by the Institute for Music in Human and Social Development (IMHSD) at the University of Edinburgh in June 2011, the theme of learning and memory was selected as the central aspect of musical experience. Symposia and posters were invited under four key topics: infants and children, adults: musicians and nonmusicians, disability and aging, and therapy and rehabilitation. Professor Richard Morris gave an opening welcome to nearly 500 international delegates, and, according to tradition, this was followed by an afternoon of methods workshops, this year on the topic of working with children. The following three days included a keynote lecture from Professor Alan Baddeley, "Human Memory;" nine platform symposia; three large and vibrant poster sessions; and a range of concerts in some of Edinburgh's most beautiful university buildings. Here, we introduce the proceedings of *The Neurosciences and Music IV: Learning and Memory*, with each section of the volume introduced by the organizers.

Methods I: Working with children—experimental methods

Katie Overy

Working with children is essential in order to improve our understanding of the development of musical skills and the potential impact of early musical experiences. Learning and memory are crucial throughout life, but particularly during early development. Noninvasive imaging techniques have been both difficult and rare in research with infants and young children, but technological advances in functional magnetic resonance imaging (fMRI), electroencephalography (EEG), and magnetoencephalography (MEG) are making such research easier. The design of age-appropriate behavioral protocols and measures are also vital, from task and stimuli design to child-friendly environments and procedures to help discourage movement during imaging. This collection of papers outlines current experimental techniques for working with children, from preterm infants to kindergarten children, and includes head-turning techniques, behavioral protocols, fMRI, and EEG.

McMahon and Lahav[5] begin with a specific discussion of the problematic noise environment of a neonatal intensive care unit (NICU), which has the potential to negatively affect preterm infant auditory development. After a detailed outline of the key milestones of auditory development *in utero*, the authors discuss the nature of auditory plasticity at critical periods, the potential impact of being deprived of maternal sounds after a premature birth, and evidence suggesting that noise exposure in the NICU can negatively influence the neurodevelopment of a child. A range of solutions is offered for reducing high noise levels and increasing exposure to natural sounds, such as offering "kangaroo care" and playing recordings of the mother's voice.

Trainor[6] continues with an in-depth discussion of current EEG and MEG methods for research with infants and young children. The paper begins by explaining the source of EEG and MEG activity, how they are measured, and the specific advantages of these techniques when working with children. Trainor then describes some of the issues and complexities that can arise, from difficulties with short attention spans and physical movements, to the changing morphology of waveforms with age, to potential artifacts in the data and how to deal with these. A series of research examples are provided, which show how different data analysis techniques have been used to explore specific questions about musical development, such as the effects of exposure to different musical timbres or the development of beat perception.

Trehub[7] presents a range of behavioral methods for working with infants, alongside key findings regarding infant music perception abilities. Trehub emphasizes that the expansion of neuroimaging technology should not replace good behavioral methods, which can offer important insights into how infants perceive and remember musical information. Trehub explains the advantages and disadvantages of different behavioral methods, regarding issues such as stimuli duration, trial numbers, and attrition rates. Infant attention is identified as the key factor that must be both attracted and observed in experimental procedures, and Trehub points to the benefits of ecologically valid stimuli such as infant-directed singing. In addition, Trehub suggests that ecologically valid infant responses are of potential importance, such as measuring rhythmic movements to music rather than looking behaviors. In conclusion, Trehub advises that future research in this area must use convergent measures and should yield conceptually as well as statistically significant findings.

Finally, Gaab and colleagues provide an extensive and invaluable survey of structural and fMRI techniques and protocols developed for infants and young children over the last 20 years.[8] Some key challenges are outlined, such as participant anxiety, movement restrictions in the scanner environment, and the availability of child-appropriate equipment and pediatric brain templates. The authors go on to review a wide and detailed range of solutions to such challenges, from preparation sessions with a mock scanner to effective data-acquisition protocols. A strong emphasis is placed on the need for young children to be comfortable, familiar with the environment, and engaged in the tasks presented. The ethical implications of such research are also considered, including the fact that research findings with infants and children can potentially influence public policy, education, and family life.

Methods II: Working with children—social, "real world" methods

Maria Majno and Nigel Osborne

A commonly held view, in many differing world cultures and at many different times, is that the experience of music has both a significant effect on an individual's mind, body, and relationships with other human beings, and plays a useful role in the social, "real world." Whereas some societies have had systems of belief that offer an accounting for this effect (e.g., the neo-Platonists in Europe,[9] or theorists such as Ibn Sina in the early medieval Islamic world[10]), the world of contemporary science has, until recently, struggled somewhat to find convincing explanations and structures for reflection.

More recently, neuroscience has begun to make major contributions in terms of hard evidence and useful understanding—something particularly welcome to educationalists, music therapists, and others who seek to develop reflection on their methodologies and to argue the case for music's social, real-world usefulness. What has certainly helped is that music is quite "neuroscience friendly." It is a rich, whole-brain, whole-body information that is highly accountable. Music is communicated mostly, but not exclusively, through sound, which is a relatively slow mechanical energy that may be measured scrupulously in frequencies of pitch and rhythm, precisely to the microsecond in duration, and captured in Fourier transforms, spectrograms, fractals, and other representations of harmonicity, rhythmicity, and turbulence.[11] This highly accountable information is processed by the fastest-firing neural system of the brain,[12] while the mechanical energy of sound is perhaps the only energy a human being may "emit" in any consciously controlled, communicative way.

The valuable insights neuroscience offers to music, and music's "friendliness" to neuroscience, point to important agendas for the future. It may be premature to speak of an "applied music neuroscience," but it is clear that many areas of social, real-world musical activity exist in which neuroscience may inform practice and where practical experience may inform neuroscience.

In the opening paper,[13] Majno discusses the psychosocial, and to some extent, political and economic significance of the "Sistema" initiative. The paper traces its progress from its origins in Venezuela, where ubiquitous centers provide irresistible musical opportunities to street children and others (including children with disabilities) as a structured alternative to the temptation of violence and crime, to the implementation of its methods in Italy and other European countries (e.g., the Raploch project in Scotland[14]), where community development and social organization have been primary objectives. Sistema is characterized by rigor, deep learning, high standards, joy, and the premise that music education should become accessible to all at no cost. The psychosocial benefits in terms of social cohesion, sense of identity, and self-respect are predicated upon issues of sense of self, empathy, synchronization, emotional intelligence, memory, cognitive development, and motor skills, all close to the domain of music neuroscience.

Uibel describes the very focused whole-child, educational, and social approach of the Musikkindergarten Berlin,[15] a project initiated by Daniel Barenboim and intended to use music not just as an "add-on," but rather as the central medium for day-to-day learning. The kindergarten provides an education *in* music, including a refreshing range of group creative and developmental activities, and *with* music, including, for example, familiar partners in drama and movement. But most radical is the work *through* music, where music leads interdisciplinary initiatives in areas such as topic-based learning, language development, and numeracy.

Overy[16] raises the thorny issue of negative musical learning experiences and contrasts the human musicality of expert instrumental performance with the more ubiquitous nonexpert human musicality of group singing. Referring to the shared affective motion experience (SAME) model of emotional responses to music,[17,18] which emphasizes motor and social aspects of musical behavior, Overy proposes that the essential and powerful nature of a musical experience lies not just in the sound, but also in the physical and human origins of the sound, creating a sense of shared experience. Referring to traditional music education methods, Overy suggests alternatives to instrumental training when investigating the nature and impact of musical learning.

Osborne considers the influence of developments in neuroscience on reflection on practice in the use of music to support children who are victims of conflict.[19] This paper locates the work within an enhanced bio-psycho-social framework. Although the standard diagnostic instruments for PTSD (post-traumatic stress disorder) are concerned with issues such as traumatic events, traumatic recall, avoidance, and hypervigilance, there are significant physiological symptoms, such as raised heart rate, respiratory problems, hyperactive or sluggish movement repertoires, and the dysregulation of endocrine systems (including the hypothalamic–pituitary–adrenal axis) and neurotransmission. Osborne further traces a circle around the bio-psycho-social framework to include aspects of communicative musicality, emotional communication, empathy, self-belief, creativity, and identity, relating each to emerging research in neuroscience. It is important for any music intervention to avoid inappropriate invasion of the physical and mental lives of traumatized children. But evidence from work in neuroscience and related disciplines with nontraumatized populations has been applied to support critical reflection on experimental methodologies seeking to use music to help regulate, for example, motor activity, the autonomic nervous system, and endocrine and respiratory systems.[20]

Symposium 1: Mechanisms of rhythm and meter learning over the life span

Erin Hannon

Dancing and moving in time with music are universal capacities that constitute a crucial component of human musical experience. Growing evidence suggests that rhythm and beat perception rely on integration of information across multiple sensory modalities and a broad network of brain regions.[21–23] A fundamental goal of research on rhythm and meter perception is to understand the conditions under which a beat or hierarchical metrical structure can be inferred from a given stimulus, the mechanisms underlying perception of a beat or meter, and the extent to which metrical perception depends on casual listening experience, formal music training, or simply emerges from the interaction between the brain and the stimulus itself. This symposium attempted to tackle these fundamental questions using various methodological approaches and technologies, and testing individuals of different ages and/or abilities.

McAuley[24] uses fMRI to examine neural correlates of beat perception. The study uses a sequence-timing paradigm that typically produces large individual differences in sensitivity to the implied beat; however, beat sensitivity declines as the stimulus is presented at slower and slower tempos. By comparing brain responses under conditions of high and low beat sensitivity, the study provides evidence for specific brain networks that are associated with beat- versus interval-based temporal processing. Honing[25] tackles the question of whether hierarchical metrical representations arise automatically in the brain or require extensive listening experience or formal music training. The paper provides a tutorial on developing optimal rhythmic stimuli and tasks that may demonstrate hierarchical beat perception through EEG by measuring a mismatch-negativity (MMN) response to stimulus omissions. It reviews evidence from EEG studies with adults and newborn infants that suggest hierarchical beat perception may be possible with minimal prior experience or training. Hannon, der Nederlanden, and Tichko [26] also explore the role of experience in meter and beat perception. They use similarity judgments to examine how well American children and adults detect beat-disrupting changes to familiar (Western) simple-meter and foreign (Balkan) complex-meter folk songs before and after at-home exposure to foreign complex-meter music. The findings indicate that listening exposure dramatically changes how children under age 10 perform in the judgment task, with accuracy improving for complex meters but declining for simple meters. By contrast, adults and older children consistently perform more accurately in the simple- than in the complex-meter conditions, and this trend is virtually unchanged by at-home listening experience. This suggests that at least some of the metrical representations that contribute to beat perception undergo slow developmental change over the course of childhood and that listening experience can fine-tune these representations during early but not late childhood or adulthood.

Together, the three papers raise important questions about the nature of beat-based perception. On the one hand, rhythmic patterns appear to activate beat-based expectancies even in highly inexperienced newborn listeners.[25] On the other hand, these expectancies are obviously influenced by listening experience and presumably by the acquisition of culture-specific musical knowledge.[26] As research continues to illuminate our understanding of the neural underpinnings of beat-based processing,[24,25] it may be possible to disentangle the various processes and mechanisms that contribute to our subjective experiences of beat and meter.

Symposium 2: Impact of musical expertise on cerebral language processing

Mathias S. Oechslin

In cognitive neuroscience, the question of whether musical expertise affects or facilitates language processing has been increasingly addressed. In this symposium, we introduced new studies that observe the influence of musical expertise on various aspects of neural language processing—making use of several brain imaging and cognitive measurements, such as event-related brain potentials (ERP), fMRI, and behavioral performance data. This spectrum of methodological approaches characterized by both time and spatial sensitivity enables a deeper understanding of musicians' neural processing of fine-grained acoustic language properties. Together, the papers in this section provide insight into language processing, ranging from basic auditory segmental speech information to suprasegmental decoding mechanisms at word and sentence level.

Disclosing cortical plasticity by focusing on basal auditory functioning, previous studies have unveiled that musical expertise alters structure and functional mechanisms of various brain areas that are involved in processing of both speech and music.[27–30] Therefore, it is reasonable to assume that neural encoding in the former domain could benefit from the expertise in the latter—even though percepts of speech and music are phenomenologically completely distinct.

Regarding basal auditory functioning, two recent studies[31,32] have demonstrated preliminary but compelling evidence that musical training induced plasticity in segmental speech processing. Performing an auditory (phoneme discrimination) categorization task, musicians and nonmusicians were asked to identify voiced and unvoiced consonant–vowel (CV) syllables with either natural or broadband noise acoustic spectra. The fMRI study revealed higher accuracy in identifying CV-syllables with manipulated spectra, flanked by a leftward hemispheric asymmetry and overall enhanced activation of the planum temporale in musicians compared to nonmusicians.[32] Moreover, using the same experimental paradigm, electrophysiological data revealed that N1 amplitudes and corresponding topography were identical in voiced and unvoiced stimuli in musicians but not in nonmusicians. Accordingly, in their paper, Meyer *et al.* comprehensively discuss the role of the planum temporale for speech processing as a function of musical expertise.[33] Taken together, these results suggest that musical expertise leads to a honed auditory capacity in spectrotemporal analysis that goes far beyond the domain of music. In this context, Besson *et al.*[34] have argued that "enhanced sensitivity to acoustic features that are common to music and speech, and that imply domain-general processes, allow musicians to construct more elaborated percepts of speech processing than nonmusicians. . .[and] facilitates stages of speech processing that are speech-specific."

Moving on along this line to the processing of linguistic structures (word segmentation level), the influence or interaction of such sensory plasticity with expertise-dependent top–down modulations (see later) is not yet clear and should be observed in more detail. However, it has been demonstrated that the nature of implicit knowledge in music and language, in general,[35] and learning of linguistic structures and musical structures, in particular, are closely related:[36] by analyzing behavioral data of a statistical learning paradigm, performance in a linguistic test has been found to be positively correlated with performances in a musical test (based on concatenated results of Refs. 37 and 38). In the latter ERP study,[38] musicians and nonmusicians were initially presented with sequences of artificial (sung) language. Then participants were asked to make familiarity ratings of musical and speech sound pairs, indicating whether the first or the second item resembled the previously presented sung sequence. Most interestingly, Francois and Schön found enhanced N400 amplitudes in musicians compared to nonmusicians, elicited by both the linguistic and the music task, concluding for musicians an advantage in stream segmentation or in lexical storage or both. The authors of this paper provide valuable insight into technical details of statistical learning paradigms.[39]

In a recent publication, Patel[40] systematically addressed the prerequisites of such possible transfer effects from musical training to neural encoding of speech. In what is formulated as the OPERA hypothesis, Patel concluded that overlapping brain networks processing shared acoustic features in speech and music are the ultimate essentials for possible transfer effects. Furthermore, to end up with a benefit in speech encoding, according to Patel, three additional conditions have to be fulfilled, including aspects of musical processing demands, training intensity, and emotional arousal. Here, Patel presents his most recent view featuring the fundamental aspects of different precision demands in music and speech processing.[41] Originally, the OPERA hypothesis mainly referenced studies that observe musical training-induced subcortical plasticity driven by descending auditory projections.[42] By analyzing such brainstem responses, it has been revealed that musical expertise yields enhanced auditory frequency representation,[43] increased accuracy in encoding linguistic pitch patterns,[44] and facilitation of hearing speech in noise.[45] Moreover, the latter acuity is supported by cognitive mechanisms, such as enhanced tonal working memory[45] and auditory attention[46] in musicians compared to nonmusicians. Accordingly, topographical ERP analyses confirmed that prefrontal responses are considerably altered (lower prefrontal auditory response variability) as a function of musical expertise during a selective auditory attention task.[46] In other words, this result signifies advantages for musical experts in situations that require sustained auditory attention. In their paper, Kraus *et al.* specifically focus on musicians' auditory advantages and the mediating role of auditory working memory.[47] In fact, previous fMRI research found that attention driven top–down modulations facilitate basal auditory speech processing.[48] This framework lucidly implies that future research is needed to disentangle the potential influence of musical expertise on interactions among structural brain characteristics, cognitive functions, and the architecture of cerebral language processing.

This ensemble of studies that addresses the same topic from different directions is clearly indicative of the increasing relevance of research on transfer effects from music to language mechanisms at several levels of complexity.

Symposium 3: Cultural neuroscience of music

Steven M. Demorest

The goal of the study of the cognitive neuroscience of music is to explain how brain structure and function interact with and shape musical thought and behavior. Research in this area has been dominated by Western thought regarding the nature of music perception and performance, a bias that has the potential to limit the scope of our theories regarding music and brain function. The implicit learning of one's culture or enculturation seems to affect virtually every area of human thought, and the relatively new field of cultural neuroscience[49] is beginning to document the important influence of culture in shaping brain function. Given that music is one of the primary agents of cultural transmission, it stands to reason that certain aspects of musical thinking are mediated by culture. This symposium explored current research in the cultural neuroscience of music and how findings in this field might clarify the role of culture in music development, cognition, and learning.

Previous research in infant music perception has identified that encultured responses to music emerge at around 12 months of age.[50] Here, Trainor *et al.* document how early musical experiences can hasten culturally biased responses in infants.[51] In a series of experiments, they explored the effect of active participation in Kindermusik classes and other formal music exposures that influence infants' behavioral and brain responses to Western rhythm, timbre, and tonality. Vuust *et al.* also present data showing learning-based changes in brain responses, but in adult populations,[52] they used an innovative MMN paradigm to differentiate the responses of performers from different genres of Western music to acoustic changes in a complex musical stimulus. Tervaniemi and her coworkers extend research on classical musicians' ERP responses to harmonic violations[53] to Finnish folk musicians,[54] whose responses reflect highly specialized sensitization to harmonic violations. Large and Almonte propose a new theory of tonality that seeks to uncover the basic neurodynamic principles that underlie tonal cognition and perception. They provide evidence that auditory neurons respond to tonal relationships in predictable ways and discuss the possibility of general neurodynamic principles that underlie all tonal cognition.[55]

Demorest and Wong both explore how implicit learning of culture influences adults' neurological responses to music of different cultures using ERP and fMRI methodologies, respectively. The first paper explores crosscultural music cognition by tracking ERP responses to melodic expectancy violations in culturally familiar and unfamiliar music.[56] As in previous research,[57] Western listeners exhibited a P600 response to scale deviations in Western melodies. They exhibited a smaller P600 when hearing similar deviations in north Indian classical melodies, which may reflect a cultural general sensitivity to statistical properties of tonal sequences or possible areas of overlap between the two tonal systems under study. In the second paper, Wong *et al.* explore fMRI responses to affective judgments across two musical cultures and how they differ between monomusicals and bimusicals.[58] In addition to confirming earlier findings of different behavioral responses,[59] they found significant group differences in the connectivity of a temporal–limbic network. This provides evidence of qualitatively different brain function as a result of exposure to two culturally distinct musical systems. The papers presented here touch on some of the central issues in cultural neuroscience, including the interaction of enculturation, formal training, and development.

Symposium 4: Memory and learning in music performance

Caroline Palmer and Peter Q. Pfordresher

The study of music performance has important implications for understanding basic mechanisms of memory and learning. In contrast to listeners, who acquire vast implicit knowledge resources accumulated during years of perceptual experience, individuals differ greatly in the explicit and implicit knowledge they have acquired for performance. These individual differences in performance knowledge leave open the possibility for studying the development of performance-specific learning and memory in cross-sectional and longitudinal contexts, with child and adult populations. In addition, music performance typically requires the execution of long, rapid sequences typically performed from memory, thus providing a rich context in which to study the development of sensorimotor learning, as performers learn to map auditory and visual feedback to specific actions. Finally, disorders that affect performers offer an excellent venue in which to study music's role in rehabilitation from sensory and motor disorders.

The first paper in this section, by Bailey and Penhune,[60] explores the acquisition of musical skills at different points in the life span to determine whether this is a "sensitive period" for musical skill acquisition. In keeping with the assumption of a sensitive period for musical development, adult musicians who began training early in life exhibited better synchronization ability than adult nonmusicians or musicians who began training later in life. Music performance is fundamentally integrative, involving the online coordination of actions with the perception of feedback from those actions.

The next two papers consider the role of perceptual feedback in performance. Pfordresher addresses how musical training influences the role of auditory feedback in nonmusicians.[61] Contrary to a purely associationist view, nonmusicians do not approach the novel task of music performance as "blank slates" but instead build on preexisting sensorimotor associations, based on their broader experience of perception/action associations, that are refined through musical training. Zimmerman and Lahav's review[62] focuses on the neural underpinnings of multisensory associations that are enhanced by musical training. They argue that the multisensory processing that is enhanced by music listening and performance underlies the success that music-related therapy has had for rehabilitation.

The fourth paper, by Palmer *et al.*,[63] discusses the time course of retrieval failures and their sensorimotor consequences during performance. They report the results of an experiment in which pianists practiced and performed novel musical pieces at fast and slow tempi. Both pitch errors and the correctly produced events immediately prior to those errors were produced with lower intensities and reduced tempo from other correctly produced tones, consistent with the view that performers covertly monitor for production errors prior to response selection.

The final paper, by Felix Strübing and Maria Herrojo Ruiz,[64] examines how sensorimotor prediction is implemented in healthy and dystonic pianists by recording neural responses to errors using event-related potentials (ERPs). Results suggested degraded error processing among dystonic patients, even preceding error production. Together, these papers intersect with the developmental, skill acquisition, and rehabilitory themes of the other papers in this volume to elucidate the neural foundations of memory for musical experience.

Symposium 5: Mind and brain in musical imagery

Andrea R. Halpern and Robert J. Zatorre

Mental imagery refers to a type of memory representation that is rich in perceptual detail. For instance, when asked to describe one's living room, most people can report the spatial layout, colors of the walls, and textures of the furniture. Auditory images also contain perceptual information. The reader is invited to imagine a piece of familiar music, perhaps "Happy Birthday." Musicians and nonmusicians alike can report auditory characteristics of tunes, such as the pitch, rhythm, and tempo.[65] Thus the experience can be quite vivid, sometimes so vivid that the auditory experience takes the form of a persistent repetitive memory, colloquially referred to as "earworms."[66] In addition to vividness, in several meaningful ways auditory imagery for music can also be said to capture qualities of the actual piece and is thus a veridical representation. For instance, people are willing to set a metronome to the tempo of an imagined familiar tune, or indicate on a keyboard the starting pitch they habitually assign to the tune, which they will replicate reliably over multiple trials.[67,68] People without absolute pitch abilities can fairly accurately report the opening pitch or key of familiar recorded music.[69,70]

In addition to behavioral research, a number of studies have documented the similarity of neural mechanisms in imagery and perception. In a prior review,[71] Zatorre and Halpern examined the substantial evidence that the secondary auditory cortex is active when people imagine music, again suggesting that we co-opt perceptual mechanisms when retrieving these vivid and veridical experiences. The review also noted that these experiences are not confusable, and that many studies find additional activation in the frontal cortex during imagery that is absent in perception, suggesting an important moderating influence of executive and memory functions on the auditory cortex.

This symposium at the conference examined new research in auditory imagery that extends earlier studies on basic aspects of auditory imagery. Among the four speakers, a number of methodological approaches were represented: behavioral studies, computer simulations, measurement of electrical activity from the brain surface (ERP), and measurement of relative blood oxygenation as a proxy for neural activity (fMRI). The variety of approaches reminds us of the importance of both "mind" and "brain" in our symposium title.

Three types of extensions to earlier work were evident among our speakers. First, several speakers presented tasks that required experimental participants to make rather fine distinctions during imagery tasks. This is perhaps most evident in Janata's paper on the accuracy with which people can make in-tune and out-of-tune judgments in imagined tones.[72] Halpern reports here on the ability to make moment-to-moment judgments on emotional aspects of imagined music using a continuous scale,[73] and Zatorre reports on neural correlates of subtle key-distance effects during mental transformations of music.[74]

Second, many of the studies discussed required fairly complex processing, particularly regarding the extent to which working memory was required for successful completion. As one example, Halpern reports a study in which people learned pairs of tunes and were required to anticipate the second pair member upon presentation of the first.[73] Keller reports studies of several aspects of musical performance that need to be maintained in parallel for successful execution.[75] Arguably, the most difficult study in the set, reported by Zatorre, was asking people to imagine tunes in reverse note order![74] This imposes a very high working memory load and was reflected in extensive frontal and parietal activations.

Third, several speakers framed their tasks in the context of the adaptive usefulness of auditory imagery. The most obvious example of this was Keller's exploration of the role of auditory imagery in synchronization of movements among musical performers.[75] Janata also suggested that auditory imagery actually enhances the fine pitch judgments required in the tasks he presented, and Halpern noted that the ability to extract emotion from imagined music adds to the enjoyment of remembered experience.[73]

Overall, the talks in the symposium showed that considerable progress has been made in extending the breadth and depth of knowledge about the quality and richness of the musical sounds we hear and manipulate in our minds, despite the essentially private nature of mental imagery.

Symposium 6: Music-induced adaptive and maladaptive brain plasticity in health and disease

Eckart Altenmüller

Emerging research over the last decade has shown that long-term music training and the associated sensorimotor skill learning can be a strong stimulant for neuroplastic changes both in the developing and adult brains, affecting both white and gray matter as well as cortical and subcortical brain structures.[76] Making music, including singing and dancing, leads to a strong coupling of perception and action mediated by sensory, motor, and multimodal brain regions and affects, either in a top–down or bottom–up fashion, important relay stations in the brainstem and thalamus.[77] Furthermore, listening to music and making music provokes motions and emotions, increases between-subject communications and interactions, and is experienced as a joyous and rewarding activity through activity changes in the amygdala, ventral striatum, and other components of the limbic system.[78] Music is a powerful tool in rehabilitation, providing an alternative entry point into compromised neural circuits due to brain damage.[79] Music thus can remediate impaired neural processes or neural connections by engaging and linking brain regions with each other that might otherwise not be linked together. The pleasurable power of listening to favorite music can even reverse maladaptive changes in the auditory cortex, causing the torturing tinnitus percept.[80] On the other hand, music-induced brain plasticity has its dark sides. Prolonged practice, high workload, overuse, and extreme demands on sensorimotor skills in professional musicians may result in a degradation of exactly these fine motor abilities, a condition termed *musician's dystonia*.[81]

The aim of this symposium was to summarize the latest research concerning the powerful impact of music on brain plasticity in both directions—adaptive and maladaptive. In the first article, Schulze and Koelsch present a review of behavioral and neuroimaging findings on similarities and differences between verbal and tonal working memory in musicians and nonmusicians.[82] They demonstrate the impact of musical training on verbal memory and its consequences for verbal learning in children and adolescents. Novel results are discussed that imply the existence of a tonal and a phonological loop in musicians, based on partly differing neural networks underlying verbal and tonal working memory. Finally, the authors propose that both verbal and tonal auditory working memory are based on the knowledge of how to produce the to-be-remembered sounds, and therefore sensorimotor representations are involved in the temporary maintenance of auditory information in working memory.

In the second article, Pantev presents a novel and promising method to reduce chronic tonal tinnitus.[83] There is evidence that maladaptive auditory cortex reorganization may contribute to the generation and maintenance of this torturing condition, which can be conceived as a permanent cortical memory trace. Because behavioral training can modify cortical organization, an approach was chosen to expose chronic tinnitus patients to self-chosen, enjoyable music, which was manipulated to contain no energy in the frequency range of the individual tinnitus, thus promoting lateral inhibition to the brain area generating tinnitus and overwriting the dysfunctional cortical memory.

The third article, authored by Zipse *et al.* from the laboratory of Schlaug, presents an impressive case report on rehabilitation of an adolescent stroke patient suffering from Broca's aphasia.[84] Using a modified version of melodic intonation therapy (MIT), the behavioral improvements were accompanied by functional MRI changes, demonstrating that the right frontal lobe takes over language functions. This case study not only provides further evidence for the effectiveness of this rehabilitation strategy, but also indicates that intensive treatment can induce functional and structural changes in a right hemisphere frontotemporal network.

Jäncke presents in his article on the dynamic audio-motor system in pianists a pilot study based on the theoretical assumption that continuous closed-loop audio–motor control could be disadvantageous for pianists.[85] He argues that the functional relationship between the intracerebral electrical activations in the auditory and premotor cortex should be rhythmically decreased and increased. To test this hypothesis, activations and connectivity of the auditory and premotor cortices were estimated using a novel method to analyze EEG time series and their causal relationship, similar to the "dynamic causal modeling" approach used in fMRI. The analysis revealed a "causal relationship" from the auditory cortex to the premotor cortex, which was considerably stronger during piano playing and weaker during rest. Interestingly, this relationship varied rhythmically during the course of piano playing, thus delivering evidence that in professional pianists, the functional coupling between the auditory and premotor cortex is instable, highly dynamic, and extremely adaptive.

Finally, Altenmüller *et al.* focus on musician's dystonia as a syndrome of dysfunctional brain plasticity.[86] This condition is characterized by a loss of fine motor control of extensively practiced and highly skilled movement patterns in professional musicians. On a neurophysiological level, this phenomenon is due to fusion of sensorimotor receptive fields in the cerebral cortex and to a lack of lateral inhibition to adjacent body parts. In their paper, the authors demonstrate that behavioral factors can trigger the manifestation of this disabling disorder. The interplay between genetic predisposition, psychological and behavioral factors, such as perfectionism, anxiety, and over-specialization, predominantly in classical reproductive musicians, is elucidated, and its important role in finally causing this neurological condition is convincingly demonstrated.

Symposium 7: The role of music in stroke rehabilitation—neural mechanisms and therapeutic techniques

Takako Fujioka and Teppo Särkämö

Neuroplasticity is a key mechanism underlying learning new skills and relearning lost skills during rehabilitation. Recent cognitive neuroscience research has shown that training in music making can enhance neural processing in sensory, motor, executive, and affective brain systems and facilitate interactions between those brain systems in healthy children and adults.[87–90] This is likely due to the fact that the massed practice associated with music making requires enormous resources within each specific brain system as well as extensive coordination between the systems. Importantly, music making also involves social interaction and induces strong emotional experiences. Even simply listening to music can have a short-term positive effect on arousal, attention, memory, and mood.[91–93]

The next question is how these findings can be incorporated and translated into the rehabilitation of the damaged brain. One of our strong allies in answering this question is the music therapy community, which has accumulated and practiced theories on how music can be used to help neurological, psychological, and functional recovery in various clinical populations, including stroke patients. Another guiding principle is the knowledge obtained from rehabilitation sciences that meaningful and relevant tasks and motivated participation in activities enhance rehabilitation outcomes in everyday life, compared to mere repetition of rote exercises. Thus, it seems that we have reached the crucial point at which each discipline can help inform the others in ascertaining how music can be used for improving the lives of patients.

How can music-related activities fit into rehabilitation? Within the framework of the International Classification of Functioning, Disability and Health (ICF) of the World Health Organization (WHO),[94] we propose that music making and/or music listening activities can be understood as an activity that influences a stroke patient's environment, which in turn interacts with behavioral activity and participation, as well at the level of brain function. For example, a music-making exercise demands not only sensorimotor but also auditory memory functions incorporated with motor planning and execution. Increasing complexity and challenge can easily be built into musical exercises, such that they are in line with the principles of effective motor learning. Also, the goal-oriented approach of music making resonates with the importance of meaningful tasks in rehabilitation. Compared with many traditional motor and cognitive rehabilitation methods, music-based activities have the advantage of being intrinsically motivating and fun as well as providing direct and natural feedback, which can be crucial for engaging patients in their rehabilitation process. At the same time, engagement in music making can become an important emotional event, often providing solace and comfort and relieving stress and anxiety, and can therefore have an important additional positive impact on quality of life.

In this section, we synthesize recent studies of musical interventions for stroke survivors. The interventions are based on existing evidence regarding underlying brain mechanisms, and outcome measures are demonstrated as functional changes at the level of behavior and the brain. These give us important insights into how the human brain in both the healthy and damaged state processes music, with its multimodal interactions. In the first article,[95] Särkämö and Soto review current evidence of the effects of music listening on emotion, cognition, and the brain; present two bodies of experimental studies showing that listening to pleasant music after a stroke can temporarily enhance visual awareness in patients suffering from unilateral spatial neglect as well as improve the recovery of memory, attention, and mood; and discuss the potential neural mechanisms underlying the rehabilitative effect of music listening. In the second article,[96] Rodriguez-Fornells, Rojo, Amengual, Ripolles, Altenmüller, and Münte review the literature concerning the application of music-supported therapy (MST) (a new motor rehabilitation method that uses musical instruments) and present experimental results indicating that MST is effective for improving the recovery of motor skills and mood after a stroke, with a direct impact on the activity and functional connectivity of the frontotemporal auditory–motor networks. MST is also discussed in the third article[97] by Fujioka, Ween, Jamali, Stuss, and Ross, which demonstrates, using MEG, that listening and tapping to a beat are both associated with the periodic modulation of beta oscillations in the brain that are typically linked to motor functions, and that MST can induce changes in the contribution of auditory and motor brain areas to the beta activity after chronic stroke. Their data also give some insight into how rhythmic auditory stimulation (RAS) using music or metronome sounds may successfully enhance rehabilitation exercises through activating the sensorimotor beta-band network.[98,99]

In the fourth article,[100] van Wijck et al. introduce the rationale and methodological basis of a novel, innovative motor intervention for stroke patients, which integrates the patient's preferred music with game technology in a rhythmic auditory cueing task for training upper limb function. Finally, Tomaino[101] describes various singing-related techniques used in clinical music therapy to enhance speech ability in nonfluent aphasic patients. Although MIT has been known as an effective singing exercise,[84,102] the paper documents how diverse the symptoms of aphasic patients can be and how emphasizing different musical features and social/motor cues in the practice can help with treating them differently.

In summary, this section highlights recent advances in the field of music therapy for stroke and provides steps toward integrating musical activities into the neurological rehabilitation of stroke patients and toward understanding how music can work in the damaged brain. Although the findings are encouraging, the field is still new, and more research is clearly needed for building a solid evidence base through randomized control trials that compare musical interventions against adequate control interventions. Through this synthesis, our hope is to inspire and stimulate future research directions with joint efforts among neuroscientists, therapists, psychologists, and engineers.

Symposium 8: Autism and music

Catherine Y. Wan

Autism spectrum disorder (ASD) is a developmental condition that affects one in 110 children, and as many as one in 70 boys.[103] In addition to social abnormalities and the presence of repetitive and stereotyped behaviors, one of the core diagnostic features of ASD is impairment of language and communication.[104] Severe deficits in communication not only diminish quality of life for affected individuals but also present a lifelong challenge for their families.

Despite their social and communication deficits, children with ASD can enjoy auditory–motor activities, such as making music, through singing or playing an instrument. In addition, such children often display enhanced music and auditory-perception abilities.[105] In the first published report of autism, Kanner[106] described the exceptional musical skills of several children, including one notable example of an 18-month-old boy who was able to discriminate among many symphonic pieces of music. Recent investigations have provided further evidence for enhanced music-perception abilities in individuals with ASD. Often described as emotionally unreachable or in their own world, these individuals are nevertheless often able to express typical affective responses to music and to derive enjoyment from music.

The paper entitled "Music: a unique window into the world of autism" presents recent advances in research examining neurobiological underpinnings of musical processing in ASD and the therapeutic potential of music making in ameliorating some of the associated deficits. Molnar-Szakacs and Heaton summarize research on the dissociation between emotional communication abilities in the musical and social domains in individuals with ASD.[107] Although individuals with ASD are often impaired in their ability to understand nonverbal expression of emotions, they can nevertheless understand simple and complex musical emotions.[17] Potential explanations for such uneven development across functional domains, as well as future directions for the study of music in autism, are discussed.

Hyde and colleagues review the behavioral and neuroimaging literature on auditory pitch and time processing in ASD.[108] Individuals with ASD generally show enhanced perceptual skills in a musical context but not in a linguistic context. These observations may be related to the locally oriented or highly focused behaviors that are common in ASD.[109] Identifying the brain–behavior relationships of auditory processing in ASD may help to identify neurobiological markers and enhanced treatment potential in ASD.

Wan *et al.* present diffusion tensor imaging data collected from a group of completely nonverbal children with ASD.[110] Abnormalities in a language-related white matter tract, the arcuate fasciculus, were investigated. This tract connects auditory and motor brain regions and is important in the mapping of sounds to articulatory actions during speech. The preliminary imaging findings reported here may explain why children with ASD fail to develop speech naturally. Furthermore, the findings complement the laboratory's ongoing treatment study of a novel intonation-based intervention (auditory–motor mapping training, AMMT), which aims to facilitate speech output in nonverbal children with ASD.[111] It is suggested that interventions that are designed to engage abnormal auditory–motor connections (such as AMMT) have the potential to effectively facilitate the development of expressive language skills.

Taken together, the three papers on music and autism review current knowledge in this emerging field, which will hopefully stimulate further research and the development of interventions for a disorder that affects the lives of a great many individuals.

Symposium 9: Learning and memory in musical disorders

Psyche Loui and Isabelle Peretz

Although music is ubiquitous across human cultures, a subset of the normal population appears to have an abnormal lack of musical ability. Increasing evidence from behavioral, neuroimaging, and genetic studies has documented difficulties in pitch perception as well as production, also known as congenital amusia or tone deafness.[112,113] Although these substantial deviations from the musical norm are fascinating to the general public, they can also be informative as a window into the psychological and neural capacities necessary for musical functioning. Furthermore, they can provide us with a model system through which to investigate capacities that, like music, might be uniquely human, such as speech and language, expectation for future events, and consciousness.

Given that musical disorders may have important implications for a broader scientific community, there is a need to have a solid understanding of their underlying cognitive processes and mechanisms. How do musical disorders inform our model of the human capacity for music? To date, tested theories of the causes of musical disorders include difficulties in fine-grained pitch discrimination,[114] spatial processing[115] (see also Ref. 116), pitch awareness,[117] disconnection of neural pathways,[118] and short-term memory limitations in pitch.[119] Despite these theories, we know relatively little about how cognitive constraints might limit musical performance. In particular, limitations on memory are only recently becoming addressed as possible explanations for musical disorders. Given that persons with congenital amusia have pitch-specific limitations on their short-term memory resources that are independent of perceptual difficulties,[120] it remains to be seen whether and to what extent these memory limitations could give rise to learning difficulties. If learning is indeed affected in people with musical disorders, then to characterize these learning deficits, and possibly to design rehabilitation strategies, we need to know about the types of information learning that are affected: whether these learning difficulties are circumscribed to music, and whether related modalities such as language learning might be affected as well.

The last section in this volume aims to address precisely these questions concerning the cognitive underpinnings of musical disorders. The section begins by exploring memory limitations in persons with congenital amusia, a topic introduced by Dalla Bella *et al.*, who outline a model of vocal production and its impairment in amusia and suggest that memory can be a source of disorders in musical production.[121] Stewart, in collaboration with Susan Anderson *et al.*, describe and present preliminary results from a musical intervention with five individuals diagnosed with congenital amusia.[122] Loui explores further the notion of learning in musical disabilities by reporting a novel investigation of rapid statistical learning abilities and its disruption in tone-deaf individuals.[123] This statistical learning approach is followed up in the last chapter by Peretz *et al.*, who report five novel experiments showing that persons with amusia can learn novel words as easily as controls, whereas they systematically fail on musical materials.[124]

Taken together, the aim of this section is to provide a comprehensive review of musical disorders, bring novel results to light regarding learning and memory and their interactions in normal and disordered brains, and ultimately to offer new perspectives toward the fundamental questions regarding neuroplasticity and the domain generality versus domain specificity of music.

Conclusion

It is evident from this introduction that the field of music neuroscience is rich and varied in ideas, approaches, implications, and potential applications. Much of the work is interdisciplinary in nature, including strong links with psychology, neurology, and clinical practice, and growing links with music performance, education, and therapy. Although the papers included here certainly reflect the current state of the art, it must also be emphasized that they represent only a small fraction of ongoing research internationally and only hint at the research being planned and prepared. The future of our understanding of the underlying neurobiology of human musical ability and behavior looks healthy, exciting, and promising, from the first milliseconds of auditory perception in infants to long-term training, individual musical preferences, and cultural diversity. We look forward with anticipation to The Neurosciences and Music V, planned for France in 2014.

Conflicts of interest

The authors declare no conflicts of interest.

References

1. Zatorre, R.J. & I. Peretz, Eds. 2001. The Biological Foundations of Music. *Ann. N.Y. Acad. Sci.* **930**.
2. Avanzini, G, C. Faienza, M. Majno & D. Minciacchi, Eds. 2003. The Neurosciences and Music. *Ann. N.Y. Acad. Sci.* **999**.
3. Avanzini, G., L. Lopez, S. Koelsch & M. Majno, Eds. 2006. The Neurosciences and Music II: From Perception to Performance. *Ann. N.Y. Acad. Sci.* **1060**.
4. Dalla Bella, S., N. Kraus, K. Overy, *et al.*, Eds. 2009. The Neurosciences and Music III: Disorders and Plasticity. *Ann. N.Y. Acad. Sci.* **1169**.
5. McMahon, E., P. Wintermark & A. Lahav. 2012. Auditory brain development in premature infants: the importance of early experience. *Ann. N.Y. Acad. Sci.* **1252:** 17–24. This volume.
6. Trehub, S.E. 2012. Behavioral methods in infancy: pitfalls of single measures. *Ann. N.Y. Acad. Sci.* **1252:** 37–42. This volume.

7. Trainor, L.J. 2012. Musical experience, plasticity, and maturation: issues in measuring developmental change using EEG and MEG. *Ann. N.Y. Acad. Sci.* **1252:** 25–36. This volume.
8. Raschle, N., J. Zuk, S. Ortiz-Mantilla, *et al.* 2012. Pediatric neuroimaging in early childhood and infancy: challenges and practical guidelines. *Ann. N.Y. Acad. Sci.* **1252:** 43–50. This volume.
9. Marenbon, J.A., Ed. 2003. *Boethius*. Oxford University Press. New York.
10. Shehadi, F. 1995. *Philosophies of Music in Medieval Islam*. Brill. Leiden, The Netherlands.
11. Katznelson, Y. 2004. *An Introduction to Harmonic Analysis*. Cambridge University Press. Cambridge, UK.
12. Bear, M.F., B.W. Connors & M.A. Paradiso. 2006. *Neuroscience: Exploring the Brain*. 3rd ed. Lippincott. Philadelphia.
13. Majno, M. 2012. From the model of *El Sistema* in Venezuela to current applications: learning and integration through collective music education. *Ann. N.Y. Acad. Sci.* **1252:** 56–64. This volume.
14. The Scottish Government. 2011. *Evaluation of the Big Noise, Sistema Scotland*. Scottish Government Social Research. ISBN 978-1-78045-099-5.
15. Uibel, S. 2012. Education through music—the model of the Musikkindergarten Berlin. *Ann. N.Y. Acad. Sci.* **1252:** 51–55. This volume.
16. Overy, K. 2012. Making music in a group: synchronization and shared experience. *Ann. N.Y. Acad. Sci.* **1252:** 65–68. This volume.
17. Molnar-Szakacs, I. & K. Overy. 2006. Music and mirror neurons: from motion to 'e'motion. *Soc. Cogn. Affect. Neurosci.* **1:** 235–241.
18. Overy, K. & I. Molnar-Szakacs. 2009. Being together in time: musical experience and the mirror neuron system. *Music Percep.* **26:** 489–503.
19. Osborne, N. 2012. Neuroscience and "real world" practice: music as a therapeutic resource for children in zones of conflict. *Ann. N.Y. Acad. Sci.* **1252:** 69–76. This volume.
20. Osborne, N. 2009. Music for children in zones of conflict and post-conflict: a psychobiological approach. In *Communicative Musicality*. S. Malloch & C. Trevarthen, Eds.: 331–356. Oxford University Press. Oxford, UK.
21. Sakai, K., O. Hikosaka, S. Miyauchi, *et al.* 1999. Neural representation of a rhythm depends on its interval ratio. *J. Neurosci.* **19:** 10074–10081.
22. Grahn, J.A. & M. Brett. 2007. Rhythm and beat perception in motor areas of the brain. *J. Cogn. Neurosci.* **19:** 893–906.
23. Penhune, V.B., R.J. Zatorre & A.C. Evans. 1998. Cerebellar contributions to motor timing: a PET study of auditory and visual rhythm discrimination. *J. Cogn. Neurosci.* **10:** 752–765.
24. McAuley, J.D., M.J. Henry & J. Tkach. 2012. Tempo mediates the involvement of motor areas in beat perception. *Ann. N.Y. Acad. Sci.* **1252:** 77–84. This volume.
25. Honing, H. 2012. Without it no music: beat induction as a fundamental musical trait. *Ann. N.Y. Acad. Sci.* **1252:** 85–91. This volume.
26. Hannon, E.E., C.M.V.B. der Nederlanden & P. Tichko. 2012. Effects of perceptual experience on children's and adults' perception of unfamiliar rhythms. *Ann. N.Y. Acad. Sci.* **1252:** 92–99. This volume.
27. Oechslin, M.S., M. Meyer & L. Jäncke. 2010. Absolute pitch–functional evidence of speech-relevant auditory acuity. *Cereb. Cortex* **20:** 447–455.
28. Schlaug, G. *et al.* 1995. In vivo evidence of structural brain asymmetry in musicians. *Science* **267:** 699–701.
29. Schön, D. *et al.* 2010. Similar cerebral networks in language, music and song perception. *NeuroImage* **51:** 450–461.
30. Schneider, P. *et al.* 2002. Morphology of Heschl's gyrus reflects enhanced activation in the auditory cortex of musicians. *Nat. Neurosci.* **5:** 688–694.
31. Ott, C.G. *et al.* 2011. Processing of voiced and unvoiced acoustic stimuli in musicians. *Front. Psychol.* **2:** 195.
32. Elmer, S., M. Meyer & L. Jäncke. 2012. Neurofunctional and behavioral correlates of phonetic and temporal categorization in musically trained and untrained subjects. *Cereb. Cortex* **22:** 650–658.
33. Meyer, M., S. Elmer & L. Jäncke. 2012. Musical expertise induces neuroplasticity of the planum temporale. *Ann. N.Y. Acad. Sci.* **1252:** 116–123. This volume.
34. Besson, M., J. Chobert & C. Marie. 2011. Transfer of training between music and speech: common processing, attention, and memory. *Front. Psychol.* **2:** 94.
35. Ettlinger, M., E.H. Margulis & P.C. Wong. 2011. Implicit memory in music and language. *Front. Psychol.* **2:** 211.
36. Schön, D. & C. Francois. 2011. Musical expertise and statistical learning of musical and linguistic structures. *Front. Psychol.* **2:** 167.
37. Francois, C. & D. Schon. 2010. Learning of musical and linguistic structures: comparing event-related potentials and behavior. *NeuroReport* **21:** 928–932.
38. Francois, C. & D. Schon. 2011. Musical expertise boosts implicit learning of both musical and linguistic structures. *Cereb. Cortex* **21:** 2357–2365.
39. François, C., B. Tillmann & D. Schön. 2012. Cognitive and methodological considerations on the effects of musical expertise on speech segmentation. *Ann. N.Y. Acad. Sci.* **1252:** 108–115. This volume.
40. Patel, A.D. 2011. Why would musical training benefit the neural encoding of speech? The OPERA hypothesis. *Front. Psychol.* **2:** 142.
41. Patel, A.D. 2012. The OPERA hypothesis: assumptions and clarifications. *Ann. N.Y. Acad. Sci.* **1252:** 124–128. This volume.
42. Kraus, N. & B. Chandrasekaran. 2010. Music training for the development of auditory skills. *Nat. Rev. Neurosci.* **11:** 599–605.
43. Musacchia, G. *et al.* 2007. Musicians have enhanced subcortical auditory and audiovisual processing of speech and music. *Proc. Natl. Acad. Sci. USA* **104:** 15894–15898.
44. Wong, P.C. *et al.* 2007. Musical experience shapes human brainstem encoding of linguistic pitch patterns. *Nat. Neurosci.* **10:** 420–422.
45. Parbery-Clark, A. *et al.* 2011. Musical experience and the aging auditory system: implications for cognitive abilities and hearing speech in noise. *PLoS ONE* **6:** e18082.
46. Strait, D.L. & N. Kraus. 2011. Can you hear me now? Musical training shapes functional brain networks for selective auditory attention and hearing speech in noise. *Front. Psychol.* **2:** 113.

47. Kraus, N., D.L. Strait & A. Parbery-Clark. 2012. Cognitive factors shape brain networks for auditory skills: spotlight on auditory working memory. *Ann. N.Y. Acad. Sci.* **1252:** 100–107. This volume.
48. Jäncke, L., S. Mirzazade & N.J. Shah. 1999. Attention modulates activity in the primary and the secondary auditory cortex: a functional magnetic resonance imaging study in human subjects. *Neurosci. Lett.* **266:** 125–128.
49. Chiao, J. & N. Ambady. 2007. Cultural neuroscience: parsing universality and diversity across levels of analysis. In *Handbook of Cultural Psychology*. S. Kitayama & D. Cohen, Eds.: 237–254. Guilford. New York.
50. Hannon, E.E. & S.E. Trehub. 2005. Metrical categories in infancy and adulthood. *Psychol. Sci.* **16:** 48–55.
51. Trainor, L.J., C. Marie, D. Gerry, *et al.* 2012. Becoming musically enculturated: effects of music classes for infants on brain and behavior. *Ann. N.Y. Acad. Sci.* **1252:** 129–138. This volume.
52. Vuust, P., E. Brattico, M. Seppänen, *et al.* 2012. Practiced musical style shapes auditory skills. *Ann. N.Y. Acad. Sci.* **1252:** 139–146. This volume.
53. Koelsch, S., B.-H. Schmidt & J. Kansok. 2002. Effects of musical expertise on the early right anterior negativity: an event-related brain potential study. *Psychophysiology* **39:** 657–663.
54. Tervaniemi, M., T. Tupala & E. Brattico. 2012. Expertise in folk music alters the brain processing of Western harmony. *Ann. N.Y. Acad. Sci.* **1252:** 147–151. This volume.
55. Large, E.W. & F.V. Almonte. 2012. Neurodynamics, tonality, and the auditory brainstem response. *Ann. N.Y. Acad. Sci.* Online only. In press.
56. Demorest, S.M. & L. Osterhout. 2012. ERP responses to cross-cultural melodic expectancy violations. *Ann. N.Y. Acad. Sci.* **1252:** 152–157. This volume.
57. Brattico E., M. Tervaniemi, R. Näätänen & I. Peretz. 2006. Musical scale properties are automatically processed in the human auditory cortex. *Brain Res.* **1117:** 162–174.
58. Wong, P.C.M., A.H.D. Chan & E.H. Margulis. 2012. Effects of mono- and bicultural experiences on auditory perception. *Ann. N.Y. Acad. Sci.* **1252:** 158–162. This volume.
59. Wong, P.C.M., A.K. Roy & E.H. Margulis. 2009. Bimusicalism: the implicit dual enculturation of cognitive and affective systems. *Music Percept.* **27:** 81–88.
60. Bailey, J.A. & V.B. Penhune. 2012. A sensitive period for musical training: contributions of age of onset and cognitive abilities. *Ann. N.Y. Acad. Sci.* **1252:** 163–170. This volume.
61. Pfordresher, P.Q. 2012. Musical training and the role of auditory feedback during performance. *Ann. N.Y. Acad. Sci.* **1252:** 171–178. This volume.
62. Zimmerman, E. & A. Lahav. 2012. The multisensory brain and its ability to learn music. *Ann. N.Y. Acad. Sci.* **1252:** 179–184. This volume.
63. Palmer, C., B. Mathias & M. Anderson. 2012. Sensorimotor mechanisms in music performance: actions that go partially wrong. *Ann. N.Y. Acad. Sci.* **1252:** 185–191. This volume.
64. Strübing, F., M. Herrojo Ruiz, H.C. Jabusch & E. Altenmüller. 2012. Error monitoring is altered in musician's dystonia: evidence from ERP-based studies. *Ann. N.Y. Acad. Sci.* **1252:** 192–199. This volume.
65. Hubbard, T.L. 2010. Auditory imagery: empirical findings. *Psychol. Bull.* **136:** 302–329.
66. Halpern, A.R. & J.C. Bartlett. 2010. Memory for melodies. In *Music Perception*. M.R. Jones, A.N. Popper & R.R. Fay, Eds. Springer-Verlag. New York.
67. Halpern, A.R. 1989. Memory for the absolute pitch of familiar songs. *Mem. Cogn.* **17:** 572–581.
68. Halpern, A.R. 1988. Perceived and imagined tempos of familiar songs. *Music Percept.* **6:** 193–202.
69. Levitin, D.J. 1994. Absolute memory for musical pitch: evidence from the production of learned melodies. *Percept. Psychophys.* **56:** 414–423.
70. Schellenberg, E.G. & S.E. Trehub. 2003. Good pitch memory is widespread. *Psychol. Sci.* **22:** 262–266.
71. Zatorre, R.J. & A.R. Halpern. 2005. Mental concerts: musical imagery and auditory cortex. *Neuron* **47:** 9–12.
72. Janata, P. 2012. Acuity of mental representations of pitch. *Ann. N.Y. Acad. Sci.* **1252:** 214–221. This volume.
73. Halpern, A.R. 2012. Dynamic aspects of musical imagery. *Ann. N.Y. Acad. Sci.* **1252:** 200–205. This volume.
74. Zatorre, R.J. 2012. Beyond auditory cortex: working with musical thoughts. *Ann. N.Y. Acad. Sci.* **1252:** 222–228. This volume.
75. Keller, P.E. 2012. Mental imagery in music performance: underlying mechanisms and potential benefits. *Ann. N.Y. Acad. Sci.* **1252:** 206–213. This volume.
76. Münte, T.F., E. Altenmüller & L. Jäncke. 2002. The musician's brain as a model of neuroplasticity. *Nat. Rev. Neurosci.* **3:** 473–478.
77. Wan, C.Y. & G. Schlaug. 2010. Music making as a tool for promoting brain plasticity across the life span. *Neuroscientist* **16:** 566–577.
78. Blood A.J. & R.J. Zatorre. 2001. Intensely pleasurable responses to music correlate with activity in brain regions implicated in reward and emotion. *Proc. Natl. Acad. Sci. USA* **98:** 11818–11823.
79. Altenmüller, E., J. Marco-Pallares, T.F. Münte & S. Schneider. 2009. Neural reorganization underlies improvement in stroke-induced motor dysfunction by music-supported therapy. *Ann N.Y. Acad. Sci.* **1169:** 395–405.
80. Okamoto, H., H. Stracke, W. Stoll & C. Pantev. 2010. Listening to tailor-made notched music reduces tinnitus loudness and tinnitus-related auditory cortex activity. *Proc. Natl. Acad. Sci. USA* **107:** 1207–1210.
81. Altenmüller, E. & H.C. Jabusch. 2010. Focal dystonia in musicians: phenomenology, pathophysiology and triggering factors. *Eur. J. Neurol.* 17(Suppl 1): 31–66.
82. Schulze, K. & S. Koelsch. 2012. Working memory for speech and music. *Ann. N.Y. Acad. Sci.* **1252:** 229–236. This volume.
83. Pantev, C., H. Okamoto & H. Teismann. 2012. Tinnitus: the dark side of the auditory cortex plasticity. *Ann. N.Y. Acad. Sci.* **1252:** 253–258. This volume.
84. Zipse, L., A. Norton, S. Marchina & G. Schlaug. 2012. When right is all that is left: plasticity of right-hemisphere tracts in a young aphasic patient. *Ann. N.Y. Acad. Sci.* **1252:** 237–245. This volume.
85. Jäncke, L. 2012. The dynamic audio-motor system in pianists. *Ann. N.Y. Acad. Sci.* **1252:** 246–252. This volume.
86. Altenmüller. E., V. Baur, A. Hofmann, *et al.* 2012. Musician's cramp as manifestation of maladaptive brain plasticity: arguments from instrumental differences. *Ann. N.Y. Acad. Sci.* **1252:** 259–265. This volume.

87. Jäncke, L. 2009. The plastic human brain. Restor. *Neurol. Neurosci.* **27:** 521–538.
88. Hannon, E.E. & L.J. Trainor. 2007. Music acquisition: effects of enculturation and formal training on development. *Trends Cogn. Sci.* **11:** 466–472.
89. Ruiz, M.H., H.C. Jabusch & E. Altenmüller. 2009. Detecting wrong notes in advance: neuronal correlates of error monitoring in pianists. *Cereb. Cortex* **19:** 2625–2639.
90. Thompson, W.F., E.G. Schellenberg & G. Husain. 2004. Decoding speech prosody: do music lessons help? *Emotion* **4:** 46–64.
91. Thompson, W.F., E.G. Schellenberg & G. Husain. 2001. Arousal, mood, and the Mozart effect. *Psychol. Sci.* **12:** 248–251.
92. Schellenberg, E.G., T. Nakata, P.G. Hunter, *et al.* 2007. Exposure to music and cognitive performance: tests of children and adults. *Psychol. Music* **35:** 5–19.
93. Greene, C.M., P. Bahri & D. Soto. 2010. Interplay between affect and arousal in recognition memory. *PLoS ONE* **5:** e11739.
94. World Health Organization (WHO). 2001. International Classification of Functioning, Disability and Health (ICF). URL http://www.who.int/classification/icf/en/ [accessed on 8 November 2011].
95. Särkämö, T. & D. Soto. 2012. Music listening after stroke: beneficial effects and potential neural mechanisms. *Ann. N.Y. Acad. Sci.* **1252:** 266–281. This volume.
96. Rodriguez-Fornells, A. *et al.* 2012. The involvement of audio–motor coupling in the music-supported therapy applied to stroke patients. *Ann. N.Y. Acad. Sci.* **1252:** 282–293. This volume.
97. Fujioka, T. *et al.* 2012. Changes in neuromagnetic beta-band oscillation after music-supported stroke rehabilitation. *Ann. N.Y. Acad. Sci.* **1252:** 294–304. This volume.
98. Thaut, M.H., G.P. Kenyon, C.P. Hurt, *et al.* 2002. Kinematic optimization of spatiotemporal patterns in paretic arm training with stroke patients. *Neuropsychologia* **40:** 1073–1081.
99. Thaut, M.H., A.K. Leins, R.R. Rice, *et al.* 2007. Rhythmic auditory stimulation improves gait more than NDT/Bobath training in near-ambulatory patients early poststroke: a single-blind, randomized trial. *Neurorehabil. Neural Repair* **21:** 455–459.
100. van Wijck, F. *et al.* 2012. Making music after stroke: using musical activities to enhance arm function. *Ann. N.Y. Acad. Sci.* **1252:** 305–311. This volume.
101. Tomaino, C.M. 2012. Effective music therapy techniques in the treatment of nonfluent aphasia. *Ann. N.Y. Acad. Sci.* **1252:** 312–317. This volume.
102. Sparks, R., N. Helm & M. Albert. 1974. Aphasia rehabilitation resulting from melodic intonation therapy. *Cortex* **10:** 303–316.
103. Autism Speaks. URL www.autismspeaks.org [accessed on 8 March 2012].
104. Tager-Flusberg, H. 2003. Language impairments in children with complex neurodevelopmental disorders: the case of autism. In *Language Competence across Populations: Toward a Definition of Specific Language Impairment*. Y. Levy & J.C. Schaeffer, Eds.: 297–321. Lawrence Erlbaum Associates. Mahway, NJ.
105. Heaton, P., L. Pring & B. Hermelin. 2001. Musical processing in high functioning children with autism. *Ann. N.Y. Acad. Sci.* **930:** 443–444.
106. Kanner, L. 1968. Autistic disturbances of affective contact. *Acta Paedopsychiatr.* **35:** 100–136.
107. Molnar-Szakacs, I. & P. Heaton. 2012. Music: a unique window into the world of autism. *Ann. N.Y. Acad. Sci.* **1252:** 318–324. This volume.
108. Ouimet, T., N.E.V. Foster, A. Tryfon & K.L. Hyde. 2012. Auditory-musical pitch processing in autism spectrum disorders: a review of behavioral and brain imaging studies. *Ann. N.Y. Acad. Sci.* **1252:** 325–331. This volume.
109. Samson, F., K.L. Hyde, A. Bertone, *et al.* 2011. Atypical processing of auditory temporal complexity in autistics. *Neuropsychologia* **49:** 546–555.
110. Wan, C.Y., S. Marchina, A. Norton & G. Schlaug. 2012. Atypical hemispheric asymmetry in the arcuate fasciculus of completely nonverbal children with autism. *Ann. N.Y. Acad. Sci.* **1252:** 332–337. This volume.
111. Wan, C.Y., L. Bazen, R. Baars, *et al.* 2011. Auditory-motor mapping training as an intervention to facilitate speech output in non-verbal children with autism: a proof of concept study. *PLoS ONE* **6:** e25505.
112. Peretz, I. 2008. Musical disorders: from behavior to genes. *Curr. Directions Psychol. Sci.* **17:** 329–333.
113. Stewart, L. 2008. Fractionating the musical mind: insights from congenital amusia. *Curr. Opin. Neurobiol.* **18:** 127–130.
114. Peretz, I. *et al.* 2002. Congenital amusia: a disorder of fine-grained pitch discrimination. *Neuron* **33:** 185–191.
115. Douglas, K.M. & D.K. Bilkey. 2007. Amusia is associated with deficits in spatial processing. *Nat. Neurosci.* **10:** 915–921.
116. Tillmann, B. *et al.* 2010. The amusic brain: lost in music, but not in space. *PLoS ONE* **5:** e10173.
117. Peretz, I. *et al.* 2009. The amusic brain: in tune, out of key, and unaware. *Brain* **132:** 1277–1286.
118. Loui, P., D. Alsop & G. Schlaug. 2009. Tone deafness: a new disconnection syndrome? *J. Neurosci.* **29:** 10215–10220.
119. Tillmann, B., K. Schulze & J.M. Foxton. 2009. Congenital amusia: a short-term memory deficit for non-verbal, but not verbal sounds. *Brain Cogn.* **71:** 259–264.
120. Williamson, V.J. & L. Stewart. 2010. Memory for pitch in congenital amusia: beyond a fine-grained pitch discrimination problem. *Memory* **18:** 657–669.
121. Dalla Bella, S., A. Tremblay-Champoux, M. Berkowska & I. Peretz. 2012. Memory disorders and vocal performance. *Ann. N.Y. Acad. Sci.* **1252:** 338–344. This volume.
122. Anderson, S., E. Himonides, K. Wise, *et al.* 2012. Is there potential for learning in amusia? A study of the effect of singing intervention in congenital amusia. *Ann. N.Y. Acad. Sci.* **1252:** 345–353. This volume.
123. Loui, P. & G. Schlaug. 2012. Impaired learning of event frequencies in tone deafness. *Ann. N.Y. Acad. Sci.* **1252:** 354–360. This volume.
124. Peretz, I., J. Saffran, D. Schön & N. Gosselin. 2012. Statistical learning of speech, not music, in congenital amusia. *Ann. N.Y. Acad. Sci.* **1252:** 361–367. This volume.

Ann. N.Y. Acad. Sci. ISSN 0077-8923

ANNALS OF THE NEW YORK ACADEMY OF SCIENCES
Issue: *The Neurosciences and Music IV: Learning and Memory*

Auditory brain development in premature infants: the importance of early experience

Erin McMahon,[1] Pia Wintermark,[2] and Amir Lahav[1,3]

[1]Department of Newborn Medicine, Brigham and Women's Hospital, Boston, Massachusetts. [2]Division of Newborn Medicine, Montreal Children's Hospital, McGill University, Montreal, Quebec, Canada. [3]Department of Pediatrics, MassGeneral Hospital for Children, Harvard Medical School, Boston, Massachusetts

Address for correspondence: Amir Lahav, ScD, PhD, The Neonatal Research Lab, Department of Newborn Medicine, 75 Francis Street, Boston, Massachusetts 02115. amir@hms.harvard.edu

Preterm infants in the neonatal intensive care unit (NICU) often close their eyes in response to bright lights, but they cannot close their ears in response to loud sounds. The sudden transition from the womb to the overly noisy world of the NICU increases the vulnerability of these high-risk newborns. There is a growing concern that the excess noise typically experienced by NICU infants disrupts their growth and development, putting them at risk for hearing, language, and cognitive disabilities. Preterm neonates are especially sensitive to noise because their auditory system is at a critical period of neurodevelopment, and they are no longer shielded by maternal tissue. This paper discusses the developmental milestones of the auditory system and suggests ways to enhance the quality control and type of sounds delivered to NICU infants. We argue that positive auditory experience is essential for early brain maturation and may be a contributing factor for healthy neurodevelopment. Further research is needed to optimize the hospital environment for preterm newborns and to increase their potential to develop into healthy children.

Keywords: auditory; brain; prematurity; newborns; noise; plasticity

Introduction

Prematurity is the leading cause of death in the first month of life and a major cause of long-term disability.[1] Advances in neonatal care have increased the survival rate of ailing premature newborn infants.[2] However, despite these advances, premature infants are still prone to developmental problems that may persist throughout their lifetime.[3–5] These include deficits in hearing, attention, memory, speech, language, social behavior, and self-regulation.[6–8] Because these problems are too often seen in neonatal intensive care unit (NICU) graduates, it is possible that the origin of these problems may be attributed to environmental factors, such as the acoustic environment in the NICU. This article is not an exhaustive review, but is intended to highlight the concept that preterm newborns are at a critical period for auditory development and therefore must be provided with appropriate auditory input and careful protection against overstimulation during their prolonged stay in the NICU. The current design of most NICUs has yet to achieve such protection.

Auditory developmental milestones

To better understand the need for auditory protection and the importance of early auditory experience, it is necessary to point out some of the most important developmental milestones of auditory development in the fetal and neonatal periods. The development of the auditory system is an elaborate process that begins very early in gestation.[9] All major structures of the ear, including the cochlea, are in place between 23 and 25 weeks gestational age (GA).[9–11] Thus, unless a congenital abnormality is present, most preterm infants can already hear when first admitted to the NICU. The human fetus can perceive and react to auditory information starting at approximately week 26 of life.[12] Beginning between 26 and 30 weeks GA, hair cells in the cochlea are fine tuned for specific frequencies and can translate vibratory acoustic stimuli into an

doi: 10.1111/j.1749-6632.2012.06445.x

electrical signal that is sent to the brainstem.[13] Beyond 30 weeks GA, the auditory system is mature enough to permit complex sounds and distinguish between different speech phonemes,[10,14,15] which is presumably the beginning of language and speech development.[16] Finally, by 35 weeks GA, auditory processing facilitates learning and memory formation.[17,18] Therefore, there is a need early on to protect preterm newborns from auditory stimuli they are not yet ready to handle.[19]

Many of the sounds that are audible in the womb are generated internally by the mother's respiration, digestion, heart rhythm, and physical movements.[20–22] Fetuses, however, can also respond to sounds originated outside of the womb, such as music and voice. These sounds stimulate the inner ear through a mechanism of bone conduction. The sound frequencies heard within the womb parallel the course of frequency development within the cochlea,[23] making the womb an optimal and sheltered environment for auditory maturation. Because maternal tissue and fluid act as filters for high-frequency sounds, the developing cochlear hair cells are protected from potentially damaging noise.[24] A gradual exposure to low-frequency sounds first permits the necessary fine tuning of the hair cells. As the auditory system matures, it can process more high-frequency patterns of human speech, such as pitch, intonation, and intensity, which provide the exogenous stimulation necessary to develop a neocortical relationship with the cochlea.[14] Being able to hear these high-frequency sounds, in fact, wires the fetus for language processing soon after it is born.

After birth, infants seem to prefer their mother's voice over an unknown female voice. This has been demonstrated in several studies showing that newborns selectively respond to their mother's speech with detectable changes in heart rate[25,26] and orienting movements toward[27] the source of the sound. The fact that newborn infants show a clear preference for their mother's voice within only hours after birth can be taken as evidence for the significance of prenatal hearing experience,[28] suggesting that auditory attention, learning, and memory begin while in the womb.

Auditory brain plasticity

Auditory brain plasticity is intimately linked with early sensory experience. Functions of the cerebral cortex and the sensory systems are mostly established during a sensitive period in the neonatal period when neural circuitry is first generated.[29] This initial neural architecture leads to different functional areas that systematically represent different environmental information, and appears to constrain further plasticity later.[30] Following this critical period, the cerebral cortex and the auditory system will only be refined and reorganized during childhood and adulthood to optimize sensory processing and adaptation to environmental changes.[31]

The development of the auditory cortex is heavily dependent on the acoustic environment, as demonstrated by both human and animal studies.[30] For example, rearing rat pups under various noise conditions significantly affects the maturation time of auditory cortical areas.[32,33] Similarly, continuous exposure to loud noise negatively affects the development of auditory-related functions, including vocal learning and spatial localization.[34–37] These animal studies demonstrate that changes in sensory input can have profound effects on the functional organization of the developing cortex. Furthermore, they implicate noise as a risk factor for abnormal sensory development, and provide a basis for our concerns regarding the adverse effects of NICU noise on neonatal brain development in humans.

Deprivation of maternal sounds

Infants born prematurely spend their first weeks and even months of life in the NICU. During this time, they are deprived of the biological maternal sounds they would otherwise be hearing *in utero*. These sounds mainly include the low-frequency bands of mother's voice and the continuous, rhythmic stimulation of the maternal heartbeat. Deprivation of these sounds when the auditory system is at a critical period for development can have a profound effect on auditory brain maturation and subsequent speech and language acquisition.[38–40]

A deprived auditory cortex cannot mature normally.[41] Evidence coming mainly from animal studies suggests that the functional development of the auditory system is largely influenced by environmental acoustic inputs in early life.[42,43] For example, depriving juvenile birds of normal auditory experience delayed the emergence of topographic brain circuitry.[44] Similarly, prolonged auditory deprivation has been shown to decrease the expression levels of selective NMDA receptors in the rat auditory cortex during early postnatal development.[45,46] In contrast, exposing infant rat pups to an enriched auditory environment—either with[47] or without[48] music stimulation—has been shown to enhance auditory discrimination and learning abilities. In addition, infant rat pups raised under sensory deafness conditions have been shown to develop abnormal synaptic morphology in the primary auditory cortex in terms of dendrite shape, length, and spine density.[49] These studies demonstrate a high inclination for auditory brain plasticity in the neonatal period.

Taken together, these animal studies lend support to our concerns about the possibly harmful auditory deprivation experienced by preterm infants, even in the seemingly quiet and protected environment of the incubator. It is, therefore, reasonable to assume that the lack of sufficient opportunities to perceive maternal speech sounds during NICU hospitalization can alter brain structure and subsequently account for some of the hearing, language, and attention deficits often seen in NICU graduates.

Effects of noise exposure

Unquestionably, the type of sounds and levels of noise typically present in the NICU are very different from those heard in the womb.[9,50] Low-frequency placental sounds in the amniotic environment are replaced by unpredictable noise coming from ventilators, cardiac monitors, infusion pumps, pagers, and alarms.[24] This accumulation of background noise,[51–53] even within the seemingly protected incubator,[54–56] well exceeds the recommended levels set by the American Academy of Pediatrics.[57] Thus, the current environment available to preterm infants during their hospitalization is not optimal.[24] NICU noise can result in detrimental health outcomes.[58–60] Loud noise can produce unwarranted physiological changes in heart rate,[61–63] blood pressure,[62] respiration, and oxygenation.[64,65] Noise also appears to cause hyperalertness, increased crying, and reduced deep sleep.[66] Our recent review paper on this topic[67] suggests that excessive exposure to noise during the neonatal period can negatively affect the cardiovascular and respiratory systems, which can increase the risk for a variety of developmental problems.

Clinical implications: what we can do

There are many things that can be done to improve the auditory environment in the NICU. Our primary target should be the quality and quantity of sounds surrounding NICU infants in order to guarantee optimal conditions for their growth and development. The two main problems that should be addressed are the high levels of noise in the NICU and the lack of meaningful acoustic input during a critical time for auditory brain development. Here, we propose a number of evidence-based suggestions that can be implemented into routine NICU care to potentially improve health outcomes of infants born prematurely.

How to address the problem of environmental noise?

Improve NICU design

The concept of individual rooms for NICU patients is still evolving.[68] However, there seems to be a consensus among both parents[69] and care givers[70] that the private-room NICU design is highly preferred. A recent model suggests that the transition from an open-bay NICU to a private-room NICU can potentially improve developmental outcomes through mediating factors such as developmental care, family-centered care, parental stress, staff behavior, and medical practices.[71,72]

Modify equipment

The default volume settings on many of the NICU's alarm systems are often unnecessarily high. Alarm volumes should be reduced, especially at night. Reducing the volume of pagers, alarms, telephones, and intercoms should improve overall noise pollution.[73] In addition, noise-making equipment, such as metal trashcans, should be replaced with quieter alternatives. Motorized paper towel dispensers should be cautiously avoided in spite of their potential to reduce infection rates.[74]

Consider a silent alarm system

Substituting audible ringtones of telephones and pagers with a silent vibration should also be considered. In this case, monitors will text events to all the nurses in the NICU pod through a wireless device rather than alarming at the infant's bedside. Previous studies taking this approach have reported positive results.[75]

Change staff behavior

It is important to educate NICU staff, including physicians, nurses, respiratory therapists, and parents, about the negative impact of noise on developmental outcomes. Increasing staff awareness can lead to significant changes in attitude and behavior, such as redirecting loud conversations away from patient areas and gently closing incubator doors. Research has shown that staff behavior alone can have an impact on the overall noise levels in the NICU, making it a more positive environment for growth and development.[72,76]

Routinely measure noise levels

Considering the evidence regarding the negative effects of noise on neonates, it is somewhat surprising that noise levels are not routinely measured in most NICUs. Noise level meters should be placed at the bedside to maintain quiet and ensure compliance with the recommendations set by the American Academy of Pediatrics.[57] Periodic monitoring of noise levels is necessary to identify new sources of environmental noise, as well as to evaluate the efficacy of noise-reduction strategies.[75]

Avoid earmuffs

While earmuffs may seem like an obvious way to filter noxious sounds, evidence for their effective use in NICU infants has been rather limited.[77] The use of ear protection in the NICU may carry risks that outweigh the benefits. The constant contact with earmuffs may present tactile overstimulation that the infant's sensory system is too immature to process.[78] In addition, earmuffs actually increase the risk for auditory deprivation by blocking the already limited human speech sounds that are available to the infant.

How to address the problem of auditory deprivation?

Provide kangaroo care

Kangaroo care is a common evidence-based method for "maternalizing" the NICU experience for preterm infants soon after birth.[79–81] During kangaroo care, the infant is placed in a supine position on the mother's (or father's) chest to have direct skin-to-skin contact.[82] The infant can then presumably hear and feel the low-frequency sounds of the maternal voice heartbeat through the skin. Kangaroo care has been associated with a decreased risk of mortality[83] and has been shown to promote maternal–infant bonding.[84,85] However, from an auditory perspective, kangaroo care also provides the infant with important opportunities to process meaningful maternal sounds that would otherwise not be available, especially in the case of low birth-weight infants who experience prolonged intubation.[81]

Play the mother's voice inside the incubator/crib

Although kangaroo care is strongly encouraged, in reality, there are times when the mother cannot be present in the NICU or the infant is too sick to be held outside the incubator. During those times, exposing the infant to audio recordings of the mother's voice will benefit both the mother and the baby: it allows the mother to be with her infant virtually, even when she is not there physically, and can provide the infant with a wide range of maternal vocalizations, including singing lullabies, reading books, and improvisational speaking. This should supply the appropriate language stimulation that is believed to promote hearing, speech, and social development. Studies have shown that preterm infants who were exposed to an audio recording of their mother's voice achieved full enteral feed quicker[86] and showed meaningful changes in heart rate[87] compared to age-matched controls receiving routine care (for review, see Ref. 88). While this approach has great potential, further research is needed to determine its optimal dose and long-term effects on child development.

Introduce vocal music

Exposing preterm infants to vocal music, such as lullabies, has been shown to increase oxygen saturations, improve weight gain, and nonnutritive sucking, and shorten overall hospital stay.[89–95] Vocal music is comprised of a large spectrum of intonations and vocalizations, both rhythmic and melodic,[96] which can provide adequate exposure to language stimuli when the live mother's voice is not available.[97] Vocal music, in particular, may also be biologically meaningful when sung by the mother. Combining live singing with kangaroo care may thus offer a perfectly viable way to warmly immerse the infant with soothing maternal sounds.[98,99] Any attempt to play other forms of music, such as purely instrumental pieces, must be taken with caution as this type of auditory stimulation cannot properly address the problem of maternal sound deprivation.

According to a recent review by Neal and Lindeke,[100] there is still controversy about the use of music as a developmental care strategy in the NICU, especially in infants <32 weeks GA. More controlled and rigorous studies are needed to definitively test the long-term effects of this therapeutic approach.

Be sensitive to the infant's state

When providing auditory stimulation in the NICU, whether it is vocal music or the mother's voice, it is critically important to pay attention to the infant's behavioral cues and modulate the stimulation accordingly. Preterm infants have limited capacity to defend themselves against sensory stimulation that is age-inappropriate with respect to duration, complexity, and intensity.[101] Infants will exhibit stress signals in autonomic, motor, and self-regulatory systems in response to irritating stimulation.[102] In accordance with the developmental care guidelines proposed by Heidi Als,[19,103] we recommend that both parents and caregivers should develop the necessary skills required to understand and respond to the infant's behavioral cues, and any automated systems that provide infants with auditory stimulation at a potentially inappropriate time should be avoided.

Carefully select audio equipment

Commercially available audio equipment, such as speakers, digital players, and cables, must be carefully tested for safety before it is used inside an incubator or a crib as these products are not typically designed for NICU use. Testing must ensure that the audio equipment delivers sound at a safe decibel level for preterm newborns, does not create electrical interference with medical equipment such as cardiac monitors and ventilators, is resistant against high temperature (~36°C) and humidity (~75%) levels typically present inside the incubator, and can be routinely cleaned with disinfectant to ensure compliance with infection control regulations. For a full description of recommended tests and audio equipment, see Ref. 104.

Conclusions

The primary auditory stimulation infants receive in the NICU is ambient noise. Thus, the prolonged hospitalization experienced by preterm infants may compromise development. This should no longer be ignored or devalued. Emphasis must be placed on making the NICU a more conducive environment for positive auditory experience. This early auditory experience occurs at the most critical period for neural wiring and subsequent neurodevelopment. Efforts should be made to envelop the preterm infants with more womb-like sounds to compensate for the loss of exposure to the maternal voice and heartbeat and to protect them from potentially adverse noise effects.

Acknowledgements

This work was supported by the following foundations: Christopher Joseph Concha, Hailey's Hope, Capita, Waterloo, Heather on Earth, Christopher Douglas Hidden Angel, and Peter and Elizabeth C. Tower, as well as by the John Alden Trust, Lifespan Healthcare, and Philips Healthcare.

Conflicts of interest

The authors declare no conflicts of interest.

References

1. Elder, D.E., A. Wong & J.M. Zuccollo. 2009. Risk factors for and timing of death of extremely preterm infants. *Aust. NZ J. Obstet. Gynaecol.* **49:** 407–410.
2. Wilson-Costello, D. 2007. Is there evidence that long-term outcomes have improved with intensive care? *Semin. Fetal Neonatal Med.* **12:** 344–354.
3. Msall, M.E. & J.J. Park. 2008. The spectrum of behavioral outcomes after extreme prematurity: regulatory, attention, social, and adaptive dimensions. *Semin. Perinatol.* **32:** 42–50.
4. Marlow, N., L. Roberts & R. Cooke. 1993. Outcome at 8 years for children with birth weights of 1250 g or less. *Arch. Dis. Child.* **68:** 286–290.
5. Marlow, N. *et al.* 2005. Neurologic and developmental disability at six years of age after extremely preterm birth. *N. Engl. J. Med.* **352:** 9–19.
6. Huang, L., K. Kaga & K. Hashimoto. 2002. Progressive hearing loss in an infant in a neonatal intensive care unit as revealed by auditory evoked brainstem responses. *Auris. Nasus. Larynx.* **29:** 187–190.
7. McCormick, M.C., K. Workman-Daniels & J. Brooks-Gunn. 1996. The behavioral and emotional well-being of school-age children with different birth weights. *Pediatrics* **97:** 18–25.
8. Xoinis, K. *et al.* 2007. Extremely low birth weight infants are at high risk for auditory neuropathy. *J. Perinatol.* **27:** 718–723.
9. Hall, J.W., 3rd. 2000. Development of the ear and hearing. *J. Perinatol.* **20:** S812–S820.
10. Cheour-Luhtanen, M. *et al.* 1996. The ontogenetically earliest discriminative response of the human brain. *Psychophysiology* **33:** 478–481.
11. Eldredge, L. & A. Salamy. 1996. Functional auditory development in preterm and full term infants. *Early Hum. Dev.* **45:** 215–228.

12. Ruben, R.J. 1992. The ontogeny of human hearing. *Acta Otolaryngol.* **112:** 192–196.
13. Querleu, D. *et al.* 1989. Hearing by the human fetus? *Semin. Perinatol.* **13:** 409–420.
14. Hepper, P., D. Scott & S. Shahidullah. 1993. Newborn and fetal response to maternal voice. *J. Reprod. Infant Psychol.* **11:** 147–153.
15. Johansson, B., E. Wedenberg & B. Westin. 1964. Measurement of tone response by the human foetus. A preliminary report. *Acta Otolaryngol.* **57:** 188–192.
16. Mehler, J. *et al.* 1988. A precursor of language acquisition in young infants. *Cognition* **29:** 143–178.
17. Birnholz, J.C. & B.R. Benacerraf. 1983. The development of human fetal hearing. *Science* **222:** 516–518.
18. Moon, C.M. & W.P. Fifer. 2000. Evidence of transnatal auditory learning. *J. Perinatol.* **20:** S37–S44.
19. Als, H. *et al.* 2005. The assessment of preterm infants' behavior (APIB): furthering the understanding and measurement of neurodevelopmental competence in preterm and full-term infants. *Ment. Retard. Dev. Disabil. Res. Rev.* **11:** 94–102.
20. Gerhardt, K.J., R.M. Abrams & C.C. Oliver. 1990. Sound environment of the fetal sheep. *Am. J. Obstet. Gynecol.* **162:** 282–287.
21. Querleu, D. *et al.* 1988. Fetal hearing. *Eur. J. Obstet. Gynecol. Reprod. Biol.* **28:** 191–212.
22. Armitage, S.E., B. A. Baldwin & M.A. Vince. 1980. The fetal sound environment of sheep. *Science* **208:** 1173–1174.
23. Fifer, W.P., Moon, C. 1988. Auditory Experience in the Fetus. In *Behavior of the Fetus.* W.P. Smotherman, S.R. Robinson, Eds: 175–188. Telford Press. Caldwell, NJ.
24. Graven, S.N. 2000. Sound and the developing infant in the NICU: conclusions and recommendations for care. *J. Perinatol.* **20:** S88–S93.
25. Ockleford, E.M. *et al.* 1988. Responses of neonates to parents' and others' voices. *Early Hum. Dev.* **18:** 27–36.
26. Kisilevsky, B.S. *et al.* 2009. Fetal sensitivity to properties of maternal speech and language. *Infant Behav. Dev.* **32:** 59–71.
27. Querleu, D. *et al.* 1984. Reaction of the newborn infant less than 2 hours after birth to the maternal voice. *J. Gynecol. Obstet. Biol. Reprod. (Paris)* **13:** 125–134.
28. DeCasper, A.J. & W.P. Fifer. 1980. Of human bonding: newborns prefer their mothers' voices. *Science* **208:** 1174–1176.
29. Sanes, D.H. & S. Bao. 2009. Tuning up the developing auditory CNS. *Curr. Opin. Neurobiol.* **19:** 188–199.
30. Dahmen, J.C. & A.J. King. 2007. Learning to hear: plasticity of auditory cortical processing. *Curr. Opin. Neurobiol.* **17:** 456–464.
31. Yan, J. 2003. Canadian Association of Neuroscience Review: development and plasticity of the auditory cortex. *Can. J. Neurol. Sci.* **30:** 189–200.
32. de Villers-Sidani, E. *et al.* 2008. Manipulating critical period closure across different sectors of the primary auditory cortex. *Nat. Neurosci.* **11:** 957–965.
33. Chang, E.F. & M.M. Merzenich. 2003. Environmental noise retards auditory cortical development. *Science* **300:** 498–502.
34. Philbin, M. K., D. D. Ballweg & L. Gray. 1994. The effect of an intensive care unit sound environment on the development of habituation in healthy avian neonates. *Dev. Psychobiol.* **27:** 11–21.
35. Marler, P. *et al.* 1973. Effects of continuous noise on avian hearing and vocal development. *Proc. Natl. Acad. Sci. USA* **70:** 1393–1396.
36. Chin, B.B. *et al.* 2002. Standardized uptake values in 2-deoxy-2-[18F]fluoro-D-glucose with positron emission tomography. Clinical significance of iterative reconstruction and segmented attenuation compared with conventional filtered back projection and measured attenuation correction. *Mol. Imaging. Biol.* **4:** 294–300.
37. Withington-Wray, D.J. *et al.* 1990. The maturation of the superior collicular map of auditory space in the guinea pig is disrupted by developmental auditory deprivation. *Eur. J. Neurosci.* **2:** 693–703.
38. Fifer, W.P. & C.M. Moon. 1994. The role of mother's voice in the organization of brain function in the newborn. *Acta Paediatr. Suppl.* **397:** 86–93.
39. Shahidullah, S. & P.G. Hepper. 1994. Frequency discrimination by the fetus. *Early Hum Dev.* **36:** 13–26.
40. deRegnier, R.A. *et al.* 2002. Influences of postconceptional age and postnatal experience on the development of auditory recognition memory in the newborn infant. *Dev. Psychobiol.* **41:** 216–225.
41. Neville, H. & D. Bavelier. 2002. Human brain plasticity: evidence from sensory deprivation and altered language experience. *Prog. Brain. Res.* **138:** 177–188.
42. Klinke, R. *et al.* 2001. Plastic changes in the auditory cortex of congenitally deaf cats following cochlear implantation. *Audiol. Neurootol.* **6:** 203–206.
43. Kral, A. *et al.* 2002. Hearing after congenital deafness: central auditory plasticity and sensory deprivation. *Cereb. Cortex* **12:** 797–807.
44. Iyengar, S. & S.W. Bottjer. 2002. The role of auditory experience in the formation of neural circuits underlying vocal learning in zebra finches. *J. Neurosci.* **22:** 946–958.
45. Lu, J. *et al.* 2008. Early auditory deprivation alters expression of NMDA receptor subunit NR1 mRNA in the rat auditory cortex. *J. Neurosci. Res.* **86:** 1290–1296.
46. Bi, C. *et al.* 2006. The effect of early auditory deprivation on the age-dependent expression pattern of NR2B mRNA in rat auditory cortex. *Brain Res.* **1110:** 30–38.
47. Xu, J. *et al.* 2009. Early auditory enrichment with music enhances auditory discrimination learning and alters NR2B protein expression in rat auditory cortex. *Behav. Brain Res.* **196:** 49–54.
48. Cai, R. *et al.* 2009. Environmental enrichment improves behavioral performance and auditory spatial representation of primary auditory cortical neurons in rat. *Neurobiol. Learn. Mem.* **91:** 366–376.
49. Bose, M. *et al.* 2010. Effect of the environment on the dendritic morphology of the rat auditory cortex. *Synapse* **64:** 97–110.
50. Abrams, R. M. & K.J. Gerhardt. 2000. The acoustic environment and physiological responses of the fetus. *J. Perinatol.* **20:** S31—S36.
51. Darcy, A.E., L.E. Hancock & E.J. Ware. 2008. A descriptive study of noise in the neonatal intensive care unit: ambient levels and perceptions of contributing factors. *Adv. Neonatal Care* **8:** S16–S26.

52. Williams, A. L., W. van Drongelen & R.E. Lasky. 2007. Noise in contemporary neonatal intensive care. *J. Acoust. Soc. Am.* **121:** 2681–2690.
53. Lasky, R.E. & A.L. Williams. 2009. Noise and light exposures for extremely low birth weight newborns during their stay in the neonatal intensive care unit. *Pediatrics* **123:** 540–546.
54. Antonucci, R., A. Porcella & V. Fanos. 2009. The infant incubator in the neonatal intensive care unit: unresolved issues and future developments. *J. Perinat. Med.* **37:** 587–598.
55. Kirchner, L. *et al.* 2011. In vitro comparison of noise levels produced by different CPAP generators. *Neonatology* **101:** 95–100.
56. Altuncu, E. *et al.* 2009. Noise levels in neonatal intensive care unit and use of sound absorbing panel in the isolette. *Int. J. Pediatr. Otorhinolaryngol.* **73:** 951–953.
57. Committee on Environmental Health. 1997. Noise: a hazard for the fetus and newborn—American Academy of Pediatrics. *Pediatrics* **100:** 724–727.
58. Brown, G. 2009. NICU noise and the preterm infant. *Neonatal Netw.* **28:** 165–173.
59. Bremmer, P., J.F. Byers & E. Kiehl. 2003. Noise and the premature infant: physiological effects and practice implications. *J. Obstet. Gynecol. Neonatal Nurs.* **32:** 447–454.
60. Blackburn, S. 1998. Environmental impact of the NICU on developmental outcomes. *J. Pediatr. Nurs.* **13:** 279–289.
61. Field, T. *et al.* 1979. Cardiac and behavioral responses to repeated tactile and auditory stimulation by preterm and term neonates. *Develop. Psychol.* **15:** 406–416.
62. Vranekovic, G. *et al.* 1974. Heart rate variability and cardiac response to an auditory stimulus. *Biol. Neonate* **24:** 66–73.
63. Zahr, L.K. & S. Balian. 1995. Responses of premature infants to routine nursing interventions and noise in the NICU. *Nurs. Res.* **44:** 179–85.
64. Johnson, A.N. 2001. Neonatal response to control of noise inside the incubator. *Pediatr. Nurs.* **27:** 600–605.
65. Wharrad, H.J. & A.C. Davis. 1997. Behavioural and autonomic responses to sound in pre-term and full-term babies. *Br. J. Audiol.* **31:** 315–329.
66. Strauch, C., S. Brandt & J. Edwards-Beckett. 1993. Implementation of a quiet hour: effect on noise levels and infant sleep states. *Neonatal. Netw.* **12:** 31–35.
67. Wachman, E.M. & A. Lahav. 2010. The effects of noise on preterm infants in the NICU. *Arch. Dis. Child Fetal. Neonatal Ed* **96:** 305–309.
68. White, R.D. 2011. The newborn intensive care unit environment of care: how we got here, where we're headed, and why. *Semin. Perinatol.* **35:** 2–7.
69. Carter, B.S., A. Carter & S. Bennett. 2008. Families' views upon experiencing change in the neonatal intensive care unit environment: from the 'baby barn' to the private room. *J. Perinatol.* **28:** 827–829.
70. Stevens, D.C. *et al.* 2010. Neonatal intensive care nursery staff perceive enhanced workplace quality with the single-family room design. *J. Perinatol.* **30:** 352–358.
71. Lester, B.M. *et al.* 2011. Infant neurobehavioral development. *Semin. Perinatol.* **35:** 8–19.
72. Philbin, M.K. & L. Gray. 2002. Changing levels of quiet in an intensive care nursery. *J. Perinatol.* **22:** 455–460.
73. Philbin, M.K. 2004. Planning the acoustic environment of a neonatal intensive care unit. *Clin. Perinatol.* **31:** 331–352, viii.
74. Brandon, D.H., D.J. Ryan & A.H. Barnes. 2008. Effect of environmental changes on noise in the NICU. *Adv Neonatal. Care* **8:** S5–S10.
75. Laudert, S. *et al.* 2007. Implementing potentially better practices to support the neurodevelopment of infants in the NICU. *J. Perinatol.* **27**(Suppl 2): S75–S93.
76. Milette, I. 2010. Decreasing noise level in our NICU: the impact of a noise awareness educational program. *Adv. Neonatal Care* **10:** 343–351.
77. Zahr, L.K. & J. de Traversay. 1995. Premature infant responses to noise reduction by earmuffs: effects on behavioral and physiologic measures. *J. Perinatol.* **15:** 448–455.
78. Aita, M. & C. Goulet. 2003. Assessment of neonatal nurses' behaviors that prevent overstimulation in preterm infants. *Intensive Crit. Care Nurs.* **19:** 109–118.
79. Ferber, S.G. & I.R. Makhoul. 2004. The effect of skin-to-skin contact (kangaroo care) shortly after birth on the neurobehavioral responses of the term newborn: a randomized, controlled trial. *Pediatrics* **113:** 858–865.
80. Charpak, N. *et al.* 2005. Kangaroo mother care: 25 years after. *Acta. Paediatr.* **94:** 514–522.
81. Gale, G., L. Franck & C. Lund. 1993. Skin-to-skin (kangaroo) holding of the intubated premature infant. *Neonatal Netw.* **12:** 49–57.
82. Anderson, G.C. 1991. Current knowledge about skin-to-skin (kangaroo) care for preterm infants. *J. Perinatol.* **11:** 216–226.
83. Conde-Agudelo, A., J.M. Belizan & J. Diaz-Rossello. 2011. Kangaroo mother care to reduce morbidity and mortality in low birthweight infants. *Cochrane Database Syst. Rev.* CD002771.
84. Tessier, R. *et al.* 1998. Kangaroo mother care and the bonding hypothesis. *Pediatrics* **102:** e17.
85. Nyqvist, K.H. *et al.* 2011. Towards universal Kangaroo mother care: recommendations and report from the First European Conference and Seventh International Workshop on Kangaroo Mother Care. *Acta Paediatr.* **99:** 820–826.
86. Krueger, C. *et al.* 2010. Maternal voice and short-term outcomes in preterm infants. *Dev. Psychobiol.* **52:** 205–212.
87. Segall, M.E. 1972. Cardiac responsivity to auditory stimulation in premature infants. *Nurs. Res.* **21:** 15–19.
88. Krueger, C. 2010. Exposure to maternal voice in preterm infants: a review. *Adv. Neonatal. Care* **10:** 13–18; quiz 19–20.
89. Cevasco, A.M. & R.E. Grant. 2005. Effects of the pacifier activated lullaby on weight gain of premature infants. *J. Music Ther.* **42:** 123–139.
90. Standley, J.M. *et al.* The effect of music reinforcement for non-nutritive sucking on nipple feeding of premature infants. *Pediatr. Nurs.* **36:** 138–145.
91. Caine, J. 1991. The effects of music on the selected stress behaviors, weight, caloric and formula intake, and length of hospital stay of premature and low birth weight neonates in a newborn intensive care unit. *J. Music Ther.* **28:** 180–192.

92. Chapman, J.S. 1975. *The relation between auditory stimulation of short gestation infants and their gross motor limb activity*. New York University. New York.
93. Coleman, J. M., R. R. Pratt, R. A. Stoddard, D. R. Gerstmann, & H. Abel. 1997. The effects of male and female singing and speaking voices on selected physiological and behavioral measures of premature infants in the intensive care unit. *Inter. J. Arts Medi.* **5:** 4–11.
94. Collins, S.K. & K. Kuck. 1991. Music therapy in the neonatal intensive care unit. *Neonatal Netw.* **9:** 23–26.
95. Malloy, G. 1979. The relationship between maternal and musical auditory stimulation and the developmental behavior of premature infants. *Birth Defects: Original Article Series* **15:** 81–98.
96. Trehub, S.E. 2001. Musical predispositions in infancy. *Ann. N.Y. Acad. Sci.* **930:** 1–16.
97. Loewy, J.V. 1995. The musical stages of speech: a developmental model of pre-verbal sound making. *Music Ther.* **13:** 47–73.
98. Lai, H.L. *et al.* 2006. Randomized controlled trial of music during kangaroo care on maternal state anxiety and preterm infants' responses. *Int. J. Nurs. Stud.* **43:** 139–46.
99. Schlez, A. *et al.* 2011. Combining kangaroo care and live harp music therapy in the neonatal intensive care unit setting. *Isr. Med. Assoc. J.* **13:** 354–358.
100. Neal, D.O. & L.L. Lindeke. 2008. Music as a nursing intervention for preterm infants in the NICU. *Neonatal Netw.* **27:** 319–327.
101. Field, T. 1990. Alleviating stress in newborn infants in the intensive care unit. *Clin. Perinatol.* **17:** 1–9.
102. Symington, A. & J. Pinelli. 2003. Developmental care for promoting development and preventing morbidity in preterm infants. *Cochrane Database Syst. Rev.* CD001814.
103. Als, H. *et al.* 1986. Individualized behavioral and environmental care for the very low birth weight preterm infant at high risk for bronchopulmonary dysplasia: neonatal intensive care unit and developmental outcome. *Pediatrics* **78:** 1123–1132.
104. Panagiotidis, J. & A. Lahav. 2010. Simulation of prenatal maternal sounds in NICU incubators: a pilot safety and feasibility study. *J. Matern Fetal Neonatal Med* **23**(Suppl 3): 106–109.

Ann. N.Y. Acad. Sci. ISSN 0077-8923

ANNALS OF THE NEW YORK ACADEMY OF SCIENCES
Issue: *The Neurosciences and Music IV: Learning and Memory*

Musical experience, plasticity, and maturation: issues in measuring developmental change using EEG and MEG

Laurel J. Trainor

Department of Psychology, Neuroscience and Behavior, McMaster Institute for Music and the Mind, McMaster University, Ontario, Canada

Address for correspondence: Laurel J. Trainor, Department of Psychology, Neuroscience and Behavior, McMaster University, 1280 Main Street West, Hamilton, ON L8S 4B2. ljt@mcmaster.ca

The neuroscientific study of musical behavior has become a significant field of research during the last decade, and reports of this research in the popular press have caught the imagination of the public. This enterprise has also made it evident that studying the development of musical behavior can make a significant contribution to important questions in the field, such as the evolutionary origins of music, cross-cultural similarity and diversity, the effects of experience on musical processing, and relations between music and other domains. Studying musical development brings a unique set of methodological issues. We discuss a select set of these related to measurement of the electroencephalogram (EEG) and magnetoencephalogram (MEG). We use specific examples from our laboratory to illustrate the types of questions that can be answered with different data analysis techniques.

Keywords: music development; EEG; MEG; artifact; oscillatory responses; beta band; gamma band

Origins of electroencephalogram and magnetoencephalogram activity

Electroencephalography (EEG) measured at the surface of the head is largely blind to action potentials and mainly reflects the summation of postsynaptic field potentials.[1–4] When a neurotransmitter acts on a single cell, it creates an electric dipole around that cell. For example, an excitatory neurotransmitter can cause current to flow into the apical dentrites of a cell, creating a net negativity outside this area of the cell. At the same time, current will flow out of the cell in the region of the cell body and basal dentrites creating a positivity outside the cell at this point. Together, these two actions create a small dipole. When many (hundreds of thousands) neurons are aligned and depolarize at the same time, the small dipoles they create will sum into a field that is large enough to measure at the surface of the head.[5–9] Because pyramidal cells tend to be aligned and perpendicular to the cortical surface, it is likely that EEG largely reflects their activity. It is important to note that to the extent that neurons are differentially excitatory and inhibitory and to the extent that they are oriented in different directions, as can happen with the folding of the cortex, their postsynaptic activity can cancel. Thus, much of the neural activity in the brain is opaque to EEG recordings.[10]

A further complication in interpreting EEG activity measured at the surface of the head is that cortical tissue is conductive and thus electrical fields will to some extent spread in all directions, blurring spatial specificity at the surface of the head. In addition, different components of the head, such as the skull and eye holes, have different effects on volume conductance, and thus dipoles originating in different brain regions will be differentially distorted. From a developmental perspective, the skull is thinner and the fontanels do not close until some months after birth, leading to age-specific differences in the distortion of the fields.[11] These factors complicate determination of the source location of EEG activity at different ages.

An electrical dipole has an associated magnetic field perpendicular to the electrical dipole, oriented according to the right-hand rule. Such magnetic fields can be measured with magnetoencephalography (MEG), which uses an array of superconducting quantum interference devices (SQIDS).[12–14] Near the surface of the head, the magnetic fields

doi: 10.1111/j.1749-6632.2012.06444.x
Ann. N.Y. Acad. Sci. 1252 (2012) 25–36 © 2012 New York Academy of Sciences.

generated by the brain are on the order of a few femtotesla, which is several orders of magnitude smaller than magnetic fields in the ambient environment, so a magnetically shielded room and active noise cancellation are necessary. Magnetic fields are not smeared by brain tissue, so determination of the brain source location of measured activity can be more precise with MEG than EEG, but also because of this, sources oriented radially to the surface of the head cannot be seen with MEG.

EEG and MEG have certain advantages over functional magnetic resonance imaging (fMRI). They have excellent temporal resolution of less than a millisecond (compared with several seconds for fMRI), so long as appropriate sampling rates are used. Unlike fMRI, both EEG and MEG are silent, which is particularly advantageous for auditory work. EEG also has the ethical advantage over fMRI of being noninvasive and virtually risk free. EEG has some advantages over MEG for developmental research. In particular, it is difficult to get infants and young children to stay still, and EEG is more tolerant of movement artifacts. Although segments of the EEG on which movement occurred might need to be eliminated (because the EEG sensors sit on the head), once movement has stopped, clean recordings can resume. With MEG, however, the child sits or lies down with his or her head in the rigid structure of the dewar containing the SQIDS, so if the head moves significantly with respect to the dewar, the MEG recordings cannot be continued. EEG is also much cheaper to purchase and operate, and portable EEG machines are becoming common, allowing easier access to special populations. On the other hand, because MEG is transparent to head tissues, it has better spatial resolution of the measured signals.

Animal studies using electrodes inserted into cortical tissue indicate that extracellular electrical field potential patterns show complex patterns of electrical sources and sinks across the six cortical layers.[7,8] In general, however, it appears that depolarizations in deeper layers with passive returns above will appear on the surface as positivities, whereas depolarizations in more superficial layers with passive returns below will appear on the surface as negativities.[5,6,8] This is important for interpreting EEG in infants and young children as the cortex matures in a layer-specific fashion. For example, although cell bodies in auditory areas are essentially all in place by birth, maturation of neurofilament expression that enables fast axon potentials and meaningful communication between neurons occurs in certain layers before others.[15] Neurofilament is expressed only in layer I at birth. Its expression in deeper cortical layers (lower III, IV, V, and VI) can be seen by 4 months and reaches adult levels by 3 to 5 years of age. Neurofilament is not expressed in upper layers (I, II, and upper III) until 5 years of age, and does not reach adult levels until about 12 years of age. Thus it would be expected that event-related potentials (ERPs) derived from EEG recordings would look very different early in development and, in particular, contain more positive components early on.[9] Indeed this is what is generally found.[16–24]

Issues in recording EEG early in development

There are two main issues specific to measuring EEG in infants and young children. First, attention spans are limited and young participants tire quickly, so experiments must be short. The number of trials needed depends on a number of factors including the size of the component of interest, the efficiency of the signal processes techniques used for analysis, and the amount of noise or artifact in the recordings. The faster the electrodes can be placed on the head and impedances checked, the more time will be left for the EEG recording. In this regard, high impedance systems involving nets of electrodes imbedded in sponges that are dipped into a conducting saline solution can be applied much more quickly than electrodes in low impedance systems requiring electrogel and abrasion of the skin. However, the former may be more subject to noise and may not be ideal for measuring small fast components such as those originating from the brain stem. Second, it can be difficult to keep young participants from moving excessively. Because muscle movements generate electrical field potentials an order of magnitude larger than potentials originating in the brain when measured at the surface of the head, the more the participant moves, the noisier will be the EEG recordings. Interestingly, young infants tend to move less than older infants, with the most challenging ages in this regard being between about 1 and 3 years. During electrode application and EEG recording, it is helpful to have one researcher whose job is to distract infants with toys, videos appropriate for infants, soap bubbles, and so on. If the infant is sitting on the parent's lap, the parent can also help by gently holding the infant's hands away from the electrodes. Figure 1 shows how to happily and efficiently place an electrode net on an infant.

Data preprocessing: dealing with artifact in developmental recordings

In addition to brain activity related to the processes of interest, measured EEG signals also contain "noise," largely in the forms of movement artifact and brain activity irrelevant to the processes of interest.[1,10] In order to see the activity of interest, typically many trials are presented and the resulting activity averaged across trials. Assuming that the timing of the noise is unrelated to that of the signal, the more the trials are averaged together, the better the signal-to-noise ratio. However, given that the amplitude of movement artifact can be an order of magnitude larger than that of the signal, additional methods are typically necessary to get a good signal-to-noise ratio within a reasonable number of trials. In the most common approach, here called conventional trial rejection (CTR), entire trials containing large amplitude artifact at any electrode are eliminated from the average. This approach works well for most adult data for which there are few trials contaminated with movement artifact. However, for data from infants or young children, CTR can result in the elimination of most of the data. A second approach in common usage is specific to the elimination of eye movements and involves modeling the dipolar sources of eye movements.[25] In addition to the recordings of experimental interest, EEG can be recorded for each subject in response to eye blinks and eye movements, and EEG sources related to eye movements can be modeled. These sources can then be eliminated subject by subject in the EEG data from the experiment of interest. Unfortunately this method does not work well with infants and young children whose eye blinks and eye movements are variable and not temporally confined.[26] There is also the problem of eliciting eye movements, and spending time recording eye movements takes time away from the recordings of experimental interest. A third approach is to perform independent component analysis (ICA) on the EEG data.[27,28] In adult data, the first few (largest) components of the analysis will be noise and these can be eliminated to reveal the signals of interest. Unfortunately, infant artifact often behaves differently. Rather than consisting of predominantly eye movements, infants can make sudden whole head movements, jaw movements, and scrunch the backs of their necks. Furthermore, such movements can cause an electrode to temporarily make a bad connection with the scalp. Fujioka *et al.* have illustrated that ICA does not typically work well with infant data.[29]

He *et al.*[16] introduced a method of independent channel rejection (ICR) whereby, if a particular electrode shows a high amplitude artifact on a particular trial, the data from that electrode are eliminated on that trial, but data from "clean" electrodes are kept. Thus, ICR assumes that electrodes can be differentially contaminated with artifact on the same trial. With this method, much less data are eliminated than with CTR, in which data from all electrodes are eliminated when there is a contaminated electrode. Mourad *et al.*[30] generalized the idea of ICR and developed the artifact blocking (AB) algorithm in which all trials are retained, but artifact is "blocked" or reduced as follows. This algorithm assumes that amplitudes greater than a certain threshold reflect artifact and attempts to reduce them toward zero by estimating a smoothing matrix that, when multiplied by the original EEG data matrix (electrode by time), creates a "clean" version of the EEG data matrix. In some regards, AB is similar to interpolation, in which an eliminated electrode is estimated from the surrounding electrodes, but it has the advantage of being completely atheoretical (does not require any knowledge of EEG conduction or brain and skull properties) and is computationally much less demanding.

Fujioka *et al.* [29] extensively compared ICR, CTR, and AB methods. They recorded real infant EEG activity in the absence of a stimulus and then repeatedly imbedded an artificially generated, and therefore known, EEG signal. They then analyzed the data using each of the three artifact rejection methods and compared the ability of each at extracting the known EEG signal. They found that ICR and AB were much better than CTR at extracting the EEG signal. Correlations between the embedded and extracted signals were much higher and residual variance much lower for ICR and AB than for CTR. Furthermore, CTR showed greater spatial distortion across the scalp than the other methods and AB showed the least spatial distortion, which

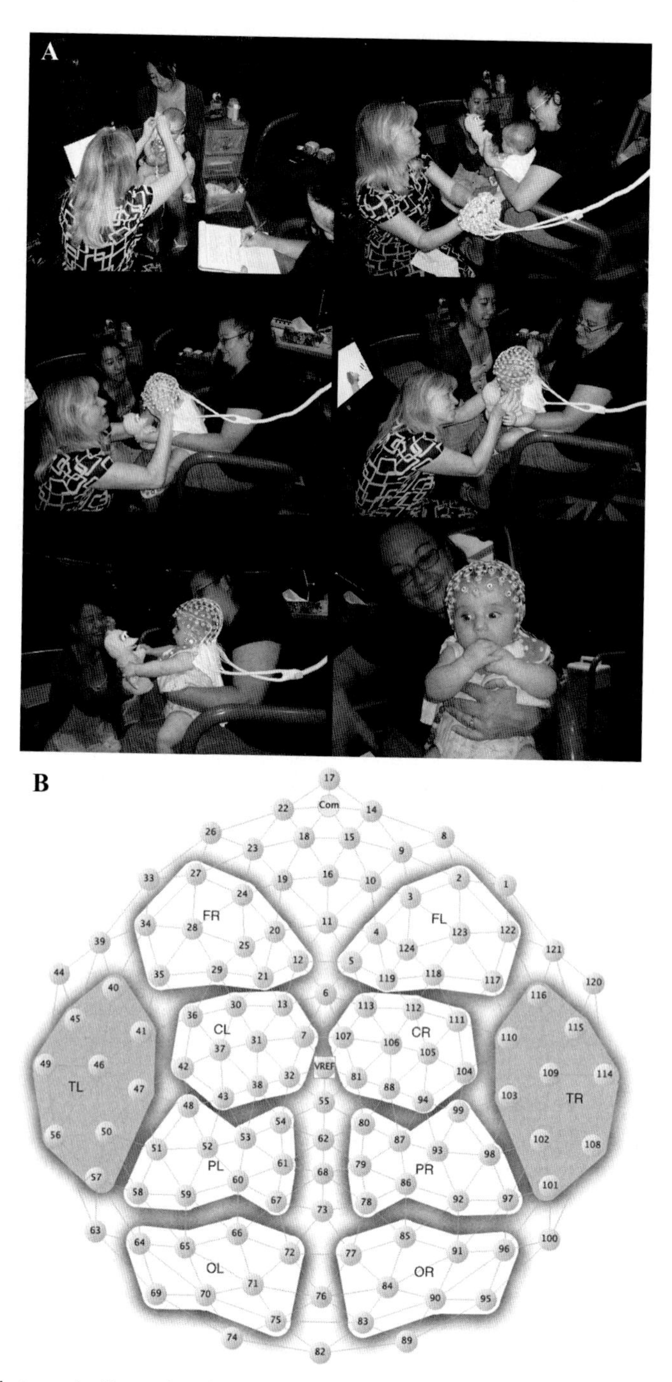

Figure 1. (A) Series of photographs illustrating placing an electrode net on an infant. The circumference of the infant's head is measured to determine net size while the mother fills out a questionnaire. Then one experimenter distracts the infant with toys while the second experimenter places the net on the infant's head and adjusts the placement of the electrodes. Once the net is on, it is quite comfortable and most infants are content. Photos by Nicole Folland. (B) For data analysis, the channels across the head can be averaged within each area to increase signal-to-noise ratio. From work by He *et al.*[16] Reprinted with permission of MIT Press Journals.

is particularly important when attempting to locate the sources of activity in the brain. To a large extent, the differences across methods appeared to be related to the amount of data that remained for analysis after artifact rejection. Thus, for adult data, AB might be less useful, but for data from infants and young children, it can offer a marked improvement over traditional methods.

Using EEG and MEG to understand musical development: examples using different data analysis techniques

Time waveforms

EEG activity reflecting the processing of a sound event is termed an event-related potential (ERP). In the time domain, ERPs consist of a series of positive or negative deflections (components) across time from the onset of the sound that reflect activity from the nuclei of the brainstem (first 15 ms after sound onset), primary auditory cortex (middle latency responses; up to 50 ms), and areas beyond (late potentials; 50 ms and later). From a developmental perspective, differences across age can be measured in all components. From a musical perspective, effects of musical training can be seen in brainstem encoding,[31] middle latency responses,[32] as well as in various components of the late potentials.[33–37] A few examples from our lab will be described to illustrate the types of questions that can be addressed by examining time waveforms.

Perhaps one of the most interesting findings in the development of auditory ERPs is that components originating in secondary auditory cortex, N1 and P2, show a very protracted developmental trajectory. Note that because N1 and P2 are processed in auditory areas around the Sylvian fissure, the fields that they produce at the surface of the head are oriented such that for N1, a negativity is seen at frontal sites in conjunction with a positivity at posterior sites, and vice versa for P2. Although obligatory responses to sound in adults, these components are so small as to be difficult to measure in children 4 to 5 years of age. With increasing age, they increase in amplitude and decrease in latency, reach a maximum amplitude around 10 to 12 years of age, and subsequently decrease in amplitude, reaching stable adult levels in the late teenage years (see Fig. 2A).[36,38,39] The development of these components appears to be affected by musical experience in that they are larger in adult musicians than nonmusicians. Furthermore, they are larger in 4- to 5-year-old children taking music lessons compared to children not taking music lessons (see Fig. 2B).[36] In sum, examining the developmental trajectories and effects of musical experience on various components in the time domain can yield valuable information about when processing develops for different musical features, and differential effects of musical experience at different ages.

Difference waves and mismatch responses

As discussed in the previous section, some EEG components that are obligatory responses to sound in adults, such as N1, are very small or nonexistent during infancy.[36,38,40,41] Interestingly, although N1 originates in superficial layers of auditory cortex, it likely reflects feedback from other cortical areas[6,8,42] and is sensitive to attentional manipulations.[43,44] On the other hand, automatic (preattentive) responses to occasional changes in an ongoing stream of sounds elicit mismatch responses that are likely processed largely within auditory cortex. In adults, such changes elicit a frontal negativity at the surface of the head between 130 ms and 250 ms accompanied by a reversal at occipital sites.[45,46] A mismatch response can also be elicited in young infants in response to occasional changes in pitch,[16,17,21,47–50] duration,[51,52] and tonal patterns.[18,53] However, in young infants, only a slow frontally positive mismatch response is typically evident.[9] The age at which the negative response emerges appears to depend to some extent on the feature that is changed. For occasional changes in pitch, at 2 months the slow positive response dominates, but the negative response can be seen at 3 months and is quite robust at 4 months (Fig. 3), whereas for temporal gap detection and changes in melodic patterns, the negative response does not emerge until later.[51,53] Interestingly, at intermediate ages, both the positive and negative mismatch responses can be seen at the same time, in the same infants, suggesting that they have different cortical origins.[16] With respect to musical training, mismatch negativity is larger in musicians than in nonmusicians for changes in melodies in transposition without accompaniment[54] and in polyphonic contexts.[55] There are few studies involving musical experience in young infants. However, one study exposed infants for 20 min a day for a week to melodies in either guitar or marimba timbre.[50] Subsequently, mismatch

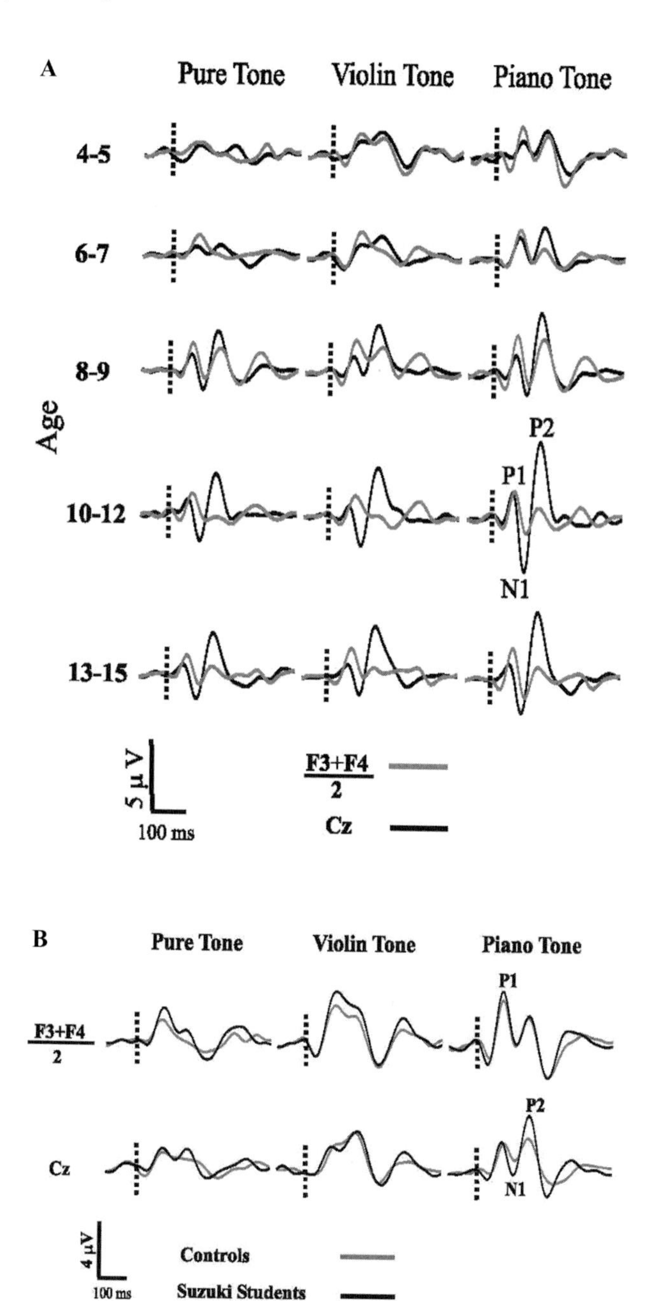

Figure 2. Development of ERPs elicited by pure, violin, and piano tones. (A) P1 reaches a maximum at 8 to 9 years of age at frontal (F2, F4) sites and diminishes thereafter. N1 reaches a maximum at 10 to 12 years of age at the vertex (Cz) and diminishes thereafter. (B) P1 and N1 are enhanced for piano tones in Suzuki piano students. The dotted vertical line represents the onset of the stimulus. From work by Shahin *et al.*[35] Adapted with permission of Wolters Kluwer/Kippincott, Williams & Wilkins.

responses were measured to quartertone changes in the pitch of a repeating tone, in one block with tones in guitar timbre and in another block with tones in marimba timbre. Differential responses were found favoring the timbre to which infants had been familiarized.

In sum, mismatch responses provide a rich context in which to measure the development of many

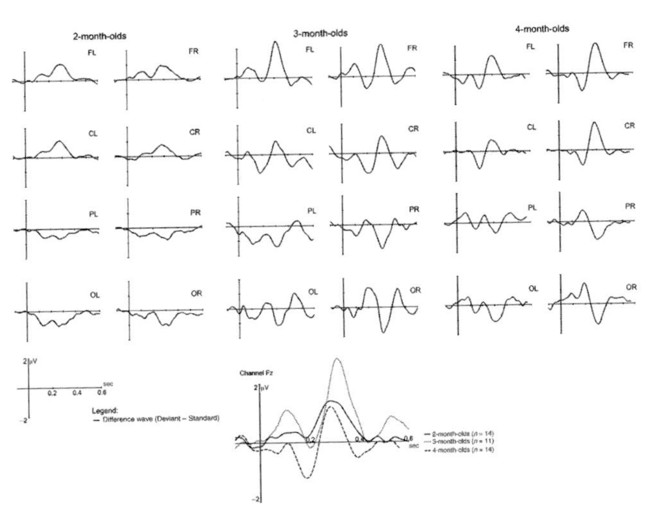

Figure 3. Development of mismatch responses to pitch change in 2-, 3-, and 4-month-old infants. Grand average difference waves are shown for each age group (filtered between 0.5 and 20 Hz), illustrating the slow positive difference wave at 2 months of age and the emergence of the mismatch negativity with increasing age. Difference waves at electrode Fz are overlaid for the three age groups at the bottom. The vertical axis represents the onset of the sound. From work by He *et al.*[16] Reprinted with permission of MIT Press Journals.

aspects of musical perception and the effects of musical experience at different ages.

Oscillatory responses

Even in the absence of specific stimulation, EEG and MEG recordings reveal ongoing oscillatory brain activity thought to reflect communication between networks of neurons. Indeed, one view is that the evoked potentials described in the last two sections reflect phase alignment of ongoing oscillatory activity that becomes entrained for analyzing a particular input.[56] Thus, changes in oscillatory rhythms in response to an auditory stimulus can reveal important aspects of stimulus processing. Oscillatory responses have been classified into five main frequency ranges, delta (0–4 Hz), theta (4–8), alpha (8–12), beta (12–30), and gamma (30–100), roughly according to proposed associated brain functions. A full discussion is beyond the scope of this paper, but can be found in recent reviews.[57,58] However, oscillatory responses change greatly over development, can be affected by attention, and can also reflect-specific effects of experience and training.[59–63] Thus we predict that they will be used increasingly in the study of musical development. Here, we briefly give examples in the alpha, beta, and gamma frequency ranges.

In the resting state, alpha is a dominant rhythm in the adult brain, and it decreases in amplitude in selective regions with stimulus presentation in different modalities.[64–65] Similar desynchronization can be seen in infancy, such that alpha-band oscillations originating from auditory cortex (termed tau) decrease in amplitude with auditory stimulus presentation.[66] However, the dominant tau frequency suppressed by sound changes with development: at 4 months of age it is 4 Hz, and by 12 months of age it is 6 Hz. Effects of experience on this development remain unknown, but further studies of the tau rhythm have the potential to increase our understanding of the development of musical sound processing.

Beta band activity has long been of interest as it is a dominant frequency in the motor system and its amplitude modulates with motor movement.[67] Recent studies indicate that the amplitude of beta activity originating in auditory cortex is modulated by the presentation of a steady beat.[68] In particular, beta amplitude decreases after each beat and rebounds before the onset of the next beat (Fig. 4). This rebound occurs across different beat tempi, indicating that the brain is predicting the timing of the next beat.[69] Applying such analyses to developmental data has the potential to further our understanding of developmental and experiential aspects of musical rhythm processing.

Gamma band activity is of interest in the auditory system as it is thought to reflect attention,

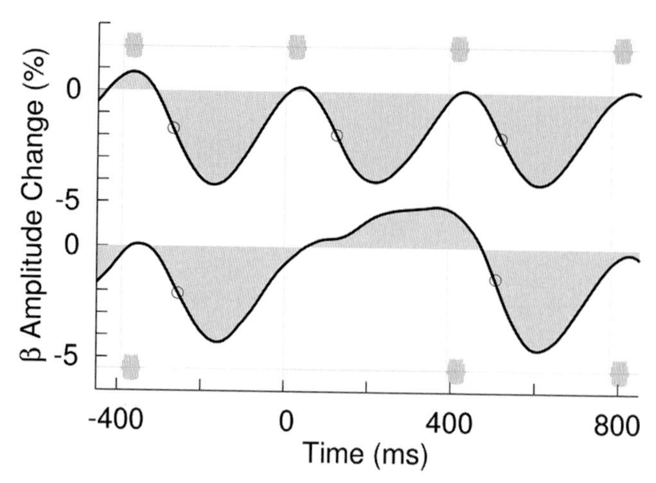

Figure 4. Beta modulation by sound presentation. The time series for event-related changes in beta (15–20 Hz) activity is shown in response to a regular stimulus sequence (top) and in response to the omission of an expected stimulus (bottom). The sound stimulus is shown above for the regular sequence and below for the omission sequence. Beta activity decreases in amplitude following stimulus onset and rebounds prior to onset of the next stimulus. From work by Fujioka *et al.*[68] Reprinted with permission from John Wiley & Sons.

anticipation, and expectation[70–72] and the binding of auditory features into a unitary percept.[73] Gamma band activity (indeed, any oscillatory activity) can be analyzed in two different categories, evoked and induced. Evoked activity is phase locked to the onset of a stimulus, and can therefore be seen by averaging the EEG or MEG signal over many trials. Induced activity, on the other hand, is also modulated by the presentation of a stimulus, but in this case it is not phase locked to the onset of the stimulus such that averaging across trials leads to canceling of the signal.[74] Thus, this type of activity must be analyzed on a trial-by-trial basis. However, because it is thought to reflect the entrainment of ongoing intrinsic brain activity with an external stimulus, it is of great interest for understanding attention and the interaction of top–down and bottom–up processes. Shahin *et al.* examined the effects of musical experience on induced gamma band activity.[61] They found that the presentation of a musical tone produced waves of increases in gamma band activity that lasted for at least half a second after stimulus onset. Furthermore, induced gamma band activity was greater in adult musicians compared to nonmusicians. Perhaps of most interest, two groups of 4-year-old children, one beginning music lessons and the other engaged in an equal amount of other extracurricular activity such as playing sports, were measured at the onset of lessons and one year later. Neither group showed any significant gamma band activity at first measurement. At the second measurement, only the group undergoing musical training showed significant gamma band activity, such that the groups did not differ at first measurement, but did differ at the second measurement (Fig. 5).

In sum, to date oscillatory activity has not been used extensively in studies of musical development. However, it holds much promise as a technique that can reveal how the brain processes music and the effects of development and experience.

Machine-learning approaches

The predominant approach to EEG and MEG data analysis is to identify features and components, study how they are affected by various manipulations, and relate them to processes of interest. Machine-learning approaches differ in that they are atheoretical and examine a vast array of data features simultaneously to determine those that best classify according to an outcome variable. They have been used with EEG data, for example, to detect seizure in infants[75] and epileptic adults,[76] and to predict responses of schizophrenic patients to different medications.[77] In terms of development, machine learning was recently applied to EEG data measured in response to musical sounds in order to predict the

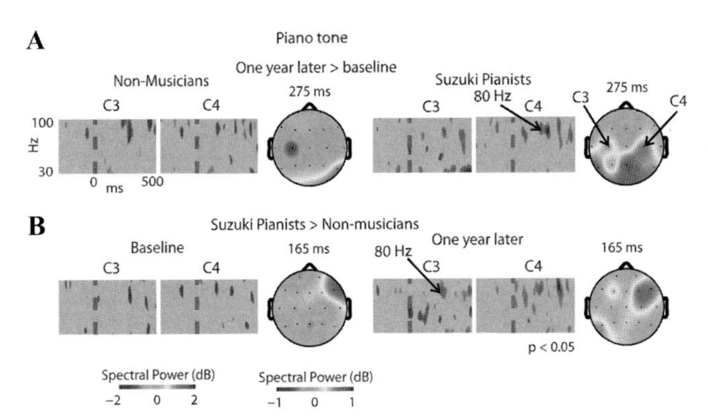

Figure 5. Development of induced gamma band activity in 4- to 5-year-old children. Spectral power of gamma band activity at central channels (C3/C4) and topographies (middle) at the peak amplitude. (A) Initial measurements contrasted with measurements one year later in nonmusician children (left) and Suzuki pianists (right) in response to piano tones. Only the Suzuki group shows evidence of gamma-band activity (periodic red signals) after one year of music lessons. (B) Contrasts between the two groups show that at the initial measurement, the groups do not differ in gamma-band activity (left) but that they do differ after one year of music lessons (right). The dotted line shows onset of the piano tones. From work by Shahin *et al.*[61] Reprinted with permission from Elsevier.

age of infants.[78] Such techniques show potential for the study of effects of musical training on brain development.

Voxel-based source waveform analysis of MEG data

As discussed previously, EEG and MEG have the advantage over fMRI of very fine temporal resolution that allows the study of oscillatory brain responses. fMRI remains superior for determining the locations of activity in the brain, but much research has been invested to develop better techniques for determining the source locations of EEG and MEG activity measured at the surface of the head. As an example, Fujioka *et al.* have used a procedure for estimating the time waveforms of MEG data in every $5 \times 5 \times 5$ mm voxel across the brain.[79] Briefly, it involves using synthetic aperture magnetometry, a spatial beamforming technique,[80] in conjunction with a common head model derived from MRI data, and applying this to time domain-averaged MEG waveforms.[81] Such techniques have successfully localized activity in auditory and motor cortices[69,82] as well as in deeper sources, including the hippocampus[69,83] and amygdala.[84]

As an example, the beta band data described previously, in which the presentation of a steady auditory beat-evoked amplitude modulations in beta-band activity in auditory areas was subject to a whole-head analysis.[69] Interestingly, even though the stimulus was auditory and there was no movement or suggestion to move, correlated modulation of activity in the beta band was seen across a wide range of motor-related areas including sensorimotor cortex, inferior frontal gyrus, supplementary motor area, and cerebellum.

At present, such techniques have not yet been applied to questions in musical development, nor have they yet been developed successfully for EEG data where activity from different sources is smeared to a greater extent at the surface of the head. But it is clear that these techniques are promising for studies of musical development and the effects of experience.

Conclusions

Collecting or analyzing developmental EEG and MEG data is not easy. However, as illustrated by the few examples from our laboratory in this paper, EEG and MEG data can contribute substantially to our understanding of musical development and the effects of musical training on brain development. Furthermore, new data analysis techniques examining oscillatory behavior, classification according to machine-learning approaches, and voxel-based source waveform extraction across the whole head offer great promise for future studies of musical development.

Acknowledgments

The writing of this paper was supported by grants from the Canadian Institutes of Health Research. I thank Andrea Unrau for comments on an earlier draft.

Conflicts of interest

The author declares no conflicts of interest.

References

1. Luck, S.J. 2005. *An Introduction to the Event-Related Potential Technique*. MIT Press. Cambridge, MA.
2. Mitzdorf, U. 1985. Current source-density method and application in cat cerebral cortex: investigation of evoked potentials and EEG phenomena. *Physiol. Rev.* **65:** 37–100.
3. Steinschneider, M. & M. Dunn. 2002. Electrophysiology in developmental neuropsychology. In *Handbook of Neuropsychology.* Vol. **7**. S.J. Segalowitz & I. Rapin, Eds.: 91–146. Elsevier. Amsterdam.
4. Vaughan Jr., H.G. & J.C. Arezzo. 1988. The neural basis of event-related potentials. In *Human Event-Related Potential.* T.W. Picton, Ed.: 45–96. Elsevier. Amsterdam.
5. Creutzfeldt, O. & J. Houchin. 1974. Neuronal basis of EEG waves. In *Handbook of Electroencephalography and Clinical Neurophysiology.* A. Remond, Ed.: 5–55. Elsevier. Amsterdam.
6. Eggermont, J.J. & C.W. Ponton. 2003. Auditory-evoked potential studies of cortical maturation in normal hearing and implanted children: correlations with changes in structure and speech perception. *Acta Otolaryngologica* **123:** 249–252.
7. Fishman, Y.I., D. Reser, J.C. Arezzo & M. Steinschneider. 1998. Pitch vs. spectral encoding of harmonic complex tones in primary auditory cortex of the awake monkey. *Brain Res.* **786:** 18–30.
8. Fishman, Y.I., D.H. Reser, J.C. Arezzo & M. Steinschneider. 2000. Complex tone processing in primary auditory cortex of the awake monkey: II. Pitch vs. critical band representation. *J. Acoust. Soc. Am.* **108:** 247–262.
9. Trainor, L.J. 2008. Event related potential measures in auditory developmental research. In *Developmental Psychophysiology: Theory, Systems and Methods.* L. Schmidt & S. Segalowitz, Eds.: 69–102. Cambridge University Press. New York.
10. Picton, T.W., O.G. Lins & M. Scherg. 1995. The recording and analysis of event-related potentials. In *Handbook of Neuropsychology*. Vol. 10. F. Boller & J. Grafman, Eds.: 3–73. Elsevier. New York.
11. Flemming, L., Y. Wang, A. Caprihan, *et al.* 2005. Evaluation of the distortion of EEG signals caused by a hole in the skull mimicking the fontanel in the skull of human neonates. *Clin. Neurophysiol.* **116:** 1141–1152.
12. Hämäläinen, M., R. Hari, R. Ilmoniemi, *et al.* 1993. Magnetoencephalography—theory, instrumentation, and applications to noninvasive studies of signal processing in the human brain. *Rev. Mod. Phys.* **65:** 413–497.
13. Hansen, P.C., M.L. Kringelbach & R. Salmelin Eds. 2010. *MEG: An Introduction to Methods.* Oxford University Press. New York.
14. Murakami, S. & Y. Okada. 2006. Contributions of principal neocortical neurons to magnetoencephalography and electroencephalography signals. *J. Physiol.* **575:** 925–936.
15. Moore, J.K. & Y.L. Guan. 2001. Cytoarchitectural and axonal maturation in human auditory cortex. *J. Assoc. Res. Oto.* **2:** 297–311.
16. He, C., L. Hotson & L.J. Trainor. 2007. Mismatch responses to pitch changes in early infancy. *J. Cogn. Neurosci.* **19:** 878–892.
17. He, C., L. Hotson & L.J. Trainor. 2009. Maturation of cortical mismatch mismatch responses to occasional pitch change in early infancy: effects of presentation rate and magnitude of change. *Neuropsychologia* **47:** 218–229.
18. He, C., L. Hotson & L.J. Trainor. 2009. Development of infant mismatch responses to auditory pattern changes between 2 and 4 months old. *J. Cogn. Neurosci.* **19:** 878–892.
19. Novak, G.P., D. Kurtzberg, J.A. Kreuzer & H.G. Vaughan Jr. 1989. Cortical responses to speech sounds and their formants in normal infants: maturational sequence and spatiotemporal analysis. *Electroen. Clin. Neuro.* **73:** 295–305.
20. Kushnerenko, E., M. Cheour, R. Čeponienė, *et al.* 2001. Central auditory processing of durational changes in complex speech patterns by newborns: an event-related brain potential study. *Dev. Neuropsychol.* **19:** 83–97.
21. Leppänen, P.H.T., K.M. Eklund & H. Lyytinen. 1997. Event-related brain potentials to change in rapidly presented acoustic stimuli in newborns. *Dev. Neuropsychol.* **13:** 175–204.
22. Molfese, D.L. & V.J. Molfese. 1985. Electrophysiological indexes of auditory-discrimination in newborn infants: the bases for predicting later language development. *Infant Behav. Dev.* **8:** 197–211.
23. Morr, M.L., V.L. Shafer, J.A. Kreuzer & D. Kurtzberg. 2002. Maturation of mismatch negativity in typically developing infants and preschool children. *Ear Hearing* **23:** 118–136.
24. Thomas, D.G. & S.M. Lykins. 1995. Event-related potential measures of 24-hour retention in 5-month-old infants. *Dev. Psychol.* **31:** 946–957.
25. Berg, P. & M. Scherg. 1991. Dipole models of eye movements and blinks. *Electroen. Clin. Neuro.* **79:** 36–44.
26. Bell, M.A & C.D. Wolfe. 2008. The use of the electroencephalogram in research on cognitive development. In *Developmental Psychophysiology: Theory, Systems, and Methods.* L.A. Schmidt & S. Segalowitz, Eds.: 150–170. Cambridge University Press. New York.
27. Makeig, S., A.J. Bell, T.P. Jung & T.J. Sejnowski. 1996. Independent component analysis of electroencephalographic data. In *Advances in Neural Information Processing Systems,* Vol. 8. D. Touretzky, M. Mozer & M. Hasselmo, Eds.: 145–151. MIT Press. Cambridge, MA.
28. Tzyy-Ping, J., S. Makeig, C. Humpheries, *et al.* 2000. Removing electroencephalographic artifacts by blind source separation. *Psychophysiology* **37:** 163–178.
29. Fujioka, T., N. Mourad, C. He & L.J. Trainor. 2011. Comparison of artifact correction methods for infant EEG applied to extraction of event-related potential signals. *Clin. Neurophysiol.* **122:** 43–51.
30. Mourad, N., J.P. Reilly, H. De Bruin, *et al.* 2007. A simple and fast algorithm for automatic suppression of high-amplitude artifacts in EEG data. In *ICASSP, IEEE International Conference on Acoustics, Speech and Signal Processing—Proceedings 1.* Honolulu, HI.

31. Kraus, N. & B. Chandrasekaran. 2010. Music training for the development of auditory skills. *Nat. Rev. Neurosci.* **11:** 599–605.
32. Schneider, P., M. Scherg, H.G. Dosch, *et al.* 2002. Morphology of Heschl's gyrus reflects enhanced activation in the auditory cortex of musicians. *Nat. Neurosci.* **5:** 688–694.
33. Pantev, C., R. Oostenveld, A. Engelien, *et al.* 1998. Increased auditory cortical representation in musicians. *Nature* **392:** 811–814.
34. Fujioka, T., B. Ross, R. Kakigi, *et al.* 2006. One year of musical training affects development of auditory cortical-evoked fields in young children. *Brain* **129:** 2593–2608.
35. Shahin, A., D. Bosnyak, L.J. Trainor & L.E. Roberts. 2003. Enhancement of neuroplastic P2 and N1c auditory evoked potentials in musicians. *J. Neurosci.* **23:** 5545–5552.
36. Shahin, A., L.E. Roberts & L.J. Trainor. 2004. Enhancement of auditory cortical development by musical experience in children. *NeuroReport* **15:** 1917–1921.
37. Kuriki, S., S. Kanda & Y. Hirata. 2006. Effects of musical experience on different components of MEG responses elicited by sequential piano tones and chords. *J. Neurosci.* **26:** 4046–4053.
38. Ponton, C.W., J.J. Eggermont, B. Kwong & M. Don. 2000. Maturation of human central auditory system activity: evidence from multi-channel evoked potentials. *Clin. Neurophysiol.* **111:** 220–236.
39. Trainor, L.J., A. Shahin & L.E. Roberts. 2003. Effects of musical training on auditory cortex in children. *Ann. N.Y. Acad. Sci.* **999:** 520–521.
40. Čeponiené, R., T. Rinne & R. Näätänen. 2002. Maturation of cortical sound processing as indexed by event-related potentials. *Clin. Neurophysiol.* **113:** 870–882.
41. Pang, E.W. & M.J. Taylor. 2000. Tracking the development of the N1 from age 3 to adulthood: an examination of speech and non-speech stimuli. *Clin. Neurophysiol.* **111:** 388–397.
42. Mitzdorf, U. 1994. Properties of cortical generators of event-related potentials. *Pharmacopsychiatry* **27:** 49–51.
43. Hyde, M. 1997. The N1 response and its applications. *Audiol. Neuro-Otol.* **2:** 281–307.
44. Woldorff, M. & S. Hillyard. 1991. Modulation of early auditory processing during selective listening to rapidly presented tones. *Electroencephalogr. Clin. Neurophysiol.* **79:** 170–191.
45. Picton, T.W., C. Alain, L. Otten, *et al.* 2000. Mismatch negativity: different water in the same river. *Audiol. Neuro-Otol.* **5:** 111–139.
46. Näätänen, R., P. Paavilainen, T. Rinne & K. Alho. 2007. The mismatch negativity (MMN) in basic research of central auditory processing: a review. *Clin. Neurophysiol.* **118:** 2544–2590.
47. Alho, K., K. Sainio, N. Sajaniemi, *et al.* 1990. Event-related brain potential of human newborns to pitch change of an acoustic stimulus. *Electroen. Clin. Neuro.* **77:** 151–155.
48. Čeponiené, R., J. Hukki, M. Cheour, *et al.* 2000. Dysfunction of the auditory cortex persists in infants with certain cleft types. *Dev. Med. Child Neurol.* **42:** 258–265.
49. Fellman, V., E. Kushnerenko, K. Mikkola, *et al.* 2004. Atypical auditory event-related potentials in preterm infants during the first year of life: a possible sign of cognitive dysfunction? *Pediatr. Res.* **56:** 291–297.
50. Trainor, L.J., K. Lee & D.J. Bosnyak. 2011. Cortical plasticity in 4-month-old infants: specific effects of experience with musical timbres. *Brain Topogr.* **24:** 192–203.
51. Trainor, L.J., M. McFadden, L. Hodgson, *et al.* 2003. Changes in auditory cortex and the development of mismatch negativity between 2 and 6 months of age. *Int. J. Psychophysiol.* **51:** 5–15.
52. Friederici, A.D., M. Friedrich & C. Weber. 2002. Neural manifestation of cognitive and precognitive mismatch detection in early infancy. *NeuroReport* **13:** 1251–1254.
53. Tew, S., T. Fujioka, C. He & L.J. Trainor. 2009. Neural representation of transposed melody in infants at 6 months of age. *Ann. N.Y. Acad. Sci.* **1169:** 287–290.
54. Fujioka, T., L.J. Trainor, B. Ross, *et al.* 2004. Musical training enchances automatic encoding of melodic contour and interval structure, *J. Cogn. Neurosci.* **16:** 1010–1021.
55. Fujioka, T., L.J. Trainor, B. Ross, *et al.* 2005. Automatic encoding of polyphonic melodies in musicians and nonmusicians. *J. Cogn. Neurosci.* **17:** 1578–1592.
56. Sauseng, P., W. Klimesch, W.R. Gruber, *et al.* 2007. Are event-related potential components generated by phase resetting of brain oscillations? A critical question. *Neuroscience* **146:** 1435–1444.
57. Baser, E. & B. Güntekin. 2008. A review of brain oscillations in cognitive disorders and the role of neurotransmitters. *Brain Res.* **1235:** 172–93.
58. Buzsáki, G. & A. Draguhn. 2004. Neuronal oscillations in cortical networks. *Science* **304:** 1926–1929.
59. Bell, M.A. 1998. The ontogeny of the EEG during infancy and childhood: implications fro cognitive development. In *Neuroimaging in Child Neuropsychiatric Disorders*. B. Garreau, Ed.: 98–111. Springer–Verlag. Berlin.
60. Marshall, P.J., Y. Bar-Haim & N.A. Fox. 2002. Development of the EEG from 5 months to 4 years of age. *Clin. Neurophysiol.* **113:** 1199–1208.
61. Shahin, A.J., L.R. Roberts, W. Chau, *et al.* 2008. Musical training leads to the development of timbre-specific gamma band activity. *NeuroImage* **41:** 113–122.
62. Shahin, A.J., L.J. Trainor, L.E. Roberts, *et al.* 2010. Development of auditory phase-locked activity for music sounds. *J. Neurophysiol.* **103:** 218–229.
63. Taylor, M.J. & T. Baldeweg. 2002. Application of EEG, ERP and intracranial recordings to the investigation of cognitive functions in children. *Develop. Sci.* **5:** 318–334.
64. Tiihonen, J., M. Kajola & R. Hari. 1989. Magnetic mu rhythm in man. *Neuroscience* **32:** 793–800.
65. Lehtela, L., R. Salmelin & R. Hari. 1997. Evidence for reactive magnetic 10-Hz rhythm in the human auditory cortex. *Neurosci. Lett.* **222:** 111–114.
66. Fujioka, T., N. Mourad & L.J. Trainor, L. J. 2011. Development of auditory-specific brain rhythms in infants. *Eur. J. Neurosci.* **33:** 521–529.
67. Pfurtscheller, G. & F.H. Lopes da Silva. 1999. Event-related EEG/MEG synchronization and desynchronization: basic principles. *Clin. Neurophysiol.* **100:** 1842–1857.
68. Fujioka, T., L.J. Trainor, E.W. Large & B. Ross. 2009. Beta and gamma rhythms in human auditory cortex during musical beat processing. *Ann. N.Y. Acad. Sci.* **1169:** 89–92.

69. Fujioka, T., L.J. Trainor, E.W. Large & B. Ross. 2012. Internalized timing of isochronous sounds is represented in neuromagnetic beta oscillations. *J. Neuro.* In press.
70. Sokolov, A., M. Pavlova, W. Lutzenberger & N. Birbaumer. 2004. Receiprocal modulation of neuromagetic induced gamma activity by attention in the human visual and auditory cortex. *NeuroImage* **22:** 521–529.
71. Snyder, J.S. & E.W. Large. 2005. Gamma-band activity reflects the metric structure of rhythmic tone sequences. *Cogn. Brain Res.* **24:** 117–126.
72. Zanto, P.Z., E.W. Large, A. Fuchs & J.A. Kelso. 2005. Gamma-band responses to perturbed auditory sequences: evidence for synchronization of perceptual processes. *Mus. Percept.* **22:** 535–552.
73. Bhattacharya, J., H. Petsche & E. Pereda. 2001. Long-range synchrony in the gamma band: role in music perception. *J. Neurosci.* **21:** 6329–6337
74. Tallon-Baudry, C. & O. Bertrand. 1999. Oscillatory gamma activity in humans and its role in object representation. *Trends Cogn. Sci.* **3:** 151–162.
75. Greene, B.R., P. De Chazal, G.B. Boylan, *et al.* 2007. Electrocardiogram based neonatal seizure detection. *IEEE Trans. Biomed. Eng.* **54:** 673–682.
76. Krajčal, V., S. Petránek, J. Mohylová, *et al.* 2007. Neonatal EEG sleep stages modeling by temporal profiles. *Computer Aided Systems Theory—Eurocast*. 195–201. Springer-Verlag, Berlin.
77. Khodayari-Rostamabad, A., G.M. Hasey, D.J. MacCrimmon, *et al.* 2010. A pilot study to determine whether machine learning methodologies using pre-treatment electroencephalography can predict the symptomatic response to clozapine therapy. *Clin. Neurophysiol.* **12:** 1998–2006.
78. Ravan, M., J.P. Reilly, L.J. Trainor & A. Khodayari-Rostamabad. 2011. A machine learning approach for distinguishing age of infants using auditory evoked potentials. *Clin. Neurophysiol.* **122:** 2139–2150.
79. Fujioka, T., B. Zendel & B. Ross. 2010. Endogenous neuromagnetic activity for mental hierarchy of timing. *J. Neuroscience* **30:** 3458–3466.
80. Robinson, S.E. & J. Vrba. 1999. Functional neuroimaging by synthetic aperture magnetometry. In *Recent Advances in Biomagnetism*. T. Yoshimoto, M. Kotani, S. Kuriki, H. Karibe & N. Nakasato, Eds.: 302–305. Tohoku UP. Sendai, Japan.
81. Cheyne, D., L. Bakhtazal & W. Gaetz. 2006. Spatiotemporal mapping of cortical activity accompanying voluntary movements using an event-related beamforming approach. *Hum. Brain Mapp.* **27:** 213–229.
82. Ross, B., J.S. Snyder, M. Aalto, *et al.* 2009. Neural encoding of sound duration persists in older adults. *NeuroImage* **47:** 678–687.
83. Riggs, L., S.N. Moses, T. Bardouille, *et al.* 2009. A complementary analytic approach to examining medial temporal lobe sources using magnetoencephalography. *NeuroImage* **45:** 627–642.
84. Cornwell, B.R., F.W. Carver, R. Coppola, *et al.* 2008. Evoked amygdala responses to negative faces revealed by adaptive MEG beamformers. *Brain Res.* **1244:** 103–112.

Ann. N.Y. Acad. Sci. ISSN 0077-8923

ANNALS OF THE NEW YORK ACADEMY OF SCIENCES
Issue: *The Neurosciences and Music IV: Learning and Memory*

Behavioral methods in infancy: pitfalls of single measures

Sandra E. Trehub[1,2]

[1]Department of Psychology, University of Toronto Mississauga, Mississauga, Ontario, Canada. [2]Département de Psychologie, Université de Montréal, Quebec, Canada

Address for correspondence: Sandra E. Trehub, Department of Psychology, University of Toronto Mississauga, Mississauga, ON, Canada L5L 1C6. sandra.trehub@utoronto.ca

This paper outlines the principal behavioral methods used to study music processing in infancy. The advantages of conditioning procedures are offset by high attrition rates and restrictions on the stimuli that can be used. The head-turn preference procedure is more user-friendly but poses greater interpretive challenges. In view of the multidimensional nature of infant attention, no single response measure, whether behavioral, physiological, or neural, can provide unambiguous information about music processing in infancy. Greater use of ecologically valid stimuli is likely to generate increased cooperation from infants and greater generality of the findings.

Keywords: behavior; methods; music; infants

Behavioral methods in infancy: pitfalls of single measures

In this golden era of the brain, behavioral measures are in danger of being relegated to secondary status, even excluded, at times, from studies of music processing in infancy. The difficulty of managing young infants' attention and arousal increases the appeal of neural measures that do not necessitate cooperation. It is fair to say that these neural measures are altering the landscape of infant music perception (see Gaab *et al.*; Trainor, this volume). Among the remarkable claims that have emerged is that beat induction is innate, as evidenced by distinctive event-related potentials in sleeping newborns to omissions of the downbeat in rhythmic music.[1] Equally remarkable is the claim that sleeping newborns detect the invariance of pitch level across changes in timbre.[2] In view of adults' difficulty with pitch matching across timbre,[3] sleeping newborns' relative ease is surprising. Unquestionably, newborn brains register these changes in stimulation, but the meaning of their responses remains to be determined. For the moment, there is no means of translating neural levels of responding to perceptual or functional levels, especially in the absence of correlated behavioral responses. So can we reasonably interpret the aforementioned findings to mean that newborns induce the beat of rhythmic musical sequences, which is to say that they abstract beat structure from complex, multidimensional auditory events? Can we also conclude that newborns perceive pitch and timbre independently, without the cross-dimension interference experienced by older listeners? These answers await corroboration from other laboratories and with additional response measures. This is not a call for the substitution of behavioral measures for neural measures but rather for their coordinated use to advance our understanding of music perception in infancy.

Conditioned responses: head-turning

Drawing attention to the interpretive problems of neural measures is not to suggest that behavioral measures are free of interpretive problems. Consider conditioning procedures, which are used widely with nonhuman species. The potent rewards that underlie the efficacy of these procedures—food for food-deprived animals or avoidance of electric shock—are obviously off limits for human infants. Instead, investigators must use relatively weak and short-lived incentives such as visual rewards.

One conditioning procedure has been used extensively with infants in the study of speech and music perception.[4,5] The conditioned head-turn

doi: 10.1111/j.1749-6632.2012.06448.x

procedure capitalizes on the propensity of infants 5 months and older to localize sound and to turn toward a sound source following the onset of sound or changes in the soundscape. Auditory changes in the environment often signal salient events that have potentially positive consequences (e.g., communication from a friend) or negative consequences (e.g., threatening sound of a foe), requiring investigation and possible action (e.g., approach, avoidance). If turning had no consequence in the experimental context (i.e., only a loudspeaker in view), infants' responses would wane or habituate after a few trials. To motivate continued responding, investigators provide brief visual rewards (e.g., moving toys, video animation) when infants turn to the loudspeaker immediately following a target sound change but at no other times. Initially, infants simply turn to investigate the change in sound (i.e., an unconditioned response). Subsequently, they turn in anticipation of the interesting visual reward (i.e., a conditioned response) if they learn the contingency between their responses and the visual events. Significantly more turns toward the sound source on trials with a change than on those without a change (responses recorded by an experimenter who is unable to hear the changes) indicate that infants detect the change in question.

One major advantage of the procedure is that it permits 20–30 observations or test trials for each infant, which is a substantial improvement over the single undifferentiated scores yielded by most other infant procedures. In principle, one could obtain discrimination estimates for individual infants. Unfortunately, the procedure has a number of limitations, including high attrition rates (as high as 50% in some studies), which can result from infants' failure to learn the contingency, or from the inability of the modest visual rewards to sustain their auditory vigilance, responsiveness, and contentment for the entire test session.[5] Because the focus is on turning to a loudspeaker at one side, trials can only be presented when infants are calm and facing directly ahead. In addition, the procedure works best with relatively brief syllable or tone sequences, precluding the use of ecologically valid speech or musical excerpts.[5] Nevertheless, the procedure has generated a number of important findings related to infants' perception of pitch and timing patterns.[6,7]

Preferential listening

Because the head-turn preference procedure, or preferential listening procedure, overcomes some limitations of the conditioned head-turn procedure, notably the high attrition rates and restrictions on stimulus duration, it has become the dominant behavioral method in infant music laboratories. Infants sit on their parent's lap with one monitor (with visual display) and loudspeaker located to their right and another (same visual display) to their left. There are no contingencies to be learned, no vigilance required, and few constraints on how little or how long infants can look. A flashing light is used to attract infants' attention to one monitor (e.g., left), at which time one musical pattern is presented until they look away. Then another flashing light attracts their attention to the other monitor (e.g., right), and a contrasting musical pattern is presented until infants look away. Over the course of several trials, which either alternate from one side to the other or occur randomly from those locations, infant looking times during the presentation of each musical pattern are tabulated. Because the stimuli are presented only when infants look at the relevant monitor, looking is essentially a proxy for listening. Significantly longer cumulative looking or listening for one of the two stimuli reveals differentiation of those stimuli. Neither the experimenter nor the mother can hear the stimuli that infants hear, which prevents mothers from exerting systematic influences on infants' behavior and also eliminates potential bias in the experimenter's judgment of gaze direction.

A number of common variations of the procedure exist, depending on the goals of the research. When the goal is to ascertain whether infants can differentiate two novel musical excerpts, the simple comparison procedure is implemented, as described previously. At other times, the question of interest is how short- or long-term exposure to a musical pattern affects infants' responsiveness to or memory for aspects of that pattern. In such instances, the presentation of the two contrasting patterns is preceded by a familiarization period, either immediately before the comparison phase or during a more extended exposure phase at home. In one study, 7-month-old infants were tested with two recordings of the same foreign lullaby (Russian, German, or Spanish) sung at different pitch levels.[8] Some infants had no prior exposure to the lullaby, but others had listened to a high- or low-pitched version of one lullaby for 5 to 10 min daily over a 2-week period. Infants with no prior exposure to the lullaby listened longer to the low-pitched version. By contrast, infants with prior exposure to the lullaby listened longer to the rendition at a novel pitch level, which indicated long-term memory for features of the original performance. Subsequent research in another laboratory indicated that infants' listening dispositions were influenced by the nature of the musical material. For example, infants listened longer to lower-pitched versions of lullabies but to higher-pitched versions of play songs, perhaps because lower pitch made the lullabies more soothing and higher pitch made play songs more playful and therefore more engaging.[9]

Perhaps the most frustrating aspect of the preferential listening procedure is that the absence of differential listening—a common occurrence, unfortunately—is uninterpretable and may not imply discrimination failure. In many cases, highly contrastive musical patterns do not result in contrastive listening times. Whereas differential listening is interpretable in some manner, its absence is both inconclusive and unpublishable.

Differential listening, which is the desirable outcome, raises other concerns. Researchers commonly interpret significantly longer listening times to one of two musical patterns as a "preference" for that pattern. As a result, infants are said to "prefer" lower-pitched lullabies,[8] higher-pitched play songs,[9] consonant over dissonant music,[10–13] the rhythms of their own musical culture,[14] novel musical patterns in some cases,[8,15] and familiar patterns in others.[16] In other domains, "preferences" for familiar or novel stimuli have been found to vary with infant age and stimulus complexity.[17] In general, however, researchers do not code infants' behavior for signs of positive or negative affect, which is problematic in view of the parallels drawn between infants' presumed preferences and adult aesthetic preferences.

Of the many studies that have used some version of the preferential listening procedure, only one systematically coded infant behavior for signs of positive or negative affect. Zentner and Kagan[13] found that infants not only looked longer but also moved less during the presentation of traditional folk melodies accompanied by parallel major and minor thirds (three- and four-semitone intervals), which sounded consonant, than during the presentation of the same melodies accompanied by parallel minor seconds (one-semitone intervals), which sounded very dissonant. Looking time and movement reduction may be complementary, both of them signaling infant attention or engagement. With respect to measures that are associated with positive or negative affect, Zentner and Kagan[13] found that eight of 32 four-month-olds fretted or turned away from the loudspeaker during the dissonant versions but not during the consonant versions, and seven infants vocalized during the consonant versions but not during the dissonant versions.

Zentner and Kagan's[13] findings are consistent with a preference for consonance. Does it follow, then, that other claims of infants' preference for consonance are well founded? Not necessarily. Zentner and Kagan[13] used small consonant and dissonant intervals, with the latter (minor seconds) sounding extremely jarring to our ears. By contrast, the other infant studies[10–12] deliberately avoided such small intervals to eliminate the confounding of interval size with degree of consonance and dissonance. Instead, they used consonant perfect fifths (seven semitones) and octaves (12 semitones) and dissonant tritones (six semitones) and minor ninths (13 semitones). In the few studies on consonance with non-human species, tests with large intervals yielded no preference for consonance,[18] but tests with small intervals yielded the previously documented preference.[19] It is possible, then, that there are inborn negative affective responses to small dissonant intervals, which result in preferential responding to small consonant intervals.

Most studies of infant music perception have used stimulus sequences consisting of pure tones or synthesized piano tones. When infants are exposed to recorded piano performances of a 20th century atonal composition (with many dissonant intervals) and a traditional 18th century tonal composition, they listen longer to the atonal selection (Plantinga and Trehub, in preparation). The distribution of listening in this case may be driven by the novelty of the atonal music relative to music that infants typically hear. Recall that infants listened significantly longer to familiar lullabies sung at a novel pitch level.[8] Familiarity is pleasing as well as comforting, but an interest in novel objects and events motivates much exploration and learning in infancy and early childhood.

Attention in infancy

Current models of infant attention illuminate problematic aspects of single measures with infant participants. Instead of considering attention as a unitary phenomenon that can be characterized by a single measure such as visual fixation, Richards, Reynolds, and Courage[20] have identified multiple facets of infant attention. For example, the onset of a novel and salient stimulus leads to stimulus orienting, which is reflected not only in visual fixation (if the stimulus has a visual component) but also in changes of posture, facial expression, limb movement, respiration, heart rate, and electroencephalographic activity. Collectively, these behavioral, physiological, and neural responses index a phase of sustained attention during which infants are engaged in heightened stimulus processing, as reflected in measures of learning, resistance to distraction, and memory.[20] With continuing exposure to the target stimulus, various behavioral, physiological, and neural indicators of infant attention or arousal return to baseline levels even though infants may continue to look at the stimulus. At this point, infants are no longer considered attentive, a view that is corroborated by various information-processing measures (e.g., learning, memory). Interpreting such instances of looking as inattention or disengagement may seem counterintuitive. That view seems plausible, however, when contemplating the daydreamers and highly focused students who populate large college classes and are often indistinguishable on the basis of their direction of gaze.

The message here is that reliance on looking measures alone, or on any single measure, results in incomplete information, at best, or, at worst, misleading information about music processing in infancy. In contrast to studies of perceptual and cognitive development in infancy, which typically involve a single-response measure, studies of social and emotional development in infancy typically include multiple measures, behavioral as well as physiological.[21,22]

Ecological validity: vocal music

With relatively few exceptions, studies of infant music perception use artificial stimuli that sacrifice ecological validity in terms of timbre (e.g., pure tones or piano tones) and expressiveness (e.g., synthesized, inexpressive performances). Such stimulus selection is guided by concerns about possible effects of extraneous variables. The assumption, generally untested, is that this sacrifice has few consequences on fundamental processes like perception and memory. There is suggestive evidence to the contrary, however. For example, after at-home exposure to a melody, infants seem to remember the original pitch level if the music was presented vocally[8] but not if it was presented instrumentally.[23] These findings are consistent with adults' detailed encoding and retention of the surface features (e.g., pitch, timbre) of familiar, ecologically valid music.[24,25] In fact, adults exhibit better memory for melodies presented vocally than for those presented instrumentally.[26]

Some investigators have attempted to examine the features of vocal music that capture and maintain infants' attention. For example, infants listen longer to audio excerpts of infant-directed singing by unfamiliar women than to noninfant-directed singing (same song and singer).[27] Moreover, infants look significantly longer and move significantly less during videos of their own mother's infant-directed songs than during videos of her infant-directed speech.[28] Face-to-face maternal singing also modulates infants' arousal, as reflected in salivary cortisol levels.[29] To date, however, there have been no attempts to examine the relative impact of familiar and unfamiliar singing voices or familiar and unfamiliar songs on infant arousal and attention.

Maternal singing is typically the primary source of music for infants. Acoustic measurements of such singing have revealed that women sing at a higher pitch level, slower tempo, and with greater dynamic variation and temporal regularity in the presence of their infant than in the infant's absence.[30–33] In everyday contexts, maternal singing is not an isolated auditory stimulus. Instead, it occurs in conjunction with maternal facial expressions and movement. For example, mothers smile and move more while singing than when talking to infants.[34] In fact, silent videos of infant-directed singing elicit longer infant looking than do comparable videos of infant-directed speech.[35]

Ecological validity: infant musical behaviors

Looking, the principal response measure in laboratory contexts, is hardly a musical behavior. Casual observations of infants in their own homes reveal responses to music, including squeals of delight, "dancing," and rudimentary singing that are rarely observable in laboratory contexts. Parent-made videos may provide clues to questions that could be addressed in the laboratory. If collected under suitable conditions, such videos may yield important information about infant musical development. One ambitious laboratory study of infants 5 to 24 months of age documented their rhythmic movement to rhythmic music and to speech by means of behavioral coding and three-dimensional motion-capture technology.[36] Interestingly, infants moved rhythmically to rhythmic patterns but not to infant- or adult-directed speech. Although such movement was not coordinated with the music, faster tempos elicited a faster tempo of movement. There was some suggestion, moreover, of greater infant smiling during fleeting periods of coordinated movement, suggesting, perhaps, a momentary surge of emotion. The coordination of movement to music unfolds gradually over the course of several years,[37] and it is influenced favorably, in the preschool period, by the presence of a social partner.[38] In sum, musical stimuli and settings that are impoverished may obscure or underestimate the impact of music on infants.

Conclusions

Over the past few decades, we have learned a great deal about infants' listening dispositions, their ability to detect changes in pitch or temporal patterning, their musical environment, and the consequences of musical exposure. Methodological improvements, both major and minor, have contributed to those gains. There have been many ingenious experiments, but the characteristic use of a single-response measure has led, at times, to interpretive difficulties. Rigor is obviously critical, but it must be achieved in ways that yield conceptually as well as statistically significant findings. The use of convergent measures may resolve some of the current interpretive problems. Finally, greater use of ecologically valid stimuli is likely to reduce attrition rates, enhance performance, and increase the generality of the findings.

Acknowledgments

Preparation of this paper was assisted by grants from the Natural Sciences and Engineering Research Council and the Social Sciences and Humanities Research Council of Canada.

Conflicts of interest

The author declares no conflicts of interest.

References

1. Winkler, I., G.P. Haden, O. Ladinig, *et al.* 2009. Newborn infants detect the beat in music. *Proc. Natl. Acad. Sci. USA* **106:** 2468–2471.
2. Hàden, G.P., G. Stefanics, M.D. Vestergaard, *et al.* 2009. Timbre-independent extraction of pitch in newborn infants. *Psychophysiology* **46:** 69–74.
3. Pitt, M.A. 1994. Perception of pitch and timbre by musically trained and untrained listeners. *J. Exp. Psychol. Hum.* **20:** 976–986.
4. Trehub, S.E. & E.E. Hannon. 2009. Conventional rhythms enhance infants' and adults' perception of musical patterns. *Cortex* **45:** 110–118.
5. Werker, J.F., L. Polka & J.E. Pegg. 1997. The conditioned head turn procedure as a method for testing infant speech perception. *Early Dev. Parent.* **6:** 171–178.
6. Trehub, S.E. 2010. In the beginning: a brief history of infant music perception. *Music Sci.* **Special Issue:** 71–87.
7. Trehub, S.E. & E.E. Hannon. 2006. Infant music perception: domain-general or domain-specific mechanisms? *Cognition* **100:** 73–99.
8. Volkova, A., S.E. Trehub & E.G. Schellenberg. 2006. Infants' memory for musical performances. *Dev. Sci.* **9:** 584–590.
9. Tsang, C. & N.J. Conrad. 2010. Does the message matter? The effect of song type on infants' pitch preference for lullabies and playsongs. *Infant Behav. Dev.* **33:** 96–100.
10. Masataka, N. 2006. Preference for consonance over dissonance by hearing newborns of deaf parents and of hearing parents. *Dev. Sci.* **9:** 46–50.
11. Trainor, L.J. & B.M. Heinmiller. 1998. The development of evaluative responses to music: infants prefer to listen to consonance over dissonance. *Infant Behav. Dev.* **21:** 77–88.
12. Trainor, L.J., C.D. Tsang & V.H.W. Cheung. 2002. Preference for sensory consonance in 2- and 4-month-old infants. *Music Percept.* **20:** 187–194.
13. Zentner, M.R. & J. Kagan. 1998. Infants' perception of consonance and dissonance in music. *Infant Behav. Dev.* **21:** 483–492.
14. Soley, G. & E.E. Hannon. 2010. Infants prefer the musical meter of their own culture: a cross-cultural comparison. *Dev. Psychol.* **46:** 286–292.
15. Hannon, E.E. & S.E. Trehub. 2005. Metrical categories in infancy and adulthood. *Psychol. Sci.* **16:** 48–55.
16. Saffran, J.R. 2003. Absolute pitch in infancy and adulthood: the role of tonal structure. *Dev. Sci.* **6:** 35–47.
17. Shinskey, J.L. & Y. Munakata. 2010. Something old, something new: a developmental transition from familiarity to novelty preferences with hidden objects. *Dev. Sci.* **13:** 378–384.
18. McDermott, J. & M. Hauser. 2004. Are consonant intervals music to their ears? Spontaneous acoustic preferences in a nonhuman primate. *Cognition* **94:** B11–B24.
19. Chiandetti, C. & G. Vallortigara. 2011. Chicks like consonant stimuli. *Psychol. Sci.* **22:** 1270–1273.

20. Richards, J.E., G.D. Reynolds & M.L. Courage. 2010. The neural basis of infant attention. *Curr. Dir. Psychol. Sci.* **19:** 41–16.
21. Condradt, E. & J. Ablow. 2010. Infant physiological response to the still-face paradigm: contributions of maternal sensitivity and infants' early regulatory behavior. *Infant Behav. Dev.* **33:** 251–265.
22. Weinberg, M.K. & E.Z. Tronick. 1996. Infant affective reactions to the resumption of maternal interaction after the still face. *Child Dev.* **67:** 905–914.
23. Plantinga, J. & L.J. Trainor. 2005. Memory for melody: infants use a relative pitch code. *Cognition* **98:** 1–11.
24. Schellenberg, E.G. & S.E. Trehub. 2003. Good pitch memory is widespread. *Psychol. Sci.* **14:** 262–266.
25. Schellenberg, E.G., P. Iverson & M.C. McKinnon. 1999. Name that tune: identifying popular recordings from brief excerpts. *Psychon. B Rev.* **6:** 641–646.
26. Weiss, M. 2011. *Vocal timbre influences memory for melodies.* Unpublished Master's thesis, University of Toronto.
27. Trainor, L.J. 1996. Infant preferences for infant-directed versus noninfant-directed play songs and lullabies. *Infant Behav. Dev.* **19:** 83–92.
28. Nakata, T. & S.E. Trehub. 2004. Infants' responsiveness to maternal speech and singing. *Infant Behav. Dev.* **27:** 455–464.
29. Shenfield, T., S.E. Trehub & T. Nakata. 2003. Maternal singing modulates infant arousal. *Psychol. Music* **31:** 365–375.
30. Nakata, T. & S.E. Trehub. 2011. Expressive timing and dynamics in infant-directed and non-infant-directed singing. *Psychomus: Music Mind Brain* **21:** 130–138.
31. Trainor, L.J., E.D. Clark, A. Huntley & B.A. Adams. 1997. The acoustic basis for infant-directed singing. *Infant Behav. Dev.* **20:** 383–396.
32. Trehub, S.E., A.M. Unyk, S.B. Kamenetsky, *et al.* 1997. Mothers' and fathers' singing to infants. *Dev. Psychol.* **33:** 500–507.
33. Trehub, S.E., A.M. Unyk & L.J. Trainor. 1993. Maternal singing in cross-cultural perspective. *Infant Behav. Dev.* **16:** 285–295.
34. Trehub, S.E., J. Plantinga, F.A. Russo. 2011. Maternal singing to infants in view or out of view. *Presented at Meetings of the Society for Research in Child Development*. Montreal, Canada.
35. Plantinga, J., S.E. Trehub & F. Russo. 2011. Multimodal aspects of maternal speech and singing. *Presented at Meetings of Neurosciences and Music IV*, Edinburgh, Scotland.
36. Zentner, M.R. & T. Eerola. 2010. Rhythmic engagement with music in infancy. *Proc. Natl. Acad. Sci. USA* **107:** 5768–5773.
37. McAuley, J.D., M.R. Jones, S. Holub, *et al.* 2006. The time of our lives: life span development of timing and event tracking. *J. Exp. Psychol. Gen.* **135:** 348–367.
38. Kirschner, S. & M. Tomasello. 2009. Joint drumming: social context facilitates synchronization in preschool children. *J. Exp. Child Psychol.* **102:** 299–314.

Ann. N.Y. Acad. Sci. ISSN 0077-8923

ANNALS OF THE NEW YORK ACADEMY OF SCIENCES
Issue: *The Neurosciences and Music IV: Learning and Memory*

Pediatric neuroimaging in early childhood and infancy: challenges and practical guidelines

Nora Raschle,[1,2] Jennifer Zuk,[1] Silvia Ortiz-Mantilla,[3] Danielle D. Sliva,[4] Angela Franceschi,[5] P. Ellen Grant,[2,4,5,6] April A. Benasich,[3] and Nadine Gaab[1,2,7]

[1]Laboratories of Cognitive Neuroscience, Division of Developmental Medicine, Children's Hospital Boston, Boston, Massachusetts. [2]Harvard Medical School, Boston, Massachusetts. [3]Center for Molecular and Behavioral Neuroscience, Rutgers, The State University of New Jersey, Newark, New Jersey. [4]Fetal-Neonatal Neuroimaging and Developmental Science Center, Children's Hospital Boston, Boston, Massachusetts. [5]Department of Radiology, Division of Neuroradiology, Children's Hospital Boston, Boston, Massachusetts. [6]Department of Medicine, Division of Newborn Medicine, Children's Hospital Boston, Boston, Massachusetts. [7]Harvard Graduate School of Education, Boston, Massachusetts

Address for correspondence: Nadine Gaab, Ph.D., Children's Hospital, Boston, Department of Medicine, Division of Developmental Medicine, Laboratories of Cognitive Neuroscience, 1 Autumn Street, Mailbox # 713, Boston, MA 02115. nadine.gaab@childrens.harvard.edu

Structural and functional magnetic resonance imaging (fMRI) has been used increasingly to investigate typical and atypical brain development. However, in contrast to studies in school-aged children and adults, MRI research in young pediatric age groups is less common. Practical and technical challenges occur when imaging infants and children, which presents clinicians and research teams with a unique set of problems. These include procedural difficulties (e.g., participant anxiety or movement restrictions), technical obstacles (e.g., availability of child-appropriate equipment or pediatric MR head coils), and the challenge of choosing the most appropriate analysis methods for pediatric imaging data. Here, we summarize and review pediatric imaging and analysis tools and present neuroimaging protocols for young nonsedated children and infants, including guidelines and procedures that have been successfully implemented in research protocols across several research sites.

Keywords: pediatric; imaging; children; magnetic resonance imaging; fMRI; MRI

Introduction

The advent of magnetic resonance imaging (MRI) has opened up new possibilities for studying human brain structure and function across the lifespan. An increase in the use of structural and functional MRI (fMRI) in infants and young children can further add to our understanding of brain development. For example, MRI research has revealed differences in brain structure and function in individuals with disabilities (e.g., dyslexia[1–3]) compared with typical controls, and working with infants and young children may unveil the developmental trajectory of such disabilities. Neuroimaging young children additionally allows for the investigation of brain plasticity during this rapid developmental period and can potentially reveal how certain perceptual, procedural, and cognitive skills, such as music perception and musical skills, develop.[4] However, in contrast to studies in school-aged children and adults, MRI research in young pediatric age groups is less common.[5] Studies involving children under the age of 6 are particularly rare, given the practical and technical challenges involved (e.g., Refs. 5, 6). Practical challenges of pediatric neuroimaging sessions include procedural difficulties (e.g., participants', anxiety or motivation, movement restriction, putting an infant to sleep in an unfamiliar environment, and parent's anxiety), technical obstacles (e.g., availability of child-appropriate equipment, masking and attenuation of scanner background noise [SBN]), as well as the challenge of choosing the most appropriate analysis methods (e.g., pediatric brain templates

doi: 10.1111/j.1749-6632.2012.06457.x

and adequate movement detection tools). In clinical populations, MRIs of infants and children are routinely obtained under sedation,[7–9] which eliminates a subset of these challenges. However, for ethical reasons, sedation is not an option for most developmental neuroimaging research. Furthermore, there is a strong push from clinicians and hospital administrators to reduce the overall need for sedation and anesthesia for cost containment and more importantly to prevent any potential negative sequelae, particularly in those receiving multiple MRI studies.

Several methods have been developed to improve an infant or child's compliance during neuroimaging sessions within the clinical (e.g., Refs. 10–12) or research setting (e.g., Refs. 13, 14). General approaches for imaging young children include play therapy,[14] behavioral training[10–12,15–17] and simulation,[18] the use of mock scanner areas,[13–19] basic relaxation,[20] and a combination of these techniques.[21] The most common practices for nonsedated newborns and infants are the natural sleep technique[22–25] and the feed and wrap procedure.[26,27] Figures S1 and S2 provide a literature overview of published protocols, guidelines, and empirical research studies using pediatric neuroimaging protocols and their sample size and success rate.

This paper summarizes successful methods for applying guidelines and recommendations on how to successfully perform neuroimaging studies in nonsedated infants and young children. Furthermore, strategies for overcoming experimental and analysis limitations and ethical implications of neuroimaging in pediatric populations are discussed.

General considerations

The age of a pediatric participant taking part in neuroimaging research should strongly influence the research protocol for pediatric imaging. Preschool-aged children may participate in structural and functional neuroimaging studies, in which the child is required to be awake and alert while performing a certain perceptual or cognitive task. Infants are usually enrolled in studies of brain structure or resting state fMRI, which are performed while the participants are asleep. fMRI techniques have been successfully applied in awake infants, but report a very high attrition rate.[28,29] Although the focus in working with preschool-aged children or older mostly lies on the child itself, the caregiver–child interaction becomes just as important during infant neuroimaging. Overall, clear communication and a child-centered, age-appropriate approaches are the fundamental aspects of pediatric neuroimaging.

Terminology

When interacting with participating children and families, it is imperative to use positive, child-friendly terminology that can be easily understood. A session may get started with "Do you know what will happen today?" All equipment should be labeled in a child-appropriate way: for example, the MRI machine may be called "brain camera," the scanner noise "camera click," and the head coil "mirror holder." Phrases like "It is really loud but won't hurt you" or "Are you doing okay inside the machine?" should be avoided. Furthermore, careful communication with all family members present during the session is just as important, if not more important, than the communication with the participants themselves to avoid parental/caregiver anxiety, which in turn impacts the behavior of the young child or infant. This involves giving the parents a clear outline of the neuroimaging session, describing each step and tool involved, reviewing safety considerations, and clarifying the research team's goal. When imaging infants, potential stress may be avoided by telling parents upfront that it is not unusual for a child to fail to fall asleep on the first visit or to wake up before the protocol is completed. A website describing each protocol step, tool, and challenge can help with the preparation process for any age group.

Environment

Scheduling an initial introductory meeting before the first neuroimaging session can help diminish both parent and child anxiety by familiarizing them with the research team and setting. Photographs, and/or a brief video overview of an actual session (either via web pages or brochures) may be combined with a tour of the actual MRI area or the mock scanner environment. Research teams working with preschoolers and older children especially benefit from the use of a mock scanner environment before or on the day of testing. It enables the research team to demonstrate the actual imaging session and may include a mock scanner with a moveable scanner bed, head coil, response tools, mirror and video system, and integrated MR sounds. Some facilities even incorporate a feedback system as part of the mock scanner area, allowing for the observation of movement and appropriate feedback to train children to lie still (e.g., Refs. 10, 17). Most importantly, the mock scanner area provides a child-friendly, appropriate preview of the actual neuroimaging session, and permits the research team to adopt a playful approach to an otherwise strictly medical topic. In addition, research teams are encouraged to have child-appropriate toys, child-sized table and chairs, snacks and drinks, and to invite participants to bring family and friends to increase children's comfort and motivation. When imaging infants, the mock scanner area is used mainly to familiarize the parents with the scanner equipment and the MRI sounds. Mock preparation further offers an opportunity to develop a relationship between the infant and the researcher, and can be used to recreate individual napping/bed-time routines within the MRI environment.

Affective state

It is crucial to recognize feelings of anxiety, frustration, or boredom and address potential issues in a child-friendly manner because children are not always able to express emotions directly. Gauging parent comfort and the level of continued consent throughout the process is equally important, particularly during neuroimaging of infants. The parent's comfort level directly influences the infant's feelings of well-being and contentment, which is critical during the natural sleep technique. Given their expertise in child development and extensive experience with children/families in the hospital environment, a child life specialist (CLS) can be an integral part of the multidisciplinary pediatric neuroimaging team.

MRI equipment

Depending on the scope of the pediatric neuroimaging session, different MRI-compatible tools may be needed. This section highlights some examples of equipment involved during neuroimaging sessions with infants and children.

Response buttons. Response tools used during the neuroimaging session should be age-appropriate in size and shape. It is recommended to position them comfortably to diminish motion when the participant attempts to reach or play with them (e.g., placing child-friendly buttons an arm-length from the participant works well).

Audio system. The frequency of sounds emitted during MR image acquisition ranges from 0–9,000 Hz[30] at intensity levels up to 115 dB with 3 tesla.[30] When working with young children who are participating in an auditory research experiment, child-friendly ear buds and/or child-sized headphones can mask the majority of SBN while allowing for the presentation of auditory stimuli. When working with infants, the goal is to decrease SBN as much as possible to avoid sleep disruption or startle. The noise can be reduced by 20–30 dB with the use of foam earplugs or industrial-grade earmuffs.[31] In addition, music or the infant's favorite lullabies can be played to mask some of the variations in the scanner noise. A steady-state sound can be an aid to sleep; however, noise that abruptly changes in sound level or intensity in a nonprogressive manner is much more likely to result in arousal and waking.

Motion attenuation. A foam mattress to line the scanner bed reduces the amount and intensity of vibration, additionally increasing comfort. For sleeping infants, a foam helmet can help reduce bone conductance, which adds sound attenuation of about 6 dB, reduces vibration, and provides additional stability.

Video/video-goggle system. For imaging older children, a video system can be used to present experimental stimuli or movies during image acquisition. However, younger children may be frightened by a goggle system, so a traditional projector–screen combination may be used.

Experimental design and image acquisition

Pediatric imaging for research purposes presents many challenges during both image acquisition and data analysis. Challenges may begin long before the first image is taken, in fact as early as during the conception and design of the research.

Experimental design

Young children often have difficulties switching between two different task instructions; however, task switching is frequently required during a traditional block or event-related design in fMRI. Our experience has shown that, if possible, it is advisable to use a block design and to separate task conditions (e.g., experimental task and control task) into two separate runs to avoid confusion.

Furthermore, the overall session length should be as short as possible and not be any longer than 90 min for a preschooler (which includes short 5–6 minute runs with breaks between each run and a total maximum of 30–40 min of actual scanning). In addition, the intense SBN, generated by the shifting of gradient coils during conventional continuous image acquisition, is one potential cause for child anxiety or discomfort. SBN affects auditory stimuli delivery,[32] leads to activation in auditory/language areas,[33,34] leads to differences in attentional demands,[35] and also influences the default mode.[36] Inhibiting the SBN during a cognitive demanding task can be quite difficult for a young child and/or a clinical population, and this should be taken into account when comparing different age groups or clinical and nonclinical populations. One way to circumvent exposure to the SBN is to use interleaved data acquisition designs, such as sparse temporal sampling,[33,34,37] if time permits.

Image acquisition

There are many challenges during image acquisition and analysis of pediatric imaging data. These barriers are most prevalent in the first three years of life, but are also present in older children and may include challenges relating to design limitations (e.g., constrained session duration in children, leading to a decrease in statistical power; and difficulty in designing task-based fMRI for infants because MR studies are most successful when the infant is imaged asleep). Challenges may also include the presence of increased movement-related artifacts or differences based on distinct anatomy (e.g., difficulties in obtaining MR head coils that provide similar fit across age, differences in brain shape and size across age, differences in brain contrast, baseline diffusivity, baseline blood flow and regional regulation, the shape of the hemodynamic response function, or default mode of brain function across age).

Many research groups are currently working on finding solutions for these challenges. For example, many collaborative groups including ours combine approaches from industry and child-life professionals in the hospital setting and further work on optimizing techniques. In parallel, technical groups are developing prospective motion mitigated sequences, and MRI coils tailored to the head size of the subject are also under development.[38,39] Development of novel pulse sequences that exploit high density–phased array coils to increase the speed of image acquisition have the potential to further increase success. Pediatric brain templates are becoming available to improve data analysis and to help account for age-related changes (e.g., Refs. 8,40,41). Furthermore, several research labs have developed techniques to better align and normalize brains that differ in size and shape. For instance, surface-based registration has been shown to provide significantly better alignment across different age groups.[42] However, these templates do not allow capture of the full range of information available, and therefore multiple sites are working on "4D" atlases that provide information on maturation of substructures.[43]

Overall, much is still unknown about the biological differences between the immature and mature brain that could influence data analysis. At each step along the way, the fundamental assumptions for any data analysis tool need to be questioned to be sure these assumptions still hold for the age of infant or child participant. As a result, a close collaboration between technical and developmental experts is needed as we continue to study these younger age groups.

Preparation and neuroimaging procedures for infants and children

Neuroimaging infants

Obtaining detailed information about the child's nap and feeding schedule is helpful in facilitating the preparation for an infant session and for deciding the best time to schedule the MRI. For some babies, and always in accordance with the parent's wishes, an earlier nap may be omitted so that the infant is sleepy upon arrival. Completing several interesting, but not too stimulating, tasks immediately before the scan can further fatigue the child. Parents are encouraged to bring anything that the infant may need for his or her bedtime routine (a familiar blanket, favorite toy, or lullaby music). Listening to a CD containing scanner background sounds (optionally overlaid with favorite lullabies) before the day of the neuroimaging session will prepare the infant for the upcoming procedure. Exposure to the scanner sounds should ideally be started at least a week before imaging. Encouraging frequent exposure to the sounds (e.g., during both awake and nap times) increases the habituation effect and may decrease startle responses during neuroimaging.

When scanning infants, replicating normal bedtime routines in the imaging suite, including bathing, feeding, dimming the lights, and playing soft lullaby music, has proven to be effective.[22] Offering a rocking chair, portacrib, bathtub, or blankets for swaddling may help parents to put the baby to sleep. Playing a CD of the MRI sequences in the room where the infant falls asleep provides a stable auditory environment throughout the process. To ease the transition between a nursery-type environment and the MRI suite, an alternative approach is to have the infant fall asleep inside the MRI suite using a MRI-compatible (inflatable) crib. Some caregivers may decide to perform their sleep routine (e.g., nursing, lullabies, etc.) inside the MRI suite or even on the MRI scanner bed. The caregiver should be allowed to try to put their child to sleep, as long as they are comfortable. The research team should leave them on their own if at all possible and not "hover." Reserving at least a two-hour time slot with the MRI allows for flexibility.

Once the infant has fallen asleep, earplugs, earmuffs ("Minimuffs," Natus Medical Inc., San Carlos, CA), and a sound attenuation helmet can be placed on the infant.[24,25] Earplugs should be placed after feeding, as having the ears blocked while nursing can make the infants uncomfortable. The infant can be slowly and gently lied on the prepared scanner bed, which should be ideally covered with a well-secured visco-elastic foam mattress. Laying a warm blanket on top of the mattress can alleviate the cool temperature in the MRI suite. The infant's head should then be positioned into the coil and the researcher can then carefully adjust the restraint straps in place to secure the infant and limit his or her movements. Once the infant has been snuggled into the space with a soft blanket and sleeps soundly, the scan can proceed.

Throughout the scan, the caregiver and a team member should stay in the imaging suite. If the infant begins to wake, MRI acquisition can be paused and an attempt made to soothe the infant back to sleep. If the infant cries or becomes distressed at any time during the scan, it is advisable to stop and immediately remove the baby from the scanner. Regardless of the success or failure of image acquisition, offering toys and onesies to infants are a nice token of appreciation and after a success, providing a CD with images of the infant brain to take home is an incentive for parents.

Neuroimaging young children

Presenting young children with a cognitive task to complete during functional imaging requires the adaptation of the tasks, instructions, and incentives used. A child-friendly theme can be useful in guiding a young child through the training and actual neuroimaging session. For example, an adventure story can motivate the child to finish all of the tasks requested. Short movies can be useful to lead into the session, engage the child, and reduce any initial anxiety. Additionally, virtual sticker charts can be used to provide feedback about the progress of the neuroimaging session and to motivate the child to complete as many sequences as possible (see also Ref. 44).

Children 4 years and older are best prepared for the fMRI session in a mock scanner area. The scope of the neuroimaging session can be explained to participants and families using applied examples and the material used during actual imaging (e.g., button boxes and headphones) can be revisited in a safe way. Once a general introduction to the MRI has been provided, the child can listen to instructions for the task and then practice while sitting on the mock scanner bed. After the child understands the task, the researcher can gradually add the use of headphones, present some prerecorded scanner sounds in the background, and have the child lie down in the mock scanner bed while simultaneously practicing the training task to replicate the actual MRI experience. If the child cannot distinguish between left and right, a response button may be labeled "monkey button" instead of "right button" and a toy animal can be placed on the appropriate side as a reminder.

One major challenge of working with young children during functional tasks is keeping movements to a minimum during image acquisition. It is advisable to discuss the impact of motion on the brain images with the child. A digital camera may be used to demonstrate the impact of movement on image quality. Print-outs of sharp and blurry pictures (e.g., of animals) or playing games, such as the game of "freeze" (challenge the child to stay still as long as possible), can be used for illustration purposes. It is best to remind the child not to speak during image acquisition because this can cause reduced task performance and data quality. Another helpful strategy is to train the child that a gentle hand press on their leg indicates there is too much movement. A member of the research team should stay in the MRI room throughout the session with the child and should implement the "hand on leg" procedure if the child demonstrates significant movement, which serves as a means of communication between the research team and the child during the experiment.

If a mock scanner is not available, game strategies such as employing play therapy techniques[45] are a recommended alternative. Providing pictures or videos of an MR scanner effectively depicts the MRI experience,[10] and desensitization to the procedure by this exposure and operant training techniques help to reduce anxiety in children.[12,17]

Researchers should keep in mind that the transition from mock training to the actual MR also brings changes to the environment (fewer toys, more medical personnel and supplies, change in temperature, etc). Thus, it is important to be attentive to the child's reaction upon entering the room and respond accordingly. Allowing the parent to accompany the child and offering comfort items may effectively smooth the transition. If a child displays resistance to any aspect of the MRI experiment, address the concerns immediately and carefully. Offering breaks during the neuroimaging session can be helpful. The researcher should take all the time necessary to make the child comfortable with the MR environment. Upon completion of image acquisition, offering incentives and a CD with images of the brain to the child is a nice reward for participation.

Ethical implication and conclusion

Here, we provide applied guidelines and recommendations on how to successfully perform neuroimaging studies on nonsedated infants and young children. These techniques have proven to be successful in a large group of infants and young children across several research sites (see Fig. S3). It is important to keep in mind that along with the advent of more accessible (pediatric) neuroimaging, ethical challenges have also arisen. Although these issues have emerged and have been discussed in detail in regard to adult neuroimaging, such ethical issues are not only present when considering pediatric neuroimaging, but also may even be magnified.[46,47] However, the anticipated benefits of using pediatric fMRI in nonclinical populations are considerable and may have far-reaching advantages including application to classroom settings.[47] Nevertheless, researchers should carefully consider what method is most appropriate for their experimental question and participant age range. Thomason has highlighted the positive experiences of pediatric research participants, but such data are not available yet for younger children.[48] Furthermore, Connors and Singh strongly emphasize that neuroimaging data are often misinterpreted by the general public and that there are a series of subtle ways in which neuroimaging data are and will affect children's lives such as the shaping of national health, strategies for criminal risk assessment, and educational practice and its overall implication for policy, education, and family life.[49] It is also important to keep ethical considerations in mind and to frequently reevaulate advances in the field of pediatric imaging in terms of its ethical implications. This will help improve existing imaging protocols and guidelines and will also facilitate development of new technological advances that will improve comfort for the children scanned and enhance data quality.

Acknowledgments

The authors thank W.C. Liu and D. Johnson from UMDNJ for their invaluable help in running nonsedated scans in infants. The authors also thank the researchers and staff at the Center for Molecular and Behavioral Neuroscience, who spent many hours designing the infant protocol and patiently acquiring scans. In particular, they thank J. Flax, S. Paterson, N. Badridze, J. Byrne, and S. Marken. The authors would also like to thank the Children's Hospital Boston MRI Team and the members of the Gaab lab who helped conduct these scans. And of course, they thank all the parents, babies, and children who worked so hard to help them acquire these important images of the developing brain. This work was supported by a grant from the Harvard Catalyst (to N.G. and E.G.) and the Santa Fe Institute Consortium (to A.A.B.), with additional funding from the Elizabeth H. Solomon Center for Neurodevelopmental Research.

Conflicts of interest

The authors declare no conflicts of interest.

Supporting information

Additional supporting information may be found in the online version of this article:

Figure S1. Overview of original research (OR) and review articles (RA) that addresses the issues of pediatric imaging and/or provide techniques for working with children within the imaging environment; in chronological order.

Figure S2. Overview of original research (OR) and review articles (RA) that addresses the issues of pediatric imaging and/or provide techniques for working with infants within the imaging environment; in chronological order.

Figure S3. Summary of the success rate of imaging infants and young children within our laboratories. Note that a subset required multiple attempts (number of imaging sessions) to successfully acquire the images. Multiple sessions markedly increased the success rate.

Please note: Wiley-Blackwell is not responsible for the content or functionality of any supporting materials supplied by the authors. Any queries (other than missing material) should be directed to the corresponding author for the article.

References

1. Richlan, F., M. Kronbichler & H. Wimmer. 2009. Functional abnormalities in the dyslexic brain: a quantitative meta-analysis of neuroimaging studies. *Hum. Brain Mapp.* **30:** 3299–3308. doi: 10.1002/hbm.20752
2. Shaywitz, S. & B. Shaywitz. 2008. Paying attention to reading: the neurobiology of reading and dyslexia. *Dev. Psychopathol.* **20:** 1329–1349. doi: 10.1017/S0954579408000631
3. Gabrieli, J.D. 2009. Dyslexia: a new synergy between education and cognitive neuroscience. *Science* **325:** 280–283. doi: 325/5938/280 [pii] 10.1126/science.1171999
4. Overy, K., A. Norton, K. Cronin, *et al.* 2005. Examining rhythm and melody processing in young children using FMRI. *Ann. N.Y. Acad. Sci.* **1060:** 210–218. doi: 1060/1/210 [pii] 10.1196/annals.1360.014
5. Bookheimer, S.Y. 2000. Methodological issues in pediatric neuroimaging. *Ment. Retard. Dev. Disabil. Res. Rev.* **6:** 161–165.
6. Poldrack, R.A., E.J. Pare-Blagoev & P.E. Grant. 2002. Pediatric functional magnetic resonance imaging: progress and challenges. *Top Magn. Reson. Imaging* **13:** 61–70.
7. Parazzini, C., C. Baldoli, G. Scotti & F. Triulzi. 2002. Terminal zones of myelination: MR evaluation of children aged 20–40 months. *AJNR Am. J. Neuroradiol.* **23:** 1669–1673.
8. Altaye, M., S.K. Holland, M. Wilke & C. Gaser. 2008. Infant brain probability templates for MRI segmentation and normalization. *NeuroImage* **43:** 721–730. doi: 10.1016/j.neuroimage.2008.07.060
9. Yamada, H. *et al.* 2000. A milestone for normal development of the infantile brain detected by functional MRI. *Neurology* **55:** 218–223.
10. Slifer, K.J. 1996. A video system to help children cooperate with motion control for radiation treatment without sedation. *J. Pediatr. Oncol. Nurs.* **13:** 91–97.
11. Slifer, K.J., J.D. Bucholtz & M.D. Cataldo. 1994. Behavioral training of motion control in young children undergoing radiation treatment without sedation. *J. Pediatr. Oncol. Nurs.* **11:** 55–63.
12. Tyc, V.L., D. Fairclough, B. Fletcher, *et al.* 1995. Children's distress during magnetic resonance imaging procedures. *Child. Health Care* **24:** 5–19.
13. Epstein, J. N. *et al.* 2007. Assessment and prevention of head motion during imaging of patients with attention deficit hyperactivity disorder. *Psychiatry Res.* **155:** 75–82. doi: S0925-4927(07)00012-1 [pii] 10.1016/j.pscychresns.2006.12.009
14. Pressdee, D., L. May, E. Eastman & D. Grier. 1997. The use of play therapy in the preparation of children undergoing MR imaging. *Clin. Radiol.* **52:** 945–947.
15. Byars, A.W. *et al.* 2002. Practical aspects of conducting large-scale functional magnetic resonance imaging studies in children. *J. Child Neurol.* **17:** 885–890.
16. Slifer, K.J., M.F. Cataldo, M. D. Cataldo, *et al.* 1993. Behavior analysis of motion control for pediatric neuroimaging. *J. Appl. Behav. Anal.* **26:** 469–470.
17. Slifer, K.J., K.L. Koontz & M.F. Cataldo. 2002. Operant-contingency-based preparation of children for functional magnetic resonance imaging. *J. Appl. Behav. Anal.* **35:** 191–194.
18. Rosenberg, D.R. *et al.* 1997. Magnetic resonance imaging of children without sedation: preparation with simulation. *J. Am. Acad. Child Adolesc. Psychiatry* **36:** 853–859.
19. de Amorim e Silva, C.J., A. Mackenzie, L. M. Hallowell, *et al.* 2006. Practice MRI: reducing the need for sedation and general anaesthesia in children undergoing MRI. *Australas. Radiol.* **50:** 319–323.
20. Lukins, R., I.G. Davan & P.D. Drummond. 1997. A cognitive behavioural approach to preventing anxiety during magnetic resonance imaging. *J. Behav. Ther. Exp. Psychiatry* **28:** 97–104. doi: S0005-7916(97)00006-2 [pii]
21. Hallowell, L.M., S.E. Stewart, E.S.C.T. de Amorim & M.R. Ditchfield. 2008. Reviewing the process of preparing children for MRI. *Pediatr. Radiol.* **38:** 271–279. doi: 10.1007/s00247-007-0704-x
22. Ortiz-Mantilla, S., M.S. Choe, J. Flax, *et al.* 2010. Associations between the size of the amygdala in infancy and language abilities during the preschool years in normally developing children. *NeuroImage* **49:** 2791–2799. doi: S1053-8119(09)01100-8 [pii] 10. 1016/j.neuroimage.2009.10.029
23. Liu, Y. *et al.* 2008. Regional homogeneity, functional connectivity and imaging markers of Alzheimer's disease: a review of resting-state fMRI studies. *Neuropsychologia* **46:** 1648–1656.
24. Paterson, S.J., N. Badridze, J.F. Flax, W.-C. Liu & A.A. Benasich. 2004. *A Method for Structural MRI Scanning of Non-Sedated Infants.* Rutgers University. Chicago.
25. Anderson, A.W. *et al.* 2001. Neonatal auditory activation detected by functional magnetic resonance imaging. *Magn. Reson. Imaging* **19:** 1–5.
26. Almli, C.R., M.J. Rivkin & R.C. McKinstry. 2007. The NIH MRI study of normal brain development (Objective-2): newborns, infants, toddlers, and preschoolers. *NeuroImage* **35:** 308–325. doi: 10.1016/j.neuroimage.2006.08.058

27. Gilmore, J.H. *et al.* 2007. Early postnatal development of corpus callosum and corticospinal white matter assessed with quantitative tractography. *AJNR Am. J. Neuroradiol.* **28:** 1789–1795. doi: 10.3174/ajnr.A0651
28. Leroy, F. *et al.* 2011. Early maturation of the linguistic dorsal pathway in human infants. *J. Neurosci.* **31:** 1500–1506. doi:10.1523/jneurosci.4141-10.2011
29. Dehaene-Lambertz, G. *et al.* 2010. Language or music, mother or Mozart? Structural and environmental influences on infants' language networks. *Brain Lang.* **114:** 53–65. doi: S0093-934X(09)00120-5 [pii] 10. 1016/j.bandl. 2009.09.003
30. Hattori, Y., H. Fukatsu & T. Ishigaki. 2007. Measurement and evaluation of the acoustic noise of a 3 Tesla MR scanner. *Nagoya J. Med. Sci.* **69:** 23–28.
31. Zahr, L.K. & J. de Traversay. 1995. Premature infant responses to noise reduction by earmuffs: effects on behavioral and physiologic measures. *J. Perinatol.* **15:** 448–455.
32. Bandettini, P.A., A. Jesmanowicz, J. Van Kylen, *et al.* 1998. Functional MRI of brain activation induced by scanner acoustic noise. *Magn. Reson. Med.* **39:** 410–416.
33. Gaab, N., J.D. Gabrieli & G.H. Glover. 2007. Assessing the influence of scanner background noise on auditory processing. I. An fMRI study comparing three experimental designs with varying degrees of scanner noise. *Hum. Brain Mapp.* **28:** 703–720.
34. Gaab, N., J.D. Gabrieli & G.H. Glover. 2007. Assessing the influence of scanner background noise on auditory processing. II. An fMRI study comparing auditory processing in the absence and presence of recorded scanner noise using a sparse design. *Hum. Brain Mapp.* **28:** 721–732. doi: 10.1002/hbm.20299
35. Tomasi, D., E.C. Caparelli, L. Chang & T. Ernst. 2005. fMRI-acoustic noise alters brain activation during working memory tasks. *NeuroImage* **27:** 377–386.
36. Benjamin, C. *et al.* 2010. The influence of rest period instructions on the default mode network. *Front. Hum. Neurosci.* **4:** 218. doi: 10.3389/fnhum.2010.00218
37. Gaab, N., C. Gaser, T. Zaehle, *et al.* 2003. Functional anatomy of pitch memory–an fMRI study with sparse temporal sampling. *NeuroImage* **19:** 1417–1426.
38. Keil, B. *et al.* 2011. Size-optimized 32-channel brain arrays for 3 T pediatric imaging. *Magn. Reson. Med.* **66:** 1777–1787. doi: 10.1002/mrm.22961
39. Brown, T.T. *et al.* 2010. Prospective motion correction of high-resolution magnetic resonance imaging data in children. *NeuroImage* **53:** 139–145. doi: S1053-8119(10)00861-X [pii] 10.1016/j.neuroimage.2010.06.017
40. Fonov, V. *et al.* 2011. Unbiased average age-appropriate atlases for pediatric studies. *NeuroImage* **54:** 313–327. doi: S1053-8119(10)01006-2 [pii] 10.1016/j.neuroimage.2010.07.033
41. Sanchez, C.E., J.E. Richards & C.R. Almli. 2012. Neurodevelopmental MRI brain templates for children from 2 weeks to 4 years of age. *Dev. Psychobiol.* **54:** 77–91. doi: 10.1002/dev.20579
42. Ghosh, S. S. *et al.* 2010. Evaluating the validity of volume-based and surface-based brain image registration for developmental cognitive neuroscience studies in children 4 to 11 years of age. *NeuroImage* **53:** 85–93. doi: S1053-8119(10)00827-X [pii] 10. 1016/j.neuroimage.2010.05.075
43. Kuklisova-Murgasova, M. *et al.* 2011. A dynamic 4D probabilistic atlas of the developing brain. *NeuroImage* **54:** 2750–2763. doi: S1053-8119(10)01311-X [pii] 10.1016/j.neuroimage.2010.10.019
44. Raschle, N.M., M. Lee, R. Buechler, *et al.* 2009. Making MR imaging child's play—pediatric neuroimaging protocol, guidelines and procedure. *J. Vis. Exp.* **29:** 1309.
45. Armstrong, T.S. & H.L. Aitken. 2000. The developing role of play preparation in paediatric anaesthesia. *Paediatr. Anaesth.* **10:** 1–4. doi: pan406 [pii]
46. Hinton, V.J. 2002. Ethics of neuroimaging in pediatric development. *Brain Cogn.* **50:** 455–468. doi: S0278262602005213 [pii]
47. Fenton, A., L. Meynell & F. Baylis. 2009. Ethical challenges and interpretive difficulties with non-clinical applications of pediatric FMRI. *Am. J. Bioeth.* **9:** 3–13. doi: 907480516 [pii] 10. 1080/15265160802617829
48. Thomason, M.E. 2009. Children in non-clinical functional magnetic resonance imaging (FMRI) studies give the scan experience a "thumbs up". *Am. J. Bioeth.* **9:** 25–27. doi: 907480617 [pii] 10. 1080/15265160802617928
49. Connors, C.M. & I. Singh. 2009. What we should really worry about in pediatric functional magnetic resonance imaging (FMRI). *Am. J. Bioeth.* **9:** 16–18. doi: 907480817 [pii] 10.1080/15265160802617944

Ann. N.Y. Acad. Sci. ISSN 0077-8923

ANNALS OF THE NEW YORK ACADEMY OF SCIENCES
Issue: *The Neurosciences and Music IV: Learning and Memory*

Education through music—the model of the Musikkindergarten Berlin

Stefanie Uibel[1,2]

[1]Institute of Occupational, Social and Environmental Medicine, Goethe University, Frankfurt am Main, Germany.
[2]Musikkindergarten Berlin, Staatsoper Unter den Linden, Berlin, Germany

Address for correspondence: Stefanie Uibel, M.D., Ph.D., Institute of Occupational, Social and Environmental Medicine, Goethe University, Theodor-Stern-Kai 7, 60590, Frankfurt am Main, Germany. uibel@med.uni-frankfurt.de

In 2005, the pianist and conductor Daniel Barenboim initiated the Musikkindergarten Berlin as the first kindergarten in which music is not only used as an occasional add-on but as the central education medium for the child every day. The skills of specially trained kindergarten teachers combined with regular visits by professional musicians of the Staatskapelle Berlin (the Berlin State Opera orchestra) form the basis of a new educational concept, in which children experience music in all its different aspects and in its unique capability as a transfer medium into all the other educational areas. In this context, the method, the aim, and the experimental ground is not only education *in* or *with* music, but *through* music. This paper provides information and examples about first-hand experiences over the last six years.

Keywords: Musikkindergarten Berlin; early childhood education; music and education; Daniel Barenboim; mimetic pedagogy

Introduction

Everything began with an idea of the pianist and conductor Daniel Barenboim, who aimed for a drastic change in children's education. For him, in connection with playing and conducting music for almost all his life, music is a fundamental human expression. A favorite story of his is that while growing up with both parents working as piano teachers, it seemed normal that every person he met would play the piano and that every time the doorbell rang someone came and played. Therefore, piano playing was a natural part of human expression, something that one does not have to learn in particular—yet learning to play *well* was important.

History and facts

In March 2005, one year after another group near Barenboim built a small music school project in Ramallah (Palestine), his wish was to have a kindergarten in Berlin where "not music education, but education of children through music" (D. Barenboim) could take place. Since then, his allies have tried to give this idea—at once natural and extraordinary—a practical translation, and with participants who give it life.

Based on a volunteer nonprofit association, in which I acted as the chairwoman, the Musikkindergarten was founded in September 2005. All members of the nonprofit association come from a wide variety of interdisciplinary backgrounds and still offer their services today for free. The kindergarten started with 21 children and 3 educators; its first stages were financed by a benefit concert of the Staatskapelle Berlin and Lang Lang under Maestro Barenboim. In 2006, the kindergarten moved into the neighborhood of the Staatsoper Unter den Linden, where it expanded. Since then, about 60 children and 9 educators share the classrooms, the gardens outside, or field and rehearsal trips. Over the years, in addition to the weekly musicians' visits, nine major projects were realized. In 2007, the nonprofit association organized an international congress under the title "Music Educates," in which the first experiences and objective observations were discussed along with presentations of other

doi: 10.1111/j.1749-6632.2011.06423.x

international projects focused on music in early childhood education. Scientific evaluation and observations have had their place since the beginning, as well as pedagogic support for the educators. The musical concept of the kindergarten was published in 2010.[1] In the near future, a practical handbook of experiences, methods, trials, and errors of the kindergarten endeavor will be completed.

What the kindergarten is, and is not

The idea of "education through music" means a general and comprehensive education of children with the help of music. Music functions as the central educational medium that transfers, into all of the main educational areas that must be covered in early childhood.

Clearly, such transfer does not happen automatically and without guidance. With the right support, for example, singing and learning to listen can assist in developing speech abilities, irrespective of cultural backgrounds; music, rhythm, and motion foster the development of motor activity; experiences with acoustic and sound phenomena build bridges to natural sciences; singing and playing are able to combine counting, recognition of structures, and social competences. Music as a medium opens up doors to all educational areas and helps children find their way into the complexity of the world.

The Musikkindergarten Berlin is *not* a kindergarten with the added musical services; it is *not* a kind of early music school; and it is *not* a training unit for little prodigies or potential future audiences. It is the daytime living space of 60 children, in which music permeates all and acts as the central educational medium.

Realization and mimetic pedagogy

The implementation of this framework is dependent on the quality of musical input, which comes from the regular and generous participation of professional musicians from the Staatskapelle Berlin. Just as important is the competence of the educational kindergarten team, whose purpose is to transfer the musical input into larger interrelations and multiple connections to other educational areas. At present, the kindergarten is comprised of about 60 children, with parents from over 23 different countries and mixed social backgrounds interacting in four mixed groups of ages two to six years.

Children imitate the good as well as the bad. The better the models and the examples, the more successful the imitation. The educational framework in the Musikkindergarten Berlin can therefore be classified as mimetic pedagogy. The combined elements of aesthetic education, play, mimesis, and experiment characterize the working space within the kindergarten.

Children relate to the world through playing, which is their primary acting mode. They learn while playing a game, having fun, and earnestly coming together.

Music is also shaped by the character of *play* it is a play with notes, the instrument, the body, as well as feelings and thoughts. One does not *work* music but *plays* music. Both the children's access to their environment and music rests upon the connecting principle of play: this tenet is the central overlap between the child's appropriation of the world and musical phenomena.

The main principles—mimetic pedagogy and an experimental mindset—are fundamental for the daily work in the Musikkindergarten. "Mimetic pedagogy" means teaching and learning by imitation. This technique operates mainly without words, but requires an accurate demonstration as well as precise observation.

An experimental attitude defines the open and progressive character of work in the kindergarten—educational ambitions should not be reached through a set of prefabricated recipes, but depend on a flexible concept of respecting and adapting the live interaction of musicians, children, and educators.

The presence of musicians in the kindergarten is achieved in different ways, such as weekly visits of one or more musicians participating in the groups, as well as small concerts in the main music room. Due to a lack of practice rooms in the main venue of the Staatskapelle, the kindergarten offers members of the orchestra space to rehearse or teach students. Thus, the work in progress can be observed in real life. In addition, over the working phases of a project, the musicians come more often to the kindergarten, and conversely the children visit musicians at the opera house.

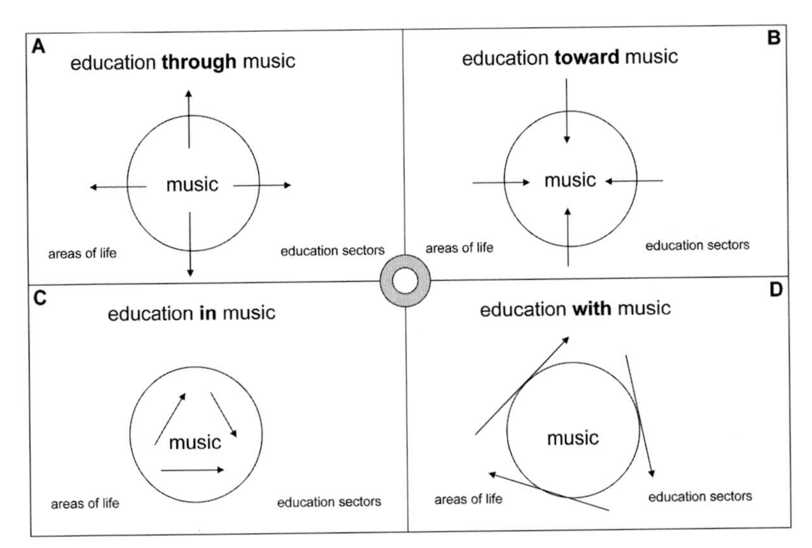

Figure 1. Relations of music and education within the kindergarten's daily work.

Education and music—relation, usage, and function

Education and music in a kindergarten setting can be combined in four ways that correspond to different functions of music (Fig. 1):

Education through music

Focusing on education *through* music (Fig. 1A), music acts as the *educational medium*, which initiates and carries the processes of how children access the world. Outside music and other aspects of life are revealed that are inherent to music.

Auditory perception and differentiation is educated by live music; communication processes are experienced and discovered in ensembles; dance and motor activity are invented according to inner motion and musical character; techniques and instrumental conditions, as well as simple physical principles of sound generation, are examined; and numerals and mathematical proportions are discovered in rhythms, songs, or compositions.

Education toward music

Education *toward* music describes the complementary way of education through music (Fig. 1B). Based on activities of daily life, the concentration is guided toward suitable music phenomena. Here, music is a planned step in the long and gradual ways of the child's learning process. Music serves as an *educational element* that enriches the child's path of world appropriation by constructing networks of aesthetic and cultural relationships.

Education in music

In the case of education *in* music, music represents the "learning object" itself, and this constitutes the more traditional field of elementary musical education (Fig. 1C).

In the domain of a kindergarten environment, this implies an introduction to basic musical features or terms like soft and loud, high and low, short and long, bright and dark. Here, contents like playing together, getting to know instruments and ways of playing them, provide the central points of interest. In this respect, music may also be considered a school subject with target-oriented teaching and remains, therefore, relatively separate from other areas of life.

Education with music

Education with music refers to a definition of music as a "didactic companion" (Fig. 1D). Nothing is developed inside the music, nothing leads to music itself, but music is used as didactic material for an extra-musical interest. Examples are the learning of short instrumental episodes, songs, or rhythmically spoken verses that mark periodic activities in the daily kindergarten routine.

When ranking these four combinations of education and music, the latter, education *with* music, is the simplest, but also the least fruitful way as it

Table 1. Educational areas in Berlin kindergartens

1 – Body, motion, and health
2 – Social and cultural environment
3 – Communication: spoken and written language, media
4 – Visual thinking, form, art
5 – Music
6 – Basic experiences in mathematics
7 – Basic experiences in natural and technical sciences

does not take into account the significance of music itself.

Education *in* music is the traditional way of subject- and goal-oriented training and is comparable with offerings like gymnastics or early Chinese language lessons—when taught at a high level, this would address the issue of music education.

Indeed, all four methods have their value, but as the basis of a wide-ranging and articulated education system education, *through* and *toward* music are essential: along those two channels, the crucial aim lies in the linkage with music of all areas of life.

Educational areas, music, and transfer

In Berlin, guidelines exist that indicate the seven main educational areas that kindergartens should receive instruction areas (Table 1). Music seems easily accessible to each of these educational fields. This can be illustrated by the example of the "Carnival of the Animals," one of our earliest projects.

Using various musical elements and pathways, children explored the aspects and behaviors of different animals: their movements, their sounds, and their character. They went to the zoo, looked at the animals, sensed them, touched some of them, and used all the impressions back in the kindergarten for further investigation: which animals live alone, which in groups, which in the sky, which in the water, and how is this possible? In an atlas, they looked up the animals' origins and natural environments and learned that in different countries, they have different names. It was easy to show, for example, that to cover the same distance, an elephant and a mouse have to take a different number of steps. The children painted, built models, invented habitat areas, and so on.

The musicians came with their instruments, played the movements, showed instruments, and performed variations, and after the children had worked at each field for weeks, some musical pieces were selected and the children and educators transposed their findings into a performance, which then was shown to the parents in a private concert. In addition, the educational and creative path itself was the goal—not a representative show of drilled children.

Experiences, results, and expectations

Visitors often are astonished when they visit because they expect a music kindergarten to be a horribly loud place, more so when they learn that all 60 children have access to all kinds of instruments the whole day. Yet the opposite is true because importance is attached to making the children attuned to their senses, music is used in a focused manner and only acquires its value in combination with its counterpart, silence.

In fact, some lessons in which no one has to teach the children with words, are very simple, yet very effective. For example, when we have soloists and tutti players, every player has to listen to the other—even two musicians have to communicate *with* each other and not *against* each other to play a song in a satisfactory way. They have to start together, find the right tempo, and stop together, and before playing, they need to tune in with the other instrument. Besides, solo parts often do alternate, so it would not be advisable to disrupt someone's solo because then others might do this to you when your own solo comes along. In all of these contexts, harmony and disharmony may be perceived in live music, thereby evoking unspoken patterns of social behavior.

Measuring the success of a kindergarten is complex and a topic that goes beyond the scope of this paper. However, evaluations from structured observations have been gathered. Independent scientists have found that, compared to age-matched children from other kindergarten, over a short period of time, children in the Musikkindergarten exhibit higher levels of social competency and communication skills. Very positive effects were also found in children with attention deficit hyperactivity disorder (ADHD) and children with other impairments in social interaction abilities.

What can be reported in parallel to scientific measurability is high satisfaction of parents and children who have attended the Musikkindergarten since 2005, also in relation to the development of their own children or siblings in other kindergartens.

Real life

As both a conclusion and a perspective, I'd like to report a short episode from the kindergarten that could be used as an example of awakened sensibilities and keen social competency or better social learning—including adults.

One day, in a regular group, a Columbian boy suddenly pretended not to understand any German and began speaking Spanish. The kindergarten educators spoke with him and asked about any problems or worries, but nothing helped. The boy remained stubbornly closed-up and still spoke Spanish; after several attempts, the educators gave up and shifted their focus on other issues. Later, another observer came into the room and witnessed an amazing situation: although all the children were talking, no one was understandable—they all had adapted to speaking Spanish—not really, of course—but they imitated the spoken sounds, syllables, and words with great amusement and considerable inventiveness. Only several minutes later did the Columbian boy suddenly start "understanding" German again.

The children simply did, through their actions, exactly what we wished them to do: they watched the scene, looked and listened carefully, and by imitation, they took the unknown language on like a new acoustic phenomenon, like a song, or simply like a speech game. It became something that with some practice, every child could do. And while interacting with each other and with the Columbian boy, they socialized and harmonized the situation unknowingly better than any adult or external person might have done—a remarkable result of education through music.

Acknowledgment

I thank all those who help to realize the Musikkindergarten Berlin through all these years, filling the idea of education through music with life every day.

Conflicts of interest

The author declares no conflicting interests.

Reference

1. Doerne, A. 2010. *Musik bildet: Der Musikkindergarten Berlin. Ein Modell*, Breitkopf & Härtel, Wiesbaden, Germany.

Ann. N.Y. Acad. Sci. ISSN 0077-8923

ANNALS OF THE NEW YORK ACADEMY OF SCIENCES
Issue: *The Neurosciences and Music IV: Learning and Memory*

From the model of *El Sistema* in Venezuela to current applications: learning and integration through collective music education

Maria Majno

Sistema delle Orchestre e Cori Giovanili e Infantili in Italia, Italy, and Fondazione Pierfranco e Luisa Mariani for Child Neurology, Milan, Italy

Address for correspondence: Maria Majno, Fondazione Mariani, Viale Bianca Maria 28, 20129 Milan, Italy. majno@fondazione-mariani.org

Over the last years, *El Sistema*—the Venezuelan project started in 1975 and now acknowledged worldwide as the most significant example of collective music education—has inspired a profusion of remarkable initiatives on all continents. From the original impulse by founder José Antonio Abreu, strong social principles of integration are combined with specific musical approaches to achieve individual empowerment as a large-scale alternative to endemic juvenile crime, counteracting the risk factors of social unease, serving as a stimulating example toward emancipation, and providing professional opportunities to the talented. Such a network, in turn, proves to be a powerful instrument of cultural progress: the tenets of "Sistema" become shared values able to foster development, reaching into issues of disability and rehabilitation. This paper presents continuities and contrasts in various ramifications of such a successful trend and outlines perspectives for further impact of this powerful transformational agent.

Keywords: *El Sistema*; Venezuela; Italy; music education; social integration; Claudio Abbado; José Antonio Abreu

Memory, this proud mother, resumed:
"My daughters are called the Muses because, all together, they have borne music. Since it recruits my nine daughters, it is the first human art: none would have access to beauty, if not through music."[1]
Learning: Homeopathy of Evil.[1]
The inspiration and rewards unleashed by music are universal benefits that must be available to all as a human right.... Best practice models exist around the world, which show how this can be achieved.[2]

Basic tonality and scoring of a dynamic "system"

Since its beginnings in 1975, the Venezuelan program, now referred to as *El Sistema*, of youth orchestras and choirs, has touched over two million children and currently involves about 400,000 youth from ages two and up,[3] in a country where the population totals about 29 million (three million of whom are in the capital, Caracas).[a] This telling proportion of the project's scope is just one measure of its importance. Over the years, the increasing success of "the most important musical initiative of our time"—a recurring definition by Sir Simon Rattle[4,5]—has spurred the Venezuelan government to invest considerable resources into the stabilization and promotion of its most celebrated cultural product. Its initiator, José Antonio Abreu,[b] a talented musician and gifted politician endowed with a unique

[a]The most current estimate, published in the *New York Times* on February 16, 2012 in an interview with Executive Director Eduardo Méndez, quotes the lower figure of 310,000 (see Ref. 11).

[b]José Antonio Abreu is acknowledged as the most significant leader in bridging music education and social progress in our time and has been given about 100 awards in the interdisciplinary facets of his activity, including a number of honorary degrees, among which a recent one in law from the State University at Milan, Italy; at the time of this writing, he is being considered for a Nobel Peace Prize nomination.

doi: 10.1111/j.1749-6632.2012.06498.x
Ann. N.Y. Acad. Sci. 1252 (2012) 56–64

vision, focused from the start on the definition of a far-reaching "system" that he consistently pursued: first with a small group of allies, then by expanding waves of influence as his country progressively realized the potential of such an endeavor.[6]

From the initial idea by Abreu and his friends, including Frank Di Polo, to bring more music where there was too little—their first rehearsal in an underground garage on February 12, 1975, has become somewhat of a legend[4]—the challenge became a concerted effort: to find an antidote to degradation and economic deprivation through an alternative offer that could be more attractive than gangs, drug-dealing, and violence.[7] The mission of *El Sistema*, under the motto "to play and to fight,"[8] is officially stated as: "To systematize music education and to promote the collective practice of music through symphony orchestras and choruses in order to help children and young people achieve their full potential and acquire values that favor their growth and have a positive impact on their lives in society."[9] Thanks to their involvement in such committed music making, children can become somebody: a main step in dignity, on a path of empowerment toward becoming an active contributor to the community.[10]

Musical goals must progress hand in hand with social aims: gradually, a still-expanding network of around 280 *Núcleos*[11]—the music schools that serve as basic units for training and performing activities—has been put in place in all Venezuelan states. It is bearing fruit with a plethora of orchestras and choirs (reportedly, 1,000 instrumental and vocal ensembles are active in the country), and a proven decrease in violent behavior and a better attitude among youth toward a realistic hope that a brighter future is at hand.[12] This connection between cultural improvement and constructive self-improvement is giving rise to more and more widespread interest with an abundance of publications—books, articles, films—posing a challenge to provide a truly current bibliography.[c]

As tangible confirmation, a number of achievements are in the public eye: the astonishing headquarters at Centro de Acción Social por la Música in Caracas (where music schools and a university are hosted under the same roof with several concert halls); the concerts of the Sinfonica Simón Bolívar (two orchestras, qualified as A and B according to seniority) and the younger but no less impressive counterpart Teresa Carreño (also launched in 2007 in addition to *La Red de Orquestas Sinfónicas Penitenciarias*, composed from detained youth); and the individual accomplishments of such prodigies as Gustavo Dudamel, Diego Matheuz, and Edicson Ruiz (whose respective careers include the positions of music director of the Los Angeles Philharmonic, music director of Teatro la Fenice in Venice, and first double-bass of the Berlin Philharmonic).[4,6] All of this confirms not only that the Venezuelan dynamic model is there to stay, but it is also ready to be exported.

Themes: exposition and developments

The approach of *El Sistema* involves a number of methodological principles primarily woven into the original form of the project in Venezuela.[4,13] A survey of such concepts and the values they encourage are useful in identifying the cardinal points essential to the orientation of other *El Sistema*–inspired programs of music education with marked social implications.

Accessibility

Accessibility implies, above all, nonselective, nonelitist criteria for admission. Music education in the Nuclei is open and free of charge; this immediately sets *El Sistema* apart. However, a flexible application of the principle has led to a modified form of optional contributions in which those who can afford more provide for those who have less or nothing. This sensible evolution, while not betraying the original tenet, provides a model for other *El Sistema*–like organizations abroad, where state support is by no means a given.

Indeed, "Culture for the poor should not mean poor culture," and along this often-quoted adage by J.A. Abreu, another challenge is that of physical accessibility of the Nuclei. Especially in remote regions, a major investment of energies and resources has to be made in order to ensure that children reach the music schools. *El Sistema*—and its derivations—must also be a system of outreach for parents and families.

[c]A comprehensive bibliography from the thesis of Esmeralda Colombo (see Ref. 7) is provided in the online materials (Appendix S1) as an appendix to this chapter.

Regularity and intensity of training

Regularity and intensity of training are indispensable ingredients when the aim is to provide a compelling alternative to dangerous occupations. Venezuelan *Núcleos* basically operate four hours seven days a week with an intensity that would be unrealistic to try and replicate under different circumstances. However, a sustained pace goes hand in hand with the tenets of discipline and commitment, which underpin the aim of fostering the sense of responsibility and autonomy in the child: "At first I did not want to do it, now I love it."[14] As an indication of how this criterion may be applied in other contexts, In Harmony in the United Kingdom (initially three selected cities for three years) offered seven hours a week,[15] and the Sistema Italia requires at the start a minimum of two hours a week,[16] as existing realities struggle to cope with the demand of gratis tuition, notwithstanding the absence of public subsidies.

Collective courses—ensemble practice

From the start, children are included in ensembles for basic instrumental and vocal training, as opposed to the traditional, conservatory-derived approach to the training of soloists, which paradoxically produce mostly unemployed musicians. Not only is ensemble playing applied on a large scale as an efficient multiplier of energies and resources, but it also provides the ideal setting to nurture a number of essential values toward better community life, such as respect, equality, sharing, cohesion, team work, and, above all, the enhancement of listening as a major constituent of understanding and cooperation.[67]

Ensemble playing also is conducive to an atmosphere of shared passion and enthusiasm (two main Abreu tenets), easily escalating into individual joy. The element of fun shines through the labels of some Nuclei and their orchestras and inspires the outlook as far as Sistema Scotland, where the overall project is named *The Big Noise*.[17]

Finally, playing together can overcome language barriers[18] and other disadvantages—an essential trait in projects addressing integration and inclusion, both within societies heavily laden with immigration and in the context of disability.

Pursuit of artistic quality and reward of excellence

It is not enough to play; one must play well—at least as well as possible. As proven by the level of Venezuelan orchestras heard on international stages, often with very young soloists, the system is also keen on developing precocious talents (Abreu himself was one) and identifies them with attentive screening while encouraging a very early start at the preschool stage. Training is carefully staggered along progressive lines—the so-called "sequencing" repertoire intertwining several methods (Kodály, Suzuki, and Dalcroze among them)—and leaves ample room for the individual teacher's initiative. The most appropriate image is that of a pyramid: with the widest possible base, it ensures a solid movement to the top level and the highest musical standards, often with poignant ambitions. "I wanted to be a professional skateboarder; now I want to be the first cellist of the Los Angeles Philharmonic;" that is hope for a member of YOLA.[19] For all of its exemplary social values, *El Sistema* and its offspring would not have sustained the world's scrutiny without its convincing artistic achievements.

Modulations: adapting to different contexts

El Sistema has given way to a string of adaptations in the different environments in which attempts to introduce similar initiatives are being pursued with more or less stable results. It is Abreu's conviction that a simple transplant "as is" would not be viable—a translation to the specificities of each context is more reasonable.[20]

Initiatives in South America bear the closest degree of analogy and are, from a historical viewpoint, also the nearest to the Venezuelan source. Most noteworthy are the results in Brazil, where a dialogue of cultures (involving European protagonists from the past and the present) mark the endeavors of the Instituto Bucarelli in Saõ Paolo[21] and the NEOJIBA network initiated in San Salvador de Bahia by pianist Ricardo Castro.[22]

The speed with which *El Sistema*–like initiatives are migrating to other continents bears witness to a growing contagion with positive outcomes. Sistema USA,[23] born as a support and advocacy organization then coming into its own with an incisive presence, is attempting to construct a census of ongoing and budding projects. The results are progressively being organized and presented in the periodical newsletter *The Ensemble,* as well as in conferences, thanks to the Abreu Fellows (a program initiated at the New England Conservatory,[24,25] as a result of the TED award to J.A. Abreu), who share their "common goal of reimagining music as a catalyst for social transformation."[26]

While looking at the awesome numbers and unprecedented achievements of *El Sistema*, followers should not be deterred by the difficulty to create a comprehensive structure respecting all principles at once on a large scale. Even with a cursory survey of what has been developed over the last 10 years in the Americas and more recently in Europe, a conspicuous effect of what may be called *micro-engagement* on a local level (akin to the model of microfinancing) is clearly notable and can play a decisive role in the kick-off of multiple smaller endeavors, with promising cumulative results.

A main distinction between the possible styles of adapting *El Sistema* to different systems are the school-based projects and projects hinging on the creation of dedicated music schools. A remarkable product of hybridization has ensued from the activities of the People's Music School in Chicago, located in an elementary school and aimed at 6- to 14-year-olds living in a strongly disadvantaged neighborhood. Thanks to the cooperation of highly qualified forces, such as the Ravinia Festival, the YOURS project is carrying on a consistent pursuit, balancing the core methodology of *El Sistema* with the need for adaptation to the American context.[7]

In newer setups, even more than in the original Venezuelan scene where state support has been the prerequisite and the operational condition from the start, eclectic alliances are the main key to sustainability. For example, in the United States, the Harmony Program in New York,[6,27] with its solidly acquired partnerships, is a bright light in a growing nationwide constellation, whereas the already-mentioned YOLA (Los Angeles Youth Orchestra) can count as an asset its proximity with the philharmonic's music director and *El Sistema's* number-one star, Gustavo Dudamel.

Possibly the most noteworthy trait in the interrelated projects on the North American continent is the capacity for cross-fertilization and the ability to transcend deep-rooted contrasts—or even conflicts (such as those with Cuba). The paradigm of Daniel Barenboim's West-Eastern Divan Orchestra,[28] reuniting forces from opposite sides of the Middle-Eastern struggle, is an inspiration toward overcoming tensions in the practice of a common goal.[29,30] Canada has also joined forces (hosting a conference at McGill University in November 2011),[31] and is concentrating its efforts on the activities of YOA, the Youth Orchestra of the Americas, whose activities are supported by world-renowned musicians and very diverse entities.[32]

In Europe, initiatives are being developed along three main channels: (1) pilot projects with close kinship to *El Sistema*, (2) national initiatives inspired by similar ideas, and (3) attempts at more concerted efforts to replicate the model on a systematic basis. Among the first initiatives is the already mentioned Big Noise–Sistema Scotland,[33] where Big Noise is the name of the youth orchestras (with their first center in Raploch), and Sistema Scotland refers to the charity supporting and organizing the project, established in 2008. A parallel effort has been cultivated by In Harmony–Sistema England,[34] where a decisive visit from the Simón Bolívar Orchestra to the UK ignited the enthusiasm of audiences and musicians alike, and spurring a key politician to earmark £1 million for the impulsive, intensive start. Three cities—Lambeth, Liverpool, and Norwich—were selected on the basis of a strictly competitive call and were methodically monitored; they now face the challenge of continuity, where sustainability must combine the forces of limited public funding with private contributions and the individual resources of musicians who believe in multiplying the effects of such significant investments.

The second group of *El Sistema*–inspired projects can be exemplified with JEKI (*jedem Kind ein Instrument* [to each child an instrument]), implemented in the German region of North-Rhine–Westphalia.[35] Again, the mission derives from an inspiration clearly related to *El Sistema*: "promoting cultural and musical education for our children, so as to encourage wider interest and enthusiasm for culture." However, in this undoubtedly affluent region (the onset of the project was connected to Ruhr's role as European Capital of Culture in 2010), it is to be expected that the social component that plays such a primary role in the Venezuelan impulse may be overshadowed by a lasting homage to the mainstream Western tradition: "the long-term goal is to improve the basic conditions for cultural education in Germany."

For a more coherent translation of *El Sistema* within Europe, the Italian Sistema delle Orchestre e dei Cori Giovanili e Infantili can be used as a paradigm. Launched by the initiative of one of the country's main musical figures, conductor Claudio Abbado—an enthusiastic supporter of the Venezuelan project since 1999—the project is rapidly taking shape by implementing a national structure with official ties to the original model (a bilateral agreement that was signed in February 2011). In parallel, a network of regional initiatives is developing, with marked features in relation to the local characteristics that are a distinguishing trait of the country's eclectic style.[36] According to the very diverse regional contexts, the focus may be directed to widespread in-school training (e.g., the Alto Adige/South Tyrol region), to the involvement of children as ambassadors of peace (Pequeñas Huellas in Piedmont), to a full-fledged, established music school renewing its course (Fiesole in Tuscany), to expanding youth ensembles (Emilia-Romagna), to pilot projects for the disadvantaged (the southern regions and a number of immigration-laden communities), or to productions aiming for higher artistic results, as in the previously mentioned pyramid of increasingly proficient ensembles, with exposure that rewards musical excellence (FuturOrchestra in Lombardy).[37]

Exports of *El Sistema* are now reaching the Far East with sustained attempts, for instance, in Korea and China; films such as *Let the Children Play*[14] bear witness to a potentially inexhaustible pool of interlocutors in economically emerging countries. Australia has also been effectively spreading its own message to nurture special musical talent,[38] and in Africa, attempts like that of K. Devroop[39,40] struggle with utmost dignity to introduce the resources of music on a smaller scale, in spite of apparent dismissal by those who downgrade this strife as an irrelevant concern where survival is at daily stake.

Counterpoint and dissonances: controversies, caveats, and hurdles

To Abreu's statement that the path of *El Sistema* concerns three spheres of the child's development—personal, family-related, and community-oriented[7]—a parallel triad can be juxtaposed of the artistic, cultural, and social aspects involved. Even if the values propagated by the *El Sistema* philosophy and its application are by now widely acknowledged, it would be unrealistic to ignore a contrasting channel in which difficulties of today echo those experienced from the start in getting the new approach recognized as a major asset for the present and the future of children. J.A. Abreu and his supporters are certainly not unfamiliar with controversy as it took the better part of 20 years to see their alternative conservatories be recognized alongside those of mainstream education; it is no accident that, in Venezuela, *El Sistema* belongs neither to the Ministry of Education nor to that of Culture, but pertains to the field of social affairs.

Having achieved official recognition of its structure, established its indispensable role to the country's cultural profile, and successfully met the challenge of artistic assessments abroad, *El Sistema* now faces the ongoing task of proving its capacity to grow beyond the striking results of its first 36 years. On an international level, a lucid assessment of the status quo emerges off and on, in the press, and might be usefully employed as constructive criticism.[41]

A first issue arises from the relationship between quantity and quality. While it is obvious that a uniformly high level may not be possible with the numbers involved in the Venezuelan project, it is also true that acceptably elevated standards should remain an important concern. Abreu has often voiced his aim to increase the pool of children up to one million, but—even if the evolution of rising support would only be welcome—overabundance may cause an actual risk of sprouting toward uncontrolled applications.

Another critical point relates to the characterization of *El Sistema* as a state régime that—contrary to the original inspiration oriented to individuals—seems to be turning lately into an increasingly patriotic affair. Revelatory symptoms are the renaming of FESNOJIV as Fundación Músical Simón Bolívar in February 2011 and the current direct dependence of *El Sistema* on the office of President Chávez, pointing to the double-edged sword of a political connotation perhaps too close for comfort.

On the artistic side, a recurring caveat is that *El Sistema* should avoid the pitfalls of a performance machine more concerned with impressive appearances than with in-depth development (for example, in the traditional encores where—after a suspenseful blackout—the explosion of the Venezuelan colors on the young musicians' jackets can become predictable when seen over and over again). Issues regarding the choices of repertoire are also crucial in the long run.

Within such a hierarchically ordered structure, born from the determined inspiration of a single far-reaching mind, it is hard to imagine how the spirit of Maestro Fundadór Abreu can be replicated, divided, and delegated without dilution of values and energies. Yet, experience has proven that precious cooperative ventures (like the instrumental patronage of the Berlin Philharmonic)[42] could benefit from more efficient management and flexible relations from the Venezuelan side, to avoid the risks of autarchy becoming self-referential.

Cadenzas: corollaries and arrows into the future

Just a few more remote steps from the model, other programs have adopted the principles of *El Sistema*, adapted its methods, and meshed its aims into their own objectives. Within this review, significant lines are being drawn to the connections with the humanitarian, political, and scientific fields.

Music as a universal language, able to transcend barriers of countries and social strata, is the ideal vehicle for conveying supranational messages. As a most poignant example of the protection of children from exploitation, the International Labour Organization's IPEC program has cultivated SCREAM (Supporting Children's Rights through Education, the Arts and the Media)[43] and produced a powerfully concise film combining music and images of almost unbearable intensity.[44] The motto "child to child solidarity against child labor" conveys very fruitful perspectives on the much-needed investment to stir empathy and complicity between the young recipients of very unevenly distributed resources.[d]

On the political front, in addition to the already mentioned West-Eastern Divan Orchestra, the role of music in conflict zones[45] is at the core of organizations like Music in Me[46] and in a growing number of networks, with increasing efforts also in systematizing knowledge, methods, and action.[47] A unique effort is being deployed by the Music Fund to ensure the availability of instruments, thanks to periodic donations organized, with impressive results, in various cities,[48,49] and through on-site training for the purpose of maintenance and repair.

Another outreach movement with distinctive social value is that of music in prisons: following the Irene Taylor Trust's commitment since 1995,[50] alliances have been established also by musicians of great caliber, such as Claudio Abbado, the Mozart Orchestra with the TAMINO project, and Salva la Musica of the Monzino Foundation.[51]

Finally, and most significantly for the field of child neurology, which provides a forum for this ongoing series of meetings and publications on the neurosciences and music, collective music education is increasingly employed in the areas of special education, rehabilitation, and music therapy. Underpinning the effectiveness of these interventions are the theories of social learning,[52] affective memory, emotional intelligence and awareness, development of motor skills, and cognitive enhancement,[53] all closely related to the domain of music neuroscience. The paradigmatic project of Manos Blancas in Venezuela, where hearing-impaired and other disabled children perform in choirs with their hands clad in white gloves, evoking the motions and emotions of music, exerts an impact beyond the immediate power of sympathy: it is an utterly practical example of integration and inclusion, now growing in numbers and spreading to other milieus. As another Sistema child simply states, "When I listen to music I hear the notes in my ear and I feel better."[14] In the area of disability, *El Sistema* is inspiring other endeavors, such as Esagramma and "Allegro Moderato," which includes members of the Sistema Italia, where severely impaired children play in ensembles with dedicated tutors in a setting where cultivating the orchestral mind[54] becomes the key to flexible yet structured modes of behavioral intervention.

Coda: a well-tempered horizon

Where is *El Sistema* heading? How is it exercising the leadership principle it so intensely vouches for? A counterpoint of Sistemas is being woven into more and more voices, as the most recent conference "*Take a Stand*" resonantly demonstrates.[55] Key figures are taking on the role of promoting ideas, multiplying projects, and fostering collaborations. The hope is that literature will also evolve toward a more scientifically minded approach.[e]

[d]Musical ensembles of the Sistema in Italy, such as Futur-Orchestra, Pequeñas Huellas, and a number of children's choirs, often dedicate their public appearances to the support of humanitarian causes.

[e]It is noteworthy that two of the most recent books about *El Sistema* (by Kaufmann & Piendl[42] and Tricia Tunstall[64]) do not include a reference list. On the other hand, more

Effects of the efforts to compensate for widespread social deprivation, and in more affluent environments the emotional disembodiment affecting many youth, are suddenly obvious and empirically proven, to the point of having become almost commonplace. However, to justify large-scale investments, it is imperative to aim for more evidence, in terms of both cognitive benefits and social improvements. Research is indeed being carried out on groups of underprivileged children (e.g., in Oregon's Head Start schools[56] or in Chicago's Hibbard Elementary School[7,57]), but coordination of efforts beyond local compilations is imperative in order to achieve a critical mass of results. A useful meta-analysis of studies could be developed in training programs like the Abreu Fellows and other advanced academic settings.

In parallel, further teamwork is needed in the pooling of resources, with the aim to optimize availability of information, accessibility of materials, exchange of experiences, and training opportunities. In today's increasing recourse to electronic media, storage and sharing have suddenly become less costly to realize and easier to promote. This kind of central support to decentralized initiatives is already in place in Venezuela:[f] other Sistemas should cooperate accordingly, as *El Sistema* officials have been expressing their willingness to contribute in this direction.

The multitude of Sistema-inspired projects involving music, psychology, pedagogy, neuroscience, politics, sociology, economics, and, respectively, musicians, scientists, thinkers, teachers, therapists, and managers provides an outstanding terrain for interdisciplinary convergence. While already on the specifically musical turf, the educational mission can create employment for instrumentalists, composers, tutors at various levels, and organizers. Yet, providing professional marketplaces is by far not the main purpose of Abreu's undertaking, and his overriding statement often recurs: "I do not just want to train better musicians—I want to form better *people.*"

In his preface to a most recent publication on *El Sistema*, Claudio Abbado elaborates: "Listening is an indispensable element in civil life, though it is often neglected. I am convinced that there is not only an aesthetic value in making music: its intrinsic beauty is the source of an intense ethical value."[58] When considering on the one hand the specificity of human culture[59] and on the other hand music's multiple roles in the course of human civilization,[60] it can be argued that the different, mostly conscious uses of music are among the most suitable traits for distinguishing humans from animals.[g] Also, the emerging subject of cultural psychology deserves to be integrated into the practices of music education.[61]

Awareness that music may be the transformational tool par excellence has gradually been reaching the public at large. While experts and policy makers strive to issue applicable guidelines as widely as possible,[62] the power of music[63] has evolved outward by becoming transformative;[64] in a recurring movement to see music as a dynamic agent of change, the growth *of* music, *in* music, and *through* music is proving to be highly sustainable also from an environmental standpoint.[65]

"For children, as for musicians, learning never ends," as Claudio Abbado likes to assert. The motto of *El Sistema* is *Tocar y Luchar* [to play and to struggle].[h] While it is established that playing is the child's work, a paraphrase might be allowed: To play and. . .to play, uniting the ideas of freedom and the discipline necessary to music making, for the benefit of children's growth, togetherness, dignity, and joy. In the words of Pliny the Younger: *Aliquando praeterea rideo, jocor, ludo, utque omnia innoxiae remissiones genera breviter amplectar, homo sum* [Moreover at times I laugh, I jest, I play; and

---

and more academic dissertations are being produced on these themes.

[f]The recently inaugurated Centro de Acción Social por la Musica includes a comprehensive library of music, documents, and audio and video recordings at the disposal of all Sistema branches.

[g]This subject was discussed in an interdisciplinary working group on "Human specificity" that operated informally under the guidance of Prof. Steven Lukes at the Berlin Wissenschaftskolleg in 2010, integrating the viewpoints of anthropology, biology, sociology, philosophy, history, neurosciences, art history, and music.

[h]The customary translation of the motto, as mentioned previously (Ref. 7), is "To Play and to Fight"; we suggest this other formulation, as it does better justice to a nonviolent alternative.

in short, to embrace all harmless recreations in one word, I am a man].[66]

Acknowledgments

The support of Fondazione Francesco Pasquinelli as partner to Il Sistema in Italy is gratefully acknowledged, as is advice from Luisa Lopez, Esmeralda Colombo, Mark Churchill, and Leonardo Panigada.

Supporting information

Additional supporting information may be found in the online version of this article.

Appendix S1. Comprehensive bibliography from the thesis of Esmeralda Colombo

Please note: Wiley-Blackwell is not responsible for the content or functionality of any supporting materials supplied by the authors. Any queries (other than missing material) should be directed to the corresponding author for the article.

Conflicts of interest

The author declares no conflicts of interest.

References

1. Serres, M. 2011. *Musique*. Le Pommier. Paris: 22. *(translation by M. Majno.)*
2. Salzburg Global Seminar. 2011. *The Value of Music: The Right To Play*. Final manifesto. URL http://www. salzburg-global.org/mediafiles/MEDIA60381.pdf [accessed on 18 February 2012].
3. Fundación Musical Simón Bolívar. URL http:// fundamusical.org.ve/es/conciertos/1171-con-dudamel-por- la-paz.html [accessed on 18 February 2012].
4. Borzacchini, C. 2010. *Venezuela en el cielo de los escenarios*. Fundación Bancaribe. Caracas, Venezuela.
5. *El Sistema. Music to Change Life*. Dir. P. Smaczny & M. Stodtmeier. Film. 2008. Euroarts. URL http://www.el-sistema-film.com/el_Sistema_The_Film.html
6. Kaufmann, M. & P. Piendl. 2011. *Das Wunder von Caracas. Wie José Antonio Abreu und El Sistema die Welt begeistern*. Irisiana. Munich, Germany.
7. Colombo, E. 2011. *Dignità sociale e rieducazione in una politica criminale "polifonica": l'esperienza di El Sistema*. Università Cattolica del Sacro Cuore. Milan, Italy.
8. *Tocar y luchar*. Dir. A. Arvelo. Film. 2005. Explorart. URL http://www.tocaryluchar.com/
9. Fundación Musical Simón Bolívar. URL http://www. fesnojiv.gob.ve/en/mission-and-vision.html [accessed on 18 February 2012].
10. TED – Technology, Entertainment, Design – Ideas Worth Spreading. Acceptance Speech by José Antonio Abreu. URL http://www.ted.com/talks/jose_abreu_on_kids_transformed_by_music.html [accessed on 18 February 2012].
11. Wakin, D.J. Fighting poverty, armed with violins. *The New York Times*, URL http://www.nytimes.com/2012/02/16/arts/music/el-sistema-venezuelas-plan-to-help-children-through-music.html?_r=1&emc=eta1 [accessed on 16 February 2012].
12. *A Slum Symphony. Allegro Crescendo*. Dir. C. Barbarossa. Film. 2010. Verve Media Company. URL http:// www.aslumsymphony.com/
13. Fundación Musical Simón Bolívar. URL http://www. fesnojiv.gob.ve/en/el-sistema-methodology.html [accessed on 18 Feburary 2012].
14. *Dudamel: Let the Children Play*. Dir. A Arvelo. Film. 2010.
15. In Harmony – Sistema England. URL http://www.ihse. org.uk/ [accessed on 18 February 2012].
16. Sistema delle Orchestre e Cori Giovanili e Infantili in Italia. URL www.federculture.it [accessed on 18 February 2012].
17. Sistema Scotland. URL http://makeabignoise.org.uk/ sistema-scotland/ [accessed on 18 February 2012].
18. Patel, A. 2008. *Music, language and the Brain*. Oxford University Press. Oxford/New York.
19. Los Angeles Philharmonic. URL http://www.laphil.com/ education/yola.cfm [accessed on 18 February 2012].
20. Fundación Musical Simón Bolívar. URL http:// www.fesnojiv.gob.ve/en/el-sistema-as-a-model.html [accessed on 18 Feburary 2012].
21. Instituto Baccarelli. URL www.institutobaccarelli. org.br/ [accessed on 18 February 2012].
22. NEOJIBA Núcleos Estaduais de Orquestras Juvenis e Infantis da Bahia. URL http://www.neojiba.org/en [accessed on 18 February 2012].
23. El Sistema USA. 2012. URL http://elsistemausa.org/.
24. New England Conservatory. URL http://necmusic. edu/abreu-fellowship [accessed on 18 February 2012].
25. Sussman, E. 2011. Music as vehicle for social change: Mark Churchill and El Sistema USA. *School Band and Orchestra*. URL http://readperiodicals.com/201101/2397594441.html#b [accessed on 1 January 2011]
26. El Sistema USA. 2012. *The Ensemble* **1:** 1.
27. Harmony Program. URL http://harmonyprogram. cuny.edu/ [accessed on 18 February 2012].
28. West-Eastern Divan Orchestra, URL http://www. west-eastern-divan.org/ [accessed on 18 February 2012].
29. The Barenboim Said Foundation, URL http://www. barenboim-said.org/en/inicio/index.html [accessed on 18 February 2012].
30. Barenboim, D. 2006. *The BBC Reith Lectures*. URL http://www.bbc.co.uk/radio4/reith2006/
31. Canadian Music Educators Association. URL http://www. cmea.ca/resources/latest-news/182-canadian-symposium-ii-on-el-sistema [accessed on 18 February 2012].
32. Youth Orchestra of the Americas. URL http:// yoa.org/ [accessed on 18 February 2012].
33. Sistema Scotland. URL http://makeabignoise.org.uk/ sistema-scotland [accessed on 18 February 2012].
34. In Harmony – Sistema England. URL http://www. ihse.org.uk/ [accessed on 18 February 2012].
35. Stiftung Jedem Kind ein Instrument. URL http://www. jedemkind.de/englisch [accessed on 18 February 2012].

36. Sistema delle Orchestre e dei Cori Giovanili e Infantili in Italia. URL www.federculture.it [accessed on 18 February 2012].
37. Radaelli, A. 2012. *La musica cambia la vita*. Feltrinelli. Milano, Italy.
38. *Mrs. Carey's Concert*. Dir. B. Connolly and Sophie Raymond. Film. 2011. Screen Australia. URL http://www.mrscareysconcert.com/
39. Devroop, K. 2009. *The Effect of Instrumental Music Instruction on Disadvantaged South African Students' Career Plans*. URL http://hdl.handle.net/10394/4244
40. Devroop, K. 2010. The South African Music Outreach Program. Presentation at the conference "Musica e Società." Scuola di Musica di Fiesole, Florence, November 2010.
41. Frei, M. 2011. Wie viel System steckt in System? Venezuela und das soziale Musikprojekt "El Sistema." *Neue Zürcher Zeitung*, 21 November 2011. URL http://www.nzz.ch/nachrichten/kultur/aktuell/wie_viel_system_steckt_im_system_1.13372944.html
42. Kaufmann, M. & P. Piendl. 2011. *Das Wunder von Caracas. Wie José Antonio Abreu und El Sistema die Welt begeistern*. Irisiana. Munich, Germany. 224–233.
43. International Labour Organization. 2012. URL http://www.ilo.org/ipec/lang--en/index.htm
44. ILO – IPEC. 2006. *Child to Child Solidarity Against Child Labour*. URL http://www.youtube.com/watch?v=0BF6y9Hfpn4
45. Osborne, N. 2012. Neuroscience and "real world" practice: music as a therapeutic resource for children in zones of conflict. *Ann. N.Y. Acad. Sci.* **1252:** 69–76. This volume.
46. Music in Me. URL http://www.musicinme.net/ [accessed on 18 February 2012].
47. Pairon, L. 2012. Practices in culture and development–A comparative study of the role of music in development projects, with a specific interest in music and conflict. PhD thesis, University of Ghent, Ghent, Belgium.
48. Music Fund. URL www.musicfund.be [accessed on 18 February 2012].
49. Milano Musica. URL www.milanomusica. org/musicfund/ [accessed on 18 February 2012].
50. Irene Taylor Trust. URL http://www.musicinprisons. org.uk/ [accessed on 18 February 2012].
51. Fondazione Antonio Carlo Monzino. URL http://www.fondazioneacmonzino.it/en/ [accessed on 18 February 2012].
52. The Royal Society, United Kingdom. 2010. Satellite meeting on social learning.
53. Hallam, S. 2011. Music education. The role of affect. In *Handbook of Music and Emotion: Theory, Research, Applications*. P.N. Juslin and J. Sloboda, Eds: 791–817. Oxford University Press. New York.
54. Sbattella, L. 2008. *La mente orchestra. Elaborazione della risonanza e autismo*. Vita e pensiero. Milano, Italy.
55. Los Angeles Philharmonic. "Take a Stand" Symposium in Los Angeles. URL http://www.laphil.com/ education/yola-symposium/index-2012.cfm [accessed on 18 February 2012].
56. Nelson, H. 2008. Effects of music training on brain and cognitive development in underprivileged 3-to 5-year old children: preliminary results. In *Learning, Arts and the Brain*, C. Asbury & B. Rich, Eds. The Dana Foundation. New York.
57. Hibbard Elementary School. URL http://www. hibbard.cps.k12.il.us/ Accessed on 18 February 2012].
58. Radaelli, A. 2012. *La musica cambia la vita*, Feltrinelli. Milano.
59. Levitin, D. 2008. *The World in Six Songs*. Penguin. New York.
60. Gazzaniga, M. 2008. *Human: The Science behind What Makes Us Unique*. HarperCollins. New York.
61. Barrett, M.S., Ed. 2010. *A Cultural Psychology of Music Education*. Oxford University Press. Oxford.
62. Salzburg Global Seminar. 2011. Instrumental value. The transformative power of music. URL http://www.salzburgglobal.org/current/News-b.cfm?IDMedia=60456
63. Mannes, E. 2011. *The Power of Music: Pioneering Discoveries in the New Science of Song*. Walker. New York.
64. Tunstall, T. 2012. *Changing Lives: Gustavo Dudamel, El Sistema and the Transformative Power of Music*. W.W. Norton. New York.
65. MUSE Musicians United to Sustain the Environment. URL http://www.musemusic.org/ et al. [accessed on 18 February 2012].
66. Pliny the Younger, circa 100 AD. Epistles, 5.III. URL http://pages.pomona.edu/~cmc24747/sources/plin_1-5.htm#book5
67. Soro, M. 2012. Democrazia della musica. Celid. Tornino, Italy.

Ann. N.Y. Acad. Sci. ISSN 0077-8923

ANNALS OF THE NEW YORK ACADEMY OF SCIENCES
Issue: *The Neurosciences and Music IV: Learning and Memory*

Making music in a group: synchronization and shared experience

Katie Overy

Institute for Music in Human and Social Development, Reid School of Music, University of Edinburgh, Edinburgh, United Kingdom

Address for correspondence: Katie Overy, IMHSD, Reid School of Music, University of Edinburgh, Alison House, 12 Nicolson Square, Edinburgh EH8 9DF, United Kingdom. k.overy@ed.ac.uk

To consider the full impact of musical learning on the brain, it is important to study the nature of everyday, non-expert forms of musical behavior alongside expert instrumental training. Such informal forms of music making tend to include social interaction, synchronization, body movements, and positive shared experiences. Here, I propose that when designing music intervention programs for scientific purposes, such features may have advantages over instrumental training, depending on the specific research aims, contexts, and measures. With reference to a selection of classroom approaches to music education and to the shared affective motion experience (SAME) model of emotional responses to music, I conclude that group learning may be particularly valuable in music pedagogy.

Keywords: music; pedagogy; group learning; social; synchronization; shared experience

Musical behavior is a fundamental part of human experience and especially important during childhood, when lullabies, nursery rhymes, and action songs provide rich and enjoyable forms of social interaction and play.[1] Educators have often suggested that musical learning during childhood can have a positive impact in other learning domains, from the high academic performance found in Hungarian singing schools in the 1960s,[2] to Swanwick's influential description of "the incredible mind-making potential of music."[3] More recently, researchers have indicated the potential impact of musical training on neural function and neural structure, from motor,[4,5] auditory,[6] and language regions,[7] to brain stem responses.[8] Such research is contributing to the growing idea that musical learning can play an important role in child development and perhaps even throughout the life span.[9,10]

In this context of potentially "recommending" music training for children, it is particularly important to note that musical learning experiences are not always positive. In experimental research, musical learning is often equated with musical instrument training—a specialized and technologically challenging form of expert human musical behavior. Although some children excel at such training, many children actually have negative experiences of learning to play a musical instrument and discontinue their lessons,[11–13] with common reasons being that they are considered "boring" or the child "dislikes practice."[14] Meanwhile, overtraining on a musical instrument can lead to problems such as repetitive strain injury[15] or focal dystonia,[16] while many musicians, both amateur and professional, experience anxiety when performing,[17,18] often relying on medication.[19,20] Even informal singing in social situations can produce severe anxiety in adults,[21] which is often anecdotally related to memories of negative experiences of childhood music lessons.[22]

At the same time, it is well established that music listening plays a strong and positive role in the daily lives of a great many individuals who do not consider themselves to be "musical," from the phonograph, to the radio, to the advent of personal, portable sound systems, and extensive personal music collections.[23,24] Music seems to be especially important to adolescents and can play a strong role in identity formation and social independence.[25–28] In addition, music is employed and enjoyed in a wide range of formal and informal social situations, such

doi: 10.1111/j.1749-6632.2012.06530.x

as the pub, nightclubs, birthday parties, weddings, rock concerts, church services, and sports events. Such "real-world" musical experiences tend to involve small or large groups of untrained "nonmusicians," usually involved in some kind of interactive movement such as clapping, foot-tapping, singing, or dancing together. The case of football fans jumping up and down together, singing can be compared with music fans jumping up and down together singing at a rock concert or, indeed, with the young children jumping up and down together to music. All three cases involve exuberant, whole-body, synchronized movement with opportunities for variation, creativity, leadership, imitation, error, and humor (and little fear of individual performance exposure).

Thus, an important question when considering music intervention programs for scientific research is how to reconcile these two types of musical behavior: nonexpert, real-world, social experiences of music, compared with expert instrumental performance. In either case, genuine musical learning and skill development is important if potential transfer effects are to be examined. This requires specific pedagogical aims and methods in addition to controlled variables. Expert instrumental training has now been directly and substantially correlated with neural differences,[29,30] presenting a very strong candidate for future research. However, there are also many other approaches to music education, many of which have been developed specifically for group learning in the classroom.[31] These range from well-established music education approaches developed in the 20th century by Kodaly,[32] Orff,[33,34] Dalcroze,[35] and Gordon,[36] to national interpretations and developments of these approaches,[37–39] traditional, intercultural approaches,[40] and methods developed in the context of psychology research with specific transfer aims in mind.[41,42] Such methods involve group learning, shared musical experiences, synchronization, imitation, and a range of other socially interactive behaviors that are common to "real-world" social musical experiences, and could perhaps be used effectively and systematically in music intervention research.

The idea of music as a shared experience is central to the SAME (shared affective motion experience) model of emotional responses to music, which proposes that auditory musical signals are heard not simply as abstract patterns of sound, but rather as a series of intentional, expressive motor acts, recruiting similar neural networks in both agent and listener.[43–46] According to this model, the synchronization of such networks between actor and perceiver (or between multiple actors or multiple perceivers) can create a sense of empathy and social bonding, which is potentially of value in educational, therapeutic, and social contexts. Research into multibrain rather than single brain conceptions of human cognition has suggested that actors and perceivers show similar neural activations during language tasks,[47] and that closer coupling of such activations correlates with increased communication.[48] It has also been shown beautifully by Kirschner and Tomasello that group musical activity with children can lead to increased cooperative behavior,[49] a finding that is comparable to studies showing that synchronization activities in adults can lead to increased cooperative behaviors.[50] In addition, there is evidence that children and adults seem to perform better when synchronizing with a human agent than with a nonhuman agent,[51,52] indicating that an important aspect of human musicality may be its capacity as a medium for social interaction, probably from infancy.[53] Such synchronized, social interaction might bring specific benefits to musical learning paradigms, for example, by improving motivation or by providing immediate temporal feedback.

The temporal, pulse-based nature of such social interaction is clearly important. The ability of humans to synchronize with a steady pulse has been shown extensively in a range of complex conditions,[54,55] with temporal prediction abilities playing an important role in synchronization accuracy.[56] It has also been shown that the auditory perception of pulse and meter engage neural motor regions, including the cerebellum,[57] basal ganglia,[58] and premotor cortex,[59] as well as the vestibular system.[60] These research findings provide strong evidence in support of theories of the embodied nature of music cognition[61] and, indeed, embodied cognition in general.[62] Evidently, musical behavior is deeply rooted in motor behavior, from vibrating vocal chords and clapping hands to expert fine motor control of musical instruments at great speeds. Even young children tend to respond very physically to a steady beat, naturally moving their bodies in approximate synchronization.[1,63] It thus seems likely that encouraging children in their natural musical movement behaviors—through clapping, marching, dancing, and singing together—might provide advantages in a musical learning context.

In conclusion, the aim of this brief paper is not to set up one kind of musical training or music education program against another, but rather to emphasize the wide range of possible approaches to musical intervention programs. In certain contexts and under certain conditions, different musical intervention paradigms may be found to have greater or lesser effects, depending on the aims and selected outcome measures. We should not ask the question, *does* music have an impact, but rather *can* specific kinds of musical experience have an impact, and *how* and *when*.[64,65] Experimentally, it is important to isolate individual variables to establish mechanisms, correlations, and causes, but pedagogically it may be crucial to combine multiple facets of musical experience, such as motivational, affective, motor, and social behaviors. When we ask children to learn music, perhaps in some cases we should encourage them to engage their entire motor systems and to jump around in groups together, rather than to sit still and develop their fine motor control on a difficult instrument. There remains a wealth of knowledge from music pedagogy, therapy, psychology, and sociology that can help us to understand the full potential of music to affect the developing brain.

Conflicts of interest

The author declares no conflicts of interest.

References

1. Campbell, P. 1998. *Songs in Their Heads.* Oxford University Press. Oxford.
2. Barkóczi, I. & C. Pléh. 1982. *Music Makes a Difference: The Effect of Kodály's Musical Training on the Psychological Development of Elementary School Children.* Kodály Institute. Kecskemét, Hungary.
3. Swanwick, K. 1998. *Music, Mind and Education.* Routledge. London.
4. Bangert, M, & G. Schlaug. 2006. Form follows function. Specialization within the specialized. *Eur. J. Neurosci.* **24:** 1832–1834.
5. Abdul-Kareem, I.A., A. Stancak, L.M. Parkes, *et al.* 2011. Plasticity of the superior and middle cerebellar peduncles in musicians revealed by quantitative analysis of volume and number of streamlines based on diffusion tensor tractography. *Cerebellum* **10:** 611–623.
6. Schneider, P., M. Scherg, H. Günter Dosch, *et al.* 2002. Morphology of Heschl's gyrus reflects enhanced activation in the auditory cortex of musicians. *Nat. Neurosci.* **5:** 688–694.
7. Sluming, V., T. Barrick, M. Howard, *et al.* 2002. Voxel-based morphometry reveals increased gray matter density in Broca's area in male symphony orchestra musicians. *NeuroImage* **17:** 1613–1622.
8. Kraus, N. & B. Chandrasekaran. 2010. Music training for the development of auditory skills. *Nat. Rev. Neurosci.* **11:** 599–605.
9. Samson, S., D. Dellacherie & H. Platel. 2009. Emotional power of music in patients with memory disorders: clinical implications of cognitive neuroscience. *Ann. N.Y. Acad. Sci.* **1169:** 245–255.
10. Parbery-Clarka, A., S. Andersona, E. Hittnera & N. Kraus. 2012. Musical experience offsets age-related delays in neural timing. *Neurobiol. Aging* [Epub ahead of print].
11. McCarthy, J.F. 1980. Individualized instruction, student achievement and dropout in an urban elementary instrumental music program. *J. Res. Music Educ.* **26:** 59–69.
12. Hallam, S. 1998. Predictors of achievement and drop out in musical instrument tuition. *Psychol. Music* **26:** 116–32.
13. Pitts, S.E., J.W. Davidson & G.E. McPherson. 2000. Models of success and failure in instrumental learning: case studies of young players in the first twenty months of learning. *Bull. Council Res. Music Educ.* **146:** 51–69.
14. Driscoll, J. 2009. 'If I play my sax my parents are nice to me': opportunity and motivation in musical instrument and singing tuition. *Music Educ. Res.* **11:** 37–55.
15. Rosety-Rodriguez, M., F.J. Ordonez, J. Farias, *et al.* 2003. The influence of the active range of movement of pianists' wrists on repetitive strain injury. *M. Eur. J. Anat.* **7:** 75–77.
16. Jabusch, H. & E. Altenmüller. 2006. Focal dystonia in musicians: from phenomenology to therapy. *Adv. Cogn. Psychol.* **2:** 207–220.
17. Hamman, D.L. 1982. An assessment of anxiety in vocal and instrumental performers. *J. Res. Music Educ.* **30:** 77.
18. Wesner, R.B., R. Noyes, Jr. & T.L. Davis. 1991. The occurrence of performance anxiety among musicians. *J. Affect. Disord.* **18:** 177–185.
19. Gates, G.A., J. Saegert, N. Wilson, *et al.* 1985. Effect of beta blockade on singing performance. *Ann. Otol. Rhinol. Laryngol.* **94**(Pt. 1): 570–4.
20. James, I. & I.B. Savage. 1984. Beneficial effect of nadolol on anxiety-induced disturbances of performance in musicians: A comparison with diazepam and placebo. Proceedings of a Symposium on the Increasing Clinical Value of Beta Blockers Focus on Nadolol. *Am. Heart J.* **108:** 1150–1155.
21. Chong, H.J. 2011. Do we all enjoy singing? A content analysis of non-vocalists' attitudes toward singing. *Arts Psychother.* **37:** 120–124.
22. Abrila, C.R. 2007. I have a voice but I just can't sing: a narrative investigation of singing and social anxiety. *Music Educ. Res.* **9:** 1–15.
23. Denora, T. 2000. *Music in Everyday Life.* Cambridge University Press. Cambridge & New York: 181.
24. Sloboda, J., A. O'Neill, A. Susan & A. Ivaldi. 2001. Functions of music in everyday life: an exploratory study using the experience sampling method. *Musicae Scientiae* **5:** 9–32.

25. Laiho, S. 2004. The psychological functions of music in the everyday life of adolescents. In *Proceedings of the 8th International Conference on Music Perception and Cognition*, August 3–7, 2004, Evanston, IL.
26. Bleich, S., D. Zillmann & J.B. Weaver. 1991. Enjoyment and consumption of defiant rock music as a function of adolescent rebelliousness. *J. Electr. Broadcast. Media* **35:** 351–366.
27. Frith, S. 1996. *Performing Rites: On the Value of Popular Music*. Oxford University Press. Oxford.
28. North, A.C., D.J. Hargreaves & S.A. O'Neill. 2000. The importance of music to adolescents. *Br. J. Educ. Psychol.* **70:** 255–272.
29. Schlaug, G. 2001. The brain of musicians: A model for functional and structural adaptation. *Ann. N.Y. Acad. Sci.* **930:** 281–299.
30. Tervaniemi, M. 2009. Musicians—same or different? *Ann. N.Y. Acad. Sci.* **1169:** 151–156.
31. Pitts, S.E. 2000. *A Century of Change in Music Education: Historical Perspectives on Contemporary Practice in British Secondary School Music*. Ashgate. Aldershot, UK.
32. Choksy, L. 1981. *The Kodaly Context*. Prentice Hall. Englewood Cliffs, NJ.
33. Hall, D. & C. Orff. 1960. *Music for Children: Teacher's Manual*. Schott Music. New York.
34. Frazee, J. & K. Kreuter. 1987. *Discovering Orff: A Curriculum for Music Teachers*. Schott Music. New York.
35. Dalcroze, E.J. 1916. *The Technique of Moving Plastic*. Durand et Cie. Paris.
36. Gordon, Edwin E. 1997. *Learning Sequences in Music: Skill, Content and Patterns: A Music Learning Theory*. GIA Publications. Chicago.
37. Stocks, M. & A. Maddocks. 1999. *Growing with Music Teacher's Book Stage 1*. Longman. London.
38. Richards, M.H. 1984. *Aesthetic Foundations for Thinking: Rethought Part 1, Experience*. The Richards Institute of Education. Bellevue, WA.
39. Geoghegan, L. 2000. *Singing Games and Rhymes for Early Years*. National Youth Choir of Scotland. A&C Black Publishers. London.
40. Kirby, J. 2009. *Teaching Taiko: Principles and Practice*. Kagemusha Taiko. Exeter, UK.
41. Bogdanowicz, M. 1999. *Metoda Dobrego Startu*. WSIP. Warszawa, Poland.
42. Overy, K. 2008. Classroom rhythm games for literacy support. In *Music and Dyslexia: A Positive Approach*. J. Westcombe, T. Miles & D. Ditchfield, Eds.: 26–44. Wiley. London.
43. Molnar-Szakacs, I. & K. Overy. 2006. Music and mirror neurons: from motion to 'e' motion. *Soc. Cogn. Affect. Neurosci.* **1:** 235–241.
44. Overy, K. & I. Molar-Szakacs. 2009. Being together in time: musical experience and the mirror neuron system. *Music Percept.* **26:** 489–504.
45. Molnar-Szakacs, I., M. Wang, E.A. Laugeson, *et al.* 2009. Autism, emotion recognition and the mirror neuron system: the case of music. *McGill J. Med.* **12:** 87–98.
46. Molnar-Szakacs, I., V. Green & K. Overy. 2012. Shared affective motion experience (same) and creative, interactive music therapy. In *Musical Imaginations: Multidisciplinary Perspectives on Creativity, Performance and Perception*. D. Hargreaves, R. MacDonald & D. Miell, Eds.: 313–331. Oxford University Press. Oxford.
47. Hasson, U., A.A. Ghazanfar, B. Galantucci, *et al.* 2012. *Brain-to-Brain Coupling: A Mechanism for Creating and Sharing a Social World*. Neuroscience Institute. Princeton, NJ.
48. Stephens, G. *et al.* 2010. Speaker-listener neural coupling underlies successful communication. *Proc. Natl. Acad. Sci. USA* **107:** 14425–14430.
49. Kirschner, S. & M. Tomasello. 2010. Joint music making promotes prosocial behavior in 4-year-old children. *Evol. Hum. Behav.* **31:** 354–364.
50. Wiltermuth, S.S. & C. Heath. 2009. Synchrony and Cooperation. *Psychol. Sci.* **20:** 1–5.
51. Kirschner, S. & M. Tomasello. 2009. Joint drumming: social context facilitates synchronization in preschool children. *J. Exp. Child Psychol.* **102:** 299–314.
52. Himberg, T. 2006. Co-operative tapping and collective timekeeping – differences of timing accuracy in duet performance with human or computer partner. In *Proceedings of the 9th International Conference on Music Perception and Cognition*. Alma Mater Studiorum University of Bologna, Bologna, Italy.
53. Malloch, S. & C. Trevarthen. 2009. *Communicative Musicality: Exploring the Basis of Human Companionship*. Oxford University Press. Oxford.
54. Large, E.W., P. Fink & J.A.S. Kelso. 2002. Tracking simple and complex sequences. *Psychol. Res.* **66:** 3–17.
55. Repp, B.H. 2005. Sensorimotor synchronization: a review of the tapping literature. *Psychon. Bull. Rev.* **12:** 969–992.
56. Pecenka, N. & P.E. Keller. 2011. The role of temporal prediction abilities in interpersonal sensorimotor synchronization. *Exp. Brain Res.* **211:** 505–515.
57. Thaut, M.H., K.M. Stephan, G. Wunderlich, *et al.* 2009. Distinct cortico-cerebellar activations in rhythmic auditory motor synchronization. *Cortex* **45:** 44–53.
58. Grahn, J.A. & M. Brett. 2007. Rhythm perception in motor areas of the brain. *J. Cogn. Neurosci.* **19:** 893–906.
59. Bengtsson, S.L., H.H. Ehrsson, T. Hashimoto, *et al.* 2009. Listening to rhythms activates motor and pre-motor cortices. *Cortex* **45:** 62–71.
60. Phillips-Silver, J. & L.J. Trainor. 2008.Vestibular influence on auditory metrical interpretation. *Brain Cogn.* **67:** 94–102.
61. Leman, M. 2007. *Embodied Music Cognition and Mediation Technology*. MIT Press. Cambridge, MA.
62. Clark, A. 1997. *Being There: Putting Brain, Body and World Together Again*. MIT Press. Cambridge, MA.
63. Eurola, T., G. Luck & P. Toiviainen. 2006. An investigation of pre-schoolers' corporeal synchronization with music. In *Proceedings of the 9th International Conference on Music Perception and Cognition*. Alma Mater Studiorum University of Bologna, Bologna, Italy.
64. Mills, J. 1998. Responses to Katie Overy's paper, "Can Music Really 'Improve' the Mind?" *Psychol. Music* **26:** 204–205.
65. Lamont, A. 1998. Responses to Katie Overy's Paper, "Can Music Really 'Improve' the Mind?" *Psychol. Music* **26:** 201–204.

Ann. N.Y. Acad. Sci. ISSN 0077-8923

ANNALS OF THE NEW YORK ACADEMY OF SCIENCES
Issue: *The Neurosciences and Music IV: Learning and Memory*

Neuroscience and "real world" practice: music as a therapeutic resource for children in zones of conflict

Nigel Osborne
Institute for Music in Human and Social Development (IMHSD), Reid School of Music, University of Edinburgh, Edinburgh, United Kingdom

Address for correspondence: Prof. Nigel Osborne, IMHSD, Reid School of Music, University of Edinburgh, Alison House, Nicolson Square, Edinburgh EH8 9DF, United Kingdom. reidprof@ed.ac.uk

Recent developments in music neuroscience are considered a source for reflection on, and evaluation and development of, musical therapeutic practice in the field, in particular, in relation to traumatized children and postconflict societies. Music neuroscience research is related to practice within a broad biopsychosocial framework. Here, examples are detailed of work from North Uganda, Palestine, and South Thailand.

Keywords: music; therapy; PTSD

Introduction

It may be premature to speak of an applied music neuroscience, but the fruits of music neuroscience research have already begun to play a role in reflection on practice in areas such as music education, community music, music therapy, and music medicine. An area of work that overlaps with all of these disciplines is the use of music as an intervention to support children in zones of conflict and postconflict.[1] Although one of the principal concerns for such children is posttraumatic stress disorder (PTSD), and the most commonly used diagnostic criteria are psychiatric (exposure to trauma, traumatic recall, avoidance, and hyperarousal—DSM-IV),[2] the symptoms of trauma are spread over a much wider range of human experience, and the potential usefulness of music is probably best considered from the perspective of a broad biopsychosocial framework (Fig. 1).

This short paper records three anecdotes of work in the field from key areas of the framework and discusses how reflection on practice, and in some cases practice itself, can be influenced by the results of research in the neuroscience of music.

Biological concerns

Peter Oloya is one of East Africa's finest sculptors, regards Damien Hirst as a major influence, and was recently artist-in-residence for Pangolin Editions near Stroud in the United Kingdom. His origins are in a small village—Lemo Bongolewich—near Kitgum in North Uganda. His childhood, however, was overshadowed by the bloody conflict between Joseph Kony's Lord's Resistance Army (LRA) and the Uganda People's Defence Force (UPDF). Between 1987 and 2007 the Luo-speaking Acholi people of North Uganda found themselves between a rock and a hard place, facing murder, rape, abductions, ethnic cleansing, and the forced recruitment of child militias and prostitutes.

It was this life experience that prompted Peter, now 33 years old, to create a nongovernmental organization (NGO) in Kitgum—AFOCOD (Art for Community Development)—using art to support the welfare of communities and children. Some of his volunteers are themselves former child soldiers. In an interview in *AfricanColours*,[3] he describes his background:

> Art is my reason for living, in fact it has given a structure to my whole life. I have lived through war

doi: 10.1111/j.1749-6632.2012.06473.x

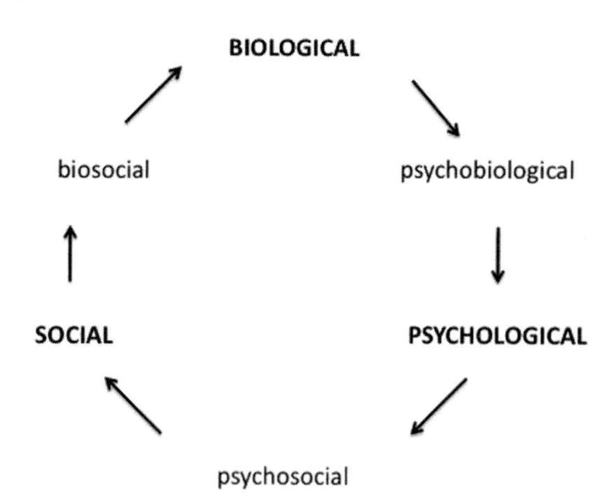

Figure 1. Proposed biopsychosocial framework for our understanding of music as a therapeutic resource.

> and experienced death throughout my early years, and I used art as a kind of therapy to help get me through those difficult times. In fact, it continues to help me to this very day. Most of the children and youths who come to us are traumatised but our training really helps with their healing process. Helping to bring some kind of normality to young, innocent lives is very important to me because they are the future and will one day help to bring peace to the country.

It is March 2007 and Peter has invited me to Kitgum to help develop musical aspects of the AFOCOD program. He believes that there is a "visceral," physical/mental engagement in music making that has an important contribution to make to the welfare of his children. We travel to Ayoma IDP camp, which lies north of Kitgum, toward the Sudan border and the "dry lands" where militias still roam. Ayoma is a densely packed maze of mud-and-wattle huts and dusty narrow passageways where children in torn clothes play with improvised toys of tin and cardboard. International NGOs have provided a few bare administrative and community buildings, and it is in a space by the community center that we decide to hold our first music workshop.

The children arrive, around 40 of them, followed by the usual gathering of curious adults. There is a specific atmosphere and feeling among groups of traumatized children that is much the same everywhere. Here, it takes the form of a restless numbness, and maybe a little fear. But I am confident I can engage the children, and I begin to play North Ugandan dances with my colleagues Robert and Hakim from the Uganda Dance Academy: first gentle rhythms and slow tempi, then faster and more vigorous music. We encourage the children to move, to clap, and to sing. We play and dance joyfully and energetically, almost to the point of clowning. It is not long before the transformation has taken place, and the children are laughing, smiling, and joining in enthusiastically. We are particularly keen that they use their voices, breathe well, and move fluently; we want them to be both energized and relaxed. It is a transformation that never fails and normally leads to requests, from both children and their caretakers, for our work to continue.

I have written elsewhere about the biological symptoms of PTSD and how music may help children deal with these problems.[1] Although we do not "apply" neuroscience in our work, we can use the research findings to help relate and connect our knowledge of the symptoms of trauma to our experience of the intervention in the field.

For example, there is a significant amount of research literature on how trauma can affect the heart, in terms of increased heart rate, blood pressure, and cardiac arrhythmias.[4–10] This may be considered alongside research from neuroscience and related disciplines that indicate that music may affect and help regulate the autonomic nervous system and related behaviors of the heart.[11–20] Finally, practitioners have anecdotal evidence from the field that their work may have a modest effect in helping regulate children's hearts and autonomic nervous systems (e.g., from discussion with general practitioners, treating traumatized children at the Centre for Children with Special Needs, Mostar, Bosnia, and Herzegovina).

Similarly, there is a small body of evidence suggesting that PTSD, and stress in general, affect and dysregulate breathing and respiratory sinus arrhythmia.[21–24] This evidence may be related to research findings that music may affect and help regulate automatic breathing and sinus arrhythmia.[25–27] Vocal music is of course directly and accountably interactive with conscious breathing,[28] and there is experience from the field of successful work supporting and improving children's breathing through singing (e.g., once again, in work with children in Bosnia and Herzegovina).[1]

PTSD is associated with dysregulation of movement repertoires, including both hyperactivity and extreme sluggishness.[29–32] The music neuroscience literature is rich in research on music and its relationship with the premotor and motor cortices, and by inference, its ability to "make us" or to "help us" move.[33] Evidence from the field, based on simple observation of changing movement repertoires, and my own experiences working with children with ADHD in Scottish special education, suggests that music may help traumatized children regulate their movements.[1,33]

Finally, there is a large body of evidence that PTSD dysregulates systems for dealing with stress and relaxation, and in particular, the hypothalamic–pituitary–adrenal (HPA) axis (e.g., Refs. 34–37). Evidence from music neuroscience and endocrinology suggests that music may help modulate and regulate these systems (e.g., Refs. 38, 39). This correlates well with anecdotal evidence—mostly observation of children's behavior—from parents and caregivers conveyed in personal communications from the field.[1]

Psychobiological and psychological concerns

It is November 2007, and we drive out to the Palestinian West Bank town of Nablus to the southeast, past the Jami' Al-Kabir Mosque to the Balata refugee camp. I am traveling with colleagues from the Palestinian Union of Social Workers and Psychologists (PUSWP) and with Sheena Boyle from the charity Children of Amal. Our objective is to support a program of therapeutic and educational work for children and to offer training to young social workers in our methods. During a previous workshop visit, journalist Donald Macintyre accompanied us and filed the following story.[40]

> The Reid Professor of Music at Edinburgh University.... is somehow managing simultaneously to play the guitar, dance, and conduct a class of 30 children in their lusty performance of a West African folk song... The electricity brought to this room would be surprising enough even if it was not happening in the heart of the West Bank's most populous—and most problematic—Palestinian refugee camp, scene of some of the worst bloodshed during the past six years of conflict.
>
> This is Balata, a stronghold of armed militancy and the target of at times almost daily Israeli incursions, where 150 Palestinians have been killed since the intifada began six years ago. It is also one of the most densely populated places on earth, home to 30,000 civilians who live in less than two square kilometres of cement-block housing packed so closely together that fat people cannot squeeze into some of the alleys between them.
>
> He first asks the children to join him in singing a melodic African chant, increasing the volume and then reducing it to a whisper. Then he has them clap in time. Then he introduces them, still clapping, to the rhythm, then, with his guitar, to the tune, and then finally—for those that need it—to the words of an old Arab song: "Aya Zeyn al-Abidin/Ir Wrd, Ir Wrd/Imfetah Baynil/Besatin" ["Zeyn al Abidin, you are like a rose that blooms in the garden."]
>
> "I have loved Arabic music for a long time, longer than you!" he tells the children through the interpreter Assim Eshtaya, 27, a school counsellor in Nablus. "I won't say more, but longer. The notes in the song we learnt come from a very old traditional Arab scale." He plays the notes on his guitar. "Now I have a proposal. Would you like to create a new song with Arabic music?" And so, with Professor Osborne allowing the children to decide the words of the song... and then to choose the melody, note by note, the song comes into being: "Dear friends, friends for ever, friends for ever," the whole group sings in Arabic.
>
> Certainly, the Balata community leaders seem delighted with the success of the project. Mrs. Boyle's July training sessions involved 90 children in three Nablus refugee camps—with others having to be turned away. "During the month, the results were clear for everyone to see," said Shaer Badawi of Balata's Yafa centre. "The children were happier."

On this visit I am keen to widen opportunities for expression and emotional communication. Back in the PUSWP headquarters in Nablus, I run training workshops for social workers involving simple musical/emotional games: we sit in a circle, half with loud, "angry" instruments (djembes, bongos, and cowbells) and half with gentle sounds (metal chimes, shakers, and crotales). A volunteer in the middle activates the different sound worlds by moving back and forth through the circle—a chance to play safely with extremes of aggressive and caressing sound. I introduce blindfold journeys through forests of musical trees, and we invent the sounds of imaginary animals to communicate in the space—what are these animals, why do they make this sound, what do they feel, what do they want to say? Most important of all, we work on co-improvisation—person-to-person exchanges of musical empathy, one of the core techniques of clinical music therapy—adapted here for more general professional use.

For practitioners, one of the most useful models for reflecting on and developing the psychological and psychobiological aspects of the work has been the theory of communicative musicality.[41] The theory has its origins, in part, in the neurosciences and related disciplines. Its basis is intersubjectivity in the phenomenological sense,[42] and it is promising that a neuroscience of intersubjectivity has recently begun to emerge.[43–45]

The empirical foundations of communicative musicality lie in the study of mother–infant vocal communication,[46–55] which has revealed the richness and significance of the responses of infants—irrespective of culture—to the "prosodic" utterances of mothers: the sharing of time and intention, emotional communication, sympathy, and movement. Research in the affective neuroscience of music opens up the possibility that this prosodic communication may be linked to the activation of a variety of cortical and subcortical neural and neuroendocrine systems.[56–69]

The theory of communicative musicality, and its associated music neuroscience research, offers to workers in the field a measure of reassurance (e.g., confirmation that changes observed in children may indeed have a basis in science), an opportunity for reflection, and the chance to advance practice. The theory further implies that through intersubjectivity, empathy, common "musical" emotions, and their neural substrates, musical activity may affect the state of mind and body of those who take part in it. From this corpus of literature and personal experience, it is reasonable to propose that children with symptoms of avoidance, numbing, or hyperarousal (DSM-IV[2]), or associated feelings of detachment, estrangement, anger, fear, lack of trust, distress, or simply unhappiness, may indeed (as they and their carers so often tell us) find a measure of physical and mental release in joyful shared experiences of musical expression.

Psychosocial, social, and biosocial concerns

It is late June 2010, and I travel to Thailand with a small group of students from Edinburgh to work on placement with colleagues from Silpakorn University, Bangkok. Anothai Nitibhon is a composer and community musician who completed her doctorate with us at the University of Edinburgh and returned to Thailand to found a new composition department and community music course at Silpakorn. I am impressed by the way their work has been embraced enthusiastically by communities in many different parts of Thailand.

Our destination is the South, and the narrow strip of Thai territory that runs between the Myanmar (Burma) border and the South China Sea down to the Kra Isthmus. These are the districts of Prachuap Khiri Khan, Chumphon, and Ranong, which have become in some ways a "no man's land." The British colonial period left large populations of ethnic Thai trapped in Burma, where they have become a significant minority with limited human rights. Over the years, there has been a slow seepage of population across the border to Thailand. There are now about 4,700 displaced Burmese Thai in the region, granted "leniency" to live in Thailand, but with no citizenship, rights, health care, employment protection, or permission to travel out of the region. Our project is to try—at their invitation—to help give the young people a voice.

Our first stop is Klong Loi, a small village that survives in dignified poverty, a bitter sweet paradise, surrounded by banana trees, long kong, and abruptly rising hillsides of tropical forest. Rachel Bradley, a postgraduate student from Edinburgh, takes up the story in her paper "No man's band" (R. Bradley, Univ. of Edinburgh, in prep.).

> In the community hall, a large open concrete platform with a stage and a roof supported by pillars to allow the breeze to blow in from the mountains, a group of teenagers (dressed like teenagers everywhere) congregates for the project. Onlookers gather as the group begins warm up games, dances and songs. The young people participate with apprehension at first, but quickly come to speak, laugh and gain confidence. . . .
>
> They are invited to talk about their lives in the displaced community. They describe what they would like to say to other teenagers living legally in Thailand, and many speak of what they would like to be when they are older. . .a painter, teacher, Thai boxer, tour guide, rapper, husband.
>
> Inspired by these conversations they split into smaller groups to compose songs with the support

of the student volunteers and their respective instruments. Participants are offered notes to choose from (from the traditional scales of Thai music); they assemble the notes of the melody one by one.. . .

The songs are a fascinating mixture of styles, including traditional, folk, pop, and rap. Interestingly, all melodies scrupulously respect the contours of the five tones of the Thai language—low, mid, high, falling, or rising—a discipline now lost in much Thai popular music. The words of the songs are touching, more a gentle *cri de coeur* than agitation propaganda.

. . .. Many times we have been hurt
Felt confused and don't know where to turn
Like the long kong fruit we come and we go
And give the long kong branch to show
That we would like to be its friend
But now we just don't see an end
So if some day you think about us
We are here, still waiting for you.

The workshop culminates in performances in the village and the local school. Our objectives here are to offer the children a positive, collective musical experience; an opportunity to build creativity, self-respect, and a sense of identity; and the chance to tell their story and to use the social power of music to raise awareness of their situation in a wider world.

There are many areas of psychosocial and social intervention among both postconflict and displaced populations where music has proven to be an agent of change.[70–72] Its potential usefulness in the psychosocial domain begins with relaxation, joy, and trust and progresses to individual creativity, self-expression, self-belief, and self-respect. At the social level, music may be effective in facilitating social communication, collective creativity, and, as in the case of Klong Loi, the expression of social identity. In situations where populations have been divided by conflict, for example, in postwar Bosnia–Hercegovina, music and the creative arts may provide a mutually acceptable meeting point in normality. It is rarely possible to "reconcile" people. But music making is a safe place for people to "be" together and rebuild trust how and when they choose. In postconflict areas, music has been effective beyond the biopsychosocial paradigm in the domains of politics and economics, as, for example, the music-based social reconstruction program in Srebrenica, or new arts-based economic initiatives in East Africa (e.g., the work of the Ruwenzori Foundation in Uganda).

All of this may seem very far from the world of music neuroscience; certainly there has been little crossover so far into postconflict social intervention. Yet, social neuroscience is a dynamically expanding field,[73–80] and an overlap of research interests with music neuroscience has already occurred—for example, in the neuroscience of intersubjectivity (see above), and in the work of Overy and Molnar–Szakacs, whose shared affective motion experience (SAME) model of emotional responses to music[81,82] proposes a shared, deep apprehension, beyond simple hearing, of the human, expressive origins of musical sound. Their suggested co-occurence of motor, emotional, and social responses to music certainly helps to explain the power and importance of group work in our activities with traumatized children. This leads the way back around the biopsychosocial circle, through issues of synchronization (e.g., in joy in making music and movement together), to the biosocial and biological starting points of this framework.

It seems clear that the circle itself may represent a continuity, synergy, and synchronicity of biological, psychological, and social concerns active in potentially therapeutic, rehabilitational, and socially reparative musical processes.

Conflicts of interest

The author declares no conflicts of interest.

References

1. Osborne, N. 2009. Music for children in zones of conflict and post-conflict: a psychobiological approach. In *Communicative Musicality*. S. Malloch & C. Trevarthen, Eds.: 331–356. Oxford University Press. New York.
2. American Psychiatric Association. 1994. *Diagnostic and Statistical Manual of Mental Disorders*. 4th Ed. APA. Washington, DC.
3. AfricanColours. 2008. Art saved my life - Peter Oloya. URL http://www.africancolours.com/african-art-news/38/uganda/art_saved_my_life_-_peter_oloya.htm [accessed on 9 February 2012].
4. Cohen, H., M. Kotler, M.A. Matar, *et al.* 1998. Analysis of heart rate variability in posttraumatic stress disorder patients in response to a trauma-related reminder. *Biol. Psychiatry* **44:** 1054–1059.
5. Cohen, H., J. Benjamin, A.B. Geva, *et al.* 2000. Autonomic dysregulation in panic disorder and in post-traumatic stress disorder: application of power spectrum analysis of heart rate variability at rest and in response to recollection of trauma or panic attacks. *Psychiatry Res.* **96:** 1–13

6. Buckley, T.C. & D.G. Kaloupek. 2001. A meta-analytic examination of basal cardiovascular activity in posttraumatic stress disorder. *Psychosom. Med.* **63:** 585–594.
7. Beckham, J.C., S.R. Vrana, J.C. Barefoot, *et al.* 2002. Magnitude and duration of cardiovascular responses to anger in Vietnam veterans with and without posttraumatic stress disorder. *J. Consult. Clin. Psychol.* **70:** 228–234.
8. Forneris, C.A., M.I. Butterfield & H.B. Bosworth. 2004. Physiological arousal among women veterans with and without posttraumatic stress disorder. *Mil. Med.* **169:** 307–312.
9. Buckley, T.C., D. Holohan, J.L. Greif, *et al.* 2004. Twenty-four-hour ambulatory assessment of heart rate and blood pressure in chronic PTSD and non-PTSD veterans. *J. Trauma. Stress* **17:** 163–171.
10. Kibler, J.L. & J.A. Lyons. 2004. Perceived coping ability mediates the relationship between PTSD severity and heart rate recovery in veterans. *J. Trauma. Stress* **17:** 23–29.
11. Updike, P.A. & D.M. Charles. 1987. Music Rx: physiological and emotional responses to taped music programs of preoperative patients awaiting plastic surgery. *Ann. Plast. Surg.* **19:** 29–33.
12. Iwanaga, M. & M. Tsukamoto. 1997. Effects of excitative and sedative music on subjective and physiological relaxation. *Percept. Mot. Skills* **85:** 287–296.
13. Byers, J.F. & K.A. Smyth. 1997. Effect of a musical intervention on noise annoyance, heart rate, and blood pressure in cardiac surgery patients. *Am. J. Crit. Care* **6:** 183–191.
14. Gerra, G., A. Zaimović, D. Franchini, *et al.* 1998. Neuroendocrine responses of healthy volunteers to 'techno-music': relationships with personality traits and emotional state. *Int. J. Psychophysiol.* **28:** 99–111
15. Cardigan, M.E., N.A. Caruso, S.M. Haldeman, *et al.* 2001. The effects of music on cardiac patients on bed rest. *Prog. Cardiovasc. Nurs.* **16:** 5–13.
16. Knight, W.E. & N.S. Rickard. 2001. Relaxing music prevents stress-induced increases in subjective anxiety, systolic blood pressure and heart rate in healthy males and females. *J. Music Ther.* **38:** 254–272.
17. Aragon, D., C. Farris & J.F. Byers. 2002. The effects of harp music in vascular and thoracic surgical patients. *Altern. Ther. Health Med.* **8:** 52–54, 56–60.
18. Mok, E. & K.Y. Wong. 2003. Effects of music on patient anxiety. *AORN J.* **77:** 396–397, 401–406, 409–410.
19. Lee, O.K., Y.F. Chung, M.F. Chan & W.M. Chan. 2005. Music and its effect on the physiological responses and anxiety levels of patients receiving mechanical ventilation: a pilot study. *J. Clin. Nurs.* **14:** 609–620.
20. Iwanaga, M., A. Kobayashi, C. Kawasaki. 2005. Heart rate variability with repetitive exposure to music. *Biol. Psychol.* **70:** 61–66.
21. Sahar, T., A.Y. Shalev & S.W. Porges. 2001. Vagal modulation of responses to mental challenge in posttraumatic stress disorder. *Biol. Psychiatry* **49:** 637–643.
22. Donker, G.A., C.J. Yzermans, P. Spreeuwenberg & J. Van der Zee. 2002. Symptom attribution after a plane crash: comparison between self-reported symptoms and GP records. *Brit. J. Gen. Pract.* **52:** 917–922.
23. Sack, M., J.W. Hopper & F. Lamprecht. 2004. Low respiratory sinus arrhythmia and prolonged psychophysiological arousal in posttraumatic stress disorder: heart rate dynamics and individual differences in arousal regulation. *Biol. Psychiatry* **55:** 284–290.
24. Nixon, R.D. & R.A. Bryant. 2005. Induced arousal and re-experiencing in acute stress disorder. *J. Anxiety Disord.* **19:** 587–594.
25. Fried, R. 1990. Integrating music in breathing training and relaxation: I. Background, rationale and relevant elements. *Biofeedback Self Regul.* **15:** 161–169.
26. Fried, R. 1990. Integrating music in breathing training and relaxation: II. Applications. *Biofeedback Self Regul.* **15:** 171–177.
27. Bernardi, L., C. Porta & P. Sleight. 2006. Cardiovascular, cerebrovascular and respiratory changes induced by different types of music in musicians and non-musicians: the importance of silence. *Heart* **92:** 445–452.
28. McCoy, S.J. 2004. *Your Voice: An Inside View—Multimedia Voice Science and Pedagogy*. Inside View. Delaware, OH.
29. Yule, W. 1994. *Posttraumatic Stress Disorder*. Plenum. New York.
30. Brent, D.A., J.A. Perper, G. Moritz, *et al.* 1995. Posttraumatic stress disorder in peers of adolescent suicide victims: predisposing factors and phenomenology. *J. Am. Acad. Child Adolesc. Psychiatry* **34:** 209–215.
31. Famularo, R., T. Fenton, R. Kinscherff & M. Augustyn. 1996. Psychiatric comorbidity in childhood post traumatic stress disorder. *Child Abuse Negl.* **20:** 953–961.
32. Adler, L.A., M. Kunz, H.C. Chua, *et al.* 2004. Attention-deficit/hyperactivity disorder in adult patients with posttraumatic stress disorder (PTSD): is ADHD a vulnerability factor? *J. Atten. Disord.* **8:** 11–16.
33. Osborne, N. 2009. Towards a chronobiology of musical rhythm. In *Communicative Musicality*. S. Malloch & C. Trevarthen, Eds.: 545–564. Oxford University Press. New York.
34. Goenjian, A.K., R. Yehuda, R.S. Pynoos, *et al.* 1996. Basal cortisol, dexamethasone suppression of cortisol and MHPG in adolescents after the 1988 earthquake in Armenia. *Am. J. Psychiatry* **153:** 929–934.
35. Yehuda, R. 2000. Neuroendocrinology. In *Post-Traumatic Stress Disorder, Diagnosis, Management and Treatment*. D. Nutt, J.R.T. Davidson & J. Zohar, Eds.: 1–260. Martin Dunitz. London.
36. Rasmusson, A.M., M. Vythilingam & C.A. Morgan. 2003. The neuroendocrinology of posttraumatic stress disorder: new directions. *CNS Spectr.* **8:** 651–656, 665–667.
37. Delahanty, D.L., N.R. Nugent, N.C. Christopher & M. Walsh. 2005. Initial urinary epinephrine and cortisol levels predict acute PTSD symptoms in child trauma victims. *Psychoneuroendocrinology* **30:** 121–128.
38. Miluk-Kolasa, B., Z. Obminski, R. Stupnicki & L. Golec. 1994. Effects of music treatment on salivary cortisol in patients exposed to pre-surgical stress. *Exp. Clin. Endocrinol.* **102:** 118–120.
39. Uedo, N., H. Ishikawa, K. Morimoto, *et al.* 2004. Reduction in salivary cortisol level by music therapy during colonoscopic examination. *Hepato-Gastroenterology* **51:** 451–453.

40. Macintyre, D. 2006. Music therapy: the pied piper of Balata. *The Independent (UK)*, October 30, 2006.
41. Malloch, S. & C. Trevarthen, Eds. 2009. *Communicative Musicality*. Oxford University Press. New York.
42. Trevarthen, C. & K.J. Aitken. 2001. Infant intersubjectivity: research, theory and clinical applications. *J. Child Psychol. Psychiatry* **42:** 3–48.
43. Dumas, G., J. Nadel, R. Soussignan, *et al.* 2010. Inter-brain sychronization during social interaction. *PloS ONE* **5:** e12166.
44. Redcay, E., D. Dodell-Feder, M.J. Pearroe, *et al.* 2010. Live face-to-face Interaction during fMRI: a new tool for social cognitive neuroscience. *Neuro Image* **50:** 1639–1647.
45. Guionnet, S., J. Nadel, E. Bertasi, *et al.* 2011. Reciprocal imitation: toward a neural basis of social interaction. *Cereb. Cortex***22:** 917–918.
46. Stern, D.N. 1974. Mother and infant at play: the dyadic interaction involving facial, vocal and gaze behaviours. In *The Effect of the Infant on its Caregiver*. M. Lewis & L.A. Rosenblum, Eds.: 187–213. Wiley. New York.
47. Alegria, J. & E. Noirot. 1978. Neonate orientation behaviour towards the human voice. *Early Hum. Dev.* **1:** 291–312.
48. Bullowa, M., Ed. 1979. *Before Speech: The Beginning of Human Communication.* Cambridge University Press. Cambridge, UK.
49. DeCasper, A.J. & W.P. Fifer. 1980. Of human bonding: newborns prefer their mothers' voices. *Science* **208:** 1174–1176.
50. Fernald, A. 1985. Four-month-old infants prefer to listen to motherese. *Infant Behav. Dev.* **8:** 181–195.
51. Papoušek, M. 1987. Models and messages in the melodies of maternal speech in tonal and non-tonal languages. *Abstr. Soc. Res. Child Dev.* **6:** 407.
52. Trehub, S.E. 1987. Infants' perception of musical patterns. *Percept. Psychophys.* **41:** 635–641.
53. Papoušek, M. 1992. Early ontogeny of vocal communication in parent-infant interactions. In *Nonverbal Vocal Communication, Comparative and Developmental Approaches.* H. Papoušek, U. Jurgens & M. Papoušek, Eds.: 230–261. Cambridge University Press. Cambridge, UK.
54. Masataka, N. 1993. Relation between pitch contour of prelinguistic vocalisations and communicative functions in Japanese infants. *Infant Behav. Dev.* **16:** 397–401.
55. Kuhl, P.K., J.E. Andruski, I.A. Chistovich, *et al.* 1997. Cross-language analysis of phonetic units in language addressed to infants. *Science* **277:** 684–686.
56. Goldstein, A. 1980. Thrills in response to music and other stimuli. *Physiol. Psychol.* **3:** 126–129.
57. Panksepp, J. 1998. The periconscious substrates of consciousness. Affective states and evolutionary origins of the SELF. *J. Conscious. Stud.* **5:** 566–582.
58. Blood, A.J., R.J. Zatorre, P. Bermudez & A.C. Evans. 1999. Emotional responses to pleasant and unpleasant music correlate with activity in paralimbic regions. *Nature Neurosci.* **2:** 322–327.
59. Pratt, R.R. & D.E. Grocke, Eds. 1999. *MusicMedicine – Music Medicine and Music Therapy: Expanding Horizons.* Vol. 3. MMB Music. St. Louis, MO.
60. Blood, A.J. & R.J. Zatorre. 2001. Intensely pleasurable responses to music correlate with activity in brain regions implicated in reward and emotion. *Proc. Natl. Acad. Sci. USA* **98:** 11818–11823.
61. Kreuz, G., S. Bongard, S. Rohrmann, *et al.* 2004. Effects of choir singing or listening on secretory immunoglobulin A, cortisol, and emotional state. *J. Behav. Med.* **27:** 623–635.
62. Stefano, G.B., W. Zhu, P. Cadet, *et al.* 2004. Music alters constitutively expressed opiate and cytokine processes in listeners. *Med. Sci. Monit.* **10:** MS18–MS27.
63. Menon, V. & D.J. Levitin. 2005. The rewards of music listening: response and physiological connectivity of the mesolimbic system. *Neuro Image* **28:** 175–184.
64. Craig, D.G. 2005. An exploratory study of changes during chills induced by music. *Musicae Scientiae* **9:** 273–287.
65. Alcaro, A., R. Huber & J. Panksepp. 2007. Behavioral functions of the mesolimbic dopaminergic system: an affective neuroethological perspective. *Brain Res. Rev.* **56:** 283–321.
66. Klockars, M. & M. Peltomaa. 2007. *Music Meets Medicine. Acta Gyllenbergiana VII.* The Signe and Ane Gyllenberg Foundation. Helsinki, Finland.
67. Grewe, O., E. Magel, R. Kopiez & E. Altenmüller. 2007. Listening to music as a re-creative process: physiological, psychological, and psychoacoustical correlates of chills and strong emotions. *Music Percept.* **24:** 297–314.
68. Guhn, M., A. Hamm & M. Zentner. 2007. Physiological and musico-acoustic correlates of the chill response. *Music Percept.* **24:** 170–180.
69. Panksepp, J. & C. Trevarthen. 2009. The neuroscience of emotion in music. In *Communicative Musicality*. S. Malloch, C. Trevarthen, Eds.: 105–146. Oxford University Press. New York.
70. Urbain, O., Ed. 2007. *Music and Conflict Transformation: Harmonies and Dissonances in Geopoilitics.* I.B. Tauris. London.
71. O'Connell, J. & S.E.-S. Castelo-Branco. 2010. *Music and Conflict*. University of Illinois Press. Champaign, IL.
72. Robertson, C. 2010. Music and conflict transformation in Bosnia: constructing and reconstructing the normal. *Music Arts Action* **2:** 2.
73. Cacioppo, J.T. 2002. Social neuroscience: understanding the pieces fosters understanding the whole and vice versa. *Am. Psychologist* **57:** 819–831.
74. Brune, M., H. Ribbert & W. Schiefenhovel. 2003. *The Social Brain: Evolution and Pathology*. Wiley & Sons. Hoboken, NJ.
75. Cacioppo, J. T. & G.G. Berntson. 2004. *Social Neuroscience: Key Readings*. Psychology Press. East Sussex, UK.
76. Wolpert, D. & C. Frith. 2004. *The Neuroscience of Social Interactions: Decoding, Influencing, and Imitating the Actions of Others*. Oxford University Press. New York.
77. Cozolino, L. 2006. *The Neuroscience of Human Relationships: Attachment and the Developing Social Brain.* W. W. Norton. New York.
78. de Haan, M. & M.R. Gunnar. 2009. *Handbook of Developmental Social Neuroscience.* Guilford Press. New York.

79. Decety, J. & W. Ickes. 2009. *The Social Neuroscience of Empathy*. MIT Press. Cambridge, MA.
80. Decety, J. & J.T. Cacioppo. 2011. *Handbook of Social Neuroscience*. Oxford University Press. New York.
81. Molnar-Szakacs, I. & K. Overy. 2006. Music and mirror neurons: from motion to 'e'motion. *Soc. Cogn. Affect. Neurosci.* **1:** 235–241.
82. Overy, K. & I. Molnar-Szakacs. 2009. Being together in time: musical experience and the mirror neuron system. *Music Percept.* **26:** 489–504.

Ann. N.Y. Acad. Sci. ISSN 0077-8923

ANNALS OF THE NEW YORK ACADEMY OF SCIENCES
Issue: *The Neurosciences and Music IV: Learning and Memory*

Tempo mediates the involvement of motor areas in beat perception

J. Devin McAuley,[1] Molly J. Henry,[2] and Jean Tkach[3]

[1]Department of Psychology, Michigan State University, East Lansing, Michigan. [2]Department of Psychology, Bowling Green State University, Bowling Green, Ohio. [3]University Hospitals, Case Western Reserve University, Cleveland, Ohio

Address for correspondence: Dr. J. Devin McAuley, Department of Psychology, Michigan State University, East Lansing, MI 48824. dmcauley@msu.edu

Increasing evidence shows that the neural circuits involved in beat perception overlap with motor circuitry even in the absence of overt movement. This study investigated effects of tempo on beat-based processing by combining functional magnetic resonance imaging with a perceptual timing paradigm where participants made simple temporal judgments about short rhythmic sequences. Of central interest were judgments about ambiguous test rhythms where the perceived direction of a timing deviation ("speeding up" vs. "slowing down") depended on the induction of an implied beat. Successful beat induction was reduced when the implied beat was at a slower tempo (1,500 ms) than when it was at a faster tempo (600 ms). Decreased beat induction was accompanied by decreased functional activity in the basal ganglia, premotor and supplementary motor regions, and thalamus. Findings support the conclusion that rhythms presented at a slow tempo reduce involvement of a striato–thalamo–cortico network in beat-based processing.

Keywords: timing; temporal processing; rhythm perception

Introduction

One central question in the neurosciences of music concerns an understanding of the neural circuitry involved in perceiving and producing musical rhythm. To date, investigations in this domain provide mounting evidence that the neural networks underpinning rhythm perception and production show substantial overlap with rhythm perception, including the perception of a periodic "pulse" or beat, recruiting many motor-related areas even in the absence of movement.[1–5] A behavioral example that highlights the close relationship between rhythm perception and production is the precise sensorimotor coordination demonstrated by performers in an orchestra or members of a dance troupe. Sensorimotor coordination is facilitated by the presence of a beat.[6–8] Moreover, rhythm perception and production demonstrate a beat-based advantage; rhythms with a strong beat have been shown to be better discriminated and reproduced than rhythms with a weak (or absent) beat.[5,9,10]

Beat perception is frequently found to be strongest (i.e., beats are most salient) for tempos around 100–120 beats per minute, which corresponds to a beat period between 500 ms and 600 milliseconds.[11] The range of tempi that afford beat perception is also limited. Musical rhythms that progress too slowly (>2 sec or so between acoustic events) lose their inherent rhythmic organization, leaving only a series of isolated events, while those that progress too rapidly (<100 ms between events) exceed perceptual limits.[11] Given that tempo is an important factor in beat perception, we were interested in the question of whether tempo mediates the involvement of motor-related circuitry in beat-based processing. This study addresses this question by combining functional magnetic resonance imaging (fMRI) with a behavioral timing paradigm previously demonstrated to reveal robust individual differences in beat perception.[12–15]

The behavioral paradigm is as follows. Participants are presented with the two types of short rhythmic sequences depicted in Figure 1: 4-tone

doi: 10.1111/j.1749-6632.2011.06433.x

Sequence Types

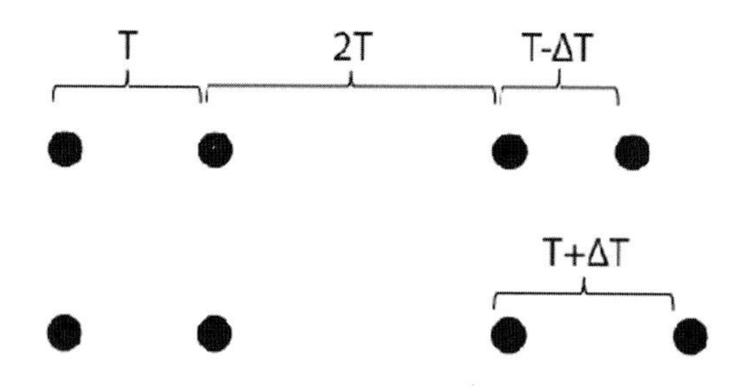

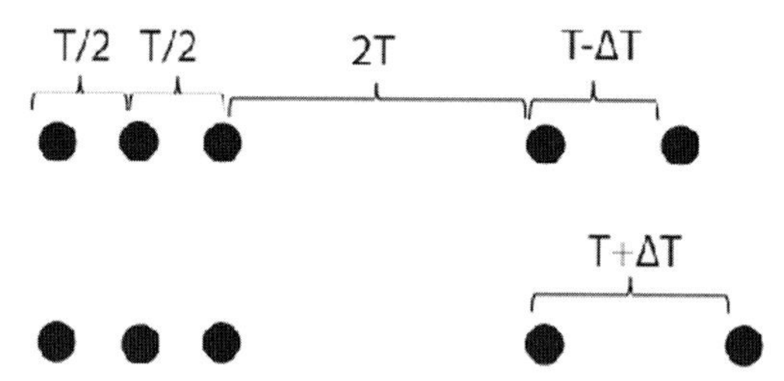

Figure 1. Diagram of stimuli and task. Four-tone (control) sequences marked out a series of three inter onset intervals: T, $2T$, and a final interval, $T \pm \Delta T$, where ΔT varies from trial to trial as a percentage deviation from T. Five-tone (test) sequences were identical to the controls except for the inclusion of an additional tone that bisects the initial inter onset interval, resulting in a series of four inter onset intervals: $T/2$, $T/2$, $2T$, and $T \pm \Delta T$. Participants' task was to judge whether each rhythm they heard was "speeding up" or "slowing down." Tempo was manipulated by varying the value of T, which was either 600 ms or 1,500 ms.

sequences and 5-tone sequences. These will be referred to hereafter as control and test sequences, respectively, for reasons that will soon become apparent. The control sequences mark out a series of three interonset intervals: T, $2T$, and a final interval, $T \pm \Delta T$, where ΔT varies from trial to trial as a percentage deviation from T. The test sequences are identical to the controls except for the inclusion of an additional tone that bisects the initial interonset interval; thus, test sequences delineate a series of four interonset intervals: $T/2$, $T/2$, $2T$, and $T \pm \Delta T$. The participants' task is to judge whether each rhythm they hear "speeds up" or "slows down" at the end.

To date, studies using this paradigm have shown that the pattern of "speeding up" and "slowing down" judgments for control sequences is very consistent and shows little variation across individuals.[12–15] Participants generally respond that control sequences "speed up" when the final interval is shorter than T (i.e., $T - \Delta T$), and "slow down" when the final interval is longer than T (i.e., $T + \Delta T$). In contrast, test sequences produce two different response patterns. Participants either respond similarly to control sequences or they judge that all test sequences slow down. The observed perceptual ambiguity likely arises because of two factors: (1) the test sequences are temporally structured so that there is an implied beat with a period equal to T, and (2) not all participants appear to perceive the implied beat. Participants who do perceive the implied beat use the same temporal referent, T, for test and control sequences, whereas those who do not perceive the implied beat use the explicit $T/2$ referent as the basis for their judgments. For these individuals, this means that all test sequences will be perceived to slow down as long as the final interval is longer than $T/2$. Thus, judgments about test sequences are of particular interest because there is a critical range of final intervals between $T/2$ and T where perception of the implied beat impacts the polarity of their judgment.

The first fMRI investigation to use this paradigm[12] provided some support that the pattern of responses to test sequences provides a valid index of beat sensitivity. Specifically, sensitivity to the implied beat of the test sequences (i.e., use of the implied referent T) was associated with functional activation in a network of motor-related areas previously linked to beat perception,[2–4] including left premotor and supplementary motor regions. Notably, individual differences in beat sensitivity predicted neural responses to both control and test sequences, suggesting that the paradigm was picking up on a general difference in rhythmic processing. That is, some individuals appeared to be processing control and test rhythms in a beat-based timing mode, while others appeared to be using an interval-based timing mode.[16,17]

A more recent fMRI investigation using the same "speeding up" versus "slowing down" paradigm examined effects of modality on beat perception, with the expectation that engagement of neural circuits in beat-based processing would be stronger when rhythms were presented in the auditory modality than in the visual modality.[14] Participants experienced control and test sequences that were marked either by brief tones or briefly presented squares. Both modalities were presented in separate blocks with the order of presented modality counterbalanced so that cross-modal order effects could be investigated. As expected, participants were more sensitive to the implied beat when test rhythms were presented in the auditory modality than in the visual modality and auditory rhythms generated larger neural responses in the basal ganglia than did visual rhythms. Visual rhythms were also found to produce a stronger sense of beat when preceded by auditory rhythms with identical temporal structure and this increase in beat perception was associated with a bilateral increase in activity in the putamen. Together, findings from these studies add to the evidence that beat perception engages a striato–thalamo–cortico network that includes the basal ganglia, thalamus, premotor, and supplementary motor areas.[2–5,12]

This study represents an extension of previous research to an investigation of effects of tempo. It is modeled on the previous fMRI investigation of effects of modality.[14] Rather than varying modality, we simply manipulated the tempo of the control and test rhythms by varying the value of T (Fig. 1); T was either 600 ms or 1,500 ms in separate blocks. Recall that for control sequences, T is the explicit referent, whereas for test sequences, T is the period of the implied beat. Based on the greater beat salience at 600 ms than at 1,500 ms, we hypothesized that slowing down the test rhythms in the 1,500 ms tempo condition should reduce the likelihood that listeners would hear the implied beat.[6,18] Moreover, if beat-based processing activates a striatal–thalamic–cortical network, then decreases in beat sensitivity should be associated with decreases in functional activity in brain regions that are part of that network.[2–5]

Methods

Participants and design

Fifteen neurologically normal, right-handed volunteers (18–52 years; $n = 12$, female) participated in return for a $25 cash payment. Participants self-reported normal hearing and varied in years of formal musical training (M = 5.5 years, SD = 4.9 years). The design of the study was a 2 (tempo, T: 600 ms, 1,500 ms) $\times 2$ (sequence type: control, test) $\times 7$ (final interval: $T \pm 4\%$, $\pm 12\%$, $\pm 20\%$, or -50%) within-subjects factorial. Participants heard control and test sequences with a variable final interval and indicated whether the sequence they heard on each trial was "speeding up" or "slowing down."

Stimuli

Figure 1 shows a diagram of the stimuli and task. Sequences were composed of 50-ms 440-Hz sine tones and were generated off-line using Praat software.[19] Control sequences consisted of four tones that marked out a series of three interonset intervals: T, $2T$, and $T \pm 4\%$, $\pm 12\%$, $\pm 20\%$, or -50%. Test sequences were identical except for the inclusion of an additional tone that bisected the initial interonset interval. Thus, test sequences consisted of five tones that marked out a series of four interonset intervals: $T/2$, $T/2$, $2T$, and $T \pm 4\%$, $\pm 12\%$, $\pm 20\%$, or -50%. The tempo factor determined the value of T and was either 600 ms or 1,500 ms.

Procedure

Participants completed two scanning sessions per tempo condition with the order of the two tempi counterbalanced across participants. Within a session, control and test sequences were randomly intermixed and the final interval varied randomly from trial to trial. Prior to the administration of each tempo condition, participants completed a familiarization block that consisted of 16 sequences (control and test) with final intervals equal to $T - 40\%$ and $T + 40\%$. No feedback was provided during familiarization, but if participants failed to reach a criterion of 75% correct for control sequences (where there was an objectively correct response), they were prompted to complete an additional familiarization block. The maximum number of familiarization blocks completed by any participant was three.

Each scanning run consisted of 48 trials with an additional nine silent trials randomly interspersed to resolve the hemodynamic response in analysis. Within a run, participants made three responses to control and test sequences for each of six final intervals ($\pm 4\%$, $\pm 12\%$, $\pm 20\%$) and 12 responses for the -50% final interval. In total, this resulted in six responses for each tempo $\times$ sequence type $\times$ final interval combination, and 24 responses to each tempo $\times$ sequence type combination for the -50% final interval. A single run lasted 6.5 or 9.5 min depending on tempo condition, and the entire experiment, including familiarization blocks and T1 image acquisition, lasted approximately 1.5 hours.

Image acquisition and preprocessing

Participants were scanned in a four-tesla Bruker Whole Body Medspec MR system using an eight-channel transmit/receive head coil. Foam pads were placed around each participant's head and supported their legs. Auditory stimuli were transmitted with equalized sound spectrum through an MR-compatible audio system (Avotec SS3100) with acoustically padded headphones to reduce fMRI acoustic noise by ~30dB. (Silent Scan; Avotec, Stuart, FL). Presentation of both auditory stimuli and visual instructions was controlled by a paradigm implemented in E-Prime 1.2 (Psychology Software Tools, Pittsburgh, PA). Visual instructions were projected to a mirror attached to the head coil. All behavioral responses were made using a MRI-compatible button box positioned under the right hand; the middle finger was used to indicate the sequence was speeding up and the index finger was used to indicate the sequence was slowing down.

Blood oxygen level–dependent (BOLD) fMRI data were obtained using a single-shot gradient-echo echo planar imaging (EPI) sequence. Each functional volume consisted of 38 contiguous axial slices, providing whole-brain coverage (TR = 2.05 s, TE = 25 ms, FOV = 240 × 240 mm, matrix size of 64 × 64). EPIs were 3 mm thick with in-plane resolution of 3.75 mm × 3.75 mm. A total of 184 and 279 fMRI volumes were collected per session (continuous acquisition) for the 600-ms tempo condition (6' 21") and the 1,500-ms tempo condition (9' 36"), respectively, including five dummy scans collected at the beginning of each session to allow magnetization to settle to a steady-state. High-resolution 1-mm isotropic T1-weighted MPRAGE anatomical images were acquired for each participant for localization and coregistration (TR = 2.5 s, TI = 1.1 s, TE = 3.52 ms, flip angle = 12°, FOV = 256 × 192 × 176 mm, and matrix = 256 × 192 × 176).

SPM5 was used for preprocessing and analysis (SPM5; Wellcome Department of Cognitive Neurology, London, UK). Images were slice-timing corrected, then realigned spatially (to correct for motion) to the image corresponding to the temporal midpoint in the series using a least-squares approach with six rigid-body parameters, and trilinear interpolation. The MPRAGE image was normalized (using affine and smoothly nonlinear transformations) to a brain template in Montreal Neurological Institute (MNI) space. Resulting normalization parameters were applied to the coregistered EPIs, and EPI images were smoothed with an 8 mm full-width half-maximum Gaussian kernel.

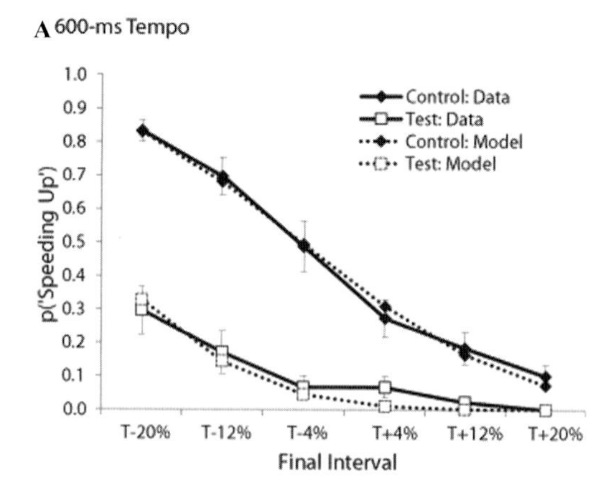

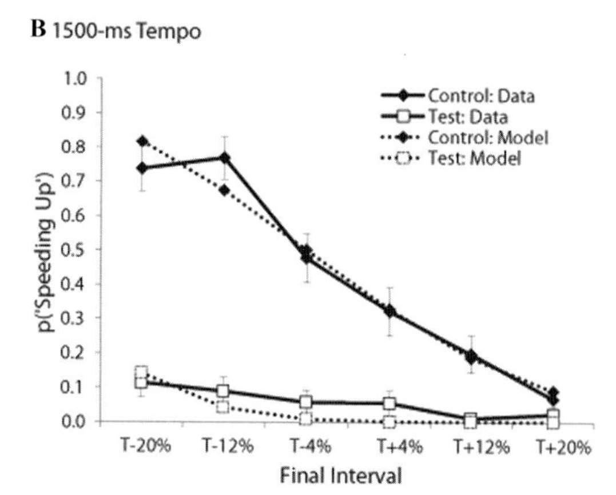

Figure 2. Proportions of speeding-up responses (solid lines) and model fits (dotted lines) for control sequences (filled markers) and test sequences (open markers) for (A) the 600-ms tempo condition; and (B) the 1,500-ms tempo condition.

Stimuli and button presses were modeled using a regressor consisting of an on-off boxcar convolved with a canonical hemodynamic response function. EPI volumes with more than 4 mm movement in any plane were included as covariates of no interest to minimize movement artifacts. Low-frequency noise was removed with a 128-second high-pass filter. Results estimated from single-subject models were entered into second-level random effects analyses for standard SPM group inference[20] to estimate effects of Tempo (600 ms vs. 1,500 ms).

Results

Behavioral results

Proportions of speeding-up responses at each final were determined for each participant separately for control and test sequences in each tempo condition. Response proportions were then fit on a by-participant basis with a signal detection model[12] to derive a beat sensitivity index, $w \in [0, 1]$. For test sequences, which have ambiguous timing, values of the beat sensitivity index, w, estimate the extent to which participants judge sequence timing using the explicit temporal referent ($T/2$) or the implied beat (which is equal to T). A value of $w = 0$ means that sequence timing judgments are based exclusively on the referent $T/2$, whereas a value of $w = 1$ means that sequence timing judgments are based exclusively on the referent T. Intermediate values of w indicate partial sensitivity to the implied beat, with larger values indicating greater beat sensitivity. For control sequences, which have unambiguous sequence timing and an explicit temporal referent equal to T, w values are expected to be $\cong 1$ for all participants.

Behavioral data and corresponding model fits are shown in Figure 2 (panel A, 600-ms tempo condition; panel B, 1,500-ms tempo condition). Mean values of the estimated beat sensitivity index, w, are shown in Figure 3. A 2 (tempo) × 2 (sequence type) × 2 (order) mixed-measures ANOVA on w revealed a main effect of tempo ($F(1,13) = 6.81$, $P < 0.05$), a main effect of sequence type ($F(1,13) = 37.36$, $P < 0.001$), and critically a tempo × sequence type interaction ($F(1,13) = 5.67$, $P < 0.05$). The main effect of order and all interactions with order were nonsignificant (all $P_s > 0.36$).

All participants responded similarly to control sequences, but demonstrated large individual differences in how they responded to test sequences. For most participants, values of w for control sequences were very close to 1.0, and did not differ significantly across tempi ($T = 600$ ms, $w = 0.94 \pm 0.02$; $T = 1{,}500$ ms, $w = 0.92 \pm 0.03$), $t(14) = 0.96$, $P = 0.36$. For test sequences, values of w were, as expected, much lower ($w = 0.47 \pm 0.08$) and more variable; w values for test sequences were close to 0.0 for some individuals, whereas for others, values of w were close to 1.0. Moreover, data were consistent with the hypothesis that sensitivity to the implied beat in test sequences would be lower for the slower tempo ($T = 600$ ms; $w = 0.59 \pm 0.09$; $T = 1{,}500$ ms; $w = 0.35 \pm 0.10$), $t(14) = 2.89$, $P < 0.01$ (one-tailed).

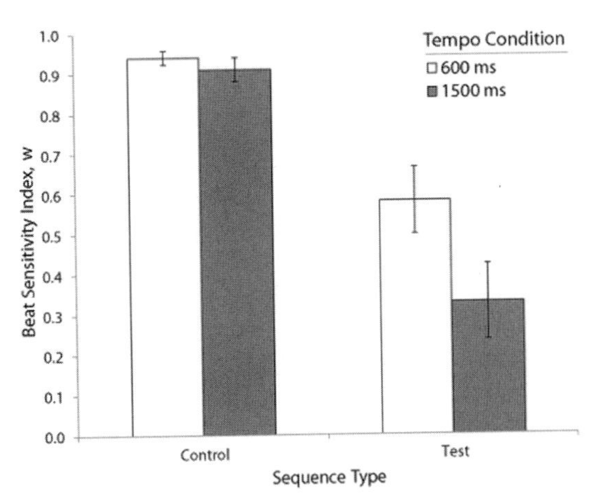

Figure 3. Beat sensitivity index, w, for control sequences (white bars) and test sequences (gray bars) shown separately for the 600-ms and 1,500-ms tempi. Values of w for control sequences were close to 1.0 and not significantly different between tempo conditions. For test sequences, the range of observed w values was much larger, and w values were significantly larger for the 600-ms tempo (indicating higher beat sensitivity) than for the 1,500-ms tempo.

We also compared temporal discrimination thresholds across tempo conditions. Relative just-noticeable differences (JNDs) were calculated for each tempo condition using the standard z-transform method.[21] A 2 (tempo) × 2 (order) mixed-measures ANOVA on relative JNDs revealed no main effects or interaction (all $P_s > 0.31$). Consistent with the scalar property of timing,[22] discrimination thresholds did not reliably differ between the 600-ms tempo condition ($M = 13.0\% \pm 0.7\%$) and the 1,500-ms tempo condition ($M = 14.1\% \pm 1.3\%$, $P = 0.41$).

fMRI results

First, we conducted a conjunction analysis of the two tempo conditions to reveal task-relevant areas. Consistent with previous neuroimaging studies of timing and temporal processing, task-related activations were observed in right superior temporal gyrus (STG),[3,23] left premotor cortex,[12,24,25] left SMA,[26,27] bilateral inferior frontal gyri,[28] bilateral middle frontal gyri,[29] left inferior parietal lobule,[30] bilateral insula,[12,28] basal ganglia (putamen and pallidum),[2,31] thalamus,[5,17] and cerebellum.[2,17] All

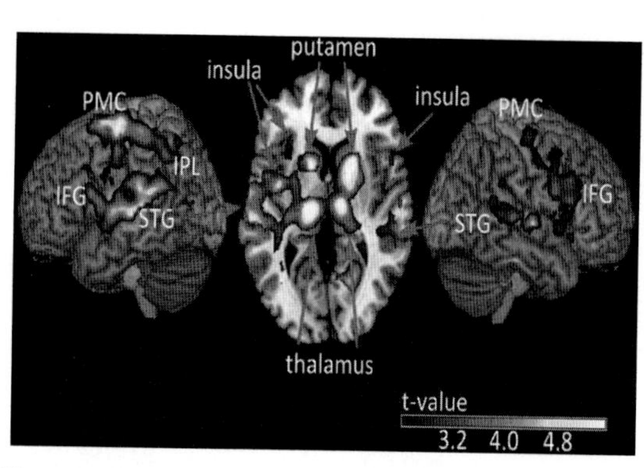

Figure 4. Areas found to be more active for the 600-ms tempo condition relative to the 1,500-ms tempo condition overlaid on a standardized brain in MNI space. PMC, premotor cortex; IFG, inferior frontal gyrus; STG, superior temporal gyrus; IPL, inferior parietal lobule.

activated clusters exceeded a whole-brain corrected threshold of P FDR < 0.05 (Tables S1 and S2).

Next, contrast images for control and test sequences estimated from single-subject analyses were entered into a second-level random effects analysis. A repeated-measures ANOVA was conducted with factors tempo (600 ms, 1,500 ms) and sequence type (control, test). Given the greater beat sensitivity in the 600-ms tempo condition than in the 1,500-ms tempo condition, of primary interest was the 600-ms tempo versus 1,500-ms tempo contrast (Fig. 4); Table S3. The results of this contrast revealed that the 600-ms tempo condition produced more activation than the 1,500-ms tempo condition in a number of brain areas including the SMA, basal ganglia (putamen and pallidum), and premotor cortices. In addition, bilateral STG, thalamus, insula, and right inferior frontal gyrus (BA 48/6) were all more active in the 600-ms tempo condition than in the 1,500-ms tempo condition at a whole-brain corrected level of significance (P FDR < 0.05). The reverse 1,500-ms tempo versus 600-ms tempo contrast did not reveal activation differences in any brain regions, even with a relaxed criterion for significance (uncorrected $P < 0.01$).

Neither the comparison between control and test sequences nor the interaction between Tempo and Sequence Type activated any brain regions at the whole brain corrected level (P FDR < 0.05).

Discussion

Behavioral results from this study yielded four main findings. First, responses to unambiguously timed control sequences produced consistent responses for most participants at both tempi. This finding is not surprising because for control sequences, the explicit temporal referent, T, was available in both tempo conditions. Second, participant responses to the ambiguously timed test rhythms revealed large individual differences. Ambiguous test rhythms were of primary interest because judgment polarity (speeding up versus slowing down) depended on the successful induction of a beat at a time level that was implied, but not explicitly marked, by the rhythms. Paralleling the pattern that we've observed previously,[12–15] some participants responded to test rhythms entirely on the basis of the explicit temporal referent, $T/2$, while others responding entirely on the basis of the implied beat. Third, there was, as expected, a reliable tendency for listeners to be less sensitive to the implied beat of the test rhythms when the rhythms were slowed down. This finding converges with recent evidence that the beat-based advantage in rhythm discrimination[2] is also reduced at slow tempi.[32] Finally, despite the difference in beat sensitivity, temporal discrimination thresholds did not vary across tempo.

fMRI results reveal a corresponding pattern of BOLD activity that supports the hypothesis that tempo mediates the engagement of motor areas in beat perception. There is mounting evidence that beat perception involves a striato–thalamo–cortico network.[2–5,12,14,17,27,33] In this study, we observed task-related activity in areas that included the basal ganglia, thalamus, premotor and supplementary motor regions, and the insula. Moreover, we found that decreases in beat sensitivity in the slow (1,500 ms) tempo condition produced corresponding reductions in functional activity in these same regions. Because there were no differences in brain activation in response to the two sequence types, the observed behavioral differences are consistent with recent proposals of distinct beat-based and interval-based timing mechanisms[12,17] and a general shift from beat-based to interval-based processing at slow tempi. In this regard, it is unlikely that the effects of tempo we observed in BOLD activity were due to differences in participants' ability to detect tempo changes (i.e., task difficulty) because temporal discrimination threshold did not reliably vary across tempo.

fMRI results also revealed an effect of tempo on neural responses in the STG. Although this finding was not necessarily expected since the amount of auditory stimulation was identical across tempo conditions, the STG has been proposed to be involved in auditory-motor transformations necessary for synchronizing rhythmic movement with an external auditory stimulus.[3] The STG also has connections with dorsal and ventral premotor cortices, as well as the insula, which have been proposed to be involved in the temporal integration of multimodal stimuli and synchrony detection;[34] previous studies have also implicated the insula in beat perception.[12]

Conclusions

Previous neuroimaging studies have revealed substantial overlap in the neural circuitry involved in rhythm perception and production, with rhythm perception recruiting many motor-related areas even in the absence of overt movement.[1–5,12,14,35] This overlap is especially apparent for rhythms that established the sense of a regular beat or pulse. This study shows that successful beat induction in short rhythmic sequences is less likely when the implied beat is at a slower tempo (1,500 ms) than when it is at a faster tempo (600 ms). Decreases in beat perception observed with the slowed rhythms are associated with decreases in functional activity in the basal ganglia, premotor and supplementary motor regions, and thalamus, supporting the conclusion that rhythms presented at a slower tempo reduce involvement of a striato–thalamo–cortico network in beat-based processing.

Conflicts of interest

The authors declare no conflicts of interest.

Supporting information

Additional supporting information may be found in the online version of this article:

Table S1. Brain areas activated in the 600-ms tempo-rest contrast.

Table S2. Brain areas activated in the 1500-ms tempo-rest contrast.

Table S3. Brain areas activated in the 600-ms tempo–1500-ms tempo contrast.

Please note: Wiley-Blackwell is not responsible for the content or functionality of any supporting materials supplied by the authors. Any queries (other than missing material) should be directed to the corresponding author for the article.

References

1. Schubotz, R.I., A.D. Friederici & D.Y. von Cramon. 2000. Time perception and motor timing: a common cortical and subcortical basis revealed by fMRI. *Neuroimage* **11:** 1–12.
2. Grahn, J.A. & M. Brett. 2007. Rhythm perception in motor areas of the brain. *J. Cogn. Neurosci.* **19:** 893–906.
3. Chen, J.L., V.B. Penhune & R.J. Zatorre. 2008. Listening to musical rhythms recruits motor regions of the brain. *Cereb. Cortex* **18:** 2844–2854.
4. Grahn, J.A. & M. Brett. 2009. Impairment of beat-based rhythm discrimination in Parkinson's disease. *Cortex* **45:** 54–61.
5. Grahn, J.A. & J.B. Rowe. 2009. Feeling the beat: premotor and striatal interactions in musicians and non-musicians during beat processing. *J. Neurosci.* **29:** 7540–7548.
6. Parncutt, R. 1994. A perceptual model of pulse salience and metrical accent in musical rhythms. *Music Percept.* **11:** 409–464.
7. Large, E.W. 2000. On synchronizing movements to music. *Hum. Mov. Sci.* **19:** 527–566.
8. Snyder, J.S. & C.L. Krumhansl. 2001. Tapping to ragtime: cues to pulse finding. *Music Percept.* **18:** 455–489.
9. Essens, P. J. & D.J. Povel. 1985. Metrical and nonmetrical representations of temporal patterns. *Percept. Psychophys.* **37:** 1–7.
10. Povel, D.J. & P.J. Essens. 1985. Perception of temporal patterns. *Music Percept.* **2:** 411–440.
11. McAuley, J.D. 2010. Tempo and Rhythm. In *Music Perception: Springer Handbook of Auditory Research.* M.R. Jones, R.R. Fay & A.N. Popper, Eds.: 165–199. Springer. New York.
12. Grahn, J.A. & J.D. McAuley. 2009. Neural bases of individual differences in beat perception. *NeuroImage* **47:** 1894–1903.
13. McAuley, J.D. & M.J. Henry. 2010. Modality effects in rhythm processing: auditory encoding of visual rhythms is neither obligatory nor automatic. *Attn. Percept. Psychophys.* **72:** 1377–1389.
14. Grahn, J.A., M.J. Henry & J.D. McAuley. 2011. FMRI investigation of cross-modal interactions in rhythm perception: audition primes vision, but not vice versa. *NeuroImage* **54:** 1231–1243.
15. Snyder, J.S., A. Pasinski & J.D. McAuley. 2011. Listening strategy modulates cognitive processing of auditory rhythms. *Psychophysiology* **48:** 198–207.
16. McAuley, J.D. & M.R. Jones. 2003. Modeling effects of rhythmic context on perceived duration: a comparison of interval and entrainment approaches to short-interval timing. *J. Exp. Psychol. Hum. Percept. Perform.* **29:** 1102–1125.
17. Teki, S., M. Grube, S. Kumar & T.D. Griffiths. 2011. Distinct neural substrates of duration-based and beat-based auditory timing. *J. Neurosci.* **31:** 3805–3812.
18. McAuley, J.D. & P. Semple. 1999. The effect of tempo and musical experience on perceived beat. *Aust. J. Psychol.* **51:** 176–187.
19. Boersma, P. & D. Weenink. 2005. Praat: doing phonetics by computer (Version 4.3.01) [Computer program]. URL http://www.praat.org/

20. Penny, W. & A.P. Holmes. 2003. Random effects analysis. In *Human Brain Function II*. R.S.J. Frackowiack, K.J. Friston, C.D. Frith, *et al.* Eds. 2nd ed. Elsevier Academic Press, San Diego.
21. MacMillan, N.A. & D.C. Creelman. 1991. *Detection Theory: A User Guide*. Cambridge University Press. New York.
22. Gibbon, J.J., R. Church & W. Meck. 1984. Scalar Timing in Memory. *Ann N.Y. Acad. Sci.* **423:** 52–57.
23. Griffiths, T.D., I. Johnsrude, J. Dean & G.G.R. Green. 1999. A common neural substrate for the analysis of pitch and duration pattern in segmented sound? *Neuroreport* **18:** 3825–3830.
24. Schubotz, R.I. & D.Y. von Cramon. 2001. Functional organization of the lateral premotor cortex: fMRI reveals different regions activated by anticipation of object properties, location and speed. *Cogn. Brain Res.* **11:** 97–112.
25. Schubotz, R.I. & D.Y. von Cramon. 2003. Functional–anatomical concepts of human premotor cortex: evidence from fMRI and PET studies. *NeuroImage* **20:** S120–S131.
26. Mayville, J.M., K.J. Jantzen, A. Fuchs, *et al.* 2002. Cortical and subcortical networks underlying syncopated and synchronized coordination revealed using fMRI. *Hum. Brain Mapp.* **17:** 214–229.
27. Macar, F., J. Coull & F. Vidal. 2006. The supplementary motor area in motor and perceptual time processing: fMRI studies. *Cogn. Process.* **7:** 89–94.
28. Platel, H., C. Price, J.C. Baron, *et al.* 1997. The structural components of music perception: a functional anatomical study. *Brain* **120:** 229–243.
29. Sakai, K., O. Hikosaki, S. Miyauchi, *et al.* 1999. Neural representation of a Rhythm depends on its Interval Ratio. *J. Neurosci.* **19:** 10074–10081.
30. Limb, C.J., S. Kemeny, E.B. Ortigoza, *et al.* 2006. Left hemispheric lateralization of brain activity during passive rhythm perception in musicians. *Anatomical Rec.* **288A:** 382–389.
31. Ferrandez, A.M., L. Hugueville, S. Lehéricy, *et al.* 2003. Basal ganglia and supplementary motor area subtend duration perception: an fMRI study. *Neuroimage* **19:** 1532–1544.
32. McAuley, J.D., K. Nave, J. Walters & A. Wiggins. 2011. Discrimination of slowed rhythms mimics beat perception impairments in Parkinson's disease. In *Proceedings of the Abstracts of the 52nd Annual Meeting of the Psychonomic Society* **16:** 52.
33. Matell, M.S. & W.H. Meck. 2004. Cortico-striatal circuits and interval timing: coincidence detection of oscillatory processes. *Cogn. Brain Res.* **21:** 139–170.
34. Lux, S., J.C. Marchall, A. Ritzla, *et al.* 2003. Neural mechanisms associated with attention to temporal synchrony versus spatial orientation. *NeuroImage.* **20:** S58–S65.
35. Chapin, H.L, T. Zanton, K.J. Jantzen, *et al.* 2010. Neural responses to complex auditory: the role of attending. *Front. Aud. Cog. Neurosci.* **1:** 244. doi:10.3389/fpsyg.2010.00224.

Ann. N.Y. Acad. Sci. ISSN 0077-8923

ANNALS OF THE NEW YORK ACADEMY OF SCIENCES
Issue: *The Neurosciences and Music IV: Learning and Memory*

Without it no music: beat induction as a fundamental musical trait

Henkjan Honing

Institute for Logic, Language and Computation, and the Cognitive Science Center Amsterdam, Universiteit van Amsterdam, Amsterdam, the Netherlands

Address for correspondence: Henkjan Honing, Institute for Logic, Language and Computation (ILLC), and the Cognitive Science Center Amsterdam (CSCA), Universiteit van Amsterdam, P.O. Box 19268, NL1000 GG Amsterdam, the Netherlands. honing@uva.nl

Beat induction (BI) is the cognitive skill that allows us to hear a regular pulse in music to which we can then synchronize. Perceiving this regularity in music allows us to dance and make music together. As such, it can be considered a fundamental musical trait that, arguably, played a decisive role in the origins of music. Furthermore, BI might be considered a spontaneously developing, domain-specific, and species-specific skill. Although both learning and perception/action coupling were shown to be relevant in its development, at least one study showed that the auditory system of a newborn is able to detect the periodicities induced by a varying rhythm. A related study with adults suggested that hierarchical representations for rhythms (meter induction) are formed automatically in the human auditory system. We will reconsider these empirical findings in the light of the question whether beat and meter induction are fundamental cognitive mechanisms.

Keywords: rhythm; meter; musicality; event-related brain potentials; attention; cognitive biology

Introduction

It seems a trivial skill: children who clap along with a song, musicians who tap their foot to the music, or a stage full of line dancers who dance in synchrony. In a way it is indeed trivial. Most people can easily pick up a regular pulse from the music or can judge whether the music speeds up or slows down. However, the realization that perceiving this regularity in music allows us to dance and make music together makes it a less trivial phenomenon. Beat induction (BI) might well be conditional to music (i.e., without it no music), and, as such, it can be considered a fundamental human trait that, arguably, must have played a decisive role in how *musicality* evolved.[1]

BI[a] has been the topic of quite a few music perception studies, mostly concerned with the theoretical and psychological aspects of this cognitive skill.[2–5] More recently, the phenomenon has attracted the interest of developmental psychologists,[6] cognitive biologists,[7] evolutionary psychologists,[1] and neuroscientists[8,9] as a skill that is fundamental to music processing.

BI has been argued to be an innate (or spontaneously developing), domain-specific, and species-specific skill.[1] However, with regard to the first issue, scientists are still divided on whether this ability develops spontaneously (emphasizing a biological basis) or whether it is learned (emphasizing a cultural basis). Some authors consider the sensitivity to beat to be acquired during the first year of life, suggesting that the ways in which babies are rocked

[a]The term *beat induction* is preferred here over *beat perception* (and synchronization) to emphasize that a beat does not always need to be physically present in order to be "perceived" (see, e.g., the section on "loud rest"). Furthermore, it stresses that beat induction is not a passive, perceptual process but an active one in which a rhythm evokes a particular regular pattern in the listener. How this process is dependent on attention and/or consciousness, and whether there might be a cognitive and neurological difference between beat induction and meter induction, are topics of current research.

doi: 10.1111/j.1749-6632.2011.06402.x

and bounced in time to music by their parents is the most important factor in developing a sense for metrical structure.[10,11] By contrast, more recent studies emphasize a biological basis, suggesting BI to be specifically tuned to music, for example, studies demonstrating that BI is already functional in young infants as well as two- to three-day-old newborns.[12,13] These recent empirical findings can be taken as a support for a genetic predisposition for BI, rather than it being a result of learning.[14]

In addition, developmental studies suggest that infants are not only sensitive to a regular pulse, but also to meter (i.e., two or more levels of pulse).[15] Thus, it is possible that humans possess some processing predisposition to extract hierarchically structured regularities from complex rhythmic patterns.[16] Research with newborns provides an appropriate context within which to understand more about these fundamental capacities.[17] However, studies addressing hierarchical perception in newborns are still underway. Hence, this review concentrates on how to study beat and meter induction using a mismatch negativity (MMN) paradigm, and addresses some open issues with regard to the cognitive and biological aspects of BI.

Before introducing the MMN paradigm, this paper begins with a theoretical music example illustrating the notion of metrical expectation.

Example: a "loud rest"

In music, an important distinction to be made is that between rhythmic pattern and metrical structure.[18] While rhythm can be characterized as the varying pattern of durations that is physically present in the music, meter involves our perception and, more importantly, anticipation and prediction of such rhythmic patterns. Meter is, as such, a cognitive phenomenon.[19]

The interaction of rhythm and meter, and the role that cognition plays in its perception and appreciation can be illustrated with the phenomenon of *syncopation*. It is often described, rather informally, as "an accent that has been moved forward," or as "a technique often used by composers to avoid regularity in rhythm by displacing an emphasis in that rhythm."[20] To illustrate this, consider the two rhythms depicted in Figure 1. Which of these is a syncopated rhythm?

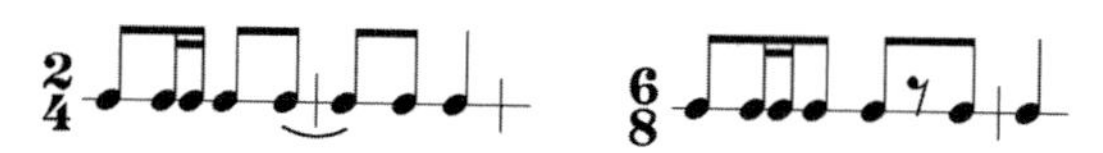

Figure 1. Which rhythm is syncopated?

A formally trained musician will easily point out the left example, guided by the slur marking a syncopation (literally a "joined beat"). However, performed by a drum computer, these notated rhythms will sound identical. Here the reader is strongly influenced by the notation. When we listen to a rhythm (even if it is simply a series of isochronous clicks, like a clock), we tend to interpret it in a metrical fashion,[21] and hear it as syncopated, or not, depending on our metric interpretation (a time signature in the notation is no guarantee that a listener will perceive the meter as such). This is illustrated by the example in Figure 2.

Western listeners tend to project a duple meter while listening to a rhythm,[22] and hence perceive a syncopation (depicted in the left panel of Fig. 2, i.e., the "loud rest" marked in gray). However, if a listener were to expect, for example, a compound meter—caused by, for instance, a different musical background or listening experience—then the syncopation will disappear altogether. It becomes a "silent rest" (see the right panel of Fig. 2).

An important insight here is that the perception of rhythm should be seen as an *interaction* between the rhythmic pattern (labeled "Rhythm" in Fig. 2) and the listener, who projects a certain meter onto it (labeled "Listener" in Fig. 2), which is induced by that very same rhythm.[23,24] We can therefore use the presence of a syncopation (or "loud rest") as evidence for the presence of a strong metric expectation (be it the result of earlier exposure to music or an inborn preference). This provides an elegant and direct method to probe metrical expectation in listeners, and is the key idea used in the experiments described below.

Using MMN to probe "loud rests"

Electrophysiological measures, such as event-related brain potentials (ERP), are a useful tool in the study of BI and the metrical encoding of rhythm, especially in examining its predictive nature. An informative component of an ERP is the MMN: a negative deflection in the brain signal that occurs if something unexpected happens while listening (even when attention is not directed to the

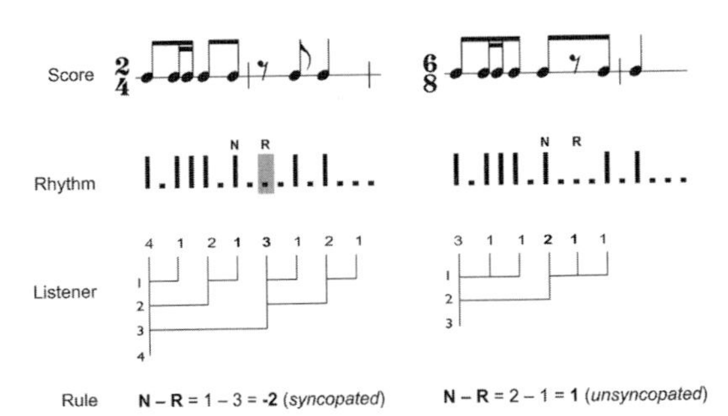

Figure 2. Two possible notations (labeled as "Score") of the same rhythm (labeled as "Rhythm"). In the left example, a metrical tree represents a duple meter, and in the right example it represents a compound meter (labeled as "Listener"). The numbers at the leaves of the metrical tree represent the theoretical metric salience (the depth of the tree at that position in the rhythm). A negative difference between the metric salience of a certain note *N* and the succeeding rest *R* indicates a syncopation. The more negative this difference, the more syncopated the note *N* or "louder" the rest *R* (adapted from Honing).[25]

rhythm).[25] This MMN is generally thought to reflect an error signal that is elicited when incoming sensory information does not match the expectations created by previous information. As such, it can be instrumental in probing a violation in a metrical expectation, such as a syncope or a "loud rest."

Characteristics of MMN

In general, an MMN is elicited when incoming stimuli mismatch the predictions produced by the neural representations of regularities extracted from the acoustic environment (e.g., pitch, duration, timbre, location).[25] Also, abstract information (i.e., one auditory feature predicting another)[26] and omissions can cause an MMN, resulting in an interpretation of the MMN as reflecting the detection of regularity violations as part of a predictive process, rather than just sample matching to sensory memory.[27] More salient deviations trigger earlier (and possibly larger amplitude) negative deflections,[28] and, as such, the MMN can be used as an index to compare metrical expectancies of different strengths. An MMN can be observed when subjects engage in a neutral primary task (e.g., watching a movie; *passive condition*) or when instructed to do an unrelated task (*unattended condition*). However, when participants focus their attention on the stimuli, the MMN is often overlapped by attention- and task-dependent ERP components, such as the P165 and the N2b.[25] This makes measuring ERP in a passive condition especially useful in studying the role of attention and consciousness in perception.[29] Finally, compared to other more recent brain imaging techniques, during measurement, ERP is more tolerant to the subject's physical movements; therefore, it is more suitable for participants whose movements are difficult to regulate and behavioral responses are limited (such as newborns).[30]

One point of concern is what to expect in the case of an omission in an acoustic signal (i.e., a silence or "rest," instead of a note)? An omission means that there is no incoming sound, and the question could then be what is linked to existing regularity representations? While this is still a topic for debate,[31] it is clear that an MMN can be elicited when the interonset intervals in a rhythm are smaller than 150 milliseconds.[32] This constraint on stimulus design has to be balanced with the absolute tempo of a rhythm, to make sure the beat occurs not too far from the preferred tempo rates.[18]

Stimuli and experimental design

An MMN is measured using an oddball design: a sound sequence in which rare sounds (deviants) are intermixed with a common sound (standard). One possible stimulus set to study metrical expectation is shown in Figure 3. It consists of eight different sound patterns, all variants of a base pattern (S1) with eight grid points. The base pattern and the four variants (containing omissions on the lowest metrical level) are "strictly metrical," that is, they contain no syncopation when interpreted in duple meter. Together, these five patterns form the set of standard patterns (S1–S5). Three deviants are constructed by omitting events on metrically salient positions in the base pattern, which lead to syncopated patterns. They are created by omitting a note on position 5 (D1), on position 3 (D2), and on position 7 (D3). According to the theoretical model described in Figure 2 (left panel), the strengths of the deviants are ordered as D1 > D2 > D3, where D1 is predicted to be the "loudest rest" or strongest syncopation, and D3 the weakest.

These stimuli were used in a pilot study (Háden, Honing, and Winkler, in preparation). In a preliminary analysis of the results, an MMN was observed for all three deviants. However, there was only a significant difference between D1 and D3, and the difference wave for D3 was close to zero (i.e., almost no MMN). These results made us wonder about

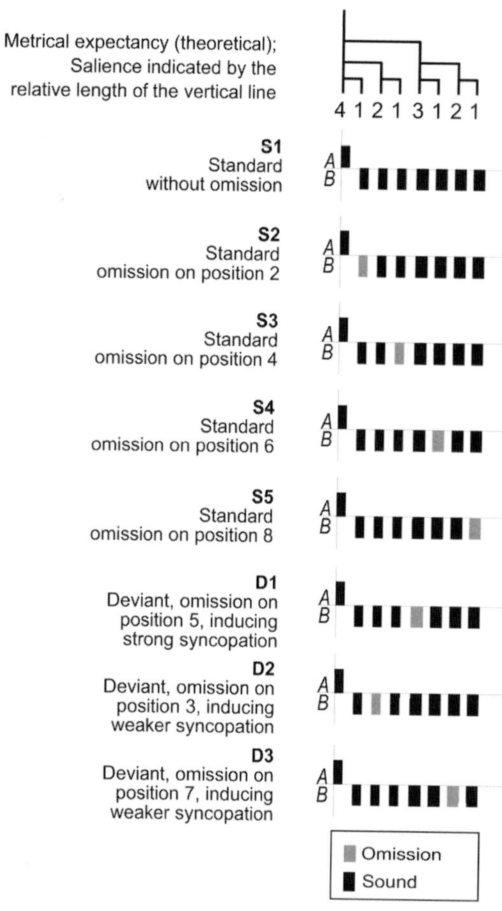

Figure 3. Stimuli as used in a pilot study on metrical expectation. S1–S4 are the standards and D1–D3 the deviants used in an oddball paradigm. Both A and B are percussive sounds, with A being higher-pitched than B, to allow phase alignment.

the relative importance of using more complex (or ecologically valid) stimuli, since in a related study, using more complex stimuli (see Fig. 4), we did get significant effects.[16] In this study, the deviant minus control (deviant – control) difference waveforms showed differences in latency, reflecting a hierarchy in violation of D1 versus D2,[28] as predicted by the theoretical model. However, an analysis of variance (ANOVA) showed a significant interaction between different attention conditions (i.e., passive vs. unattended) and casts some doubt on whether the extraction of a metrical hierarchy is fully automatic.[33]

A similar design using the same stimuli but with only one deviant D1 (because of time constraints) was used with newborns.[13] In that study, we could show that the electrical brain responses elicited by the standard and deviant – control patterns were very similar to each other, whereas the response to D1 differed significantly, providing evidence that (sleeping) newborns can sense the beat. So it appears that the capability of predicting the beat in rhythmic sound sequences is already functional at birth.

Discussion

These studies suggest BI (but possibly not meter induction) to be an automatic process outside of the focus of attention, and they provide evidence that BI is shared among adults and newborns, as such supporting a biological basis.

But how sure can we be that finding an MMN is indeed an evidence for beat and/or meter induction? Are alternative explanations possible?

Because the deviant – control in these studies did not elicit an MMN, we can be sure that the MMN is not a result of the acoustic qualities of the D-pattern *per se*. Furthermore, the response is not simply a result of detecting omissions, otherwise it would show up in the other silent locations as well. Also, the response is not caused by separate representations formed by the three streams (hi-hat, snare, and bass; see Fig. 4): only omissions of the downbeat within the rhythmic context elicited this response (as was checked in a separate experiment with adults).[13]

This leaves the fact that for the deviants, two instead of one element was deleted from the base pattern (see Fig. 4). Although a deviant on its own (deviant – control condition) did not elicit an MMN, along with the fact that all omissions are psychically identical (150 ms of silence) and all stimuli were normalized with respect to amplitude, we can not completely rule out a contribution to the MMN resulting from this manipulation. Introducing an additional pattern with an omission on position 6 (cf. Fig. 4), and including it either as S5 or as D3 in the stimulus set, should reveal this[34] (Bouwer, Háden, van Zuijen, and Honing, in preparation).

Finally, one could wonder to what extent BI is a domain-specific phenomenon, and, as such, represents a predisposition for music. Or is BI a particular instantiation of a general tendency of the brain to recognize mismatches in acoustic signals, including spoken language? However, if such an effect would be found in language, it does not rule out the interpretation that it draws from a fundamental musical trait.[24] So, for now, in the absence of empirical evidence, the domain-specific hypothesis is as likely as the domain-general hypothesis.

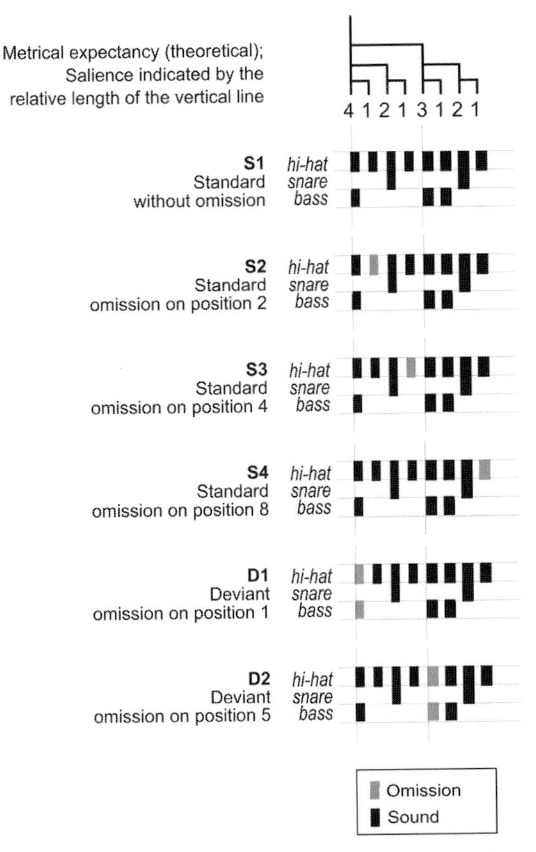

Figure 4. Stimuli as used in an adult and newborn study on metrical expectation. S1–S4 are the standards and D1 and D2 the deviants used in an oddball paradigm. The different percussion sounds are marked as hi-hat, snare, and bass.

Is hierarchy in rhythm innate, learned, or emergent?

As mentioned before, developmental studies suggest that infants are not only sensitive to a regular pulse, but also to meter. While BI requires the length of the full cycle (period) and its onset (phase) to be represented in the brain, it is also possible that newborn infants form an abstract mental representation of the base pattern, for instance, by learning the probabilities of each event in the varying rhythmic pattern or, alternatively, inducing multiple levels of beat. This would allow them not only to sense the beat, but also to build a hierarchical representation of the rhythm (meter induction). It would predict a difference in MMN latency (and possibly amplitude) for a D1 versus a D2, as has been demonstrated in adults.[16] This exciting possibility is an issue for further research. Together with the ongoing work on beat versus meter induction and the role of attention, it will help to address the question whether these hierarchical representations are innate (or at least active at day one), emergent (are they a structural property of the stimuli?), explicitly learned (as a result of musical training), or implicitly learned (as a result of exposure, however brief, to Western music).

Is BI species specific?

As discussed elsewhere,[1] BI might be considered a spontaneously developing, domain-specific, and species-specific skill. With regard to the first aspect, the newborn study provides one single piece of evidence suggesting such early bias. With regard to the second aspect, convincing evidence is still lacking, although it was recently argued that BI does not play a role (or is even avoided) in spoken language.[35] With regard to the final aspect, it was recently suggested that we might share BI with a selected group of bird species,[36,37] and not with a more closely related species such as nonhuman primates.[38] This is surprising when one assumes a close mapping between specific genotypes and specific cognitive traits. However, more and more studies show that genetically distantly related species can show similar cognitive skills, skills that more genetically closely related species fail to show.[39] This offers a rich basis for comparative studies of this specific cognitive function.

Most existing animal studies have used behavioral methods to probe the presence (or absence) of BI, such as tapping tasks[39] or measuring head bobs.[38] It might well be that if more direct electrophysiological measures are used (such as analogs of the MMN in several species),[40] nonhuman primates might indeed also show BI. This is a topic of current research (Honing, *et al.*, in preparation).

Conclusion

BI has been argued to be a spontaneously developing, domain-specific, and species-specific skill.[1,35] Although both (culture-specific) learning and perception/action coupling are relevant in development,[10,11] at least one study shows that the auditory system of a newborn is sensitive to periodicities induced by a varying rhythm. Although learning by movement is probably important, the newborn auditory system is apparently sensitive to periodicities and develops expectations about when a new cycle should start. This result is fully compatible with the notion that BI is innate. However, it is still an open question whether this regularity detection in newborns is restricted to beat only, or whether it can be hierarchical, either as a (statistically) learned structural property of the stimulus or by inducing multiple levels of periodicity. Finally, with regard to the domain specificity and species specificity of BI, convincing evidence is still lacking and both of these aspects are the topics of current research.

Acknowledgments

The author is supported by the Hendrik Muller chair designated on behalf of the Royal Netherlands Academy of Arts and Sciences (KNAW) and is a member of the Research Priority Area "Brain & Cognition" at the University of Amsterdam. Fleur Bouwer and Gábor Háden are thanked for comments on an earlier version of this manuscript.

Conflicts of interest

The author declares no conflicts of interest.

References

1. Honing, H. & A. Ploeger. Cognition and the evolution of music: pitfalls and prospects. *Topics Cogn. Sci.* In press.
2. Povel, D.J. & P. Essens. 1985. Perception of temporal patterns. *Music Percept.* **2:** 411–440.
3. Desain, P. & H. Honing. 1999. Computational models of beat induction: the rule-based approach. *J. New Music Res.* **28:** 29–42.
4. Large, E.W. & M.R. Jones. 1999. The dynamics of attending: how people track time-varying events. *Psychol. Rev.* **10:** 119–159.
5. McAuley, J.D., M.R. Jones, S. Holub, *et al.* 2006. The time of our lives: life span development of timing and event tracking. *J. Exp. Psychol. Gen.* **135:** 348–367.
6. Hannon, E.E. & S.E. Trehub. 2005. Metrical categories in infancy and adulthood. *Psychol. Sci.* **16:** 48–55.
7. Fitch, W.T. 2006. The biology and evolution of music: a comparative perspective. *Cognition* **100:** 173.
8. Grahn, J.A. & M. Brett. 2007. Rhythm and beat perception in motor areas of the brain. *J. Cogn. Neurosci.* **19:** 893–906.
9. Grube, M., F.E. Cooper, P.F. Chinnery & T.D. Griffiths. 2010. Dissociation of duration-based and beat-based auditory timing in cerebellar degeneration. *Proc. Natl. Acad. Sci. USA* **107:** 11597–11601.
10. Trehub, S.E. & E.E. Hannon. 2006. Infant music perception: domain-general or domain-specific mechanisms? *Cognition* **100:** 73–99.
11. Phillips-Silver, J. & L.J. Trainor. 2005. Feeling the beat: movement influences infants' rhythm perception. *Science* **308:** 1430.
12. Zentner, M. & T. Eerola. 2010. Rhythmic engagement with music in infancy. *Proc. Natl. Acad. Sci. USA* **107:** 5768–5773.
13. Winkler, I., G. Haden, O. Ladinig, *et al.* 2009. Newborn infants detect the beat in music. *Proc. Natl. Acad. Sci. USA* **106:** 2468–2471.
14. Honing, H., O. Ladinig, I. Winkler & G. Háden. 2009. Is beat induction innate or learned? Probing emergent meter perception in adults and newborns using event-related brain potentials (ERP). *Ann. N.Y. Acad. Sci.* **1169:** 93–96.
15. Hannon, E.E. & S.P. Johnson. 2005. Infants use meter to categorize rhythms and melodies: implications for musical structure learning. *Cogn. Psychol.* **50:** 354–377.
16. Ladinig, O., H. Honing, G. Háden & I. Winkler. 2009. Probing attentive and pre-attentive emergent meter in adult listeners with no extensive music training. *Music Percept.* **26:** 377–386.
17. Winkler, I., E. Kushnerenko, J. Horváth, *et al.* 2003. Newborn infants can organize the auditory world. *Proc. Natl. Acad. Sci. USA* **100:** 1182–1185.
18. Honing, H. Structure and interpretation of rhythm and timing in music. In *Psychology of Music*. 3rd ed. D. Deutsch, Ed. Academic Press. London. In press.
19. Longuet-Higgins, H.C. & C.S. Lee. 1984. The rhythmic interpretation of monophonic music. *Music Percept.* **1:** 424–441.
20. Oxford Music Online. 2011. Entry 'Syncopation'. Available at: http://www.oxfordmusiconline.com/subscriber/article/grove/music/27263.
21. Brochard, R., D. Abecasis, D. Potter, *et al.* 2003. The "ticktock" of our internal clock: direct brain evidence of subjective accents in isochronous sequences. *Psychol. Sci.* **14:** 362–366.
22. Drake, C. & D. Bertrand. 2001. The quest for universals in temporal processing in music. *Ann. N.Y. Acad. Sci.* **930:** 17–27.
23. Fitch, W.T. & A.J. Rosenfeld. 2007. Perception and production of syncopated rhythms. *Music Percept.* **25:** 43–58.
24. Honing, H. 2011. *The Illiterate Listener. On Music Cognition, Musicality and Methodology*. Amsterdam University Press. Amsterdam.
25. Winkler, I. 2007. Interpreting the mismatch negativity (MMN). *J. Psychophysiol.* **21:** 147–163.
26. Paavilainen, P., P. Arajärvi & R. Takegata. 2007. Preattentive detection of nonsalient contingencies between auditory features. *Neuroreport* **18:** 159–163.
27. Bendixen, A., E. Schröger & I. Winkler. 2009. I heard that coming: event-related potential evidence for stimulus-driven prediction in the auditory system. *J. Neurosci.* **29:** 8447–8451.
28. Schröger, E. & I. Winkler. 1995. Presentation rate and magnitude of stimulus deviance effects on human pre-attentive change detection. *Neurosci. Lett.* **193:** 185–188.
29. Näätänen, R., T. Kujala & I. Winkler. 2010. Auditory processing that leads to conscious perception: a unique window to central auditory processing opened by the mismatch negativity and related responses. *Psychophysiology* **48:** 4–22.
30. Gaab, D. (this volume). Current fMRI methods with children. *Ann. N.Y. Acad. Sci.*
31. May, P.J.C. & H. Tiitinen. 2010. Mismatch negativity (MMN), the deviance-elicited auditory deflection, explained. *Psychophysiology* **47:** 66–122.
32. Yabe, H., M. Tervaniemi, K. Reinikainen & R. Näätänen. 1997. Temporal window of integration revealed by MMN to sound omission. *NeuroReport* **8:** 1971–1974.

33. Ladinig, O., H. Honing, G. Háden & I. Winkler. 2011. Erratum to probing attentive and pre-attentive emergent meter in adult listeners with no extensive music training. *Music Percept.* **26:** 444.
34. Honing, H. & F. Bouwer. 2011. Is hierarchy in rhythm perception consciously learned? In *Proceedings of the Rhythm Perception and Production Workshop.* Max Planck Institute for Human Cognitive and Brain Sciences. Leipzig.
35. Patel, A.D. 2008. *Music, Language, and the Brain.* Oxford University Press. Oxford.
36. Fitch, W.T. 2009. Biology of music: another one bites the dust. *Curr. Biol.* **19:** 403–404.
37. Patel, A.D., J.R. Iversen, M.R. Bregman & I. Schulz. 2009. Studying synchronization to a musical beat in nonhuman animals. *Ann. N.Y. Acad. Sci.* **1169:** 459–469.
38. Zarco, W., H. Merchant, L. Prado & J.C. Mendez. 2009. Subsecond timing in primates: comparison of interval production between human subjects and rhesus monkeys. *J. Neurophysiol.* **102:** 3191–3202.
39. De Waal, F.B.M. 2009. Darwin's last laugh. *Nature* **460:** 175.
40. Nelken, I. & N. Ulanovsky. 2007. Mismatch negativity and stimulus-specific adaptation in animal models. *J. Psychophysiol.* **21:** 214–223.

Ann. N.Y. Acad. Sci. ISSN 0077-8923

ANNALS OF THE NEW YORK ACADEMY OF SCIENCES
Issue: *The Neurosciences and Music IV: Learning and Memory*

Effects of perceptual experience on children's and adults' perception of unfamiliar rhythms

Erin E. Hannon, Christina M. Vanden Bosch der Nederlanden, and Parker Tichko
Department of Psychology, University of Nevada, Las Vegas, Nevada

Address for correspondence: Dr. Erin E. Hannon, Department of Psychology, University of Nevada, Las Vegas, 4505 Maryland Parkway #455030, Las Vegas, NV 89154. erin.hannon@unlv.edu

Rhythm and meter are fundamental components of music that are universal yet also culture specific. Although simple, isochronous meters are preferred and more readily discriminated than highly complex, nonisochronous meters, moderately complex nonisochronous meters do not pose a problem for listeners who are exposed to them from a young age. The present work uses a behavioral task to examine the ease with which listeners of various ages acquire knowledge of unfamiliar metrical structures from passive exposure. We examined perception of familiar (Western) rhythms with an isochronous meter and unfamiliar (Balkan) rhythms with a nonisochronous meter. We compared discrimination by American children (5 to 11 years) and adults before and after a 2-week period of at-home listening to nonisochronous meter music from Bulgaria. During the first session, listeners of all ages exhibited superior discrimination of isochronous than in nonisochronous melodies. Across sessions, this asymmetry declined for young children but not for older children and adults.

Keywords: perceptual learning; rhythm perception; development; enculturation

Introduction

Rhythm and meter are essential, universal components of music that vary across cultures. Just as a child must learn his or her native language without formal instruction, so too must listeners acquire implicit knowledge of the rules that govern the music of their culture, including rules about musical temporal structure.[1,2] Recent studies suggest that newborns and young infants are sensitive to beat and rhythm,[3–6] yet culture-specific perceptual experience clearly plays a crucial role in shaping rhythm and meter perception and production.[2,7–12] Despite evidence of cultural influence, relatively little is currently known about the developmental trajectory of culture-specific musical rhythm and meter processing.

A widely held assumption by educators, policymakers, and parents is that when it comes to learning music, earlier is better. Early experience has profound effects on human development,[13,14] and age-related declines in learning have been observed in language and other domains,[15–17] yet surprisingly few studies have directly addressed the question of whether younger learners have an advantage over older learners in the context of acquiring musical knowledge. Most studies address this question in the context of formal music training. For example, beginning music lessons prior to the age of 7 is thought to be essential for acquiring *absolute pitch*, the ability to name a pitch class in the absence of a reference.[18,19] Likewise, anatomical and functional brain enhancements observed among musicians are negatively correlated with the age at which musicians began taking music lessons.[20–23] One problem with interpreting such findings, however, is that onset of music training is usually confounded with total amount of training (in years) because individuals who began music lessons at a young age tend to have more years of music training overall than those who began later, particularly among college-age samples. However, recent evidence suggests that even when total amount of music training is controlled, performance on rhythm perception and synchronization is better among musicians who began lessons prior to the age of 7 than those who began after that age.[24,25]

doi: 10.1111/j.1749-6632.2012.06466.x
Ann. N.Y. Acad. Sci. 1252 (2012) 92–99

The above findings have potentially important implications because they suggest that, like language, there may be an advantage to acquiring musical expertise early in development. However, even if individuals have the same amount of formal training in years, other factors may contribute to age-related learning differences. For example, music lessons for young children may differ in content, structure, and appeal when compared with lessons for older children. Practice patterns may also differ across younger and older learners. Thus, given the complexity and diversity of music training experiences, it is nearly impossible for researchers to ensure that all participants in a study receive the same amount and type of formal training. Even if we observe age-related decline in the capacity to learn to play an instrument, this does not necessarily imply that the same decline characterizes informal learning in the context of everyday music listening and enculturation. As an informal, implicit process, the latter type of learning is perhaps more comparable to language acquisition, yet little is known about age-of-acquisition effects in the context of musical enculturation.

The present experiment examines how passive exposure to music from a foreign culture influences perception of rhythm and meter at different ages. Although young listeners can infer the underlying beat from a rhythmic pattern,[4–6] they nevertheless appear to acquire hierarchical metrical representations or categories that influence beat induction in a top–down fashion.[26] Metrical structures vary cross-culturally, and Western music in particular tends to contain isochronous beats at multiple levels and durations that stand in simple 2:1 or 1:1 ratios. It is perhaps for this reason that Western listeners have difficulty perceiving, remembering, and producing rhythms that contain complex duration ratios and fail to conform to an isochronous beat.[7,26–31] By contrast, music from various regions of the world (the Balkans, South Asia, Africa, South America) can contain nonisochronous meter with alternating long and short durations having 3:2 ratios.[32] Accordingly, listeners from these cultures perceive isochronous and nonisochronous meters similarly as long as the structures are culturally familiar (S. Ullal, E.E. Hannon & J.S. Snyder, in prep.).[7,33–34]

Among Western listeners, biases toward isochronous meter appear to emerge during infancy, as shown by the finding that American infants readily distinguish folk melodies with either type of meter at 6 months of age, but by 12 months they only discriminate melodies with isochronous meter.[7,8] Interestingly, this developmental decline in perception of nonisochronous meter can be prevented if 12-month-olds are given 2 weeks of at-home exposure to nonisochronous Balkan folk music for 10 min per day, after which they show robust discrimination of nonisochronous melodies.[8] By contrast, the same amount of exposure has minimal effect on the performance of adults, who continue to exhibit a strong bias toward isochronous meters and simple rhythmic ratios even after exposure.[8] Thus, the capacity to learn from everyday music exposure appears to change between infancy and adulthood, perhaps in tandem with musical enculturation. As culture-specific musical knowledge develops, listeners may acquire increasingly robust representations that are decreasingly susceptible to modification.[2] The present study therefore adapted the at-home exposure paradigm described previously[8] to examine how readily children of different ages and adults learn from passive exposure to unfamiliar (Balkan) nonisochronous meters. Listeners between the ages of five and adulthood participated in a perceptual judgment task with both isochronous and nonisochronous folk melodies before and after at-home exposure to recordings of nonisochronous, Balkan folk music. This allowed us to directly and precisely manipulate informal exposure to music to examine potential age-of-acquisition effects in music learning.

Methods

Participants

Five groups of participants were approximately of ages 5, 7, 9, 11, or 18+ years. There were 24 5-year-olds (11 female; $M_{age} = 5.2$, age range: 4.5–5.8), 23 7-year-olds (8 female; $M_{age} = 6.99$, age range: 6–7.8), 24 9-year-olds (11 female; $M_{age} = 8.92$, age range: 8–9.8), 26 11-year-olds (15 female; $M_{age} = 11.3$, age range: 10–13), and 23 adults (14 female; $M_{age} = 21.3$, age range: 18–36). Families of child participants volunteered in response to letters distributed in the community and received a toy after each testing session, and adult participants received course credit for their participation. All participants had normal hearing, no history of hearing problems, did not suffer from a cold or illness on the day of testing, and spoke English fluently. Some participants were fluent in a second language but were bilingual from infancy ($n = 10$) or acquired English prior to age five ($n = 10$). Forty-one participants (30 children) reported having had some formal music training, which ranged from zero to 11 years ($M = 0.80$). An additional 25 participants (24 children) reported having had formal dance training, which ranged from zero to 7 years ($M = 0.43$). No participants had ever visited the Balkan peninsula, and none reported being familiar with music, dances, or languages from that region.

Only the participants who tested in both sessions were included in the final sample. An additional 13 participants were run but excluded from the final sample because of equipment failure ($n = 5$ children), insufficient exposure period ($n = 1$ child), or because the participant gave the same response throughout the entire experiment ($n = 7$ children, 1 adult). We also established an inclusion threshold to exclude children who clearly did not understand the task and/or failed to follow instructions. We reasoned that if children understood the task, they should have little difficulty correctly distinguishing perfectly accurate from highly disrupted renditions of a previously presented melody. The final sample therefore excluded participants who gave higher similarity ratings to the *severely disrupted* test stimulus than to the *unaltered* test stimulus in the isochronous conditions of the initial session (see Fig. 2 and stimulus description in the next section). This led to the exclusion of 33 children (12 5-year-olds, 10 7-year-olds, 4 9-year-olds, and 7 11-year-olds), roughly 21% of the total sample.

Stimuli

The similarity judgment task was identical across the two test sessions. Two blocks of each meter were presented, for a total of four blocks per session. All stimuli were generated using a MIDI sequencer (Digital Performer) and converted to AIFF using the Apple QuickTime synthesizer (Apple, Inc., Cupertino, CA). Each block began with a familiarization stimulus followed by four test stimuli (identical to those used previously, Refs. 7 and 8). Four traditional Balkan folk songs, each eight measures in duration, were used as familiarization stimuli, two originally scored in 4/4 isochronous meter and the other two in 7/8 nonisochronous meter.[35] Each familiarization stimulus had two melodic instruments

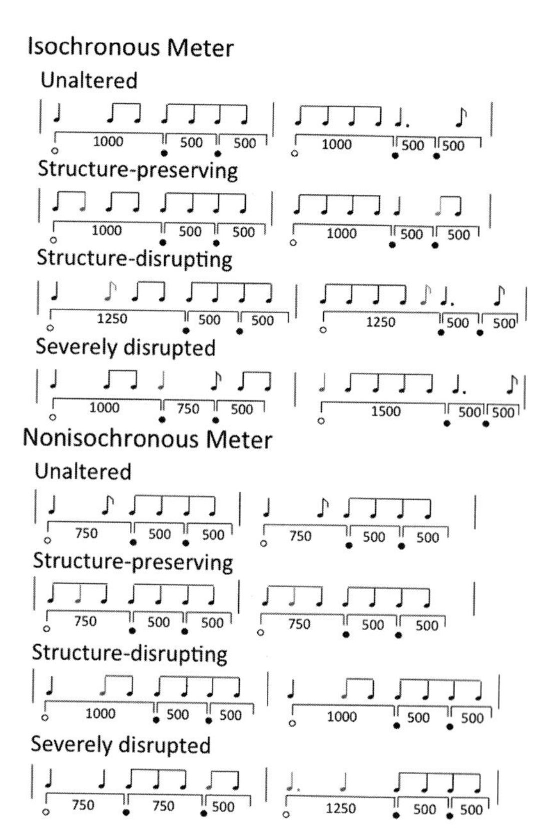

Figure 1. Examples of the four types of test stimuli (unaltered, structure-preserving, structure-disrupting, and severely disrupted), presented separately for each type of meter (isochronous and nonisochronous). Added notes are depicted in gray.

playing in unison or thirds, a harmonic instrument, and a fourth percussion instrument presenting a repeating long–short–short or short–short–long pattern. Drum patterns with isochronous meter alternated between 1,000 msec and 500 msec intervals, yielding a 2:1 ratio, whereas nonisochronous meter drum patterns alternated between 750 msec and 500 msec, yielding a 3:2 ratio (see "unaltered" stimuli in Fig. 1). All familiarization stimuli were accompanied by a dynamic cartoon display of a tiger holding a guitar and swaying back and forth.

For each familiarization stimulus, four test renditions were created, two that preserved the original meter and drum pattern of the familiarization stimulus, and two that disrupted its meter and drum pattern (see Fig. 1). Test stimuli contained only the drum pattern and one melodic instrument (piano). Thus, even the unaltered test stimulus had novel instrument timbre and texture relative to familiarization, though it preserved the exact pitch and rhythm of the original. Both "structure-preserving" and "structure-disrupting" test stimuli contained an extra 250-msec eighth note inserted into every measure. Structure-preserving stimuli reduced adjacent note durations to maintain the original metrical structure and drum pattern, whereas structure-disrupting stimuli left adjacent durations unchanged, which lengthened a drum interval and therefore disrupted the meter. The severely disrupted stimulus contained many extra notes that were 250–500 msec in duration, inserted pseudorandomly one to three times per measure. Test stimuli thus presented both obvious disruptions as well as more subtle disruptions. All auditory test stimuli were randomly paired with a video of one of five animated sheep. Each sheep was paired with each test stimulus, creating four possible sheep–stimulus pairings per stimulus. For use in a brief practice block preceding testing, an additional set of familiarization and test stimuli were created from the song "Mary Had a Little Lamb," with familiarization and test renditions created as described previously.

For the at-home listening phase of the study, a compact disc (CD) was created that contained five recordings of nonisochronous dance music from Macedonia, Bulgaria, or Bosnia used in prior work.[8] The CD was 10 min long and contained none of the folk songs used during test sessions. Audio CDs were burned using iTunes on a Macintosh computer (Apple, Inc., Cupertino, CA).

For the second test session, a brief recognition test was prepared to determine whether participants followed the at-home CD listening instructions. Fifteen 20-sec excerpts were presented in random order. Five were targets (clips from the take-home CD), five were nontargets drawn from the same Balkan artists with nonisochronous meters, and five were nontargets drawn from folk music recordings from other regions of the world. Thus, nontargets were highly confusable with targets, making it difficult for participants to succeed on the recognition test without listening at home to the audio CD.

Apparatus and procedure

Each testing session included four blocks (one per familiarization song), plus a practice block at the beginning of the session. Within each block, participants first heard a familiarization stimulus for 2 minutes, followed by four "test" renditions of the same song. Participants rated how similar each test stimulus was to the original familiarization stimulus on a scale of 1 (very similar) to 5 (very different). Sheep–stimulus pairings, block order, and test stimulus order were counterbalanced between subjects.

Participants were tested alone on a Mac Mini computer (Apple, Inc., Cupertino, CA) running PsyScope[36] and equipped with an ioLab USB response box and two desktop computer speakers. Responses were collected using a rectangular game board with a row of six colored squares, the leftmost of which contained a picture of the tiger which accompanied the familiarization song. Five laminated game pieces were created (e.g., rocket, star, flower), which participants placed on the game board to indicate their rating. The game board thus served as a tangible representation of the five-point similarity rating scale with maximal similarity on the left (near the tiger) and dissimilarity on the right (far away from the tiger). The experimenter gave each participant verbal instructions based on the following: "A tiger will play you the first song. Listen carefully! His song is the best song. His friends are each going to try to play the same song, but they can't all play it like the tiger. Use the scale to show how close each animal gets to the tiger's song. If the animal's version is very close to the tiger's, put your game piece next to the tiger. If the animal's version is wrong, put your game piece far away from the tiger." For any given trial, the child placed his or her game piece on the game board and the experimenter entered the child's response using the response box.

Each test session was preceded by a brief practice block, during which participants were given practice using the rating scale. During the test phase of the practice block, each test stimulus was presented twice—first with explicit feedback about the similarity of each stimulus relative to familiarization and a second time without feedback. Participants could repeat the familiarization block until they felt comfortable with the task.

Participants were asked to return for a second test session 12–15 days later. During the interim, they were asked to listen at home twice daily to the audio CD using a provided log sheet to indicate listening days and times. At the end of the second test session, participants were given the recognition test by means of PsyScope,[36] which presented targets and nontargets in random order. Participants indicated whether each excerpt had or had not been on the take-home audio CD. During the final test session participants completed a questionnaire assessing their musical and cultural background.

Results

Similarity ratings

Accurate performance in the similarity judgment task entailed giving higher dissimilarity ratings to structure-disrupting than to structure-preserving test stimuli, so difference scores were calculated as a measure of accuracy.[8] For each block we subtracted ratings of test stimuli that preserved the meter (the mean of unaltered and structure-preserving ratings) from ratings of test stimuli that disrupted the meter (mean of structure-disrupting and severely disrupted ratings). We combined ratings of both subtle and more obvious disruptions because we wanted to ensure the task was easy enough for the younger age groups. We expected overall accuracy to increase with age regardless of meter.

Session 1

As shown in Figure 2A, older participants were generally more accurate than younger participants, consistent with our prediction that younger children would find the task more challenging. Moreover, all groups exhibited higher accuracy in isochronous rather than nonisochronous conditions, consistent with prior work.[7,8] These trends were confirmed by a two-way mixed design analysis of variance (ANOVA) with factors of meter (within subjects) and age group (between subjects), which revealed a main effect of age group, $F(4,115) = 8.322$, $P < 0.001$, meter, $F(1,115) = 37.037$, $P < 0.001$, and a significant interaction between meter and age group, $F(4,115) = 2.76$, $P < 0.05$. Bonferroni-corrected t-tests revealed that accuracy was higher in isochronous than in nonisochronous conditions for 5-year-olds, $t(23) = 3.807$, $P < 0.001$, 7-year-olds, $t(22) = 2.59$, $P < 0.05$, and adults, $t(23) = 3.029$, $P < 0.01$, and marginally significant for 9-year-olds, $t(23) = 1.892$, $P = 0.07$, and 11-year-olds, $t(25) = 1.681$, $P = 0.10$ (see Fig. 2A).

Session 2

Performance in session 2 differed markedly from session 1, particularly for the younger groups. The two-way meter by age group ANOVA revealed a main effect of age group, $F(4,115) = 16.058$, $P < 0.001$, but only a marginally significant main effect of meter, $F(1,115) = 3.42$, $P = 0.07$, and a significant interaction between meter and age group, $F(4,115) = 4.245$, $P < 0.01$. Unlike during session 1, where an advantage for isochronous-meter stimuli was apparent at all ages, during session 2, this advantage was evident only among adult participants, $t(23) = 3.79$, $P < 0.01$, and to an extent among 11-year-olds, although the advantage was not significant in this group, $t(25) = 1.602$, $P = 0.12$. Seven- and 9-year-olds performed comparably in isochronous and nonisochronous conditions during session 2, $t(23) < 0.30$, $P > 0.76$, and 5-year-olds performed more accurately in the nonisochronous condition, although this trend was not significant, $t(23) = 1.27$, $P = 0.217$ (see Fig. 2B).

Changes across sessions

An overall three-way mixed-design ANOVA with factors of meter, session, and age group confirmed that there was a significant three-way interaction,

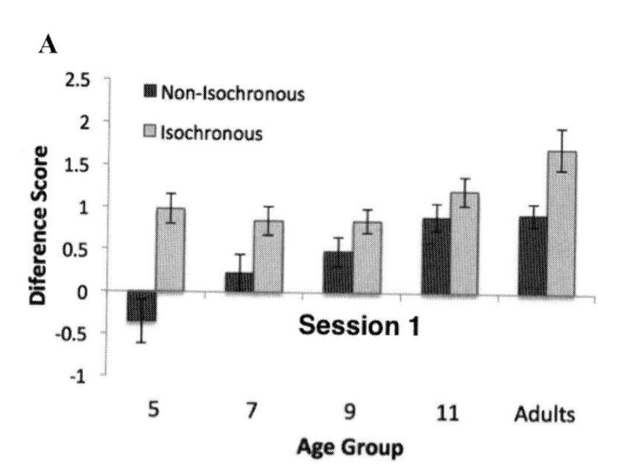

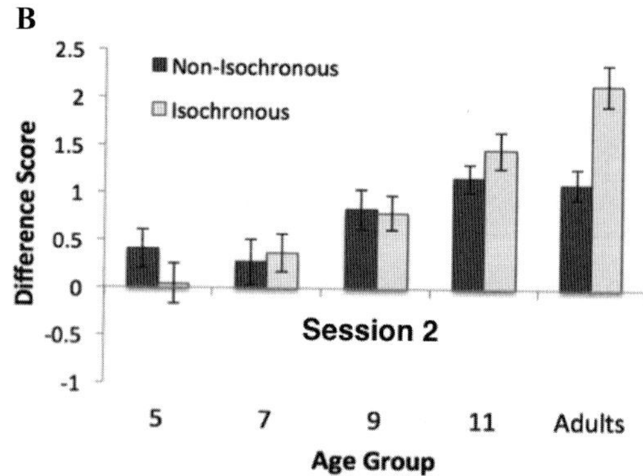

Figure 2. Difference scores (calculated by subtracting ratings of structure-preserving from structure-disrupting stimuli) provide an index of accuracy in the isochronous and nonisochronous conditions at session 1 (A) and session 2 (B), across age groups.

$F(4,115) = 4.612$, $P < 0.01$. Across the two sessions, performance in the nonisochronous condition improved significantly among the 5-year-olds, $t(23) = 2.413$, $P = 0.024$, but not among 7-year-olds, $t(23) = 0.14$, $P = 0.891$, 9-year-olds, $t(23) = 1.474$, $P = 0.154$, 11-year-olds, $t(25) = 1.289$, $P = 0.21$, or adults, $t(22) = 0.904$, $P = 0.14$. By contrast, performance in the isochronous condition actually decreased among 5-year-olds, $t(23) = -3.909$, $P = 0.001$, and 7-year-olds, $t(23) = -2.15$, $P = 0.043$, and remained unchanged or slightly improved among 9-year-olds, $t(23) = -0.229$, $P = 0.821$, 11-year-olds, $t(25) = 1.197$, $P = 0.242$, and adults, $t(22) = 1.613$, $P = 0.121$.

It is clear from the robust main effects of age group that the overall task, regardless of meter, was less challenging for older participants, which precludes direct comparison of accuracy across age groups. However, because the same participants were tested in both meter conditions, it is reasonable to compare relative accuracy across metrical conditions as an index of enculturation. Because processing advantages for isochronous over nonisochronous meters appear to be acquired through culture-specific experience (S. Ullal, E.E. Hannon & J.S. Snyder, in prep.),[7,8,33–34] we reason that asymmetries in performance across the two meter conditions should reflect the extent to which individuals are enculturated to Western music. We therefore computed "enculturation" scores for each participant at each session by subtracting accuracy in the nonisochronous condition from accuracy in the isochronous condition, with positive values indicating an advantage for Western isochronous meter (see Fig. 3).

Enculturation scores were analyzed with a two-way mixed-design ANOVA with factors of session (within-subjects) and age group (between-subjects). There was a significant main effect of age group, $F(4,115) = 2.456$, $P < 0.05$, session, $F(1,115) = 9.195$, $P < 0.01$, and a significant interaction between age group and session, $F(4,115) = 4.612$, $P < 0.01$. There was a tendency for overall enculturation to increase with age ($M_{age5} = 0.48$, SD $= 0.32$; $M_{age7} = 0.35$, SD $= 0.28$; $M_{age9} = 0.17$, SD $= 0.18$; $M_{age11} = 0.30$, SD $= 0.18$; $M_{adult} = 0.91$, SD $= 0.26$), and enculturation scores were generally higher during session 1 ($M = 0.68$, SD $= 0.24$) than during session 2 ($M = 0.20$, SD $= 0.25$). Separate one-way ANOVAs for each session revealed effects of age during both session 1, $F(4,115) = 2.76$, $P < 0.05$, and session 2, $F(4,115) = 4.245$, $P < 0.01$, but the nature of age trends differed for the two sessions. In session 1, enculturation scores were positive for all age groups, and planned Bonferroni comparisons revealed that only 5-year-olds had significantly higher scores than 11-year-olds. By contrast, in session 2, enculturation scores were 0 for 5-, 7-, and 9-year-olds, whereas 11-year-olds and adults had positive scores (adult scores were significantly higher than that for 5- and 9-year-olds, $P < 0.05$, and marginally higher than that for 7-year-olds, $P = 0.10$). Thus, across the two sessions enculturation changed most for the youngest participants.

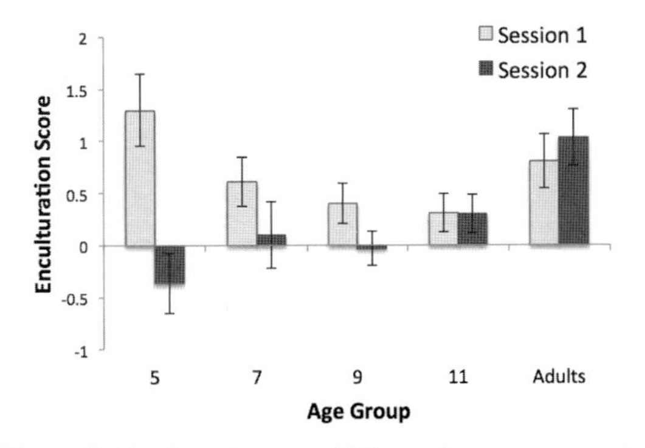

Figure 3. Enculturation scores (difference between accuracy in isochronous and nonisochronous conditions) at both sessions as a function of age.

Recognition scores

We wanted to ensure that observed age differences did not arise because of age-related changes in compliance with the at-home listening regimen. Because we could not precisely control at-home exposure, the recognition test served as an objective measure of familiarity with the CD. If 5-year-olds' improvement in the nonisochronous meter condition arose because they listened more frequently or with more concentration than did other age groups, they should also have higher recognition scores. The recognition test contained 10 nontargets and five targets; so to minimize response bias, recognition scores were calculated as the mean of percent hits and false alarms. For all age groups, recognition scores (percent correct) were above chance ($M_{age5} = 0.79$, SD $= 0.14$; $M_{age7} = 0.86$, SD $= 0.17$; $M_{age9} = 0.85$, SD $= 0.14$; $M_{age11} = 0.84$, SD $= 0.15$; $M_{adult} = 0.84$, SD $= 0.14$), and a one-way ANOVA revealed that recognition accuracy did not vary by age, $F(4,112) = 0.843$, $P > 0.50$. It was therefore unlikely that younger children listened more or were more compliant than older children or adults.

Conclusions

The present findings suggest that incidental exposure to music with foreign, nonisochronous meters has different effects on younger than older listeners. Specifically, passive at-home exposure to Balkan folk music with nonisochronous meter gave rise to dramatic changes in the perception of nonisochronous and isochronous meter among 5-year-olds and to some extent among 7-year-olds, but minimal effects among 11-year-olds and adults. It was also shown that at-home exposure eliminated a previously observed bias toward isochronous meters among children of age 9 and below that. By contrast, 11-year-olds and adults consistently performed more accurately in the isochronous than in the nonisochronous conditions across sessions, regardless of exposure. Thus, greater susceptibility to perceptual experience, previously documented among 12-month-old infants,[8] appears to extend to the first 5 years of childhood, and culture-specific representations may continue to undergo development up until age 11.

It is important to acknowledge limitations that should be addressed in future work. By providing all participants with the same CD, the present experiment manipulated musical experience in a much more controlled fashion than has been possible in prior work.[18–25] However, it was impossible to ensure that all subjects received the same amount and quality of exposure to the CD. The recognition test provided an important index of listening quality, but future work, perhaps in a laboratory setting, could attempt to more closely monitor exposure to ensure maximal compliance.

A second limitation arose from using the same task with all participants. Younger children were at a disadvantage relative to older children and adults who presumably have more developed working memory, cognitive control, attention span, and superior task comprehension. We felt that using the same task across ages was a more conservative approach than reducing task difficulty for younger groups, given the possibility that younger learners might outperform older learners. The fact that the youngest children showed the greatest gains in performance despite their overall poorer performance provides strong support for the claim that they learn more readily. It would nevertheless be helpful in future work to equate task difficulty across groups to ensure that it does not interact with learning outcomes.

Finally, the study could be improved by including a control group of listeners who receive no at-home listening experience. Because exposure to isochronous music remained constant across sessions, we expected no change in performance in the isochronous condition and thus considered it a within-subjects control. However, we found that 5- and 7-year-olds' performance in the isochronous condition diminished over the two sessions. This outcome might be expected if young children have weak representations of isochronous meter that are readily reorganized through exposure to nonisochronous meter. It is thus essential to determine whether declines in performance arose due to exposure or due to repeated testing.

In conclusion, the present work provides an initial glimpse into the fundamental question of how and when culture-specific musical knowledge is acquired over the course of child development. Such research is essential for understanding musical development and the mechanisms underlying music learning. The work also adds to a growing body of evidence suggesting that age-of-acquisition effects characterize music learning just as they do language learning, and that such effects arise in the context of formal music training as well as informal musical enculturation.

Conflicts of interest

The authors declare no conflicts of interest.

References

1. Hannon, E.E. & L.J. Trainor. 2007. Music acquisition: effects of enculturation and formal training on development. *Trends Cogn. Sci.* **11:** 466–472.
2. Hannon, E.E. 2010. Music enculturation: how young listeners construct musical knowledge through perceptual experience. In *Neoconstructivism: The New Science of Cognitive Development*. S. Johnson, Ed.: 132–158. Oxford University Press. New York.
3. Demany, L., B. McKenzie & E. Vurpillot. 1977. Rhythm perception in early infancy. *Nature* **266:** 718–719.
4. Hannon, E.E. & S.P. Johnson. 2005. Infants use meter to categorize rhythms and melodies: implications for musical structure learning. *Cogn. Psychol.* **50:** 354–377.
5. Winkler, I., G.P. Haden, O. Ladinig, *et al.* 2009. Newborn infants detect the beat in music. *Proc. Natl. Acad. Sci. USA* **106:** 2468–2471.

6. Phillips-Silver, J. & L.J. Trainor. 2005. Feeling the beat: movement influences infant rhythm perception. *Science* **308:** 1430.
7. Hannon, E.E. & S.E. Trehub. 2005. Metrical categories in infancy and adulthood. *Psychol. Sci.* **16:** 48–55.
8. Hannon, E.E. & S.E. Trehub. 2005. Tuning in to rhythms: infants learn more readily than adults. *Proc. Natl. Acad. Sci. USA* **102:** 12639–12643.
9. Hannon, E.E., G. Soley & R.S. Levine. 2011. Constraints on infants' musical rhythm perception: effects of interval ratio complexity and enculturation. *Dev. Sci.* **14:** 865–872.
10. Soley, G. & E.E. Hannon. 2010. Infants prefer the musical meter of their own culture: a cross-cultural comparison. *Dev. Psychol.* **46:** 286–292.
11. Drake, C. & J.B. El Heni. 2003. Synchronizing with music: intercultural differences. *Ann. N.Y. Acad. Sci.* **999:** 429–437.
12. Gerry, D.W., A.L. Faux & L.J. Trainor. 2010. Effects of Kindermusik training on infants' rhythmic enculturation. *Dev. Sci.* **13:** 545–551.
13. Meltzoff, A.N., P.K. Kuhl, J. Movellan & T.J. Sejnowski. 2009. Foundations for a new science of learning. *Science* **325:** 284.
14. Knudsen, E.I., J.T. Heckman, J.L. Cameron & J.P. Shonkoff. 2006. Economic, neurobiological, and behavioral perspectives on building America's future workforce. *Proc. Natl. Acad. Sci. USA* **103:** 10155–10162.
15. Hernandez, A.E. & P. Li. 2007. Age of acquisition: its neural and computational mechanisms. *Psychol. Bull.* **133:** 638–650.
16. Flege, J.E., G. Yeni-Komshian & S. Liu. 1999. Age constraints on second language learning. *J. Mem. Lang.* **41:** 78–104.
17. Johnson, J.J. & E.L. Newport. 1989. Critical period effects in second language learning: the influence of maturational state on the acquisition of English as a second language. *Cogn. Psychol.* **1:** 60–99.
18. Deutsch, D., T. Henthorn, E. Marvin & H. Xu. 2006. Absolute pitch among American and Chinese conservatory students: prevalence differences, and evidence for a speech-related critical period. *J. Acoust. Soc. Am.* **119:** 719–722.
19. Takeuchi, A. & S. Hulse. 1993. Absolute pitch. *Psychol. Bull.* **113:** 345–361.
20. Amunts, K., G. Schlaug, L. Jäncke, *et al.* 1997. Motor cortex and hand motor skills: structural compliance in the human brain. *Hum. Brain Mapp.* **5:** 206–215.
21. Elbert, T., C. Pantev, C. Weinbruch, *et al.* 1995. Increased cortical representation of the fingers of the left hand in string players. *Science* **270:** 305–307.
22. Pantev, C., R. Oostenveld, A. Engelien, *et al.* 1998. Increased auditory cortical representations in musicians. *Nature* **392:** 811–814.
23. Schlaug, G., L. Jäncke, Y. Huang & H. Steinmetz. 1995. In vivo evidence of structural brain asymmetry in musicians. *Science* **267:** 699–701.
24. Watanabe, D., T. Savion-Lemieux & V.B. Penhune. 2007. The effect of early musical training on adult motor performance: evidence for a sensitive period in motor learning. *Exp. Brain Res.* **176:** 332–340.
25. Bailey, J.A. & V.B. Penhune. 2010. Rhythm synchronization performance and auditory working memory in early- and late-trained musicians. *Exp. Brain Res.* **204:** 91–101.
26. Desain, P. & H. Honing. 2003. The formation of rhythmic categories and metric priming. *Perception* **32:** 341–365.
27. Essens, P. 1986. Hierarchical organization of temporal patterns. *Percept. Psychophys.* **40:** 69–73.
28. Essens, P. & D. Povel. 1985. Metrical and nonmetrical representations of temporal patterns. *Percept. Psychophys.* **37:** 1–7.
29. Fraisse, P. 1982. Rhythm and tempo. In *The Psychology of Music*. D. Deutsch, Ed.: 149–180. Academic Press. New York.
30. Repp, B.H., J. London & P.E. Keller. 2005. Production and synchronization of uneven rhythms at fast tempi. *Music Percept.* **23:** 61–78.
31. Snyder, J.S., E.E. Hannon, E.W. Large & M.H. Christiansen. 2006. Synchronization and continuation tapping to complex meters. *Music Percept.* **24:** 135–146.
32. London, J. 2004. *Hearing in Time: Psychological Aspects of Musical Meter.* Oxford University Press. New York.
33. Soley, G. & E.E. Hannon. 2010. Infants prefer the musical meter of their own culture: a cross-cultural comparison. *Dev. Psychol.* **46:** 286–292.
34. Hannon, E.E., G. Soley & S. Ullal. In press. Familiarity overrides simplicity in rhythmic pattern perception: a cross-cultural examination of American and Turkish listeners *J. Exp. Psychol. Hum. Percept. Perf*.
35. Geisler, R. 1989. *The Bulgarian and Yugoslav Collections.* The Village and Early Music Society. Grass Valley, CA.
36. Cohen, J.D., B. MacWhinney, M. Flatt & J. Provost. 1993. Psyscope: a new graphic interactive environment for designing psychology experiments. *Behav. Res. Methods Instrum. Comput.* **25:** 257–271.

Ann. N.Y. Acad. Sci. ISSN 0077-8923

ANNALS OF THE NEW YORK ACADEMY OF SCIENCES
Issue: *The Neurosciences and Music IV: Learning and Memory*

Cognitive factors shape brain networks for auditory skills: spotlight on auditory working memory

Nina Kraus,[1,2,3,4] Dana L. Strait,[1,2] and Alexandra Parbery-Clark[1,3]

[1]Auditory Neuroscience Laboratory, [2]Institute for Neuroscience, [3]Department of Communication Sciences, [4]Departments of Neurobiology and Physiology, Otolaryngology, Northwestern University, Evanston, Illinois

Address for correspondence: Nina Kraus, Frances Searle Building, 2240 Campus Drive, Evanston, IL 60208. nkraus@northwestern.edu; www.brainvolts.northwestern.edu

Musicians benefit from real-life advantages, such as a greater ability to hear speech in noise and to remember sounds, although the biological mechanisms driving such advantages remain undetermined. Furthermore, the extent to which these advantages are a consequence of musical training or innate characteristics that predispose a given individual to pursue music training is often debated. Here, we examine biological underpinnings of musicians' auditory advantages and the mediating role of auditory working memory. Results from our laboratory are presented within a framework that emphasizes auditory working memory as a major factor in the neural processing of sound. Within this framework, we provide evidence for music training as a contributing source of these abilities.

Keywords: hearing in noise; auditory working memory; experience-dependent plasticity; brainstem

Introduction

Listening to and understanding speech is an extraordinarily complex task involving a vast array of sensory and cognitive processes. The acoustic complexity of speech makes it particularly vulnerable to masking by other environmental sounds. Still, for the normal system, understanding speech in noise is something that our intricately tuned auditory system routinely accomplishes. Humans often face situations in which background noise impairs speech perception, yet musicians are less impeded by noise than the rest of us.[1–3] In standardized testing circumstances, accomplished musicians perform better in understanding speech in noise than their age- and hearing-matched peers. Aside from the implications that this result promotes regarding common music/speech physiological mechanisms, it also opens the question of the route by which musical training affords speech-in-noise processing advantages. Patel's OPERA hypothesis,[4] which is reviewed later, outlines conditions necessary for the successful transfer of learning from music to language domains. It addresses the convergence of the many levels of processing that music and speech share, and how musical training can enable us to capitalize on this overlap to enhance language function.

Hearing speech in noise

The ability to successfully listen to speech in noisy backgrounds involves both sensory- and cognitive-based skills. At the sensory end of the continuum, the auditory system must lock on to the target speech signal while excluding competing voices and ambient noise. This is accomplished by organizing disparate, overlapping auditory inputs into different streams by using grouping strategies based on shared characteristics such as location and acoustical similarity.[5] The relative stability of voice pitch, or fundamental frequency, over time in the course of running speech gives a speech stream an identity and aids in grouping it separately from other voices, even those close in pitch to the voice of interest.[6,7] In addition to voicing, other signal-based cues, such as timing, harmonics, and location, aid in group formation of speech.[8] At the cognitive end of the spectrum, a listener's attention and working

doi: 10.1111/j.1749-6632.2012.06463.x

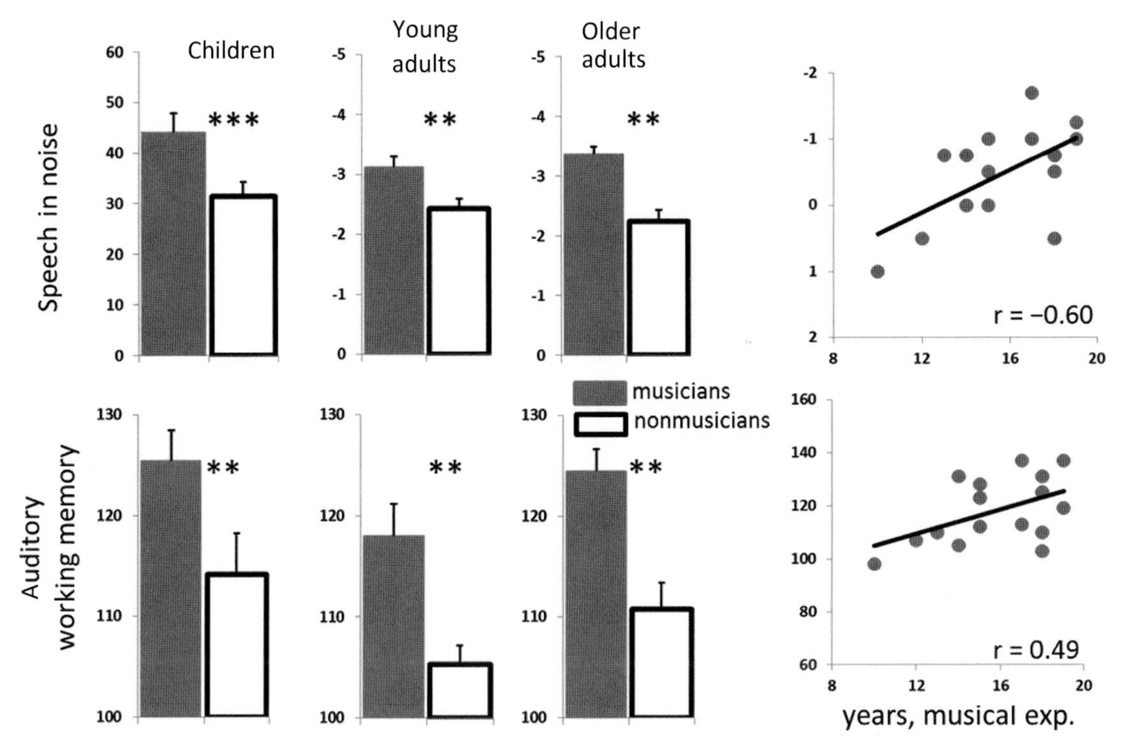

Figure 1. First three columns: musicians (solid, red) perform better than nonmusicians (open, black) in both hearing-in-noise ability (top) and auditory working memory (bottom). This is true for school-age children (left, age range 7–13; musician $n = 15$, nonmusician $n = 16$), young adults (center, age range 18–30; musician $n = 16$, nonmusician $n = 15$), and older adults (right, age range 45–65; musician $n = 18$, nonmusician $n = 19$). In all age ranges, groups were otherwise matched in IQ and audiometric thresholds. Child speech-in-noise scores are expressed in percentiles; young and older adult speech-in-noise scores expressed in threshold signal-to-noise levels (dB). Auditory working memory expressed as standardized scores. $^{**}P < 0.01$; $^{***}P < 0.001$. Right column: hearing-in-noise ability and working memory skill vary as a function of years of musical experience in young adults. These relationships also hold for children and older adults. Child, young adult, and older adult data adapted from Strait *et al.*[10] and Parbery-Clark *et al.*,[1,3] respectively.

memory skills, as well as knowledge about the world, are used to their utmost in the pursuit of a conversation in noise.[8,9] The better one's working memory and attention skills, the better the ability to hear speech in noise.[1]

Musicians, speech-in-noise perception, and auditory cognitive skills

In studies involving participants of all ages, our group is investigating the advantage musical training affords to hearing speech in noise (Fig. 1, top row). Both younger[1] and older[3] adult musicians outperform nonmusicians on standardized measures of speech-in-noise perception. School-age children (8 to 12 years) with musical backgrounds,[10] despite having enjoyed considerably fewer years of training than adult musicians, similarly outperform their peers. In these same age groups, auditory working memory has proven to be better in musicians (Fig. 1, bottom row).[1,3,10] Advantages may emerge at even younger ages, such as in preschool-age children embarking in Suzuki-Orff music training.[11] Indeed, auditory working memory and speech-in-noise perceptual abilities are correlated in all age groups and—relevant to the nature/nurture question discussed below—both track with years of musical experience (Fig. 1, right column). Although not reviewed here, musicians have demonstrated superior auditory attention skills as well.[12,13] In both children and adults, the cognitive enhancements exhibited in musicians tend to be auditory-domain specific.[1,12,14,15] It is noteworthy that rhythm skill is linked to auditory working memory and attention, and the ability of the nervous system to enhance stimulus regularities.[16]

The speech-in-noise advantage in musicians, given the importance of stream formation discussed previously, is unsurprising. A musician's auditory system is constantly tuned in to complex auditory streams, and the ability to separate and organize them is crucial to musical performance. This ability, in turn, translates to auditory perceptual advantages in other domains such as speech. The memory advantage in musicians, which we are not the first to report,[14,17] is postulated to have a basis in functional cortical activation: in a pitch memory task, nonmusicians rely more on auditory sensory areas, while musicians rely on areas of cortex more devoted to short-term memory.[18] The fact that musicians have an edge over nonmusicians in auditory memory is not entirely surprising given that much of music training involves memorization and short-term memory manipulations, such as those involved in working out a tough passage by listening to it, holding it in memory, and repeatedly executing the motor complexities of playing it. Improvisation also exercises memory, as the hook must be held in memory in order to successfully execute an improvisational flight. In addition, auditory memory is involved in the learning of notes and auditory patterns, instrumental fingerings and tuning, and remembering lyrics. Attention is likewise strongly involved in focusing on musical notation; sounds; body control, for example, fingering; and on tempi and dynamics of other musicians that you are playing with. This is akin to following a conversation in a noisy environment—memory of, and attention to, what was said a few seconds before is crucial to allowing you to make sense of what is being said at this moment.

The role of music training and a neurophysiological approach

As we have seen, musicians are better at speech-in-noise perception and auditory working memory than nonmusicians. We are working to identify the biological bases for this advantage. This section summarizes recent findings and posits a model of sensory–cognitive reciprocity accounting for the connections among neural processing, cognitive abilities, and the ability to decode speech in noisy backgrounds.

The auditory brainstem is a hub of sensory–cognitive interactions.[19] Once thought to be passive relay stations between the cochlea and the cortex, subcortical nuclei such as inferior colliculus are now understood to be highly reciprocally connected with cortical areas, affected by cognitive and emotional influences, and plastic in their response properties.[20–22] This plasticity can be wrought over different time scales—from online processing to weeks-long training to lifelong skill learning—and is accomplished via the massive efferent auditory connections that are active even out to the peripheral extreme—the hair cells of the cochlea.[23]

Our neurophysiological approach—recording auditory brainstem responses to complex sounds, the cABR—is not capable of arbitrating between bottom-up– and top-down–mediated plasticity—in other words, whether neurophysiological patterns visible in musician subcortical responses originated with hypertuned attention and memory or with locally sharpened response properties in the sensory structures. Our approach, however, provides a window into the functioning of a subcortical auditory system that is strongly affected by *both* sensory and cognitive influences. Therefore, it offers a powerful measure of the sensory–cognitive auditory system.

Accessing biology in humans

There is a long list of literature reporting biological changes following pervasive musical experience.[13,18,24–36] To better arm the reader to interpret the findings of the particular biological measure presented here, we need to say a few words about what a brainstem response to a complex sound looks like and some of the ways it can be analyzed. When stimulating with a complex sound, such as a speech syllable, the response of the auditory brainstem measured at the scalp is strikingly similar to the stimulating sound. In fact, a digitized cABR recording, played through a speaker, sounds very much like the original evoking stimulus. In the time domain, neural firing to transient events such as syllable onsets and offsets is readily visible in the response, as are the responses to the periodic voicing cycles of the vowel. In the frequency domain, a mirroring of spectral peaks is apparent, albeit with the auditory system's low-pass characteristic affecting higher-frequency amplitudes in the response. Unlike cortical responses that provide an abstract representation of the stimulus, the fidelity to the stimulus of the cABR and the resulting morphological richness lend themselves to a host of signal-processing techniques that permit examination of how stimulus attributes are biologically transduced.[37] In this review, the processing approaches used include correlation of the response to the stimulus; measuring noise-induced shifts in response timing; analyzing the frequency content of the response, especially the harmonics of voice pitch; and quantifying the timing/phase differences arising from frequency glides in consonant sounds.

Music training, speech in noise, working memory, and biological processing

The advantages that musicians have in hearing speech in noise and in cognitive processes, such as auditory memory, were examined with respect to the speech-evoked brainstem response. Using this approach enables us to determine the biological processing differences between musicians and nonmusicians and whether these differences relate to speech-in-noise perception and working memory. *Musicians*, in our studies, are defined as individuals who began their music training before the age of 9 years and have been playing at minimum three times weekly up to the time of enrollment in the study. *Speech-in-noise* tests require participants to repeat sentences that they hear in varying amounts of background noise until a threshold signal-to-noise ratio is determined.[38,39] Standardized *auditory working memory* tasks require the participant to remember, manipulate (e.g., reorder), and recite lists of words, numbers, or sentences.[40]

Response fidelity

The extent to which the nervous system generates a response that resembles the incoming sound reveals the fidelity with which the nervous system encodes sound. Correlating a digitized cABR waveform to the digitized stimulus waveform is one way to quantify this fidelity. The extent of similarity between the sound "da" and its evoked brainstem response is correlated with the ability to hear speech in noise on a standardized test (Fig. 2, top left), and this measure of biological processing, in turn, correlates with auditory working memory (Fig. 2, top center). Both child and adult musicians have responses that more

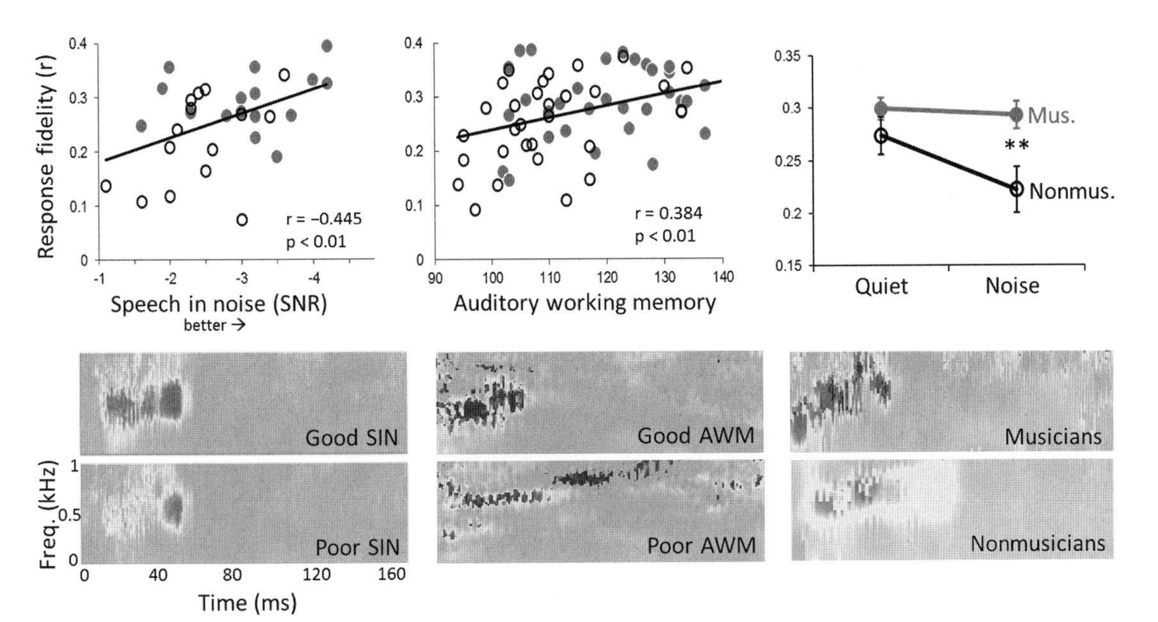

Figure 2. Two measures of neural processing, response correlation to stimulus (top row) and cross-phaseograms of responses to stop consonants ba and ga (bottom row), reveal biological underpinnings of behavior and experience. Left column: biological processing both correlates with speech-in-noise perceptual ability and reveals differences between good and poor speech-in-noise perceivers. Center column: auditory working memory patterns similarly with neural processing. In the two scatterplots, solid symbols are musicians, open symbols are nonmusicians. Right: musician and nonmusician groups also have different biological processing patterns, particularly when the evoking stimulus is masked by noise. To interpret the phaseograms, green indicates no phase differences between the two responses; warm colors, as seen in the 20–60 msec region in all cases, signify faster neural processing of ga than ba. This is the expected pattern based on the frequency content of the two syllables. $^{**}P < 0.01$.

highly reflect the acoustic properties of the evoking stimulus compared to their age-, IQ, and hearing-matched nonmusician peers (Fig. 2, top right).[10,42]

Noise-induced response delay

Background noise delays the timing of neural processing. It is thought that a better-tuned afferent auditory system will result in a response that is less delayed. To assess timing delays brought about by noise, we can either measure the timing of discrete response peaks or use cross-correlation procedures to compare responses to the same stimulus when presented in a quiet versus a noisy background.[41] The extent of the response shift between quiet and noisy backgrounds can serve as a metric of processing integrity. Indeed, in young adult musicians, the delay incurred by background noise is smaller than in otherwise-matched nonmusicians.[10,42] An almost identical pattern is seen in children who are either good or poor speech-in-noise perceivers.[43] In all of these populations, working memory is strongly correlated with both speech-in-noise perception and the degree of noise-induced cABR timing shift.

Response spectrum

The frequency composition of speech and music is preserved in the neural response, and the spectrum of the response yields a rich source of information regarding the encoding of a sound's frequency composition. The encoding of the harmonics, in particular, reveals a musician/nonmusician distinction. To a "da," the neural encoding of the syllable's harmonics is enhanced in musicians both when presented in quiet and in a noisy background.[10,42] This is seen both in children and young adults, and in both groups the extent of the harmonic enhancement is correlated with auditory working memory. We have also found that response spectrum depends on stimulus context, and the extent to which context (e.g., regular vs. random stimulus presentation) enhances the response spectrum is correlated with music and language abilities in children[16] and years of musical experience in adults.[44]

Timing

Speech and music are spectrotemporally dynamic signals and the processing of their rapid changes requires precise neural timing. Precise neural timing is essential for effective auditory-based communication, and timing breaks down with certain communication disorders.[45–47] Subtle differences between stimuli, such as the frequency content of a formant transition differentiating two stop consonants, result in quantifiable timing differences in the response. In addition to measuring the timing of discrete peaks, it is also possible to use cross-spectrum techniques to compute phase differences between two responses.[48] The cross-phaseogram produces a color representation of subtle timing differences between a pair of biological responses that may be difficult to quantify in the time-domain waveforms.[41,48] An example is the differentiation of responses evoked by differing stop consonants, such as /b/ and /d/. We have synthesized a trio of such sounds, ba, da, ga, such that they differ acoustically only in a subtle difference in the frequency sweep of the second formant. Although their time-domain responses are very similar, the differences in biological processing among them are readily apparent through the use of the cross-spectrum technique. Figure 2, bottom left, demonstrates that a poor speech-in-noise–perceiving cohort of subjects has a smaller phase difference between ba and ga than a good speech-in-noise–perceiving cohort (warm colors in the 0–60 msec range of the phaseograms represent neural differentiation between two sounds). Likewise, this timing precision is enhanced in good performers on a task of auditory working memory and in musicians (Fig. 2, bottom, center and right).

It must be mentioned that the complex auditory brainstem response is not monolithic in its response properties. The focus of this paper is to highlight that it differs between groups and relates to other phenomena, so a reader might have an impression that the response is depressed as a whole in poor speech-in-noise perceivers or nonmusicians. However, individual properties of the cABR are quite separable,[49] and there are many aspects of the response that do not differ between these groups. Nevertheless, in a variety of populations we have evidence of a three-way relationship between the subcortical processing of complex sounds, auditory working memory, and the ability to hear speech in noise.

How do we know that musical experience was the driving force behind improved outcomes on skills such as speech-in-noise perception and enhancements in subcortical auditory processing? Might it be the case that people with enhanced auditory processing and other skills are more inclined to persevere with music education? Evidence for "nurture" in this question comes from three sources:

longitudinal studies of individuals as they undergo musical training (e.g., Schlaug *et al.*[50]), cross-sectional studies of musicians with a range of years of experience (e.g., Forgeard *et al.*[51] and Fig. 1, right column[1]), and findings that musicians' brains react preferentially to their own instruments.[52–55] Research using these designs provides examples of anatomical, physiological, and behavioral development that coincide with degree of musical experience.[56–59] It is unlikely that such correlations would arise from preexisting conditions influencing the pursuit of musical expertise.

Discussion

Musical experience is a driving force in shaping biological responses to sound and, as we have seen, the benefits afforded by music transfer to other realms of auditory processing, including speech.[26,34,60,61] The recently proposed OPERA hypothesis[4] describes mechanisms by which music can lead to generalized learning. Among them is the *overlap* in biological resources available for the processing of these two types of sounds, along with the greater demands of *precision* that music, relative to speech, puts on these shared resources. Furthermore, music practice and performance elicits strong *emotions*. Emotion, in particular, is a strong driving force behind auditory learning in animals.[62,63] Finally, the *repetition* and the cognitive demands required by intensive music practice, such as *attention* and working memory, initiate enhanced cortical plasticity, which in turn strengthens subcortical circuitry and tunes the afferent system for signal processing of incoming speech.

Here, we present a model that describes the cortical, subcortical, and emotional mechanisms that interact to affect speech processing and how this reciprocally interactive network is influenced by musical experience (Fig. 3).[16] The relationships between musical skill and hearing speech in noise are no doubt mediated by cognitive factors such as memory, and the subcortical response patterns tying music, speech perception, and auditory memory together suggest a corticofugal-mediated shaping of sensory function. Auditory working memory and hearing in noise are intrinsically linked, and biological processing of sound, accessed by cABR, provides a biological basis for that link. It appears that cognitive function, such as working memory, is a force that drives the biological representation of sound.

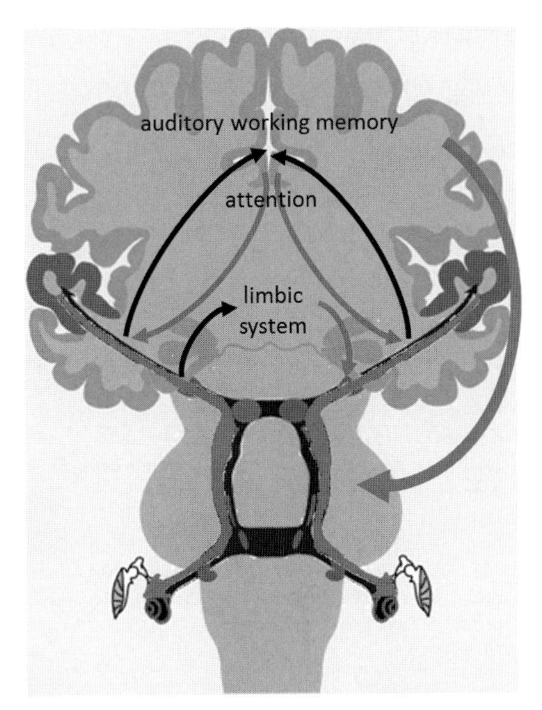

Figure 3. The afferent auditory pathway, from cochlea to cortex, is complemented by descending projections originating in brain regions responsible for executive and limbic functions. These corticofugal connections sharpen auditory processing. Auditory working memory, in particular, is stronger in musicians and drives strengthened auditory processing as well as perceptual benefits for following conversations in noise.

We propose that music training first drives cognitive enhancement that, in turn, shapes the nervous system's response to sound. Music training as a means of augmenting corticofugal auditory networks has the potential to enhance everyday communication.

In the last decade, we have moved away from the classical view of hearing as a one-way street from the cochlea to higher brain centers in the cortex. It is now accepted that cognition, once thought to play no role in hearing, has a dramatic influence on hearing and subsequent communication. Our approach—the convergent study of cognition, perception, and biological processing—is one means of understanding the mechanistic bases of cognition's role in auditory processing.

Acknowledgments

Supported by NSF Grants BCS-0921275, BCS-1057556, and SMA-1015614, NIH Grant F31DC-011457, and the Grammy Foundation.

Conflicts of interest

The authors declare no conflicts of interest.

References

1. Parbery-Clark, A., E. Skoe, C. Lam & N. Kraus. 2009. Musician enhancement for speech-in-noise. *Ear Hear.* **30:** 653–661.
2. Zendel, B.R. & C. Alain. 2012. Musicians experience less age-related decline in central auditory processing. *Psychol. Aging.* doi: 10.1037/a0024816
3. Parbery-Clark, A., D.L. Strait, S. Anderson, *et al.* 2011. Musical experience and the aging auditory system: implications for cognitive abilities and hearing speech in noise. *PLoS One* **6:** e18082.
4. Patel, A.D. 2011. Why would musical training benefit the neural encoding of speech? The OPERA hypothesis. *Front. Psychol.* **2:** 142.
5. Bregman, A.S. 1990. *Auditory Scene Analysis: The Perceptual Organization of Sound.* MIT Press. Cambridge, MA.
6. Brokx, J.P.L. & S.G. Nooteboom. 1982. Intonation and the perceptual separation of simultaneous voices. *J. Phonetics* **10:** 23–36.
7. Darwin, C.J. 2005. Pitch and auditory grouping. In *Pitch: Neural Coding and Perception*, Vol. 24. C.J. Plack, R.R. Fay & A.J. Oxenham, Eds.: 278–305. Springer. New York.
8. Shinn-Cunningham, B.G. & V. Best. 2008. Selective attention in normal and impaired hearing. *Trends Amplif.* **12:** 283–299.
9. Heinrich, A., B.A. Schneider & F.I. Craik. 2008. Investigating the influence of continuous babble on auditory short-term memory performance. *Q. J. Exp. Psychol.* **61:** 735–751.
10. Strait, D.L., A. Parbery-Clark, E. Hittner & N. Kraus. 2011. Music lessons in early childhood enhance subcortical speech encoding in challenging listening environments. *The Neurosciences and Music IV*, Edinburgh, Scotland.
11. Meyer, M., S. Elmer, M. Ringli, *et al.* 2011. Long-term exposure to music enhances the sensitivity of the auditory system in children. *Eur. J. Neurosci.* **34:** 755–765.
12. Strait, D.L., N. Kraus, A. Parbery-Clark & R. Ashley. 2010. Musical experience shapes top-down auditory mechanisms: evidence from masking and auditory attention performance. *Hear. Res.* **261:** 22–29.
13. Strait, D.L. & N. Kraus. 2011. Can you hear me now? Musical training shapes functional brain networks for selective auditory attention and hearing speech in noise. *Front. Psychol.* **2:** 113.
14. Chan, A.S., Y.C. Ho & M.C. Cheung. 1998. Music training improves verbal memory. *Nature* **396:** 128.
15. Ho, Y.C., M.C. Cheung & A.S. Chan. 2003. Music training improves verbal but not visual memory: cross-sectional and longitudinal explorations in children. *Neuropsychology* **17:** 439–450.
16. Strait, D.L., J. Hornickel & N. Kraus. 2011. Subcortical processing of speech regularities underlies reading and music aptitude in children. *Behav. Brain Funct.* **7:** 44.
17. Jakobson, L.S., S.T. Lewycky, A.R. Kilgour & B.M. Stoesz. 2008. Memory for verbal and visual material in highly trained musicians. *Music Percept.* **26:** 41–55.
18. Gaab, N. & G. Schlaug. 2003. The effect of musicianship on pitch memory in performance matched groups. *NeuroReport* **14:** 2291–2295.
19. Winer, J.A. 2006. Decoding the auditory corticofugal systems (corrected version of original publication, vol. 207, 2005). *Hear. Res.* **212:** 1–8.
20. Bajo, V.M., F.R. Nodal, D.R. Moore & A.J. King. 2010. The descending corticocollicular pathway mediates learning-induced auditory plasticity. *Nat. Neurosci.* **13:** 253–260.
21. Gao, E.Q. & N. Suga. 2000. Experience-dependent plasticity in the auditory cortex and the inferior colliculus of bats: role of the corticofugal system. *Proc. Natl. Acad. Sci. USA* **97:** 8081–8086.
22. Marsh, R.A., Z.M. Fuzessery, C.D. Grose & J.J. Wenstrup. 2002. Projection to the inferior colliculus from the basal nucleus of the amygdala. *J. Neurosci.* **22:** 10449–10460.
23. de Boer, J. & A.R.D. Thornton. 2007. Effect of subject task on contralateral suppression of click evoked otoacoustic emissions. *Hear. Res.* **233:** 117–123.
24. Bangert, M. & G. Schlaug. 2006. Specialization of the specialized in features of external human brain morphology. *Eur. J. Neurosci.* **24:** 1832–1834.
25. Parbery-Clark, A., S. Anderson & E. Hittner. 2012. Musical experience offsets age-related delays in neural timing. *Neurobiol. Aging.* doi: 10.1016/j.neurobiolaging.2011.12.015
26. Schlaug, G. 2001. The brain of musicians. A model for functional and structural adaptation. *Ann. N.Y. Acad. Sci.* **930:** 281–299.
27. Hutchinson, S.L., H.L. Lee, N. Gaab & G. Schlaug. 2003. Cerebellar volume of musicians. *Cereb. Cortex* **13:** 943–949.
28. Koelsch, S., E. Schroger & M. Tervaniemi. 2000. Superior pre-attentive and attentive processing of auditory information in musicians: an MMN study. *J. Psychophysiol.* **14:** 64–65.
29. Tervaniemi, M., V. Just, S. Koelsch, *et al.* 2005. Pitch discrimination accuracy in musicians vs nonmusicians: an event-related potential and behavioral study. *Exp. Brain Res.* **161:** 1–10.
30. Zatorre, R.J., J.L. Chen & V.B. Penhune. 2007. When the brain plays music: auditory-motor interactions in music perception and production. *Nat. Rev. Neurosci.* **8:** 547–558.
31. Fujioka, T., B. Ross, R. Kakigi, *et al.* 2006. One year of musical training affects development of auditory cortical-evoked fields in young children *Brain* **129:** 2593–2608.
32. Hannon, E.E. & L.J. Trainor. 2007. Music acquisition: effects of enculturation and formal training on development. *Trends Cogn. Sci.* **11:** 466–472.
33. Hyde, K.L., J. Lerch, A. Norton, *et al.* 2009. Musical training shapes structural brain development. *J. Neurosci.* **29:** 3019–3025.
34. Schlaug, G., A. Norton, K. Overy & E. Winner. 2005. Effects of music training on the child's brain and cognitive development. *Ann. N.Y. Acad. Sci.* **1060:** 219–230.
35. Shahin, A., L.E. Roberts & L.J. Trainor. 2004. Enhancement of auditory cortical development by musical experience in children. *NeuroReport* **15:** 1917–1921.
36. Kraus, N. & B. Chandrasekaran. 2010. Music training for the development of auditory skills *Nat. Rev. Neurosci.* **11:** 599–605.

37. Skoe, E. & N. Kraus. 2010. Auditory brainstem response to complex sounds: a tutorial. *Ear Hear.* **31:** 302–324.
38. Killion, M.C., P.A. Niquette, G.I. Gudmundsen, *et al.* 2004. Development of a quick speech-in-noise test for measuring signal-to-noise ratio loss in normal-hearing and hearing-impaired listeners. *J. Acoust. Soc. Am.* **116:** 2395–2405.
39. Nilsson, M., S.D. Soli & J.A. Sullivan. 1994. Development of the hearing in noise test for the measurement of speech reception thresholds in quiet and in noise. *J. Acoust. Soc. Am.* **95:** 1085–1099.
40. Woodcock, R.W., K.S. McGrew & N. Mather. 2001. *Woodcock-Johnson Psycho-educational Battery*. 3rd ed., Vol. 3. Riverside. Itasca, IL.
41. Tierney, A., A. Parbery-Clark, E. Skoe & N. Kraus. 2011. Frequency-dependent effects of background noise on subcortical response timing. Hear. Res. **282:** 145–150.
42. Parbery-Clark, A., E. Skoe & N. Kraus. 2009. Musical experience limits the degradative effects of background noise on the neural processing of sound. *J. Neurosci.* **29:** 14100–14107.
43. Anderson, S., E. Skoe, B. Chandrasekaran & N. Kraus. 2010. Neural timing is linked to speech perception in noise. *J. Neurosci.* **30:** 4922–4926.
44. Parbery-Clark, A., D.L. Strait & N. Kraus. 2011. Context-dependent encoding in the auditory brainstem subserves enhanced speech-in-noise perception in musicians. *Neuropsychologia* **49:** 3338–3345.
45. Tallal, P. 1981. Language disabilities in children: perceptual correlates. *Int. J. Pediatr. Otorhinolaryngol.* **3:** 1–13.
46. Tallal, P. & N. Gaab. 2006. Dynamic auditory processing, musical experience and language development. *Trends Neurosci.* **29:** 382–390.
47. Gaab, N., J.D.E. Gabrieli, G.K. Deutsch, *et al.* 2007. Neural correlates of rapid auditory processing are disrupted in children with developmental dyslexia and ameliorated with training: an fMRI study. *Restor. Neurol. Neurosci.* **25:** 295–310.
48. Skoe, E., T. Nicol & N. Kraus. 2011. Cross-phaseogram: objective neural index of speech sound differentiation. *J. Neurosci. Methods* **196:** 308–317.
49. Kraus, N. & T.G. Nicol. 2005. Brainstem origins for cortical 'what' and 'where' pathways in the auditory system. *Trends Neurosci.* **28:** 176–181.
50. Schlaug, G., M. Forgeard, L. Zhu, *et al.* 2009. Training-induced neuroplasticity in young children. *Ann. N.Y. Acad. Sci.* **1169:** 205–208.
51. Forgeard, M., E. Winner, A. Norton & G. Schlaug. 2008. Practicing a musical instrument in childhood is associated with enhanced verbal ability and nonverbal reasoning. *PLoS One* **3:** e3566.
52. Strait, D.L., K. Chan, R. Ashley & N. Kraus. 2012. Specialization among the specialized: auditory brainstem function is tuned in to timbre. *Cortex.***48:** 360–362.
53. Pantev, C., L.E. Roberts, M. Schulz, *et al.* 2001. Timbre-specific enhancement of auditory cortical representations in musicians. *NeuroReport* **12:** 169–174.
54. Margulis, E.H., L.M. Mlsna, A.K. Uppunda, *et al.* 2009. Selective neurophysiologic responses to music in instrumentalists with different listening biographies. *Hum. Brain Mapp.* **30:** 267–275.
55. Shahin, A.J., L.E. Roberts, W. Chau, *et al.* 2008. Music training leads to the development of timbre-specific gamma band activity. *NeuroImage* **41:** 113–122.
56. Gaser, C. & G. Schlaug. 2003. Brain structures differ between musicians and nonmusicians. *J. Neurosci.* **23:** 9240–9245.
57. Schneider, P., M. Scherg, H.G. Dosch, *et al.* 2002. Morphology of Heschl's gyrus reflects enhanced activation in the auditory cortex of musicians. *Nat. Neurosci.* **5:** 688–694.
58. Musacchia, G., M. Sams, E. Skoe & N. Kraus. 2007. Musicians have enhanced subcortical auditory and audiovisual processing of speech and music. *Proc. Natl. Acad. Sci. USA* **104:** 15894–15898.
59. Lee, K.M., E. Skoe, N. Kraus & R. Ashley. 2009. Selective subcortical enhancement of musical intervals in musicians. *J. Neurosci.* **29:** 5832–5840.
60. Bidelman, G.M. & A. Krishnan. 2010. Effects of reverberation on brainstem representation of speech in musicians and nonmusicians. *Brain Res.* **1355:** 112–125.
61. Bidelman, G.M., J.T. Gandour & A. Krishnan. 2011. Cross-domain effects of music and language experience on the representation of pitch in the human auditory brainstem. *J. Cogn. Neurosci.* **23:** 425–434.
62. Weinberger, N.M. 2007. Auditory associative memory and representational plasticity in the primary auditory cortex. *Hear. Res.* **229:** 54–68.
63. Kilgard, M.P. & M.M. Merzenich. 1998. Cortical map reorganization enabled by nucleus basalis activity. *Science* **279:** 1714–1718.

Ann. N.Y. Acad. Sci. ISSN 0077-8923

ANNALS OF THE NEW YORK ACADEMY OF SCIENCES
Issue: *The Neurosciences and Music IV: Learning and Memory*

Cognitive and methodological considerations on the effects of musical expertise on speech segmentation

Clément François,[1] Barbara Tillmann,[2] and Daniele Schön[3]

[1]INCM-CNRS UMR6193 and Aix-Marseille University, Marseille, France. [2]CNRS UMR5292, INSERM U1028, Lyon Neuroscience Research Center, Auditory Cognition and Psychoacoustics Team, Lyon, France and Université de Lyon, France. [3]INS, INSERM and Aix-Marseille University, Marseille, France

Address for correspondence: Daniele Schön, INS-INSERM-Aix-Marseille University, Service Neurophysiologie Clinique—CHU Timone, Rue Saint Pierre-13385 Marseille Cedex 5, Marseille, France 13385. schon@incm.cnrs-mrs.fr

Both speech and music are constituted by sequences of sound elements that unfold in time and require listeners to engage cognitive functions such as sequencing, attention, and memory. We recently ran a set of experiments with the aim of testing the effect of musical expertise on a rather high cognitive function: speech segmentation. Here, we will present the main concepts underlying the investigation of speech segmentation as well as its link to music and musical expertise. Interestingly, our results seem to show that musical training and expertise have effects on brain plasticity that may go beyond primary regions. Moreover, to facilitate and improve future research in this domain, we will here describe several delicate methodological precautions that need to be taken into account (e.g., the choice of stimuli, participants, data analyses). Finally, we will give some possible future directions to better understand the impact that music may have on speech processing.

Keywords: speech segmentation; musical expertise; transfer of training

An introduction to statistical learning

Learning the mother tongue or a second language is a rather long process that goes through several dependent phases. Because word boundaries are not systematically flagged by acoustic cues, such as stresses or pauses, one important step in language learning is the ability to extract words that unfold in time. Many studies inspired by the seminal work of Saffran *et al.* showed the importance of the statistical structure of the speech stream for an efficient segmentation.[1–3] Indeed, in a speech stream, within-word syllables tend to be associated more often than between-words syllables. The importance of these statistics (called conditional or transitional probabilities) has been shown in adults, infants, and neonates.[4–7] The experimental paradigm consists of a familiarization (learning) phase followed by a test. During the familiarization phase, participants listen to several minutes of a statistically structured, continuous flow of artificial syllables without any acoustic cues at word boundaries. The test phase depends on the participant population (infants, adults). In the case of adult participants, the test is a two-alternative forced choice procedure (AFC) and participants have to choose, in each trial, between a word that was part of the language and a word built with similar syllables, but that was not part of the language (henceforth, a partial word). Above-chance performance suggests participants' ability to segment the auditory (speech) stream on the basis of transitional probabilities. This ability has been also demonstrated for nonlinguistic auditory sequences, such as sequences of pitches[4,8,9] or of instrumental timbres,[10] thus suggesting that this type of learning is not specific to speech. Finally, this ability has also been shown for sequences of visual stimuli (shapes or movements) and sequences of tactile stimulations,[11] suggesting that this type of learning is at work in several modalities.[12–14]

Song is particularly well suited to better understand the relationship between music and language processing. Indeed, the segmentation of linguistic and nonlinguistic inputs has been studied separately

doi: 10.1111/j.1749-6632.2011.06395.x

(with different tasks and different participants), thus rendering it difficult to draw any clear conclusion on the nonspecificity of the learning processes for language and music. Recently, we compared speech segmentation of an artificial sung language (with speech and music combined in the same signal) to speech segmentation of a spoken language. Speech segmentation was better when sung than when spoken, possibly due to structural and motivational properties of music.[15] In work by François and Schön,[16] participants (nonmusicians) were exposed to the sung speech stream, but then tested separately on linguistic and musical dimensions of the sung language (both behavioral and EEG responses were recorded at test). Performance was above chance for the linguistic test, but at chance level in the musical test. By contrast to this behavioral measure, the analysis of the event-related potentials (ERPs) revealed in both linguistic and musical tests a similar frontocentral late negative component that was larger for nonfamiliar (partial words/sequences) than for familiar items (words/sequences). In agreement with previous findings,[3,8,17] we interpreted this late negative component as an index for the search of memory traces that have been shaped during learning (i.e., in the familiarization phase). In another study, François and Schön[18] compared sung speech segmentation in nonmusicians and professional musicians. While the behavioral results did not show a clear effect of expertise, ERP data showed a larger late negative component for musicians than for nonmusicians, in both language and musical tests.

Finally, to show that music training was the cause of this difference (and not prior differences of the participants in the two groups), François *et al.*[19] conducted a longitudinal study spanning over two years using a test–training–retest procedure with a pseudorandom assignment of children to two different artistic training programs. Children followed a training on either music or painting, and were tested on their ability to extract words from a continuous flow of syllables (the linguistic test). Both behavioral and electrophysiological measures showed a greater improvement in speech segmentation across testing sessions in the music group compared to the painting group. Taken together, these findings point to a benefit of musical expertise and musical training for both speech and music segmentation.

Why would speech segmentation benefit from musical expertise?

In 2007, Hickok and Poeppel proposed a dual-route model of speech processing,[20] which includes dorsal and ventral pathways. The dorsal pathway acts as a sensorimotor interface aiming at mapping the phonologico-acoustic representations of speech sounds to articulatory representations. The ventral pathway acts as a lexico-conceptual interface aiming at mapping the phonologico-acoustic representations to lexico-conceptual information. Based on a set of studies combining behavioral, EEG, and fMRI techniques, Rodriguez-Fornells *et al.*[3] have recently adapted and completed this model for speech segmentation. They described a large cortical network involved in speech segmentation comprising the posterior part of the superior temporal gyrus (STG) and the premotor cortex (PMC) connected via the arcuate fasciculus.

Within this framework, several nonexclusive hypotheses can explain the differences in speech segmentation between musicians and nonmusicians. First, musicians might benefit from more efficient sound encoding at subcortical and cortical levels. Indeed, previous studies have shown functional differences between musicians' and nonmusicians' encoding of musical and linguistic sounds.[21] In our experiment,[18] we also found differences due to expertise in rather early auditory components, such as the N1, the P2, and the MMN, all possibly generated in the primary auditory cortex and the planum temporale.[22,23] Thus, early differences in the functioning of the brainstem and the auditory cortex might explain the musical expertise effect, notably with musical training leading to a reorganization of auditory neurons along the auditory dorsal pathway, thus facilitating the subsequent processing steps in the STG and the left inferior frontal gyrus (IFG).[24]

Second, an alternative explanation could be that musicians may have a more developed and efficient dorsal pathway than nonmusicians. Its implication for statistical learning has been suggested by a recent study demonstrating that white matter integrity in the vicinity of the left IFG-predicted performance in artificial grammar learning.[25] Similarly, in a speech segmentation experiment combining fMRI and EEG measures, Cunillera *et al.*[26] observed a frontocentral late negative component (at word onsets) that increased over the learning period and that had its generator around the PMC and the left IFG. Taken together with the observation of increased gray matter density and volume in the left IFG for musicians in comparison to nonmusicians,[27] these results suggest that musicians' advantage in speech segmentation might be related to enhanced involvement of the premotor brain areas in comparison to nonmusicians.

Third, we might hypothesize that musical training improves the connectivity between these two subsystems. This hypothesis can be integrated in the recent proposition that sound acts as a scaffolding framework for cognitive sequencing,[28] also supporting how to process and interpret sequential and temporal information in the environment. It is suggested that sound and speech processing has an additional unspecified influence on the development of general cognitive sequencing abilities (also in other modalities). While Conway *et al.*[28] find support for their hypothesis in the consequences of auditory deprivation on domain-general sequencing (i.e., impaired implicit learning for visual, nonlinguistic regularities in deaf children), additional support can be found in research investigating the effect of musical expertise, that increased training on sound analyses and enhanced sound exposure should lead to improved sequencing also in nonmusical tasks (such as speech segmentation).

Methodological considerations

The comparison of musicians and nonmusicians

When comparing the performance of musicians and nonmusicians, one has to keep in mind some general difficulties and restrictions to conduct proper experiments testing for the effect of musical practice. Indeed, most studies comparing musicians and nonmusicians use a cross-sectional approach: a group with 10 years of musical practice is compared to a group without musical practice possibly comparable in age, handedness, sex, and level of education. These studies are very instructive, but always fall in the criticism of noncontrolled, preexisting genetic or other factors explaining the observed between-group differences. These issues can be ruled out with a longitudinal approach testing naive (with respect to music) participants enrolled in a test–training–retest procedure, while still controlling for several standardized neuropsychological tests and several socioeconomic variables. After the first testing session, this specific procedure requires a pseudorandom assignment of participants to either a musical group or a control group. This procedure also encounters some difficulties, like the critical choice of the activity proposed to the control group and the repetition of the tests, which renders the test explicit. This is particularly delicate when interpreting speech segmentation as an implicit learning process (cf. section "Measuring learning: from explicit to implicit measures").

The choice of the instructions

The instructions before the learning phase can become, in our opinion, important because they determine to which extent learning may take place implicitly or not. For instance, in some studies, participants are explicitly told about the presentation of an invented "nonsense" language, or even to look for the words embedded in the speech stream—for example, Saffran *et al.*: "Figure out where the words begin and end."[29] Another study presents the isolated words (visually) before the familiarization phase in order to maximize learning.[17] By contrast, several studies, like ours, do not give any explicit information before the familiarization phase; participants are asked to carefully listen to a continuous stream of syllables for several minutes.

Measuring learning: from explicit to implicit measures

Another methodological difficulty refers to the implicit nature of the tests and, of course, the implicit nature of the learning process. Studies investigating speech segmentation mostly refer to "statistical learning" rather than "implicit learning,"[30] which leads to the absence of testing whether the learned knowledge is implicit. The domain thus adapts a definition of implicit learning as the incidental nature of the acquisition process and without the intention to learn (at least for some of the instructions, see above), rather than extending it to the implicitness of the acquired knowledge.[31,32]

The standard behavioral test used in this domain can be criticized for theoretical purposes in the field of implicit learning. The AFC procedure requires participants to make an explicit judgment on the two presented items without feedback ("Indicate which of the two strings sounded more like a word from the language you have heard before."). However, the representations of the segmented items might be weak and might vanish rapidly in time, probably also due to the interference caused by the presentation of nonfamiliar items. Thus, there is a need to promote implicit measures, like ERPs, which do not necessary require an overt behavioral response (e.g., analysis of ERPs during the familiarization phase). The priming paradigm could also be adapted to the testing of speech segmentation as it does not require telling participants about the structure of the material and as it has already been successfully applied to artificial grammar learning.[33] Similarly, the serial reaction time procedure can also be used as a promising implicit measure, as introduced by Hunt and Aslin[12] for the visual modality with three element units: over exposure, response times decreased for elements within the units, compared to response times for elements crossing unit boundaries.

Another relevant point to consider is the use of repeated tests in within-subject designs, such as for test–training–retest designs or designs comparing different types of languages or studying the transfer from one language structure to another (see, for example, work by Omigie and Stewart[34]). Indeed, when interpreting results and comparing them to other results reported in the literature, one needs to be aware of the fact that once tested with a typical AFC task, learning of a successive language structure will be even less implicit, insofar as subjects are then aware that the stream contains words and possibly of the structure of the stream (e.g., trysillabic words). One possibility to at least blur the cues for potential strategies gained in the first test phase and then orienting the perception of the second exposure phase is the use of test items of different lengths (bi- and trisyllabic items), even though only trisyllabic words are relevant.[35]

The control of the stimuli (units and stream)

When preparing an artificial language, great care must be taken about the choice of the elements, the definition of the units, and their chaining in the stream. The first choice is related to the choice of the phonemes. For instance, a language containing consonants t, d, b, p, and vowels i and u will be more difficult to segment than a language containing t, s, m, p, and a and o, due to the greater phonetic proximity in the former compared to the latter. Also, when planning to acquire EEG, one may want to use consonants with a short and similar attack time (e.g., plosives) in order to have clearer ERPs (N1-P2 complex) to the onset of syllables/words.[3]

Another important parameter is the number of words to be used to build the stream (typically between four and six). Because the stream is typically generated by a pseudorandom concatenation of the words (no repetition of the same word twice in row), using a very small number of words has a direct impact on the transitional probabilities at word boundaries. In the extreme case of using two words (e.g., bada-tibu), the transitional probability at word boundaries (da-ti, bu-ba) would be identical to the transitional probabilities between the two words. This might be particularly relevant when using partial words in the test phase (in particular, for the AFC test) because, when using a few words, TPs for the partial words will be very close to TP for words (identical in the case of two words). Thus, one may want to calculate TPs for words and partial words before running the experiment. Also, note that while most studies used only trisyllabic words, some studies started to use words of different lengths aiming to get closer to real-world language learning situations.[35]

Another parameter that needs to be taken into account is the ratio between the duration of the familiarization phase and the number of words, weighted by the number of syllables building a word. These choices can vary considerably across the published experiments, going from a 2-min stream using four bisyllabic words (in babies[1]) up to a 21-min stream using six trisyllabic words (in adults[29]). One may also note that while behavioral experiments often use rather long and boring familiarization durations, EEG analysis seems to show that learning may take place in the very first minutes.[26,36] This suggests that the AFC might not be a sensitive measure of implicit learning, as already discussed above (cf. section on "Measuring learning: from explicit to implicit measures").

Another decision to be made concerns the definition of the 2-AFC test phase: participants have to choose between one item that is a word and the other item that is not. For the latter one, three types are commonly used: nonwords for which syllables are arranged in an order that has never occurred in the familiarization phase; partial words for which two of the three syllables have appeared in that order in the familiarization phase, but the association with the third syllable has never been heard; and partial words that have been heard in the familiarization phase, thus containing the boundary between two words (e.g., da-ti in the two-word language example given above). Of course, the choice of one of the three test item types influences the interpretation in terms of learning processes. When using nonwords, participants may simply rely on the detection of a new transition between two syllables, while the use of partial words discards this possibility and is thus more informative for statistical learning (see Ref. 37 for a more complete discussion on this issue).

For the speech material preparation, another delicate point that is rarely addressed in the literature is the synthesis procedure used to generate the stream. Indeed, because acoustic cues need to be controlled, a voice synthesizer is used to generate speech (even though cross-splicing techniques might represent an alternative method). However, speech synthesis is far from being a simple affair, and it is thus important to carefully check the acoustic features of the generated stream. This is even more important when using a sung language, insofar as there might be unexpected interactions between vowels and pitch, resulting in perceptual accents or lengthening as well as in more or less clearly pronounced consonants.

The importance of a careful selection in the material construction is not restricted to verbal material only, but also affects nonverbal segmentation tasks. In the case of music, for instance, tone triplets may pop out within the stream (reinforcing or obscuring boundaries). When tone sequences (or sung sequences) are used, special care must be taken in the choice of the pitches (and pitch classes), their potential link with musical structures (e.g., tonality), but also the defined interval sizes and interval directions between adjacent tones (e.g., jumps, reversals) either within units or across unit boundaries (see also Ref. 4). When musical timbers are used, care must be taken regarding acoustic similarity (or more generally surface features) for the definition of units and boundaries (see Ref. 10). It is worth underlining that these indications only concern the choice of the various features in relation to the definition of units and their boundaries in the stream. Nonetheless, implicit learning can be studied with materials varying in distance and similarity with real-life materials (e.g., Loui *et al.*, showing artificial grammar learning with a new musical scale[38]).

Regarding the selection of elements and units, some experimental controls have been proposed (though not systematically applied yet) to ensure that above-chance performance is not due to some preference bias of syllable (or tone) combinations in statistical units or other unrelated influences. Saffran *et al.*[4] proposed to define two language systems (L1, L2) that were built in such a way that L1 consists of partial units of L2, and vice versa. This construction allows using the same test phase for two participant groups (having been exposed to one or the other language), and thus avoids a confound between learning and other influences (e.g., perceptual biases, preferences). Reber and Perruchet[39] suggested using multiple implementations, where all languages have the same statistical structures, but are instantiated by different syllable attributions (see also Ref. 40).

Data analyses

In several statistical learning studies, *t*-tests are used to make statistical inferences on whether learning has taken place or not. In particular, group performance is compared to chance level. While the two-sample *t*-test is not appropriate because a "chance" sample has no variance, one should use a one-sample *t*-test, setting up a normal distribution with a mean specified by the null hypothesis (here, $P = 0.5$). Also, nonparametric tests, such as the one-sample Wilcoxon signed rank test, might be more appropriate. Another (nonexclusive) approach is to have either a between-subject design directly comparing results of two or more samples/populations or a within-subject design directly comparing, for instance, learning of the linguistic dimension and the musical dimension, or learning before and after music training (however, see above for comments on repeated-measure designs). Finally, a solution at the individual subject level might be to use binomial tests to estimate the exact probability for each individual. Then, one may run second-order statistics on the *P* values, although this might lead to a very conservative test (with 36 trials the individual threshold for significance would be 24 correct responses, $P = 0.0326$).

Another type of analysis consists of taking into account the differences between items. Because each item has average transitional probabilities between syllables (TPs), it is possible to test whether items with high TPs have been learned better than items with low TPs. With this aim, one may either run a nonparametric ANOVA with items as a factor, or contrast directly high versus low TPs items. These analyses should point out the importance of TPs by showing poor performance for low TP items and good performance for high TP items.[29]

To better define the role of TPs in the learning process, one could also take into account the distinctiveness between the two items in each trial. This is of specific interest when using partial words that have been heard during the familiarization phase (i.e., TPs > 0) and that by consequence may also have a sort of prelexical representation competing with words. The underlying idea is that trials with a high TP contrast between words and partial words (i.e., very different TPs) should be easier and thus show better performance than trials with a low TP contrast (i.e., very close TPs).

Finally, another interesting way to model data is, in our view, taking into account *time*. As we discussed above, the test is rather explicit and can lead to interference or learning effects along the testing session (typically 36 trials). Thus, modeling time in the statistical analyses by comparing performance in the first half of the test to performances in the second half of the test can be very instructive in this respect. For instance, Rohrmeier *et al.*[41] used an artificial grammar paradigm with sequences of tones and showed increasing performance along the testing session for the untrained group while the trained group showed decreasing performance along the test.

Speech segmentation and musical expertise: perspectives and future directions

This last section presents possible future directions in the field of musical training and/or practice-shaped brain plasticity and statistical learning, keeping in mind that the most important point, in our view, is understanding what aspects of music training (and its consequences on the brain) might contribute to beneficial effects for speech segmentation and to what extent. Several nonexclusive directions might be interesting in this respect. For instance, aiming to determine the relevant processes, one could study the effect of different types of musical trainings, such as rhythmic training versus pitch/tonal training, by comparing, for example, a group of drummers to a group of singers.

Another promising perspective will be investigating the relation between perceptual and productive musical skills and their interaction with speech segmentation processes. To address this aim, psychophysical measures of perceptual and productive skills could be correlated with statistical learning performance as well as with EEG or fMRI markers of speech segmentation collected on the same participants. Furthermore, because musicians' advantage could be in part explained by better preattentive sound processing, other experiments could record both pre-attentive and attentive subcortical (e.g., frequency-following responses) and cortical auditory responses (e.g., MMN and P300) and correlate these measures with behavioral performance and psychophysical tests.

Another possible direction is to test interactions between musical expertise and different acoustic and linguistic properties of the speech segmentation input. For instance, it has been shown that lexical stress (including changes in pitch and timing) and subliminal gaps at word boundaries facilitate speech segmentation (e.g., Refs. 36, 41, and 42). Thus, it could be interesting to see to what extent musical practice modifies the sensitivity to these acoustic cues on both pitch and time dimensions.

On a more linguistic side, Tyler and Cutler[43] compared English, French, and Dutch speakers in an artificial language–learning paradigm and reported an interaction of participants' mother tongue and the influence of vowel lengthening and pitch movement on speech segmentation performance. Similarly, a speech stream respecting the native phonotactic rules results in better segmentation than a "foreign" speech stream.[44] These manipulations of the stream could shed new light on the extent to which musical practice may affect language universals and/or language-specific knowledge.

Finally, a very promising direction will be to investigate learning processes *per se*, and to study the familiarization phase using electrophysiological measures (EEG/MEG). This approach has the double advantage of an implicit measure of learning as well as of giving access to the learning curve over time. Interestingly, some studies have reported different learning curves as a function of behavioral performance, contrasting EEGs of high versus low learners.[8,26,36,45] For the present aim, advances in signal-processing methods, such as time-frequency analyses[46] and frequency-tagging analyses,[47] will allow for a better understanding of the cortical processes participating in speech segmentation and will probably turn out to be highly informative when applied to investigate the bases of the differences between musicians and nonmusicians in speech segmentation.

Acknowledgments

Preparation of this paper was supported by a grant from the French National Research Agency to D. Schön and B. Tillmann (ANR Blanc DMBB #NT09˙520631). C. François was a Ph.D. student supported by the ANR-Neuro (#024-01).

Conflicts of interest

The authors declare no conflicts of interest.

References

1. Saffran, J.R., R.N. Aslin & E.L. Newport. 1996. Statistical learning by 8-month old infants. *Science* **274:** 1926–1928.
2. Kuhl, P.K. 2004. Early language acquisition: cracking the speech code. *Nat. Rev. Neurosci.* **5:** 831–843.
3. Rodriguez-Fornells, A., T. Cunillera, A. Mestress-Misse & R. De Diego Balaguer. 2009. Neurophysiological mechanisms involved in language learning in adults. *Phil. Trans. Roy. Soc. B.* **364:** 3711–3735.
4. Saffran, J.R., E. Johnson, R. Aslin & E. Newport. 1999. Statistical learning of tone sequences by human infants and adults. *Cognition* **70:** 27–52.
5. Aslin, R.N., J.R. Saffran & E. Newport. 1998. Computation of conditional probability statistics by 8-month-old infants. *Psychol. Sci.* **9:** 321–324.
6. Gervain, J., F. Macagno, S. Cogoi, *et al.* 2008. The neonate brain detects speech structure. *Proc. of Natl. Acad. Sci.* **105:** 14222–14227.
7. Teinonen, T., V. Fellman, R. Näätänen, *et al.* 2009. Statistical language learning in neonates revealed by event-related brain potentials. *BMC Neurosci.* **13:** 10–21.
8. Abla, D., K. Katahira & K. Okanoya. 2008. On-line assessment of statistical learning by event related potentials. *J. Cogn. Neuro.* **20:** 952–964.
9. Kudo, N., Y. Nonada, N. Mizuno, *et al.* 2011. On-line statistical segmentation of non-speech auditory stream in neonates as demonstrated by event-related brain potentials. *Develop. Sci.* **14:** 1100–1106.
10. Tillmann, B. & S. McAdams. 2004. Implicit learning of musical timbre sequences: statistical regularities confronted with acoustical (dis)-similarities. *J. Exp. Psychol. Learn. Mem. Cogn.* **30:** 1131–1142.
11. Conway, M.C. & M. Christiansen. 2005. Modality-constrained statistical learning of tactile, visual, and auditory sequences. *J. Exp. Psychol. Learn Mem. Cogn.* **31:** 24–39.
12. Hunt, R.H. & R.N. Aslin. 2001. Statistical learning in a serial reaction time task: access to separable statistical cues by individual learners. *J. Exp. Psychol. Gen.* **130:** 658–680.
13. Fiser, J. & R.N. Aslin. 2002. Statistical learning of higher-order temporal structure from visual shape sequences. *J. Exp. Psychol. Learn Mem. Cogn.* **28:** 458–467.
14. Fiser, J. & R.N. Aslin. 2005. Encoding multielement scenes: statistical learning of visual feature hierarchies. *J. Exp. Psychol. Gen.* **134:** 521–537.
15. Schön, D., M. Boyer, S. Moreno, *et al.* 2008. Song as an aid for language acquisition. *Cognition* **106:** 975–983.
16. François, C. & D. Schön. 2010. Learning of musical and linguistic structures: comparing event-related potentials and behavior. *Neuroreport* **21:** 928–932.
17. Sanders, L.D., E.L. Newport & H.J. Neville. 2002. Segmenting nonsense: an event-related-potential index of perceived onsets in continuous speech. *Nat. Neurosci.* **5:** 700–703.
18. François, C. & D. Schön. 2011. Musical expertise boosts implicit learning of both musical and linguistic structures. *Cereb. Cortex.* **21:** 2357–2365.
19. François, C.J. Chobert, M. Besson & D. Schön. Submitted. Music training for the development of language acquisition. *Cereb. Cortex*
20. Hickok, G. & D. Poeppel. 2007. The cortical organization of speech processing. *Nat. Rev. Neurosci.* **8:** 393–402.
21. Kraus, N. & B. Chandrasekaran. 2010. Music training for the development of auditory skills. *Nat. Rev. Neurosci.* **11:** 599–605.
22. Godey, B., D. Schwartz, J.B. De Graaf, *et al.* 2001. Neuromagnetic source localization of auditory evoked fields and intracerebral evoked potentials: a comparison of data in the same patients. *Clin. Neurophysiol.* **112:** 1850–1859.
23. Molholm, S., S. Martinez, W. Ritter, *et al.* 2005. The neural circuitry of pre-attentive auditory change detection an fMRI study of pitch and duration mismatch negativity generators. *Cereb. Cortex* **15:** 545–551.
24. Saur, D., B.W. Kreher, S. Schnell, *et al.* 2008. Ventral and dorsal pathways for language. *Proc. Natl. Acad. of Sci.* **105:** 18035–18040.
25. Flöel, A., M.H. De Vries, J. Scholz, *et al.* 2009. White matter integrity in the vicinity of Broca's area predicts grammar learning success. *Neuroimage* **47:** 1974–1981.
26. Cunillera, T., E. Càmara, J.M. Toro, *et al.* 2009. Time course and functional neuroanatomy of speech segmentation in adults. *Neuroimage* **48:** 541–553.
27. Sluming, V., T. Barrick, M. Howard, *et al.* 2002. Voxel-based morphometry reveals increased gray matter density in Broca's area in male symphony orchestra musicians. *Neuroimage* **17:** 1613–1622.
28. Conway, C., D. Pisoni & W. Kronenberger. 2009. The importance of sound for cognitive sequencing abilities: the auditory scaffolding hypothesis. *Curr. Direct. Psychol. Sci.* **18:** 275–279.
29. Saffran, J.R., E.L. Newport & R.N. Aslin. 1996. Word segmentation: the role of distributional cues. *J. Memory Lang* **35:** 606–621.
30. Perruchet, P. & S. Pacton. 2006. Implicit learning and statistical learning: one phenomenon, two approaches. *Trends Cogn. Sci.* **10:** 233–238.
31. Perruchet, P. 2008. Implicit learning. In *Cognitive Psychology of Memory. Vol.2 of Learning and Memory: A Comprehensive Reference.* J. Byrne, Ed.: 597–621. Elsevier. Oxford.
32. Berry, D.C. & Z. Dienes. 1993. Implicit learning: theoretical and empirical issues. *Essays in Cognitive Psychology.* Lawrence Erlbaum Associates, Inc. Hillsdale, NJ, England.

33. Tillmann, B. & B. Poulin-Charonnat. 2010. Auditory expectations for newly acquired structures. *Quart. J. Exp. Psychol.* **63:** 1646–1664.
34. Omigie, D. & L. Stewart. 2011. Preserved statistical learning of tonal and linguistic material in congenital amusia. *Front. Aud. Cong. Neurosci.* **2:** 1–11.
35. Perruchet, P. & B. Tillmann. 2010. Exploiting multiple sources of information in learning an artificial language: human data and modelling. *Cogn. Sci.* **34:** 255–285.
36. De Diego Balaguer, R., J.M. Toro, A. Rodriguez-Fornells & A.C. Bachoud-Lévi. 2007. Different neurophysiological mechanisms underlying word and rule extraction from speech. *PLoS One* **2:** e1175.
37. Mirman, D., K. Graf Estes & J.S. Magnuson. 2010. Computational modeling of statistical learning: effects of transitional probability versus frequency and links to word learning. *Infancy* **15:** 471–486.
38. Loui, P., D.L. Wessel & C.L. Hudson Kam. 2010. Humans rapidly learn grmmatical structure in a new musical scale. *Music Percept.* **27:** 377–388.
39. Reber, R. & P. Perruchet. 2003. The use of control groups in artificial grammar learning. *Quart. J. Exp. Psychol.* **56A:** 97–115.
40. Perruchet, P., M.D. Tyler, N. Galland & R. Peereman. 2004. Learning nonadjacent dependencies: no need for algebraic-like computations. *J. Exp. Psychol. Gen.* **133:** 573–583.
41. Rohrmeier, M., P. Rebuschat & I. Cross. 2011. Incidental and online learning of melodic structure. *Conscious. Cogn.* **20:** 214–222.
42. Cunillera, T., J.M. Toro, N. Sebastián-Gallés & A. Rodríguez-Fornells. 2006. The effects of stress and statistical cues on continuous speech segmentation: an event-related brain potential study. *Brain Res.* **1123:** 168–178.
43. Tyler, M. & A. Cutler. 2009. Cross-language differences in cue use for speech segmentation. *J. Acoust. Soc. Am.* **126:** 367–376.
44. Cunillera, T., A. Gomila & A. Rodríguez-Fornells. 2008. Beneficial effects of word final stress in segmenting a new language: evidence from ERPs. *BMC Neurosci.* **9:** 23.
45. Mersad, K. & T. Nazzi. 2011. Transitional Probabilities and positional frequency phonotactics in a hierarchical model of speech segmentation. *Memory Cogn.* **39:** 1085–1093.
46. De Diego Balaguer, R., L. Fuentemilla & A. Rodríguez-Fornells. 2011. Brain dynamics sustaining rapid rule extraction from speech. *J. Cogn. Neurosci.* **23:** 3105–3120.
47. Buiatti, M., M. Peña & G. Dehaene-Lambertz. 2009. Investigating the neural correlates of continuous speech computation with frequency-tagged neuroelectric responses. *Neuroimage* **44:** 509–519.

Ann. N.Y. Acad. Sci. ISSN 0077-8923

ANNALS OF THE NEW YORK ACADEMY OF SCIENCES
Issue: *The Neurosciences and Music IV: Learning and Memory*

Musical expertise induces neuroplasticity of the planum temporale

Martin Meyer,[1,2] Stefan Elmer,[3] and Lutz Jäncke[2,3]

[1]Research Unit for Neuroplasticity and Learning in the Healthy Aging Brain, University of Zurich, Zurich, Switzerland. [2]Center for Integrative Human Physiology, University of Zurich, Zurich, Switzerland. [3]Division of Neuropsychology, University of Zurich, Zurich, Switzerland

Address for correspondence: Martin Meyer, Institute for Psychology, University of Zurich, Sumatrastrasse 30, CH-8006 Zurich, Switzerland. mmeyer@access.uzh.ch

The present manuscript summarizes and discusses the implications of recent neuroimaging studies, which have investigated the relationship between musical expertise and structural, as well as functional, changes in an auditory-related association cortex, namely, the planum temporale (PT). Since the bilateral PT is known to serve as a spectrotemporal processor that supports perception of acoustic modulations in both speech and music, it comes as no surprise that musical expertise corresponds to functional sensitivity and neuroanatomical changes in cortical architecture. In this context, we focus on the following question: To what extent does musical expertise affect the functioning of the left and right plana temporalia? We discuss the relationship between behavioral, hemodynamic, and neuroanatomical data obtained from musicians in light of maturational and developmental issues. In particular, we introduce two studies of our group that show to what extent brains of musicians are more proficient in phonetic task performance.

Keywords: transfer effects; hemispheric lateralization; planum temporale; auditory neuroplasticity; voice onset time; musical expertise

Introduction

Brain function and anatomy in musicians is an insightful topic for studying the mechanisms of neuroplasticity.[1–3] As compared to nonmusicians, musicians perform better on elementary auditory and motor tasks, thereby, demonstrating the superiority of their auditory and motor systems.[4–10] It has been shown that musical training also facilitates the processing of linguistic and metalinguistic information, in particular, prosody.[11–14] These transfer effects provide striking evidence for the view that music and language share at least partially overlapping neural underpinnings.[15–17] Beyond the primary motor cortex, it has been evidenced that regions accommodating the auditory system and territories that are involved in audio–motor coupling support receptive and expressive functions of both speech and music processing.[18,19] The planum temporale (PT), a part of the auditory-related cortex, plays a key role in subserving auditory functions that underlie music and speech processing.

The PT: neuroanatomy and function

The PT is situated caudally to the transverse temporal gyrus in the left and right hemisphere.[20] It covers the posterior surface of the supratemporal plane[21] and demonstrates a strong structural leftward asymmetry that is independent of biological sex, handedness, and language lateralization.[22–24] Developmental malformations or lack of asymmetry of the PT have been previously associated with neurological dysfunctions, such as developmental stuttering, developmental dyslexia, and psychiatric disorders.[25–28] However, these observations are only somewhat supported by previous research endeavors because the dysfunction of one sole area does not sufficiently account for the complexity of symptoms associated with most developmental disorders. Various researchers have suggested that the PT coincides, at least partially, with the Wernicke area in the left hemisphere. Since the notion of "Wernicke area" is a neuroanatomically ill-defined concept and because there is still disagreement as regards the

doi: 10.1111/j.1749-6632.2012.06450.x
Ann. N.Y. Acad. Sci. 1252 (2012) 116–123

precise location and boundaries of this area,[29] we prefer to refrain completely from using this term in the present publication.

Once the morphological leftward asymmetry of the PT was discovered it was enthusiastically dubbed the "temporal speech region."[30] This may be due to it being located caudally to the primary and secondary auditory cortex, thereby suggesting the functionally and structurally dense intertwinement between the supratemporal auditory fields. Following this line of reasoning, it may be postulated that inflowing speech signals enter the primary auditory cortex bilaterally and are redirected to the left planum temporale for further analysis at higher linguistic levels. This argumentation had long been the dominant model of the brain–language relationship until several empirical observations called this view into question. A seminal study by Schlaug *et al.*,[31] however, provided evidence for a general association of auditory functions with the PT. The authors observed a significantly stronger leftward asymmetry of the PT surface in musicians with absolute pitch compared to musicians with relative pitch and nonmusicians. Based on these findings they proposed that the PT should not be considered a "speech center" or a cortical module that selectively performs linguistic tasks. Meanwhile, it is undisputed that the PT is involved in numerous auditory,[32] audio-visual,[33] purely visual,[34] sensory-motor integration,[35] and even linguistic functions.[36] In a highly cited metaanalysis,[37] researchers investigated the recruitment of the bilateral PT in a number of various tasks; they concluded that the PT should be primarily considered a spectro-temporal "computational hub." In other words, the neural ensembles that make up both the left and the right PT are preferentially driven by spectro-temporal modulations of the acoustic signal. Once the acoustic signal has passed the primary and secondary auditory cortices it enters the auditory-related cortex in the bilateral PT in order to undergo further steps of spectro-temporal signal decoding. Even though the PT is no longer considered the sole speech region in humans, novel striking evidence for a robust relationship between anatomical and functional data related to speech processing became evident.[38] The correspondence between functional and structural leftward asymmetry implicates that some aspects of speech perception are mediated via the left PT.[36,39–41]

The PT as a fundamental processor for speech

To more thoroughly explore the role of the PT in speech perception, Jäncke *et al.* performed a series of functional magnetic resonance imaging (fMRI) studies.[40] In these studies participants were presented with the following acoustic stimuli: consonant–vowel (CV) syllables, tones, white noise, and vowels. The results of this investigation indicated a specialization of the left PT for decoding phonetic features, namely, the voice-onset time (VOT). The VOT can be defined as the temporal interval between the release of a stop consonant and the beginning of the vibration of the vocal folds. Thus, the left PT is preferentially driven by rapidly changing acoustic cues. The identification of such cues is essential for speech processing, for example, perceiving the difference between the syllables /da/ and /ta/. Therefore, subtle temporal modulations of the acoustic signal could be considered the foundation of speech processing. This observation is in accordance with previous clinical studies that observed deficient temporal decoding in patients with acquired aphasia following left hemisphere impairment.[42] Zaehle *et al.*[41] have performed a follow-up fMRI study during which volunteers listened to acoustic stimuli that were either tied up with linguistic content (CV syllables) or not (acoustic noise with varying brief silent periods that correspond to the varying VOT in syllables). The data analysis revealed left-sided responses in the primary and posterior association cortex (PT) during the perception of rapidly changing temporal information in speech and nonspeech sounds. Another fMRI study by Meyer *et al.*[43] provided further corroboration for the role the left posterior auditory-related cortex in brief auditory events. The confluent results of these studies support the central statement of a neurobiological framework, namely the "asymmetric sampling in time" (AST) hypothesis.[44] According to this framework, the auditory association cortex, presumably the PT, is asymmetrically driven by temporal modulations in acoustic signals. While the left PT is more strongly involved in decoding (sub)segmental rapidly acoustic changing cues (ca. 40 Hz), the contralateral area primarily supports the processing of suprasegmental (slowly) changing acoustic cues (ca. 4 Hz). In linguistic terms, this means that the analysis of fine-grained acoustic information, which is important for phonetic perception, is a left-dominant function. In addition, the right auditory-related cortex mediates acoustic patterns that span over several speech units; these suprasegmental patterns are essential for decoding prosodic and rhythmic information. Two previous fMRI studies have provided support for the latter proposal in that they have shown the specific involvement of the right PT during the processing of speech melody[45] and speech rhythm.[46]

One may wonder how this knowledge fits into the frequently sketched notion that "language" per se resides in the brain's "left hemisphere" and that "music" *per se* is situated in the "right hemisphere." First and foremost, the two hemispheres do not work in distinct isolation no matter what task they are currently engaged in. It is not plausible to assume that music or speech perception is a function that merely resides in one sole hemisphere. It must be emphasized that even asymmetrically dominant tasks are partially supported by the contralateral hemisphere. Second, it does not make any sense to consider one hemisphere as a distinct entity. The cerebral hemisphere is made up of gray matter portions and white matter fiber tracts that form cortical ribbons and subcortical nuclei. Some of these cortical and subcortical regions form functional circuits that preferentially support sensory and cognitive functions. Thus, it is not accurate to use the term "hemisphere" in a global sense in the context of functional neuroanatomy of language and music. In terms of language, it has been shown that expressive and receptive language functions populate the banks of the left Sylvian fissure (sulcus lateralis) and are situated along the smaller ascending and horizontal rami that branch off from the Sylvian fissure.[47] Even though the left perisylvian region is the dominant region for language processing,[48,49] it must not be ignored that the contralateral cortex also mediates linguistic and paralinguistic functions.[50–52] Subcortical nuclei, such as the neostriatum,[53] and the massive white matter tracts[54–59] complement the language network. Interestingly, expressive and receptive musical functions have been shown to recruit perisylvian areas that overlap with regions involved in supporting linguistic functions at the elementary level.[60] Nevertheless, as regards music, a moderate dominance of right supratemporal sections has been noted.[61] Third, it is important to understand that music and language cannot be considered monolithic entities, but should rather be decomposed into different subcomponents or more fine-grained levels of processing.[15] Sub- and suprasegmental signal modulations available in speech and music are crucial for the processing of acoustic qualities, namely, melodic contour, timbre, phrasing, and rhythm. Whether the left or the right auditory cortex is more strongly involved in the processing of speech and music is not dependent on the domain of the signal. This can be shown by simply exemplifying what happens when individuals hear sung speech, for example, an aria. It comes as no surprise that both the left and the right perisylvian cortex are involved in the processing of an aria.[18] According to the aforementioned "AST" framework, temporal modulations in the acoustic signal patterns should be considered relevant for the issue of hemispheric lateralization. Thus, the AST framework accounts not only for an improved understanding of the speech–brain relationship, but also better elucidates the music–brain relationship. Since suprasegmental modulations of the melodic contour represent the constitutive hallmark in music, a functional rightward asymmetry of the processing of acoustical features that span over several acoustic segments can be expected.[62]

A myriad of studies published in the last decade have shown that individuals who were musically trained or who are professional musicians process musical (and speech) signals differently than individuals who lack this expertise.[1,2,63] In the context of the present topic, one fMRI study by Ohnishi *et al.* is of particular interest.[64] According to this study, the left PT is more strongly involved when highly skilled musicians simply listen to a piece of piano music; however, untrained participants exhibited an increased rightward lateralization of brain responses in the PT. This result demonstrates neural plasticity in the posterior auditory-related cortex, which has been induced by musical expertise. It is possible that highly proficient musicians scan the incoming acoustic signal with higher temporal resolution in order to process the music in a more fine-grained mode, which permits the analysis of subtle information beyond the sole appreciation of melody. This reasoning is in line with the central tenant of the AST model, which states that any acoustic signal is processed via two temporal resolution frames. Due to the neuroplasticity of the auditory system, individual hearing acuity and musical expertise may affect the standard mode of the "division of labor" between the two hemispheres. In musicians, the development of exceptional hearing skills has formed both the left and right auditory-related cortex so that decoding of rapidly changing acoustic information in music becomes just as important as the processing of the melodic contour. The musicians who partook in the study of Ohnishi *et al.*[64]—unlike the nonmusicians—may have more carefully computed the subtlety of the music's "phonology," and they may have benefited from the enhanced fine-grained auditory analysis that is mediated by the left PT. Indeed, there is empirical evidence demonstrating neuroplastic effects in the left PT and in the adjacent auditory regions in musicians' brains,[65] and especially in absolute pitch possessors.[66,67]

Having emphasized the relationship between musical expertise, enhanced auditory skills, and neuroplastic changes of the PT, one may wonder whether this amalgamation may also lead to specific conditions for speech perception in musicians. To address this issue, we performed two studies (S. Elmer, J. Hänggi, M. Meyer & L. Jäncke, in preparation).[63] In the first investigation we used fMRI and scanned musicians and nonmusicians who were presented with the same rapidly changing speech and nonspeech stimuli as used by Zaehle *et al.*,[41] that is, voiceless /ka/ and voiced /da/ German CV syllables, as well as stimuli that were synthesized from the two CV syllables resulting in white noise analogs of speech with reduced spectrum, but preserved temporal structure. All participants were asked to perform a phonetic categorization task, namely to decide whether a heard stimulus belongs to the voiced category or to the voiceless category. The results of this categorization task demonstrated convincingly that musicians are better at decoding subtle temporal acoustic cues irrespective of hearing speech or noisy analogs with reduced acoustic spectrum. It should be noted that the musicians were significantly superior to the nonmusicians. In terms of functional brain imaging, the musicians—in agreement with our predictions—demonstrated magnified responses in the left PT when they heard and then categorized the speech syllables and the noisy sounds.

Furthermore, we observed an intriguing positive correlation between hemodynamic parameters for the two noisy conditions and behavioral data. The better the behavioral performance in the phonetic categorization task, the higher the strength of the fMRI responses. Moreover, a nonparametric correlational analysis within the group of musicians revealed a relationship between the estimated number of hours of individual musical training across the life span and BOLD response in the left PT. Thus, this study provides several interesting results, three of which we shall now focus on. First, from the positive correlation between performance data and hemodynamic signal strength we learn that an increase of brain responses in the PT appears to reflect competence ("working better") rather than incompetence ("working harder"). Secondly, since the PT was the sole brain region displaying the patterns described above, one may reason that the left PT acts as a primary processor of rapidly changing acoustic cues in speech and nonspeech signals. Third, the correlation seen between the amount of training and the strength of brain responses clearly supports the view that the musicians' brain is a result of learning-induced neuroplasticity, rather than the product of connate talent. In a second study, we investigated the relationship between the anatomical architecture of the left PT and performance in the phonetic categorization task (S. Elmer, J. Hänggi, M. Meyer & L. Jäncke, in preparation). As predicted, the musicians demonstrated an enlargement of the left PT surface area relative to nonmusicians. Furthermore, when only considering the relationship between neuranatomical architecture and behavior in the musicians' group we noted a positive correlation between the left PT surface area and behavioral accuracy in categorizing spectrally reduced stimuli. Consequently, this additional analysis lends further credence to the argument that function and structure of the left PT may evolve as a function of long-term musical training. These neuroplastic changes may not only affect musical skills, but they may also impact elementary auditory functions that could be considered a major foundation of speech.

(Comparative) neuroanatomy

One may argue that the exceptional leftward structural PT asymmetry observed in musicians[31] reflects an anatomical disposition for musical expertise. In other words, one may proclaim that the changes in the left PT do not occur as a function of long-term musical training but already exist in human newborns. Along this line, the larger the left PT in terms of surface or volume, the more likely it is that an individual becomes a proficient musician. However, based on sole neuroanatomical asymmetry it is not possible to predict musical proficiency. In fact, the leftward asymmetry of the PT is already manifest during the 29th to 31st week of gestation.[68,69] According to one recent *in utero* investigation of temporal lobe asymmetry, structural brain asymmetry is determined early in infancy and appears to be related to asymmetric hemispheric gene expression as early as the 12th week of gestation.[70] However, as emphasized by Steinmetz,[69] one must be aware that the brain of an unborn human child matures and develops as a function of both genetic blueprints and prenatal sensory stimulation, such as hearing voices, music, or any other sound. Thus, determining to which extent genes, environment, or the interaction between genes and external stimulation affects PT architecture is almost impossible. Secondly, when reasoning about PT asymmetry, one must keep in mind that differential anatomical parameters, namely cortical volume, cortical surface, and cortical thickness may be linked to various maturational and developmental issues (M. Meyer, L. Jäncke, S. Hirsiger & J. Hänggi, under review).[71] Notwithstanding these discrepancies, the majority of studies in the context of *in vivo* morphometry consider cortical surface area to be the most reliable correspondent to functional issues. The compilation of data reviewed by Shapleske *et al.*[21] suggests that the PT surface area most strongly accounts for the structural asymmetry not only in musicians, but also in the majority of individuals (~78%) who have not received musical training. Thus, it is obvious that a leftward structural PT surface asymmetry exists in almost all human individuals and that this asymmetry may reflect the importance of phonetic competence as a key skill for speech comprehension. Recent cytoarchitectonic and imaging studies in humans demonstrated that cortical columns in the left PT area are more widely spaced apart, less densely interconnected; however, they appear to be more strongly connected via large-scale white matter tracts that results in a larger surface or volume.[50,72] Musical training alters this relationship because it increases the asymmetry by enhancing the PT surface, thereby providing more computational resources for a fine-grained auditory analysis of acoustic signals, which in turn is beneficial for phonetic competence. This dynamic interplay between behavior and structural and anatomical asymmetry explains the aforementioned results provided by Elmer *et al.* (S. Elmer, J. Hänggi, M. Meyer & L. Jäncke, in preparation).[63]

When comparing this evidence with reports of PT asymmetry in great apes, one notes that these animals demonstrate a human-like anatomical macroscopic pattern of the PT.[73,74] How can these observations be reconciled with the notion of PT (asymmetry) as a decisive morphological device for phonetic competence in humans? As outlined above, phonetic competence is based on the skill to decode fine-grained rapidly changing acoustic information at the subsegmental level. Akin to humans, auditory perception in great apes is reliant upon the same elementary mechanisms. Monkeys[75] and other mammals, for example rodents,[76] have demonstrated similar leftward hemispheric preferences when presented with rapidly changing acoustic cues. At least with respect to humans and great apes it cannot be disputed that PT surface asymmetry and specific aspects of categorial perception of fine-grained acoustic signal cues are closely linked. However, this link alone is not a sufficient explanation for why humans possess the ability to use language. Even though it appears that humans and apes share the biological foundation for the elementary levels of speech decoding, it cannot be ignored that the discrimination of phonemes and syllables and the identification of these stimuli as linguistically relevant information is unique to humans. Cytoarchitectonic studies may provide important insight. Similar to the macroscopic leftward PT asymmetry, research pertaining to cytoarchitecture of the posterior auditory region has shown leftward asymmetry.[77–80] In particular, microscopic parameters, such as minicolumn spacing, the peripheral neuropil, and the spacing density of neurons within columns, also demonstrate a leftward asymmetry in humans,[81] which was absent in chimpanzees and rhesus monkeys.[82] Presently, it is difficult to reason about the implications of these studies. Perhaps the specific architecture of the left posterior auditory association cortex with distant microcolumns is advantageous for the rapid and efficient recognition of temporal patterns in acoustic signals that convey linguistically relevant information, namely phonetic cues. In any case these findings support the view that the human brain does not contain a particular cortical area that houses the faculty of language. Instead, it is likely that the evolution of language went along with a "rewiring" of already existing areas and the establishment of large-scale neural connections. Whether the evolution of music may have reinforced these neuroanatomical modifications cannot be reliably evaluated based on the present data at hand.

Conclusions

The present article reviews several studies that have provided behavioral, functional brain imaging, and neuroanatomical evidence for a specification of the left PT for fine-grained auditory analysis. While this specification has long been considered a marker for the left hemisphere's dominance for language functions, it is now evident that more universal auditory and less language-specific preferences may account for this asymmetry. In particular, recent studies conducted by our research team that investigated the relationship between anatomical and hemodynamic parameters, as well as phonetic categorization skills in proficient musicians have demonstrated a dense relationship between musical experience, performance in phonetic processing, and leftward asymmetry of PT functional neuroanatomy. Thus, the macro- and microscopic modifications of the PT that have been evidently induced by musical experience could be considered an interesting demonstration for long-term neuroplasticity and the potentially beneficial effects of musical training for elementary speech mechanisms.

Acknowledgments

This research was supported by the Swiss National Foundation (Grant no. 320030-120661 to MM) and by the "Fonds zur Förderung des akademischen Nachwuchses" (FAN) des "Zürcher Universitätsvereins" (ZUNIV). We are indebted to Nathalie Giroud, Sarah McCourt-Meyer, Angela Müller, and Mathias Oechslin for helpful comments on an earlier version of this manuscript.

Conflicts of interest

The authors declare no conflicts of interest.

References

1. Jäncke, L. 2009. The plastic human brain. *Restor. Neurol. Neurosci.* **27:** 521–538.
2. Münte, T.F., E. Altenmüller & L. Jäncke. 2002. The musician's brain as a model of neuroplasticity. *Nat. Neurosci.* **3:** 473–478.
3. Wu, C.Y. & G. Schlaug. 2010. Music making as a tool for promoting brain plasticity across the life span. *Neuroscientist* **16:** 1123–1130.
4. Tervaniemi, M., V. Just, S. Koelsch, *et al.* 2005. Pitch discrimination accuracy in musicians vs nonmusicians: an event-related potential and behavioral study. *Exp. Brain Res.* **161:** 1–10.
5. Fujioka, T., L.J. Trainor, B. Ross, *et al.* 2004. Musical training enhances automatic encoding of melodic contour and interval structure. *J. Cog. Neurosci.* **16:** 1010–1021.
6. Meyer, M., S. Baumann & L. Jöncke. 2006. Electrical brain imaging reveals spatio-temporal dynamics of timbre perception in humans. *NeuroImage* **32:** 1510–1532.
7. Jäncke, L., G. Schlaug & H. Steinmetz. 1997. Hand skill asymmetry in professional musicians. *Brain Cogn.* **34:** 424–432.
8. Jäncke, L., S. Baumann, S. Koeneke, *et al.*, 2006. Neural control of playing a reversed piano: empirical evidence for an unusual cortical organization of musical functions. *NeuroReport* **17:** 447–451.
9. Meyer, M., S. Elmer, M. Ringli, *et al.* 2011. Long-term exposure to music enhances the sensitivity of the auditory system in children. *Eur. J. Neurosci.* **34:** 755–765.
10. Musacchia, G., M. Sams, E. Skoe & N. Kraus. 2007. Musicians have enhanced subcortical auditory and audiovisual processing of speech and music. *Proc. Natl. Acad. Sci. USA* **104:** 15894–15898.
11. Moreno, S., C. Marques, A. Santos, *et al.* 2009. Musical training influences linguistic abilities in 8-year-old children: more evidence for brain plasticity. *Cereb. Cortex* **19:** 712–723.
12. Marques, C., S. Moreno, S.L. Castro & M. Besson. 2007. Musicians detect pitch violations in a foreign language better than nonmusicians: behavioral and electrophysiological evidence. *J. Cog. Neurosci.* **19:** 1453–1463.
13. Francois, C. & D. Schon. 2011. Musical expertise boosts implicit learning of both musical and linguistic structures. *Cereb. Cortex* **21:** 2357–2365.
14. Oechslin, M., M. Meyer & L. Jäncke. 2010. Absolute pitch – functional evidence of speech relevant auditory acuity. *Cereb. Cortex* **20:** 447–455.
15. Besson, M. & D. Schon. 2001. Comparison between language and music. *Ann. N.Y. Acad. Sci.* **930:** 232–258.
16. Patel, A.D. 2003. Language, music, syntax and the brain. *Nat. Rev. Neurosci.* **6:** 674–680.
17. Patel, A.D. 2011. Why would musical training benefit the neural encoding of speech? The OPERA hypothesis. *Folia Phoniatr.* **2:** 142.
18. Kleber, B., N. Birbaumer, R. Veit, *et al.* 2007. Overt and imagined singing of an Italian aria. *NeuroImage* **36:** 889–900.
19. Tervaniemi, M., A. Castaneda, M. Knoll & M. Uther. 2006. Sound processing in amateur musicians and nonmusicians: event-related potential and behavioral indices. *NeuroReport* **17:** 1225–1228.
20. Westbury, C.F., R.J. Zatorre & A.C. Evans. 1999. Quantifying variability in the planum temporale: a probability map. *Cereb. Cortex* **9:** 392–405.
21. Shapleske, J., S.L. Rossell, P.W.R. Woodruff & A.S. David. 1999. The planum temporale: a systematic, quantitative

review of its structural, functional and clinical significance. *Brain Res. Rev.* **29:** 26–49.
22. Dos Santos Sequeira, S., W. Woerner, C. Walter, *et al.* 2006. Handedness, dichotic-listening ear advantage, and gender effects on planum temporale asymmetry: a volumetric investigation using structural magnetic resonance imaging. *Neuropsychologia* **44:** 622–636.
23. Dorsaint-Pierre, R., V.B. Penhune, K.E. Watkins, *et al.* 2006. Asymmetries of the planum temporale and Heschl's gyrus: relationship to language lateralization. *Brain* **129:** 1164–1176.
24. Sommer, I.E., M. Aleman, A. Somers, *et al.* 2008. Sex differences in handedness, asymmetry of the planum temporale and functional language lateralization. *Brain Res.* **1206:** 76–88.
25. Foundas, A.L., A.M. Bollich, J. Feldman, *et al.* 2004. Aberrant auditory processing and atypical planum temporale in developmental stuttering. *Neurology* **63:** 1640–1646.
26. Galaburda, A.M., G.F. Sherman, G.D. Rosen, *et al.* 1985. Developmental dyslexia: four consecutive patients with cortical anomalies. *Ann. Neurol.* **18:** 222–233.
27. Heiervang, E., K. Hugdahl, H. Steinmetz, *et al.* 2000. Planum temporale, planum parietale and dichotic listening in dyslexia. *Neuropsychologia* **38:** 1704–1713.
28. Petty, R.G., P.E. Barta, G.D. Pearlson, *et al.* 1995. Reversal of asymmetry of the planum temporale in schizophrenia. *Am. J. Psychiatr.* **152:** 715–721.
29. Bogen., J.E. & G.M. Bogen. 1976. Wernicke's region—Where is it? *Ann. N.Y. Acad. Sci.* **280:** 834–843.
30. Galaburda, A., F. Sanides & N. Geschwind. 1978. Human brain. Cytoarchitectonic left–right asymmetries in the temporal speech region. *Arch. Neurol.* **35:** 812–817.
31. Schlaug, G., L. Jäncke, Y. Huang & H. Steinmetz. 1995. In vivo evidence of structural brain asymmetry in musicians. *Science* **267:** 699–701.
32. Mustovic, H., K. Scheffler, F. Di Salle, *et al.* 2003. Temporal integration of sequential auditory events: silent period in sound pattern activates human planum temporale. *NeuroImage* **20:** 429–434.
33. Pekkola, J., V. Ojanen T., Auhi T.,*et al.* 2006. Attention to visual speech gestures enhances hemodynamic activity in the left planum temporale. *Hum. Brain Mapp.* **27:** 471–477.
34. Petitto, L.A., R.J. Zatorre, K. Gauna, *et al.* 2000. Speech-like cerebral activity in profoundly deaf people processing signed languages: implications or the neural basis of human language. *Proc. Natl. Acad. Sci. USA* **97:** 13961–13966.
35. Isenberg, A.L., K.I. Vaden, K. Saberi, *et al.* 2011. Functionally distinct regions for spatial processing and sensory motor integration in the planum temporale. *Hum. Brain Mapp*, doi:10.1177/1073858407305726.
36. Xu, Y., J. Gandour, T. Talavage, *et al.* 2006. Activation of the left planum temporale in pitch processing is shaped by language experience. *Hum. Brain Mapp.* **27:** 173–183.
37. Griffiths, T.D. & J.D. Warren 2002. The planum temporale as a computational hub. *Trends Neurosci.* **25:** 348–353.
38. Josse, G., B. Mazoyer., F. Crivello & N. Tzourio-Mazoyer 2003. Left planum temporale: an anatomical marker of left hemispheric specialization for language comprehension. *Cogn. Brain Res.* **18:** 1–14.
39. Binder, J.R., J.A. Frost, T.A. Hammeke, *et al.* 1996. Function of the left planum temporale in auditory and linguistic processing. *Brain* **119:** 1239–1254.
40. Jäncke, L., T. Wüstenberg, H. Scheich & H.-J. Heinze. 2002. Phonetic perception and the temporal cortex. *NeuroImage* **15:** 733–746.
41. Zaehle, T., T. Wüstenberg, M. Meyer & L. Jäncke. 2004. Evidence for rapid auditory perception as the foundation of speech processing—a sparse temporal sampling fMRI study. *Eur. J. Neurosci.* **20:** 2447–2456.
42. Efron, R. 1963. Temporal perception, aphasia amd d'eja vu. *Brain* **86:** 403–424.
43. Meyer, M., S. Zysset, D.Y. von Cramon & K. Alter. 2005. Distinct fMRI responses to laughter, speech, and sounds along the human perisylvian cortex. *Cogn. Brain Res.* **24:** 291–306.
44. Poeppel, D. 2003. The analysis of speech in different temporal integration windows: cerebral lateralization as 'asymmetric sampling in time'. *Speech Commun.* **41:** 245–255.
45. Meyer, M., K. Steinhauer, K. Alter, *et al.* 2004. Brain activity varies with modulation of dynamic pitch variance in sentence melody. *Brain Lang.* **89:** 277–289.
46. Geiser, E., T. Zaehle, L. Jäncke & M. Meyer. 2008. The neural correlate of speech rhythm as evidenced by metrical speech processing: an fMRI study. *J. Cogn. Neurosci.* **20:** 541–552.
47. Vigneau, M., V. Beaucousin, P.Y. Hervé, *et al.* 2006. Meta-analyzing left hemisphere language areas: phonology, semantics, and sentence processing. *NeuroImage* **30:** 1414–1432.
48. Kell, C.A., B. Morillon, F. Kouneiher & A.-L. Giraud 2011. Lateralization of speech production in sensory cortices—a possible sensory origin of cerebral left dominance for speech. *Cereb. Cortex* **21:** 932–937.
49. Tzourio-Mazoyer, N., G. Simon, F. Crivello, *et al.* 2010. Effect of familial sinistrality on planum temporale surface and brain tissue asymmetries. *Cereb. Cortex* **20:** 1476–1485.
50. Jung-Beeman, M. 2005. Bilateral brain processes for comprehending natural language. *Trends Cogn. Sci.* **9:** 512–518.
51. Perani, D., M.C. Saccuman, P. Scifo, *et al.* 2011. Neural language networks at birth. *Proc. Natl. Acad. Sci. USA* **108:** 16056–16061.
52. Vigneau, M., V. Beaucousin, P.Y. Hervé, *et al.* 2011. What is right-hemisphere contribution to phonological, lexico-semantic, and sentence processing? Insights from meta-analysis. *NeuroImage* **54:** 577–593.
53. Kotz, S.A. & M. Schwartze. 2010. Cortical speech processing unplugged: a timely subcortico-cortical framework. *Trends Cogn. Sci.* **14:** 392–399.
54. Elmer, S., J. Hänggi, M. Meyer & L. Jäncke. 2011. Differential language expertise related to white matter architecture in regions subserving sensory-motor coupling, articulation, and interhemispheric transfer. *Hum. Brain Mapp.* **32:** 2064–2074.
55. Frey, S., J.S.W. Campbell, G.B. Pike & M. Petrides. 2008. Dissociating the human language pathways with high angular resolution diffusion fiber tractography. *J. Neurosci.* **28:** 11435–11444.
56. Friederici, A.D. 2009. Pathways to language: fiber tracts in the human brain. *Trends Cogn. Sci.* **13:** 175–181.
57. Glasser, M.F. & J.K. Rilling. 2008. DTI tractography of the human brain's language pathways. *Cereb. Cortex* **18:** 2471–2482.

58. Makris, N. & D.N. Pandya. 2009. The extreme capsule in humans and rethinking of the language circuitry. *Brain Struct. Funct.* **213:** 343–358.
59. Upadhyay, J., K. Hallock, M. Ducros, *et al.* 2008. Diffusion tensor spectroscopy and imaging of the arcuate fasciculus. *NeuroImage* **39:** 1–6.
60. Brown, S., M.J. Martinez & L.M. Parsons. 2006. Music and language side by side in the brain: a PET study of the generation of melodies and sentences. *Eur. J. Neurosci.* **23:** 2791–2803.
61. Janata, P., J.L. Birk, J.D. Van Horn, *et al.* 2002. The cortical topography of tonal structures underlying Western music. *Science* **298:** 2167–2170.
62. Boemio, A., A. Fromm, A. Braun & D. Poeppel 2005. Hierarchical and asymmetric temporal sensitivity in human auditory cortices. *Nat. Neurosci.* **8:** 389–395.
63. Elmer, S., M. Meyer & L. Jöncke. 2011. Neurofunctional and behavioral correlates of phonetic and temporal categorization in musically trained and untrained subjects. *Cereb. Cortex*, doi:10.1093/cercor/bhr142.
64. Ohnishi, T., H. Matsuda, T. Asada, *et al.* 2001. Functional anatomy of musical perception in musicians. *Cereb. Cortex* **11:** 754–760.
65. Bermudez, P., J.P. Lerch, A.C. Evans & R.J. Zatorre. 2009. Neuroanatomical correlates of musicianship as revealed by cortical thickness and voxel-based morphometry. *Cereb. Cortex* **19:** 1583–1596.
66. Hirata, Y., S. Kuriki & C. Pantev. 1999. Musicians with absolute pitch show distinct neural activities in the auditory cortex. *NeuroReport* **10:** 999–1002.
67. Wilson, D., S.J. Lusher, C.Y. Wan, *et al.* 2009. The neurocognitive components of pitch processing: insights from absolute pitch. *Cereb. Cortex* **19:** 724–732.
68. Dubois, J., M. Benders, F. Lazeyras, *et al.* 2010. Structural asymmetries of perisylvian regions in the preterm newborn. *NeuroImage* **52:** 32–42.
69. Steinmetz, H. 1996. Structure, function and cerebral asymmetry: in vivo-morphometry of the planum temporale. *Neurosci. Biobehav. Rev.* **20:** 587–591.
70. Kasprian, G., G. Langs, P.C. Brugger, *et al.* 2011. The prenatal origin of hemispheric asymmetry: an in utero neuroimaging study. *Cereb. Cortex* **21:** 1076–1083.
71. Panizzon, M.S., C. Fennema-Notestine, L.T. Eyler, *et al.* 2009. Distinct genetic influences on cortical surface area and cortical thickness. *Cereb. Cortex* **19:** 2728–2735.
72. Sigalovsky, I.S., B. Fischl & J.R. Melcher. 2006. Mapping an intrinsic MR property of gray matter in auditory cortex of living humans: a possible marker for primary cortex and hemispheric differences. *NeuroImage* **32:** 1524–1537.
73. Cantalupo, C., D.L. Pilcher & W.D. Hopkins. 2003. Are planum temporale and Sylvian fissure asymmetries directly related? A MRI study in great apes. *Neuropsychologia* **41:** 1975–1981.
74. Gannon, P.J., R.L. Holloway, D.C. Broadfield & A.R. Braun. 1998. Asymmetry of the planum temporale: human-like pattern of Wernicke's brain language area homolog. *Science* **279:** 220–222.
75. Steinschneider, M., I.O. Volkov, Y.I. Fishman, *et al.* 2005. Intracortical responses in human and monkey primary auditory cortex support a temporal processing mechanism for encoding of the voice onset time phonetic parameter. *Cereb. Cortex* **15:** 170–186.
76. Wetzel, W., F.W. Ohl & H. Scheich. 2008. Global versus local processing of frequency-modulated tones in gerbils: an animal model of lateralized auditory cortex functions. *Proc. Natl. Acad. Sci. USA* **105:** 6753–6758.
77. Chance, S.A., M.F. Casanova, A.E. Switala & T.J. Crow. 2006. Minicolumnar structure in Heschl's gyrus and planum temporale: asymmetries in relation to sex and fiber number. *Neuroscience* **143:** 1041–1050.
78. Galuske, R.A.W., W. Schlote, H. Bratzke & W. Singer. 2000. Interhemispheric asymmetries of the modular structure in human temporal cortex. *Science* **289:** 1946–1949.
79. Hutsler, J. & R.A.W. Galuske. 2003. The specialized structure of human language cortex: pyramidal cell size asymmetries within auditory and language-associated regions of the temporal lobes. *Brain Lang.* **86:** 226–242.
80. Hutsler, J.J. 2003. Hemispheric asymmetries in cerebral cortical networks. *Trends Neurosci.* **28:** 429–435.
81. Buxhoeveden, D.P. & M. Casanova 2000. Comparative lateralisation patterns in the language area of human, chimpanzee, and rhesus monkey brains. *Laterality* **5:** 315–330.
82. Buxhoeveden, D.P., A.E. Switala, M. Litaker, *et al.* 2001. Lateralization of minicolums in human planum temporale is absent in nonhuman primate cortex. *Brain Behav. Evol.* **57:** 349–358.

Ann. N.Y. Acad. Sci. ISSN 0077-8923

ANNALS OF THE NEW YORK ACADEMY OF SCIENCES
Issue: *The Neurosciences and Music IV: Learning and Memory*

The OPERA hypothesis: assumptions and clarifications

Aniruddh D. Patel
The Neurosciences Institute, San Diego, California

Address for correspondence: Aniruddh D. Patel, The Neurosciences Institute, 10640 John Jay Hopkins Dr., San Diego, CA 92121. apatel@nsi.edu

Recent research suggests that musical training enhances the neural encoding of speech. Why would musical training have this effect? The OPERA hypothesis proposes an answer on the basis of the idea that musical training demands greater precision in certain aspects of auditory processing than does ordinary speech perception. This paper presents two assumptions underlying this idea, as well as two clarifications, and suggests directions for future research.

Keywords: musical training; speech perception; neural plasticity; neural encoding; pitch perception

Introduction

Recent neuroscientific research has revealed that musically trained individuals show enhanced neural encoding of speech sounds in the auditory brainstem relative to musically untrained individuals.[1] One possible source of this difference is innate differences in the structure and function of the auditory system between those who pursue versus who do not pursue musical training. Another possible source is experience-dependent neural plasticity because of musical training. A role for plasticity is suggested by the repeated finding that the degree of enhancement correlates with the number of years of musical training. This plasticity may be driven by the extensive corticofugal pathways that originate in the auditory cortex and project to subcortical auditory regions.[2]

Firm evidence that musical training causes enhancements in brainstem speech encoding awaits controlled studies in which individuals are randomly assigned to musical training versus some other enjoyable artistic training (such as painting), and brainstem processing of speech is measured before and after training.[3,4] Hopefully such research will be conducted soon. It seems very likely that such studies will demonstrate a causal role for musical training in enhancing the quality of brainstem speech sound encoding. This will be an important result because the quality of brainstem speech sound encoding is correlated with real-world language skills such as reading ability and hearing speech in noise.[1] This, in turn, suggests that musical training may be useful for improving these skills in normal individuals or in those who suffer from developmental problems with these skills.

If musical training causes enhancements in subcortical speech encoding, this raises a fundamental question: Why would musical training have this effect? After all, musical instruments (such as guitars or trumpets) sound very different from human voices. Furthermore, spoken sentences are acoustically very complex, and many musical instruments produce sound patterns that are simpler by comparison (e.g., consider the acoustics of a flute melody vs. a speaking voice). Why would learning to produce and perceive relatively simple acoustic patterns improve the brain's processing of complex acoustic patterns?

One possible explanation is offered by the OPERA hypothesis.[5] OPERA argues that musical training enhances the neural encoding of speech when five conditions are met: overlap, precision, emotion, repetition, and attention. Together, these conditions drive adaptive neural plasticity in auditory processing networks, leading them to function with higher precision than needed for ordinary speech communication. Yet, because speech shares these networks with music, speech processing may benefit.

doi: 10.1111/j.1749-6632.2011.06426.x
Ann. N.Y. Acad. Sci. 1252 (2012) 124–128

Of these five conditions, the primary novel idea is that musical training influences speech processing because it demands greater precision than speech perception in certain aspects of auditory processing. This paper focuses on this issue of precision, discussing two assumptions underlying OPERA that were not made explicit in the original statement of the hypothesis, as well as two clarifications. This paper focuses on pitch processing, an important part of both speech and music.[6] The auditory system has neural mechanisms for extracting pitch from complex sounds,[7] and it is very likely that common brainstem circuits are involved in pitch processing in speech and music (satisfying the "overlap" condition of OPERA).

Before turning to the assumptions and clarifications mentioned above, it is worth reviewing the idea that musical training demands greater precision than speech perception in terms of pitch processing. One way to approach this issue is to ask: "How important is hearing fine pitch distinctions for speech comprehension versus for musical training?"

If a person cannot hear fine distinctions in pitch, chances are that he or she will still comprehend the basic meaning of spoken sentences. This is because speech has many redundant cues to meaning. For example, if a listener doesn't perceive the pitch rise at the end of "Is it your birthday?," he or she will still know the sentence is a question by virtue of word choice ("Is it...") and grammar. Indeed, it has recently been shown that native speakers of Mandarin Chinese (a tone language) can understand sentences with no pitch variation (i.e., native listeners find monotone Mandarin sentences just as intelligible as natural sentences, when heard in quiet).[8] Presumably, listeners used the remaining phonetic information and their knowledge of Mandarin to guide their perception in a way that allowed them to infer which words were being said. (Similar experiments have also been performed in English.[9]) The larger point is that understanding the basic meaning of sentences does not require ability to hear fine distinctions in pitch because of the redundancy of spoken language (this redundancy likely includes top-down semantic and syntactic cues to word meaning). Of course, in natural language, pitch variation does convey a variety of information such as prosodic focus, phrase boundaries, emotional tone, attitude, and so forth. It seems, however, that deriving the basic semantic meaning of a sentence does not seem to require high-precision pitch encoding by the brain.

In contrast, musical training does demand high-precision pitch encoding. This is because musical training involves constantly monitoring whether one is in tune and playing the right notes. To do this, a performer needs to hear subtle distinctions in pitch (in Western music, neighboring musical pitches such as B vs. B-flat differ by just ~6% in pitch). According to OPERA, this high-precision demand on pitch encoding sets the stage for musical training to sharpen pitch processing in the auditory system. The remaining three conditions of OPERA (emotion, repetition, and attention) facilitate neural plasticity. According to OPERA, when all five conditions are met, musical training drives subcortical pitch-encoding networks to function with higher precision than needed for ordinary speech processing. However, because speech and music share subcortical pitch-processing circuits, speech processing benefits. For example, higher precision encoding of voice pitch may enhance speech perception in noise[10,11] because natural modulations in voice pitch contribute to enhanced intelligibility of speech in noise.[8,9] Higher precision encoding of voice pitch may also contribute to the superior performance of musicians in recognizing affective prosody in spoken language.[12]

With this background in place, it is now possible to discuss certain assumptions and clarifications concerning the "precision" component of OPERA, focusing on pitch processing.

Assumption 1: the neural encoding of pitch and spectral shape is distinct

Speech perception relies on the neural encoding of several aspects of complex acoustic signals. Important aspects include waveform periodicity (related to pitch), the distribution of acoustic energy across the frequency spectrum, or spectral shape (related to formant patterns and phoneme identity), and overall amplitude envelope (related to syllable structure and prominence). In claiming that music demands greater precision in pitch encoding than does speech, it is crucial to distinguish pitch from spectral shape. Speech perception requires sensitivity to spectral shape and how it changes over time (indeed, dynamic spectral shape information is both necessary and sufficient for speech perception even in the absence of any pitch information).[13,14] If the neural mechanisms for encoding pitch and spectral shape were identical, then it would not make sense to suggest that music demands greater auditory encoding precision than speech because speech places fairly high demands on the encoding of spectral shape (e.g., for detecting formants and their movements). In other words, the idea that music demands higher precision pitch processing than speech depends on the assumption that the neural encoding of pitch and of spectral shape rely on distinct brain mechanisms. For example, the encoding of low-frequency pitches (such as those used in speech and music) may rely heavily on temporal patterns of activity in auditory neurons, ultimately rooted in phase-locking of action potentials to periodicities in the acoustic waveform.[15] In contrast, encoding of spectral shape may rely on spatial patterns of activity across auditory tonotopic maps that occur in multiple subcortical and cortical regions. (As an aside, it is interesting to note that humans can readily understand whispered speech, which contains no acoustic periodicity and no clear pitch information, but which does contain rapid spectral-shape changes because of acoustic filtering by vocal tract articulators. Hence, the auditory system can do spectral shape encoding in the absence of any clear pitch information.)

Assumption 2: the normal auditory system tolerates "good enough" processing

If high-precision pitch encoding benefits speech perception (e.g., perhaps by enhancing speech perception in noise and sensitivity to emotional prosody), then why doesn't our day-to-day use of speech result in high-precision pitch encoding? OPERA assumes that the precision of pitch encoding demanded by speech perception or any other kind of auditory perception (e.g., perception of environmental sounds) reflects what is good enough to get the job done. In ordinary speech perception, the primary job is understanding the semantic content of utterances in the ecological contexts in which spoken language most often takes place (i.e., in face-to-face conversations in relatively quiet settings). If this is being done satisfactorily, then there is no "upward pressure" on the auditory system to enhance the precision of pitch encoding. According to this view, individuals can vary widely in their pitch-encoding precision (perhaps due to inborn neuroanatomical differences) and still engage successfully in ordinary speech communication because of the many redundant cues to meaning in spoken language, as discussed previously. Crucially, however, when it comes to musical training, the range of encoding precision that is "good enough to get the job done" is narrower and skewed to the high end. This is because musical training involves carefully regulating and monitoring the pitches produced. Presumably, one cannot do this job with an auditory system that does imprecise pitch encoding. This sets the stage for experience-dependent plasticity to enhance pitch encoding by the brain.

Clarification 1: musical training and speech perception—apples and oranges?

OPERA is concerned with explaining why musical training benefits the neural encoding of speech. OPERA compares the auditory encoding demands made by musical training and by speech perception. These are very different activities. Most notably, musical training involves sound production, whereas speech perception does not. Hence, does OPERA compare apples and oranges, that is, things that should not be compared? The crucial point is that musical training also involves auditory training because it involves careful monitoring of what one produces. As mentioned earlier, musicians must constantly listen to themselves to judge whether they are in tune and are playing the right notes, and these judgments require hearing subtle distinctions in pitch. Early in one's musical training, feedback about whether one is in tune or playing the right notes may come from an external source (such as a music teacher), but as one progresses, self-monitoring becomes essential, and production and perception become tightly intertwined. Thus, in discussing the auditory encoding demands made by musical training versus by speech perception, one is really talking about the demands made by music perception (in the context of performance training) versus speech perception.

Clarification 2: musical performance training versus ear training

Musical training does not always involve sound production. Ear training is also part of music education, for example, learning to distinguish different rhythms, pitch intervals, and cadence types. In principle, if such ear training requires making subtle pitch distinctions and also involves positive emotion, extensive repetition, and focused attention, then purely aural training in music could enhance the neural encoding of pitch in speech according to the principles of OPERA. However, in practice, it may be much easier to meet the conditions of OPERA in performance-based musical training than in purely perception-based training. For example, one major difference between performance-based training and ear training is that the former is often social, involving group music making. This can heighten emotion and attention because of the psychological impact of engaging in a coordinated, pleasing group activity. Furthermore, sensorimotor training may drive neural plasticity more strongly than purely sensory training.[16]

Looking ahead: beyond pitch

Aside from pitch, are there other auditory attributes for which musical training might demand higher precision encoding than speech perception? Of course, what constitutes "musical training" depends on the instrument and the type of music being studied. For example, there are many percussion instruments and musical styles that place a high premium on rhythm and timing. How do rhythmic music and speech compare in terms of the demands placed on accurate encoding of timing information? Small temporal differences can be quite important in speech. For example, small differences in voice-onset-time help distinguish stop consonants, such as /b/ versus /p/, and small differences in the rise time of amplitude envelopes help distinguish stop consonants from glides, such as /b/ versus /w/. However, just as with pitch, the critical question is "How important is hearing fine timing distinctions for speech comprehension versus for musical training?" It may be that because of the redundancy of spoken language, if a person cannot hear fine temporal distinctions, he or she will still comprehend the basic meaning of sentences. In contrast, musical training on rhythmic instruments may require hearing fine timing distinctions, for example, as part of determining whether a rhythm is being played correctly or whether one is in synchrony with other members of a group.

More broadly, empirical work is needed to determine which auditory attributes shared by speech and music are subject to higher precision-encoding demands in the musical domain. Such information can help guide hypothesis-driven explorations of music as a tool for enhancing or rehabilitating certain speech perception abilities.

Acknowledgments

I thank D. Robert Ladd, LouAnn Gerken, Devin McAuley, and Dan Levitin for insightful comments on the original OPERA paper, which helped inspire this update, and Mathias Oechslin, Nina Kraus, Trent Nicol, and Alexandra Parbery-Clark for comments on this paper. This work was supported by Neurosciences Research Foundation as part of its program on music and the brain at the Neurosciences Institute, where ADP is the Esther J. Burnham Senior Fellow.

Conflicts of interest

The author declares no conflict of interest.

References

1. Kraus, N. & B. Chandrasekaran. 2010. Music training for the development of auditory skills. *Nat. Rev. Neurosci.* **11:** 599–605.
2. Patel, A.D. & J.R. Iversen. 2007. The linguistic benefits of musical abilities. *Trends Cognitive Sci.* **11:** 369–372.
3. Song, J. H., E. Skoe, P. C. Wong & N. Kraus. 2008. Plasticity in the adult human auditory brainstemfollowing short-term linguistic training. *J. Cognitive Neurosci.* **10:** 1892–1902.
4. Moreno, S., C. Marques, A. Santos, *et al.* 2009. Musical training influences linguistic abilities in 8-year-old children: more evidence for brain plasticity. *Cerebral Cortex* **19:** 712–723.
5. Patel, A.D. 2011. Why would musical training benefit the neural encoding of speech? The OPERA hypothesis. *Front. Psychol.* **2:** 142. doi:10.3389/fpsyg.2011.00142
6. Patel, A.D. 2008. *Music, Language, and the Brain.* Oxford University Press. New York.
7. McDermott, J.H. & A.J. Oxenham. 2008. Music perception, pitch, and the auditory system. *Curr. Opin. Neurobiol.* **18:** 452–463.
8. Patel, A.D., Y. Xu & B. Wang. 2010. The role of F0 variation in the intelligibility of Mandarin sentences. In: *Proceedings of Speech Prosody 2010*, Chicago, IL.
9. Binns, C. & J.F. Culling 2008. The role of fundamental frequency contours in the perception of speech against interfering speech. *J. Acoust. Soc. Am.* **122:** 1765–1776.
10. Parbery-Clark, A., E. Skoe & N. Kraus 2009. Musical experience limits the degradative effects of background noise on the neural processing of sound. *J. Neurosci.* **29:** 14100–14107.
11. Parbery-Clark, A., D.L. Strait, S. Anderson, *et al.* 2011. Musical training and the aging auditory system: implications for cognitive abilities and hearing speech in noise. *PLoS ONE* **6:** e18082. doi: 10.1371/journal.pone.0018082
12. Lima, C.F. & S.L. Castro 2011. Speaking to the trained ear: musical expertise enhances the recognition of emotions in speech prosody. *Emotion* **11:** 1021–1031.

13. Drullman, R., J.M. Festen & R. Plomp. 1994. Effect of temporal envelope smearing on speech reception. *J. Acoust. Soc. Am.* **95:** 1053–1064.
14. Shannon, R.V., F.G. Zeng, V. Kamath, *et al.* 1995. Speech recognition with primarily temporal cues. *Science* **270:** 303–304.
15. Cariani, P.A. & B. Delgutte 1996. Neural correlates of the pitch of complex tones. I. Pitch and pitch salience. *J. Neurophysiol.* **76:** 1698–1716.
16. Lappe, C., S.C. Herholz, L.J. Trainor & C. Pantev. 2008. Cortical plasticity induced by short-term unimodal and multimodal musical training. *J. Neurosci.* **8:** 9632–9639.

Ann. N.Y. Acad. Sci. ISSN 0077-8923

ANNALS OF THE NEW YORK ACADEMY OF SCIENCES
Issue: *The Neurosciences and Music IV: Learning and Memory*

Becoming musically enculturated: effects of music classes for infants on brain and behavior

Laurel J. Trainor,[1,2] Céline Marie,[1] David Gerry,[1] Elaine Whiskin,[1] and Andrea Unrau[1]

[1]McMaster Institute for Music and the Mind, Department of Psychology, Neuroscience & Behaviour, McMaster University, Hamilton, Ontario, Canada. [2]Rotman Research Institute, Baycrest Hospital, Toronto, Ontario, Canada

Address for correspondence: Laurel J. Trainor, Director, McMaster Institute for Music and the Mind, Department of Psychology, Neuroscience & Behaviour, McMaster University, Hamilton, ON L8S 4K1, Canada. LJT@mcmaster.ca

Musical enculturation is a complex, multifaceted process that includes the development of perceptual processing specialized for the pitch and rhythmic structures of the musical system in the culture, understanding of esthetic and expressive norms, and learning the pragmatic uses of music in different social situations. Here, we summarize the results of a study in which 6-month-old Western infants were randomly assigned to 6 months of either an active participatory music class or a class in which they experienced music passively while playing. Active music participation resulted in earlier enculturation to Western tonal pitch structure, larger and/or earlier brain responses to musical tones, and a more positive social trajectory. Furthermore, the data suggest that early exposure to cultural norms of musical expression leads to early preferences for those norms. We conclude that musical enculturation begins in infancy and that active participatory music making in a positive social setting accelerates enculturation.

Keywords: enculturation; music acquisition; brain development; EEG; learning; infancy; tonality; esthetics

Introduction

Music is a cornerstone of human culture, found in everyday life in all societies and used to mark important occasions such as weddings and funerals. The ability of music to engender a common emotion and engage people in cooperative social behavior likely explains why music is ubiquitous in religious rituals, sporting events, cooperative work engagement, and interactions between caregivers and infants.[1–9] Part of the power of music lies in the fact that it is generative; that is, it does not consist of a fixed set of vocal signals, but rather a structural framework in which an unlimited number of musical compositions are possible. Furthermore, although there are biological constraints that shape the space of possible musical structures,[10,11] the particular structures used vary from musical system to musical system, and even within a musical system, these structures evolve from generation to generation. Thus, for young infants, learning to become full participants in their culture necessarily involves becoming sensitive to, and specialized for, the particular musical system of that culture.

Enculturation takes place on many levels. Musical systems differ structurally in how they organize pitch space, the musical scales they use, whether and how they define harmony, and what rhythmic and metrical structures they employ. In this regard, many studies indicate that even Western adults with no formal musical training have become enculturated listeners through everyday exposure to Western music.[12–20] Musical systems also differ in terms of how musical esthetics and expressivity are conveyed; for example, timbral voice qualities considered pleasing in one culture may be aversive in another. Performance group structures also differ, from a predominance of solo to antiphonal to chorus styles.[21] Finally, there are cultural rules as to who performs what music when, including songs for males or females or children and special songs for religious leaders.[22]

The vast majority of developmental research on musical enculturation concerns infants and children learning Western tonal and rhythmic structure.[11,23–32] In this regard, Hannon and her colleagues have shown that young Western infants are

doi: 10.1111/j.1749-6632.2012.06462.x

able to process both simple and complex metrical structures found in music around the world, but become specialized for the simple metrical structures predominant in Western music by 12 months of age.[23,24,33] Similarly, in contrast to Western adults, young infants are not yet sensitive to Western tonal pitch structure, processing equally well wrong notes that go outside the key of a melody and wrong notes that are consistent with the key and implied harmony of a melody.[34] Interestingly, studies in preschool children suggest that music lessons accelerate musical acquisition as measured both behaviorally[35] and with brain imaging techniques such as electroencephalography (EEG), magnetoencephalography (MEG), and functional magnetic resonance imaging (fMRI).[36–38] Little work has been done in the infancy period. However, one study suggests that metrical specialization can be slowed by exposure to foreign musical systems around 12 months of age.[33] There is also evidence that participation in Kindermusik classes for infants and parents can accelerate specialization for Western meters in seven-month-old infants.[39] In the pitch domain, one study indicates that controlled listening to melodies in either marimba or guitar timbre at four months of age strengthens brain responses to the exposed timbre,[40] but it remains an open question as to whether musical training in infancy can accelerate enculturation to Western tonality and Western esthetic values related to musical expression.

Experiment on the effects of active musical experience in infancy on enculturation to Western music

Gerry, Unrau, and Trainor investigated the effects of music classes for infants and parents on enculturation to Western music.[41] Their findings on enculturation to Western tonality and the social implications of this enculturation are summarized below. In the present paper, we also present new data from this large project on the effects of these classes on esthetic responses and brain development.

Study background

We hypothesized that musical enculturation takes place through social interaction and participation in music making. To test this idea, we randomly assigned infants at 6 months of age to participate for 6 months in one of two types of weekly hour-long music classes for infants and parents. A total of 38 infants completed the musical training, as defined by attending at least 75% of the classes. After this participation, we measured their sensitivity to Western tonality, their esthetic preferences, their brain responses to musical sounds, and their social development. We also measured brain responses and social development at the beginning of the classes. We did not expect to see sensitivity to Western tonality or esthetic preferences at 6 months of age, so these were only measured at 12 months. The classes took place at Ontario Early Years Centers, which are government-sponsored drop-in centers for preschool children and their families. In the active classes, a Suzuki-philosophy approach was used in which the teacher engaged parents and infants in a curriculum that focused on movement, singing, playing percussion instruments, and building a repertoire of lullabies and action songs.[42] The classes emphasized musical expression, listening in order to play a percussion instrument or sing at the correct time, repetition of the repertoire, and encouraging parents to develop an awareness of their infants' responses. Parents were encouraged to play a CD at home that included the songs learned in class. In the "passive classes," parents and infants listened to a rotation of CDs from the Baby Einstein (The Walt Disney Co., Burbank, CA) series while the teacher encouraged play and interaction at art, book, ball, block, and stacking cup play stations. These CDs consist of synthesized classical music, rendered without musical expression, but nonetheless, as the name indicates, marketed as a tool to help make your baby more intelligent. Parents were encouraged to take home a different Baby Einstein CD each week to listen to at home. Thus, the passive classes were matched as much as possible to the active classes in terms of amount of musical stimulation, motivation, and social interaction. Details concerning the classes can be found in Gerry *et al.*[41] The classes took place at two different centers, one in a lower socioeconomic status (SES) area and the other in a middle-class SES area, such that each center had one active and one passive class. Teachers of the classes were blind as to the content and hypotheses of the tests given to infants, and experimenters conducting the tests were blind as to whether each infant was in the group participating in passive or active classes.

Enculturation to Western tonality

As reported by Gerry and colleagues,[41] sensitivity to Western tonality was measured by examining infants' preferences for two versions of a sonatina by Thomas Atwood (1765–1838). The *tonal* version was presented in G major as written by Atwood. The *atonal* version had additional accidentals added, such that it alternated between G major and G-flat major every beat and therefore had no feeling of a tonal center. This modification maintained the rhythm, phrasing, and melodic contours of the original piece as well as the amount of sensory consonance (every chord maintained its identity, for example as major or minor, because the manipulation simply transposed the whole chord down by a semitone or not).

Infants' preferences for these two versions were measured in a head-turn preference procedure in which the experimenter sat across from the infant, who sat on his or her parent's lap. The parent and the experimenter wore headphones and listened to masking music so that they were unaware of what the infant was hearing and could not influence the results. The experimenter signaled to the computer through a button box to begin a trial when the infant was facing forward. The computer initiated a trial by flashing a light on one side of the infant, illuminating an interesting toy. When the infant looked at the toy, the light remained on and one of the two versions of the sonatina began playing. The light remained on and the music played until the infant looked away, signaling the end of the trial. The next trial was identical except that it occurred on the opposite side of the infant and involved the other version of the sonatina. Which version was played on the first trial and which version was played on which side were counterbalanced across infants. Trials of the two versions alternated until 20 trials had been completed. In this way, the infant controlled how long they listened to each version. The dependent measure of infants' tonality preference was the proportion of the total time listening (listening times to tonal plus atonal versions) spent listening to the tonal version.

The final sample consisted of 20 infants in the active classes group and 10 in the passive classes group. An additional 20 infants were tested in a no-training age-matched control group in order to verify the results in the passive classes group. Interestingly, a higher proportion of parents completed the active compared to passive training classes with their infants, measured by attending at least 75% of the classes. Furthermore, no infants in the active group, but four in the passive group, had to be eliminated for fussing during the preference procedure or producing anomalous data. For details see Gerry *et al.*[41] Analysis of the preference data indicated that the groups differed significantly and that those in the active classes group showed a preference for the tonal version, whereas those in the passive classes and no-training groups showed no significant preferences.[41]

These results clearly show that some sensitivity to Western tonality is possible by 12 months of age. Furthermore, they indicate that active musical participation involving social interaction between infants, their parents, and others in the group promotes earlier enculturation to the pitch structure of music.

Esthetic enculturation to Western classical music

Enculturation to stylistic norms of expressive performance is an important aspect of musical development, but one that has been little studied. There are many aspects of esthetics, and infant listeners have limited attention spans in which to be tested, so this first study could only scratch the surface of this question. We decided to test whether infants in the active and passive classes groups would show differential responding to musically expressive versus synthesized versions of a classical Western piece from the Romantic period, in which there is a large scope for expression during performance. Two versions of the opening section of Chopin's Waltz in A-flat, op. 69, No. 1, were used in this experiment (Fig. 1A). The first version was played in an expressive manner on acoustic piano by Dinu Lipatti, considered to be one of the world's greatest interpreters of Chopin. The second version was synthesized using Cakewalk in Midi Chorus timbre, with no timbral or dynamic variation, no expressive timing, and no velocity contour. Due to dynamic variation within the Lipatti performance, dynamics in both versions were normalized in order to obtain dynamic uniformity. Average tempo was matched between the two versions. These stimuli can be heard in the supporting information at

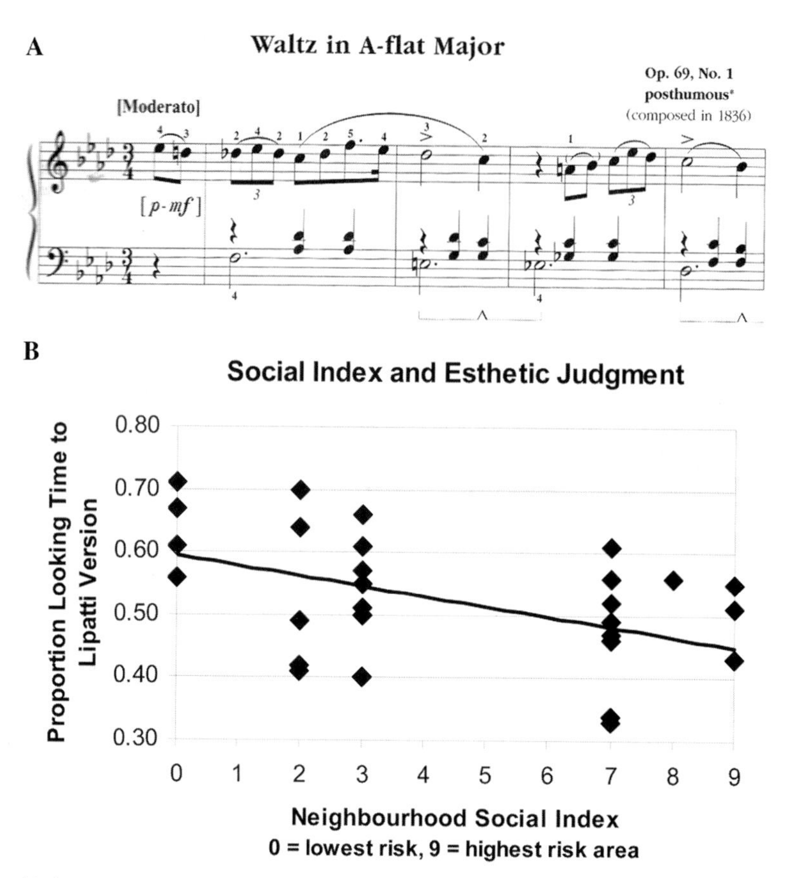

Figure 1. Relationship between esthetic preferences and SES vulnerability. (A) The opening of the Chopin Waltz in A-flat, op. 69, No. 1, from the Romantic period of the Western classical music repertoire. Two versions of the waltz were used, one performed by Dinu Lipatti on acoustic piano and the other synthesized with no dynamic variation and no expressive timing. The two versions can be heard in the supporting information at http://www.psychology.mcmaster.ca/ljt/estheticstimuli.htm. (B) Correlations between SES vulnerability and degree of each individual infant's preference for the Lipatti compared to the synthesized version. SES vulnerability is a composite score based on the area where the infant resides. It can be seen that lower SES vulnerability is associated with greater preference for the Dinu Lipatti version, consistent with greater exposure to acoustic classical music in higher-SES homes.

http://www.psychology.mcmaster.ca/ljt/estheticstimuli.htm.

Infants' preferences for the two versions were tested in the preferential-looking paradigm described in the last section. Each infant completed 20 trials. The sample had 20 infants from the active classes and 14 from the passive classes group. Of these, two infants did not complete the testing due to fussing or crying and two were eliminated as outliers (total looking time more than 2 SD from the mean), leaving a final sample of 20 infants from the active classes and 10 from the passive classes. A t-test with proportion listening time to the Lipatti version (time listening to Lipatti/time listening to Lipatti plus synthesized version) as the dependent measure revealed no significant difference between the groups ($P > 0.55$). Furthermore, a single-group t-test revealed that there was no significant preference for either version (not significantly different from 0.5, representing the expected value for chance performance, $P > 0.30$; mean proportion $= 0.52$, SD $= 0.10$). Thus, no evidence was found for an effect of active versus passive musical experience on esthetic preferences for two highly contrastive versions of a Chopin waltz at 12 months of age. This is perhaps not surprising as neither class type was exposed to expressive exemplars of this type from the Western classical corpus within the classes.

We conducted a further analysis based on SES vulnerability. A neighborhood social index (NSI) is available for the Hamilton, Ontario area.[43,44] It rates each area on nine subscales, with a total score of 0, representing lowest vulnerability, and 9 the highest vulnerability for young children. This score takes into account the percentage of unemployed residents, low-income families, income derived from government transfer payments, residents without high-school diplomas, recent immigrants, residents who do not speak English or French, residents who do not own their own home, resident mobility, and single-parent families (all compared to national averages). From the home address of each participating family, we obtained the SES vulnerability score for each infant. Classifying infants as low or high vulnerability according to the NSI scores resulted in 10/20 infants in the active classes and 4/10 infants in the passive classes being classified as residing in high-vulnerability neighborhoods. Interestingly, NSI scores were negatively correlated with preference for the expressive Lipatti version ($r = -0.48$, $n = 30$, $P = 0.008$; Fig. 1B), and this correlation remained significant for each group separately. Thus, infants from higher-SES neighborhoods tended to prefer the expressive Lipatti version.

While it is difficult to determine causality from correlational data, this result is consistent with higher-SES families exposing infants to more acoustic classical music in their homes compared to lower-SES families. If this interpretation is correct, it indicates that enculturation to esthetic norms of expression in musical performances begins early in development through exposure to exemplars from the musical culture.

Effects of musical enculturation on brain development

Musical training has been shown to affect brain development in preschool children,[36–38] but previous studies have not examined this question in infants. To determine whether participation in the active classes set infants on a different trajectory of brain development compared to participation in the passive classes, we measured EEG while infants listened to a repeating standard piano tone (C_5, 523 Hz) through a speaker (custom-built Westsun Jason Sound, Mississauga, Ontario, Canada) at 70 dB(A) over a noise floor of 29 dB(A) measured at the location of the infant's head. On 10% of repetitions (deviants), the pitch was changed by one semitone ($C\#_5$, 554 Hz), although responses to these trials are not reported (see below). The piano tones were 300 msec in duration with stimulus-onset asynchronies of 400 msec and infants were recorded for up to 20 min, as long as they did not fuss. Recordings were made at 6 months of age at the onset of the music classes and again at 12 months of age, at the end of the classes.

EEG was recorded at a sampling rate of 1,000 Hz from a 124-channel HydroCel GSN net (Electrical Geodesics, Eugene, OR) referenced to the vertex. The impedance of all electrodes was below 50 KΩ during the recording. EEG data were bandpass filtered between 2 and 18 Hz (roll-off = 12 dB/octave) using EEprobe software. The sampling rate was modified to 200 Hz in order to run the Artifact Blocking program in Matlab.[45,46] Recordings were rereferenced off-line using an average reference and then segmented into 500-msec epochs (−100 to 400 msec relative to stimulus onset) to create event-related potential (ERP) waveforms. Standard and deviant trials were averaged separately for each electrode for each infant at each age (6, 12 months). Standard trials immediately following deviant trials were excluded from the average. To increase signal-to-noise ratios, groups of electrodes were averaged together to form 10 scalp regions covering left and right frontal, central, parietal, occipital, and temporal regions (FL, FR, CL, CR, PL, PR, OL, OR, TL, TR), following He *et al.*[47] As the amplitude of the ERPs was almost flat at parietal regions, as expected from the scalp distribution of ERPs in response to sound,[48] these regions (PL and PR) were eliminated from further analysis. Unfortunately, the deviant waveforms were too noisy (there were too few trials) to analyze. The initial sample contained 24 infants who completed testing at both 6 months and 12 months. An additional three infants (two from the active and one from the passive classes) had to be eliminated because their data contained too few trials to analyze and/or their data contained too much movement artifact at one of the test sessions (6, 12 months), leaving a final sample of 14 infants from the active classes group and seven infants from the passive classes group.

As can be seen in Figure 2A, the ERPs to standards were dominated by a large positivity peaking around 175 msec at 6 months and 155 msec at 12 months after stimulus onset. This is consistent

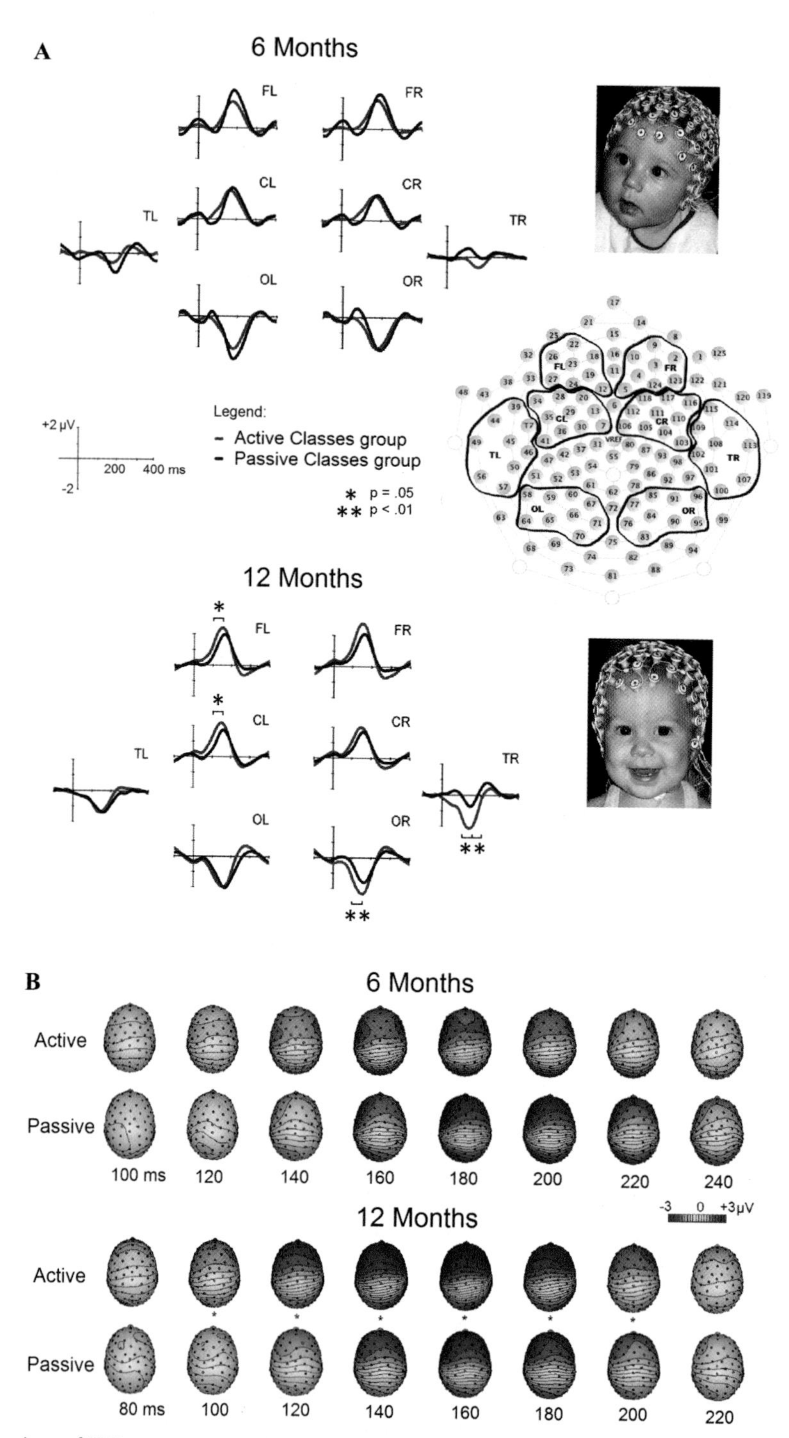

Figure 2. Comparison of ERP responses to a piano tone between infants in the active and passive classes groups. (A) Grand average ERP waveforms to standard piano tones comparing the results for infants in the active and passive classes groups at 6 months (upper) and 12 months (lower) of age. The *y*-axis represents the onset of the piano tone stimulus. Scalp regions at which the two groups differed significantly between 100 and 150 msec and between 150 and 200 msec after stimulus onset are shown. The electrodes included in the eight scalp regions used for analysis (following He *et al.*[47]) are shown to the right, with illustrations of infants in the 124-channel HydroCel GSN nets. (B) Topographic head maps for the active and passive classes groups at 6 and 12 months of age every 20 msec, illustrating the time course of the positivity across age and group.

with the previous ERP reports of infants' responses to piano tones (as reviewed by Trainor[49]). Topographic voltage maps were calculated every 20 msec so that the evolution of the amplitude and the spatial distribution of this component could be seen across age and type of training (Fig. 2B). The precise peak of the positivity was sometimes difficult to measure in individual infants at 6 and 12 months of age. Therefore, in order to analyze the differences between the active classes and passive classes groups, we computed mean amplitudes in two time windows, 100–150 msec and 150–200 msec after stimulus onset for each infant for each age for each region as the dependent measure. Because the number of infants was different in the two groups (14 and 7), the nonparametric Mann–Whitney U-test was used rather than an analysis of variance (ANOVA) approach.

At 6 months of age, before the training, the ERPs elicited by the piano tones were not significantly different between the active and passive classes groups in either the 100–150 msec or 150–200 msec time windows at any region (Fig. 2A). After 6 months of training, when infants were 12 months of age, infants in the active classes group showed a larger positivity between 100 and 150 msec compared to infants in the passive classes group in the left frontal (1.99 μV vs. 1.28 μV; $P = 0.05$) and left central (1.88 μV vs. 1.10 μV; $P = 0.05$) regions, and a larger negativity in the right occipital (–1.87 μV vs. –0.96 μV; $P = 0.007$) and right temporal (–1.85 μV vs. –0.40 μV; $P = 0.001$) regions. Between 150 and 200 msec, infants in the active classes group showed a larger negativity compared to infants in the passive classes group in the right temporal region (–1.46 μV vs. –0.25 μV; $P = 0.006$).

These results revealed no differences between ERP responses to piano tones between the active and passive classes groups before training began, but significantly different responses after 6 months of music classes. Although the latency difference is not possible to examine statistically, the plots in Figure 2 indicate that the large positive component might be both larger and earlier in the active classes group compared to passive classes group after the 6 months of musical training. This difference indicates that tone processing was more advanced for those in the active classes in that responses were faster and/or involved more synchronous neural firing. Because infants were randomly assigned to groups, the differences between the groups can be attributed to differences related to the training they received. Furthermore, although the sample size is relatively small, the fact that there were no differences at 6 months but robust differences at 12 months again suggests that the ERP differences were due to the differences in the music classes that the groups received. Although we cannot determine precisely which aspect of the active musical classes led to advanced brain development for musical processing, good candidates include active participation in music making, involvement of the parent in musical interaction with their infant, modeling of good singing by the teacher, routine and repetition of musical materials, and encouraging infants to attend to the music in a social setting.

Social consequences of musical enculturation

Because making music with other people leads to entrainment, social cooperation, and prosocial behavior in older children and adults,[1–7,9] Gerry *et al.*[41] hypothesized that infants in the active classes might show advanced social development compared to infants in the passive classes. They administered several subscales of the Infant Behavior Questionnaire (IBQ)[50,51] and indeed found that although infants in the active and passive classes showed no significant differences at 6 months of age, at 12 months, after 6 months of classes, those in the active classes showed less distress to limitations, less distress when confronted with novel stimuli, more smiling and laughter, and easier soothability compared to those in the passive classes.

The IBQ is a parent report measure, so it is not clear from these results whether the infants in the two groups were actually different, whether they were perceived to be different by their parents, or both. In a sense, it does not matter which interpretation is correct because it is clear that parents in the active classes developed better social interactions with their infants and rated their infants more positively compared to parents in the passive group. With social interaction, positive feelings on the part of the parent are likely to result in positive interactions with the infant, which likely lead to better outcomes for the infant. Reciprocally, more positive responses from infants likely lead to more positive responses from the parent, which again feed back to better outcomes for the infant. The important result is that the active classes led to more positive parent–infant social interaction compared to the passive classes.

Conclusions

Many different musical systems are used around the world with unique pitch and rhythmic structures, devices for expressive performance, and rules of pragmatic use. Enculturation involves the development of perceptual processing that is specialized for the particular pitch and rhythmic structures of the musical system used in the culture, familiarity with esthetic and expressive norms, and learning what music is used pragmatically in different social situations. Here, we explored the beginnings of some of these processes in Western infants by comparing the effects of 6 months of music classes beginning at 6 months of age that either emphasized active musical participation or passive exposure to synthesized music. The results indicate that active participation leads to earlier enculturation to tonal pitch structure. Furthermore, we found suggestive evidence that exposure to Western classical music leads to earlier sensitivity to esthetic norms for musical expression in this genre. Intriguingly, the social context of the musical experience appears to be crucial, such that infants in the active classes, in which parents and infants participated in active music making together, showed more positive social developmental trajectories compared to infants in the passive classes. Finally, the results indicate that differences between these two groups can also be measured at the brain level, with larger and/or earlier ERP responses to musical sounds in the active classes group compared to passive classes group evident after but not before participation in the classes. Previous studies in older children have indicated that musical lessons are associated with more advanced brain responses,[36–38] but the present study is the first to show effects of musical training on sound processing in the brain during the first year after birth.

In this initial study, there were a number of differences between the active classes and passive classes, so we cannot be sure as to which features of the active classes were most crucial for promoting musical enculturation. The most obvious candidate is active music making in a social context involving infants and parents. Other features of the active classes may have contributed as well, however, such as the use of live, expressive singing as modeled by the teacher and parents, the high degree of repetition that enabled parents to learn the songs and feel comfortable singing them at home with their infants, the more formal class routine that may have directed infants' attention to important features of the music, and the encouragement for parents to observe their infants' behaviors and progress over the course of the classes. In any case, the results clearly indicate that infants can benefit from participatory early musical classes and that musical enculturation begins in early infancy.

Acknowledgments

This research was supported by grants from the Grammy Foundation, the Canadian Institutes of Health Research (CIHR), and the Natural Sciences and Engineering Research Council of Canada (NSERC). We thank summer NSERC students Kristen Tonus, Adrienne Cheung, and Kathleen Lee for help with data reduction and analysis, and Heather Huck, Grace Liao, Nazanin Rajabi, and Lily Zhou for help testing the participants.

Supporting information

Additional supporting information may be found in the online version of this article:

The stimuli used to test infants' understanding of norms of expression in Western classical music from the Romantic period. Dynamics in both versions below were normalized and average tempo was matched in order to obtain uniformity across the stimuli.

Sound example 1. Chopin's Waltz in A-flat, op. 69, no. 1, as played by Dinu Lipatti, considered to be one of the world's greatest interpreters of Chopin.

Sound example 2. Chopin's Waltz in A-flat, op. 69, no. 1, as synthesized using Cakewalk midi chorus timbre, with no timbral or dynamic variation, no expressive timing, and no velocity contour.

Please note: Wiley-Blackwell is not responsible for the content or functionality of any supporting materials supplied by the authors. Any queries (other than missing material) should be directed to the corresponding author for the article.

Conflicts of interest

The authors declare no conflicts of interest.

References

1. Bispham, J. 2006. Rhythm in music: what is it? Who has it? And why? *Music Percept.* **24:** 125–134.
2. Brown, S. & U. Volgsten. 2006. *Music and Manipulation: On the Social Uses and Social Control of Music*. Berghahn Books. New York.
3. Hove, M.J. & J.L. Risen. 2009. It's all in the timing: interpersonal synchrony increases affiliation. *Soc. Cogn.* **27:** 949–961.
4. Kirschner, S. & M. Tomasello. 2009. Joint drumming: social context facilitates synchronization in preschool children. *J. Exp. Child Psychol.* **102:** 299–314.
5. Kirschner, S. & M. Tomasello. 2010. Joint music making promotes prosocial behavior in 4-year-old children. *Evol. Hum. Behav.* **31:** 354–364.
6. McNeill, W. 1995. *Keeping Together in Time: Dance and Drill in Human History*. Harvard University Press. Cambridge, MA.
7. Merker, B. 2000. Synchronous chorusing and human origins. In *The Origins of Music*. N.L. Wallin, B. Merker & S. Brown, Eds.: 315–327. MIT Press. Cambridge, MA.
8. Trehub, S.E. & L.J. Trainor. 1998. Singing to infants: lullabies and playsongs. *Adv. Infancy Res.* **12:** 43–77.
9. Wiltermuth, S.S. & C. Heath. 2009. Synchrony and cooperation. *Psychol. Sci.* **20:** 1–5.
10. Huron, D. 2001. Tone and voice: a derivation of the rules of voice-leading from perceptual principles. *Music Percept.* **19:** 1–64.
11. Trehub, S.E. 2005. Developmental and applied perspectives on music. *Ann. N.Y. Acad. Sci.* **1060:** 1–4.
12. Bigand, E. & M. Pineau. 1997. Global context effects on musical expectancy. *Percept. Psychophys.* **59:** 1098–1107.
13. Bigand, E. & B. Poulin-Charronnat. 2006. Are we "experienced listeners"? A review of the musical capacities that do not depend on formal musical training. *Cognition* **100:** 100–130.
14. Janata, P., J.L. Birk, J.D. Van Horn, *et al.* 2002. The cortical topography of tonal structures underlying western music. *Science* **298:** 2167–2170.
15. Koelsch, S., T. Gunter, E. Schröger & A.D. Friederici. 2003. Processing tonal modulations: an ERP study. *J. Cogn. Neurosci.* **13:** 520–541.
16. Peretz, I. 2006. The nature of music from a biological perspective. *Cognition* **100:** 1–32.
17. Tillmann, B., E. Bigand, N. Escoffier & P. Lalitte. 2006. The influence of musical relatedness on timbre discrimination. *Eur. J. Cogn. Psychol.* **18:** 343–358.
18. Tillmann, B., E. Bigand & F. Madurell. 1998. Local versus global processing of harmonic cadences in the solution of musical puzzles. *Psychol. Res.* **61:** 157–174.
19. Trainor, L.J., K.L. McDonald & C. Alain. 2002. Automatic and controlled processing of melodic contour and interval information measured by electrical brain activity. *J. Cogn. Neurosci.* **14:** 430–442.
20. Trainor, L.J. & S.E. Trehub. 1994. Key membership and implied harmony in Western tonal music: developmental perspectives. *Percept. Psychophys.* **56:** 125–132.
21. Lomax, A. 1968. *Folk Song Style and Culture*. Transaction Books. New Brunswick, NJ.
22. Merriam, A.P. 1964. *The Anthropology of Music*. Northwestern University Press. Evanston, IL.
23. Hannon, E.E. & L.J. Trainor. 2007. Music acquisition: effects of enculturation and formal training on development. *Trends Cogn. Sci.* **11:** 466–472.
24. Hannon, E.E. & S.E. Trehub. 2005. Metrical categories in infancy and adulthood. *Psychol. Sci.* **16:** 48–55.
25. Trainor, L.J. 2005. Are there critical periods for music development? *Dev. Neuropsychol.* **46:** 262–278.
26. Trainor, L.J. 2008. The neural roots of music. *Nature* **453:** 598–599.
27. Trainor, L.J. & K.A. Corrigall. 2010. Music acquisition and effects of musical experience. In *Springer Handbook of Auditory Research: Music Perception.* M. Riess-Jones & R.R. Fay, Eds.: 89–128. Springer. Heidelberg, Germany.
28. Trainor, L.J. & A.J. Unrau. 2012. Development of pitch and music perception. In *Springer Handbook of Auditory Research: Human Auditory Development*, Vol. 42. L.A. Werner, *et al.*, Eds.: 223–251. Springer. New York.
29. Trehub, S.E. 2001. Musical predispositions in infancy. *Ann. N.Y. Acad. Sci.* **930:** 1–16.
30. Trehub, S.E. 2003. The developmental origins of musicality. *Nat. Neurosci.* **6:** 669–673.
31. Trehub, S.E. 2003. Toward a developmental psychology of music. *Ann. N.Y. Acad. Sci.* **999:** 402–413.
32. Trehub, S.E. 2009. Music lessons from infants. In *Oxford Handbook of Music Psychology*. S. Hallam, I. Cross & M. Thaut, Eds.: 229–234. Oxford University Press. Oxford.
33. Hannon, E.E. & S.E. Trehub. 2005. Tuning in to rhythms: infants learn more readily than adults. *Proc. Natl. Acad. Sci. USA* **102:** 12639–12643.
34. Trainor, L.J. & S.E. Trehub. 1992. A comparison of infants' and adults' sensitivity to Western musical structure. *J. Exp. Psychol. Hum. Percept. Perform.* **18:** 394–402.
35. Corrigall, K.A. & L.J. Trainor. 2010. Musical enculturation in preschool children: acquisition of key and harmonic knowledge. *Music Percept.* **28:** 195–200.
36. Fujioka, T., B. Ross, R. Kakigi, *et al.* 2006. One year of musical training affects development of auditory cortical-evoked fields in young children. *Brain* **129:** 2593–2608.
37. Shahin, A.J., L.J. Roberts, W. Chau, *et al.* 2008. Musical training leads to the development of timbre-specific gamma band activity. *NeuroImage* **41:** 113–122.
38. Schlaug, G., M. Forgeard, L. Zhu, *et al.* 2009. Training-induced neuroplasticity in young children. *Ann. N.Y. Acad. Sci.* **1169:** 205–208.
39. Gerry, D.W., A.L. Faux & L.J. Trainor. 2010. Effects of Kindermusik training on infants' rhythmic enculturation. *Dev. Sci.* **13:** 545–551.
40. Trainor, L.J., K. Lee & D.J. Bosnyak. 2011. Cortical plasticity in 4-month-old infants: specific effects of experience with musical timbres. *Brain Topogr.* **24:** 192–203.
41. Gerry, D.W., A. Unrau & L.J. Trainor. In press. Active music classes in infancy enhance musical, communicative and social development. *Developmental Sci.*
42. Jones, D. 2004. Suzuki early childhood education. *Am. Suzuki J.* **36:** 32–38.

43. Connor, S. 2001. *Understanding the Early Years: Early Childhood Development in North York*. Publications Office, Human Resources Development Canada. Canada.
44. Janus, M., S. Brinkman, E. Duku, *et al.* 2007. *The Early Development Instrument: Population-based Measure for Communities: a Handbook on Development, Properties and Use*. Publications Office, Library and Archives Canada. Canada.
45. Mourad, N., J.P. Reilly, H. De Bruin, *et al.* 2007. A simple and fast algorithm for automatic suppression of high-amplitude artifacts in EEG data. *ICASSP, IEEE International Conference on Acoustics, Speech and Signal Processing—Proceedings*, Vol. 1. Honolulu, HI.
46. Fujioka, T., N. Mourad, C. He & L.J. Trainor. 2011. Comparison of artifact correction methods for infant EEG applied to extraction of event-related potential signals. *Clin. Neurophysiol.* **122:** 43–51.
47. He, C., L. Hotson & L.J. Trainor. 2007. Mismatch responses to pitch changes in early infancy. *J. Cogn. Neurosci.* **19:** 878–892.
48. He, C., L. Hotson & L.J. Trainor. 2009. Maturation of cortical mismatch responses to occasional pitch change in early infancy: effects of presentation rate and magnitude of change. *Neuropsychologia* **47:** 218–229.
49. Trainor, L.J. 2008. Event related potential measures in auditory developmental research. In *Developmental Psychophysiology: Theory, Systems and Methods*. L. Schmidt & S. Segalowitz, Eds.: 69–102. Cambridge University Press. New York.
50. Gartstein, M.A. & M.K. Rothbart. 2003. Studying infant temperament via the Revised Infant Behavior Questionnaire. *Infant Behav. Dev.* **26:** 64–86.
51. Parade, S.H. & E.M. Leerkes. 2008. The reliability and validity of the Infant Behavior Questionnaire—revised. *Infant Behav. Dev.* **31:** 637–646.

Ann. N.Y. Acad. Sci. ISSN 0077-8923

ANNALS OF THE NEW YORK ACADEMY OF SCIENCES
Issue: *The Neurosciences and Music IV: Learning and Memory*

Practiced musical style shapes auditory skills

Peter Vuust,[1,2,*] Elvira Brattico,[3,4,*] Miia Seppänen,[3,4] Risto Näätänen,[1,3,5] and Mari Tervaniemi[3,4]

[1]Center of Functionally Integrative Neuroscience, Aarhus University, Denmark. [2]The Royal Academy of Music, Aarhus, Denmark. [3]Cognitive Brain Research Unit, Institute of Behavioral Sciences, Faculty of Behavioral Sciences, University of Helsinki, Finland. [4]Finnish Centre of Excellence in Interdisciplinary Music Research, University of Jyväskylä, Finland. [5]Department of Psychology, University of Tartu, Tartu, Estonia

Address for correspondence: Peter Vuust, Centre of Functionally Integrative Neuroscience, Aarhus University Hospital, Norrebrogade 44, Building 14A 8000 Aarhus C, Denmark. pv@pet.auh.dk

Musicians' processing of sounds depends highly on instrument, performance practice, and level of expertise. Here, we measured the mismatch negativity (MMN), a preattentive brain response, to six types of musical feature change in musicians playing three distinct styles of music (classical, jazz, and rock/pop) and in nonmusicians using a novel, fast, and musical sounding multifeature MMN paradigm. We found MMN to all six deviants, showing that MMN paradigms can be adapted to resemble a musical context. Furthermore, we found that jazz musicians had larger MMN amplitude than all other experimental groups across all sound features, indicating greater overall sensitivity to auditory outliers. Furthermore, we observed a tendency toward shorter latency of the MMN to all feature changes in jazz musicians compared to band musicians. These findings indicate that the characteristics of the style of music played by musicians influence their perceptual skills and the brain processing of sound features embedded in music.

Keywords: mismatch negativity (MMN); EEG; musicians; multifeature MMN paradigm; musical style; learning

Introduction

Learning to play music at a professional level requires years of targeted training and dedication to music. The study of how musicians' brains evolve through daily training has recently emerged as an effective way of gaining insight into changes of the human brain during development and training.[1–4] Mismatch negativity (MMN) studies have consistently revealed neural differences in early sound processing between people with different musical backgrounds. The stimuli used in these studies, however, have often been far from musical sounding, hours long, and very repetitive, making the experiments less ecologically valid. Furthermore, MMN studies and studies of musical abilities in cognitive neuroscience, in general, have mainly been confined to cross-sectional studies of musicians versus nonmusicians (NMs), treating musicians as a unified group.[5]

This paper poses the questions: Can the MMN paradigms be adapted to resemble a musical context while keeping the experimental duration contained, and will they reveal differences in sound-related brain activity between different types of musicians?

The MMN, as measured with electroencephalography (EEG) or magnetoencephalography (MEG) with subjects' attention diverted from the stimuli, is a preattentive brain response mainly originating from the auditory cortices at around 100–200 ms after a change in sound features, such as pitch, timbre, and intensity.[6–8] Its amplitude and latency depends on change magnitude such that larger feature changes yield larger and faster MMNs.[9] The MMN is considered a candidate index of auditory capabilities. This has been substantiated by studies showing that the amplitude and latency of the MMN depend highly on instrument, practice strategies, and on the level of expertise. Musicians who need to intone while playing their instrument, such as

*Both the authors contributed equally.

doi: 10.1111/j.1749-6632.2011.06409.x

violinists, display a greater sensitivity to small differences in pitch compared to NMs,[10] singers respond with a stronger MMN than instrumentalists to small pitch changes,[11] conductors process spatial sound information more accurately than professional pianists and NMs,[12] and rhythmically skilled jazz musicians (JMs) respond to rhythmic deviations with a stronger, more left-lateralized and faster MMN than NMs.[13,14] However, to uncover fine-grained processing differences between musicians' and NMs' MMN responses, the musical context in which the feature change is placed is crucial. When presented with simple sinusoidal tones with greater mistuning instead of fine-grained differences, violinists were not superior to NMs in discriminating pitch changes.[10] A similar lack of MMN differences was obtained when comparing processing of isolated infrequent sinusoidal tones, harmonic sounds, or of infrequent minor chords within a sequence of major chords in different kinds of musicians against NMs.[2,15–17] Therefore, we needed to investigate stimuli consisting of realistic, complex musical material.

The classical MMN paradigms, most often consisting of simple auditory patterns with a single deviant embedded in the repetitive sequence, do not resemble typical music as such. Nevertheless, there is a prevalence of repetitive patterns in real music.[18] To simulate the patterns in real music, there are two characteristic aspects that our stimulus would need to contain. First, patterns in music often reflect harmonic progressions or at least motion between different parts of the musical form. For example, the bass ostinato accompanied by repeated drum rhythms and grooves, which is extensively used in contemporary Western music styles such as rock, soul, and Latin music, is usually transposed according to a chord scheme and often varies slightly to reflect the chord quality between, for example, major and minor chords.[19] Similarly, in classical music, the so-called Alberti bass, which is a musical accompaniment encountered in Mozart's sonatas or Beethoven's rondos and later adopted with variations in other contemporary musical genres,[20] reflect the underlying harmonic scheme. This accompaniment features an arpeggio-like texture that is used to provide the harmonic background for a melody.

Second, to reflect real music, we need the stimulus to embed more than one type of sound deviant into music with alternating pitches. This is typical for patterns in more complex music as exemplified by the short excerpt from Herbie Hancock's piano solo "All of You," as played on the Miles Davis record *Four and More,* and transcribed in Figure 1. It shows how different musical and auditory features, such as pitch, rhythm, and intensity, create intertwining patterns embedded within the musical phrases.[22]

Recently, Näätänen *et al.* developed a novel, multifeature MMN paradigm,[23] lasting only 15 minutes, in which several types of acoustic changes are presented in the same repetitive sound sequence. This technique allows for several MMNs to be independently elicited for different auditory attributes within the same sequence in a very short time. Importantly, no difference was observed between the MMNs recorded using the new paradigm and the ones obtained in the traditional longer oddball paradigm in which only one feature is repeated and changes are randomly interspersed. In this study we could, therefore, accommodate the two characteristics of musical patterning by combining the fast multifeature MMN-paradigm with an Alberti bass sequence moving around between different major and minor chords. Using this musical multifeature paradigm, we sought to test differences among musicians playing different styles of Western music, specifically among classical musicians (CMs), JMs, and pop/rock musicians.

There are broad differences between these musical genres not only regarding the listening experience, but also in relation to how they are taught and learned. JMs learn and perform music to a great extent by using the ear,[24,25] and separate ear training classes are taught at all the primary jazz schools around the world. Furthermore, jazz music in its modern form is characterized by complex chord changes, rich harmonies, and challenging rhythmic structures, such as polyrhythms,[22,26,27] which place great demands on listeners' and performers' theoretical and ear-training skills. At the other end of the spectrum, the teaching tradition within classical music focuses less on learning by ear training, taking notated music as a starting point, even though some schools such as the Suzuki method[28] teach music by ear in the early years of childhood.

In this study, we applied the new fast musical multifeature MMN paradigm with CMs, JMs, band musicians (BMs), and NMs. In this paradigm, six types

Figure 1. Example of patterns in "real" music. Transcription of measures 20–24 in Herbie Hancock's piano solo (upper system) and drum accompaniment (bottom system) on "All of You," from the record *Four and More.*[21] The example shows how patterns are woven into each other in real music. The patterns include simple patterns in rhythm, intensity, pitch height, and more abstract musical patterns, such as rising and falling sequences of thirds.

of acoustic changes (pitch mistuning, intensity, timbre, sound-source location slide, and rhythm) relevant for musical processing[16,29–31] in different musical genres are presented in the same sound sequence, lasting in total about 15 minutes. We hypothesized that this musically adapted MMN paradigm would elicit MMNs even in NMs. Furthermore, we expected to find enhanced processing of auditory features embedded in this more musical context in JMs compared to other groups.

Materials and methods

Participants

Eleven NMs (4 women, 7 men), 10 JMs (1 woman, 9 men), 7 CMs (5 women, 2 men), and 14 BMs (4 women, 10 men) gave informed consent and participated in the experiment. NMs were university students with minimal musical background and with occasional or no playing. JMs and CMs were defined as having formal education in music academies or conservatories mainly concentrated on the jazz or the classical genre, or earning their living by teaching or performing classical or jazz music. In turn, the main criteria for classifying musicians as BMs was their active playing and regular performing in a rock or pop band; formal education in a pop/jazz conservatory was not considered as crucial because many BMs learn to play in informal situations or are self-taught.

The participants did not differ in age between groups (mean age, JMs: 34 ± 9 years; CMs: 32 ± 9 years; BMs: 28 ± 7 years; and NMs: 26 ± 1 years), in onset age of playing music (CMs: 8 ± 3 SD years; JMs: 11 ± 3 SD years, BMs: 11 ± 5 SD years), but BMs chose their main instrument later than CMs and JMs (playing experience with their main instrument: CMs: 23 ± 7 SD; JMs: 23 ± 10 SD; BMs: 12 ± 4 SD). From the groups of CMs and JMs, the

Stimulus (Alberti Bass)

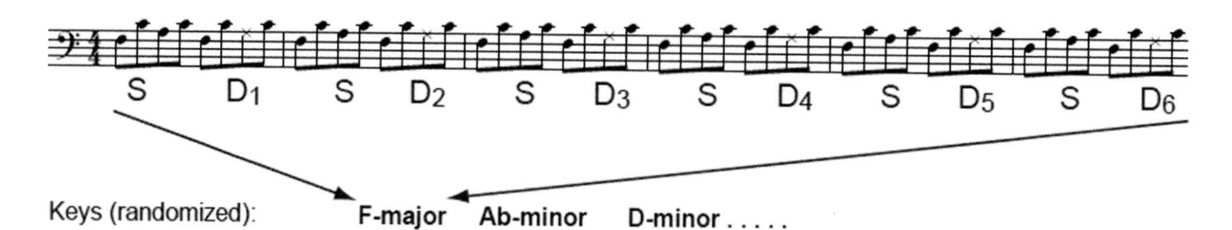

S: Standard
D1: Pitch-deviant: 24 cents lower
D2: Timbre-deviant: filtered, having an 'old time radio' effect
D3: Location-deviant: slightly shifted to the left
D4: Intensity-deviant: 6 dB reduction
D5: Slide-deviant: sliding up from a whole note below
D6: Rhythm-deviant: 30 ms earlier

Figure 2. Stimulus: Alberti bass played with piano sounds. Patterns alternated between standard sequence and a deviant sequence, periodically transposed to different keys and or modality with an interval of six bars. Each tone was 200 ms in duration, having an SOA of 5 ms, yielding a tempo of around 146 beats per minute. Comparisons were made between the third note of the standard sequence and the third note of the deviant sequence. The pitch deviant was created mistuning the third tone by 24 cents, tuned downward in the major mode, upward in the minor mode. For all analyses, major and minor pitch deviations were pooled. The rhythm deviant was created by anticipating the third note by 30 ms, compared with when it was expected. The timbre used the "old-time radio" effect provided with Adobe Audition using a four-channel parametric equalizer. The location deviant was generated by decreasing the amplitude of the right channel up to 10 dB, perceptually resulting in a sound coming slightly from a location left of center. The loudness deviant was made by reducing the original intensity by 6 dB and the pitch-slide deviant by sliding up to the standard from two semitones below. Sounds were amplitude normalized. Each tone was in stereo, 44,100 in sample frequency, and 200 ms in duration, having an Interstimulus interval (ISI) of 5 milliseconds.

Advanced Measure of Musical Audiation (AMMA) test data of two participants were missing in each, and of one participant from the BMs.

Stimuli and procedure

The musical, multifeature MMN paradigm[32] was based on the "optimal" paradigm presented in Näätänen *et al.*,[23] although more complex and musically enriched. The "optimal" paradigm, in which a "standard" simple tone is presented once after each "deviant" tone makes it possible to record event-related potentials (ERP) responses for many auditory feature deviations with an equally good signal-to-noise ratio as with the traditional oddball paradigms. Here, we presented auditory deviants occurring as the third note within an Alberti bass sequence consisting of four notes (Fig. 2).

To simulate the way the Alberti bass occurs in music, we changed the key every sixth measure, allowing for six different types of deviants (pitch mistuning, intensity, timbre, sound-source location, and rhythm) to appear exactly once in each key in a randomized order. The order of the 24 possible keys (12 major and 12 minor) was pseudorandomized so that each key appeared once for every 24 transpositions. The sequences were programmed in Cubase using a standard piano sound, and the deviants were created in Adobe Audition.

Participants passively heard auditory sequences through headphones for 20 min while watching a silenced document film. Before EEG recording, participants answered a background questionnaire consisting of questions about their musical knowledge. After the EEG recording, an AMMA musicality test was conducted to obtain an additional behavioral measure of the musical skills of the subjects.[33] The results of these tests and its correlation with MMN will be reported elsewhere.[32]

EEG recording and data analysis

The EEG was recorded with a BioSemi ActiveTwo system (BioSemi B.V., Amsterdam, Netherlands) using a 64-channel cap based on a 10/5 system with an active electrode sampling rate during recordings of a maximum of 2,048, down-sampled into 512 Hz with BDFDecimator (BioSemi B.V., Amsterdam, Netherlands). Double-sided adhesive electrode rings were used to attach the electrodes to the mastoids behind auricles electro-oculography (EOG; below the lower eyelid of the right eye), and the reference electrode to the nose.

The EEG was offline filtered (bandpass 1–30 Hz). Further, epochs of 100 ms prestimulus (used as baseline) and 400 ms poststimulus periods were separately averaged for the six types of deviant stimuli in each condition and for the standard stimuli, divided into six groups preceding each deviant type. Epochs including an EEG or EOG change exceeding ± 100 μV for more than four isolated channels were omitted from the averaging. Isolated channels with exceeding range were interpolated. A few contaminated channels were discarded in some subjects.

First, the MMN peak latencies were measured from the most negative peak at the frontal electrode approximating the Fz in the 10–20 system and the most positive peaks at the mastoid electrodes (LM and RM) occurring at the 100–200 ms poststimulus period. The mean MMN amplitudes were calculated as the average voltage at the 40-ms period centered at the individual peak latencies measured from Fz. Subsequently, to delineate the MMN and to include also the values from the mastoids, the standard stimulus (the third note of the standard Alberti pattern) ERPs were subtracted from the corresponding deviant-stimulus ERPs of the same sequence, and hence re-referenced to the average values measured from both mastoid channels. This procedure resulted in six different waveforms per subject that was further analyzed with respect to MMN amplitudes, latencies, and scalp distribution as earlier.

A repeated-measures ANOVA was calculated with respect to the MMN amplitude on a subset of electrodes (F3, F4, C3, C4, P3, and P4) for feature (pitch, timbre, intensity, slide, sound-source location, and rhythm), frontality (frontal, central, parietal), and laterality (left, right) as dependent variables, and group (JM, BM, CM, and NM) as an independent variable. MMN latencies were tested similarly but only from values obtained at Fz electrode. Results are reported with Greenhouse–Geisser corrected test values with the uncorrected degrees of freedom. LSD *post hoc* tests were used to test the direction of the effects obtained in the repeated-measures ANOVA. Because the groups were not perfectly matched, we used gender and age as covariates in all the repeated measures ANOVAs to account for the group differences unrelated to the variables of interest.

Results

As illustrated in Figure 3, the musical, multifeature paradigm produced MMNs for all six feature deviations even in NMs, as demonstrated by the significant differences between the mean amplitudes to deviant versus standard stimuli recorded at Fz ($P < 0.05$). MMN amplitudes also differed according to the feature, with a significant main effect of Feature ($F_{5,180} = 2.6$, $P = 0.03$). The largest MMN was elicited by the localization deviant ($P < 0.0001$ compared with all other deviants in LSD post hoc tests) and the smallest MMNs by the intensity and rhythm deviants ($P < 0.0001$ in LSD post hoc tests).

There was a significant main effect of group on the MMN amplitudes ($F_{3,36} = 5.3$, $P = 0.004$, Fig. 3). JMs had the largest MMNs across deviants compared with all other groups ($P < 0.01$ in post hoc LSD tests).

The distribution of the MMN over the scalp was also modulated by musicianship (interaction group × frontality: $F_{6,72} = 3.9$, $P = 0.002$; group × feature × frontality × laterality: $F_{30,360} = 1.6$, $P = 0.02$). Separate ANOVAs for the group factor showed that for JMs only there was a significant interaction feature × frontality ($F_{10,70} = 3.5$, $P = 0.02$) due to marginally significant effects of frontality for pitch and location, whereas for CMs, only there was a significant interaction feature × laterality ($F_{5,20} = 3.9$, $P = 0.01$) deriving from the left-lateralized MMN to timbre deviants (main effect of laterality: $F_{1,4} = 16.5$, $P = 0.02$).

We also found a marginal difference in latency of the MMN between groups ($F_{3,36} = 2.4$, $P = 0.09$), caused by JMs having slightly shorter MMN latencies than BMs.

Discussion

Using a novel musical, multifeature MMN paradigm, we have found reliable MMNs even in NMs to six different sound deviants embedded in a musical sounding structure, showing that the MMN paradigm can be adapted to reflect processing of real music. Furthermore, we found differences in MMN to these deviants among musicians playing different types of music. In particular, we obtained a larger MMN amplitude in JMs as compared with

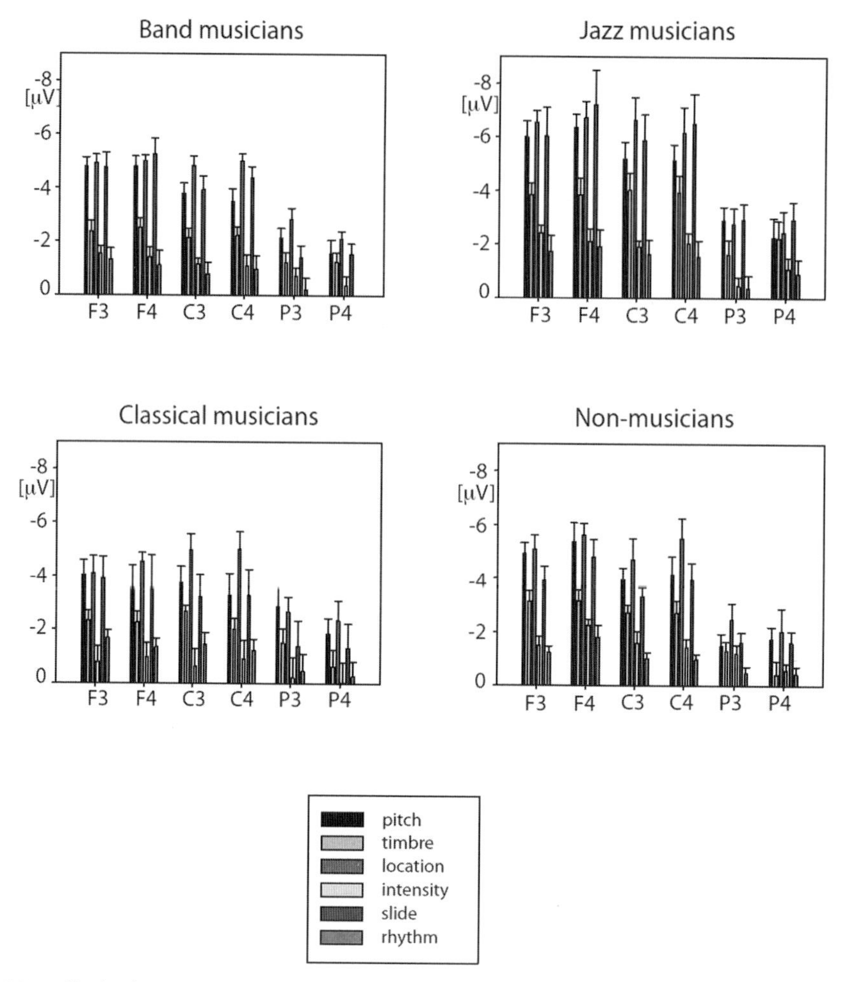

Figure 3. MMN amplitudes for each deviation, group, and electrode.

CMs, BMs, and NMs across the six different sound features and further, a tendency for a shorter MMN latency in JMs compared to BMs. These results suggest that the style/genre of music that professional musicians are engaged with influences, at least partially, early auditory skills.

The MMNs obtained in relation to the auditory deviants in our musical multifeature paradigm show that it is possible to develop highly controlled brain-measuring paradigms approaching a musical context. We managed to incorporate two traits typical of real music by combining the multifeature paradigm formulated by Näätänen *et al.* with the Alberti bass, an accompaniment commonly used in the Western musical culture in both classical and improvisational music genres providing a harmonic background for a melody. First, this approach allowed for the stimulus to simulate chord progressions that could have been found in, for example, jazz from the modal period in which random chord plateaus were frequent.[34] An example is the composition "Sketches of Spain" from Miles Davis' *Kind of Blue*,[35] in which there is no apparent relation between different plateaus of modalities. Second, we were able to present a number of deviants embedded in a relatively complex musical structure lasting overall less than 20 minutes. This is a musical development of the original MMN paradigms, in which only one deviant is presented, and of Näätänen's multifeature MMN paradigm,[23] in which several deviants are presented in a sequence of identical notes (standards). Our musical multifeature paradigm, however, admittedly lacks both melody and intentionality. Future studies should be conducted to determine if adding melody and potentially a composer's touch would still allow for preattentive processing that can be detected with MMN.

This paradigm opens a novel ecological method of comparing MMNs in musicians from different musical genres. This is important because complexity in certain instances is crucial for disclosing fine-grained auditory processing differences among participants from various musical backgrounds.[10,36–38] Here, we show, for the first time, differences in preattentive brain responses between CMs, JMs, and BMs to a range of deviants embedded in continuous streams of music-like material.

The main finding of this study, that JMs overall have larger MMNs to the six sound deviants than all other groups, may be the result of two factors. First, JMs learn and perform music to a great extent by ear.[24,25] Furthermore, jazz music is characterized by complex chord changes, rich harmonies, and challenging rhythmic material, which may boost these performers' theoretical and ear-training skills. In comparison, even though BMs play primarily pop/rock and learn music primarily by ear, they are typically exposed to less-challenging musical material, especially with regard to harmonic content. Classical music, on the other hand, is typically practiced and learned from notated music, such that there is less focus on ear-training skills. Hence, the combination of typical learning strategies, with a strong focus on ear-training skills, taught in separate classes and included in the instrumental training, and the acoustic and the music theoretical complexity inherent to the style/genre, may account for the fact that JMs had greater overall strength of the preattentive ERP amplitude to music-related auditory deviants.

Even though the present data indicate that stylistic features and learning of a style/genre of music influences auditory brain processing, it is not possible to rule out the possibility of innate differences among musician groups. A recent study found personality differences between CMs and jazz/rock musicians, suggesting that musicians partly choose musical careers on the basis of personality traits.[39] Also, it may be that people with good ear-training skills are attracted to music that involves complex harmonies and places strong demands on exact ear-training skills, such as jazz music.

With these limitations in mind, the musical, multifeature paradigm presents itself as a possible ecological objective measure of auditory skills relevant to music perception because MMNs are preattentively elicited with no behavioral task, and are correlated with individual behavioral measures and musical expertise.[13,30,40,41] This could potentially be of particular interest to future studies on music training and also on individuals with congenital amusia. Moreover, in the case that future developments of the ERP method will reach sensitivity and reliability even at the individual level, it may be possible to draw individual multiattribute "profiles" of sound-discrimination abilities.

Acknowledgments

The authors would like to thank Mr. Enrico Glerean and Mr. Lauri Janhunen, B.A., for their assistance with collecting and preparing the data.

Conflicts of interest

The authors declare no conflicts of interest.

References

1. Schlaug, G., L. Jäncke, Y. Huang & H. Steinmetz. 1995. In vivo evidence of structural brain asymmetry in musicians. *Science* **267:** 699–701.
2. Fujioka, T., B. Ross, R. Kakigi, *et al.* 2006. One year of musical training affects development of auditory cortical-evoked fields in young children. *Brain* **129:** 2593–2608.
3. Chakravarty, M.M. & P. Vuust. 2009. Musical morphology. *Ann. N.Y. Acad. Sci.* **1169:** 79–83.
4. Tervaniemi, M., T. Tupala & E. Brattico. 2011. Expertise in folk music alters the brain processing of Western harmony. *Ann. N.Y. Acad. Sci.*, this issue.
5. Tervaniemi, M. 2009. Musicians–same or different? *Ann. N.Y. Acad. Sci.* **1169:** 151–156.
6. Näätänen, R., P. Paavilainen, T. Rinne & K. Alho. 2007. The mismatch negativity (MMN) in basic research of central auditory processing: a review. *Clin. Neurophysiol.* **118:** 2544–2590.
7. Näätänen, R., M. Tervaniemi, E. Sussman, *et al.* 2001. "Primitive intelligence" in the auditory cortex. *Trends Neurosci.* **24:** 283–288.
8. Näätänen, R. & I. Winkler. 1999. The concept of auditory stimulus representation in cognitive neuroscience. *Psychol. Bull.* **125:** 826–859.
9. Näätänen, R., P. Paavilainen, K. Alho, *et al.* 1987. The mismatch negativity to intensity changes in an auditory stimulus sequence. *Electroencephalogr. Clin. Neurophysiol. Suppl.* **40:** 125–131.
10. Koelsch, S., E. Schröger & M. Tervaniemi. 1999. Superior pre-attentive auditory processing in musicians. *Neuroreport* **10:** 1309–1313.

11. Nikjeh, D.A., J.J. Lister & S.A. Frisch. 2008. Hearing of note: an electrophysiologic and psychoacoustic comparison of pitch discrimination between vocal and instrumental musicians. *Psychophysiology* **45:** 994–1007.
12. Münte, T.F., C. Kohlmetz, W. Nager, *et al.* 2001. Neuroperception. Superior auditory spatial tuning in conductors. *Nature* **409:** 580.
13. Vuust, P. *et al.* 2005. To musicians, the message is in the meter. *Neuroimage* **24:** 560–564.
14. Vuust, P., L. Østergaard, K.J. Pallesen, *et al.* 2009. Predictive coding of music. *Cortex* **45:** 80–92.
15. Brattico, E., R. Näätänen & M. Tervaniemi. 2001. Context effects on pitch perception in musicians and non-musicians: evidence from event-related-potential recordings. *Music Percept.* **19:** 199–222.
16. Brattico, E. *et al.* 2009. Neural discrimination of nonprototypical chords in music experts and laymen: an MEG study. *J. Cogn. Neurosci.* **21:** 2230–2244.
17. Tervaniemi, M., V. Just, S. Koelsch, *et al.* 2005. Pitch discrimination accuracy in musicians vs non-musicians: an event-related potential and behavioral study. *Exp. Brain Res.* **161:** 1–10.
18. Krumhansl, C.L. 1990. *Cognitive Foundations of Musical Pitch.* Oxford University Press. Oxford, UK.
19. Butler, M. 2006. *Unlocking the Groove: Rhythm, Meter, and Musical Design in Electronic Dance Music.* Indiana University Press. Bloomington, IN.
20. Fuller, D. 2010. Alberti bass. *Grove Music Online.*
21. Davis, M. 1964. *The Complete Concert: Four and More & My Funny Valentine.* Columbia Records (Audio CD).
22. Vuust, P. & A. Roepstorff. 2008. Listen up! Polyrhythms in brain and music. *Cogn. Semiotics* **3:** 131–159.
23. Näätänen, R., S. Pakarinen, T. Rinne & R. Takegata. 2004. The mismatch negativity (MMN): towards the optimal paradigm. *Clinical Neurophysiology* **115:** 140-144.
24. Monson, I. 1997. *Saying Something: Jazz Improvisation and Interaction.* (Chicago Studies in Ethnomusicology), The University of Chicago Press. Chicago.
25. Berliner, P.F. 1994. *Thinking in Jazz: The Infinite Art of Improvisation.* The University of Chicago Press. Chicago.
26. Vuust, P., A. Roepstorff, M. Wallentin, *et al.* 2006. It don't mean a thing. . .Keeping the rhythm during polyrhythmic tension, activates language areas (BA47). *Neuroimage* **31:** 832–841.
27. Vuust, P., M. Wallentin, K. Mouridsen, *et al.* 2011. Tapping polyrhythms in music activates language areas. *Neurosci. Lett.* **494:** 211–216.
28. Galvao, A. 1999. Kinaesthesia and instrumental music instruction: some implications. *Psychol. Music* **27:** 129–137.
29. Vuust, P., L. Ostergaard, K. J. Pallesen, *et al.* 2009. Predictive coding of music–brain responses to rhythmic incongruity. *Cortex* **45:** 80–92.
30. Pantev, C. *et al.* 2003. Music and learning-induced cortical plasticity. *Ann. N.Y. Acad. Sci.* **999:** 438–450.
31. Tervaniemi, M., A. Castaneda, M. Knoll & M. Uther. 2006. Sound processing in amateur musicians and non-musicians: event-related potential and behavioral indices. *Neuroreport* **17:** 1225–1228.
32. Vuust, P. *et al.* 2011. New fast mismatch negativity paradigm for determining the neural prerequisites for musical ability. *Cortex* **47:** 1091–1098.
33. Gordon, E.E. 1989. *Manual for the Advanced Measures of Music Audiation.* GIA Publications. Chicago.
34. Vuust, P. 2000. *Polyrhythm and Metre in Modern Jazz–A Study of the Miles Davis' Quintet of the 1960'ies (Danish).* Royal Academy of Music. Aarhus, Denmark.
35. Davis, M. 1959. *Kind of Blue.* Columbia Records (Audio CD).
36. Huotilainen, M., V. Putkinen & M. Tervaniemi. 2009. Brain research reveals automatic musical memory functions in children. *Ann. N.Y. Acad. Sci.* **1169:** 178–181.
37. Seppänen, M., E. Brattico & M. Tervaniemi. 2007. Practice strategies of musicians modulate neural processing and the learning of sound-patterns. *Neurobiol. Learn. Mem.* **87:** 236–247.
38. Wallentin, M., A.H. Nielsen, M. Friis-Olivarius, *et al.* 2010. The Musical Ear Test, a new reliable test for measuring musical competence. *Learn. Individ. Differ.* **20:** 188–196.
39. Vuust, P. *et al.* 2009. Personality influences career choice: sensation seeking in professional musicians. *Music Educ. Res.* **12:** 219–230.
40. Nikjeh, D. A., J.J. Lister & S. A. Frisch. 2009. The relationship between pitch discrimination and vocal production: comparison of vocal and instrumental musicians. *J. Acoust. Soc. Am.* **125:** 328–338.
41. Rüsseler, J., E. Altenmüller, W. Nager, *et al.* 2001. Event-related brain potentials to sound omissions differ in musicians and non-musicians. *Neurosci. Lett.* **308:** 33–36.

Ann. N.Y. Acad. Sci. ISSN 0077-8923

ANNALS OF THE NEW YORK ACADEMY OF SCIENCES
Issue: *The Neurosciences and Music IV: Learning and Memory*

Expertise in folk music alters the brain processing of Western harmony

M. Tervaniemi,[1,2] T. Tupala,[1,2] and E. Brattico[1,2]

[1]Cognitive Brain Research Unit, Cognitive Science, Institute of Behavioral Sciences, University of Helsinki, Helsinki, Finland.
[2]Center of Excellence in Interdisciplinary Music Research, University of Jyväskylä, Jyväskylä, Finland

Address for correspondence: Mari Tervaniemi, Cognitive Brain Research Unit, Institute of Behavioural Sciences, P.O. Box 9, 00014 University of Helsinki, Helsinki, Finland. mari.tervaniemi@helsinki.fi

In various paradigms of modern neurosciences of music, experts of Western classical music have displayed superior brain architecture when compared with individuals without explicit training in music. In this paper, we show that chord violations embedded in musical cadences were neurally processed in a facilitated manner also by musicians trained in Finnish folk music. This result, obtained by using early right anterior negativity (ERAN) as an index of harmony processing, suggests that tonal processing is advanced in folk musicians by their long-term exposure to both Western and non-Western music.

Keywords: musical expertise; auditory event-related potentials; learning; music; education

Introduction

Since the origins of the study of the neurosciences of music, one of the most prominent research areas has been the neural basis of musical expertise. Most empirical endeavors of this tradition have focused on comparing adult musicians with explicit training in classical Western music with adults who lack formal training in music. Remarkably, both functional and structural differences between these groups of participants have been obtained. In most occasions, musicians have shown faster and/or enhanced brain activity and, furthermore, have had larger brain areas devoted to music at various parts of the brain.[1–4] These differences have not only been visible at the cortical areas underlying primary auditory and somato-motor functions, but have been traced back at several crossmodal functions at cortical and subcortical areas and in several neural tracts, for example, connecting the two hemispheres.

Recently, however, it has been acknowledged that not all music and musicians originate from the classical genre: musicians active in rock[5] and jazz[6,7] have also been studied using brain imaging methodology. Interestingly, their neurocognitive profiles as determined by auditory event-related potentials (ERPs) and magnetic event-related fields (ERFs) seem to differ from those observed in classical musicians. For instance, compared with nonmusicians, rock musicians display higher sensitivity to changes in sound intensity and location than in sound frequency or duration.[5] Jazz musicians, in turn, have highly advanced left-lateralized processing of rhythmic changes[6] and enhanced discriminative brain processes in general.[7]

The current project was established to explore the neurocognitive profiles of folk musicians. Similar to folk music in many other parts of the world, Finnish folk music is aurally transmitted and makes use of improvisation and melodic variations. There are two major traditions: first, the older, less Western (*runo*) tradition with typical five-tone scales, various musical meters (e.g., 5/4, 4/4, 9/4), and emphasis on singing; and, second, the younger (*pelimanni*) tradition with diatonic scale and its major and minor modes, newer instruments (e.g., violin, clarinet, harmonium), and song and dance forms with regular meters originating from or influenced by Western tonal music. Nowadays, folk musicians are involved in well-structured musical training offered by public music schools (childhood, adolescence) and music academies (adult musicians), where they

doi: 10.1111/j.1749-6632.2011.06428.x

familiarize themselves with the Finnish traditions, contemporary world music, and, more or less, with folk music traditions of other cultures. Thus, folk musicians offer an ideal possibility to investigate the auditory neurocognition, which in their special case is a result of the interplay between various musical systems, both from the Western tonal tradition and from other non-Western musical cultures.

As a probe to auditory neurocognition of folk musicians, we used an ERP component called early right anterior negativity (ERAN). It is elicited by violations of chord cadences, and occurs 150–250 ms after the onset of a chord violation.[8] The ERAN is elicited by harmonically incongruous chords, such as the Neapolitan chord,[9,10] or double dominant chords (DD), at the tonic position[11,12] or at the dominant position.[13,14] Of particular interest in the current context is the sensitivity of ERAN to reflect the degree of encountered harmonic violation. Its time course (latency) of occurrence and strength (amplitude) sensitively reflect the degree of harmonic appropriateness.[14,15] As Leino *et al.*[13] and Garza Villarreal *et al.*[14] have shown, the positioning of the Neapolitan chord after a tonic or a dominant chord within the cadence is sensitively reflected in the ERAN attributes. Therefore, we also varied the position of the chord violation. Our working hypothesis was that the explicit training of folk musicians in other non-Western musical systems would be reflected by reduced ERAN amplitude (which is a special index of Western tonal system), without being affected by the chord position.

Methods

Stimuli

We created the seven-chord cadences by using digitally produced piano chords. The Neapolitan was placed at the third, fifth, or seventh position, each with 25% probability. Thus, only 25% of all cadences were without the Neapolitan chord. The Neapolitan chords at the third and seventh positions, which follow a dominant chord, thus produce a rare, highly incongruous succession according to Western tonal music. In contrast, the Neapolitan chord at the fifth position, which follows a tonic chord, results in a milder incongruity according to the Western musical system. The chords with organ timbre were included in about 8% of the cadences, and they were similar to the other chords in all aspects except timbre. These deviant chords were never presented at positions three, five, or seven.

The sounds were digitally edited to be of equal length and intensity. The first six chords of the cadence were 600 ms in duration, and the last one was 1,200 ms. The chord cadences were transposed to 12 frequency levels, corresponding to the twelve pitch classes of the chromatic scale.

Participants and procedure

Twenty-two healthy young adults with normal hearing aged 20 to 35 years old participated in the experiments. Of them, 11 were folk musicians who studied in a music academy or university (4 males; $M = 27.7$ years) and 11 were nonmusicians who had studied or practiced music for no more than two years and lessons had finished over 10 years before the experiment (2 males; $M = 25.3$ years). Their participation was compensated for with movie tickets.

During the experiment, participants were instructed to press a button whenever they heard a chord with an organ-like timbre. Thus, the task was "semiattentive," making it possible to compare the results to those of previous ERAN experiments:[8] the participants did not have a primary task outside the auditory modality but did not focus their attention to the Neapolitan chords either.

The sounds were presented with Presentation software at 50 dB above individually determined hearing thresholds in the beginning of the experimental session. The sounds were presented via headphones while the participant was sitting in a comfortable chair in an electronically and acoustically shielded room.

Data recording and analysis

The EEG was recorded with a 64 Ag–Cl active electrode cap using BioSemi ActiView equipment (4096 Hz downsampled to 512 Hz offline; band-pass 0.16–100 Hz). Additional electrodes were attached to the mastoids, under the right eye, and on the nose. The data were analyzed in a Matlab environment. The data were filtered offline using a 1–30 Hz band-pass filter and baseline set up to the mean 100-ms precadence interval. The analysis epoch lasted from 100 ms prestimulus to 600 ms poststimulus. When there were EEG epochs containing a deflection exceeding ± 100 μV on at least eight channels, the epoch was rejected from further processing.

The data were rereferenced to the mean value recorded from the two mastoid channels. Based on visual inspection of the ERP responses, we set up two time windows for analysis, the first at 100–230 ms and the second at 230–350 ms. Peak latencies in these two time windows were computed from Cz. Thereafter, mean amplitudes were calculated averaging on a 20 ms window around the grand-average peaks defined for each group and stimulus condition separately for each electrode or based on Cz. We extracted mean amplitudes from the following electrodes: F1, F3, F5, FC1, FC3, FC5, C1, C3, C5, CP1, CP3, CP5, P1, P3, P5, F2, F4, F6, FC2, FC4, FC6, C2, C4, C6, CP2, CP4, CP6, P2, P4, and P6.

To test the presence of differences between groups and stimulus conditions, we performed repeated-measures ANOVA with group (folk musicians, nonmusicians) as the between-subjects factor, and chord position (third, fifth, seventh), laterality (left, right), frontality (frontal, fronto-central, central, parieto-central, parietal), and electrodes, which included in each region of interest (three levels) as the within-subject factors. In the current report, however, we concentrate on results with group and chord position alone. The findings on laterality and frontality will be reported elsewhere.

Results

Early negativity (100–230 ms)

As illustrated by Figure 1, the Neapolitan chords at the three chord positions elicited a systematically larger ERAN amplitude in folk musicians as compared with nonmusicians ($F[1,20] = 10.96$, $P < 0.001$). There were no significant effects of chord positions or interactions including group as factor ($F < 1$).

Late negativity and positivity (230–350 ms)

As illustrated by Figure 1, the late ERAN around 285 ms was modulated by group and chord positions. The main effect of Chord position ($F[2,40] = 36.6$, $P < 0.001$) and subsequent tests indicated that ERP in response to the Neapolitan chord at the fifth position in the cadence was more negative than the ERP at the two other positions ($P < 0.001$). Importantly, while the groups did not differ in the overall ERP amplitudes across chord positions ($F < 1$), they did differ in how they processed chords based on their positions. This was evidenced by the significant interaction group × chord position ($F[2,40] = 3.7$, $P = 0.03$).

Separate ANOVAs for the chord position revealed main effects of group for the fifth ($F[1,40] = 3.9$, $P = 0.06$) and seventh chord positions ($F[1,40] = 5.7$, $P = 0.03$), whereas no significant group effect was obtained for the chord at the third position ($F < 1$). These findings resulted from an enhanced ERAN in folk musicians to the Neapolitan chord at the fifth position only (particularly in the frontal area, as evidenced by separate ANOVAs), and from an enhanced P3a at the ending seventh position. In contrast, the P3a in response to the third position did not differentiate folk musicians from nonmusicians.

Discussion

In this paper, our aim was to explore whether expertise in Finnish folk music is reflected in the auditory neurocognition, particularly in terms of ERAN response evoked by Neapolitan chords. According to our data, the earliest group difference between folk musicians and non-musically trained individuals was observable at around 170 ms: ERAN was overall enhanced in folk musicians when compared with nonmusicians. Following this early ERAN, the stronger chord violations in the final chord position elicited a P3a that was enhanced for folk musicians. The milder chord violation in the fifth position evoked a prominent late ERAN in folk musicians compared

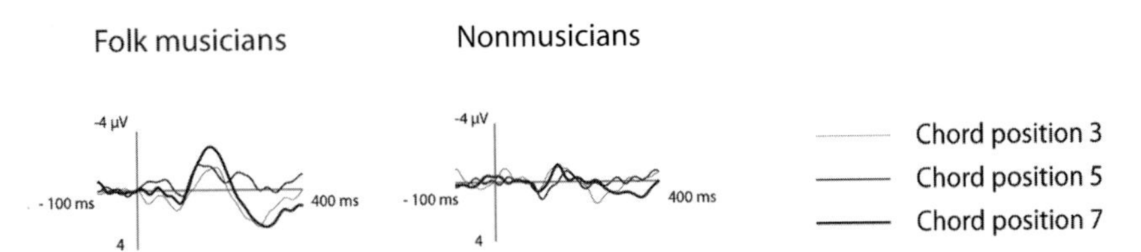

Figure 1. ERAN elicited by chord violations as indicated by thin (position 3), medium thick (position 5), and thick (position 7) lines (Cz electrode). Early ERAN was systematically larger in folk musicians than in nonmusicians. It was followed by late ERAN (position 5) and P3a (position 7) in folk musicians only.

with nonmusicians. The latest P3a and ERAN responses were thus sensitive to the strength of the encountered chord violation. We interpret these findings as showing a culture-dependent neuroplasticity of brain responses to musical rule violations.

The early ERAN was enhanced in folk musicians as compared to nonmusicians, irrespective of the position of the Neapolitan chord in the cadence. This finding replicates and generalizes to folk musicians what was observed in classically trained musicians.[16,17] Hence, our original hypothesis about the reduced ERAN in folk musicians was not supported. This hypothesis was based on an assumption that since ERAN probes expectancy violations, particularly in the context of Western tonal system, folk musicians would not display enhanced neural representations of these violations as a consequence of their expertise in various (mainly non-Western) musical systems. Instead, the result of enhanced ERAN, discrepant from our expectations, may have resulted from lessons in a classical music instrument during the childhood of some folk musicians. In addition, unavoidably, all members of modern society receive exposure to Western music in everyday life. So, these two factors together increased (rather than decreased) their observed ERAN and the underlying neural representations of the chord succession rules of Western tonal music. Moreover, this ERAN increment was present equally in chord violations, independent of the chord position.

In contrast, the positive response following the early ERAN at about 285 ms was sensitively modulated by both group and chord position in the cadence. It was elicited in all subjects in response to the strong incongruity at the third and seventh positions of the cadence, and enhanced only in the seventh position in folk musicians compared to nonmusicians. Our interpretation is that, based on its latency, polarity, and topography, this response can be labeled as a P3a response that reflects an involuntary switch of attention toward the Neapolitan chords.[18] In the current context, the P3a elicitation is a novel finding and, because of group comparison design and P3a sensitivity to the expertise of the participants in folk music, a highly interesting one. Its elicitation can be interpreted to reflect an intriguing sensitivity of folk musicians to target their attention toward chords that are highly discrepant from the most congruous chords. We can suppose that the enhanced P3a response to the ending chord of the cadence in folk musicians stems from their robust expectations for the tonic chord, unfulfilled because of the presentation of the Neapolitan.

Late ERAN-resembling response of negative polarity to the chord at the fifth position appears to be enhanced also in folk musicians. Its presence might be attributed to the possibility of the Neapolitan to occur after a tonic chord, as it happened in the fifth position. In a cadence constructed according to the most traditional rules of the Western music theory, it is a rare but feasible alternative. Our interpretation is that folk musicians, thanks to their implicit and, in some cases, explicit training in musical systems, which follow the Western music theory (even if it has merged with their expertise in other harmonic systems), are sensitive to those intrinsic regularities of Western music. If a Neapolitan chord generates a "wrong" chordal succession within a cadence, their attention is automatically shifted toward such a chord reflected by P3a. However, if it follows a chord producing some expectation for the Neapolitan, only a late ERAN-kind of negativity is elicited. In our study, folk musicians tended to have superior ERAN to the milder incongruity of the Neapolitan chord at the fifth position, showing sensitivity toward chord progression following the rules of Western tonal music.

In summary, the current findings indicate a highly specialized neurocognitive ability of folk musicians that is activated when harmonic encoding is requested. Not only did these musicians display generally enhanced early ERAN, but also subsequent P3a and late-ERAN, which were evoked in a very sensitive manner with regard to the chord positions. These data offer further support for the existence of different profiles that adult musicians might have as a function of their long and profound training within a given genre and thus emphasize the need for taking the personal history and enculturation of the subjects into close account in neuroscientific studies.

Acknowledgments

The authors would like to thank MMus. Leena Joutsenlahti, MMus. Kirsi Ojala, and Dr. Juha Ojala for influential discussions on the practices of music making among Finnish folk musicians.

Conflicts of interest

The authors declare no conflicts of interest.

References

1. Münte, T.F., E. Altenmüller & L. Jäncke. 2002. The musician's brain as a model of neuroplasticity. *Nat. Neurosci.* **3:** 473–478.
2. Jäncke, L. 2009. The plastic human brain. *Restor. Neurol. Neurosci.* **27:** 521–538.
3. Tervaniemi, M. 2009. Musicians – same or different? *Ann. N.Y. Acad. Sci.* **1169:** 151–156.
4. Tervaniemi, M. In press. Musicianship—how and where in the brain? In *Musical Imaginations: Multidisciplinary Perspectives on Creativity, Performance and Perception.* D. Hargreaves, D. Miell & R. MacDonald, Eds: 285–295. Oxford University Press. Oxford.
5. Tervaniemi, M., A. Castaneda, M. Knoll & M. Uther. 2006. Sound processing in amateur musicians and non-musicians: event-related potential and behavioral indices. *NeuroReport* **17:** 1225–1228.
6. Vuust, P., K.J. Pallesen, C. Bailey, *et al.* 2005. To musicians, the message is in the meter: pre-attentive neuronal responses to incongruent rhythm are left-lateralized in musicians. *NeuroImage* **24:** 560–564.
7. Vuust, P., E. Brattico, M. Seppänen, *et al.* 2012. Practiced musical style shapes auditory skills. *Ann. N.Y. Acad. Sci.* **1252:** 139–146 (this issue).
8. Koelsch, S. 2009. Music-syntactic processing and auditory memory: similarities and differences between ERAN and MMN. *Psychophysiology* **48:** 179–190.
9. Koelsch S., T.C. Gunter, A.D. Friederici & E. Schröger. 2000. Brain indices of music processing: "Non-musicians" are musical. *J. Cogn. Neurosci.* **12:** 520–541.
10. Loui, P., T. Grent-'t-Jong, D. Torpey & M. Woldorff. 2005. Effects of attention on the neural processing of harmonic syntax in Western music. *Cogn. Brain Res.* **25:** 678–687.
11. Koelsch S., S. Jentschke, D. Sammler & D. Mietchen. 2007. Untangling syntactic and sensory processing: an ERP study of music perception. *Psychophysiology* **44:** 476–490.
12. Sammler, D., S. Koelsch & A.D. Friederici. 2011. Are left fronto-temporal brain areas a prerequisite for normal music-syntactic processing? *Cortex* **47:** 659–673.
13. Leino, S., E. Brattico, M. Tervaniemi & P. Vuust. 2007. Representation of harmony rules in the human brain: further evidence from event-related potentials. *Brain Res.* **1142:** 169–177.
14. Garza Villarreal, E.A., E. Brattico, S. Leino, *et al.* 2011. Distinct neural responses to chord violations: a multiple source analysis study. *Brain Res.* **1389:** 103–114.
15. Koelsch S., S. Kilches, N. Steinbeis & S. Schelinski. 2008. Effects of unexpected chords and of performer's expression on brain responses and electrodermal activity. *PLoS One* **3:** e2631.
16. Koelsch, S., B.-H. Schmidt & J. Kansok. 2002. Effects of musical expertise on the early right anterior negativity: an event-related brain potential study. *Psychophysiology* **39:** 657–663.
17. Müller, M., L. Hofel, E. Brattico & T. Jacobsen. 2010. Aesthetic judgments of music in experts and laypersons–an ERP study. *Int. J. Psychophysiol.* **76:** 40–51.
18. Escera, C., K. Alho, R.E. Schröger & I. Winkler. 2000. Involuntary attention and distractibility as evaluated with event-related brain potentials. *Audiol. Neurootol.* **5:** 151–166.

Ann. N.Y. Acad. Sci. ISSN 0077-8923

ANNALS OF THE NEW YORK ACADEMY OF SCIENCES
Issue: *The Neurosciences and Music IV: Learning and Memory*

ERP responses to cross-cultural melodic expectancy violations

Steven M. Demorest[1] and Lee Osterhout[2]

[1]Laboratory for Music Cognition, Culture and Learning, School of Music, [2]Cognitive Neuroscience of Language Lab, Department of Psychology, University of Washington, Seattle, Washington

Address for correspondence: Steven M. Demorest, School of Music, University of Washington, Box 353450, Seattle, WA 98195-3450. demorest@uw.edu

In this preliminary study, we measured event-related potentials (ERPs) to melodic expectancy violations in a cross-cultural context. Subjects ($n = 10$) were college-age students born and raised in the United States. Subjects heard 30 short melodies based in the Western folk tradition and 30 from North Indian classical music. Each melody was presented in its original and deviation form, and subjects were asked to judge the congruence of the melody. Results indicated that subjects found the Indian melodies less congruous overall and were less sensitive to deviations in the Indian melody condition. ERP data were partly consistent with the behavioral data with significant P600 responses to deviations in both cultural conditions, but less robust in the Indian context. Results are interpreted in light of previous research on listeners' abilities to generate expectancies in unfamiliar cultures and the possibility of overlap in the scale systems influencing the findings.

Keywords: culture; melodic expectancy; event-related potential; music cognition

Introduction

Music, like language, is a universal phenomenon occurring in virtually every known human culture. Humans seem predisposed to process musical information and can make relatively sophisticated musical judgments from an early age[1] and without formal training.[2–4] Unlike language, music's abstract and nonreferential nature can lead to a perception that musical understanding easily crosses cultural boundaries where language cannot. Unlike language, music's abstract and nonreferential nature can lead to a perception that musical understanding easily crosses cultural boundaries where language cannot. While the presence of music may be a cultural universal, even a cursory overview of the world's diverse musical practices suggest that musical understanding is not. Even a shared definition of the word *music* across cultures is difficult, if not impossible to find.[5] Within a given culture, what functions as music is often highly contextualized, adhering to a specific set of conventions particular to that culture. Consequently, exposure to those cultural conventions is likely to shape an individual's understanding of music at a cognitive level.

Research into cross-cultural music cognition has found evidence for both the universality and particularity of musical thought and behavior. Early research into the cross-cultural perception of tonal hierarchies in music of India,[6] Indonesia,[7] China,[8] Finnish folk music,[9] and Sami yoiks[10] found that cultural outsiders could in fact derive tonal hierarchy profiles and melodic expectations from minimal exposure to another culture's music, but that their judgments revealed some biases of their home culture and a lack of sensitivity to some cultural conventions. A more recent study by Curtis and Bharucha[11] sought to expose listeners' cultural biases more explicitly. Using a recognition memory paradigm, they presented listeners with melodies based on either a Western or Indian scale but with one scale tone missing. After hearing each melody, they played a single tone and asked listeners to identify whether they had heard that tone played in the melody. Western listeners were more likely to "remember" hearing a tone from the Western scale even when the melody was played in an Indian

doi: 10.1111/j.1749-6632.2012.06464.x

scale context. Research into the role of culture in memory has revealed a consistent in-culture bias for recognition memory for listeners from multiple cultures[12–15] suggesting that prior exposure shapes our ability to process and retain musical structure. Bimusical individuals, on the other hand, perform equally well on memory tasks in both their "home" cultures, further strengthening the argument that enculturation shapes music cognition.[15]

When individuals listen to a piece of music, they do not just passively take in information from the piece, they use that information to make predictions about what will come next. Those predictions or *expectancies* are based in part on the content of what they are listening to and in part on their prior experiences with music.[16] The more familiar a particular style of music, the stronger the predictions are regarding what comes next. Expectancy plays a crucial role in the ability to understand the structure of music and derive enjoyment from it. If the music does not make sense, then there is no clear sense of what is coming next and it is unlikely to hold one's interest. Conversely, if a piece of music is too familiar or repetitive, interest also wanes. Emotional responses to music are often a result of the tension between what is expected and what actually happens.[17,18]

When our expectations for a piece of music are violated, it initiates an event-related electrical response in the brain. The nature of the brain response depends on the nature of the deviation. Previous research with Western music has documented that listeners respond to an out-of-key note with a late positive (P300 or P600) event-related potential (ERP) component when they are actively engaged in listening.[19–21] An ERP study by Brattico *et al.*,[21] featured both passive and active music listening tasks. In the active task, subjects were asked to respond to out-of-tune or out-of-key notes that were inserted in random spots intact melodies. They found that listeners responded behaviorally by rating deviant versions of the melodies as significantly less congruous than originals. Neurologically, subjects responded to both types of deviations with a P600 event related potential. In all of the ERP studies mentioned thus far, listeners were detecting deviations to melodic expectations formed within a culturally familiar context. Because different cultures have different tonal systems, it is not clear whether listeners can detect melodic expectancy violations outside of a familiar cultural context and whether such deviations would elicit a similar brain response.

Two previous cross-cultural ERP studies have explored listeners' ability to detect out-of-key notes in a scale context using an oddball paradigm.[22,23] A study by Bischoff Reninger *et al.*[23] explored whether Western encultured listeners could detect out-of-key notes within a Javanese *pélog* scale. They had two groups of Western-born musicians, one group with no exposure to Javanese music, and one that had taken lessons in Javanese music at a U.S. university. They found that both subject groups could detect scale deviations in both conditions both behaviorally and with a P300 response to deviant tones, suggesting that mere exposure to a standard allows listeners to construct expectancies for out-of-culture examples. However, listeners with no exposure to Javanese music exhibited smaller P300 responses than those with exposure to the other culture. In both prior studies, scale tones were presented in an oddball paradigm that focused on detection of change from a standard. The judgments do not represent the contextualized tonality judgments that are typical of most music listening. To test the influence of enculturation on top–down processes of music cognition, judgments would need to be contextualized in a melodic framework requiring the inference of underlying scale components for expectancy formation. We applied the procedures of Brattico *et al.*[21] measuring both behavioral and brain responses to out-of-key notes in a melodic context, but using melodies from two different cultures as the stimuli. The purpose of this initial study was to explore the influence of cultural background on U.S.-born listeners' ability to respond to expectancy violations in intact melodies from two cultural contexts. The hypotheses were:

(1) Participants will rate North Indian melodies in any condition as less congruous than Western melodies.
(2) Participants will respond to out-of-key deviations in Western melodies with a P600 ERP response.
(3) Participants will not show a P600 response to similar deviations in North Indian music.

Methods

Participants for the study were college-aged adults drawn from a university subject pool. The mean age of the participants was 20.2 years with a range of 19–22 years and the average amount of musical training was 6.7 years with a range of 0–13 years. Participants had to have been born and raised in the United States and could not be of Indian descent.

All participants listened to synthesized presentations of 30 European folk song excerpts and 30 excerpts from the *alap* portion of different North Indian *ragas*. All melodies were heard in their original form and in deviation form (120 total). Deviations for both sets of melodies occurred on a strong beat after at least one measure of music. Deviations were created by changing a target note by one semitone. The direction of the change was initially determined at random, though modifications were made if the direction of change moved to another note within the same scale or *thaat*. The target notes were purposely not tonic pitches or notes that had occurred with great frequency to minimize alternate cues regarding scale membership.

Original melodies and deviations were heard in two blocks of 30 melodies for each culture. Within each block all 30 melodies from each culture were heard, 15 in deviation form and 15 in original form. The order of the melodies within each block, the order of blocks within each culture, and the order in which each culture was presented to participants were randomized into four distinct presentation orders.

Continuous electroencephalogram (EEG) was recorded from 29 scalp electrodes referenced to the left mastoid, using a bioamplifier (SA Instrumentation Co., Encinitias, CA) (bandpass 0.01–100 Hz, 3 dB cut-off) with a sampling rate of 200 Hz. Electrodes were positioned at extended 10–20 locations at midline, medial lateral, and lateral locations. Two additional electrodes were used to monitor for horizontal and vertical eye movements. Averaged ERPs were calculated off-line from trials free of artifact. Of particular interest was the P600 effect. Therefore, ERPs were quantified as mean amplitudes within a 500–800 ms window, which represents the typical temporal window for the P600.

Participants were first briefly familiarized with the apparatus and the judgment task. They then heard the melodies over speakers at a range of 62–66 dB SPL, with an average peak level of approximately 65 dB SPL. The speakers were positioned on either side of a computer monitor that provided a fixation point during listening and instructions to rate the melodies after each hearing. Subjects pressed a button after rating each melody to hear the next melody.

Results

Hypothesis 1 predicted that participants would find Indian music significantly less congruous than Western music. We conducted a repeated measures analysis of variance on participants' ratings of congruousness across the two cultural styles (Western/Indian) and the two melody conditions (original/deviation). The analysis revealed a significant main effect for culture with Indian music rated as significantly less congruous than Western ($F[1, 9] = 26.04$, $P < 0.001$) (Fig. 1). There was a significant main effect for condition as well with deviation melodies rated as less congruous than the original versions ($F[1, 9] = 27.97$, $P < 0.001$), however a significant condition by culture interaction ($F[1, 9] = 15.17$, $P < 0.01$) suggests that participants were less sensitive to deviations in the culturally unfamiliar melody context.

We also predicted that our subjects would respond to melodic deviations in Western music by exhibiting the same late positive component found in earlier studies, but would not exhibit a similar sensitivity in the Indian melody context. An analysis of the ERP data found a significant main effect for condition for the Western melodies ($F[1, 9] = 18.09$, $P < 0.01$), as well as a significant condition by midline interaction ($F[2, 18] = 5.69$, $P < 0.05$), suggesting a widely distributed P600 response with the strongest activation in the posterior regions. There was no corresponding main effect of condition for Indian music, but there was a significant condition by midline interaction ($F[2, 18] = 5.26$, $P < 0.05$), indicating that subjects did respond to the Indian deviations, but their response was less robust. Figure 2 shows the difference waves for the Western versus Indian deviation conditions across the two cultures, which visually demonstrate the difference in the magnitude of the responses.

Discussion

The behavioral results show that listeners are less sensitive to melodic expectancy violations in the music of unfamiliar cultures compared to their own culture. The ERP data were more mixed, with subjects exhibiting a late positive component

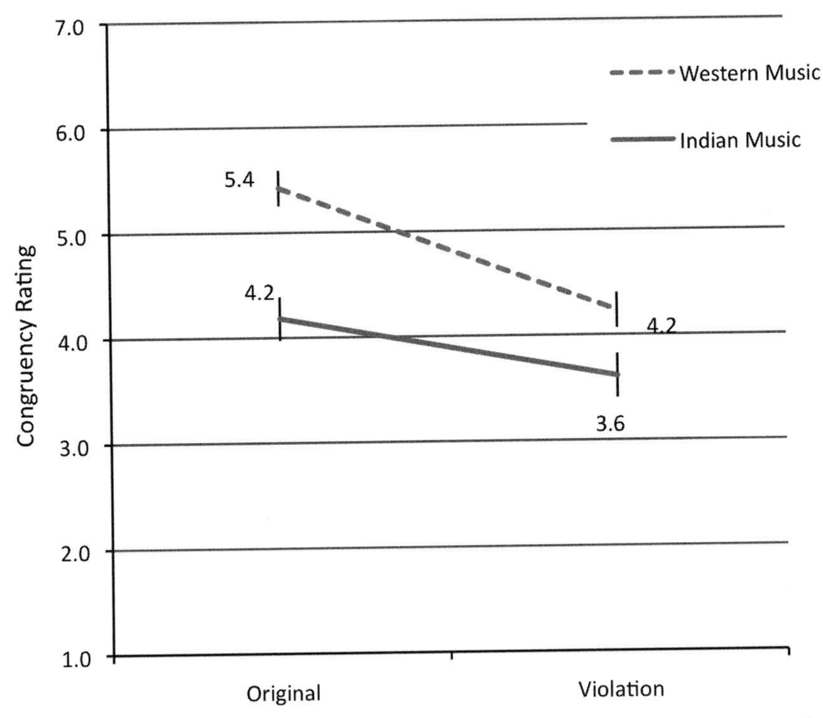

Figure 1. Participants' ratings of congruousness on a seven-point Likert scale across culture and melody condition.

in response to deviations in both cultural conditions, though less robust in the culturally unfamiliar context. This finding was similar to the weaker P300 response of out-of-culture listeners to scale deviations in a previous study.[23] The results provide some support for the idea that listeners can, with minimal exposure, internalize certain patterns in culturally unfamiliar music that allow them to respond to stimulus changes. This is consistent with findings from tonal hierarchy research that listeners can respond to the distribution of tones in culturally unfamiliar music[6–10] and with the aforementioned ERP study of cross-cultural scale perception.[23] However, we are cautious about endorsing such an interpretation because there are some possible confounds that may have influenced these results.

We predicted that Western listeners would not respond to melodic expectancy violations in culturally unfamiliar music. However, to the extent that the out-of-culture stimuli contained elements of Western musical structures, listeners might apply their in-culture schema to the interpretation of out-of-culture music. This misapplication of encultured schemas to out-of-culture music has been demonstrated in both behavior[10,11] and brain[13,24] responses for previous cross-cultural studies. Although Indian *ragas* are not scale-based in the same way that Western melodies are, they do represent underlying collections of pitches or *thaats*. North Indian music has 32 possible scales that are called *thaats* (though only 10 are in frequent use). Some of these *thaats* have tone sets that correspond to Western major and minor scales. For example, the *Bilawal thaat* corresponds to the Western major mode. To the extent that the Indian stimuli activated subjects' Western schemas, our listeners may have responded to a deviation, but of a Western rather than Indian melodic structure. Subsequent research needs to clarify the extent of this theoretical intersection between the two systems and either choose *ragas* based in *thaats* that do not resemble Western scales or include the possible intersection as a variable in the analysis.

Because Indian listeners were not included in the initial phase of this exploratory study, we cannot determine conclusively how well the notion of melodic expectancy "violation" was operationalized for North Indian music. Although the stimuli were based on existing *ragas*, and the deviations were

Figure 2. ERP difference waves in response to deviations in Western and Indian music listening conditions demonstrating a late positive component but with a different magnitude and distribution by condition.

created by an in-culture colleague, the limitations of ERP may have interfered with results. ERP stimuli typically last for a few seconds because subjects need to listen to a large number of stimuli for signal averaging purposes. On the other hand, the typical time scale of an Indian *raga* unfolds over minutes not seconds. Given that North Indian musical structure is melody based, it is unclear how long a *raga* needs to be heard to generate strong expectancies with in-culture listeners. These questions should be resolved in the next phase of our investigation by conducting a fully comparative ERP study involving the music and participants from both cultures.

The interaction of cultural background with basic cognitive processes like expectancy formation and memory is an important area for future research in the cognitive neuroscience of music.[25] Results can tell us something about the role of culture in shaping music cognition and the nature of musical learning. Once we understand differences between in-culture and out-of-culture responses to music, we can begin to track an individual's learning. Such tracking has already been accomplished in second language learning, finding that even short-term intensive exposure can change subjects' brain responses to a new language.[26] In music, such information could help educators to determine what types of experiences contribute to better learning of unfamiliar musical systems.

From a theoretical standpoint, ERP can be a powerful tool to test theories of expectancy formation. For example, researchers have developed some statistical profiles of Western melodic movement based on large databases of songs.[17] Such profiles should predict how strongly listeners respond to deviations of cultural rules, predictions that can be tested using ERP. With such tools, we might be able to quantify the "cultural distance" between two musical styles in a way that could predict how easily members of one group could learn the music of the other.

The issues encountered in this study also reveal some of the potential challenges of cross-cultural comparative research. When theories such as melodic expectancy have their genesis in Western notions of scale and key, it can be difficult to realize those concepts authentically in non-Western musical styles. In order for any theory of music cognition to apply broadly to human music cognition, it must operate within these different cultural contexts. Cross-cultural research has the potential to clarify our understanding of human music cognition in ways that transcend individual cultural conventions.

Acknowledgments

The authors are grateful to Professor Ramesh Gangolli for his work in creating the Indian stimuli and consulting on aspects of North Indian music. We also thank Dr. Judy McLaughlin and Geoffrey Valentine for their expert assistance in data collection and analysis.

Conflicts of interest

The authors declare no conflicts of interest.

References

1. Trehub, S.E. 2001. Musical predispositions in infancy. *Ann. N.Y. Acad. Sci.* **930:** 1–16.
2. Koelsch, S., T. Gunter, A.D. Friederici & E. Schröger . 2000. Brain indices of music processing: "nonmusicians" are musical. *J. Cogn. Neurosci.* **20:** 520–541.
3. Bigand, E. 2003. More about the musical expertise of musically untrained listeners. *Ann. N.Y. Acad. Sci.* **999:** 304–312.
4. Bigand, E. & B. Poulin-Charronnat. 2006. Are we "experienced listeners"? A review of the musical capacities that do not depend on formal musical training. *Cognition* **100:** 100–130.
5. Cross, I. 2008. Musicality and the human capacity for cultures. *Musicae Scientiae, Special Issue: Narrative in Music and Interaction* **12:** 147–167.
6. Castellano, M.A., J.J. Bharucha & C.L. Krumhansl. 1984. Tonal hierarchies in the music of north India. *J. Exp. Psychol. Gen.* **113:** 394–412.
7. Kessler, E.J., C. Hansen & R.N. Shepard. 1984. Tonal schemata in the perception of music in Bali and the West. *Music Percept.* **2:** 131–165.
8. Krumhansl, C.L. 1995. Music psychology and music theory: problems and prospects. *Music Theor. Spectr.* **17:** 53–80.
9. Krumhansl, C.L., J. Louhivuori, P. Toiviainen, *et al.* 1999. Melodic expectation in Finnish spiritual folk hymns: convergence of statistical, behavioral, and computational approaches. *Music Percept.* **17:** 151–195.
10. Krumhansl, C. L., P. Toivanen, T. Eerola, *et al.* 2000. Cross-cultural music cognition: cognitive methodology applied to North Sami Yoiks. *Cognition* **76:** 13–58.
11. Curtis, M.E. & J.J. Bharucha. 2009. Memory and musical expectation for tones in cultural context. *Music Percept.* **26:** 365–375.
12. Demorest, S.M., S.J. Morrison, M.N. Beken & D. Jungbluth. 2008. Lost in translation: an enculturation effect in music memory performance. *Music Percept.* **25:** 213–223.
13. Demorest, S.M., S.J. Morrison, M.N. Beken, *et al.* 2010. Music comprehension among western and Turkish listeners: FMRI investigation of an enculturation effect. *Soc. Cogn. Affect Neurosci.* **5:** 282–291.
14. Morrison, S.J., S.M. Demorest & L.A. Stambaugh. 2008. Enculturation effects in music cognition: the role of age and music complexity. *J. Res. Music Educ.* **56:** 118–129.
15. Wong, P.C.M., A.K. Roy & E.H. Margulis. 2009. Bimusicalism: the implicit dual enculturation of cognitive and affective systems. *Music Percept.* **27:** 81–88.
16. Vuust, P., L. Østergaard, K.J. Pallesen, *et al.* 2009. Predictive coding of music. *Cortex* **45:** 80-92.
17. Huron, D. 2006. *Sweet Anticipation: Music and the Psychology of Expectation.* The MIT Press. Cambridge, MA.
18. Meyer, L.B. 1956. *Emotion and Meaning in Music.* University of Chicago Press. Chicago, IL.
19. Besson, M., F. Faita & J. Requin. 1994. Brain waves associated with musical incongruities differ for musicians and non-musicians. *Neurosci. Lett.* **168:** 101–105.
20. Besson, M. & F. Faita. 1995. An event-related potential (ERP) study of musical expectancy: comparison of musicians with nonmusicians. *J. Exp. Psychol. Hum. Percept. Perform.* **21:** 1278–1296.
21. Brattico, E., M. Tervaniemi, R. Näätänen & I. Peretz. 2006. Musical scale properties are automatically processed in the human auditory cortex. *Brain Res.* **1117:** 162–174.
22. Neuhaus, C. 2003. Perceiving musical scale structures. A cross-cultural event-related brain potentials study. *Ann. N.Y. Acad. Sci.* **999:** 184–188.
23. Bischoff Renninger, L., M.P. Wilson & E. Donchin. 2006. The processing of pitch and scale: an ERP study of musicians trained outside of the Western musical system. *Empir. Music Rev.* **1:** 185–197.
24. Nan, Y., T.R. Knosche & A.D. Friederici. 2006. The perception of musical phrase structure: a cross-cultural ERP study. *Brain Res.* **1094:** 179–191.
25. Morrison, S.J. & S.M. Demorest. 2009. Cultural constraints on music perception and cognition. *Prog. Brain Res.* **178:** 67–77.
26. McLaughlin, J., L. Osterhout, & A. Kim. 2004. Neural correlates of second-language word learning: minimal instruction produces rapid change. *Nat. Neurosci.* **7:** 703–704.

Ann. N.Y. Acad. Sci. ISSN 0077-8923

ANNALS OF THE NEW YORK ACADEMY OF SCIENCES
Issue: *The Neurosciences and Music IV: Learning and Memory*

Effects of mono- and bicultural experiences on auditory perception

Patrick C.M. Wong,[1,2,3] Alice H.D. Chan,[4] and Elizabeth H. Margulis[5]

[1]The Roxelyn and Richard Pepper Department of Communication Sciences and Disorders, Northwestern University, Evanston, Illinois. [2]Department of Otolaryngology—Head and Neck Surgery, Northwestern University, Chicago, Illinois. [3]Hugh Knowles Center for Clinical and Basic Science in Hearing and Its Disorders, Northwestern University, Evanston, Illinois. [4]Division of Linguistics and Multilingual Studies, Nanyang Technological University, Singapore. [5]Department of Music, University of Arkansas, Fayetteville, Arkansas

Address for correspondence: Patrick C.M. Wong, The Roxelyn and Richard Pepper Department of Communication Sciences and Disorders, Northwestern University, 2240 Campus Drive, Evanston, IL 60208. pwong@northwestern.edu

The auditory system functions in the context of everyday life and the cultural environment in which we live. Although cultural-invariant, universal principles certainly contribute to sound processing, cultural factors play a role as well. In this review paper, we discuss two potential sources of cultural influence on auditory perception. We term the first type bottom–up, and use it to refer to the way that increased exposure to particular kinds of sound could shape our auditory and auditory–neural responses. The second type we term top–down, and use it to refer to the way our cultural upbringing broadly shapes how we think, which may in turn have an impact on how we perceive the world. An important consideration regarding cultural influences is that many individuals grow up with exposure to environmental stimulations of more than one culture. In our discussion, we will consider both mono- and bicultural experiences.

Keywords: music perception; bimusicalism; culture; fMRI

Introduction

Auditory events are heard within the confines of a cultural environment and are often produced by individuals of a particular culture. In the auditory domains of speech and music, culture plays an important role in both perception and production.

Research on the behavioral and neural sciences of auditory processing has focused on identifying universal principles, such as areas of the brain that are associated with musical syntactic processing,[1] auditory object recognition,[2,3] speech,[4] as well as the ways spoken language and sound perception can be impaired due to brain injuries[5] and cognitive decline.[6,7] This line of work has informed us about the biobehavioral mechanisms of sound processing.

In contrast to work that examines the universal principles of human perception, cultural psychologists have devoted much effort to identifying cultural differences.[8,9] For example, it has been found that individuals from East Asian and North American cultures might focus on the foreground and background aspects of the same visual image to a different extent.[10] Much of this work on cultural differences focuses on individuals who identify themselves as belonging predominately to one culture.[11]

In our view, culture may provide two types of influences to our (auditory) perceptual system. First, culture may provide a bottom–up influence on our brain and behavioral responses to specific types of sounds. Just as exposure to the sound of our own language conditions the way we respond to it and to other languages,[12] exposure to the sound of our culture's music may condition the way we respond to it and to other musical systems. Second, and less discussed, culture may provide a top–down cultural influence on auditory processing. It is possible that cultural upbringing shapes our general thinking style, which may in turn affect how we perceive sound in our environment. This top–down,

doi: 10.1111/j.1749-6632.2012.06407.x
Ann. N.Y. Acad. Sci. 1252 (2012) 158–162

culture-to-perception influence is in line with research on visual perception that attributes cultural group differences in image viewing to top–down processes.[10]

In our discussion later, we will review studies that provide support to both types of cultural influences. In addition to discussing individuals who have been exposed to and brought up in one culture, we will also discuss the effects of bicultural experiences on our auditory system, especially pertaining to the perception of music.

Bottom–up cultural effects: behavioral evidence

In our earlier work, we were interested in examining the behavioral responses to in-culture and other-culture music by Western and Indian (South Asian) listeners who reported having been predominately exposed to the music of only one culture.[13,14] In addition, we were interested in the responses of listeners who were "bimusical;" that is, listeners who grew up listening to music from both Western and Indian cultures and reported to continue to listen to music of both cultures. All listeners performed a cognitive recognition memory experiment and an affective tension judgment experiment. Monocultural Indian listeners participated in the experiments in the state of Bihar, India. Monocultural Western listeners and bicultural Indian–Western listeners participated in the experiments in our laboratory in the United States. All participants reported having less than three years of formal musical training.

In the recognition memory experiment, all three groups of listeners were exposed to 30-sec excerpts of Western music (symphonies by J. Stammitz and G.B. Sammartini) and Indian music (compositions by N. Banerjee and U.R. Skhan) in the exposure phase. In the testing phase, they heard musical clips familiar from the exposure phase, in addition to new musical clips that they had not previously heard. They were asked whether they had heard the clips during testing. Similar to previous face perception,[15] voice perception,[16–18] and music perception[19–21] studies, we found an in-culture advantage. Monocultural Indian and Western listeners remembered music of their own culture (Indian and Western, respectively) more accurately than music of the other culture. Interestingly, bicultural listeners remembered both types of music equally well.

In the tension judgment experiment, the three groups of listeners listened to 10- to 18-sec Western and Indian melodies matched for tempo, tonic pitch, and timbre (half were played on the piano, half on the sitar) and were asked to indicate the tension they felt by using a dial, after previous studies.[22,23] Similar to the recognition memory experiment, we found clear cultural differentiation for the monocultural groups. Although the Indian listeners judged the Western music to be tenser than the Indian music, the reverse was true for the Western listeners. As in the recognition memory experiment, we found no differences between the two types of music in tension judgment by the bicultural group.

Taken together, these two behavioral experiments suggest that even in listeners who have relatively little formal musical training, sensitivity to musical culture can be measured behaviorally. Note that in this case, cultural sensitivity is likely a reflection of prolonged exposure to one or more musical cultures. Although cognitive and affective consequences of both mono- and bicultural musical experiences can be revealed in these tasks, it is not known from these experiments the neural consequences of mono- and bicultural musical experiences. We examined this topic in a recent functional magnetic resonance imaging (fMRI) study.[24]

Bimusicalism and the brain

To investigate the neural consequences of bimusicalism, we conducted an fMRI study examining cerebral hemodynamic responses to Indian and Western music by monocultural Western listeners and bicultural Indian–Western listeners while they performed a tension judgment task using the stimuli and experimental procedures described earlier.[13] To accommodate the scanning environment, tension responses were collected using a response box rather than a dial as in previous studies.

We were particularly interested in evaluating possible extensions of two hypotheses in bilingualism[25] to bimusicalism. The first is the fractional hypothesis, which postulates that two separate and isolable linguistic systems exist in the bilingual brain. In the holistic hypothesis, qualitative differences between the mono- and bilingual brains are expected. Extending to bimusicalism, the fractional hypothesis would predict that when listening to Western music, both monocultural Western listeners and bicultural Indian–Western listeners would show

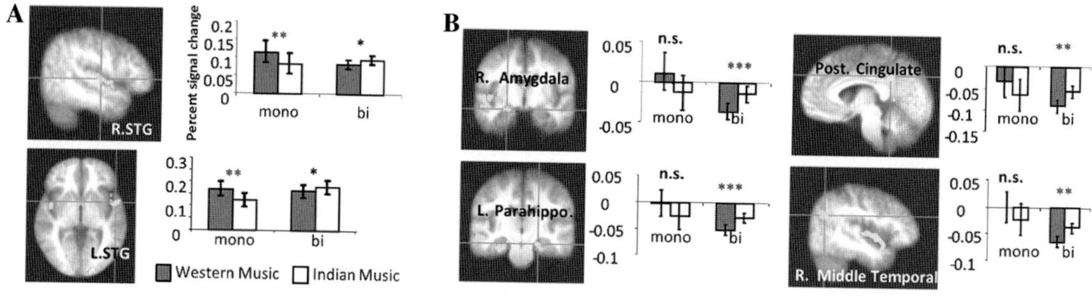

Figure 1. Functional MRI results. Whereas monocultural listeners used the perceptual areas in the bilateral superior temporal gyrus (A) to differentiate the two types of music (Western and Indian music), bicultural listeners differentiated the two types of music in the limbic areas (B). R.STG, right superior temporal gyrus; L.STG, left superior temporal gyrus; R. Amygdala, right amygdala/parahippocampal gyrus; L. Parahippo, left parahippocampal gyrus; Post. Cingulate, posterior cingulate; and R. Middle Temporal, right middle temporal gyrus.

similar brain responses. On the other hand, the holistic hypothesis would predict differences across these two groups as they listen to Western music, which is "native" to both groups. Some of the qualitative differences could be associated with the ways that the listeners' cognitive and affective systems are engaged at the regional and network brain levels, as well as the relationships among brain systems and behaviors.

In terms of listeners' tension responses to music, we replicated our earlier study,[13] which showed monocultural Western listeners tended to report higher tension ratings for Indian than Western music, whereas the bicultural Indian–Western listeners perceived no such difference in tension. In terms of regional brain activities, we found significant group (monocultural Western vs. bicultural Indian–Western) by music (Western vs. Indian) interactions across several brain regions (Fig. 1). Most notably, monocultural listeners were more likely to differentiate Western and Indian music reflected in activation in auditory temporal regions (Fig. 1A), potentially an effect due to prolonged increased exposure to one kind of music (perceptual learning). Bicultural listeners differentiated the two types of music predominately in limbic regions (Fig. 1B), suggesting the potential engagement of affective processes.[26] We also saw differences at the network level. Using structural equation modeling (SEM), we found stronger brain connectivity across numerous temporal-limbic pathways in the bicultural relative to the monocultural listeners. In addition, the strengths of connectivity in these paths differed between the two music conditions in the bicultural listeners, whereas monocultural listeners showed a similar strength of connectivity across the two music conditions.

Besides examining regional and network neural characteristics of our mono- and bicultural listeners, we also examined the relationships among degree of music exposure, affective responses (tension rating), and activation in various affective and auditory brain regions using SEM. Whereas tension ratings were driven predominately by amygdala activities in the monocultural listeners, in the bicultural listeners, these affective responses represented an amalgamation of limbic and temporal brain influences and the degree of music exposure, suggesting more complex behavioral–neural relationships in the bicultural listeners.

Whereas our earlier behavioral study found that bicultural musical experiences can be measured by examining both cognitive and affective processing,[13] our recent fMRI study suggests that mono- and bicultural music listening experiences can drive qualitative neural differences. These results provide support for the holistic hypothesis.

Potential top–down effects

Although interesting, our studies on music culture and the brain may only speak to one potential source of influence of culture on our auditory system, namely bottom–up, exposure-driven influences. To everyday music listeners who had little formal musical training, music exposure could reflect bottom–up inputs, and their brain responses are likely results of years of musical exposure. In addition to this bottom–up effect, culture might potentially have a top–down effect on auditory processing.

Increasingly, cultural psychologists have been interested in the impact of cultural upbringing on relatively basic perceptual processes.[9] These culture-to-perception studies primarily focus on visual perception. In general, researchers in this field argue that cultural upbringing could shape thought patterns, which could in turn influence basic perceptual processes. For example, it is possible that growing up in an East Asian culture shapes thinking style to be more collectivistic, which could in turn result in East Asians focusing on both the background and foreground of a visual image. On the other hand, growing up in North America could shape thought patterns to be more individualistic, which could then lead to a focus on the foreground when viewing the same visual image. In fact, several visual perception studies using different paradigms have found such results. For example, in eye tracking experiments, Chua *et al.*[10] found a larger number of fixations on the background of visual images in East Asian participants relative to their American participants. Kitayama *et al.*[27,28] asked American (Western) and Japanese (East Asian) participants to perform a drawing task. After being presented with a square containing a line printed in the middle (e.g., a line of 3 cm that is printed in a square with a height of 9 cm), subjects were then asked to draw a line in a new square with either the same or different dimensions as the first. In the relative task, subjects were asked to draw a line of the same relative length (according to the proportions of the first line and square combination) in a new square with different dimensions. For instance, in the example given, if the subject was then presented with a new square with a height of 12 cm, the correct length of the line to be drawn would be 4 cm. On the other hand, in the absolute task, subjects were asked to draw a line of the exact same length as that they had seen in the previously shown square. In this example, the correct length to be drawn would be 3 cm, no matter how big the new square was. The relative task required subjects to focus on the relationship between the context and the target whereas the absolute task required subjects to ignore the context. Indeed, Kitayama *et al.* found significant group differences in their performance in such a task. Specifically, Japanese participants made fewer errors in the relative task, which involved making relative judgment of the line based on the length of the surrounding square.

Researchers in this line of work generally argue that the group differences observed are due to top–down, culture-to-perception effects, relating to how cultural upbringing shapes thought processes and ultimately changes how individuals of different cultures view the same image. However, a critical test of this potential top–down effect would be to examine potential cultural differences in another sensory domain (e.g., auditory perception). A domain-general pattern of results would provide stronger support for the top–down nature of cultural influences.

We have recently embarked on experiments that examine a potential influence of culture on auditory perception after the aforementioned experiments in visual perception. Thus far, we have examined monocultural East Asian and monocultural European American participants and found encouraging results that are similar to the visual perception research we discussed earlier. In the future, we plan to examine bicultural participants as they perform the same visual and auditory tasks.

Conclusions

For individuals with a functional auditory system, sounds are almost ubiquitous. Very often, what determines the kind of sounds that we hear is our cultural environment. We have reviewed experiments that investigate the potential influence of culture on auditory perception and discussed potential bottom–up and top–down influences. These influences affect not only listeners with monocultural experiences but also bicultural experiences, although potentially in qualitatively different manners.

Acknowledgments

This work is supported by Grants from the National Science Foundation (BCS-1125144) awarded to P.W. and A.C., the National Institutes of Health (R01DC008333 & K02AG035382) awarded to P.W., and the College of Humanities, Arts, and Social Sciences, Nanyang Technological University (M58100050) awarded to A.C.

Conflicts of interest

The authors declare no conflicts of interest.

References

1. Patel, A.D., E. Gibson, J. Ratner, *et al.* 1998. Processing syntactic relations in language and music: an event-related potential study. *J. Cogn. Neurosci.* **10:** 717–733.
2. Warrier, C., P. Wong, V. Penhune, *et al.* 2009. Relating structure to function: Heschl's gyrus and acoustic processing. *J. Neurosci.* **29:** 61–69.
3. Leaver, A.M. & J.P. Rauschecker. 2010. Cortical representation of natural complex sounds: effects of acoustic features and auditory object category. *J. Neurosci.* **30:** 7604–7612.
4. Binder, J.R., J.A. Frost, T.A. Hammeke, *et al.* 2000. Human temporal lobe activation by speech and nonspeech sounds. *Cereb. Cortex* **10:** 512–528.
5. Peach, R.K. & P.C.M. Wong. 2004. Integrating the message level into treatment for agrammatism using story retelling. *Aphasiology* **18:** 429–441.
6. Wong, P.C.M., M. Ettlinger, J.P. Sheppard, *et al.* 2010. Neuroanatomical characteristics and speech perception in noise in older adults. *Ear Hear* **31:** 471–479.
7. Wong, P.C.M., J.X. Jin, G.M. Gunasekera, *et al.* 2009. Aging and cortical mechanisms of speech perception in noise. *Neuropsychologia* **47:** 693–703.
8. Nisbett, R.E. & Y. Miyamoto. 2005. The influence of culture: holistic versus analytic perception. *Trends Cogn. Sci.* **9:** 467–473.
9. Nisbett, R.E. & T. Masuda. 2003. Culture and point of view. *Proc. Natl. Acad. Sci. USA* **100:** 11163–11170.
10. Chua, H.F., J.E. Boland & R.E. Nisbett. 2005. Cultural variation in eye movements during scene perception. *Proc. Natl. Acad. Sci. USA* **102:** 12629–12633.
11. Chiao, J.Y., T. Iidaka, H.L. Gordon, *et al.* 2008. Cultural specificity in amygdala response to fear faces. *J. Cogn. Neurosci.* **20:** 2167–2174.
12. Kuhl, P.K., K.A. Williams, F. Lacerda, *et al.* 1992. Linguistic experience alters phonetic perception in infants by 6 months of age. *Science* **255:** 606–608.
13. Wong, P.C.M., A.K. Roy & E.H. Margulis. 2009. Bimusicalism: the implicit dual enculturation of cognitive and affective systems. *Music Percept.* **27:** 81–88.
14. Wong, P.C.M., T.K. Perrachione & E.H. Margulis. 2009. Effects of asymmetric cultural experiences on the auditory pathway: evidence from music. *Ann. N.Y. Acad. Sci.* **1169:** 157–163.
15. Bruce, V. & A. Young. 1986. Understanding face recognition. *Br. J. Psychol.* **77:** 305–327.
16. Perrachione, T.K. & P.C. Wong. 2007. Learning to recognize speakers of a non-native language: implications for the functional organization of human auditory cortex. *Neuropsychologia* **45:** 1899–1910.
17. Perrachione, T.K., J.B. Pierrehumbert & P.C.M. Wong. 2009. Differential neural contributions to native- and foreign-language talker identification. *J. Exp. Psychol. Human* **25:** 1950–1960.
18. Perrachione, T.K., J.Y. Chiao & P.C.M. Wong. 2010. Assymetric cultural effects on perceptual expertise underlie an own-race bias for voices. *Cognition* **114:** 42–55.
19. Demorest, S.M., S.J. Morrison, M.N. Beken, *et al.* 2008. Lost in translation: an enculturation effect in music memory performance. *Music Percept.* **25:** 213–223.
20. Morrison, S.J., S.M. Demorest, E.H. Aylward, *et al.* 2003. FMRI investigation of cross-cultural music comprehension. *Neuroimage* **20:** 378–384.
21. Morrison, S.J., S.M. Demorest & L.A. Stambaugh. 2008. Enculturation effects in music cognition: the role of age and music complexity. *J. Res. Music Educ.* **56:** 118–129.
22. Lerdahl, F. & C.L. Krumhansl. 2007. Modeling tonal tension. *Music Percept.* **24:** 329–366.
23. Margulis, E.H. 2007. Silences in music are musical not silent: an exploratory study of the effect of context on musical pauses. *Music Percept.* **24:** 485–506.
24. Wong, P.C.M., A.H.D. Chan, A. Roy, *et al.* 2011. The bimusical brain is not two monomusical brains in one: Evidence from musical affective processing. *J. Cogn. Neurosci.* **23:** 4082–4093.
25. Grosjean, F. 1989. Neurolinguists, beware! The bilingual is not two monolinguals in one person. *Brain Lang.* **36:** 3–15.
26. Blood, A.J. & R.J. Zatorre. 2001. Intensely pleasurable responses to music correlate with activity in brain regions implicated in reward and emotion. *Proc. Natl. Acad. Sci. USA* **98:** 11818–11823.
27. Kitayama, S., H. Park, A.T. Sevincer, *et al.* 2009. A cultural task analysis of implicit independence: comparing North America, Western Europe, and East Asia. *J. Pers. Soc .Psychol.* **97:** 236–255.
28. Kitayama, S., S. Duffy, T. Kawamura, *et al.* 2003. Perceiving an object and its context in different cultures: a cultural look at new look. *Psychol. Sci.* **14:** 201–206.

Ann. N.Y. Acad. Sci. ISSN 0077-8923

ANNALS OF THE NEW YORK ACADEMY OF SCIENCES
Issue: *The Neurosciences and Music IV: Learning and Memory*

A sensitive period for musical training: contributions of age of onset and cognitive abilities

Jennifer Bailey and Virginia B. Penhune
Psychology Department, Concordia University, Montreal, Quebec, Canada

Address for correspondence: Jennifer Bailey, Concordia University – Psychology, 7141 Sherbrooke Street West, Montreal, Quebec H4R 1B6, Canada. j.anne.bailey@gmail.com

The experiences we engage in during childhood can stay with us well into our adult years. The idea of a sensitive period—a window during maturation when our brains are most influenced by behavior—has been proposed. Work from our laboratory has shown that early-trained musicians (ET) performed better on visual-motor and auditory-motor synchronization tasks than late-trained musicians (LT), even when matched for total musical experience. Although the groups of musicians showed no cognitive differences, working memory scores correlated with task performance. In this study, we have replicated these findings in a larger sample of musicians and included a group of highly educated nonmusicians (NM). Participants performed six woodblock rhythms of varying levels of metrical complexity and completed cognitive subtests measuring verbal abilities, working memory, and pattern recognition. Working memory scores correlated with task performance across all three groups. Interestingly, verbal abilities were stronger among the NM, while nonverbal abilities were stronger among musicians. These findings are discussed in context of the sensitive period hypothesis as well as the debate surrounding cognitive differences between musicians and NM.

Keywords: sensitive period; musicians; cognitive abilities

Introduction

The plastic changes that occur in each of our brains as we mature are the result of an interaction between maturational changes and experience. A fascinating example of this interaction is a "sensitive period"—a window of time during development when brain systems are more susceptible to the influence of experience or stimulation. In our lab, we have used trained musicians to study possible sensitive period effects. In these studies, musicians who began training before age 7 demonstrated enhanced rhythm synchronization performance compared to musicians who began their training later in development, when matched for total musical experience.[1,2] In addition, although these two groups of highly trained musicians did not differ on global cognitive variables, individual working memory scores predicted synchronization performance. In this study, the sample size has been increased and a nonmusician's (NM) group has been added to further elucidate the association between working memory, musical training, and task performance. Including a group of highly educated NM also provides insight into possible cognitive differences between musicians and NM.

As our knowledge about brain plasticity evolves, evidence for sensitive periods related to the acquisition of a variety of skills increases. The idea of a sensitive period may have gained most widespread attention through the results of a number of studies showing that second-language proficiency is greater in individuals who were exposed to the second language before age 11–13.[3,4] Recent evidence using neuroimaging techniques also supports the idea that the sensory systems have developmental windows of time during which they are most sensitive to stimulation. Differences in occipital recruitment for nonvisual functions between congenitally blind individuals and those who acquired blindness later in

doi: 10.1111/j.1749-6632.2011.06434.x

development suggest that the visual system also has a developmental window during which it is most responsive to stimulation.[5] Cochlear implantation studies suggest that the auditory system is more responsive the earlier these devices are implanted.[6,7] Studies have reported differences in brain structure between early-trained (ET) and late-trained (LT) musicians and have associated these differences with the extent of musical experience.[8–10] However, an important addition to the investigation of a sensitive period is the matching paradigm developed in our laboratory.[2] When ET and LT musicians are matched for musical experience (e.g., years of formal instruction, years of playing, current hours of practice), the general effects associated with musical experience are controlled and the age at which they began their musical training is isolated as the variable of interest.

Evidence from previous studies in our lab supports the idea of a sensitive period among musicians, even when cognitive abilities are considered. ET musicians (those who began before age 7) outperformed LT musicians (those who began after age 7) on an auditory–motor synchronization task, as well as a visual–motor synchronization task, when matched for total musical experience.[1,2] The two groups did not differ on cognitive measures such as vocabulary (VC), matrix reasoning (MR), digit-span (DS), and letter–number sequencing (LN).[1,11,12] However, working memory scores predicted performance on the rhythm synchronization task across both groups of musicians. A regression analysis revealed that after controlling for working memory, group membership still accounted for variance in task performance. These results suggest that a musician's working memory and age of starting musical training both contributed to their ability to perform the rhythm synchronization task.

This study aims to replicate our previous findings in a larger sample of musicians, and shed light on the debate surrounding cognitive differences between musicians and NM. Although cognitive differences between musicians and NM have been reported, there is controversy in the literature over how or why these differences emerge.[13,14] Studies have used child samples to examine the interaction between music lessons and cognitive and brain development.[15,16] Using an adult sample complements studies of children by allowing us to test whether differences associated with musical training persist into adulthood, especially because we are comparing musicians to a group of highly educated NM. In addition, using a group of adult musicians with extensive but variable lengths of musical training allows us to investigate the nature of the association between music lessons and cognitive abilities.

Method

Participants

Fifty neurologically healthy individuals between the ages of 18 and 36 ($M = 25.5$ years old, SD $= 4.6$) participated in this study. Participants were screened for significant head injuries, history of neurological disease, or medication that could affect task performance. Of the 50 participants, 30 were highly trained and currently practicing musicians and 20 were NM (less than three years of musical experience). The musical training and experience of each participant was determined through a Musical Experience Questionnaire (MEQ) that was developed within our laboratory.[1] The MEQ quantifies the amount of instrumental, vocal, and dance training an individual has received; at what age this training occurred; and the amount of time currently dedicated to practicing on a weekly basis. All musicians had extensive musical experience ($M = 16.4$ years; SD $= 4.4$). Musicians were classified as ET ($n = 15$) or LT ($n = 15$) musicians, based on their MEQ data. Those who began their musical experience prior to or at the age of 7 were placed in the ET group and those who began after the age of 7 were classified as LT. The age of 7 was chosen based on previous studies.[1,2,7] The two groups were matched on years of musical experience, years of formal training, and hours of current practice. All participants gave informed consent and the Concordia University Research Ethics Committee had approved the protocol.

Stimuli

The rhythm task used in this study consisted of six woodblock rhythms of varying difficulty based on their metrical structure.[17,18] Each test rhythm consisted of 11 woodblock notes and had a total duration of six seconds. These rhythms differed in their temporal structure, such that the inter onset intervals between musical notes varied, but not the duration of the notes themselves. More specifically, each rhythm was made up of five eighth notes (each 250 ms), three quarter notes (each 500 ms), one

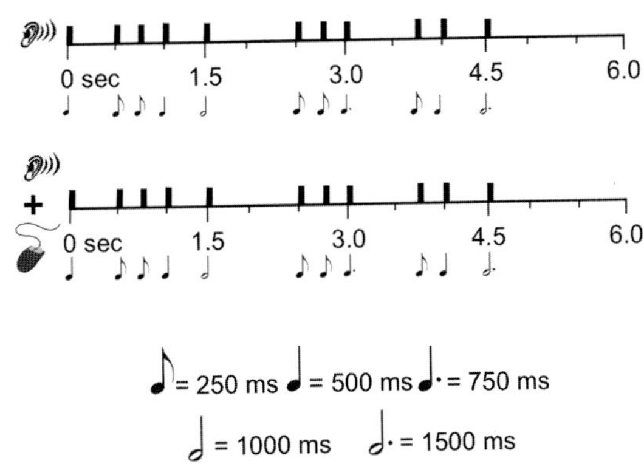

Figure 1. Illustration of the rhythm task. Participants were exposed to six rhythms presented in random order for approximately two 12-min blocks. Two different rhythms of each rhythmic complexity were used. Each trial consisted of a listening component followed by a listening and tapping component.

dotted quarter note (750 ms), one half note (1000 ms), and one dotted half note (1,500 ms). Manipulation of the temporal structure of the notes resulted in progressively more complex and less metrically structured rhythms. For a more detailed description of this task and the metrical complexity manipulation, please see Bailey and Penhune.[1]

Participants completed the DS and LN subtests from the Wechsler Adult Intelligence Scale–III (WAIS) and the VC and MR subtests from the Wechsler Abbreviated Scale of Intelligence (WASI).[10,11] The DS requires individuals to recall strings of numbers, and the LN requires individuals to recall and mentally manipulate strings of letters and numbers. Both of these subtests tap into working memory abilities; however, LN imposes a heavier load on working memory, while DS consists of a rote auditory memory recall section in addition to a mental manipulation section. The VC assesses an individual's ability to orally define words and the MR assesses nonverbal reasoning and visual pattern recognition abilities. Both VC and MR are strongly correlated with global IQ, although they assess different types of intelligence.

Procedure

During the rhythm task, participants alternated between listening and tapping along while each rhythm played twice (Fig. 1). Participants were instructed to tap as accurately as possible with the rhythm as it played during the tapping repetition. Two very basic practice rhythms were administered to familiarize participants with the task. Each rhythm presented in a counterbalanced fashion six times over approximately 12 min in each block and participants performed two blocks. Once participants had completed the first block of the task, they were asked to perform DS. Participants then performed the second block of the rhythm synchronization task, followed by VC, LN, and finally, MR.

Measures

Musical information was quantified for each participant in terms of years of experience, years of formal training, and hours of current weekly practice using the MEQ.[1] Individual cognitive abilities were measured using the four chosen cognitive subtests (DS, LN, VC, and MR). Results were scored according to standard procedure. Performance on the rhythm synchronization task was measured using three dependent variables: percent correct (PC), asynchrony (ASYN), and inter-tap interval (ITI) deviation. A tap was considered correct if it was made within half of

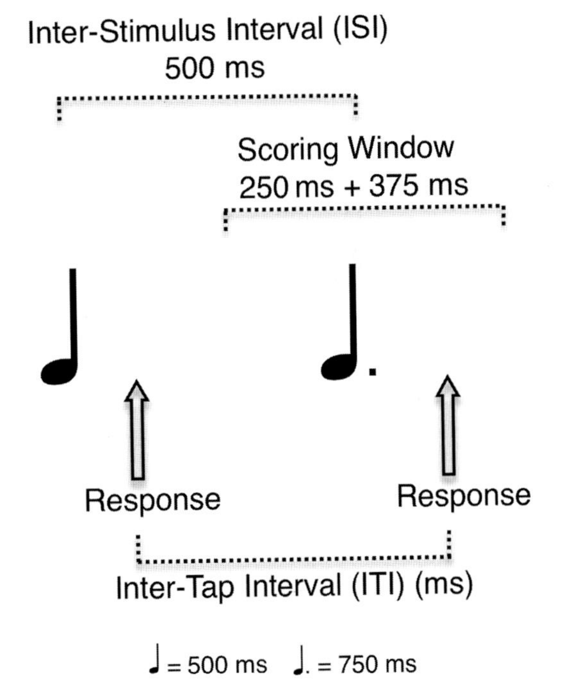

Figure 2. Illustration of the scoring method used to evaluate rhythm task performance. A response was scored correctly if the mouse tap was made within half of the onset-to-onset interval before and after a woodblock note. Asynchrony was measured as the difference between each woodblock note and the participant's response. ITI deviation was calculated as a ratio of the ITI and the ISI subtracted from 1.

the onset-to-onset interval before or after a woodblock note (Fig. 2). The ASYN measure was defined as the absolute value of temporal difference between the onset of each woodblock note and the associated mouse key press. The ITI deviation measure indicated the extent of deviation of the participant's tap interval from the actual interval between each pair of woodblock notes. It was calculated by dividing the interval between each pair of the participant's taps by the interval between each corresponding pair of woodblock notes in the rhythms and subtracting this ratio from a value of one. This measure is indicative of how well participants reproduced the temporal structure of the rhythms.

Data analysis

To compare rhythm synchronization across the three groups, a repeated-measures analysis of variance (ANOVA) for each of the dependent variables was conducted, with group as the between-subjects factor and rhythm type as the within-subjects factor. Pair-wise comparisons for between group differences were analyzed using least significant differences (LSD) correction for multiple comparisons. The result of our matching procedure was evaluated using *t*-test analyses for years of musical experience, years of formal training, and hours of current practice among the musicians. Group differences on the cognitive subtests were assessed using a one-way ANOVA for each cognitive variable with group as the between-subjects factor. Pair-wise comparisons were conducted using an LSD correction for multiple comparisons. The relationships among cognitive measures, musical experience variables, and task performance were examined using one-tailed Pearson correlation analyses. Raw scores on the cognitive subtests were used to correlate with performance measures and scaled scores were used when comparing the three groups on the cognitive measures. However, results were consistent whether raw or scaled scores were used in the analyses.

Based on a previously observed relationship between individual working memory abilities and task performance among musicians, a hierarchical regression analysis was conducted with all three groups in order to assess whether the observed group difference persists after individual working memory scores are considered.[1] A model was created with total ITI deviation as the dependent measure and both group and working memory as predictors. A composite score for each participant's working memory ability was created by summing their LN and DS scores and used in the regression analysis.

Results

Group comparisons of musical and cognitive measures between the ET and LT musicians confirmed that the two groups were well matched in terms of years of musical experience, years of formal training, and hours of current practice (Table 1). The one-way ANOVA revealed no significant differences in DS or LN scores between groups, although statistical trends toward a main effect of group on MR and VC were observed (Fig. 3). Pair-wise comparisons revealed that the NM VC scores were higher than the ET ($P = 0.026$), and the MR scores of the LT were higher than those of the NM ($P = 0.017$). Scaled scores were used for these analyses.

Behavioral measures

The ANOVA comparing accuracy (PC) of the rhythm reproduction task across the three groups showed a significant main effect of group ($F[2, 47] = 3.99$, $P < 0.05$; Fig. 4A). Pair-wise comparisons using an LSD correction revealed differences between the ET and NM ($P < 0.01$). These results confirm that all three groups were performing the

Table 1. Group demographics of musical experience variables

Group	Age (years)	Age of onset (years)	Formal training (years)	Musical experience (years)	Current practice (years)
Early-trained	23.47 (±3.85)	5.87 (±1.19)	11.73 (±3.97)	16.87 (±4.10)	15.23 (±9.97)
Late-trained	26.60 (±5.22)	10.47 (±2.03)	10.03 (±4.39)	15.90 (±4.74)	14.43 (±7.80)
t-values	−1.87	−7.57*	1.11	0.60	0.25

Note: Standard deviation values are in brackets.
*P value < 0.001.

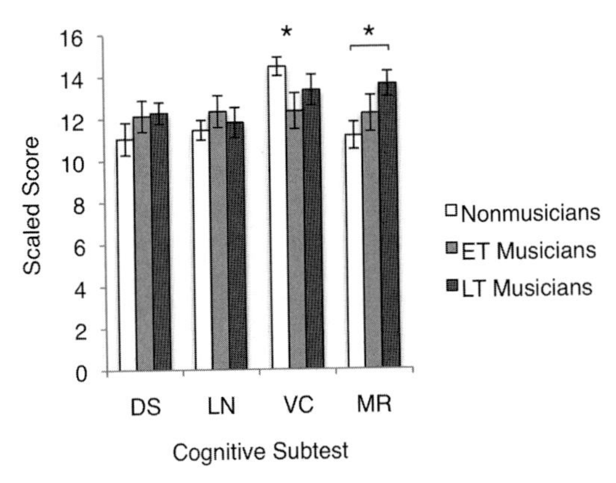

Figure 3. Group mean cognitive scaled scores. DS, digit-span; LN, letter–number sequencing; VC, vocabulary; MR, matrix reasoning. No group differences were observed on the two measures of working memory (DS and LN); however, statistical trends toward group differences were observed on VC ($P = 0.078$) and MR ($P = 0.055$). Pair-wise comparisons revealed specific group differences ($^*P < 0.05$).

task correctly overall and the mean performance values were in the expected order (i.e., ET > LT > NM).

The ANOVA comparing performance on the synchronization measure (ASYN) across the three groups revealed a similar pattern of results, such that there was a main effect of group ($F\,[2, 47] = 16.76$, $P < 0.001$; Fig. 4B). Pair-wise comparisons using an LSD correction revealed lower ASYN scores for the ET and LT when compared to the NM ($P < 0.001$ for both comparisons). In addition, the ET group was better able to synchronize their responses than the LT musician group as revealed by lower ASYN scores ($P = 0.05$). These results suggest that the group differences were heightened on this more sensitive performance measure compared to our more global measure of accuracy (PC).

Consistent with the other performance measures, the ANOVA comparing reproduction of the temporal structure of the rhythms using our ITI measure of deviation across the three groups showed a significant main effect of group ($F[2, 47] = 20.30$, $P < 0.001$; Fig. 4C). Pair-wise comparisons using an LSD correction revealed a similar pattern of results as on the ASYN measure such: The ET had lower deviation scores than the LT ($P < 0.05$) and both musician groups had lower deviation scores than the NM ($P < 0.001$ for both comparisons).

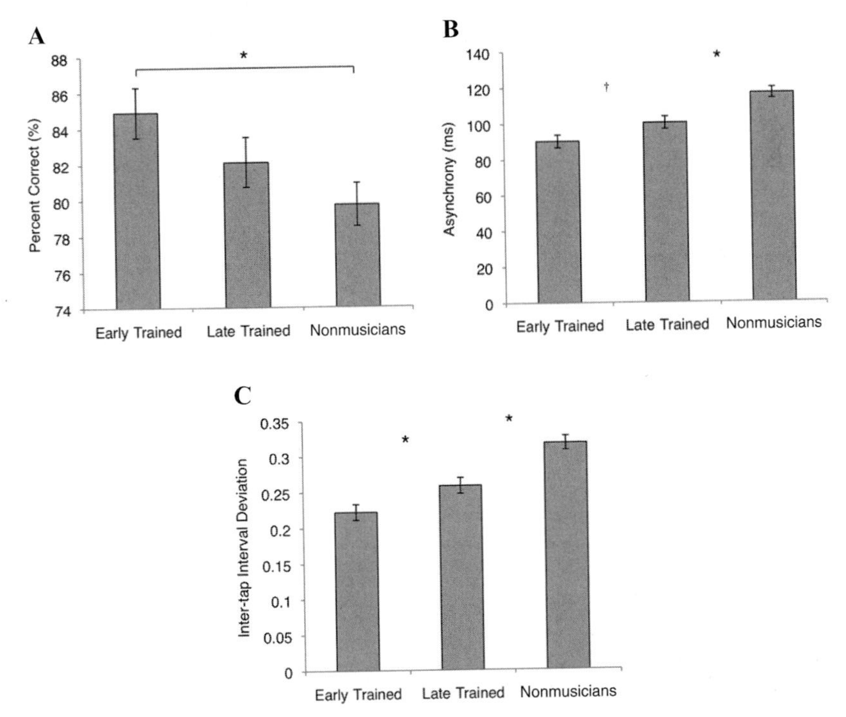

Figure 4. Task performance results for all three groups: (A) percent correct; (B) asynchronization; and (C) inter-tap interval deviation. Repeated measures ANOVA for each performance measure revealed a significant main effect of group, and pair-wise comparisons revealed specific group differences ($^*P < 0.05$, $\dagger = 0.05$). Standard error bars have been used.

Table 2. Correlation results between cognitive scores and task performance measures

Performance measure	Digit-span	Letter–number sequencing	Matrix reasoning	Vocabulary
Percent correct	0.275***	0.360*	0.147	−0.072
Asynchrony	−0.258***	−0.307*	−0.262***	0.269***
Inter-tap interval deviation	−0.378**	−0.340*	−0.339*	0.187

Note: Raw scores were used for the cognitive measures.
* P values < 0.05.
** P values < 0.01.
*** P values < 0.08 but greater than 0.05.

These results further illustrate that as the measure of performance increased in sensitivity to temporal aspects of the rhythms, the observed group differences were heightened.

Correlations

In order to examine the relationship between task performance and cognitive variables across the three groups, raw scores for PC, ASYN, and ITI were correlated with raw scores for VC, MR, DS, and LN (Table 2). Both working memory measures (DS and LN) correlated significantly with the three performance measures (PC, ASYN, and ITI) in the expected directions, confirming that the rhythm reproduction task implicates the use of working memory. Surprisingly, VC correlated with the synchronization measure such that higher VC scores were associated with poorer performance of the rhythm task. In addition, MR correlated with both synchronization and ITI measures of task performance. These results were likely driven by the group differences observed on these cognitive scores and this will be addressed in the discussion section.

Results from the correlational analyses between the behavioral measures and musical variables in the musicians (Table 3) reveal a significant association between years of formal training and ITI deviation ($r = -0.367$, $P < 0.05$). In addition, age of onset showed a significant relationship with ASYN and ITI, as well as a relationship trend toward significance with PC. In order to examine the association among years of formal training, cognitive scores, and task performance, correlations were performed between years of formal training and each cognitive measure. This set of analyses revealed an association trend toward significance between years of formal training and DS ($r = 0.342$, $P = 0.06$); however, no significant associations with LN, VC, or MR.

Regression analysis

In order to determine if the amount of variance in ITI deviation during task performance accounted for by group was above and beyond what was explained by individual working memory abilities, a hierarchical regression analysis was conducted using the three groups (Table 4). These results confirmed that, while individual working

Table 3. Correlation results between musical experience and task performance measures

Performance measure	Age of onset (years)	Formal training (years)	Musical experience (years)	Current practice (hours)
Percent correct	−0.352***	0.010	0.141	−0.052
Asynchrony	0.459*	−0.214	−0.139	−0.079
Inter-tap interval deviation	0.509**	−0.367*	−0.095	0.046

Note: Raw scores were used for the cognitive measures.
* P values < 0.05.
** P values < 0.01.
*** P values < 0.08 but greater than 0.05.

Table 4. Regression analysis results examining the predictive value of group membership above and beyond working memory to task performance

	R^2	β	R^2 change	F
Step 1	0.165		0.165	9.45
Working memory		−0.406*		
Step 2	0.538			27.336
Working memory		−0.293*		
group		0.621*	0.373	

Note: A working memory composite score was used for this analysis comprised of individual raw DS and LN scores.
*P values < 0.01.

memory abilities were predictive of task performance, group membership accounted for additional portions of the variance in ITI deviation scores.

Discussion

These findings replicate our previous findings but in a larger sample, and provide further evidence for a sensitive period for musical training that may have a specific impact on sensorimotor synchronization abilities. In this study, the ET musicians were better able to reproduce the rhythms than the LT musicians, even after controlling for years of formal training, playing experience, and current hours of practice. In addition, the two musician groups did not differ on the four cognitive measures. In other words, this observed group difference on task performance could not be attributed to differences in musical experience or cognitive ability, but to the developmental window during which musical training began. As expected, NM rhythm synchronization abilities were inferior to both musician groups. Although there were no differences in working memory performance across the three groups, individual working memory scores correlated with task performance, suggesting similar reliance on working memory resources for all groups. In further support of the sensitive period hypothesis, the regression results suggest that even after controlling for individual working memory scores, group membership still predicted a significant amount of variance in task performance. This reinforces the idea that musical training, especially early musical training, improves task performance above and beyond the contribution of working memory abilities.

In addition to the differences between ET and LT musicians on the rhythm task, we observed differences in performance on global cognitive variables between musicians and NM. Specifically, the NM obtained higher VC scores, while the musician groups had higher MR scores. These findings are interesting and can shed light on the types of cognitive effects associated with musical training. One hypothesis is that music lessons benefit the underlying cognitive abilities that are measured by MR, and that in contrast, NM are exercising their verbal intelligence via other avenues. If this were the case, one would expect the length of musical training (i.e., years of formal training) to be correlated with MR scores among the musician group, but it is not. Alternatively, one could hypothesize that individuals with strong visual-spatial organization skills are inclined to take up music lessons, and those with strong verbal abilities are likely to take up other NM activities. If this were true, then no relationship between length of musical training and MR would be present, yet group differences would persist between musicians and NM. The current data supports this assumption. The more general question of what is driving cognitive differences between musicians and NM is an area of controversy. Recently, Schellenberg and Peretz proposed that the observed association between music lessons and cognition may be mediated by executive function, although a more recent publication by Schellenberg failed to report convincing evidence that this was the case.[12,13] In our sample, a weak association between working memory, a component of executive function, and years of formal training was observed among the musicians; however, their scores were not higher than the NM, suggesting that if years of formal training impacts working memory, it does not do so above and beyond other nonmusical activities in which NM engage. Other factors, such as socioeconomic status or the family environment, may contribute to the differences between the two groups. Both our musicians and NM were either in the process of completing an undergraduate degree or had obtained one, and some were pursuing higher-level education. Thus, in these highly educated samples, any enhanced cognitive abilities observed in musicians over NM are likely to be a combination of innate predisposition and effects associated with exercising the abilities implicated in music lessons during development. Similarly, NM are likely predisposed to engage in other nonmusical activities and exercise other abilities during their development.

In summary, this study adds to the growing literature in support of a sensitive period for sensorimotor-integration abilities among musicians and considers NM as a comparison sample. Any differences in brain structure between early and LT musicians associated with these enhanced synchronization abilities have yet to be explored. The results from this study also add to the evidence that musicians and NM possess different cognitive strengths, even in a sample of highly educated adults. However, the exact contributions of innate predisposition and the influence of training remain unknown.

Conflicts of interest

The authors declare no conflicts of interest.

References

1. Bailey, J.A., & V.B. Penhune. 2010. Rhythm synchronization performance and auditory working memory in early- and late-trained musicians. *Exp. Brain Res.* **204:** 91–101.
2. Watanabe, D., T. Savion-Lemieux & V. Penhune. 2007. The effect of early musical training on adult motor performance: evidence for a sensitive period in motor learning. *Exp. Brain Res.* **176:** 332–340.
3. Johnson, J.S. & E.L. Newport. 1989. Critical periods effects in second language learning: the influence of maturational state on the acquisition of English as a second language. *Cogn. Psychol.* **21:** 60–99.
4. Weber-Fox, C. & H.J. Neville. 2001. Sensitive periods differentiating processing of open- and closed-class words: an ERP study of bilinguals. *J. Speech Lang. Hear. Res.* **44:** 1338–1353.
5. Voss, P., F. Gougoux, R. Zatorre, *et al.* 2008. Differential occipital responses in early- and late-blind individuals during a sound-source discrimination task. *NeuroImage* **40:** 746–758.
6. Kral, A., R. Hartmann, J. Tillein, *et al.* 2001. Delayed maturation and sensitive periods in the auditory cortex. *Audio. Neuro-Otol.* **6:** 346–362.
7. Sharma, A., P. Gilley, M. Dorman & R. Baldwin. 2007. Deprivation-induced cortical reorganization in children with cochlear implants. *Int. J. Audiol.* **46:** 494–499.
8. Schlaug, G., L. Jäncke, Y. Huang, *et al.* 1995. Increased corpus callosum size in musicians. *Neuropsychologia* **33:** 1047–1055.
9. Bengtsson, S.L., Z. Nagy, S. Skare, *et al.* 2005. Extensive piano practicing has regionally specific effects on white matter development. *Nat. Neurosci.* **8:** 1148–1150.
10. Imfeld, A., M. Oechslin, M. Meyer, *et al.* 2009. White matter plasticity in the corticospinal tract of musicians: a diffusion tensor imaging study. *NeuroImage* **46:** 600–607.
11. Wechsler, D. 1997. *Wechsler Adult Intelligence Scale.* 3rd ed. Psychological Corporation. San Antonio, TX.
12. Wechsler, D. 1999. *Wechsler Abbreviated Scale of Intelligence.* Psychological Corporation. San Antonio, TX.
13. Schellenberg, E. & I. Peretz. 2008. Music, language and cognition: unresolved issues. *Trends Cogn. Sci.* **12:** 45–46.
14. Schellenberg, E.G. 2011. Examining the association between music lessons and intelligence. *Br. J. Psychol. (London, England: 1953)* **102:** 283–302.
15. Hyde, K.L., J. Lerch, A. Norton, *et al.* 2009. Musical training shapes structural brain development. *J. Neurosci.* **29:** 3019–3025.
16. Schellenberg, E.G. 2006. Long-term positive associations between music lessons and IQ. *J. Educ. Psychol.* **98:** 457–468.
17. Essens, P. 1995. Structuring temporal sequences: comparison of models and factors of complexity. *Percep. Psychophys.* **57:** 519–532.
18. Essens, P. & D. Povel. 1985. Metrical and nonmetrical representations of temporal patterns. *Percep. Psychophys.* **37:** 1–7.

Ann. N.Y. Acad. Sci. ISSN 0077-8923

ANNALS OF THE NEW YORK ACADEMY OF SCIENCES
Issue: *The Neurosciences and Music IV: Learning and Memory*

Musical training and the role of auditory feedback during performance

Peter Q. Pfordresher
University at Buffalo, State University of New York, Buffalo, New York

Address for correspondence: Peter Q. Pfordresher, Department of Psychology, 355 Park Hall, University at Buffalo, Buffalo, NY 14260. pqp@buffalo.edu

Recent research has shown that music training enhances music-related sensorimotor associations, such as the relationship between a key press on the keyboard and its associated musical pitch (auditory feedback). Such results suggest that the role of auditory feedback in performance may be based on learned associations that are task specific. Here, results from various studies will be presented that suggest that the real state of affairs is more complex. Several recent studies have shown similar effects of altered auditory feedback during piano performance for pianists and individuals with no piano training. Other recent research suggests dramatic differences between pianists and nonmusicians concerning the influence of auditory feedback on melody switching that suggest greater influence of auditory feedback among nonmusicians than pianists. Taken together, results suggest that musical training refines preexisting sensorimotor associations.

Keywords: auditory feedback; musical expertise; sensorimotor coordination; sequence production

Introduction

The performing musician is, in a sense, also a member of the audience insofar as he or she perceives the music being performed in the form of auditory feedback. A compelling question thus concerns the role of auditory feedback for the performer. Whereas earlier accounts proposed that performers use auditory feedback for error monitoring,[1–3] more recent data suggest a limited role for auditory feedback.[4–6] Existing data from the domain of music performance suggest that performers are sensitive both to the way in which auditory feedback is coordinated with actions in terms of the timing of pitch onsets, and to the way that action sequences are coordinated with pitch patterns.[7–11]

Most studies concerning the role of auditory feedback have adopted the altered auditory feedback (AAF) paradigm, in which participants produce an action sequence although hearing auditory feedback that differs from what one would usually expect. Effects of such alterations on the planning and execution of musical sequences can be used to determine how and to what degree performers rely on auditory feedback. In my lab, we address the role of auditory feedback in music performance among both musically trained and untrained populations, by incorporating simplified melodies and forms of music notation that are easily learnable by nonmusicians.[9] In addition to the AAF paradigm, other paradigms can also be used to assess the role of auditory feedback during music performance, which are discussed here.

In this paper, I consider how musical expertise influences the role of auditory feedback during music performance. Music performance offers an excellent context in which to explore questions related to expertise, given large individual differences. Furthermore, despite such large-scale differences in expertise, most humans possess elaborate implicit knowledge of the rules that guide musical pattern formation.[12,13] As such, I review here the results of several recent studies in which participants of varying musical experience produce short musical sequences on a keyboard, in experiments designed to test the kind of associations that these participants

doi: 10.1111/j.1749-6632.2011.06408.x

have between their actions and resulting pitches. I focus specifically on associations with respect to pitch content, rather than synchronization between actions and sounds (as in traditional delayed auditory feedback), because pitch-based associations are most likely to be shaped by experience.

Three views of the learning process

Before discussing experimental effects of training on the role of auditory feedback during performance, I first consider three canonical hypotheses concerning the relationship one might expect, based on long-standing trends in research on perception and action.

Hypothesis 1: strict associationism

The first hypothesis follows from a standard associationist view, first articulated in psychology in the "serial chaining" hypothesis of William James,[14] and expanded on in the "closed loop" theory of motor learning.[15] According to this view, the novice has no associations between actions and pitches before training. Over the course of training, these associations form and become solidified. Thus, whereas an

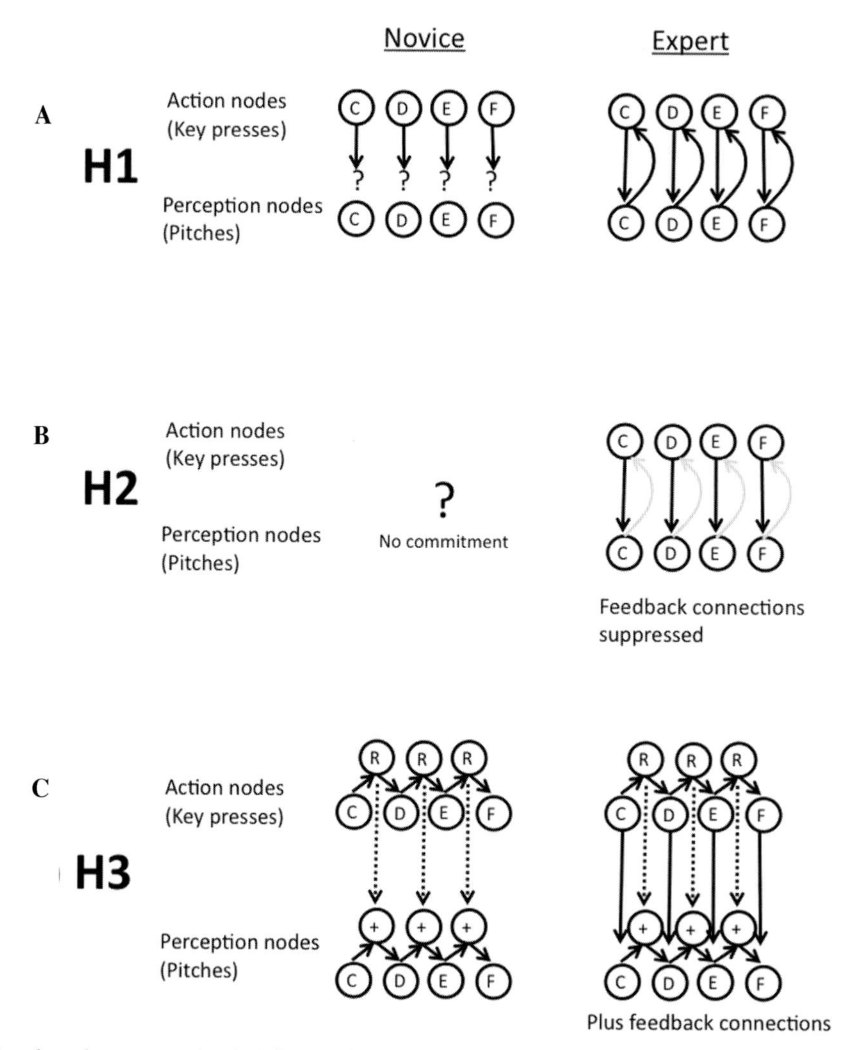

Figure 1. Three hypotheses concerning the influence of musical training (novice vs. expert) on neural associations between action plans and auditory feedback. Planned actions (key presses) are represented as nodes (circles) with the intended key press denoted by the letter name of the pitch associated with that key. Perception nodes denote the perceived pitch category. Solid arrows indicate zero-order associations between planned action and perceived events; dotted lines indicate first-order associations. Question marks indicate absent associations, and gray bars indicate suppressed associations. Text has further details.

expert should be sensitive to relationships between actions and auditory feedback, a novice should not. Figure 1A illustrates such an architecture, adopting the framework of a neural network in which action "nodes" that represent target key presses may be associated with perceptual "nodes" that represent pitch events. Before training, action nodes are not systematically associated with perceptual nodes. Training leads to bidirectional connections, such that perceptual feedback confirms planned actions when feedback is appropriate (the "perceptual trace" in closed-loop theory). This approach predicts sensitivity to alterations of auditory feedback among experts but not among novices.

Hypothesis 2: motor schema formation

A second hypothesis follows from the framework of motor programming and schema theory,[16] as well as problems in relating a strict associationist view to skilled performance.[17] According to this view, experts develop internalized schemas that allow them to perform without the use of feedback. In Figure 1B, this prediction for experts is represented by a set of unidirectional perception action connections leading from planning nodes to feedback nodes, whereas feedback connections to planning are suppressed. Indeed, the ability to ignore potentially interfering auditory events is critical for expert performers, who often have to perform in ensembles in which sounds made by other musicians may cause interference. This approach does not make an explicit commitment to the kind of connections that may exist among novices. However, it is clear that any sensitivity to auditory feedback would be restricted to novices.

Hypothesis 3: hierarchical shared representations

A third hypothesis adopts a more hierarchical view, inspired by evidence that motor planning adopts higher order retrieval structures.[8,18–20] In keeping with other views of action planning,[21] the hierarchy proposed here is based on transitions among events in a sequence, as opposed to dominance hierarchies that have been used to characterize musical schemas.[13,22,23] In Figure 1C, a two-level hierarchy is shown for action in which movements from C to F are linked by nodes that encode movement transitions between action nodes. In Figure 1C these higher order transitional nodes encode rightward movement between adjacent key presses, though other transitions are possible. Commensurate transitional associations among perception nodes are represented as upwards pitch motion. According to this perspective, the novice is sensitive to certain aspects of perception/action relationships, but only those aspects that are general to a wide range of behaviors. Thus, with respect to the piano, the novice may be sensitive to the mapping between movement transitions on the keyboard and concurrent patterns of pitch motion, though not to specific pitch class–piano key relationships. With training, one retains these general-purpose associations although building more refined associations with music-specific characteristics of auditory feedback.

Results from different experimental paradigms

Here, I consider relevant data concerning the effect of musical training on the role of auditory feedback during performance. All paradigms summarized here concern associations between produced action and concurrent sequences of musical pitches (auditory feedback). However, these paradigms differ with respect to how such associations are manifested. My division of results across paradigms is done in part because each paradigm, on the surface, appears to support a different hypothesis described above.

Effects of sensorimotor associations on subsequent unimodal processing

Various studies have examined how the action pitch associations formed in musical training influence the subsequent planning of actions or the processing of pitch information. What is common across these paradigms is that the researcher measures how binding across modalities (perception and action) affect subsequent processing of one modality on its own. Moreover, a common theme across these studies is that results tend to support the associationist view (hypothesis 1) given above.

Keller and Koch[24] explored how musical training influenced sensitivity to perception/action associations with respect to the initial planning of motor sequences. They incorporated a proto-musical task in which participants tapped sequences on vertically arranged metal plates. The authors varied the vertical mapping of actions to pitch height, such that the mapping could be compatible or incompatible. Sequences were only three taps long and the authors were not concerned with effects of these alterations on production of the three taps. They were instead interested in how quickly participants could prepare a subsequent movement sequence based on the action/effect contingencies they experienced in a block of trials. Musicians' performance deteriorated when the mapping was incompatible. By contrast, nonmusicians seemed to be insensitive to the mapping, in keeping with hypothesis 1.

Further support for hypothesis 1 has been found in neuroimaging studies that have investigated how musical training may lead to motor associations during the perception of musical pitch. In one study, Bangert and Altenmüller[25] used slow-moving frequencies in the electroencephalography (EEG) signal to identify brain areas that reflect audiomotor associations after training. In their task, nonmusicians were trained to produce musical sequences, although hearing auditory feedback that was either mapped appropriately with keys on a keyboard, or had an unpredictable mapping. Though both novice groups were able to learn musical sequences, subsequent perceptual responses of the brain included motor activity (in the inferior frontal gyrus) only for the group that had experienced reliable mapping between actions and pitch. In a related fMRI study by Lahav *et al.*,[26] novices likewise learned to play musical sequences (with normal mapping of pitch), and then afterwards listened to the melodies they had learned to play or melodies comprising different pitch classes. Similar to Bangert and Altenmüller, Lahav *et al.* found that training led to activations in the inferior frontal gyrus, only when participants heard the melodies they had learned to play.

Taken together, these effects are in agreement with the associationist hypothesis 1. It is furthermore significant that these results may reflect associations that are localized at the inferior frontal gyrus, given the proposal that this area of the brain plays a role in the mirroring of perception and action.[27]

Disruptive effects of AAF during performance

Now we turn to the way in which pianists and nonmusicians respond to AAF when feedback pitches are altered but feedback events are presented in synchrony with key presses. The primary difference between the AAF paradigm and the paradigms discussed in the previous section is that for AAF tasks, one is interested in the effect of altered feedback on ongoing production. As such, the AAF paradigm speaks to the effects of sensorimotor coordination on concurrent production, whereas paradigms discussed earlier reflect generalization of sensorimotor coordination to subsequent processing.

To date, several studies have addressed the effects of AAF for pitch.[7,9,11,28,29] In general, the most disruptive alteration to pitch is the *serial shift* of pitch, in which feedback events originate from serial positions at a constant lag (or lead) relative to the current position. For instance, in the lag-1 serial shift (results of which I report here), every key press generates the pitch associated with the immediately preceding key press. Note that disruption from this condition would necessarily reflect action pitch-based associations, given that sounds are always synchronized with key presses.

The effects of serially shifted AAF have been explored among pianists and nonmusicians. Figure 2A shows error rates (which typically reflect disruption from alterations to pitch) among a group of participants with no musical training, as well as a group of participants with at least six years of private piano lessons. The data shown here are pooled across five experiments from a previously published paper.[9] Both groups performed single-voice melodies with the right hand, although pianists performed 12-note melodies that required changes in hand position, and nonmusicians performed 8-note melodies with fixed finger–key relationships. Melodies were varied to match difficulty to skill level; the success of this matching is reflected in similar error rates for the normal feedback condition. Both groups made more errors during serially shifted AAF than during normal feedback. Thus nonmusicians, as well as pianists, were disrupted by alterations to pitch. At the same time, pianists made significantly more errors when hearing serially shifted feedback than nonmusicians. Figure 2B illustrates the change in the amount of disruption from serial shifts when treating musical training on a continuum. As can be seen, increase in the years of training are associated with greater disruption from serial shifts among individuals reporting at least one year of private piano lessons.

Two important implications arise from these results. First, nonmusicians do experience disruption when hearing altered feedback pitches. This result is critical because it suggests that sensorimotor associations during music performance may not be due

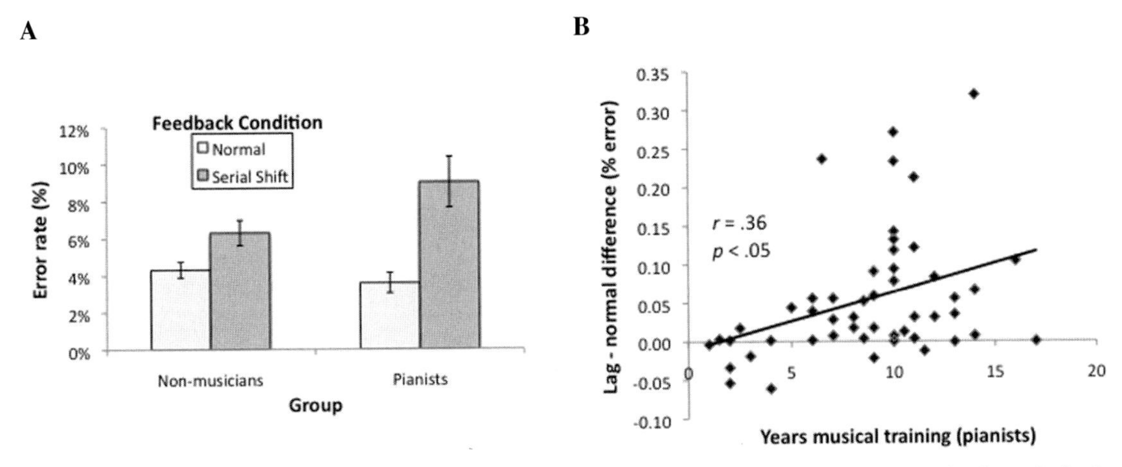

Figure 2. (A) Mean error rates for the sequencing of pitch events during piano performance as a function of auditory feedback condition and musical training. Error bars represent one standard error of the mean. (B) Scatterplot relating years of total musical training on any instrument to the disruptive effect of serial shifts on error rates (difference in error rates across serially shifted and normal feedback conditions) among pianists.

simply to learned task-specific associations. That is, nonmusicians may transfer other kinds of associations to the novel task of performing a melody on the piano. Further support for this idea comes from recent evidence that nonmusicians may have associations between spatial location and pitch (the SMARC effect, Ref. 30). Moreover, recent fMRI data suggests that the brain's responses to serially shifted auditory feedback are not localized in the inferior frontal gyrus, but instead involve a network of centers including the cerebellum, thalamus, and the anterior cingulate cortex.[31] These areas are typically associated with error monitoring[32] and the use of internal models during production.[33]

The second important implication is that sensitivity to AAF, though present among novices, may be enhanced during skill acquisition. The fact that pianists experience more disruption than nonmusicians may reflect the fact that pianists are sensitive to a broader range of hierarchical levels when relating actions to sound than are nonpianists, in keeping with hypothesis 3 (see Fig. 1C). In support of this view, we have found that pianists are also more sensitive to random alterations of feedback pitch than are nonmusicians, who show no disruption from this manipulation.[9] In addition, pianists are disrupted by serial shifts of a melody that is a melodic variation of the melody they play, suggesting generalization. Nonmusicians are not.[11]

Taken together, these results support the hierarchical perspective of hypothesis 3 described above. Both pianists and nonmusicians are disrupted by AAF, although the basis for disruption among nonmusicians may be different than among trained pianists.

Melody switching

The third research paradigm described here combines characteristics of both paradigms described earlier. In this paradigm, participants learn two different melodies and on each trial perform one of them although hearing either normal feedback—the melody they are playing—or feedback from the other melody they learned, referred to as alternate feedback. While playing and experiencing one of these feedback conditions, participants hear a randomly positioned auditory response cue (a single tone). This cue may signal participants to switch from the melody they are playing to the alternate melody, or to continue their current melody depending on the timbre.

We recently reported results for a group of 10 nonpianists, some of whom had modest amounts of musical training on instruments other than the piano.[28] Surprisingly, nonmusicians showed no effect of auditory feedback during trials in which the response cue signaled a switch, but did show a feedback effect during trials in which the response cue was irrelevant. In such trials, nonmusicians paused after hearing the irrelevant cue (leading to a lengthened interonset interval) although experiencing alternate auditory feedback, but not when they heard

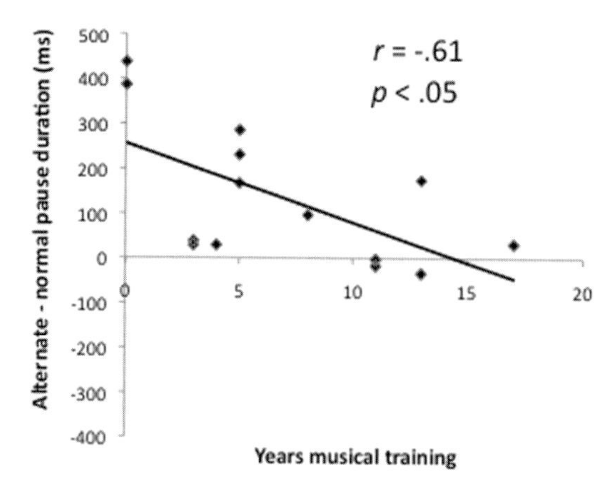

Figure 3. Scatterplot relating years of total musical training (abscissa) to the effect of hearing alternate feedback on the duration of a pause that follows an irrelevant response cue (the ordinate) in a melody-switching task.

feedback from the performed melody. These results suggested that participants had difficulty withholding a switch response when auditory feedback activated pitch events associated with the switch.

We also collected data from five pianists (not reported in the original paper). Surprisingly, pianists did not show any effects of cue type or auditory feedback. Furthermore, the effect of alternate feedback diminished as a function of musical experience, as shown in Figure 3 (which includes the sample from the published paper plus the additional pianists). Whereas alternate feedback leads to long pauses after the irrelevant response cue for nonmusicians (even greater than 400 ms), more musically trained participants are better able to withhold the switching response even with alternate auditory feedback. This is initially puzzling as the result seems to support hypothesis 2. In fact, the effect of training on the role of auditory feedback appears to be the exact opposite of what we have found for the effects of AAF (see Fig. 3), and is clearly distinct from the effects of auditory feedback on subsequent unimodal processing.

Resolving a puzzle

The results provided here are puzzling in that each paradigm seems to support a different hypothesis regarding the effects of learning on sensorimotor associations. It is of course unlikely that all three learning mechanisms operate in parallel. The most plausible resolution would involve an approach that modifies one of the existing hypotheses. I here propose that a modification of hypothesis 3 (hierarchical shared representations) holds the greatest promise. Hypothesis 3 is the only candidate that suggests that all individuals are sensitive to perception/action relationships during music performance, based on general principles of sensorimotor coordination, with the nature of these associations being modified through experience.

Hypothesis 3 clearly accounts for the effects of AAF, but why do results from other paradigms differ? A particularly informative feature of the paradigms presented here has to do with whether the effects of auditory feedback emerge in generalizations from one performance context to a subsequent task or not. Such generalization is present when one examines the influence of auditory feedback on subsequent unimodal processing, or the effects of auditory feedback on the ability to switch from one melody to another within a trial. Results from these paradigms suggest *qualitative* differences across training groups. When one assesses the effect of altered feedback on concurrent production, the qualitative effects on performance (a significant increase in error rates) is there for both untrained and trained performers, even though disruption varies in magnitude (a quantitative, rather than qualitative, difference).

Thus, the effects of AAF on concurrent production may reflect basic properties of the perception/action system. Specifically, the effects of altered feedback may be based primarily on one's sensitivity to the degree of coordination between patterns of movement through space, and any pattern of perceptual events that coincides with these movement patterns. Given this interpretation it is not surprising that disruption from alterations of pitch during music performance are based primarily on how the pitch contour pattern is related to the pattern of movements during production, more so than whether the performer does or does not hear the specific pitch that is associated with a single key.[9,11]

By contrast, tasks that reveal differences across pianists and nonmusicians may reflect the way in which musical training enhances one's hierarchical representation of music, beyond levels summarized in Figure 1C. Earlier research concerning music perception suggested that musically trained individuals might have more refined hierarchical representations for pitch[34] and for timing.[23] Furthermore, in transfer of learning tests that involve music performance, pianists have been found to generalize across sequences based on abstract structural characteristics.[35,36] Thus it is possible that pianists may differ from nonmusicians with respect to their tendency to generalize relationships in one trial to a subsequent trial.

However, the melody-switching paradigm is different in that "alternate" feedback is not disruptive. In fact, in that study all melodies comprised the same metrical structure. Given the lack of disruption in this task, the challenge becomes evaluating the response cue, which is facilitated by ignoring auditory feedback. Here is where one aspect of hypothesis 2 emerges. When feedback is coordinated with respect to metrical pattern, pianists are able to focus on the task-switching component. Pianists are good at this and so any influence of alternate feedback is diluted by their enhanced ability to generalize from one sequence to the other when switching. Because of their enhanced memory capacity, pianists can have access to two sequences at the same time, and readily switch between the two. Nonmusicians, however, may have difficulty sustaining two sequence representations in memory, and so the effect of alternate feedback is more apparent.

Conclusion: toward a new model of learning

These results suggest a new framework for understanding the way in which enhanced music performance skill influences sensorimotor associations associated with that skill. First, with training, musicians build sensitivity to action-effect relationships across multiple hierarchical scales. Nonmusicians begin with sensitivity to coarse-grained and easily generalizable aspects of perception/action coordination, specifically the way in which spatial transitions on the keyboard map on to changes in pitch height. With skill, these associations may spread to both lower and higher levels. This leads to both costs and benefits. Whereas switching between two melodies may become an easier task, given the richer action hierarchies one has stored, a greater tendency to be disrupted by incompatible relationships may emerge.

Results reported here come exclusively from paradigms that involve piano performance, and thus our focus on perception/action relationships has been on relationships between patterns of movement in space (the keyboard) and variations in pitch. For performance on other instruments, and in singing, such spatial mapping is either indirect (as in the trumpet) or absent (the voice). Nevertheless, we suspect that similar kinds of mappings will emerge in these contexts, although the parameters may vary (for example, laryngeal tension may substitute spatial movement). For the present, this claim remains a hypothesis for the future.

Acknowledgment

The research summarized here was supported in part by NSF Grants BCS-0344892, BCS-0704516, and BCS-0642592.

Conflicts of interest

The author declares no conflicts of interest.

References

1. Lee, B.S. 1950. Effects of delayed speech feedback. *J. Acoust. Soc. Am.* **22:** 824–826.
2. Black, J.W. 1951. The effect of dleayed side-tone upon vocal rate and intensity. *J. Speech Disord.* **16:** 56–60.
3. Chase, R.A. 1965. An information-flow model of the organization of motor activity. I: transduction, transmission and central control of sensory information. *J. Nerv. Ment. Dis.* **140:** 239–251.
4. Howell, P. 2001. A model of timing interference to speech control in normal and altered listening conditions applied to the treatment of stuttering. In *Speech Motor Control in Normal and Disordered Speech*. B. Maassen, *et al.*, Eds.: 291–294. Uttgeverij Vantilt. Nijmegen, the Netherlands.
5. Howell, P. 2004. Assesment of some contemporary theories of stuttering that aply to spontaneous speech. *Contemp. Issues Commun. Sci. Disord.* **39:** 122–139.
6. Howell, P., D.J. Powell & I. Khan. 1983. Amplitude contour of the delayed signal and interference in delayed auditory feedback tasks. *J. Exp. Psychol. Hum. Percept. Perform.* **9:** 772–784.
7. Finney, S.A. 1997. Auditory feedback and musical keyboard performance. *Music Percept.* **15:** 153–174.
8. Pfordresher, P.Q. 2003. Auditory feedback in music performance: evidence for a dissociation of sequencing and timing. *J. Exp. Psychol. Hum. Percept. Perform.* **29:** 949–964.
9. Pfordresher, P.Q. 2005. Auditory feedback in music performance: the role of melodic structure and musical skill. *J. Exp. Psychol. Hum. Percept. Perform.* **31:** 1331–1345.
10. Pfordresher, P.Q. 2006. Coordination of perception and action in music performance. *Adv. Cogn. Psychol. Music Perform.* **2:** 183–198.
11. Pfordresher, P.Q. 2008. Auditory feedback in music performance: the role of transition-based similarity. *J. Exp. Psychol. Hum. Percept. Perform.* **34:** 708–725.
12. Bigand, E. & B. Poulin-Charronnat. 2006. Are we "experienced listeners"? A review of the musical capacities that

do not depend on formal musical training. *Cognition* **100:** 100–130.
13. Tillmann, B., J.J. Bharucha & E. Bigand. 2000. Implicit learning of tonality: a self-organizing approach. *Psychol. Rev.* **107:** 885–913.
14. James, W. 1890. *The Principles of Psychology*, Vol. 2, Holt. New York.
15. Adams, J.A. 1971. A closed-loop theory of motor learning. *J. Mot. Behav.* **3:** 111–149.
16. Schmidt, R.A. 1975. A schema theory of discrete motor skill learning. *Psychol. Rev.* **82:** 225–260.
17. Lashley, K. 1951. The problem of serial order in behavior. In *Cerebral Mechanisms in Behavior*. L.A. Jeffress, Ed.: 112–136. Wiley. New York.
18. Williamon, A. & E. Valentine. 2002. The role of retrieval structures in memorizing music. *Cogn. Psychol.* **44:** 1–32.
19. Klapp, S.T. & R.J. Jagacinski. 2011. Gestalt principles in the control of motor action. *Psychol. Bull.* **137:** 443–462.
20. Palmer, C. & C. van de Sande. 1995. Range of planning in music performance. *J. Exp. Psychol. Hum. Percept. Perform.* **21:** 947–962.
21. Rosenbaum, D.A., S.B. Kenny & M.A. Derr. 1983. Hierarchical control of rapid movement sequences. *J. Exp. Psychol. Hum. Percept. Perform.* **9:** 86–102.
22. Krumhansl, C.L. 1979. Psychological representation of musical pitch in a tonal context. *Cogn. Psychol.* **11:** 346–374.
23. Palmer, C. & C.L. Krumhansl. 1990. Mental representations for musical meter. *J. Exp. Psychol. Hum. Percept. Perform.* **16:** 728–741.
24. Keller, P.E. & I. Koch. 2008. Action planning in sequential skills: relations to music performance. *Q. J. Exp. Psychol.* **61:** 275–291.
25. Bangert, M. & E.O. Altenmüller. 2003. Mapping perception to action in piano practice: a longitudinal DC-EEG study. *BMC Neurosci.* **4:** 26.
26. Lahav, A., E. Saltzman & G. Schlaug. 2007. Action representation of sound: audiomotor recognition network while listening to newly acquired actions. *J. Neurosci.* **27:** 308–314.
27. Rizzolatti, G. & L. Craighero. 2004. The mirror-neuron system. *Annu. Rev. Neurosci.* **27:** 169–192.
28. Pfordresher, P.Q. *et al.* 2011. Activation of learned action sequences by auditory feedback. *Psychon. Bull. Rev.* **18:** 544–549.
29. Pfordresher, P.Q. & C. Palmer. 2006. Effects of hearing the past, present, or future during music performance. *Percept. Psychophys.* **68:** 362–376.
30. Rusconi, E. *et al.* 2006. Spatial representation of pitch height: the SMARC effect. *Cognition* **99:** 113–129.
31. Pfordresher, P.Q. *et al.* 2010. Neural stratification of sequencing and timing? An fMRI study. *J. Cogn. Neurosci., Suppl.*, 190.
32. Carter, C.S., *et al.* 1998. Anterior cingulate cortex, error detection, and the online monitoring of performance. *Science* **280:** 747–749.
33. Wolpert, D.M., R.C. Miall & M. Kawato. 1998. Internal models in the cerebellum. *Trends Cogn. Sci.* **2:** 338–347.
34. Krumhansl, C.L. & R.N. Shepard. 1979. Quantification of the hierarchy of tonal functions within a diatonic context. *J. Exp. Psychol. Hum. Percept. Perform.* **5:** 579–594.
35. Palmer, C. & R.K. Meyer. 2000. Conceptual and motor learning in music performance. *Psychol. Sci.* **11:** 63–68.
36. Meyer, R.K. & C. Palmer. 2003. Temporal and motor transfer in music performance. *Music Percept.* **21:** 81–104.

Ann. N.Y. Acad. Sci. ISSN 0077-8923

ANNALS OF THE NEW YORK ACADEMY OF SCIENCES
Issue: *The Neurosciences and Music IV: Learning and Memory*

The multisensory brain and its ability to learn music

Emily Zimmerman[1,2] and Amir Lahav[1,2]

[1]Department of Newborn Medicine, Brigham and Women's Hospital, Boston, Massachusetts. [2]Harvard Medical School, Boston, Massachusetts

Address for correspondence: Amir Lahav, Sc.D., Ph.D., Department of Newborn Medicine, The Neonatal Research Lab, 75 Francis Street, Boston, MA 02115. amir@hms.harvard.edu

Playing a musical instrument requires a complex skill set that depends on the brain's ability to quickly integrate information from multiple senses. It has been well documented that intensive musical training alters brain structure and function within and across multisensory brain regions, supporting the experience-dependent plasticity model. Here, we argue that this experience-dependent plasticity occurs because of the multisensory nature of the brain and may be an important contributing factor to musical learning. This review highlights key multisensory regions within the brain and discusses their role in the context of music learning and rehabilitation.

Keywords: brain; multisensory; learning; music

Introduction

We live in a multisensory world, and as a result, the brain is equipped with many multisensory areas. Traditionally, the brain was thought to contain several modality-specific, or unisensory, regions that could receive input from one of the primary senses. However, with advancing research it has become clear that many of the unisensory brain regions are actually multisensory in nature. For example, visual and somatosensory inputs have been observed in the auditory cortex,[1–12] auditory and somatosensory inputs have been observed in the visual cortex,[13–17] and visual and auditory inputs have been observed in the somatosensory cortex.[18,19] Therefore, the brain is composed of several multisensory regions, making it possible to handle the sensory challenges of our multisensory world.

The human brain depends highly on sensory input. When a sensory signal is received, receptor organs create neural impulses that are sent along unisensory pathways into subcortical and cortical structures. Typically, when the signals reach the cortex they remain unisensory. However, there are specialized areas within the brain that integrate and store information from different sensory modalities. These multisensory brain areas receive inputs through several unisensory cortical pathways, including auditory, visual, and somatosensory circuits.[20]

Musical training and the multisensory brain

Multisensory brain regions are shaped by specialized sensory experiences. One example of a specialized sensory experience is the process of learning to play a musical instrument. Initially, musical training takes advantage of the innate multisensory capabilities of the brain; however, as skillful expertise develops, further plasticity and neural specificity occurs. Previous studies have shown structural and functional brain differences in auditory,[21–24] sensorimotor,[25–27] and multimodal integration areas[26,28–30] in musicians compared to nonmusicians. Later in the review, we will argue that it is indeed the multimodal integration areas that play an advantageous role in music learning, performance, and rehabilitation.

In addition, brain plasticity seems to be highly sensitive to the multisensory conditions under which training was obtained. For instance, it has been shown that trained violin players have increased somatosensory representation of their left hand,[25] trained trumpet players have enhanced interactions between auditory and somatosensory inputs to the lips,[31] and trained pianists have increased activation in presupplementary motor cortex and lateral dorsal premotor cortex.[32] Taken together, these studies show that musical training can result in plasticity-induced cortical changes to further optimize the multisensory demands required for music performance.

Is there an advantage to having multisensory regions in the brain?

When multiple senses are stimulated at the same time, termed multimodal, the result can be advantageous. For example, when a violinist is playing, you can not only hear the music, but can also see the violinist bowing against the strings. Because these stimuli (i.e., sound/sight of violin) originate from the same event, the resultant combination is synergistic and can significantly exceed the response from either sensory input alone or their arithmetic sum.[33–40] Often, the neural responses can exceed the sum of the responses to each sense alone by approximately 1,000%.[36,41] This type of additive neural response derived from multisensory regions is beneficial for learning music and may be a key factor for the effective use of music in physical rehabilitation.

Multisensory cortical brain regions associated with music learning

Although the brain contains a vast amount of multisensory areas, several distinct cortical brain regions have been particularly implicated in the music domain. These regions mainly include the superior temporal sulcus (STS), the intraparietal sulcus (IPS), and the prefrontal cortex (PFC). As we discuss each brain region, we will first introduce the multisensory nature of the region and then highlight its role in the context of music learning and performance.

STS

The STS is the sulcus separating the superior temporal gyrus from the middle temporal gyrus in the temporal lobe. Studies have shown that approximately 36–38%[42,43] of the anterior STS neurons and 12% of the caudal STS neurons are multimodal.[44] Barraclough *et al.*[45] have shown that the same STS neurons that code the sight of actions can also code the sound of those actions. They found that 23% of neurons responsive to the sight of an action were also responsive to the sound of that action.[45] Thus, neurons in the STS seem to form multisensory representations of observed actions.

Functional magnetic resonance imaging (fMRI) studies have shown activation of the STS during various musical tasks, such as the processing of melody,[46] pitch,[47] and timbre changes,[48] perception of musical chords,[49] recognition of musical memories,[50] and during singing.[50] This region seems to be activated by the fine structures evident within musical pieces.

IPS

The IPS is located on the lateral surface of the parietal lobe and is thought to be related to perceptual-motor coordination. Three subareas within the IPS are involved in multisensory integration. These areas include the lateral intraparietal region, which receives sensory signals for eye positioning and auditory signals;[51] the ventral intraparietal region, which contains neurons that can respond to visual, auditory, somatosensory, and vestibular to create bimodal and trimodal receptive fields;[52–55] and the temporoparietal junction, which is thought to contain many multimodal representations of space.[56]

Schulze *et al.*[57] found large IPS activation during auditory working memory in musicians compared to nonmusicians. Another study by Lahav *et al.*[58] has shown activation in the IPS[59] during motionless listening to a rehearsed musical piece five days after learning. IPS activation has also been shown during temporal reversal of a melody—where musicians were presented with the first few notes of a familiar tune, or its title, followed by a string of notes in a reversed order. These studies implicate the IPS as a brain area involved in auditory memory circuitry—a crucial component for music learning.

PFC

The PFC plays a role in temporal integration and receives projections from both the auditory and visual cortices.[60–62] This widespread connectivity makes the PFC a prime location for multisensory integration. This region has been of particular interest in the past 15 years with the discovery of mirror neurons. A mirror neuron is a type of neuron that discharges both when an individual performs an action and when they observe another individual performing a similar action.[63,64] Mirror neurons likely play a large role in music integration and learning. Buccino *et al.*[65] scanned nonmusicians during the following tasks: guitar chords being played, a pause after the model observation, execution of the observed chords, and then rest. They found that the basic circuit underlying imitation learning consisted of the IPL, posterior aspect of the inferior frontal gyrus, and the mirror neurons in PFC.[65]

Leaver *et al.*[66] examined the predictive or "anticipatory imagery" of music at various stages using fMRI. They found that anticipatory imagery in silence for highly familiar music activated rostral prefrontal cortex and premotor areas. Not only is the prefrontal cortex important for musical prediction, it is also activated during error correction during musical performance.[67] In addition, the prefrontal cortex has been shown to be activated in pitch processing,[68,69] temporal modulations in rhythmic stimuli,[70] musical chord violations,[71] recognition of tonal structures,[72] and emotional responses to music.[73,74] It is clear that the prefrontal cortex is a necessity for music because planning, execution, and emotion are all vital aspects for a successful musical performance.

The three multisensory regions discussed previously are crucial for musical training and performance. For example, music is not attained by merely activating the motor cortex to press down a piano key with your finger; it requires the fine structure and specificity afforded by multisensory regions. A musical piece must contain pitch, timing, and timbre, which are all housed in these multisensory brain regions. We hypothesize that music requires more than an integrated sensorimotor system; it requires an integrated multisensory system. A prime model for this hypothesis is a nonmusician, who may have a well-intact sensorimotor system but still lacks the fine structure of a good musician. It may be, in fact, that these nonmusicians have abnormal neuronal structure, characterized by reduced gray matter volume or impaired white matter connectivity in the three multisensory regions discussed previously. Future studies are encouraged to examine this hypothesis in more detail.

Implications for music learning

Is the multisensory learning environment (e.g., vision, audition, and touch) indeed beneficial in the initial process of learning to play a musical instrument? The usage of nonmusicians as study subjects can provide investigators with a unique experimental model to answer this question. For example, Eldridge *et al.*[75] examined the effect of unisensory versus multisensory feedback on the ability to code and recognize pitch information. In this study, two groups of nonmusicians were trained to play a piano piece by ear. One group received uninterrupted audiovisual feedback, whereas the other group could hear but not see their hands on the keyboard (audio alone). This study found that multimodal stimulation increased learning and encouraged the use of audiovisual (rather than audio or visual alone) in learning new skills.[75] Similarly, Pantev *et al.*[76] found that multimodal sensorimotor-auditory training in nonmusicians resulted in greater plastic changes in auditory cortex than auditory training alone. Music is a multimodal experience and when one of the modalities is occluded—it can come at the cost of learning. Thus, when it comes to music learning, receiving feedback from multiple senses is probably better than receiving feedback from one.

Although this review focuses primarily on cortical multisensory regions in the context of music learning, there are also several subcortical processes that are attributed to musical training (for a review, see Ref. 77). Kraus *et al.*[78] point to pitch, timbre, and timing as having subcortical populations that can be enhanced through musical training. Similarly, other studies have shown that musicians have been shown to have enhanced electrophysiological responses to melodic contour and interval information[79] and enhanced evoked potentials in response to pitch changes during speech processing[80–82] in the auditory brainstem. As the focus on subcortical structures in musical training increases, it will likely reveal vast multisensory regions and cortico–subcortical connectivity that further integrate and evolve during musical training.

Implications for rehabilitation

Music is often a highly motivating and pleasurable form of therapy that can have global benefits on the brain (for review, see Ref. 83). Previous studies have shown that rapid plastic adaptation because of music performance is not restricted to motor areas but also incorporates auditory and auditory-sensorimotor circuitry.[84–86] Thus, the multisensory nature of music making and listening concurrently stimulates multiple systems within the brain. This type of multimodal stimulation is believed to facilitate crosstalk and connectivity between key regions of the brain, which may be particularly beneficial for neurologically impaired patients.

Schneider *et al.*[86] provided 20 stroke patients with a musical training program (15 × /3 weeks) and compared them to 20 stroke patients who did not receive the program. They found that the stroke patients who received the musical training program had significant improvements in speed, precision, and smoothness of movements, as well as in everyday activities compared to the control group. The reason music is so powerful is because it requires integration of multisensory and motor inputs as well as the precise monitoring of the resultant motor performance by multiple feedback mechanisms.[87] Therefore, if all possible connections and networks are being targeted, not only does the likelihood for rehabilitation increase, but the ability to create new connections through experience-dependent plasticity also increases.

The engagement of multiple multisensory and motor regions activated by playing music can have beneficial effects on the physiological and psychological health of individuals.[83] In addition, playing music in rehabilitation can improve attention,[88–90] emotion,[91] cognition,[92] behavior,[93,94] and communication[95] skills (for a review on this topic, see Ref. 83). It is clear that the use of music for neurologically impaired patients is a favorable clinical option.

Conclusion

These multisensory areas enable musicians to integrate complex sensory inputs and the resultant motor outputs necessary to play music. This paper argues that the use of multisensory feedback during musical training should be highly preferred because of the vast interconnectivity within and between multisensory areas, which allows for music-induced brain plasticity to occur. This plasticity may be particularly important for neurologically impaired patients. Thus, the implementation of multisensory music-based therapies in physical rehabilitation should be further encouraged.

Conflicts of interest

The authors declare no conflicts of interest.

References

1. Schroeder, C.E. & J.J. Foxe. 2002. The timing and laminar profile of converging inputs to multisensory areas of the macaque neocortex. *Brain Res. Cogn. Brain Res.* **14:** 187–198.
2. Calvert, G.A. *et al.* 1997. Activation of auditory cortex during silent lipreading. *Science* **276:** 593–596.
3. Foxe, J.J. *et al.* 2000. Multisensory auditory-somatosensory interactions in early cortical processing revealed by high-density electrical mapping. *Brain Res. Cogn. Brain Res.* **10:** 77–83.
4. Giard, M.H. & F. Peronnet. 1999. Auditory-visual integration during multimodal object recognition in humans: a behavioral and electrophysiological study. *J. Cogn. Neurosci.* **11:** 473–490.
5. Molholm, S. *et al.* 2002. Multisensory auditory-visual interactions during early sensory processing in humans: a high-density electrical mapping study. *Brain Res. Cogn. Brain Res.* **14:** 115–128.
6. Besle, J. *et al.* 2004. Bimodal speech: early suppressive visual effects in human auditory cortex. *Eur. J. Neurosci.* **20:** 2225–2234.
7. Gobbele, R. *et al.* 2003. Activation of the human posterior parietal and temporoparietal cortices during audiotactile interaction. *NeuroImage* **20:** 503–511.
8. Lutkenhoner, B. *et al.* 2002. Magnetoencephalographic correlates of audiotactile interaction. *NeuroImage* **15:** 509–522.
9. Ghazanfar, A.A. *et al.* 2005. Multisensory integration of dynamic faces and voices in rhesus monkey auditory cortex. *J. Neurosci.* **25:** 5004–5012.
10. Schroeder, C.E. *et al.* 2001. Somatosensory input to auditory association cortex in the macaque monkey. *J. Neurophysiol.* **85:** 1322–1327.
11. Fu, K.M. *et al.* 2003. Auditory cortical neurons respond to somatosensory stimulation. *J. Neurosci.* **23:** 7510–7515.
12. Fu, K.M. *et al.* 2004. Timing and laminar profile of eye-position effects on auditory responses in primate auditory cortex. *J. Neurophysiol.* **92:** 3522–3531.
13. Morrell, F. 1972. Visual system's view of acoustic space. *Nature* **238:** 44–46.
14. Hagen, M.C. *et al.* 2002. Tactile motion activates the human middle temporal/V5 (MT/V5) complex. *Eur. J. Neurosci.* **16:** 957–964.
15. Amedi, A. *et al.* 2001. Visuo-haptic object-related activation in the ventral visual pathway. *Nat. Neurosci.* **4:** 324–330.
16. James, T.W. *et al.* 2002. Haptic study of three-dimensional objects activates extrastriate visual areas. *Neuropsychologia* **40:** 1706–1714.
17. Pietrini, P. *et al.* 2004. Beyond sensory images: object-based representation in the human ventral pathway. *Proc. Natl. Acad. Sci. USA* **101:** 5658–5663.
18. Zhou, Y.D. & J.M. Fuster. 2004. Somatosensory cell response to an auditory cue in a haptic memory task. *Behav. Brain Res.* **153:** 573–578.
19. Zhou, Y.D. & J.M. Fuster. 2000. Visuo-tactile cross-modal associations in cortical somatosensory cells. *Proc. Natl. Acad. Sci. USA* **97:** 9777–9782.
20. Stein, B.A. & M.A. Meredith. 1993. *The Merging of the Senses.* MIT Press. Cambridge, MA.
21. Bermudez, P. & R.J. Zatorre. 2005. Differences in gray matter between musicians and nonmusicians. *Ann. N.Y. Acad. Sci.* **1060:** 395–399.

22. Lappe, C. *et al.* 2008. Cortical plasticity induced by short-term unimodal and multimodal musical training. *J. Neurosci.* **28:** 9632–9639.
23. Pantev, C. *et al.* 1998. Increased auditory cortical representation in musicians. *Nature* **392:** 811–814.
24. Zatorre, R.J. 1998. Functional specialization of human auditory cortex for musical processing. *Brain* **121**(Pt 10): 1817–1818.
25. Elbert, T. *et al.* 1995. Increased cortical representation of the fingers of the left hand in string players. *Science* **270:** 305–307.
26. Gaser, C. & G. Schlaug. 2003. Gray matter differences between musicians and nonmusicians. *Ann. N.Y. Acad. Sci.* **999:** 514–517.
27. Hund-Georgiadis, M. & D.Y. von Cramon. 1999. Motor-learning-related changes in piano players and nonmusicians revealed by functional magnetic-resonance signals. *Exp. Brain Res.* **125:** 417–425.
28. Bangert, M. & G. Schlaug. 2006. Specialization of the specialized in features of external human brain morphology. *Eur. J. Neurosci.* **24:** 1832–1834.
29. Sluming, V. *et al.* 2007. Broca's area supports enhanced visuospatial cognition in orchestral musicians. *J. Neurosci.* **27:** 3799–3806.
30. Zatorre, R.J., J.L. Chen & V.B. Penhune. 2007. When the brain plays music: auditory-motor interactions in music perception and production. *Nat. Rev. Neurosci.* **8:** 547–558.
31. Schulz, M., B. Ross & C. Pantev. 2003. Evidence for training-induced crossmodal reorganization of cortical functions in trumpet players. *Neuroreport* **14:** 157–161.
32. Baumann, S. *et al.* 2007. A network for audio-motor coordination in skilled pianists and nonmusicians. *Brain Res.* **1161:** 65–78.
33. Diedrich, A. & H. Colonius. 2004. *Modeling the Time Course of Multisensory Interaction in the Manual and Saccadic Responses.* MIT Press. Cambridge, MA.
34. Jiang, W. *et al.* 2001. Two cortical areas mediate multisensory integration in superior colliculus neurons. *J. Neurophysiol.* **85:** 506–522.
35. King, A.J. & A.R. Palmer. 1985. Integration of visual and auditory information in bimodal neurones in the guinea-pig superior colliculus. *Exp. Brain Res.* **60:** 492–500.
36. Meredith, M.A. & B.E. Stein. 1986. Spatial factors determine the activity of multisensory neurons in cat superior colliculus. *Brain Res.* **365:** 350–354.
37. Meredith, M.A. & B.E. Stein. 1983. Interactions among converging sensory inputs in the superior colliculus. *Science* **221:** 389–391.
38. Peck, C.K. 1987. Visual-auditory interactions in cat superior colliculus: their role in the control of gaze. *Brain Res.* **420:** 162–166.
39. Perrault, Jr., T.J. *et al.* 2003. Neuron-specific response characteristics predict the magnitude of multisensory integration. *J. Neurophysiol.* **90:** 4022–4026.
40. Stanford, T.R., S. Quessy & B.E. Stein. 2005. Evaluating the operations underlying multisensory integration in the cat superior colliculus. *J. Neurosci.* **25:** 6499–6508.
41. Wallace, M.T. & B.E. Stein. 1997. Development of multisensory neurons and multisensory integration in cat superior colliculus. *J. Neurosci.* **17:** 2429–2444.
42. Bruce, C., R. Desimone & C.G. Gross. 1981. Visual properties of neurons in a polysensory area in superior temporal sulcus of the macaque. *J. Neurophysiol.* **46:** 369–384.
43. Benevento, L.A. *et al.* 1977. Auditory–visual interaction in single cells in the cortex of the superior temporal sulcus and the orbital frontal cortex of the macaque monkey. *Exp. Neurol.* **57:** 849–872.
44. Hikosaka, K. 1993. The polysensory region in the anterior bank of the caudal superior temporal sulcus of the macaque monkey. *Biomed. Res. (Tokyo)* **14:** 41–45.
45. Barraclough, N.E. *et al.* 2005. Integration of visual and auditory information by superior temporal sulcus neurons responsive to the sight of actions. *J. Cogn. Neurosci.* **17:** 377–391.
46. Lee, Y.S. *et al.* 2011. Investigation of melodic contour processing in the brain using multivariate pattern-based fMRI. *NeuroImage* **57:** 293–300.
47. Stewart, L. *et al.* 2008. fMRI evidence for a cortical hierarchy of pitch pattern processing. *PLoS One* **3:** e1470.
48. Menon, V. *et al.* 2002. Neural correlates of timbre change in harmonic sounds. *NeuroImage* **17:** 1742–1754.
49. Klein, M.E. & R.J. Zatorre. 2011. A role for the right superior temporal sulcus in categorical perception of musical chords. *Neuropsychologia* **49:** 878–887.
50. Peretz, I. *et al.* 2009. Music lexical networks: the cortical organization of music recognition. *Ann. N.Y. Acad. Sci.* **1169:** 256–265.
51. Andersen, R.A. *et al.* 1997. Multimodal representation of space in the posterior parietal cortex and its use in planning movements. *Annu. Rev. Neurosci.* **20:** 303–330.
52. Avillac, M. *et al.* 2005. Reference frames for representing visual and tactile locations in parietal cortex. *Nat. Neurosci.* **8:** 941–949.
53. Bremmer, F. *et al.* 2002. Visual-vestibular interactive responses in the macaque ventral intraparietal area (VIP). *Eur. J. Neurosci.* **16:** 1569–1586.
54. Duhamel, J.R., C.L. Colby & M.E. Goldberg. 1998. Ventral intraparietal area of the macaque: congruent visual and somatic response properties. *J. Neurophysiol.* **79:** 126–136.
55. Schlack, A. *et al.* 2005. Multisensory space representations in the macaque ventral intraparietal area. *J. Neurosci.* **25:** 4616–4625.
56. Leinonen, L. 1980. Functional properties of neurones in the posterior part of area 7 in awake monkey. *Acta. Physiol. Scand.* **108:** 301–308.
57. Schulze, K., K. Mueller & S. Koelsch. 2011. Neural correlates of strategy use during auditory working memory in musicians and nonmusicians. *Eur. J. Neurosci.* **33:** 189–196.
58. Lahav, A., E. Saltzman & G. Schlaug. 2007. Action representation of sound: audiomotor recognition network while listening to newly acquired actions. *J. Neurosci.* **27:** 308–314.
59. Zatorre, R.J., A.R. Halpern & M. Bouffard. 2010. Mental reversal of imagined melodies: a role for the posterior parietal cortex. *J. Cogn. Neurosci.* **22:** 775–789.
60. Gaffan, D. & S. Harrison. 1991. Auditory-visual associations, hemispheric specialization and temporal-frontal interaction in the rhesus monkey. *Brain* **114**(Pt 5): 2133–2144.

61. Fuster, J.M., M. Bodner & J.K. Kroger. 2000. Cross-modal and cross-temporal association in neurons of frontal cortex. *Nature* **405:** 347–351.
62. Romanski, L.M., J.F. Bates & P.S. Goldman-Rakic. 1999. Auditory belt and parabelt projections to the prefrontal cortex in the rhesus monkey. *J. Comp. Neurol.* **403:** 141–157.
63. Gallese, V. *et al.* 1996. Action recognition in the premotor cortex. *Brain* **119**(Pt 2): 593–609.
64. Rizzolatti, G. *et al.* 1996. Premotor cortex and the recognition of motor actions. *Brain Res. Cogn. Brain Res.* **3:** 131–141.
65. Buccino, G. *et al.* 2004. Neural circuits involved in the recognition of actions performed by nonconspecifics: an FMRI study. *J. Cogn. Neurosci.* **16:** 114–126.
66. Leaver, A. M. *et al.* 2009. Brain activation during anticipation of sound sequences. *J. Neurosci.* **29:** 2477–2485.
67. Ruiz, M.H. *et al.* EEG oscillatory patterns are associated with error prediction during music performance and are altered in musician's dystonia. *NeuroImage* **55:** 1791–1803.
68. Wilson, S.J. *et al.* 2009. The neurocognitive components of pitch processing: insights from absolute pitch. *Cereb. Cortex* **19:** 724–732.
69. Zatorre, R.J., A.C. Evans & E. Meyer. 1994. Neural mechanisms underlying melodic perception and memory for pitch. *J. Neurosci.* **14:** 1908–1919.
70. Thaut, M.H. 2003. Neural basis of rhythmic timing networks in the human brain. *Ann. N.Y. Acad. Sci.* **999:** 364–373.
71. Garza Villarreal, E.A. *et al.* 2011. Distinct neural responses to chord violations: a multiple source analysis study. *Brain Res.* **1389:** 103–114.
72. Janata, P. *et al.* 2002. The cortical topography of tonal structures underlying Western music. *Science* **298:** 2167–2170.
73. Blood, A.J. & R.J. Zatorre. 2001. Intensely pleasurable responses to music correlate with activity in brain regions implicated in reward and emotion. *Proc. Natl. Acad. Sci. USA* **98:** 11818–11823.
74. Omar, R. *et al.* 2011. The structural neuroanatomy of music emotion recognition: evidence from frontotemporal lobar degeneration. *NeuroImage* **56:** 1814–1821.
75. Eldridge, M.S. & A. Lahav. 2010. Seeing what you hear: visual feedback improves pitch recognition. *Eur. J. Cogn. Psychol.* **23:** 1–14.
76. Pantev, C. *et al.* 2009. Auditory-somatosensory integration and cortical plasticity in musical training. *Ann. N.Y. Acad. Sci.* **1169:** 143–150.
77. Kraus, N. & B. Chandrasekaran. 2010. Music training for the development of auditory skills. *Nat. Rev. Neurosci.* **11:** 599–605.
78. Kraus, N. *et al.* 2009. Experience-induced malleability in neural encoding of pitch, timbre, and timing. *Ann. N.Y. Acad. Sci.* **1169:** 543–557.
79. Lee, K.M. *et al.* 2009. Selective subcortical enhancement of musical intervals in musicians. *J. Neurosci.* **29:** 5832–5840.
80. Magne, C., D. Schon & M. Besson. 2006. Musician children detect pitch violations in both music and language better than nonmusician children: behavioral and electrophysiological approaches. *J. Cogn. Neurosci.* **18:** 199–211.
81. Besson, M. *et al.* 2007. Influence of musical expertise and musical training on pitch processing in music and language. *Restor. Neurol. Neurosci.* **25:** 399–410.
82. Musacchia, G. *et al.* 2007. Musicians have enhanced subcortical auditory and audiovisual processing of speech and music. *Proc. Natl. Acad. Sci. USA* **104:** 15894–15898.
83. Koelsch, S. 2009. A neuroscientific perspective on music therapy. *Ann. N.Y. Acad. Sci.* **1169:** 374–384.
84. Bangert, M. & E.O. Altenmüller. 2003. Mapping perception to action in piano practice: a longitudinal DC-EEG study. *BMC Neurosci.* **4:** 26.
85. Bangert, M. *et al.* 2006. Shared networks for auditory and motor processing in professional pianists: evidence from fMRI conjunction. *NeuroImage* **30:** 917–926.
86. Schneider, S. *et al.* 2007. Using musical instruments to improve motor skill recovery following a stroke. *J. Neurol.* **254:** 1339–1346.
87. Altenmüller, E. 2008. Neurology of musical performance. *Clin. Med.* **8:** 410–413.
88. Koelsch, S. 2009. Music-syntactic processing and auditory memory: similarities and differences between ERAN and MMN. *Psychophysiology* **46:** 179–190.
89. Nelson, A. *et al.* 2008. The impact of music on hypermetabolism in critical illness. *Curr. Opin. Clin. Nutr. Metab. Care* **11:** 790–794.
90. Klassen, J.A. *et al.* 2008. Music for pain and anxiety in children undergoing medical procedures: a systematic review of randomized controlled trials. *Ambul. Pediatr.* **8:** 117–128.
91. Thaut, M.H. *et al.* 2009. Neurologic music therapy improves executive function and emotional adjustment in traumatic brain injury rehabilitation. *Ann. N.Y. Acad. Sci.* **1169:** 406–416.
92. Gerdner, L.A. & E.A. Swanson. 1993. Effects of individualized music on confused and agitated elderly patients. *Arch. Psychiatr. Nurs.* **7:** 284–291.
93. Schlaug, G., S. Marchina & A. Norton. 2009. Evidence for plasticity in white-matter tracts of patients with chronic Broca's aphasia undergoing intense intonation-based speech therapy. *Ann. N.Y. Acad. Sci.* **1169:** 385–394.
94. Altenmüller, E. *et al.* 2009. Neural reorganization underlies improvement in stroke-induced motor dysfunction by music-supported therapy. *Ann. N.Y. Acad. Sci.* **1169:** 395–405.
95. Hillecke, T., A. Nickel & H.V. Bolay. 2005. Scientific perspectives on music therapy. *Ann. N.Y. Acad. Sci.* **1060:** 271–282.

Ann. N.Y. Acad. Sci. ISSN 0077-8923

ANNALS OF THE NEW YORK ACADEMY OF SCIENCES
Issue: *The Neurosciences and Music IV: Learning and Memory*

Sensorimotor mechanisms in music performance: actions that go partially wrong

Caroline Palmer, Brian Mathias, and Maxwell Anderson
Department of Psychology, McGill University, Montreal, Quebec, Canada

Address for correspondence: Caroline Palmer, Department of Psychology, McGill University, 1205 Dr. Penfield Ave., Montreal QC H3A 1B1, Canada. Caroline.palmer@mcgill.ca

Even expert musicians make errors occasionally, and overt responses that are correct may be accompanied by partial-error behavior that can be indicative of online error detection processes. We compare pianists' production of correct pitches, pitch errors, and partial errors (correct pitches with incorrect force or duration) by examining events prior to errors. Errors tended to be produced with slower durations and softer intensities (associated with force reduction) than correct events. In addition, pre-error events tended to have durations and intensities that fell between those of errors and correct responses, presumably due to response competition with upcoming errors that resulted in partial-error outcomes. These findings support the inference that partial information about upcoming (planned) sequence events is used to guide current responses, consistent with cascade models of activation during sequence production.

Keywords: conflict monitoring; mismatch detection; executive control; partial errors

Introduction

Even the best musicians make errors occasionally: tones whose sensory outcomes do not resemble the performers' expected or intended tones. Errors can arise from many sources, including inattention, lack of skill, or reading failures; we focus here on the occasional unambiguous pitch error when a pianist strikes the wrong key, which can arise even in well-practiced performance. Studies of piano performance indicate a 2–3% error rate in performance of familiar pieces that have been practiced for many hours,[1] similar to error rates reported in other well-learned tasks.[2] One intriguing finding across many domains is that error outcomes resemble events intended for elsewhere in a sequence: for example, a sightreading error that slips the eye because the error fits well within the musical context,[3] or an exchange of order between neighboring elements of an action sequence.[4]

Several studies suggest that performance errors influence the production of surrounding events. The altered force or timing with which those surrounding events are produced has led them to be termed "partial errors," which have been defined in simple response tasks as error tendencies (such as a faulty timing or response force) that do not result in a full-fledged inaccurate response (such as pressing the wrong key).[5] These partial errors are particularly interesting because they are often accompanied by electromyographic or electroencephalographic signals prior to their production that indicate some awareness that an error has been planned prior to its execution.[6,7] Explanations of how errors and partial errors arise, and what they reveal about real-time processing mechanisms, fall into two primary camps. One explanation is that error detection is the product of a mismatch comparator,[7,8] in which a representation of a correct response is compared with a representation of an actual response. Sufficient mismatch between these representations triggers the comparator to generate an error signal, which is thought to generate an error-related negativity that often precedes the error outcome.[6,7] An alternate explanation is the conflict monitoring hypothesis,[9] in which two or more events prepared for execution are compared; the comparison reflects the degree of conflict between these events, which generates a larger signal, the greater the conflict.

doi: 10.1111/j.1749-6632.2011.06427.x

Possible functions of an error-related negativity include immediate error correction (within-trial), as have been reported for fast error corrections[10] and for partial (inhibited) errors.[5] Only the error mismatch theory is consistent with the immediate error correction function. Another possible function is strategic adjustments between trials, as seen in post-error slowing[10] or in conflict adaptation.[11] The conflict monitoring theories are most consistent with this function. Unfortunately, the timing of the event-related negativity (ERN) is not always consistent with post-error slowing or error correction,[6,12] and larger ERN amplitudes have sometimes been noted for lower conflict than for higher conflict conditions.[13]

Several researchers have proposed that errors and, to some extent, partial errors arise from mismatches to a predictive sensorimotor consequence;[6,8] as a response unfolds, the predicted sensory outcome of the action is compared with the actual sensory input. This predicted outcome is thought to include stimulus features and kinesthetic properties, and deviations from the predictions are used to correct movements. Similar internal prediction models have been proposed for motor control tasks,[14] and for visual and kinesthetic feedback in aimed movements.[15] Most of these tasks entail single arm or hand movements, in which the sensory consequences do not overlap with preceding or successive outcomes that are typical of the relatively fast tone sequences that musicians must execute.

We describe a study of the occasional errors that arise in piano performance, typified by production of very fast series of individual hand and finger movements whose sensory/kinesthetic properties overlap in time. Music performance provides an excellent ground for testing the adaptive nature of the brain in response to occasional pitch errors that arise when fingers strike the wrong keys. We test whether information associated with error outcomes is used for online corrections, and we examine the events preceding errors for signs of partial errors that have been reported in piano performance.[16,17] Pianists performed two-part musical pieces at fast tempi after producing a note-perfect performance of the pieces at a slower tempo. The musical contexts in which pianists performed were examined for their influence on the types of errors that occurred.

Method

Participants and materials

Twenty-four adult pianists from the Montreal community participated in the study; all participants had at least seven years of piano training (M = 16.2 years). Every participant reported playing the instrument regularly; none reported any hearing problems. They performed eight novel pieces, composed in 4/4 meter and consisting of 33 isochronous sixteenth notes in two parts (one part controlled by each hand). Half of the pieces were in a major key and half were in a minor key, and all conformed to conventions of Western polyphonic music.

Design and procedure

Each participant performed four musical pieces twice per block in each of four experimental blocks ($4 \times 2 \times 4$), with the musical notation in view. Two pieces were performed at a medium tempo (225 ms per sixteenth-note interonset interval [IOI]), and the other two at a fast tempo (187.5 ms per sixteenth-note IOI). Each piece was performed twice at one of the two experimental tempi within each trial. The ordering of pieces and assignment of tempo to piece were counterbalanced across subjects.

At the start of the session, participants practiced each piece until an error-free performance was achieved at a slow tempo (429 ms per sixteenth-note IOI). This criterion ensured that errors that occurred at faster tempi were not due to sight-reading failures, errors of perception, or incorrect learning of the musical pieces. Pianists were instructed to perform at the tempo indicated by the metronome throughout the trial and to perform without correcting any errors. Additional (unrelated) pieces were performed either before or after the performances of the pieces evaluated in the study.

Errors in pitch accuracy were identified by computer comparison of pianists' performances with the information in the notated musical score.[18,19] Although pianists were instructed at the beginning of the session not to stop and correct any errors that they produced, they occasionally corrected some of the errors and thus the error was interrupted and could not be coded unambiguously. These correction errors were excluded from all analyses (less than 1.2% of total errors).

Results

The mean pitch error rate across all performances was 0.12 (mean error rate in medium tempo performance = 0.092; fast performance = 0.144). Pitch errors were first compared with correct tones in terms of their duration, measured by IOIs (shown in Fig. 1). To evaluate influences of errors on the performance of events that preceded them, correct tones were divided into two groups for analysis. The first group consisted of every correct event that was produced immediately before an error, referred to as pre-error tones, and the second group consisted of every correct event that was not followed by an error, referred to as correct tones. IOIs were determined from the onset of a particular pitch event to the onset of the subsequent pitch that was produced by the same hand of the pianist.

A repeated measures analysis of variance (ANOVA) on the mean event IOIs in each performance by tempo and event type indicated a significant main effect of tempo, $F(1, 23) = 367.68$, $P < 0.001$. As expected, event IOIs in the medium condition ($M = 227$ ms) were significantly greater than event IOIs in the fast condition ($M = 195$ ms) and closely matched the prescribed tempi of 225 ms IOI in the medium condition and 187.5 ms IOI in the fast condition. Also, there was a significant effect of event type on produced IOI, $F(2, 46) = 3.25$, $P < 0.05$, and no interaction of tempo with event type. *Post hoc* comparisons (Tukey HSD = 5.22, $\alpha = 0.05$) indicated that mean IOIs of errors were significantly greater than IOIs of correct tones, but mean IOIs of pre-errors did not differ significantly from those of correct tones or error tones. Thus, the accuracy of performance pitches influenced the relative length of time that the pitches were sounded, from fastest (correct) performance to slowest (errorful) performance. Pitch errors were performed less accurately in time relative to the metronome tempo than were correct events.

We further analyzed whether error tones differed from correctly performed tones in terms of the intensity with which they were produced. Figure 2 shows mean intensities of correctly performed tones, pitch errors, and pre-error correct tones. The repeated measures ANOVA on event intensities by tempo and event type indicated a significant main effect of tempo, $F(1, 23) = 6.44$,

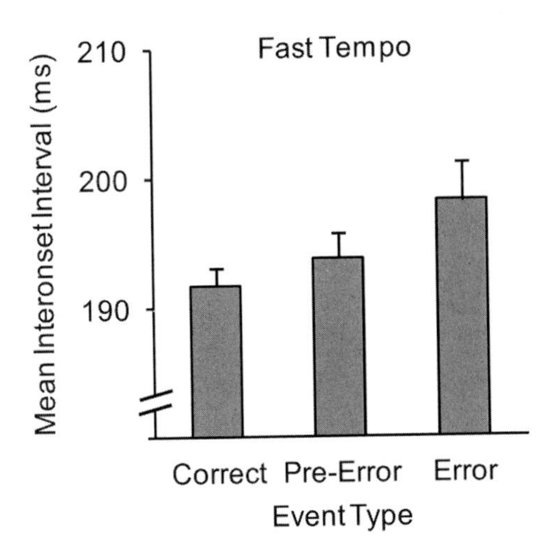

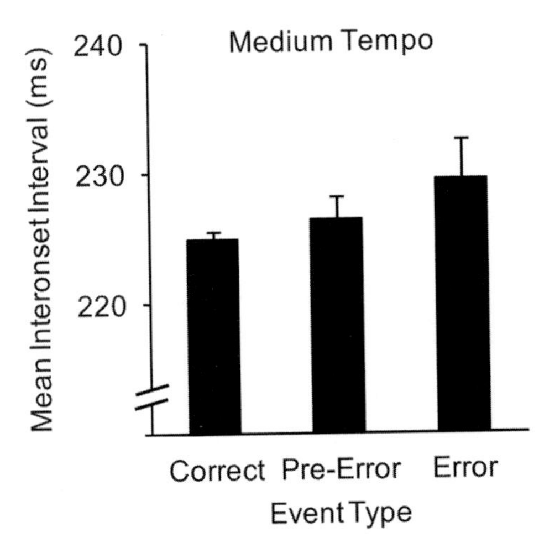

Figure 1. Mean IOIs of correctly performed pitches that were not immediately followed by an error, correct pitches produced one tone before errors, and pitch errors. Fast tempo condition on top; medium tempo condition on bottom.

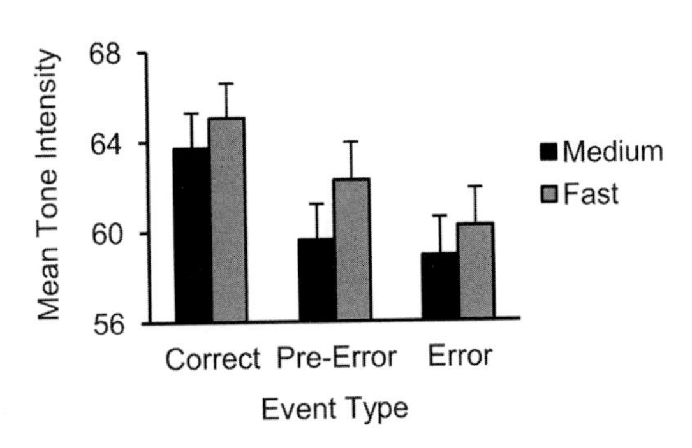

Figure 2. Mean intensities (MIDI velocity values) of correctly performed pitches that were not immediately followed by an error, correct pitches produced immediately before an error, and pitch errors for the medium- and fast-tempo conditions.

Figure 3. One of the notated stimulus pieces with predicted metrical accent strengths according to a 4-tier metrical grid. Events aligned with higher tiers are more strongly accented.

$P < 0.05$. Similar to previous findings,[20] pitches in the fast tempo condition ($M = 62.5$) were produced with significantly more force (yielding greater tone intensities) than pitches in the medium tempo condition ($M = 60.8$). There was also a significant effect of event type on production intensity, $F(2, 46) = 31.52$, $P < 0.001$. *Post hoc* comparisons (Tukey HSD = 1.50, $\alpha = 0.05$) indicated that, on average, correct tones were produced with greater intensity than both pre-error tones and error tones, while pre-error and error tones did not differ in intensity. Thus, the locus of the error effect extended to the amount of force with which performers produced pre-error tones. There was no significant interaction of tempo and event type.

The position of sequential events within a piece's metrical framework may influence the intensity with which errors and correct tones are produced. To examine metrical effects on tone intensities, correct and error tones were coded for metrical accent strength according to a 4-tier metrical grid, consistent with the time signature and smallest notated duration in the musical score of the pieces, as shown in Figure 3 for one piece.[21]

It is possible that online adjustments to the execution of pre-error tones reflect preparation specific to the pre-error position, in addition to (or in place of) preparation of the upcoming error. For example, the metrical accent strength of an error and of the tone preceding it may influence the production of both the error and the tone preceding it. To test this possibility, we compared mean tone intensities for correct events, pre-error events, and errors within each performance by the metrical accent level with which they were aligned, as prescribed in Figure 3. Figure 4 shows the mean intensities of correct, pre-error, and error tones as a function of the metrical accent strength with which those tones aligned in the musical score.

A repeated measures ANOVA on performance intensities by tempo, event type, and metrical accent strength (1–4) indicated again the significant main effects of event type, $F(2, 46) = 22.84$, $P < 0.001$, and tempo, $F(1, 23) = 5.24$, $P < 0.05$, such that correct events and events in fast-tempo performances were played with greater intensity on average. There was also a significant main effect of accent, $F(3, 69) = 53.72$, $P < 0.001$. Tones aligned with stronger metrical accents (metrical tiers 3 and 4) were produced with significantly greater intensity than tones that were aligned with weaker metrical accents (tiers 1 and 2) (Tukey HSD = 1.94, $\alpha = 0.05$), similar to other reports.[22] In addition, tones were produced with significantly greater intensity when they aligned with metrical accent tier 2 than when aligned with tier 1 (Tukey HSD = 1.94, $\alpha = 0.05$). There were no interactions of metrical accent with other variables.

If pre-error events are distinguished from correct events due to a temporal overlap of processing with errors, they may inherit the properties of the errors they precede. To test this possibility, we reexamined the intensities of pre-error tones in terms of the metrical accent strength of the error they preceded (as opposed to their own metrical accent strength shown in Fig. 4). Figure 5 shows the intensity of pre-error tones as indexed by the metrical accent strength of the error they precede; this pattern differs distinctly from the intensities of the pre-error tones as indexed by their own metrical accent strength (Fig. 4, middle) as well as by the metrical accent strength of correct tones and error tones. Thus, pre-error tones, while reflecting some properties of the errors they precede (weaker intensities), retain their

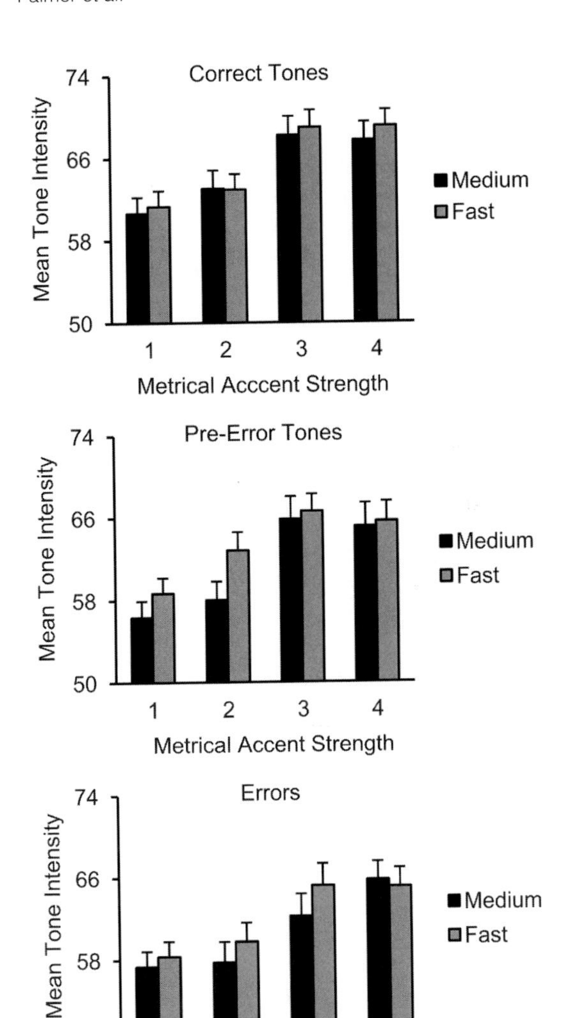

Figure 4. MIDI values as a function of metrical accent strength and tempo. Correct tone intensities on top; pre-error tones in middle; and error tones on bottom.

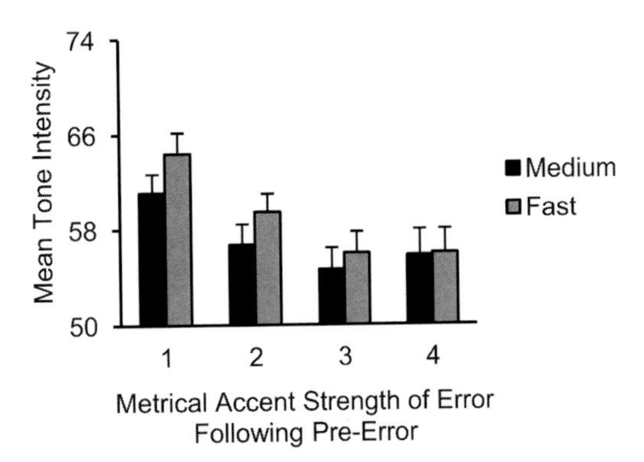

Figure 5. MIDI values of pre-error tones as a function of tempo and metrical accent strength of the following error.

own metrical accent strength and are not completely defined by the error.

Overall, meter influenced how forcefully tones were produced irrespective of their accuracy: pre-error tones inherited the metrical strength of their intended positions, as opposed to that of the error they preceded. This finding suggests that the online correction interpretation, consistent with the mismatch detection hypothesis, does not extend to all aspects of pre-error tones. Thus, the produced intensity of pre-error tones was influenced by both the position of the tones relative to upcoming errors, and by the metrical hierarchy with which the pre-error tones aligned.

Discussion

Pianists' errors produced during speeded performance documented pre-error intensity reductions arising from the reduced force with which pianists' fingers struck the keys, consistent with predictions of an error monitoring hypothesis. Pre-error events indicated some evidence of slowing as well, though not as large as found in previous literature, perhaps due to the task demands to perform at a fixed tempo. These findings extend comparisons of correct events and errors[16,17] to pre-error events which, like partial errors, inherit some unintended (incorrect) properties of the errors they anticipate (reduced tempo, decreased intensity) but retain their own pitch features in performance. The current findings also extend the previous findings to performances at different tempi; the rate at which pre-error events are produced is modulated by both the intended rate and by the upcoming error, features that are typical of partial errors.

Coles *et al.*[5] proposed that subthreshold motor activity associated with partial errors may affect the way in which correct responses are produced, and thus hypothesized that the processing system noticed partial errors and adjusts its behavior. The partial errors noted in the current study were not uniquely predictive of upcoming errors; that is, the distributions of timing and intensity features of partial errors overlapped with the same properties of error events; this is likely due, we believe, to the rapid tempi at which the sequences were performed, yielding overlapping time-courses of processing.

Pianists' pitch errors and the partial errors that precede them may be distinguished by electrophysiological markers of ERNs. Partial errors observed in button-pressing tasks[13,23] tend to elicit ERNs whose amplitudes[6] and latencies[24] are associated with reduced force during response production. If the ERN signals error correction, then ERNs that occur sooner before an error outcome should be more likely to be associated with a partial error (and reduced response force) than those that occur later.[25] The reduced force that we observed for pre-error events was suggestive of a corrective signal that precedes error production by at least 180 ms; this timecourse is consistent with Ruiz and colleagues'[17] findings that pre-error slowing was detected at least 125 ms before pianists' pitch errors.

Pre-error events inherited some properties of upcoming outcomes while retaining other unique (correct) features, consistent with the fact that many errors in music performance arise from response competition among intended sequence events, called contextual errors.[19] The idea that response competition effects are graded (not all or none) fits with musicians' task demands to produce a sequence of pitches, many of which repeat in the sequence and, therefore, recur as viable responses. Given a musical context of speeded sequences of repeating responses, it is reasonable to expect that the time-courses of errors and pre-error events overlap; thus, it is possible that pre-errors resemble errors in part because their time-course overlaps with that of planned error actions, as well as because pre-errors have registered as too slow or quiet (intensity) in their own right. The fact that we were able to demonstrate the same patterns of pre-error tone properties (slowing and reduced force) in fast and medium performances suggests that the properties shared between errors and pre-error tones are not solely due to their overlapping time-courses.

Pianists' increased performance tempi yielded higher intensities (faster finger forces) and shorter interonset intervals for all tones (correct events, errors, and pre-error events). The performance tempi were both fairly fast and elicited relatively high error rates. Thus, perhaps it is not surprising that the pre-errors did not change their features across the similar performance tempi. A wider range of performance tempi may provide a more sensitive test of the overlapping influences of pre-error and error tones.

The piano performances of two-handed novel musical pieces elicited more errors, and thus more pre-error events, compared with previous studies of piano performance.[16,17] Ruiz and colleagues[17] studied fast performances of single-handed musical excerpts, which tended to elicit fewer planning demands and less slowing in tempo attributable to musical difficulty.[18] Maidhof *et al.*[16] used scales and octaves, which also may not have taxed pianists as much as the novel musical pieces used here. Thus, some experimental differences may have accounted for why so many partial errors resulted in this study. It would be revealing to examine whether partial errors that retain their own properties elicit a smaller ERN than those that inherit most characteristics of the error they anticipate.

Acknowledgments

This work was supported by Canada Research Chair and NSERC Grant 298173 to the first author, and a Tomlinson Fellowship to the second author.

Conflicts of interest

The authors declare no conflicts of interest.

References

1. Palmer, C. & C. Drake. 1997. Monitoring and planning capacities in the performance of musical skills. *Can. J. Exp. Psychol.* **51:** 369–384.
2. Rumelhart, D.E. & D.A. Norman. 1982. Simulating a skilled typist: a study of skilled cognitive-motor performance. *Cogn. Sci.* **6:** 1–36.
3. Sloboda, J.A. 1977. Phrase units as determinants of visual processing in music reading. *Br. J. Psychol.* **68:** 117–124.
4. Norman, D.A. 1981. Categorization of action slips. *Psychol. Rev.* **88:** 1–15.
5. Coles, M.G.H., M.K. Scheffers & L. Fournier. 1995. Where did you go wrong? Errors, partial errors, and the nature of human information processing. *Acta Psychol.* **90:** 129–144.
6. Gehring, W.J., B. Goss, M.G.H. Coles, *et al.* 1993. A neural system for error detection and compensation. *Psychol. Sci.* **4:** 385–390.
7. Coles, M.G.H., M.K. Sheffers & C.B. Holroyd. 2001. Why is there an ERN/Ne on correct trials? Response representations, stimulus-related components, and the theory of error-processing. *Biol. Psychol.* **56:** 173–189.
8. Falkenstein, M., J. Hohnsbein & J. Hoormann. 1991. Effects of cross-modal divided attention on late ERP components. II. Error processing in choice reaction tasks. *Electroencephalogr. Clin. Neurophysiol.* **78:** 447–455.
9. Botvinick, M., T. Braver, D. Barch, *et al.* 2001. Conflict monitoring and cognitive control. *Psychol. Rev.* **108:** 624–652.
10. Rabbitt, P.M.A. (1966). Errors and error-correction in choice-response tasks. *J. Exp. Psychol.* **71:** 264–272.

11. Gratton, G., M.G.H. Coles & E. Donchin. 1992. Optimizing the use of information: strategic control of activation and responses. *J. Exp. Psychol.: Gen.* **121:** 480–506.
12. Gehring, W.J. & D.E. Fencsik. 2001. Functions of the medial frontal cortex in the processing of conflict and errors. *J. Neurosci.* **21:** 9430–9437.
13. Burle, B., C. Roger, S. Allain, *et al.* 2008. Error negativity does not reflect conflict: a reappraisal of conflict monitoring and anterior cingulate cortex activity. *J. Cogn. Neurosci.* **20:** 1637–1655.
14. Wolpert D.M., Z. Ghahramani & M.I. Jordan. 1995. An internal model for sensorimotor integration. *Science* **269:** 1880–1882.
15. Meyer, D.E., R.A. Abrams, S. Kornblum, *et al.* 1988. Optimality in human motor performance: ideal control of rapid aimed movements. *Psychol. Rev.* **95:** 340–370.
16. Maidhof C., M. Rieger, W. Prinz & S. Koelsch. 2009. Nobody is perfect: ERP effects prior to performance errors in musicians indicate fast monitoring processes. *PLoS One* **4:** e5032.
17. Ruiz, M.H., H.C. Jabusch & E. Altenmüller. 2009. Detecting wrong notes in advance: neuronal correlates of error monitoring in pianists. *Cereb. Cortex* **19:** 2625–2639.
18. Palmer, C. & C. van de Sande. 1995. Range of planning in music performance. *J. Exp. Psychol.: Hum. Percept. Perform.* **21:** 947–962.
19. Palmer, C. & P.Q. Pfordresher. 2003. Incremental planning in sequence production. *Psychol. Rev.* **110:** 683–712.
20. Dalla Bella, S. & C. Palmer. 2011. Rate effects on timing, key velocity, and finger kinematics in piano performance. *PLoS One* **6:** e20518.
21. Lerdahl, F. & R. Jackendoff. 1983. *A Generative Theory of Tonal Music.* MIT Press. Cambridge, UK.
22. Palmer, C. 1996. On the assignment of structure in music performance. *Music Percept.* **14:** 21–54.
23. Coles, M.G.H., G. Gratton, T.R. Bashore, *et al.* 1985. A psychophysiological investigation of the continuous-flow model of human information-processing. *J. Exp. Psychol.: Hum. Percept. Perform.* **11:** 529–533.
24. Carbonnell, L. & M. Falkenstein. 2006. Does the error negativity reflect the degree of response conflict? *Brain Res.* **109:** 124–130.
25. Gehring, W.J., Y. Liu, J.M. Orr & J. Carp. 2011. The error-related negativity (ERN/Ne). In *Oxford Handbook of Event-Related Potential Components.* S.J. Luck & E. Kappeman, Eds.: in press. Oxford University Press. New York.

Ann. N.Y. Acad. Sci. ISSN 0077-8923

ANNALS OF THE NEW YORK ACADEMY OF SCIENCES
Issue: *The Neurosciences and Music IV: Learning and Memory*

Error monitoring is altered in musician's dystonia: evidence from ERP-based studies

Felix Strübing,[1] María Herrojo Ruiz,[1,2] Hans Christian Jabusch,[1,3] and Eckart Altenmüller[1]
[1]Institute of Music Physiology and Musicians' Medicine, Hannover University of Music, Drama and Media, Hannover, Germany. [2]Department of Neurology, Charité University of Medicine, Berlin, Germany. [3]Institute of Musicians' Medicine, Dresden University of Music "Carl Maria von Weber," Dresden, Germany

Address for correspondence: Eckart Altenmüller, Institute of Music Physiology and Musicians' Medicine, Hannover University of Music, Drama and Media, Emmichplatz 1, Hannover 30175, Germany. altenmueller@hmt-hannover.de

Musician's dystonia (MD) is a task-specific movement disorder characterized by a loss of voluntary motor control in highly trained movements like piano playing. Its underlying pathophysiology is defined by deficient functioning of neural pathways at different levels of the central nervous system. However, a few studies have examined the brain responses associated with executive functions such as error monitoring in MD. We recorded the electroencephalogram (EEG) in professional pianists during the performance of memorized music sequences at fast tempi. Event-related potentials (ERPs) locked to pitch errors were investigated in MD and a control group. In MD patients, significantly larger error-related brain responses before and following errors were observed as compared with healthy pianists. Our results suggest that in MD, the generalized degraded neural activity at all levels of the central nervous system is manifested in specific neural correlates of the executive functions that monitor an overlearned sensorimotor performance.

Keywords: error monitoring; dystonia; music performance

Introduction

Musician's dystonia (MD) is a focal task-specific dystonia (FTSD), which is defined by involuntary and dysfunctional movement patterns when playing a musical instrument. It affects one out of a 100 professional musicians and is therefore the most common of all dystonias in a specific population.[1] As underlying pathology, external triggering factors, such as overuse and biomechanical constraints, seem to contribute importantly. Furthermore, recent studies have shown that impaired sensorimotor integration and decreased inhibition on all levels of the sensorimotor pathways play important roles in its manifestation.[2–4] Deficient inhibition also leads to hyperactive basal ganglia pathways.[5,6]

It has been shown that disabilities involving basal-ganglia dysfunction such as Gilles de la Tourette syndrome (TS), obsessive-compulsive disorder (OCD), and Parkinson's disease (PD) can result in different modulation of event-related potentials (ERPs). Therefore, the goal of our study was to further investigate the pathology of musicians' dystonia by means of evoked potentials in an ecological design involving frontal executive functions and error monitoring in professional pianists suffering from dystonia.[7–10]

With respect to error monitoring in general, recent electroencephalogram (EEG) studies have shown several error-specific ERP components. The most relevant are the so-called error-related negativity (ERN or Ne) and the error-related positivity (Pe).[11,12] An ERN is characterized by its negative deflection of voltage between 50 and 100 ms after the actual error commitment, regardless of the type of task error.[11,12] The most recent evidence demonstrates that the posterior cingulate cortex is the generator of the ERN and that it is functionally related to the dorsal anterior cingulate cortex.[13] Evidence suggests that the ERN mainly reflects earlier stages of error detection, although its amplitude may be influenced by the affective or emotional significance of the error.[14–16] Another specific error-related component is the Pe, the maximum of which is usually

doi: 10.1111/j.1749-6632.2011.06417.x
Ann. N.Y. Acad. Sci. 1252 (2012) 192–199

at parietal scalp electrodes between 200 and 500 ms after an incorrect action. It has been suggested that the Pe is influenced by the subjective error perception.[17,18]

Playing the piano at a professional level is exceptionally well suited to study error monitoring because it is an extremely demanding task, requiring the highest spatio-temporal precision of complex movements under the unyielding control of the auditory system. Because in classical reproductive music "correct" notes have to be played and wrong notes have to be avoided, piano playing has recently been used in several EEG studies to investigate error-monitoring processes.[19–22] A very interesting and novel finding in this context was an early ERP component termed pre-ERN.[20,21] The pre-ERN shows maximum negative deflection at about 70–30 ms before note onset. Its neural generator was localized in the rostral ACC.[20] The pre-ERN was interpreted as a neural correlate of error prediction signals in overlearned performance.

The objective of this study was to take advantage of the expertise of highly trained classical pianists and compare electrophysiological correlates of error monitoring in pianists with MD and healthy pianists. Our hypothesis was that deficient inhibition and disturbed basal ganglia loops would be reflected in altered error-related brain potentials.

Materials and methods

Participants

Twelve professional pianists took part in the study, six of whom were MD patients (four males; age range, 28–52 years; mean, 40 years; SD, 10 years) and six were healthy control subjects (four males; age range, 26–44 years; mean, 35 years; SD, 7 years). All patients were right handed according to the Edinburgh Handedness Scale and reported normal hearing. Informed consent was obtained from each participant.[23] The study received approval by the local ethics committee of Hannover and the patient data are presented in Table 1.

Before participating in the study, all patients were examined by a movement disorders specialist (E.A., the senior author) to confirm the diagnosis of MD, based on a neurological examination and visual inspection while they played the piano. In all patients, solely the left hand was affected. Secondary dystonia and genetic forms of dystonia were excluded by a laboratory test and clinical examinations. In all participants, motor control at the piano was assessed by musical instruments digital interface (MIDI)-based scale analysis previously reported as a valid tool for this purpose.[24]

Stimulus material

The musical stimuli consisted of six sequences taken from the right-hand parts of Preludes V, VI, and X of *The Well-Tempered Clavier* (Part 1) by Johann Sebastian Bach and the Piano Sonata No. 52 in E Flat Major by Joseph Haydn, described previously by Herrojo Ruiz *et al.*[20] We chose these pieces because they consist mainly of 16th notes and therefore provide homogeneous stimulus material of the same duration. The tempo for each piece was chosen so that the interonset interval (IOI; time between two consecutive onset keypresses) was 125 ms, which resulted in fast tempi even for

Table 1. MD and healthy pianists did not differ in the accumulated practice time

Patient	Sex	Age	Affected digits of the left hand (in descending degree of impairment)	Year of MD manifestation	Last therapy	Accumulated practice time (h)
Pat_1	Female	52	2	1992	Botox (9 years since last injection)	26,645
Pat_2	Male	51	4, 5, 1, 2, 3	2004	–	92,892
Pat_3	Male	49	3	1995	Botox (7 years since last injection)	62,962
Pat_4	Male	40	2	1996	–	36,135
Pat_5	Male	39	2, 4	1996	Botox (6 months after last injection)	27,922
Pat_6	Female	29	2	2004	–	37,595

professional musicians. This was necessary in order to induce a higher error-production rate. All participants were instructed to learn and rehearse the sequences using a metronome, though. The score of the pieces is depicted in Figure 1.

Experimental design

The experiment was carried out in a dimly lit and acoustically shielded room, where participants sat comfortably in front of a digital piano (Wersi Digital Piano CT2) in an armchair, with their left forearms resting on the armrest. For artifact reduction interfering with the EEG recording, participants were told to also let their right forearm rest on a movable armrest attached to a sled-type device, which had been constructed specifically for that purpose. In order to prevent pianists from tracking their finger movement with their eyes, the surface of the keyboard was covered with a black board, which still made effortless playing possible.

Before the experimental session we checked that every participant was able to perform the pieces from memory and according to the score at the desired tempi. Every pianist was instructed to perform the score from beginning to the end without stopping to correct errors, but they were unaware of our interest in error-specific data.

The experimental session consisted of 10 trials per type of sequence, which were presented randomly. The 60 sequences they played in total amounted to 40 min of continuous performance.

EEG recording and preprocessing

Continuous EEG signals were recorded from 35 electrodes placed on the scalp according to the extended 0–20 system, linked mastoids as references. To monitor blinks and eye movements, we additionally recorded a transversal electrooculogram. EEG signals were digitized at a sampling frequency of 500 Hz and impedance was kept below 5 kΩ. The upper cutoff was 100 Hz (software by NeuroScan Inc., Herndon, VA). Note onsets, visual trigger stimuli, and metronome beats were automatically recorded with markers within the continuous EEG file. Performance was recorded as a MIDI file using a standard MIDI sequencer program.

For filtering and processing the continuous EEG files, we used the EEGLAB MATLAB® Toolbox v. 7.2.11.22bb.[25] After data acquisition, we applied a 0.5–35 Hz band-pass finite impulse response (FIR) filter to remove linear trends and muscle artifacts. We then performed a wavelet-enhanced independent component analysis (wICA)[26] after first computing the ICA components with the FastICA algorithm[27] to clean data from artifacts, such as blinks and eye movements. The use of ICA for artifact removal has been reported to constitute a loss of neural activity since the rejected components usually contain not only artifacts, but also neuronal

Figure 1. Examples of musical stimuli. The first bars of the six musical sequences are illustrated. Pieces 1 and 2 were taken from Prelude V of *The Well Tempered Clavier* (Part 1) by Johann Sebastian Bach. Pieces 3 and 4 were adapted from Prelude VI and piece 6 from Prelude X. The fifth sequence was adapted from the Piano Sonata No. 52 in E Flat Major by Joseph Haydn. The tempi as were given in the experiment are indicated: metronome 120 for quarter note and 160 for triplet of eighth notes. In all cases, the IOI was 125 milliseconds.

activity. wICA solves this problem by using wavelet thresholding, which filters out the artifacts by their specific time–frequency properties.[20] Any artifacts that could still reside in the epoched EEG file were manually subtracted. Data were epoched into two conditions representing correct and erroneous notes with a time window time-locked from 300 ms before note onset to 500 ms after note onset. The baseline was set from 300 to 150 ms before the actual keystroke.

Data analysis

Errors were defined as wrong pitches. Correct notes were defined as keypresses correct in pitch and timing, according to the given tempi. To discriminate errors from correct notes, we used an algorithm developed in MATLAB that compared each MIDI performance with its template in pitch.[20] Only incorrect or correct notes that were preceded or followed by three correct notes entered the analysis. Additional criteria for accepting a correct or erroneous note as events for further analysis were as follows: First, the time interval between keypress and key release was not accepted above 150 ms in order to avoid overlapping of auditory processing from two simultaneous notes. Second, in correct pitch notes, we set a strict timing criterion to make sure that there were no deviations in timing, which could lead to neural processing related to timing monitoring. Specifically, the IOI of correct pitch notes was accepted when in the range of 120–130 milliseconds. In pitch errors, the timing criterion was set for the range of 100–300 milliseconds. We did not set a stricter timing criterion for pitch errors because performance errors typically lead to slowing in the next event (post-error slowing) and can also be associated with pre-error slowing or speeding phenomena.[11,12,20] All errors that appeared in at least seven out of 10 trials of each type, which therefore could be related to learning errors, were removed from the analysis.

Statistical analysis

Analysis of the behavioral data was performed by means of nonparametric permutation tests.[28] For statistical assessment of the ERPs, the waveforms were first averaged across trials in each subject and condition (error, correct) and in five medial scalp electrodes (Fz, FCz, Cz, CPz, Pz). Before investigating the between-group differences, we first conducted a two-sample t-test (errors versus correct notes) in the patient and control groups separately. The analysis focused on two different relevant time windows that were based on visual inspection: (i) the pre-ERN time window (70–20 ms prior to note onset) and (ii) the Pe time window (230–270 ms after note onset).

Based on the findings from previous studies, only anterior or posterior electrodes were selected for the analysis of the pre-ERN or Pe.[20,21] Because the topographic maximum for the ERN is acknowledged to spread across all medial electrodes, we used all medial electrodes to compute the statistics.[29]

The between-group statistics were computed using two-way analyses of variance (ANOVAs) with the factors group (patients versus controls) and condition (erroneous versus correct notes). Multiple comparisons were corrected by controlling the false discovery rate (FDR) at level $q = 0.05$.[30]

Results

Performance analysis

MD and healthy pianists did not differ in the accumulated practice time ($P = 0.83$; Table 1 in MD; healthy pianists had an accumulated practice time between 25,000 and 78,110 h; mean, 44,147 h) or in average age ($P = 0.077$). As a reliable parameter for the assessment of motor control dysfunction, we investigated the mean standard deviation of the IOI (mSD-IOI) in both groups based on a scale analysis procedure previously described.[24] As expected, this value differed significantly between groups in the affected left hand ($P = 0.000001$; mean mSD-IOI, 20 ms [SD, 3 ms] in patients, 11 ms [SD, 1 ms] in healthy pianists). Both groups did not differ in the average of isolated errors ($P = 0.72$; 80 [SD, 40] in healthy pianists, 70 [SD, 30] in MD). Furthermore, we observed a reduced loudness (MIDI velocity) in incorrect notes compared to correct notes in both groups ($P = 0.001$ in MD, $P = 0.04$ in healthy pianists), although there was no difference between groups ($P > 0.05$). Additionally, pre- and posterror slowing (~150 ms) was found in both groups in the IOI between incorrect note and the neighboring event note, which differed significantly from the mean IOI in the performance ($P = 0.001$). Again, no difference between groups was found.

ERP analysis

A graphical overview of the ERP components can be found in Figure 4. Control subjects and patients

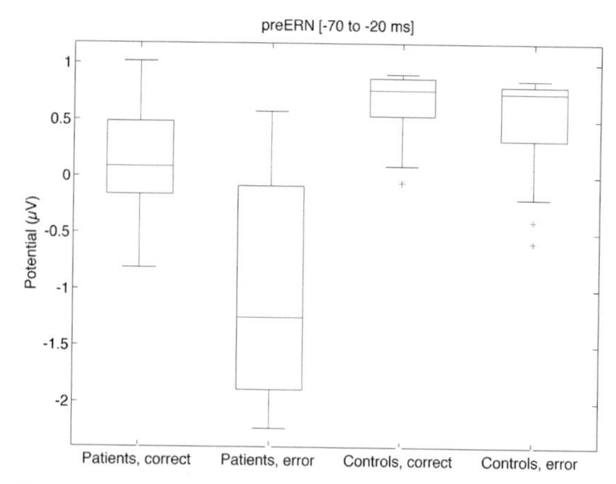

Figure 2. pre-ERN component overview. This box-plot graph was calculated for the pre-ERN time interval ranging from 70 to 20 ms before note onset. Patients and controls showed larger negative deflections for erroneous keypresses compared to correct keypresses.

showed larger negative oscillations in the pre-ERN time interval (70–20 ms before note onset) for incorrect notes (Fig. 2). Later, a positive amplitude modulation was observed between 230 and 270 ms when comparing incorrect to correct notes (Fig. 3). This could be related to the Pe. An ERN component, which would typically peak at around 50–100 ms after keypress, could not be observed.

Within-group ERP analysis

In the control group, we found a significant difference between errors and correct notes for the pre-ERN ($P = 0.0003$) as well as for the Pe ($P = 0.02$). Similarly, in the patient group we found significant differences between errors and correct notes for the pre-ERN ($P < 0.0001$) and the Pe ($P < 0.0001$). These results confirmed the presence of pre-ERN and Pe in both experimental groups.

Between-group data analysis

We found significant effects (i) in the pre-ERN time window for the factors group ($P < 0.0001$) and condition ($P = 0.0002$), as well as for their interaction ($P < 0.0001$). Contrasts revealed that the main effect for factor condition resulted from larger negative deflections of the pre-ERN in the patient group than in the control group. (ii) In the Pe time window, we obtained significant main effects for the factors group ($P = 0.0002$) and condition ($P < 0.0001$). Error minus correct contrasts revealed that the Pe in the control group was characterized by smaller positive amplitude deflections than in the patient group. Therefore, both groups differed in the pre-ERN and Pe components due to larger pre-ERN and Pe deflections in the MD group.

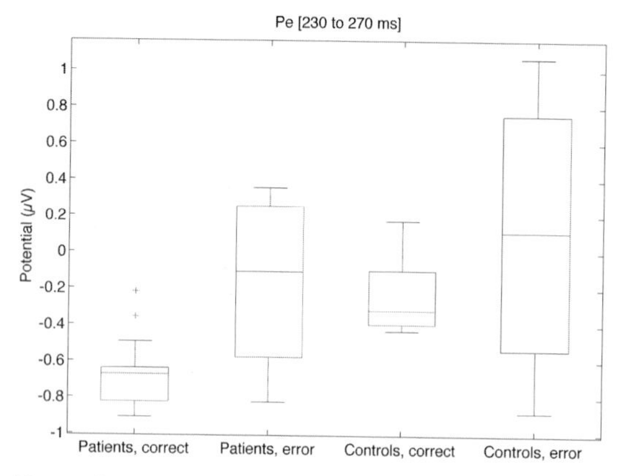

Figure 3. Pe component overview. This box-plot graph was calculated for the Pe time interval ranging from 220 to 270 ms after note onset. As compared to correct notes, the positive deflection is larger for erroneous keypresses in both groups.

Discussion

The current study investigated the error-related potentials in professional pianists suffering from dystonia and in a healthy group. When comparing correct and erroneous notes, we could show two specific error-related potentials within each group, that is, pre-ERN and Pe. Both components were significant within each group. These results are in line with the previous studies that used a similar experimental paradigm, and therefore confirm the pre-ERN and Pe as key ERP components reflecting error-monitoring processes during piano performance.[20,21]

As it has been suggested, the pre-ERN might reflect a predictive error signal triggered by an internal forward model or by a reward estimation system, which anticipates the lack of reward. In both cases, the pre-ERN can be interpreted as a neural signal reflecting the mismatch between a planned keypress and the predicted erroneous outcome.[20–22] Furthermore, its neural generator lies in the rostral ACC, a region associated with detecting motivational errors.[31,32] Interestingly, our between-group analysis depicted a larger pre-ERN in MD patients. This could be the result of higher motivational modulation of the predictive error signal in MD: patients might place greater importance on pressing

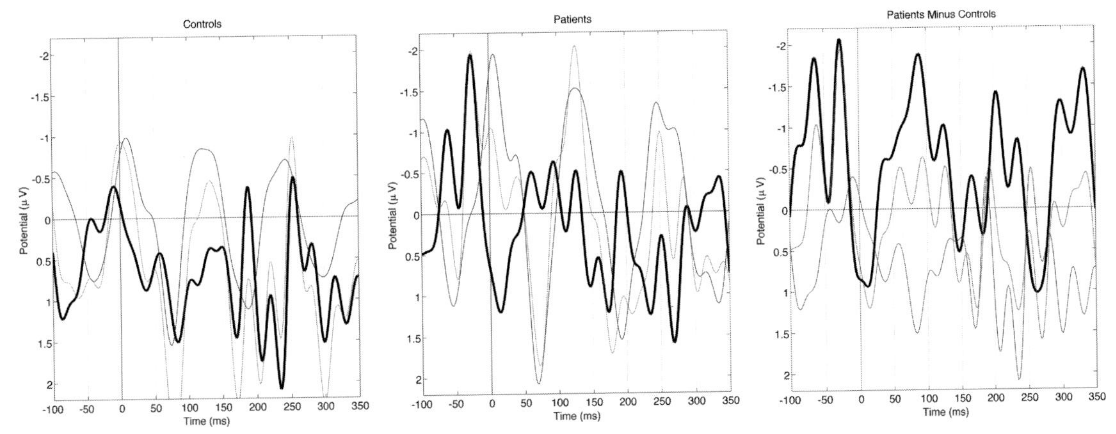

Figure 4. Grand-averaged EEG overview. The grand-averaged EEG files across patients and healthy controls are illustrated at electrode position Fz. The bold line in controls and patients represents the difference between incorrect (red line, dashed) and correct (black line, solid) notes. In the group comparison (Patients Minus Controls), differences from patients (violet, dashed line) and controls (blue line, solid) were subtracted again, resulting in the between-group difference (bold line).

the right note. Therefore, it is plausible to assume that the mismatch between the predicted error and the planned correct event is more enhanced.

In fact, findings of recent studies in which the relation between psychological factors and the development of MD has been examined could support the thesis of a higher motivational impact of errors on MD patients. Not only do MD patients tend to have a more distinct sense of perfectionism, it is also assumed that MD is accompanied by higher levels of anxiety and neuroticism.[33,34] Thus, the abnormal cortico-thalamic neural activity found in MD may influence limbic loops, resulting in both altered motor and affective processing.[35]

These results converge with the second outcome of our study: a larger Pe following errors in pianists with MD than in healthy pianists. This ERP has been suggested to signal conscious error recognition as well as evaluation, and it might even reflect response strategy adaptation.[18,36,37] Enhanced Pe amplitudes have also been found in children suffering from OCD.[38] The authors conclude that the emotional impact of an error might be reflected by the Pe such that the greater the emotional significance of the error, the higher the Pe.

If taken into consideration that the Pe in this paradigm also seems to be generated by rostral ACC areas and that the rostral ACC is associated with affective processing, a higher Pe in MD patients may suggest enhanced conscious and emotional error evaluation.[10,17,20,39]

Other studies have shown that patients with dystonia suffer from higher intensity of obsessive-compulsive symptoms.[40,41] Cavallaro and Bihari concluded that OCD and FD might indeed share a common pathologic background indicating basal-ganglia dysfunction.[42,43]

There is also support for a reduced pallidal inhibition of the thalamus in dystonia, which results in the over-activity of medial and prefrontal cortical areas.[44] In MD, this could be expressed by an excessive activation of sensorimotor cortical areas during skilled movements of the affected hand. In addition, it could be speculated that the phenomenon of enhanced pallidal disinhibition of the thalamus in dystonia leads to altered error signals projected from the output of the basal ganglia to the posterior frontomedial cortex.[44]

Further studies have shown evidence for hyperactive basal-ganglia signaling in FTSD.[5] It has been hypothesized that the integration of prefrontal and motor fronto-thalamico-striato-cortical circuits provides contextual information to the motor ACC to enable their function in performance monitoring.[45]

We did not observe an ERN component in our experiment in both healthy pianists and patients. Most studies focusing on the ERN use flanker or Stroop tasks that participants are not trained for, whereas the pianists in our study were all professionally trained pianists and the performance was overlearned. Their experience provided them with a fast-functioning, internal self-monitoring system during the memorized performance, which enables an earlier detection of the errors. In contrast, participants in speeded choice reaction tasks are not trained for the task and, in addition, no planning in terms of a memory representation of the task is possible. These key differences in the population and in the task can account for the lack of ERN in the present study as well as in related piano performance studies.[20,21]

This study provides evidence to support the idea that the degraded neural activity in the basal-ganglia-thalamocortical loops might interact with error-monitoring processes associated with the basal ganglia, which results in altered error-monitoring mechanisms in focal dystonia.[22]

Notwithstanding the altered ERP patterns observed during error monitoring in MD, at the behavioral level no differences in performance between patients (performing with the healthy hand) and healthy pianists were observed. This outcome suggests that it is necessary to look further into the interaction between behavior and brain activity during action monitoring in MD. Only then can the implications of the reported altered error-related ERPs in MD be better understood.

Acknowledgment

The authors want to acknowledge the help of Michael Großbach, who helped out with the EEG recording in some experimental sessions.

Conflicts of interest

The authors declare no conflicts of interest.

References

1. Altenmüller, E. & H.C. Jabusch. 2010. Focal dystonia in musicians: phenomenology, pathophysiology and triggering factors. *Eur. J. Neurol.* **17**(Suppl. 1): 31–36.
2. Ruiz, M.H., P. Senghaas, M. Grossbach, *et al.* 2009. Defective inhibition and inter-regional phase synchronization in pianists with musician's dystonia: an EEG study. *Hum. Brain Mapp.* **30:** 2689–2700.
3. Stinear, C.M. & W.D. Byblow. 2004. Impaired modulation of intracortical inhibition in focal hand dystonia. *Cereb. Cortex* **14:** 555–561.
4. Lin, P.T. & M. Hallett. 2009. The pathophysiology of focal hand dystonia. *J. Hand Ther.* **22:** 109–113; quiz 114.
5. Preibisch, C., D. Berg, E. Hofmann, *et al.* 2001. Cerebral activation patterns in patients with writer's cramp: a functional magnetic resonance imaging study. *J. Neurol.* **248:** 10–17.
6. Naumann, M., W. Pirker, K. Reiners, *et al.* 1998. Imaging the pre- and postsynaptic side of striatal dopaminergic synapses in idiopathic cervical dystonia: a SPECT study using [123I] epidepride and [123I] beta-CIT. *Mov. Disord.* **13:** 319–323.
7. Greenberg, B.D., U. Ziemann, G. Cora-Locatelli, *et al.* 2000. Altered cortical excitability in obsessive-compulsive disorder. *Neurology* **54:** 142–147.
8. Beste, C., C. Saft, J. Andrich, *et al.* 2006. Error processing in Huntington's disease. *PLoS One* **1:** e86.
9. Johannes, S., B.M. Wieringa, W. Nager, *et al.* 2002. Excessive action monitoring in Tourette syndrome. *J. Neurol.* **249:** 961–966.
10. Ito, J. & J. Kitagawa. 2006. Performance monitoring and error processing during a lexical decision task in patients with Parkinson's disease. *J. Geriatr. Psychiatry Neurol.* **19:** 46–54.
11. van Veen, V. & C.S. Carter. 2006. Error detection, correction, and prevention in the brain: a brief review of data and theories. *Clin. EEG Neurosci.* **37:** 330–335.
12. Gehring, W.J., B. Goss, M.G. Coles, *et al.* 1993. A neural system for error-detection and compensation. *Psychol. Sci.* **4:** 385–390.
13. Agam, Y., M.S. Hamalainen, A.K. Lee, *et al.* 2011. Multimodal neuroimaging dissociates hemodynamic and electrophysiological correlates of error processing. *Proc. Natl. Acad. Sci. USA* **108:** 17556–17561.
14. Menon, V., N.E. Adleman, C.D. White, *et al.* 2001. Error-related brain activation during a Go/NoGo response inhibition task. *Hum. Brain Mapp.* **12:** 131–143.
15. Kiehl, K.A., P.F. Liddle & J.B. Hopfinger. 2000. Error processing and the rostral anterior cingulate: an event-related fMRI study. *Psychophysiology* **37:** 216–223.
16. Garavan, H., T.J. Ross, K. Murphy, *et al.* 2002. Dissociable executive functions in the dynamic control of behavior: inhibition, error detection, and correction. *Neuroimage* **17:** 1820–1829.
17. Herrmann, M.J., J. Rommler, A.C. Ehlis, *et al.* 2004. Source localization (LORETA) of the error-related-negativity (ERN/Ne) and positivity (Pe). *Brain Res. Cogn. Brain Res.* **20:** 294–299.
18. Nieuwenhuis, S., K.R. Ridderinkhof, J. Blom, *et al.* 2001. Error-related brain potentials are differentially related to awareness of response errors: evidence from an antisaccade task. *Psychophysiology* **38:** 752–760.
19. Katahira, K., D. Abla, S. Masuda & K. Okanoya. 2008. Feedback-based error monitoring processes during musical performance: an ERP study. *Neurosci. Res.* **61:** 120–128.
20. Ruiz, M.H., H.C. Jabusch & E. Altenmüller. 2009. Detecting wrong notes in advance: neuronal correlates of error monitoring in pianists. *Cereb. Cortex* **19:** 2625–2639.
21. Maidhof, C., M. Rieger, W. Prinz & S. Koelsch. 2009. Nobody is perfect: ERP effects prior to performance errors in musicians indicate fast monitoring processes. *PLoS One* **4:** e5032.
22. Ruiz, M.H., F. Strubing, H.C. Jabusch & E. Altenmüller. 2011. EEG oscillatory patterns are associated with error prediction during music performance and are altered in musician's dystonia. *Neuroimage* **55:** 1791–1803.
23. Oldfield, R. 1971. The assesment and analysis of handedness: the Edinburgh inventory. *Neuropsychologia* **9:** 97–113.

24. Jabusch, H.C., H. Vauth & E. Altenmüller. 2004. Quantification of focal dystonia in pianists using scale analysis. *Mov. Disord.* **19:** 171–180.
25. Delorme, A. & S. Makeig. 2004. EEGLAB: an open source toolbox for analysis of single-trial EEG dynamics including independent component analysis. *J. Neurosci. Methods* **134:** 9–21.
26. Castellanos, N.P. & V.A. Makarov. 2006. Recovering EEG brain signals: artifact suppression with wavelet enhanced independent component analysis. *J. Neurosci. Methods* **158:** 300–312.
27. Hyvarinen, A. & E. Oja. 2000. Independent component analysis: algorithms and applications. *Neural Netw.* **13:** 411–430.
28. Good, P. 2005. *Permutation, Parametric and Bootsrap Tests of Hypotheses*. Springer Verlag. New York.
29. Hajcak, G., J.S. Moser, N. Yeung & R.F. Simons. 2005. On the ERN and the significance of errors. *Psychophysiology* **42:** 151–160.
30. Benjamini, Y., A.M. Krieger & D. Yekutieli. 2006. Adaptive linear step-up procedures that control the false discovery rate. *Biometrika* **93:** 491–507.
31. Dunning, J.P. & G. Hajcak. 2007. Error-related negativities elicited by monetary loss and cues that predict loss. *Neuroreport* **18:** 1875–1878.
32. Luu, P., P. Collins & D.M. Tucker. 2000. Mood, personality, and self-monitoring: negative affect and emotionality in relation to frontal lobe mechanisms of error monitoring. *J. Exp. Psychol. Gen.* **129:** 43–60.
33. Jabusch, H.C., S.V. Muller & E. Altenmüller. 2004. Anxiety in musicians with focal dystonia and those with chronic pain. *Mov. Disord.* **19:** 1169–1175.
34. Enders, L., J.T. Spector, E. Altenmüller, *et al.* 2011. Musician's dystonia and comorbid anxiety: two sides of one coin? *Mov. Disord.* **26:** 539–542.
35. Lencer, R., S. Steinlechner, J. Stahlberg, *et al.* 2009. Primary focal dystonia: evidence for distinct neuropsychiatric and personality profiles. *J. Neurol. Neurosurg. Psychiatry* **80:** 1176–1179.
36. van Veen, V. & C.S. Carter. 2002. The anterior cingulate as a conflict monitor: fMRI and ERP studies. *Physiol. Behav.* **77:** 477–482.
37. Falkenstein, M., J. Hoormann, S. Christ & J. Hohnsbein. 2000. ERP components on reaction errors and their functional significance: a tutorial. *Biol. Psychol.* **51:** 87–107.
38. Santesso, D.L., S.J. Segalowitz & L.A. Schmidt. 2006. Error-related electrocortical responses are enhanced in children with obsessive-compulsive behaviors. *Dev. Neuropsychol.* **29:** 431–445.
39. Van Veen, V. & C.S. Carter. 2002. The timing of action-monitoring processes in the anterior cingulate cortex. *J. Cogn. Neurosci.* **14:** 593–602.
40. Bugalho, P., B. Correa, J. Guimaraes & M. Xavier. 2008. Set-shifting and behavioral dysfunction in primary focal dystonia. *Mov. Disord.* **23:** 200–206.
41. Kubota, Y., T. Murai, T. Okada, *et al.* 2001. Obsessive-compulsive characteristics in patients with writer's cramp. *J. Neurol. Neurosurg. Psychiatry* **71:** 413–414.
42. Cavallaro, R., G. Galardi, M.C. Cavallini, *et al.* 2002. Obsessive compulsive disorder among idiopathic focal dystonia patients: an epidemiological and family study. *Biol. Psychiatry* **52:** 356–361.
43. Bihari, K., J.L. Hill & D.L. Murphy. 1992. Obsessive-compulsive characteristics in patients with idiopathic spasmodic torticollis. *Psychiatry Res.* **42:** 267–272.
44. Berardelli, A., J.C. Rothwell, M. Hallett, *et al.* 1998. The pathophysiology of primary dystonia. *Brain* **121:** 1195–1212.
45. Ullsperger, M. & D.Y. von Cramon. 2006. The role of intact frontostriatal circuits in error processing. *J. Cogn. Neurosci.* **18:** 651–664.

Ann. N.Y. Acad. Sci. ISSN 0077-8923

ANNALS OF THE NEW YORK ACADEMY OF SCIENCES
Issue: *The Neurosciences and Music IV: Learning and Memory*

Dynamic aspects of musical imagery

Andrea R. Halpern

Psychology Department, Bucknell University, Lewisberg, Pennsylvania

Address for correspondence: Andrea R. Halpern, Psychology Department, Bucknell University, Lewisburg, PA 17837. ahalpern@bucknell.edu

Auditory imagery can represent many aspects of music, such as the starting pitches of a tune or the instrument that typically plays it. In this paper, I concentrate on more dynamic, or time-sensitive aspects of musical imagery, as demonstrated in two recently published studies. The first was a behavioral study that examined the ability to make emotional judgments about both heard and imagined music in real time. The second was a neuroimaging study on the neural correlates of anticipating an upcoming tune, after hearing a cue tune. That study found activation of several sequence-learning brain areas, some of which varied with the vividness of the anticipated musical memory. Both studies speak to the ways in which musical imagery allows us to judge temporally changing aspects of the represented musical experience. These judgments can be quite precise, despite the complexity of generating the rich internal representations of imagery.

Keywords: auditory imagery; memory; emotion; anticipation; music

Introduction

The introduction to this symposium referred to some of the characteristics of mental imagery as memory representations. People report these experiences as capturing perceptual aspects of the event or object, but in addition to the phenomenology, numerous reports suggest that people can make accurate decisions about objects or experiences that are only imagined.[1,2] Although auditory imagery is studied less often than visual, people commonly claim that they can recollect music in such a way as to simulate the actual hearing of the tune. For instance, they can reproduce timing and pitch relationships as well as evoke the sound qualities of musical instruments (reviewed in Ref. 3).

The relationship between heard and imagined sounds is of interest for both theoretical and applied reasons. The theoretical interest lies in the ways in which mental representation both preserves but also changes experience. Memory is a dynamic process, and memory researchers have long since abandoned thinking of memory as an actual audio or video recorder. Not even "flashbulb" memories are veridical,[4] and many psychologists caution against the reification of eyewitness testimony in legal systems.[5] However, as noted previously, the extent to which imagery representations maintain aspects of experience that are not easily verbalizable or even obviously codable suggests dissociations between the content and labeling of memories. For instance, I tested people's abilities to reproduce the opening pitch of familiar songs, like "Happy Birthday." Although people differed in what pitch they chose, individuals were quite reliable in reproducing or choosing the same opening pitch for the tune on two occasions several days apart.[6] But none of these people had absolute pitch ability; that is, none could name the notes they were selecting, thus they apparently were accessing a perceptual code.

Understanding auditory imagery is also useful in several applied domains. Consider mental practice. Musicians can suffer from crippling dystonias or other medical conditions that can shorten a career.[7] Relieving some of the motor stress by engaging in mental practice (which involves both auditory and motor imagery) could help alleviate some of these concerns, and indeed mental practice is recommended for even healthy musicians or athletes when physical practice is not practical. Knowing how

doi: 10.1111/j.1749-6632.2011.06442.x

imagery processes work could lead to recommendations about how to implement these strategies, as well as offering guidance to teachers for students who have more and less vivid imagery.[8] In addition, the emerging field of neural prosthetics could benefit from understanding the neural underpinnings of these vivid mental representations.[9]

My focus in this symposium was on two aspects of auditory imagery that could be considered *dynamic* in the sense that the imagery representation is required to be updated continuously to serve the purpose at hand. I use this term to contrast with the more static representations required in some of my previous auditory imagery studies. For instance, in the study on mental pitches mentioned earlier,[6] the pitch selection does not require any processing past the initial memory retrieval. Likewise, comparing the sound of imagined musical instruments requires a memory retrieval of the sound of a violin or saxophone, and maintenance long enough to generate a similarity rating, but no additional operations.[10] However, to accomplish a wider variety of tasks, it is likely we use dynamic imagery processes to track time-varying information, make novel or creative judgments, generate predictions, or to represent multiple aspects of information simultaneously.

I will illustrate dynamic auditory imagery processes with two studies that I was recently involved with. In the first case, Lucas, Schubert, and I[11] examined how well people could extract emotional judgments of music in real time as they listened to, and then imagined, a familiar piece of music. In this case, even the perceptual judgment is dynamic in that listeners must monitor the piece as well as make a corresponding judgment in continuous, real time. Dynamic imagery processes are even more important when the piece is being generated internally.

In the second case, I will describe a study carried out in conjunction with Leaver and Rauschecker[12] that looked at people's ability to predict an upcoming melody based on having heard a melody previously associated with it. We call this "anticipatory imagery," in that the imagery is evoked only in response to the cue, but then must be retrieved in a prospective manner, as the second melody, in this case, is not actually played. The analogy from real life would be anticipating the next track on one's favorite album or the next movement in a familiar symphony as the previous segment ends. In fact, the first experiment in that paper used favorite CDs as stimuli. However, I was only involved in the second study that used paired associate learning of tunes, so I will be describing that one. And I will be making a link between these, as it is entirely possible that the tracking displayed in the emotion judgments was facilitated by anticipation.

During talks, I sometimes ask the audience if they ever deliberately imagine music so as to regulate their moods; I usually get wide agreement with this statement. Even persistent musical memories, or "earworms," seem to consist of preferred and pleasant music.[13] So it would not be surprising if people could extract emotional judgments from imagined music, at least on a global basis. What was less evident to us was whether emotional judgments could be made in real time as the music was actually proceeding (either real or imagined). Some preliminary evidence by Schubert *et al.*[14] suggested a mixed answer. They asked a pianist to listen to a recording of himself and make continuous judgments of valence and arousal in the piece. He then did the same thing as he imagined the piece. The two response profiles were similar, but the pianist began lagging in the responses to the imagined relative to the heard piece. This is consistent with imagery being a fairly expensive cognitive process in that it requires full consciousness and responses in typical mental imagery experiments are often rather slow compared to other kinds of tasks. However, we did not know if these results would generalize beyond a case study.

We therefore asked undergraduate students to participate in two tracking studies. We tested musicians (with at least 8 years of private instrument playing, averaging about 10.5 years) partly to insure that our participants would have the fine motor control needed for continuous response, but also because of the music we used. After considerable discussion, we decided that classical music would be a good genre to test because we needed pieces that varied on the selected dimensions of valence, arousal, and emotionality over the time span of about one minute. We decided that music from genres more familiar to nonmusicians, such as pop, rock, and movie scores, tend not to vary as much over short durations. Furthermore, we needed the pieces to be highly familiar to our listeners for the imagery condition, which again pointed to students with extensive musical training.

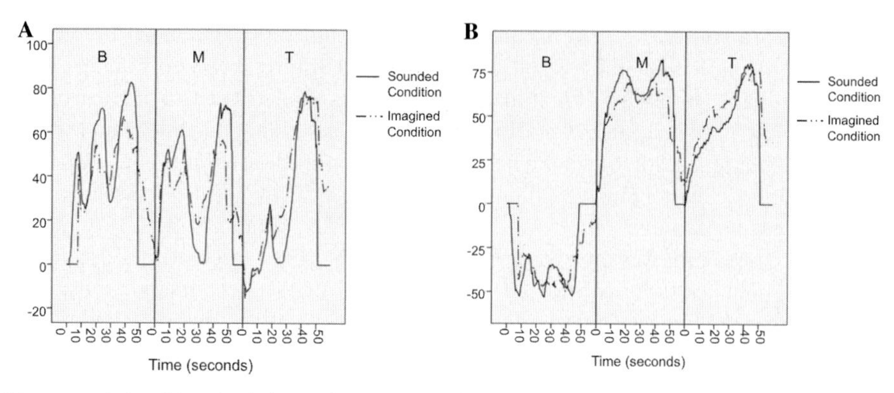

Figure 1. Mean sounded and imagined time-series responses. (A) Mean arousal response for Beethoven, B; Mozart, M; and Tchaikovsky, T. (B) Mean valence response for Beethoven, B; Mozart, M; and Tchaikovsky, T. Figure reprinted with permission from Ref. 11.

We selected approximately the first minute of Beethoven's Symphony no. 5 in C minor," op. 67, Tchaikovsky's "Waltz of the Flowers" from the *Nutcracker Suite*, op. 71, and the "Allegro" from Mozart's *Serenade in G*, K. 525 (*Eine Kleine Nachtmusik*). These were very familiar to our participant pool, had the requisite changes in emotional characteristics in their opening sections, and were rather different from each other. For instance, the Beethoven piece was in a minor key and the others in major. [The figures refer to these pieces by the first letter of the composers: B, T, and M.]

We trained the participants to use a continuous response recorder. As the piece played (or as they imagined it), they moved a mouse on a display; mouse positions were recorded twice per second. In the first study, the task was to track valence on the x-axis simultaneously with arousal on the y-axis. Valence was explained as the range of negative to positive emotion and arousal as the range of sleepy to excited emotion. Experiment 2 used the single dimension of emotionality (in the range of high to low). Several trials of practice preceded the real trials. In the sounded condition, the piece was played over speakers. In the imagined condition, the participants heard the first few notes of the to-be-imagined piece and also could refer to a musical score. We analyzed data from 17 participants in experiment 1, and 11 participants (4 from the previous study) in experiment 2, but I will discuss the results together. We also gave participants a tapping synchronization task, wherein they heard two measures of an isochronous beat at 160 bpm, then had to keep the beat steady without the cue for 40 measures.

Figures 1 and 2 show the averaged response profiles for the three dimensions: the solid line indicates the mouse position over time in the sounded condition and the dashed line in the imagined condition. As is evident, the response profiles on average were very similar in both conditions on all three dimensions, both in pattern (the same peaks and valleys) and timing. Quantitative analysis showed that the two profiles were highly correlated in most participants. A cross-correlation function analysis allowed us to determine the lead or lag at which the highest profile correlations were obtained. We found that depending on the dimension, 85–100% of the analysis showed significant correlation at some lead or lag; and the modal lead–lag was 0 samples. That is, most people most of the time tracked the emotional profile similarly, in both pattern and timing, in the sounded and imagined conditions. The small leads and lags shown by some people in some conditions were actually within the small error range on the tapping synchronization task. We also related many of the peaks and valleys in the profiles to predesignated "turning points" in the music. For instance, we predicted an increase in arousal when loudness increased in the music. Over the eight identified turning points across the three pieces, 84% of the participants responded in the predicted way in their tracking responses.

Thus, we concluded that musicians, at least, could indeed make temporally fine-grained judgments. In addition, the emotional message of music can be

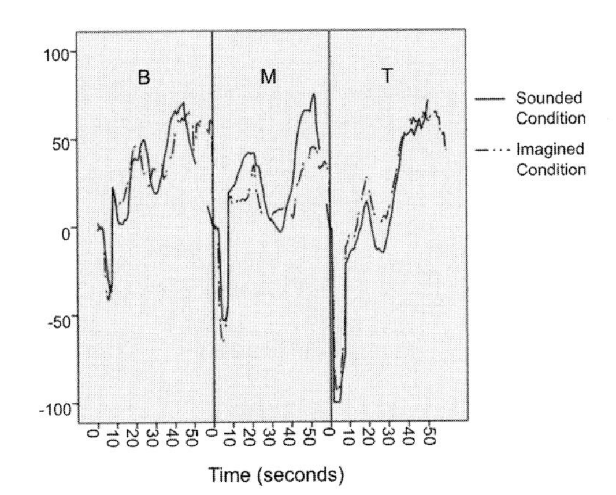

Figure 2. Mean emotionality response for Beethoven, B; Mozart, M; and Tchaikovsky, T. Figure reprinted with permission from Ref. 11.

decoded, and presumably enjoyed in real time, even in a mental representation. In that context, it is interesting to hear the words of Romel Joseph, a violinist buried in the rubble of his collapsed music school for 18 hours before being rescued from the 2010 Haiti earthquake. He later related how running through familiar pieces in his mind helped him cope: "For example, if I perform the Franck sonata, which is [sic] 35 minutes long in my honors recital at Juilliard, then I would bring myself to that time. That allows me ... to mentally take myself out of the space where I was..."[15]

In this study, we were somewhat surprised that the tracking in the imagined condition did not slow down. Although carried out before the one just described, the other study I am using to illustrate dynamic auditory imagery could help explain the outcome. Perhaps the participants were anticipating the notes in the imagined music, which counteracted what would otherwise have been a slowdown in response because of the exigencies of generating the auditory image.

As mentioned earlier, Leaver, Rauschecker, and I built on a prior study in Rauschecker's lab that examined neural correlates of anticipation of the next track in familiar CDs.[12] To reduce individual differences and to gain more experimental control, we taught participants seven otherwise unrelated pairs of unfamiliar but melodious tunes. The training session involved repeated study of a pair, with the goal that when presented with the first member of the pair, everyone could conjure up the image of the second member of the pair. The scanning session involved some additional training before entering the fMRI scanner, and after the first run, we asked them to overtrain by hearing the pairs for about 10 additional minutes. We tested 10 people who had a minimum of two years of musical training (averaging about 6.5 years).

Although in the scanner, trials were of three types: silent baseline, anticipatory imagery trials in which

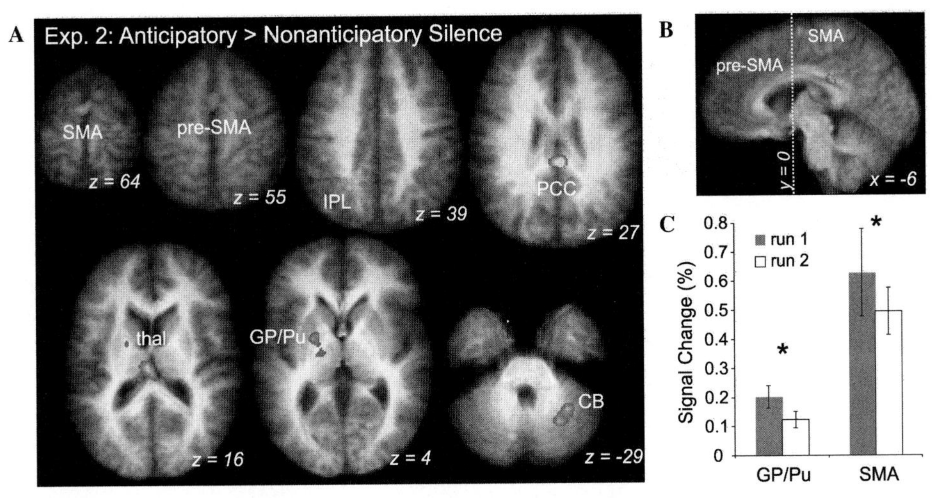

Figure 3. (A) Significant activation associated with anticipatory imagery (AS > NS). Areas include cerebellum, globus pallidus/putamen (GP/Pu), thalamus, posterior cingulate cortex (PCC), presupplementary motor area (pre-SMA), and SMA proper. (B) Sagittal view of medial frontal activation. Dotted line indicates Talairach coordinate axis, $y = 0$, separates pre-SMA and SMA proper. (C) ROI analysis reveals percentage signal change differences in the AS conditions between run 1 (shaded) and run 2 (white) ($^{*}P < 0.05$). ROIs were defined by analysis shown in A. Error bars indicate SEM. Figure reprinted with permission from Ref. 12.

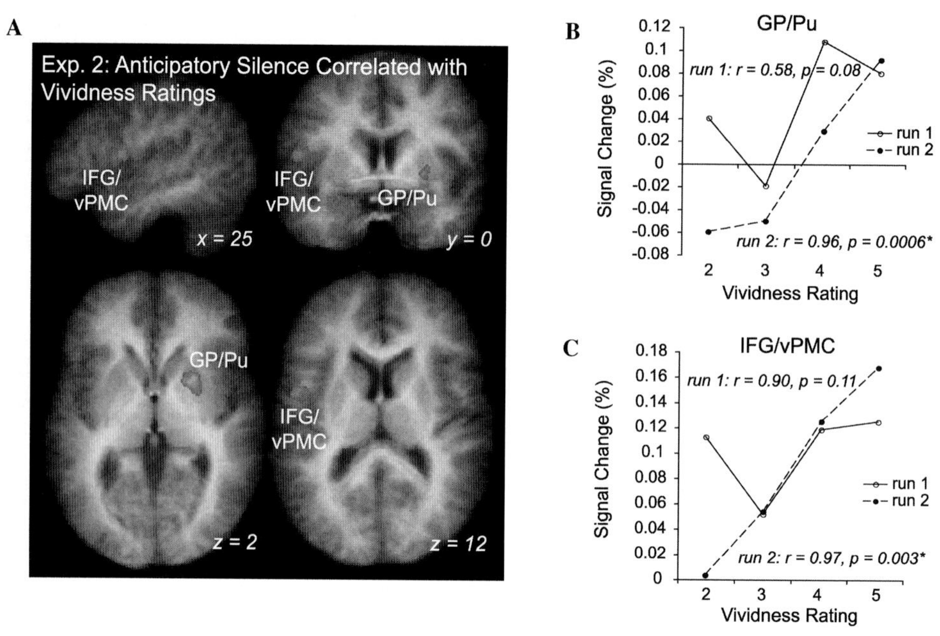

Figure 4. (A) Areas for which BOLD signal is correlated with vividness rating in the familiar silence condition (GP/Pu, inferior frontal gyrus, ventral premotor cortex (IFG/vPMC)). (B, C) Graphs show correlations between percentage signal change and vividness ratings for ROIs resulting from the parametrically weighted GLM analysis in A. Figure reprinted with permission from Ref. 12.

the first member of a learned melody pair was presented and people were asked to imagine the associated tune, and nonanticipatory trials. On these trials, a novel tune was presented so that participants could not imagine any associated tune. The acquisition of the fMRI data occurred in the silence after the tune had been presented. The silence after a learned tune ought to have engendered anticipation of the next tune, hence we called these anticipatory silences (AS). In these trials, we asked people to rate the vividness of their imagery for the anticipated tune, on a 1 (no image) to 5 (very vivid) scale. The silence after a novel tune should not have engendered any particular auditory imagery, hence nonanticipatory silence (NS). For details of the scanning parameters and analyses, see Ref. 12.

Figure 3 shows the contrast between AS and NS trials. A number of brain areas were more active in the anticipatory response, and of particular interest were several areas normally associated with motor sequence learning, such as the supplementary motor areas (SMA and pre-SMA) and the globus pallidus and putamen in the basal ganglia. These two areas showed decreased signal after overtraining, as shown on the right side of the figure. Figure 4 shows two brain areas where activity in AS trials was directly correlated with the vividness ratings: the basal ganglia areas just mentioned and a ventral premotor area. As seen in the graphs, correlations were strongly positive and even linear in the second run (correlations in the first run were positive but not significantly so). That is, for every increase in vividness ratings, a proportional increase only in these two areas occurred in the hemodynamic response of AS trials.

This experiment showed how the brain response changes as anticipatory imagery becomes stronger. When people know fairly well what tune will follow a cue tune (i.e., during the first run), sequence learning areas are fairly active. They also rate the vividness of the upcoming target tune as being moderately vivid (a mean of 3.62 of 5). But when the associations are recent and overlearned, less activity is shown in several areas, presumably because the associations have already been formed. Vividness ratings go up (a mean of 4.01), and the relationship between the behavioral rating and signal strength in two areas associated with the task becomes impressively fine tuned. Both types of findings speak to the dynamic aspects of auditory imagery: the neural locus reflects what happens as the anticipatory imagery becomes a more accurate predictor of the upcoming information. Anticipation is likely one way that musicians calibrate their timing when playing in an ensemble;[16] here we see that even small amounts of training can yield some dramatic neural changes associated with that mechanism.

From both sets of studies, we can draw some conclusions about both the construct of auditory imagery and the methods for studying it. First, it is possible to devise tasks that externalize the essentially private experience of auditory imagery. Furthermore, we can capture time-locked aspects of auditory imagery, which is important for studying imagery experiences that unfold over time. The more obvious example of this was the ability of the participants in the emotion study to track, with exquisite temporal resolution, the moment-to-moment changes in rather complex imagined music. But even the simpler task of anticipating one melody after hearing another one requires temporal tracking.

We also saw that the dynamic aspect of imagery can in some respects be reflected in short-term neural changes. The additional training in the Leaver *et al.* study[12] comprised a mere 10 minutes. Nevertheless, this was enough to engender clear differences in neural response—in this case the reduction of response as the anticipation became more entrenched.

Finally, we saw remarkable correspondences in self-report measures and neural patterns. As Zatorre reports in another paper in this Symposium, self-report on an auditory imagery vividness scale predicts response in auditory cortex and intraparietal sulcus during mental musical transformation tasks. We might think of this as a relationship between a *trait* measure of auditory imagery and neural response, where overall stronger ability to imagine sounds predicts more activity. The trial-by-trial vividness measure in the anticipatory imagery study reflects more a *state* measure as participants judged each particular instance of anticipation. We cannot disentangle causality; we do not know if the ratings engender the neural changes or vice versa. However, we can say that the essentially linear relationship between self-reported vividness and neural activity suggests that these reports are psychologically robust. These results should encourage cognitive neuroscientists to include both objective and subjective behavioral measures as they probe complex mental phenomena.

Acknowledgments

The research reported in Ref. 12 was supported by the National Science Foundation.

Conflicts of interest

The author declares no conflicts of interest.

References

1. Kosslyn, S.M. 1980. *Image and Mind*. Harvard University Press. Cambridge, MA.
2. Finke, R.A. 1985. Theories relating mental imagery to perception. *Psychol. Bull.* **98:** 236–259.
3. Hubbard, T.L. 2010. Auditory imagery: empirical findings. *Psychol. Bull.* **136:** 302–329.
4. Talarico, J.M. & D.C. Rubin. 2003. Confidence, not consistency, characterizes flashbulb memories. *Psychol. Sci.* **14:** 455–461.
5. Wells, G.L., A. Memon & S.D. Penrod. 2006. Eyewitness evidence: improving its probative value. *Psychol. Sci. Pub. Int.* **7:** 45–75.
6. Halpern, A.R. 1989. Memory for the absolute pitch of familiar songs. *Mem. Cognit.* **17:** 572–581.
7. Altenmüller, E. & H.-C. Jabusch. 2010. Focal dystonia in musicians: phenomenology, pathophysiology and triggering factors. *Eur. J. Neurol.* **17:** 31–36.
8. Highben, Z. & C. Palmer. 2004. Effects of auditory and motor mental practice in memorized piano performance. *B. Coun. Res. Music Ed.* **159:** 58–65.
9. Nicolelis, M.A.L. 2001. Actions from thoughts. *Nature* **409:** 403–407.
10. Halpern, A.R., R.J. Zatorre, M. Bouffard & J. Johnson. 2004. Behavioral and neural correlates of perceived and imagined musical timbre. *Neuropsychologia* **42:** 1281–1292.
11. Lucas, B.L., E. Schubert & A.R. Halpern. 2010. Perception of emotion in sounded and imagined music. *Music Percept.* **27:** 399–412.
12. Leaver, A.M., J. Van Lare, B. Zielinski, *et al.* 2009. Brain activation during anticipation of sound sequences. *J. Neurosci.* **29:** 2477–2485.
13. Halpern, A.R. & J.C. Bartlett. 2011. The persistence of musical memories: a descriptive study of earworms. *Music Percept.* **28:** 425–431.
14. Schubert, E., P. Evans & J. Rink. 2006. Emotion in real and imagined music: same or different? In *Proceedings of the Ninth International Conference on Music Perception and Cognition*. M. Baroni, A.R. Addessi, R. Caterina & M. Costa, Eds.: 810—814. Bologna, Italy.
15. Turpin, C. (Producer). 2010. Wife, school lost in quake, violinist vows to rebuild. *All Things Considered* [audio podcast]. Available at: http://www.npr.org/2010/01/23/122900781/wife-school-lost-in-quake-violinist-vows-to-rebuild. Accessed January 23, 2010.
16. Keller, P.E. & M. Appel. 2010. Individual differences, auditory imagery, and the coordination of body movements and sounds in musical ensembles. *Music Percept.* **28:** 27–46.

Ann. N.Y. Acad. Sci. ISSN 0077-8923

ANNALS OF THE NEW YORK ACADEMY OF SCIENCES
Issue: *The Neurosciences and Music IV: Learning and Memory*

Mental imagery in music performance: underlying mechanisms and potential benefits

Peter E. Keller[1,2]

[1]Music Cognition & Action Group, Max Planck Institute for Human Cognitive and Brain Sciences, Leipzig, Germany. [2]MARCS Auditory Laboratories, University of Western Sydney, Australia

Address for correspondence: Peter Keller, Max Planck Institute for Human Cognitive and Brain Sciences, Stephanstr. 1a, 04103 Leipzig, Germany. keller@cbs.mpg.de

This paper examines the role of mental imagery in music performance. Self-reports by musicians, and various other sources of anecdotal evidence, suggest that covert auditory, motor, and/or visual imagery facilitate multiple aspects of music performance. The cognitive and motor mechanisms that underlie such imagery include working memory, action simulation, and internal models. Together these mechanisms support the generation of anticipatory images that enable thorough action planning and movement execution that is characterized by efficiency, temporal precision, and biomechanical economy. In ensemble performance, anticipatory imagery may facilitate interpersonal coordination by enhancing online predictions about others' action timing. Overlap in brain regions subserving auditory imagery and temporal prediction is consistent with this view. It is concluded that individual differences in anticipatory imagery may be a source of variation in expressive performance excellence and the quality of ensemble cohesion. Engaging in effortful musical imagery is therefore justified when artistic perfection is the goal.

Keywords: auditory imagery; music performance; temporal prediction; action simulation; internal models; individual differences

Introduction

Picture a musician striding onto stage and then, after settling into the optimal position, pausing momentarily with eyes closed in silent concentration before producing a sound. This paper addresses the question of whether mental imagery serves some useful function in solo and ensemble music performance, and what that function (or functions) may be. In doing so, we will touch upon topics related to different modes and modalities in which imagery may take place, anecdotal evidence for the use of imagery by performers, putative cognitive and motor mechanisms that underlie such imagery, as well as where these processes may take place in the brain.

Musical imagery is assumed to be a multimodal process by which an individual generates the mental experience of auditory features of musical sounds, and/or visual, proprioceptive, kinesthetic, and tactile properties of music-related movements, that are not (or not yet) necessarily present in the physical world. Such mental images may be generated through either deliberate thought or automatic responses to endogenous and exogenous cues. A large part of the following discussion concentrates on auditory imagery because it appears to be prominent in the phenomenology of performing musicians.

Imagery modes and modalities

Although topics related to musical imagery have occupied researchers for some time,[1–3] little scientific work has dealt specifically with the role of imagery in music performance. This may be due to challenges associated with isolating the effects of auditory imagery on behavior and brain processes in the presence of exogenous auditory stimulation, as well as due to the threat of movement artifacts in recordings of brain activity. It is also possible that the need for performance studies has been partially obviated by the fact that imagining music can itself be considered a form of performance, albeit covert.[4]

doi: 10.1111/j.1749-6632.2011.06439.x

Research on musical imagery has nevertheless been concerned with issues that are pertinent to overt music performance.

A central theme in this work concerns similarities and differences in how structural and temporal properties of sound (e.g., pitch, duration, rhythm, tempo, timbre, and loudness) are represented during auditory imagery and auditory perception.[2] Related research has sought to investigate the degree to which auditory imagery and perception are equivalent in terms of their neural correlates and their effects on behavior.[5,6] A panoply of psychophysical and neuroscientific techniques have been employed to this end, the latter including electroencephalography (EEG),[7,8] magnetoencephalography (MEG),[9,10] functional magnetic resonance imaging (fMRI),[11,12] and positron emission tomography (PET).[13] Evidence gathered with these tools converges on the conclusion that musical imagery involves the interplay of brain regions implicated in auditory and motor processing.[14,15] Exciting new developments have shown, moreover, that it is possible not only to detect whether an individual is engaging in imagery, but also to decode patterns of brain activity associated with the time course of imagining specific musical pieces and rhythmic structures.[16–18]

Another prominent research theme concerns the relationship between auditory imagery and more general brain functions subserving attention, memory, and the prediction of future events.[7,9,19,20] Just as these brain functions differ between people, individual differences—mainly related to musical experience—have been observed in the vividness of auditory images and the potency of their effects on skilled behavior.[21–24] Finally, a growing body of research focuses upon manifestations of auditory imagery in everyday life, both in specialist populations—for example, the musings of music students[25] and the hallucinations that plague schizophrenics[26]—and in regular folk who, for example, experience spontaneous auditory imagery in the form of tunes getting "stuck in the head."[27]

The way in which mental imagery is used by music performers can take several forms.[28] These include mental practice away from the instrument,[29] the silent reading of musical scores (as when conductors and instrumentalists prepare for performance, which requires an advanced skill that is referred to as "notational audiation"),[30] and thinking of the ideal sound during performance.[31,32] Empirical research

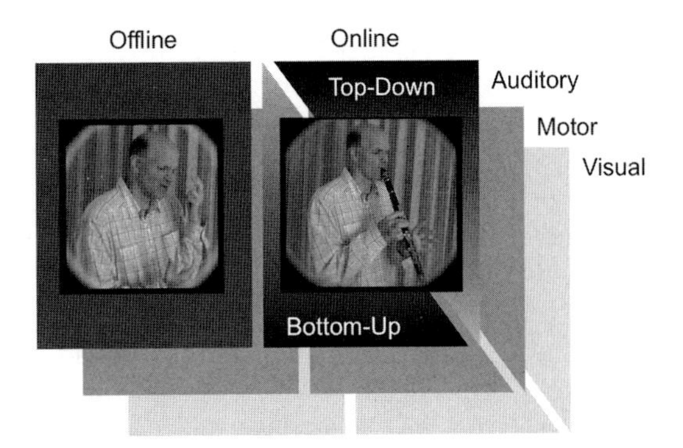

Figure 1. Taxonomy of musical imagery modes (offline, online) and modalities (auditory, motor, visual). The photographs illustrate how, in the offline mode, the performer imagines the ideal sound away from the instrument, while in the online mode, he imagines the ideal sound while playing (in this case with unconventional fingering, thus demonstrating his virtuosity).

has by and large confirmed the effectiveness of these real-world practices. For instance, one study[23] found a positive correlation between pianists' auditory imagery abilities and success at learning novel piano pieces from notation in the absence of auditory feedback.

The previous examples suggest a distinction between the use of mental imagery prior to performance (i.e., offline) and during performance (online). A taxonomy of imagery modes and modalities is shown in Figure 1. Online imagery during performance, furthermore, may proceed via a top–down route—in which the performer deliberately (and possibly effortfully) generates mental images of action goals—and/or a bottom–up route, in which expectancies based on perceptual input automatically trigger mental images. These different modes of imagery may also take place in different modalities—auditory, motor (proprioceptive, kinesthetic, and tactile), and visual—depending on the performer's goals and strategies, as well as the multisensory nature of the context in which the performance takes place (e.g., opera stage vs. isolated recording studio).

Anecdotal evidence for the importance of imagery in musical excellence

Several sources of anecdotal evidence exist that show how covert imagery affects overt musical performance. Some sources are subtle and indirect. For example, Robert Schumann's *Humoreske*, op. 20 for solo piano, contains a third stave sandwiched between the conventional treble and bass staves in one section. A lyrical melodic line is notated on this extra stave, with the mysterious marking *innere Stimme* (inner voice). This inner voice, which apparently represents Schumann's future wife Clara Wieck singing one of her own compositions, is intended to be only imagined.[33] Presumably, doing so affects the character of the parts actually played by the pianist.

There are also numerous more direct anecdotes about the use of mental imagery by virtuoso performers. In one of these, the legendary pianist Artur Rubinstein is sitting on a train with a score for César Franck's Symphonic Variations. Apparently, he learned the entire piece via notational audiation *en route* to the concert.[34]

Less spectacular, but highly informative, accounts of the use of imagery to achieve performance excellence are provided in a study of self-reports by members of the Chicago Symphony brass section.[32] One musician states, "If I don't hear it [the ideal sound] or conceptualize it in my brain, there's no way I'm going to get it" (Ref. 32, p. 146). Another alludes to the use of auditory imagery during private practice for ensemble performance, claiming that "the sound of what is going on in the rest of the orchestra is always in my imagination... You're hearing the whole picture..." (Ref. 32, pp. 145–146). This quote is noteworthy because it implies that imagining the sound of others' parts when practicing one's own part ultimately assists in achieving a cohesive ensemble sound.

Musicians' intuitions about the beneficial effects of imagery on performance are crystallized in self-help books and how-to manuals that address the process of achieving excellence as an instrumentalist.

One influential book in this mold, entitled *The Inner Game of Music*,[35] is remarkable in the sense that it resonates strongly with psychological principles concerning intentional action. For instance, note the correspondence between concepts expressed by bassist Barry Green in this book and William James in his *Principles of Psychology*.[36] The musician states, "When you can hold the sound and pitch of the music clearly in your head... performing it accurately becomes easier. Your body has a sense of its goal" (Ref. 35, p. 75). This echoes James's statement illustrating the operation of his ideo-motor principle (i.e., the notion that actions are triggered automatically by the anticipation of their intended effects): "The marksman ends by thinking only of the exact position of the goal, the singer only of the perfect sound..." (Ref. 36, p. 774). Furthermore, with respect to the benefits of imagery in terms of promoting automaticity in motor control, the musician writes, "Effectively, you are playing a duet between the music in your head and the music you are performing. Any notes you play that don't correspond to your imagined sense of the music stand out, and your nervous system is able to make instant, unconscious adjustments" (Ref. 35, p. 75). The psychologist clearly agrees: "We are then aware of nothing between the conception and the execution. All sorts of neuromuscular processes come between... but we know absolutely nothing of them. We think the act, and it is done" (Ref. 36, p. 790). James' words are fitting to usher us into the realm of cognitive and motor mechanisms that support imagery.

Cognitive/motor mechanisms underlying imagery during performance

Musical imagery relies, in one way or another, on cognitive processes that act upon memory representations. Working memory is involved to the extent that musical imagery requires mental representations of information related to specific rhythmic, pitch, timbral, and/or intensity patterns to be accessed, temporarily maintained, and manipulated in accordance with the demands of the task at hand.[37–39] Two additional mechanisms—which are intimately linked to motor control—are also likely to be relevant to the use of imagery in music performance: action simulation and internal models.

Action simulation occurs when sensorimotor brain processes that resemble those associated with executing an action are engaged in the absence of overt movement.[40–42] Such covert activity may be triggered by observing or imagining an action or its effects,[43,44] for example, tones in the case of music.[45] This triggering is mediated by experience-based associations between sensory and motor processes.[15] Brain activations indicative of musical action simulation are hence especially strong in individuals who have had the opportunity to learn associations between movements involved in playing an instrument and the ensuing auditory effects.[46–49]

Internal models constitute another mechanism that relies on experience-based learning. The idea

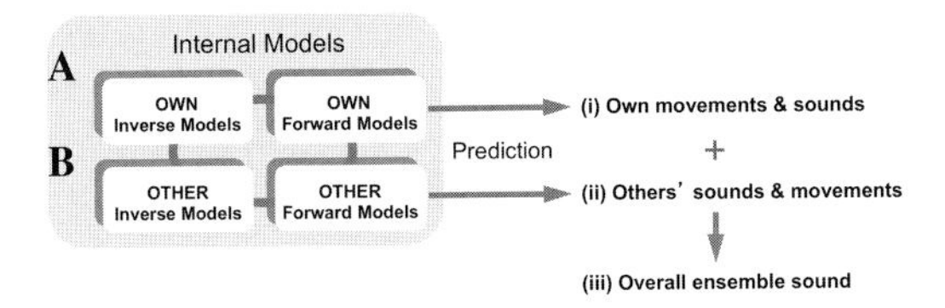

Figure 2. Forward and inverse internal models representing sensorimotor transformations related to (A) one's own actions and (B) others' actions. Together, these models generate predictions that can facilitate (i) anticipatory control of a performer's movements and sounds; (ii) anticipation of coperformers' sounds and movements; and (iii) anticipation and partial control of the overall ensemble sound.

behind these models, put forth by researchers in the field of computational movement neuroscience, is that sensorimotor transformations between bodily states and events in the immediate environment are represented in the brain.[50] There are two types of internal models (see Fig. 2A), both purportedly residing in the cerebellum and communicating with other brain regions.[51] *Forward models* represent the causal relationship between motor commands and sensory experiences related to their effects on the body and environment. *Inverse models* represent transformations from desired action outcomes (sounds, in the case of music) to the motor commands that give rise to these outcomes. Forward and inverse models increase the efficiency of motor control by running slightly ahead of action execution, thereby allowing movement errors to be anticipated and corrected in advance.[52,53] This notion is compatible with so-called "predictive coding" approaches to perception and action,[54,55] which have been applied to musical skills.[56]

Action simulation during music performance entails running internal models that trigger auditory and motor images of one's own upcoming actions.[57] Thus, anticipatory imagery facilitates the planning and execution of musical actions.[58] This type of imagery is a top–down controlled process to the extent that the performance goal—a representation of the ideal sound—is kept active in working memory.

Benefits of anticipatory auditory imagery

A series of studies investigating the role of anticipatory auditory imagery in musical action planning and execution has revealed several potential functional benefits. First, anticipatory auditory imagery assists in selecting which movements to produce,[6] for example, in which sequential order to strike piano keys. Second, such imagery promotes efficient (rapid) movement by enabling thorough action preplanning.[59] Third, it facilitates timing accuracy by optimizing movement kinematics.[60] Finally, anticipatory auditory imagery allows for economical force control by reducing the performer's reliance upon tactile feedback.[61]

The results of other work indicate that the benefits of anticipatory auditory imagery may extend to musical ensemble performance. The temporal precision that characterizes successful ensemble coordination requires performers to predict each others' actions. It has been proposed that these predictions are generated by a second class of "socially endowed" internal models (see Fig. 2B) that serve to simulate coperformers' actions (*cf.* Ref. 62) and to generate predictions about the overall ensemble sound.[57]

In a preliminary investigation of this process, pianists were asked to record one part from several duets and then, months later, to play the complementary part in synchrony with either their own or others' recordings.[63] Synchronization was most precise when pianists played with their own recordings. This finding suggests that the pianists predicted the timing of sounds in the recordings by simulating the performances online, as the match between simulated event timing and actual timing in a complementary part is presumably best when both are products of the same cognitive/motor system. Whether this online action simulation of the other part involved auditory imagery was not addressed.

A subsequent study, however, identified a link between auditory imagery and the quality of interpersonal coordination in musical ensembles.[58] Fourteen pianists were invited to the lab, first in pairs, to perform piano duos while their movements (keystroke timing and anterior–posterior body sway) were recorded, and then individually to perform a task that assessed the vividness of imagery for upcoming sounds in a paradigm that required the production of rhythmic sequences with or without auditory feedback (see Ref. 61). It was found that individual differences in anticipatory auditory imagery were correlated with the degree of synchrony in the duos. Specifically, asynchronies between movements of pianists within duos—at the level of both keystrokes and body sway—decreased with increasing scores on the imagery assessment task. Using imagery to predict the time course of others' actions may seem like an excessively effortful solution to the problem of ensemble coordination, but, as noted earlier, some musicians claim to imagine their coperformers' parts even while practicing their own part in private. If such imagery skills are practiced, then why should they be eschewed during performance?

A series of studies conducted in my lab by Nadine Pecenka has sought to gain a better understanding of the relationship between auditory imagery, temporal prediction, and sensorimotor synchronization. In a first step, separate measures of prediction tendencies and imagery skills were obtained for a large group of individuals (the majority of whom were musically trained).[64,65]

Prediction tendencies were indexed by a task that required finger tapping with auditory pacing signals that contained tempo changes. The degree to which each individual predicted upcoming tempo changes was estimated by computing the cross-correlation between the individual's intertap intervals and the pacing signal's interonset intervals at different lags: The lag-0 cross-correlation is high to the extent that a person is able to predict interonset intervals, while the lag-1 cross-correlation is high to the extent that he or she tracks the tempo changes (see also Refs. 66 and 67). It turns out that people vary widely in their prediction abilities, and that this variation is positively correlated with amount of musical training.

Temporal imagery skills were assessed in the same individuals using a perceptual judgment task. This task required participants to mentally continue a tempo change in a short auditory sequence with a gap, and then to judge whether a probe tone occurred early or late relative to the imagined continuation. It was found that imagery thresholds derived from this task (where low values indicate good performance) were correlated with individuals' prediction scores, and with their accuracy on various sensorimotor synchronization tasks (employing isochronous and tempo-changing pacing signals). Thus, individual differences in auditory imagery ability were related to temporal prediction and sensorimotor synchronization skills.

Next, the ecological validity of the previously mentioned relationship between temporal prediction ability and sensorimotor synchronization skills was tested in an interpersonal coordination task.[68] The same participants were invited back to the lab in pairs, and were asked to tap in synchrony with one another at a moderate, regular tempo. Taps triggered distinctive percussion sounds. A crucial aspect of the experimental design was that the individuals were paired in such a way that they formed three types of dyads: individuals with high prediction tendencies were paired with other high predictors, individuals with low prediction tendencies were paired with other low predictors ("trackers"), and predictors were paired with trackers in mixed dyads. The main result was that interpersonal coordination was most accurate in dyads comprised of predictors.

Neural correlates of imagery-based temporal prediction

The foregoing behavioral evidence is consistent with the notion that sensorimotor synchronization in musical contexts is facilitated by temporal prediction mechanisms that involve auditory imagery. This raises the question of whether there is evidence that these processes are mediated by common brain regions. Previous studies have shown overlap in brain areas involved in imagery and serial prediction for a variety of tasks, including those that require judgments about whether the structure of an ongoing pitch or rhythmic sequence is violated.[69] A meta-analysis has situated this overlap in the (inferior ventral) lateral premotor cortex.[14] Other work has found that real and imagined rhythmic coordination of movement with auditory pacing sequences recruit similar brain regions, including the premotor cortex, supplementary motor area, superior temporal gyrus, basal ganglia, and cerebellum.[70] The relevance of additional cortical and subcortical structures has been highlighted by the results of studies concerned with various forms of auditory imagery, particularly those targeting anticipatory processes[12] and complex sensorimotor transformations,[71] as well as in studies of the role of internal models in temporal prediction.[72]

A recent fMRI experiment aimed to identify the specific brain regions that mediate online temporal prediction during sensorimotor synchronization.[73] The question of interest was whether these regions would overlap with those activated in brain imaging studies of auditory imagery. The behavioral task involved finger tapping in synchrony with tempo-changing pacing signals under three conditions that varied in terms of concurrent working memory demands. In one condition, participants tapped while watching a stream of novel objects; in a more difficult condition, participants counted the number of consecutively repeated objects; and in the most difficult condition, they counted objects repeated after an intervening item. A parametric analysis revealed a network of brain regions in which activity decreased as a function of decrements in the degree of temporal prediction across the three conditions. This network spanned areas that other work has found to be implicated in auditory imagery and auditory attention (e.g., superior/middle temporal gyrus and inferior frontal gyrus), internal models (cerebellum), and processes subserving sensorimotor integration and sensorimotor transformations (sensorimotor cortex). These results provide evidence that auditory imagery and temporal prediction may be linked through brain regions that subserve multiple modalities and levels of processing, and that these links may support anticipatory action control during the synchronization of movements with externally controlled sound sequences.

Conclusions

Mental imagery facilitates multiple aspects of music performance. The deliberate use of anticipatory auditory (and/or motor and visual) imagery during performance may assist in planning and executing one's own actions—with potential beneficial effects on the control of parameters such as timing, intensity, articulation, and intonation—and in predicting others' actions with a view to optimizing ensemble coordination. Individual differences in anticipatory imagery may, therefore, be a source of differences in expressive performance capabilities and in the quality of ensemble cohesion. Although mental imagery during music performance may be effortful, it is justifiable when artistic perfection is the goal.

Conflicts of interest

The author declares no conflicts of interest.

References

1. Godøy, R.I. & H. Jørgensen. 2001. *Musical Imagery*. Swets & Zeitlinger. Lisse, the Netherlands.
2. Hubbard, T.L. 2010. Auditory imagery: empirical findings. *Psychol. Bull.* **36:** 302–329.
3. Reisberg, D. 1992. *Auditory Imagery*. Erlbaum. Hillsdale, NJ.
4. Zatorre, R.J. & A.R. Halpern. 2005. Mental concerts: musical imagery and auditory cortex. *Neuron* **47:** 9–12.
5. Halpern, A.R. 1992. Musical aspects of auditory imagery. In *Auditory Imagery*. D. Reisberg, Ed.: 1–27. Erlbaum. Hillsdale, NJ.
6. Keller, P.E. & I. Koch. 2006a. Exogenous and endogenous response priming with auditory stimuli. *Adv. Cogn. Psychol.* **2:** 269–276. [www.ac-psych.org]
7. Janata, P. 2001. Brain electrical activity evoked by mental formation of auditory expectations and images. *Brain Topogr.* **13:** 169–193.
8. Schaefer, R.S. 2011. *Measuring the Mind's Ear: EEG of Music Imagery*. Radboud University Nijmegen. Nijmegen, the Netherlands.
9. Tian, X. & D. Poeppel. 2010. Mental imagery of speech and movement implicates the dynamics of internal forward models. *Front. Psychol.* **1:** 166. doi:10.3389/ fpsyg.2010.00166
10. Herholz, S.C., C. Lappe, A. Knief & C. Pantev. 2008. Neural basis of music imagery and the effect of musical expertise. *Eur. J. Neurosci.* **28:** 2352–2360.
11. Kraemer, D.J.M., C.N. Macrae, A.E. Green & W.M. Kelly. 2005. Sound of silence activates auditory cortex. *Nature* **434:** 158.
12. Leaver, A.M., J. Van Lare, B. Zielinski, *et al.* 2009. Brain activation during anticipation of sound sequences. *J. Neurosci.* **29:** 2477–2485.
13. Zatorre, R.J., A.R. Halpern, D.W. Perry, *et al.* 1996. Hearing in the mind's ear: a PET investigation of musical imagery and perception. *J. Cogn. Neurosci.* **8:** 29–46.
14. Schubotz, R.I. 2007. Prediction of external events with our motor system: towards a new framework. *Trends Cogn. Sci.* **11:** 211–218.
15. Zatorre, R.J., J.L. Chen & V.B. Penhune. 2007. When the brain plays music. Auditory-motor interactions in music perception and production. *Nat. Rev. Neurosci.* **8:** 547–558.
16. Iversen, J.R., B.H. Repp & A.D. Patel. 2009. Top-down control of rhythm perception modulates early auditory responses. *Ann. N.Y. Acad. Sci.* **1169:** 58–73.
17. Nozaradan, S., I. Peretz, M. Missal & A. Mouraux. 2011. Tagging the neuronal entrainment to beat and meter. *J. Neurosci.* **31:** 10234–10240.
18. Schaefer, R.S., P. Desain & P. Suppes. 2009. Structural decomposition of EEG signatures of melodic processing. *Biol. Psychol.* **82:** 253–259.
19. Rauschecker, J.P. & S. Scott. 2009. Maps and streams in the auditory cortex. *Nat. Neurosci.* **12:** 718–724.
20. Voisin, J., A. Bidet-Caulet, O. Bertrand & P. Fonlupt. 2006. Listening in silence activates auditory areas: a functional magnetic resonance imaging study. *J. Neurosci.* **26:** 273–278.
21. Aleman, A., M.R. Nieuwenstein, K.B.E. Boecker & E.H.F. de Hann. 2000. Music training and mental imagery ability. *Neuropsychologia* **38:** 1664–1668.
22. Brodsky, W., A. Henik, B. Rubinstein & M. Zorman. 2003. Auditory imagery from musical notation in expert musicians. *Percept. Psychophys.* **65:** 602–612.
23. Highben Z. & C. Palmer. 2004. Effects of auditory and motor mental practice in memorized piano performance. *B. Coun. Res. Music Ed.* **159:** 58–65.
24. Janata, P. & K. Paroo. 2006. Acuity of auditory images in pitch and time. *Percept. Psychophys.* **68:** 829–844.
25. Bailes, F. 2007. The prevalence and nature of imagined music in the everyday lives of music students. *Psychol. Music* **35:** 555–570.
26. Hugdahl, K., E.M. Løberg, K. Specht, *et al.* 2008. Auditory hallucinations in schizophrenia: the role of cognitive,

brain structural and genetic disturbances in the left temporal lobe. *Front. Hum. Neurosci.* **1:** 6. doi:10.3389/neuro.09.006. 2007

27. Halpern, A.R. & J.C. Bartlett. 2011.The persistence of musical memories: a descriptive study of earworms. *Music Percept.* **28:** 425–443.
28. Repp, B.H. 2001. Expressive timing in the mind's ear. In *Elements of Musical Imagery.* R.I. Godøy & H. Jørgensen, Eds.: 185–200. Swets & Zeitlinger. Lisse, the Netherlands.
29. Johnson, R. 2011. Musical tempo stability in mental practice: a comparison of motor and non-motor imagery techniques. *Res. Stud. Music Edu.* **33:** 3–30.
30. Brodsky, W., Y. Kessler, B.S. Rubinstein, *et al.* 2008. The mental representation of music notation: notational audiation. *J. Exp. Psychol. Human* **34:** 427–445.
31. Bailes, F. 2009. Translating the musical image: case studies of expert musicians. In *Sounds in Translation: Intersections of Music, Technology and Society.* A. Chan & A. Noble, Eds.: 41–59. Australian National University E Press. Canberra, Australia.
32. Trusheim, W.H. 1993. Audiation and mental imagery: implications for artistic performance. *Q. J. Music Teach. Learn.* **2:** 139–147.
33. Ostwald, P.F. 2010. *Schumann: The Inner Voices of a Musical Genius.* Northeastern University Press. Boston.
34. Rubinstein, A. 1980. *My Many Years.* Jonathan Cape. London.
35. Green, B. & W.T. Gallwey. 1986. *The Inner Game of Music.* Doubleday. New York.
36. James, W. 1890. *Principles of Psychology.* Holt. New York.
37. Baddeley, A.D. & R.H. Logie. 1992. Auditory imagery and working memory. In *Auditory Imagery.* D. Reisberg, Ed.: 179–197. Erlbaum. Hillsdale, NJ.
38. Deutsch, D. 1975. The organization of short term memory for a single acoustic attribute. In *Short Term Memory.* D. Deutsch & J.A. Deutsch, Eds.: 107–151. Academic Press. New York.
39. Kalakoski, V. 2001. Musical imagery and working memory. In *Musical Imagery.* R.I. Godøy & H. Jørgensen, Eds.: 43–55. Swets & Zeitlinger. Lisse, the Netherlands.
40. Decety, J. & J. Grezes. 2006. The power of simulation: imagining one's own and other's behaviour. *Cogn. Brain Res.* **1079:** 4–14.
41. Gallese, V., C. Keysers & G. Rizzolatti. 2004. A unifying view of the basis of social cognition. *Trends Cogn. Sci.* **8:** 396–403.
42. Jeannerod, M. 2006. *Motor Cognition: What Actions Tell the Self.* Oxford University Press. Oxford, UK.
43. Sebanz, N. & G. Knoblich. 2009. Prediction in joint action: what, when, and where. *Top. Cogn. Sci.* **1:** 353–367.
44. Wilson, M. & G. Knoblich. 2005. The case for motor involvement in perceiving conspecifics. *Psychol. Bull.* **131:** 460–473.
45. Repp, B.H. & G. Knoblich. 2004. Perceiving action identity: how pianists recognize their own performances. *Psychol. Sci.* **15:** 604–609.
46. Bangert, M., T. Peschel, G. Schlaug, *et al.* 2006. Shared networks for auditory and motor processing in professional pianists: evidence from fMRI conjunction. *NeuroImage* **30:** 917–926.
47. Baumann, S., S. Koeneke, M. Meyer, *et al.* 2005. A network for sensory-motor integration: what happens in the auditory cortex during piano playing without acoustic feedback? *Ann. N.Y. Acad. Sci.* **1060:** 186–188.
48. Haueisen, J. & T.R. Knösche. 2001. Involuntary motor activity in pianists evoked by music perception. *J. Cogn. Neurosci.* **13:** 786–792.
49. Lahav, A., E. Saltzman & G. Schlaug. 2007. Action representation of sound: audiomotor recognition network while listening to newly acquired actions. *J. Neurosci.* **27:** 308–314.
50. Wolpert, D.M., R.C. Miall & M. Kawato. 1998. Internal models in the cerebellum. *Trends Cogn. Sci.* **2:** 338–347.
51. Ito, M. 2008. Control of mental activities by internal models in the cerebellum. *Nat. Rev. Neurosci.* **9:** 304–313.
52. Rauschecker, J.P. 2011. An expanded role for the dorsal auditory pathway in sensorimotor control and integration. *Hearing Res.* **271:** 16–25.
53. Wolpert, D.M. & Z. Ghahramani. 2000. Computational principles of movement neuroscience. *Nat. Neurosci.* **3:** 1212–1217.
54. Friston, K.J., J. Daunizeau, J.M. Kilner & S.J. Kiebel. 2010. Action and behavior: a free-energy formulation. *Biol. Cybern.* **102:** 227–260.
55. Friston, K.J. & S.J. Kiebel. 2009. Predictive coding under the free-energy principle. *Philos. T. Roy. Soc. B.* **364:** 1211–1221.
56. Vuust, P., L. Ostergaard, K.J. Pallesen, *et al.* 2009. Predictive coding of music: brain responses to rhythmic incongruity. *Cortex* **45:** 80–92.
57. Keller, P.E. 2008. Joint action in music performance. In *Enacting Intersubjectivity: A Cognitive and Social Perspective to the Study of Interactions.* F. Morganti, A. Carassa & G. Riva, Eds.: 205–221. IOS Press. Amsterdam.
58. Keller, P.E. & M. Appel. 2010. Individual differences, auditory imagery, and the coordination of body movements and sounds in musical ensembles. *Music Percept.* **28:** 27–46.
59. Keller, P.E. & I. Koch. 2008. Action planning in sequential skills: relations to music performance. *Q. J. Exp. Psychol.* **61:** 275–291.
60. Keller, P.E. & I. Koch. 2006. The planning and execution of short auditory sequences. *Psychon. B. Rev.* **13:** 711–716.
61. Keller, P.E., S. Dalla Bella & I. Koch. 2010. Auditory imagery shapes movement timing and kinematics: evidence from a musical task. *J. Exp. Psychol. Human* **36:** 508–513.
62. Wolpert, D.M., K. Doya & M. Kawato. 2003. A unifying computational framework for motor control and social interaction. *Philos. T. Roy. Soc. B.* **358:** 593–602.
63. Keller, P.E., G. Knoblich & B.H. Repp. 2007. Pianists duet better when they play with themselves: on the possible role of action simulation in synchronization. *Conscious. Cogn.* **16:** 102–111.
64. Pecenka, N. & P.E. Keller. 2009. Auditory pitch imagery and its relationship to musical synchronization. *Ann. N. Y. Acad. Sci.* **1169:** 282–286.
65. Pecenka, N. & P.E. Keller. 2009. The relationship between auditory imagery and musical synchronization abilities in musicians. In *Proceedings of the 7th Triennial Conference of European Society for the Cognitive Sciences of Music ESCOM 2009.* J. Louhivuori, T. Eerola, S. Saarikallio, T. Himberg & P.S. Eerola, Eds.: 409–414. University of Jyväskylä. Jyväskylä, Finland.
66. Rankin, S.K., E.W. Large & P.W. Fink. 2009. Fractal tempo fluctuation and pulse prediction. *Music Percept.* **26:** 401–413.

67. Repp, B.H. 2002. The embodiment of musical structure: effects of musical context on sensorimotor synchronization with complex timing patterns. In *Common Mechanisms in Perception and Action: Attention and Performance XIX*. W. Prinz & B. Hommel, Eds.: 245–265. Oxford University Press. Oxford, UK.
68. Pecenka, N. & P.E. Keller, 2011. The role of temporal prediction abilities in interpersonal sensorimotor synchronization. *Exp. Brain Res.* **211:** 505–515.
69. Schubotz, R.I. & D.Y. von Cramon. 2003. Functional-anatomical concepts on human premotor cortex: evidence from fMRI and PET studies. *NeuroImage* **20:** S120–S131.
70. Oullier, O., K.J. Jantzen, F.L. Steinberg & J.A.S. Kelso. 2005. Neural substrates of real and imagined sensorimotor coordination. *Cereb. Cortex* **15:** 975–985.
71. Zatorre, R.J., A.R. Halpern & M. Bouffard. 2010. Mental reversal of imagined melodies: a role for the posterior parietal cortex. *J. Cogn. Neurosci.* **22:** 775–789.
72. Coull, J.T., R.K. Cheng & W.H. Meck 2011. Neuroanatomical and neurochemical substrates of timing. *Neuropsychopharmacol.* **36:** 3–25.
73. Pecenka, N., A. Engel, M.T. Fairhurst & P.E. Keller. 2011. Neural correlates of auditory temporal predictions during sensorimotor synchronization. *Poster presented at The Neurosciences and Music IV: Learning and Memory*. University of Edinburgh. Edinburgh, UK.

Ann. N.Y. Acad. Sci. ISSN 0077-8923

ANNALS OF THE NEW YORK ACADEMY OF SCIENCES
Issue: *The Neurosciences and Music IV: Learning and Memory*

Acuity of mental representations of pitch

Petr Janata
Center for Mind and Brain and Department of Psychology, University of California, Davis, California

Address for correspondence: Petr Janata, Center for Mind and Brain, University of California, 267 Cousteau Pl., Davis, CA 95618. pjanata@ucdavis.edu

Singing in one's mind or forming expectations about upcoming notes both require that mental images of one or more pitches will be generated. As with other musical abilities, the acuity with which such images are formed might be expected to vary across individuals and may depend on musical training. Results from several behavioral tasks involving intonation judgments indicate that multiple memory systems contribute to the formation of accurate mental images for pitch, and that the functionality of each is affected by musical training. Electrophysiological measures indicate that the ability to form accurate mental images is associated with greater engagement of auditory areas and associated error-detection circuitry when listeners imagine ascending scales and make intonation judgments about target notes. A view of auditory mental images is espoused in which unified mental image representations are distributed across multiple brain areas. Each brain area helps shape the acuity of the unified representation based on current behavioral demands and past experience.

Keywords: imagery; music; attention; discrimination; event-related potentials

Introduction

Our internal auditory worlds are vibrant. From inner speech to anticipating the upcoming word or note in a familiar song that we are listening to, our minds maintain an "auditory" world that exists both independently and in interaction with the auditory world around us.

Before proceeding to discuss the acuity of mental auditory images, it is useful to discuss what is meant by an auditory image or mental representation of a sound. I find it most useful to start with a broad definition, such as any sound that you hear playing in your mind. A fundamental distinction can be drawn between external (physical) sources of mental representations of sound and internal (mental) sources of those representations. A slightly more specific definition can be fashioned at the neural level such that the neural activity associated with representations of external sources is contemporaneous with the external source and derives from activity ascending the auditory pathway from the ear, whereas representations deriving from internal sources need not satisfy either of those constraints. I use the term *auditory image* synonymously with mental representation of a sound.

Expectation, covert orientation of attention, and imagery

The focus of this paper is on the acuity of auditory images of pitch that are formed and maintained in a number of different task contexts. I argue that it is parsimonious to consider a unified concept of auditory image across task contexts rather than separate types of auditory images for each of the different types of tasks that in some way instantiate and cause listeners to maintain or manipulate auditory images (Fig. 1).[1] Examples of different types of tasks in which auditory images must be maintained are target detection tasks, auditory working memory tasks, and traditional imagery tasks, such as imagining a familiar melody. While the informational sources from which the auditory image derives may be different in each of those tasks, the auditory image nonetheless becomes available for comparison with mental representations of sensory auditory input so that the listener can decide whether the two mental representations are the same. One might argue that an auditory image should necessarily be associated with a single brain region or population of neurons within that brain region. In such a view, there would be a different auditory image associated with each

doi: 10.1111/j.1749-6632.2011.06441.x

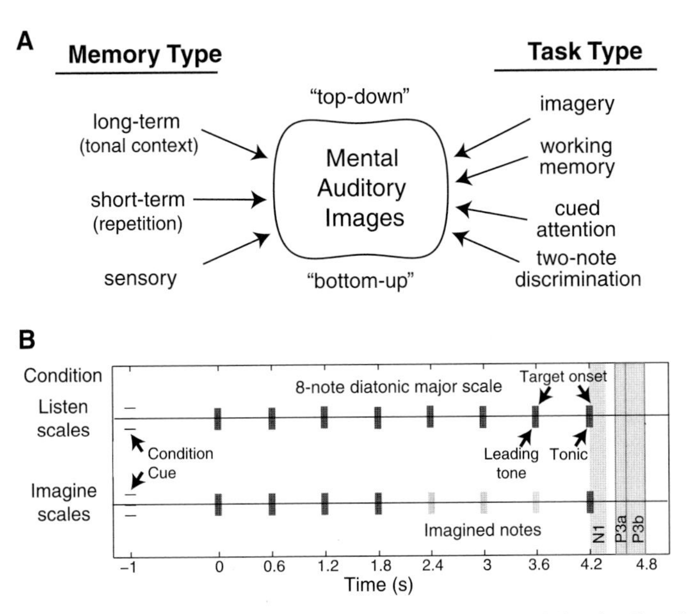

Figure 1. (A) A schematic view of the idea that mental auditory images might be regarded as distributed representations that are subject to influence by different types of memory processes associated with different brain regions as a function of the demands of any particular task. This view is distinct from one in which different types of auditory images are associated with each of the different task types. (B) Diagram of the cued-attention (listen scales) and imagery tasks (imagine scales) used to examine acuity of target tones. Adapted with permission from Ref. 1 (with permission from the Acoustic Society of America) and Ref. 14 (with permission from Elsevier).

successive population of neurons that participates in the representation/maintenance of the auditory image. Thus, even though behavioral responses are made as a consequence of comparing mental and sensory representations in the task types mentioned previously, implying a single locus for the auditory image, the possibility of different task-specific sources of the mental auditory images would imply that different auditory images are maintained in different areas of the brain, before being broadcast to the brain area in which the mental and sensory representations are compared. A contrasting view is one in which the neural representation of a single auditory image is delocalized, that is, distributed across multiple brain areas. In this view, each task-relevant brain area may assist in maintaining the image, and the shaping of the image by any of those areas can influence the neural representation at the other areas. I elaborate this idea for the case of pitch.

Comparisons of pitch acuity across task types

I consider three types of tasks that vary in the degree of sensory support that they provide for the auditory image that is to be compared with a probe note so that an intonation judgment can be made about the probe note. Extensive reports of experiments using these tasks have been presented elsewhere.[1,2] The tasks are as follows.

(1) A basic *two-tone discrimination task* in which two pairs of two 250 ms complex harmonic tones are presented in quick succession (600 ms stimulus-onset asynchrony) and the listener must indicate in which of the two pairs the two tones differ. This task provides the greatest degree of sensory support, as the exact information that is to be compared is maintained in sensory memory for a brief period of time. While this task would not be considered an imagery task in the traditional sense of imagining a melody, there is nonetheless a requirement to maintain a mental image of a just-heard sound in memory.
(2) A *cued-attention task* in which an expectation for a probe note is elicited by the preceding notes of an ascending major scale. Two scales are heard on each trial, and participants must indicate which of the two scales contained the mistuned terminal probe note. This task

provides an intermediate degree of sensory support in the sense that although a note corresponding to the probe note has not yet been heard—that is, there is no direct sensory representation of it—the sequence of notes and associated sensory representations leading up to it is highly predictive and can potentially support the formation of an accurate mental image to coincide with the arrival of the probe tone. While this task is not a traditional imagery task, there is nevertheless a requirement to construct a mental image of a sound that has not yet been heard.

(3) An *imagery task* that is similar to the cued-attention task, but differs in that only the initial four notes of the major scale are heard and the rest must be imagined, so that a probe note that occurs at the time it would have occurred, had the entire scale been played, can be judged. This task provides very little sensory support for the auditory image.

The cued-attention and imagery tasks can be manipulated further by varying the number of keys that the scales are presented in and by varying the terminal probe note about which the intonation judgment is made. Holding the key constant from trial to trial allows a short-term absolute memory for the target notes to build up. Placement of the probe note at the leading tone (an unstable scale position) versus the tonic (the most stable scale position) allows an estimate to be made of the influence of long-term memory for tonal knowledge on the momentary auditory image.

These brief descriptions of the task types and additional mnemonic manipulations hopefully serve to illustrate the difficulty in specifying precisely what the criteria are for labeling a process as *auditory imagery* and the elements that are being operated on as *auditory images*.

When the tasks are structured as two-alternative forced choice (2AFC) tasks, the exact amount of mistuning of the probe notes is varied from trial to trial so that each listener's intonation threshold can be determined for each task type. Not surprisingly, the smallest thresholds are obtained in the two-tone discrimination task. Interestingly, thresholds in the cued-attention task remain comparable to two-tone thresholds for many, but not all listeners (Fig. 2A). Moreover, when the key in which the scale is played is varied across trials, thresholds tend to become worse. These effects are further exacerbated in the imagery task when sensory support is further removed, even though some individuals

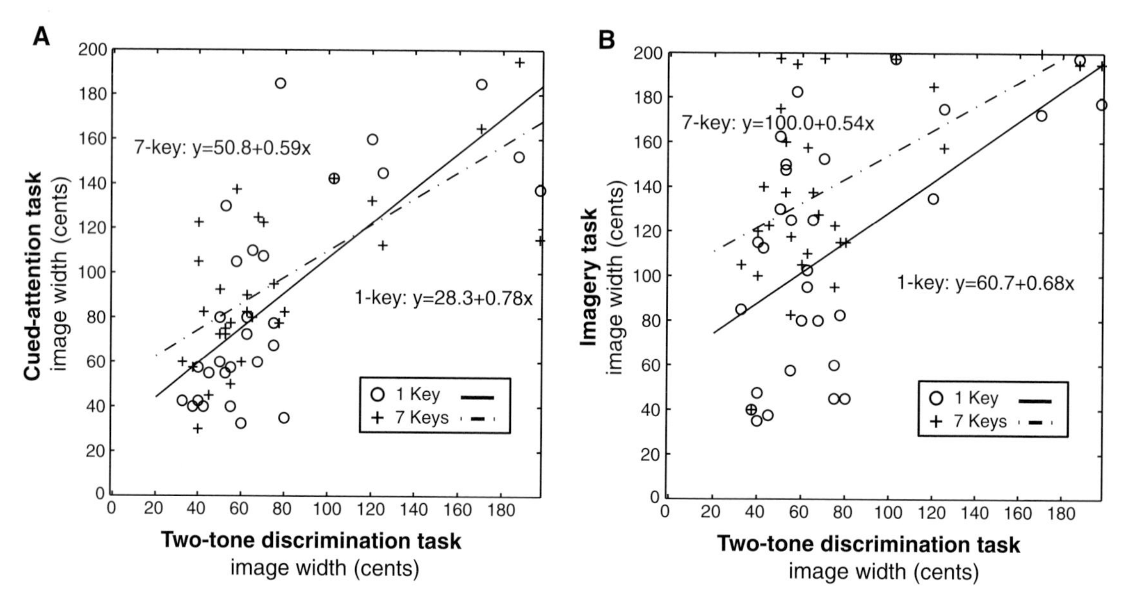

Figure 2. Correlations between auditory image widths in a two-tone discrimination task and image widths for the last note of an ascending major scale in a cued-attention task (A) and imagery task (B). The influence of short-term memory build-up was assessed by contrasting performance on blocks of trials in which the key was held constant and blocks in which the key varied randomly among seven keys. Reprinted with permission from Ref. 1.

perform as well in the imagery task as they do in the two-tone discrimination task (Fig. 2B). These observations indicate that auditory images are shaped by multiple mechanisms, that comparable quality of the image can be obtained whether it arises directly from sensory input or from internal sources, and that individuals vary considerably in the degree to which they utilize these different mechanisms when forming and maintaining auditory images.[1,2] Although considerable variance in pitch image acuity is observed in individuals with no musical training, with some untrained individuals exhibiting pitch images that are as precise as those of trained individuations, overall, musical training significantly improves acuity in both cued-attention and imagery conditions.[1,2]

Electrophysiological measures of image acuity

The decisions made by listeners in the behavioral experiments summarized earlier depend on the comparison of an internally maintained auditory image (top–down) with incoming sensory information (bottom–up). The suggestion that internal auditory images might be maintained across a number of brain areas implies that comparisons between top–down and bottom–up representations could also transpire at and across multiple brain areas. Particular attention has been paid to the auditory cortex. Mismatch negativity (MMN) experiments have demonstrated a role of the auditory cortex in the detection of acoustic deviants outside the focus of attention,[3] though auditory representations in earlier stages of the ascending auditory pathway have been shown to be modulated by attention or the preceding stimulus context.[4] Here, I also focus on the auditory cortex, or, more specifically, electrophysiological signatures of secondary auditory cortex processing with a particular emphasis on the N1 component of the auditory evoked potential.[5,6]

Both attentional cuing and auditory imagery influence activity in auditory cortical areas as demonstrated by electrophysiological, positron emission tomography (PET), and functional magnetic resonance imaging (fMRI) studies.[7–11] Imagining notes in a melody establishes a voltage distribution across the scalp with a topography that resembles that of the N1, and can be modulated focally in time to appear as an evoked potential.[12] MMN responses have also been observed in response to imagined notes in familiar melodies.[13]

The N1

Utilizing the intonation judgment tasks described earlier, we demonstrated that N1 amplitude can serve as a marker of auditory image acuity.[14] The amplitude of the N1 depends on the rate at which successive tones are presented, and decreases rapidly when multiple tones are presented.[15] If the tones cease, the amplitude of the N1 in response to an eventual tone is large again. What might happen to the amplitude of the N1 to a probe tone given the task of imagining an ascending scale? If the notes in an ascending scale are imagined successfully and each of those imagery events activates the auditory cortex, the N1 to the probe note should be small, as if each of the intervening imagined notes had actually been heard. By contrast, if no imagery task is performed, or if a participant who is supposed to engage in imagining the sequence of tones does not do so successfully, the N1 amplitude in response to a probe tone about which a judgment must be made should be large because it is effectively the first input to the auditory cortex in several seconds. Figures 3A and 3B illustrate this effect. Those individuals who succeeded in forming accurate auditory images showed smaller N1 amplitudes than those who did not, suggesting that by virtue of imagining the tones leading up to the probe tone they instantiated representations of those tones in the auditory cortex.

These observations parallel the findings of studies in which the conceptual framework has centered on the orienting of attention. Visual and auditory tasks in which expectations are manipulated in time and another feature dimension, for example, space or pitch, result in faster reaction times and reduced N1 amplitudes when expectations are met.[16] Utilizing a very similar paradigm, in which all the notes of an ascending scale across two octaves were heard except for the two notes preceding the final tonic, Lange[16] observed reductions in N1 amplitude as temporal and pitch expectations were met. Although no explicit instruction was provided to imagine the missing notes, the results illustrate that regularity in the stimulus structure implicitly facilitates the instantiation of a mental pitch representation at a specific moment in time that can be compared with the incoming sensory representation.[17,18]

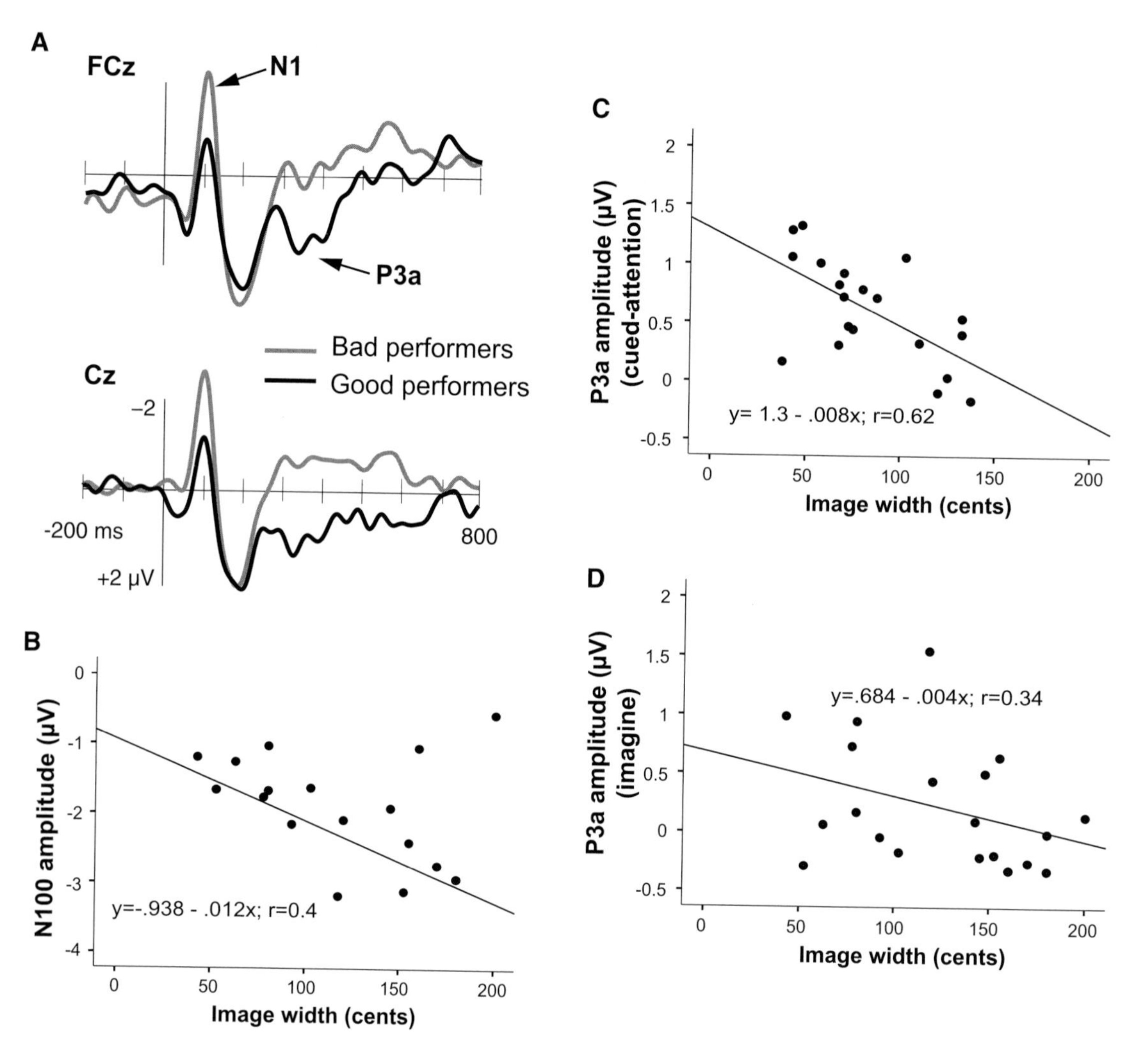

Figure 3. Event-related potential (ERP) indexes of auditory image acuity. (A) Listeners with poor pitch image acuity (bad performers) exhibited a larger N1 component of the auditory-evoked potential, indicative of a failure to imagine the notes leading up to the final note properly. (B) N1 amplitudes increased (became more negative) as image acuity decreased. The P3a component, a measure of deviance detection, was larger in those listeners who formed more accurate images in the cued-attention (C) and imagery conditions (D). Adapted from Ref. 14 with permission.

Auditory images within a perception–action cycle framework

I suggested earlier that the role of establishing and maintaining auditory images might be shared across multiple brain areas depending on the particular task context. Why, however, should such sharing take place and what would the other areas be that participate in the shaping of the representation? Paralleling the work of others, I believe it is useful to think of auditory images in terms of a perception–action cycle.[19] In simplest terms, the perception side of the perception–action cycle might be associated with representing bottom–up sensory information, while the action side mediates the projection of top–down information, not only to motor areas that drive effectors, but also to sensory areas in the form of "efference-copy" or expectations more generally.[20–22] The concept of cyclicality becomes pertinent whenever a sequence of events/actions comes into play, in that an image/action sets the stage for an upcoming image/action, often shaped by the interaction of the image with sensory input. Musical examples include both covert and overt singing of a melody with or without actual accompaniment, or adjusting intonation across time while singing a single note.

P3a

The P3a is an event-related potential (ERP) component that appears as a frontocentral positivity along the midline in response to novel auditory stimuli.[23] The P3a can even appear in response to musical events, for example, chords, that are incongruent with a highly expected and imagined chord.[24] I believe the P3a response observed in the cued-attention and imagery tasks may reflect activity of an incongruity detector within the perception–action cycle. As with the N1, its amplitude was related to image acuity in both the cued-attention and imagery conditions: listeners with more precise mental images had a larger P3a amplitude (Fig. 3C and D). Note, that in contrast to the reduction of N1 amplitude for the probe note, presumably due to activation of the auditory cortex by the preceding imagined notes, the P3a increased in amplitude, consistent with its interpretation as a marker of deviance detection. Attributing to the P3a, a relationship to the perception–action cycle stems from its focal frontocentral midline topography with a presumed source in the anterior cingulate cortex,[25] the ACCs known role in conflict and action monitoring[26,27] and auditory target detection.[23,25,28] However, given the location of the probes at the end of a note sequence (chain of perception–action cycle events), it cannot be ruled out that the P3a simply reflects the conscious marking of a probe note as a deviant note for a subsequent response, rather than an error signal that would influence the production of the next note in a sequence.

Is the concept of a delocalized auditory image useful?

Arguing for a view of auditory images as neural representations that span several brain areas involved to varying degrees in perception and action is somewhat perilous because it risks making the concept of an auditory mental representation unspecific and difficult to identify in the brain. Many might argue that it is easier to regard the representations of auditory objects in terms of discrete processing stages that are accessible to top–down influences and conscious awareness to varying degrees.[29] For example, in the context of attention-orienting paradigms, one can ask in what part of the brain is activity modulated when an expectation is established for a particular auditory feature. However, the processes and focal representations in the attention/expectation case are traditionally viewed as different than the distributed representations maintained during an auditory working memory task across the brain areas that constitute the phonological loop. In attention/expectation paradigms, the emphasis is on a single locus in feature space at a certain point in time, with an emphasis on perception, whereas in the working memory case there is an emphasis on the maintenance of multiple objects over time across both sensory and motor areas. Moreover, in the context of natural behaviors, such as controlling intonation of a pitch while singing, or singing along with a piece of familiar music in one's mind, it is very difficult to specify where the sensory representation at one moment in time stops and the motor representation of the next moment in time begins. In this regard it seems more parsimonious to consider a delocalized sensorimotor representation that is then subject to influences from and accessible to multiple other brain regions as a function of the broader behavioral context in which the sensorimotor representation is being created. A delocalized representation should not imply a lack of constraints, however.

Anatomical connections impose constraints on the activity in the connected areas insofar as the pattern of activity in one area will be correlated with the pattern of activity in the other. In the case of auditory images, the arcuate fasciculus is a connection that binds sensory (lateral temporal) and motor (lateral prefrontal) brain areas and supports verbal and presumably tonal working memory.[30,31] It is also associated with the ability to perceive pitch sequences, such that individuals with poor ability to discriminate pitch contours (amusics) have a thinner arcuate fasciculus in the right hemisphere.[32] Interestingly, amusics show a dissociation in their representations of pitch contours, with more accurate, albeit not fully tuned, representations evident if they are probed with a production task instead of a perception task.[33] This observation serves as a cautionary note that the inferred quality of a mental representation may depend on the behavioral context in which it is probed.

If a task involves a decision, the accuracy of which depends on pitch image quality, the behavioral and neural acuity measures simply reflect the multiple influences that have combined to shape the top–down and bottom–up representations up to the point that they interact. However, if the behavioral or neural measure probes the quality of the auditory image at some other location in the brain, the inferred quality may be different. For example, take two individuals who are asked to imagine one of the songs of their favorite band that they heard the previous evening at a concert they attended together. They are both performing the same imagery task. Both of them might recall the same song equally well in terms of the melodic contour and the words if probed with a production task, whereas only one of them might be able to determine accurately whether sounded probe tones are "in tune" with the imagined notes. Rather than considering that one of these individuals deploys more mental auditory images than the other during the imagery task, it may make sense to consider mental auditory images as unified mental constructs. However, the exact appearance of the image at any point in the brain nonetheless depends on the functional perspective of the brain area engaging the construct and the individual's success in engaging the functions of that brain area. Whether such a view is useful from an experimentalist's perspective remains to be seen. One advantage of it is that it diminishes the drive to consider every different type of task that in some way manipulates an auditory representation to be giving rise to a different type of represented entity.

Conflicts of interest

The author declares no conflicts of interest.

References

1. Navarro Cebrian, A. & P. Janata. 2010. Influences of multiple memory systems on auditory mental image acuity. *J. Acoust. Soc. Am.* **127:** 3189–3202.
2. Janata, P. & K. Paroo. 2006. Acuity of auditory images in pitch and time. *Percept. Psychophys.* **68:** 829–844.
3. Näätänen, R. 1992. *Attention and Brain Function.* Lawrence Erlbaum Associates. Hillsdale, NJ.
4. Suga, N. *et al.* 2000. The corticofugal system for hearing: recent progress. *Proc. Natl. Acad. Sci. USA* **97:** 11807–11814.
5. Näätänen, R. & T. Picton. 1987. The N1 wave of the human electric and magnetic response to sound: a review and an analysis of the component structure. *Psychophysiology* **24:** 375–425.
6. Pantev, C. *et al.* 1995. Specific tonotopic organizations of different areas of the human auditory cortex revealed by simultaneous magnetic and electric recordings. *Electroencephalogr. Clin. Neurophysiol.* **94:** 26–40.
7. Halpern, A.R. & R.J. Zatorre. 1999. When that tune runs through your head: a PET investigation of auditory imagery for familiar melodies. *Cereb. Cortex.* **9:** 697–704.
8. Jäncke, L., S. Mirzazade & N.J. Shah. 1999. Attention modulates activity in the primary and the secondary auditory cortex: a functional magnetic resonance imaging study in human subjects. *Neurosci. Lett.* **266:** 125–128.
9. Woldorff, M.G. & S.A. Hillyard. 1991. Modulation of early auditory processing during selective listening to rapidly presented tones. *Electroencephalogr. Clin. Neurophysiol.* **79:** 170–191.
10. Woldorff, M.G. *et al.* 1993. Modulation of early sensory processing in human auditory cortex during auditory selective attention. *Proc. Natl. Acad. Sci. USA* **90:** 8722–8726.
11. Zatorre, R.J. *et al.* 1996. Hearing in the mind's ear: a PET investigation of musical imagery and perception. *J. Cogn. Neurosci.* **8:** 29–46.
12. Janata, P. 2001. Brain electrical activity evoked by mental formation of auditory expectations and images. *Brain Topogr.* **13:** 169–193.
13. Herholz, S.C. *et al.* 2008. Neural basis of music imagery and the effect of musical expertise. *Eur. J. Neurosci.* **28:** 2352–2360.
14. Navarro Cebrian, A & P. Janata. 2010. Electrophysiological correlates of accurate mental image formation in auditory perception and imagery tasks. *Brain Res.* **1342:** 39–54.
15. Woods, D.L. & R. Elmasian. 1986. The habituation of event-related potentials to speech sounds and tones. *Electroencephalogr. Clin. Neurophysiol.* **65:** 447–459.
16. Lange, K. 2009. Brain correlates of early auditory processing are attenuated by expectations for time and pitch. *Brain Cogn.* **69:** 127–137.
17. Barnes, R. & M.R. Jones. 2000. Expectancy, attention, and time. *Cogn. Psychol.* **41:** 254–311.
18. Jones, M.R. *et al.* 2002. Temporal aspects of stimulus-driven attending in dynamic arrays. *Psychol. Sci.* **13:** 313–319.
19. Fuster, J.M. 2001. The prefrontal cortex: an update—time is of the essence. *Neuron.* **30:** 319–333.
20. Haggard, P. & B. Whitford. 2004. Supplementary motor area provides an efferent signal for sensory suppression. *Cognit. Brain Res.* **19:** 52–58.
21. Baess, P., T. Jacobsen & E. Schroeger. 2008. Suppression of the auditory N1 event-related potential component with unpredictable self-initiated tones: evidence for internal forward models with dynamic stimulation. *Int. J. Psychophysiol.* **70:** 137–143.
22. Lange, K. 2011. The reduced N1 to self-generated tones: an effect of temporal predictability? *Psychophysiology* **48:** 1088–1095.
23. Polich, J. 2007. Updating P300: an integrative theory of P3a and P3b. *Clin. Neurophysiol.* **118:** 2128–2148.
24. Janata, P. 1995. ERP measures assay the degree of expectancy violation of harmonic contexts in music. *J. Cogn. Neurosci.* **7:** 153–164.
25. Crottaz-Herbette, S. & V. Menon. 2006. Where and when the anterior cingulate cortex modulates attentional response: combined fMRI and ERP evidence. *J. Cogn. Neurosci.* **18:** 766–780.
26. Gehring, W.J. & R.T. Knight. 2000. Prefrontal-cingulate interactions in action monitoring. *Nat. Neurosci.* **3:** 516–520.
27. Carter, C.S. & V. van Veen. 2007. Anterior cingulate cortex and conflict detection: an update of theory and data. *Cognit. Affect. Behav. Neurosci.* **7:** 367–379.

28. Kiehl, K.A. *et al.* 2001. Neural sources involved in auditory target detection and novelty processing: an event-related fMRI study. *Psychophysiology.* **38:** 133–142.
29. Näätänen, R. & I. Winkler. 1999. The concept of auditory stimulus representation in cognitive neuroscience. *Psychol. Bull.* **125:** 826–859.
30. Catani, M. & M. Mesulam. 2008. The arcuate fasciculus and the disconnection theme in language and aphasia: history and current state. *Cortex* **44:** 953–961.
31. Koelsch, S. *et al.* 2009. Functional architecture of verbal and tonal working memory: an fMRI study. *Hum. Brain Mapp.* **30:** 859–873.
32. Loui, P., D. Alsop & G. Schlaug. 2009. Tone deafness: a new disconnection syndrome? *J. Neurosci.* **29:** 10215–10220.
33. Loui, P. *et al.* 2008. Action-perception mismatch in tone-deafness. *Curr. Biol.* **18:** R331–R332.

Ann. N.Y. Acad. Sci. ISSN 0077-8923

ANNALS OF THE NEW YORK ACADEMY OF SCIENCES
Issue: *The Neurosciences and Music IV: Learning and Memory*

Beyond auditory cortex: working with musical thoughts

Robert J. Zatorre

McGill University, Montreal, Quebec, Canada

Address for correspondence: Robert J. Zatorre, Montreal Neurological Institute, 3801 University, Montreal, Quebec H4A 2B4, Canada. robert.zatorre@mcgill.ca

Musical imagery is associated with neural activity in auditory cortex, but prior studies have not examined musical imagery tasks requiring mental transformations. This paper describes functional magnetic resonance imaging (fMRI) studies requiring manipulation of musical information. In one set of experiments, listeners were asked to mentally reverse a familiar tune when presented backwards. This manipulation consistently elicits neural activity in the intraparietal sulcus (IPS). Separate experiments requiring judgments about melodies that have been transposed from one musical key to another also elicit IPS activation. Conjunction analyses indicate that the same portions of the IPS are recruited in both tasks. The findings suggest that the dorsal pathway of auditory processing is involved in the manipulation and transformation of auditory information, as has also been shown for visuomotor and visuospatial tasks. As such, it provides a substrate for the creation of new mental representations that are based on manipulation of previously experienced sensory events.

Keywords: musical imagery; mental transformations; intraparietal sulcus; fMRI

Introduction

Mental imagery takes many forms and can have various functions. Perhaps the most familiar form of imagery involves evocation of previously experienced information. This aspect of imagery is most closely tied to memory recall, and can serve to enhance information retrieval. But from the earliest cognitive studies of imagery it became evident that imagery could also involve a component of manipulation or modification. For example, one of the clearest demonstrations of visual imagery required volunteers to mentally transform viewed objects from one, seen orientation to another, imagined one.[1] This ability to transform an internal representation to arrive at an answer to a question, or to solve a problem, raises many questions about the mechanisms by which such processes take place, and of their functional significance. In the domain of music, a number of studies have pointed to the utility of auditory imagery to accomplish specific musical goals: for example, in the study of a written score,[2] as a way to ensure accurate intonation during a performance,[3] or as an aid to learning.[4] There is in fact good evidence that mental practice—which no doubt includes more than just an auditory imagery component—can be beneficial to musicians,[5,6] and can even result in changes in cortical functional organization.[7] These studies confirm the utility (and hence the psychological reality) of musical imagery for musicians and performers, but do not clarify to what extent the imagery required for the task may involve manipulation of existing representations as opposed to evocation of those memory traces. Yet it seems clear that in order for creativity to exist, it would be necessary for musicians, especially composers or improvisers, to have the ability to recombine or juxtapose previously experienced musical events into novel combinations.

Experimental studies of musical imagery have tended to focus more on its perceptual aspect, rather than on active manipulation of information, in contrast to studies in the visual domain (e.g., Ref. 8). For example, tasks typically used to study musical imagery require a volunteer to imagine a familiar tune and make a judgment about it;[9] such procedures tap into retrieval and experiential aspects of the imagery process fairly well, but do not require

doi: 10.1111/j.1749-6632.2011.06437.x

anything beyond that. More recently, investigators have become interested in how musical imagery can be used in a more active way (see the other contributions in this volume for examples of these approaches). This idea leads to the studies that I will describe in this paper, which focus on the neural substrates of mental manipulation of musical information.

Previous research into the neural basis of musical imagery has primarily been carried out in the context of the more passive tasks just alluded to. The aim has mostly been to identify the neural structures that are implicated in people's ability to imagine music, usually well-known tunes. Among the first experimental approaches to this phenomenon was a study in which Andrea Halpern and I investigated the ability of people with unilateral temporal-lobe excisions to make judgments about the pitch of imagined versus heard melodies.[10] We found that patients with removals within the right temporal lobe performed more poorly on both perception and imagery tasks than did those with left temporal excision or controls. The "common fate" of the two tasks paralleled findings in the visual imagery domain in which damage to visual cortex led to deficits in both visual imagery and perception.[11] Furthermore, the result was consistent with many previous experiments with this patient population, indicating that damage to auditory cortical regions—especially those within the right temporal lobe—resulted in tonal processing deficits.[12]

A series of functional neuroimaging studies using positron emission tomography (PET), functional magnetic resonance imaging (fMRI), and magnetoencephalography (MEG) followed that consistently implicate auditory cortex in a variety of musical imagery tasks. We and others have now documented increases in neural activity in auditory cortical regions while volunteers perform imagery tasks that include judging the pitch change of two syllables within an imagined tune;[13] imagining the continuation of a familiar melody when cued with its opening tones;[14] comparing the similarity of two imagined instrumental timbres;[15] imagining a familiar tune during gaps in its presentation;[16] and judging if a sounded tone is a correct continuation of an imagined melody.[17] The important contribution of auditory cortex to musical imagery is thus well established, but the question remains about how this system might be implicated in more active forms of imagery, and whether additional neural resources might be needed for tasks that require more than evocation of a previously experienced auditory event.

Clues to this question arise from two sources: first, from a consideration of the neural pathways associated with processing of auditory information, and second, from the literature on manipulation and transformation of perceptual information in nonauditory domains. A large body of neurophysiological and neuroanatomical studies in monkeys, coupled with functional imaging and other types of studies in humans, have led to the view that there are (at least) two processing pathways originating in auditory cortex. One is more ventrally directed, leading along the superior and middle temporal gyri with eventual targets inferior frontal cortex; and another is more dorsally directed, going to parietal, premotor, and ultimately dorsolateral frontal cortices.[18] The functional significance of these processing streams has often been discussed in terms of spatial versus nonspatial processing, or in terms of language-specific processes.[19] The relevant concept here, however, is that there are hierarchically organized information-processing loops that might be expected to play a role in tasks requiring processing of auditory representations, such as those that might be involved in active imagery. This idea, in turn, meshes well with a wealth of evidence from other domains in which the dorsal stream, specifically regions within the posterior parietal cortex, is implicated in tasks requiring manipulation of information. For example, parietal cortex, especially the intraparietal sulcus (IPS), is known to be recruited by visual mental rotation tasks,[20] as well as visuomotor tasks,[21] and tasks requiring manipulation (as opposed to monitoring) of items in working memory.[22] There is scant evidence for an involvement of these areas in auditory tasks, although some studies have reported relevant findings.[23] Furthermore, the posterior parietal cortex receives inputs from auditory cortex in the temporal lobe.[24,25] It is thus reasonable to investigate whether active musical imagery tasks might also involve portions of these same networks.

The foregoing gives the background to a recent fMRI study from our lab[26] in which we sought to create a musical imagery task that would involve a manipulation component, and not simply evocation as had been the case in prior studies. Our

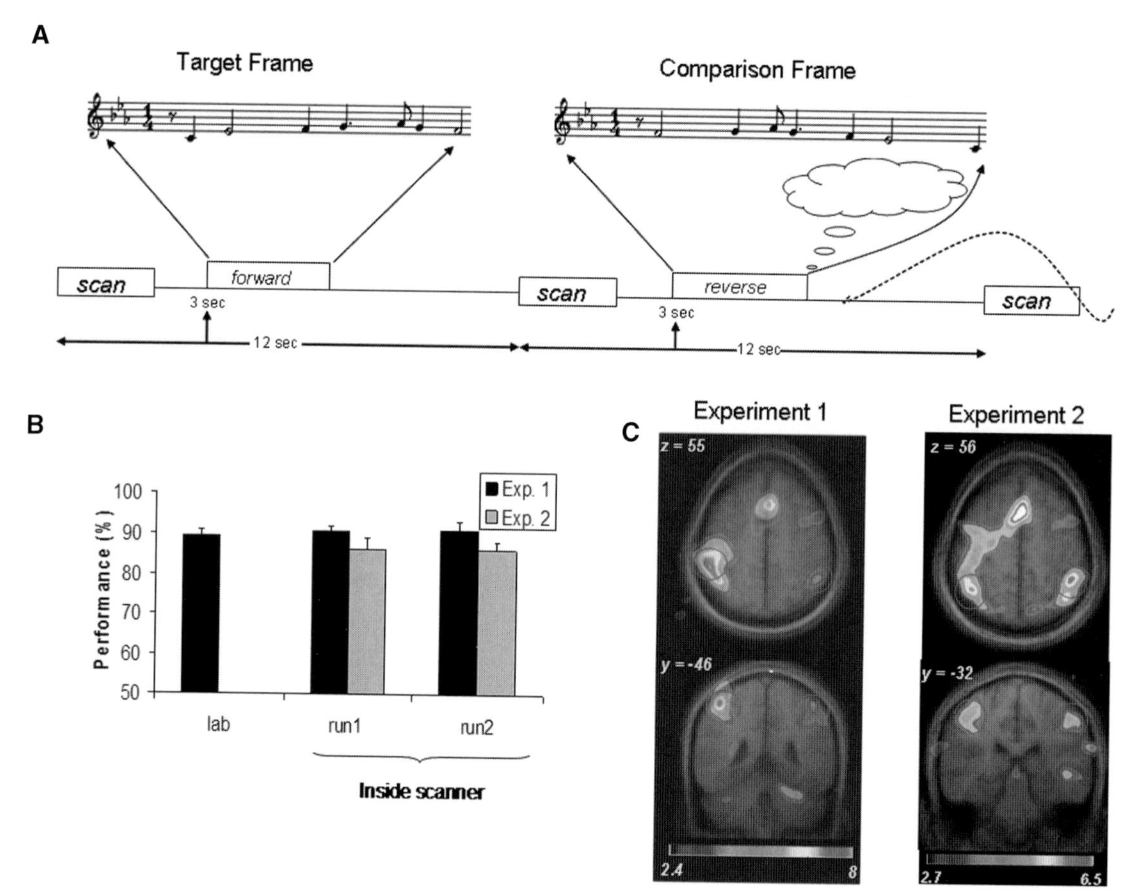

Figure 1. (A) Timeline of a trial in the fMRI mental reversal experiment (experiment 1 of Ref. 26). Each trial contained two 12-sec frames. The target stimulus (in this example, "Greensleeves") was sounded in the first frame, and was followed in the second frame by a comparison stimulus that was either a true or incorrect temporal reversal of that melody. The mental event of interest (depicted by the "cloud") presumably occurs sometime between the end of the reversed melody presentation and the scan. The dotted line illustrates the presumed hemodynamic function associated with performing the mental reversal task. The fMRI volume acquisitions were clustered as shown within the sparse-sampling paradigm so as to maximize the likelihood of capturing the BOLD signal associated with the mental reversal. Experiment 2 was similar but only a visual tune title was given in the first frame. (B) Behavioral performance on the tasks in experiments 1 and 2, collected first in the laboratory, and subsequently during fMRI scanning. Note the high levels of performance. (C) fMRI results from the two experiments. The top (horizontal) and bottom (coronal) sections illustrate BOLD signal increases within the intraparietal sulcus (circled) when contrasting the reverse condition to a control condition containing matched acoustical stimulation.

prediction was that there would be activation of auditory regions to a greater extent when manipulation was required than when it was not, but we also expected that extra-auditory regions might also be recruited. We developed a task requiring temporal reversal, or reordering of tones in time from front to back(Fig. 1A). This task is artificial in the sense that the circumstances where a listener would need to perform a mental reversal are essentially nil in normal musical listening (although retrograde permutations are occasionally used as a compositional device). Our aim in any case was not to mimic normal listening, but to create a controlled situation requiring the mental reorganization of auditory material. This task is arguably a good way to achieve this goal, and also is similar to some classic tasks used in neuropsychology (e.g., digit repetition backwards).

Because identification of a tune presented in reverse order would prove too difficult for most people, we opted for a comparison task in which a familiar tune is first presented in its normal form, and is then followed by a reversed form; this latter may or may not be an exact reversal, and the listener's task is to mentally reorder the second pattern to determine whether it matches the target tune or not. On those trials in which the second pattern was not identical to the first, any changed notes were chosen from the same key and from the same range as those in the target tune in order not to provide any obvious cues. This is a mental manipulation that musicians can perform, but it can still be argued that it could be carried out without any mental reversal because one could, in principle, create some sort of inventory of tones present in the two stimuli and then compare them without needing to reorder. We judged this highly unlikely based on our own intuitions, but as a test of this potential issue (and to satisfy some pesky reviewers) we carried out a control behavioral task in which the incorrect note in the reversed melody was systematically varied in position. We reasoned that if reversal were in fact taking place as we claimed, it should take longer to judge the incorrect item when it occurred at the beginning than at the end. That is, if the reversed tune is represented by tones 6, 5, 4, 3, 2, 1, and note 5 is incorrect, it will take longer to judge than if note 2 is incorrect, assuming that the listeners are following the instructions and mentally replaying the tones in the normal 1–6 order. This is indeed what we observed, with response latencies differing by close to 200 ms in the two conditions. These type of chronometric behavioral data are quite important in validating the task, and fits with other behavioral studies both in the auditory[27] and visual[28] domains, indicating that the intended mental transformation was taking place.

Having developed the task, we were now in a position to examine the pattern of brain activity associated with the mental reversal. We selected from among a group of musically trained listeners 12 individuals who on average were able to perform the task at between 80 and 90% correct (Fig. 1B). The fMRI paradigm was set up so that we could analyze the activity after the target tune separately from the brain activity associated with the comparison tune—the one requiring reversal. We also implemented a control stimulus that was acoustically similar to the tunes. When we contrasted the reversal condition to either the control or forward conditions we observed increases in blood-oxygenation signal in a number of cortical and subcortical regions. The most prominent of these included the IPS, dorsolateral and ventrolateral frontal cortex, and the anterior cingulate (Fig. 1C). We did not observe the expected recruitment of auditory cortical areas, as had been seen in prior studies, although when we looked for blood oxygen level–dependent (BOLD) activity that correlated with individual differences in imagery vividness, as measured via an off-line questionnaire, we did see a significant correlation in the right planum temporale. One reason for the lack of auditory cortex activity in the contrast analysis is that the scanning sequence was set up to pick up the brain activity from imagery but not from the sounded stimulus; to do so necessitated placing the acquisition rather late in the trial, perhaps resulting in the wrong timing to detect the (very likely weak) auditory cortex activity. To remedy this situation, we carried out a second experiment in which the trial time was shortened; in addition, we removed the sounded stimulus from the "forward" condition replacing it by the title of the to-be-imagined tune. This way we avoided any potential contamination of the BOLD signal in auditory cortex to the real stimulus as opposed to the imagined one. The results mirrored the data from the first experiment in that similar areas of parietal and frontal cortex were activated; in addition, we did observe some BOLD activity in the right superior temporal sulcus area, albeit weak in magnitude.

The clearest finding in this study was the recruitment of dorsal-pathway structures in the reversal task. In both experiments, the findings consistently indicated large changes in BOLD signal in the IPS, along with dorsolateral frontal and anterior cingulate regions. The expectation that we would observe enhanced auditory cortex activation was not borne out very clearly, although there was some evidence in experiment 2, and also in experiment 1 in relation to individual variability in auditory imagery vividness. Although most prior auditory imagery studies have reported clear auditory cortex recruitment, as mentioned above, this is not universally the case.[29] Although this issue is not resolved to our satisfaction, we can put it aside for the moment and turn our attention instead to the robust finding of IPS activation. In part, this change of focus was motivated by an unrelated series of experiments being carried out at the same time in our lab, which were initially motivated by different questions. We had been interested for some time in the processing of melodies outside the context of imagery, and had recently been pursuing behavioral probes that would be sensitive to individual differences in the ability to

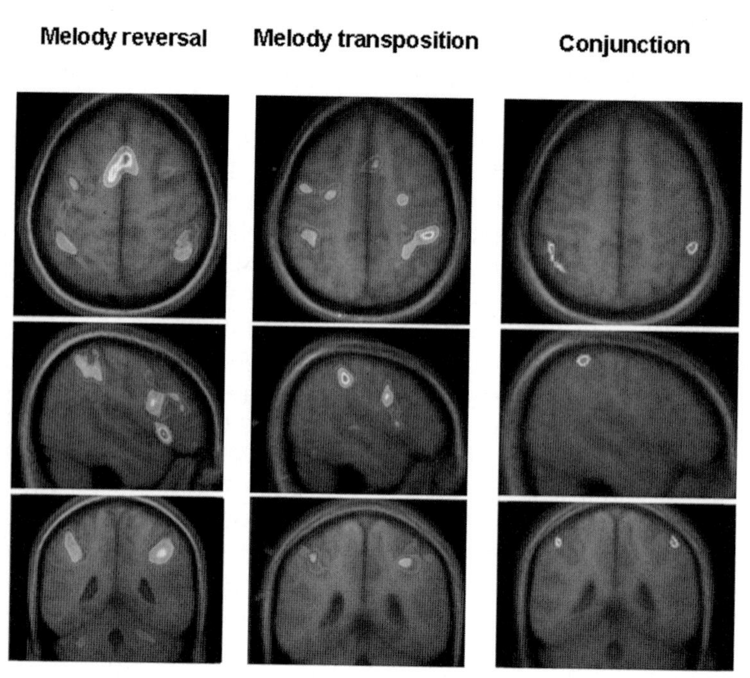

Figure 2. Comparison of melody reversal and melody transposition tasks. Each column shows, from top to bottom, horizontal, sagittal, and coronal sections through the IPS region. The left column shows BOLD signal increases in a contrast of reverse versus forward conditions; the middle column shows BOLD signal increases in a contrast of transposed versus untransposed conditions; the right column shows the conjunction analysis of the data from the first two columns. Note that the principal region showing significant conjunction is within the IPS.

encode and recognize melodic patterns. To accomplish this aim we developed a discrimination task using novel, unfamiliar melodies, in which the second item in a discrimination pair is either transposed to a different musical key or not.[30] The listener's instruction is to determine whether the two patterns are identical or if there is a single changed tone in the second item. On trials in which the second melody is transposed relative to the first, the task would then require that the pitch intervals between successive tones be abstracted because the absolute pitch values would all be different. This task met several criteria relevant to our needs because (1) it is sensitive to musical training, but does not require musical training for successful execution (indeed, nonmusicians find the concept intuitive even if they also often find the task itself challenging); and (2) it elicits a wide range of scores from near-chance to near-perfect. The latter feature was important to us because we were attempting to capture the population variance in this aspect of auditory processing.

When we compared brain activity measured with fMRI in the transposition condition to the non-transposition condition, we observed strong activation within the IPS.[30] This activity was also stronger during transposition than it was during other, control tasks with similar cognitive demands (including, in particular, working memory load), such as a rhythm discrimination task and a phoneme discrimination task. None of these control tasks required anything akin to the transposition. The argument that IPS activity is directly linked to the specific demands of transposition was strengthened by an additional finding: when we ran an analysis taking each individual's behavioral score as a regressor and then looking throughout the entire brain volume for voxels whose activity was predictive of success, we observed that the peak response was located in the right IPS. In other words, IPS activity was directly linked to performance on the transposition task.

If the two tasks we have studied, musical transposition and musical reversal, are indeed related to similar underlying processes, then we should be able to demonstrate that the same subregion within the IPS is in fact involved. One simple way to do this is to superimpose the findings from the two experiments, to see whether there is overlap. A conjunction analysis allowed us to do just that, and we did observe significant conjunction in a number of voxels within the IPS when comparing the images derived from the reversal and transposition tasks (Fig. 2). This finding is suggestive but not sufficient to demonstrate that the same specific subregion is in fact involved. To do so, we need to demonstrate that there is overlap in individual brains, else the effect could be attributed to averaging and smoothing artifacts; this could only be done if the same people were tested with the two tasks. Moreover, the two tasks in question had not been designed to be compared because they used quite different materials (familiar vs. unfamiliar tunes; different timbres; different durations), and also different control conditions. In order to allow a direct comparison, we therefore implemented a new study in which identical stimulus materials were used for both a reversal and a transposition task, allowing us to test these on the same individuals. As expected, each task yielded strong activation within the IPS, thus replicating each of the two prior studies. More importantly, when we conducted a group conjunction analysis we found overlapping voxels across the two tasks. Most critically of all, we conducted separate conjunctions in each individual data set without any spatial smoothing, and this confirmed that there was significant overlap in nine out of 10 individual brains. We are confident, therefore, in the conclusion that these two tasks share an underlying neural substrate.

This conclusion raises a further question: what do these two tasks have in common that they should recruit similar neural structures? At first glance these sets of findings might seem to be so disparate as to be unrelated. In fact, the IPS has been implicated in a wide range of tasks,[21] and it is therefore reasonable to assume that there is a wide range of processes that take place within this complex cortical region. This is no doubt partly the case given that there are gradients within the IPS in terms of its anatomical features, such as its connectivity.[31] Yet, the various tasks that have been linked to the posterior parietal cortex in general, and the IPS in particular, do share some underlying computational features. In some general sense, they can all be said to involve transformations of some kind, often from one reference frame to another. This is the most accepted model in the visuomotor domain, for example, where work from both monkeys and humans indicates that the IPS is a critical link in a network involved in operations such as eye movements to a target, reaching, and grasping.[31,32] On a cognitive level, it has already been remarked that the IPS is important for visual mental rotation[20] and for working memory tasks requiring manipulation[22] as opposed to monitoring.

We would propose that just as the quite varied visuospatial, visuomotor, and cognitive operations mentioned previously all require some kind of transformation, this is also the case for musical reversal and transposition. In both cases, it is the relationship between the individual elements (tones) that must be abstracted, rather than their absolute values (temporal order or pitches) in order for the transformation to be applied (reordering in time, or raising/lowering in pitch). These operations bear a formal similarity, we argue, with the required computations in rotating an object in visual space, or even reaching a target, in that sensory information has to be represented in a sufficiently abstract form to allow the required action. Although the IPS has traditionally been viewed as an interface for visual inputs, it does receive inputs from many modalities, including auditory in both monkeys[24] and humans.[25] It is therefore well-situated to be involved in carrying out the type of transformation operations we have discussed here.

To come back to the topic of imagery, then, what have we learned? The main conclusion we draw is that, in order to understand more active aspects of imagery, we must move beyond neural representations that involve auditory cortex alone. Instead, we propose that the sensory-motor pathways that are critical for other aspects of auditory processing, especially the dorsal pathway, are critical to the ability to work with musical thoughts. The interaction between sensory representations within auditory cortical areas and the manipulation mechanisms involving parietal (and frontal) cortices are the substrate that allows for representations of previous events to be generated internally, and then manipulated to create novel structures. In this way of thinking, then, we may have the beginnings of a model to explain some aspects of creative thinking.

Acknowledgments

The research described in this paper was supported by funding from the Canadian Institutes of Health Research.

Conflicts of interest

The author declares no conflicts of interest.

References

1. Shepard, R.N. & J. Metzler. 1971. Mental rotation of three-dimensional objects. *Science* **171:** 791–793.
2. Mountain, R. 2001. Composers & imagery: myths & realities. In *Musical Imagery*. R.I. Godøy & H. Jorgensen, Eds.: 271–288. Routledge. Florence, KY.
3. Trusheim, W.H. 1991. Audiation and mental imagery: implications for artistic performance. *Q. J. Music Teach. Learn.* **2:** 139–147.
4. Highben, Z. & C. Palmer. 2004. Effects of auditory and motor mental practice in memorized piano performance. *Bull. Council Res. Music Ed.* **159:** 58–65.
5. Coffman, D.D. 1990. Effects of mental practice, physical practice, and knowledge of results on piano performance. *J. Res. Music Ed.* **38:** 187–196.
6. Theiler, A.M. & L.G. Lipppman. 1995. Effects of mental practice and modeling on guitar and vocal performance. *J. General Psychol.* **122:** 329–343.
7. Pascual-Leone, A. 2003. The brain that plays music and is changed by it. In *The Cognitive Neuroscience of Music*. I. Peretz & R. Zatorre, Eds.: 396–412. Oxford University Press. Oxford.
8. Kozhevnikov, M., S. Kosslyn & J. Shepard. 2005. Spatial vs. object visualizers: characterization of visual cognitive style. *Memory Cognit*. **33:** 710–726.
9. Halpern, A.R. 1992. Musical aspects of auditory imagery. In *Auditory Imagery*. D. Reisberg, Ed.: 1–27. Lawrence Erlbaum. Hillsdale, NJ.
10. Zatorre, R.J. & A.R. Halpern. 1993. Effect of unilateral temporal-lobe excision on perception and imagery of songs. *Neuropsychologia* **31:** 221–232.
11. Farah, M.J. 1988. Is visual imagery really visual? Overlooked evidence from neuropsychology. *Psychol. Rev.* **95:** 307–317.
12. Stewart, L. *et al.* 2006. Music and the brain: disorders of musical listening. *Brain* **129:** 2533–2553.
13. Zatorre, R.J. *et al.* 1996. Hearing in the mind's ear: a PET investigation of musical imagery and perception. *J. Cogn. Neurosci.* **8:** 29–46.
14. Halpern, A.R. & R.J. Zatorre. 1999. When that tune runs through your head: a PET investigation of auditory imagery for familiar melodies. *Cereb. Cortex.* **9:** 697–704.
15. Halpern, A.R. *et al.* 2004. Behavioral and neural correlates of perceived and imagined musical timbre. *Neuropsychologia* **42:** 1281–1292.
16. Kraemer, D.J.M. *et al.* 2005. Musical imagery: sound of silence activates auditory cortex. *Nature* **434:** 158.
17. Herholz, S.C. *et al.* 2008. Neural basis of music imagery and the effect of musical expertise. *Eur. J. Neurosci.* **28:** 2352–2360.
18. Rauschecker, J.P. & S.K. Scott. 2009. Maps and streams in the auditory cortex: nonhuman primates illuminate human speech processing. *Nat. Neurosci.* **12:** 718–724.
19. Hickok, G. & D. Poeppel. 2004. Dorsal and ventral streams: a framework for understanding aspects of the functional anatomy of language. *Cognition* **92:** 67–99.
20. Zacks, J.M. 2008. Neuroimaging studies of mental rotation: a meta-analysis and review. *J. Cogn. Neurosci.* **20:** 1–19.
21. Culham, J.C. & K.F. Valyear. 2006. Human parietal cortex in action. *Curr. Opin. Neurobiol.* **16:** 205–212.
22. Champod, A.S. & M. Petrides. 2007. Dissociable roles of the posterior parietal and the prefrontal cortex in manipulation and monitoring processes. *Proc. Nat. Acad. Sci.* **104:** 14837–14842.
23. Rudner, M., J. Rönnberg & K. Hugdahl. 2005. Reversing spoken items: mind twisting not tongue twisting. *Brain Lang.* **92:** 78–90.
24. Lewis, J.W. & D.C. Van Essen. 2000. Corticocortical connections of visual, sensorimotor, and multimodal processing areas in the parietal lobe of the macaque monkey. *J. Comp. Neurol.* **428:** 112–137.
25. Frey, S. *et al.* 2008. Dissociating the human language pathways with high angular resolution diffusion fiber tractography. *J. Neurosci.* **28:** 11435–11444.
26. Zatorre, R.J., A.R. Halpern & M. Bouffard. 2010. Mental reversal of imagined melodies: a role for the posterior parietal cortex. *J. Cogn. Neurosci.* **22:** 775–789.
27. Halpern, A.R. 1988. Mental scanning in auditory imagery for tunes. *J. Exp. Psychol: Learn. Memory Cognit.* **14:** 434–443.
28. Kosslyn, S., W.L. Thompson & G. Ganis. 2006. *The Case for Mental Imagery*. Oxford University Press. Oxford.
29. Leaver, A.M. *et al.* 2009. Brain activation during anticipation of sound sequences. *J. Neurosci.* **29:** 2477–2485.
30. Foster, N.E.V. & R.J. Zatorre. 2009. A role for the intraparietal sulcus in transforming musical pitch information. *Cereb. Cortex* **20:** 1350–1359.
31. Grefkes, C. & G.R. Fink. 2005. The functional organization of the intraparietal sulcus in humans and monkeys. *J. Anatomy* **207:** 3–17.
32. Husain, M. & P. Nachev. 2007. Space and the parietal cortex. *Trends Cogn. Sci.* **11:** 30–36.

Ann. N.Y. Acad. Sci. ISSN 0077-8923

ANNALS OF THE NEW YORK ACADEMY OF SCIENCES
Issue: *The Neurosciences and Music IV: Learning and Memory*

Working memory for speech and music

Katrin Schulze[1] and Stefan Koelsch[2]

[1]Developmental Cognitive Neuroscience Unit, UCL Institute of Child Health, London, United Kingdom. [2]Cluster "Languages of Emotion," Freie Universität Berlin, Berlin, Germany

Address for correspondence: Katrin Schulze, Developmental Cognitive Neuroscience Unit, UCL Institute of Child Health, 30 Guilford Street, London WC1N 1EH, UK. kschulze@ich.ucl.ac.uk

The present paper reviews behavioral and neuroimaging findings on similarities and differences between verbal and tonal working memory (WM), the influence of musical training, and the effect of strategy use on WM for tones. Whereas several studies demonstrate an overlap of core structures (Broca's area, premotor cortex, inferior parietal lobule), preliminary findings are discussed that imply, if confirmed, the existence of a tonal and a phonological loop in musicians. This conclusion is based on the findings of partly differing neural networks underlying verbal and tonal WM in musicians, suggesting that functional plasticity has been induced by musical training. We further propose a strong link between production and auditory WM: data indicate that both verbal and tonal auditory WM are based on the knowledge of how to produce the to-be-remembered sounds and, therefore, that sensorimotor representations are involved in the temporary maintenance of auditory information in WM.

Keywords: auditory working memory; verbal; tonal; musical expertise; strategy

Introduction

Working memory (WM) describes a brain system responsible for temporary storage and simultaneous manipulation of information,[1–3] which is critical for higher cognitive functions such as planning, problem solving, and reasoning, but also for understanding or appreciating speech and music.

The present paper is based on the WM model developed by Baddeley and Hitch,[3] which assumes an attentional control system (the "central executive") that operates in conjunction with two subsidiary systems: the visuospatial sketchpad and the phonological loop. The visuospatial sketchpad processes and stores visual and spatial information; the phonological loop represents verbal short-term memory (STM). The mutual interaction between long-term memory (LTM) and WM was recognized by the introduction of a fourth component to the model: the episodic buffer. This limited capacity system is assumed to bind information from the subsidiary systems, store information in a multimodal code, and enable the interaction between WM and LTM.[4]

We acknowledge that there are many other STM or WM models (for an overview, see Refs. 5, 6–11). Our paper, however, is theoretically embedded in the highly influential Baddeley and Hitch[3] WM model because studies exploring the question whether WM for music and language differs have been primarily based on this WM model.[3,12–18] Furthermore, although parts of this model are still discussed, no other model of verbal WM is as well investigated, developed, and accepted as the phonological loop.[19]

The terms STM and WM have not been used consistently in the literature.[20] One possibility is to use the term STM to refer to the simple temporary storage of information, and WM to refer to the maintenance and manipulation of information.[5,20] Often it is not well defined whether a task needs additional processing and/or manipulation;[a] therefore, no distinction has been made between STM and WM in the present paper, but we will refer only to WM instead. In addition, we will use the term

[a]For example, if participants in an auditory WM experiment have to decide whether one test stimulus was presented previously during a sequence consisting of several stimuli, it is not known whether or how much manipulation in addition to simply storing and rehearsing the auditory sequence is required.

doi: 10.1111/j.1749-6632.2012.06447.x

auditory WM to describe WM processes for verbal or nonverbal stimuli that were presented auditorily. Finally, it should be noted that whereas verbal WM experiments used both recall and recognition tasks, studies exploring WM for tone material were relying on recognition tasks (but see Ref. 12 for a recall task for musical stimuli).

Behavioral studies

WM for verbal information

Baddeley and Hitch[3] suggested a multicomponent WM model in which verbal information is processed by a phonological loop. This component can be further subdivided into a passive storage component (phonological storage) and an active rehearsal mechanism (articulatory rehearsal process). It is assumed that the passive storage component can store auditory or speech-based information for a few seconds.[2,8] If the information has to be maintained for longer, the articulatory rehearsal process can rehearse the verbal information, a process comparable to subvocal speech. The articulatory rehearsal can be interrupted by articulatory suppression,[1,2] which usually involves overt articulation (for example, Refs. 21 and 22–26), preventing the articulatory rehearsal mechanism to subvocally rehearse verbal material and thus reducing the verbal WM function. The word length effect, on the other hand, refers to the phenomenon that participants show a greater memory span[22] and a superior recognition accuracy[21] for short words than for long words. The effect of articulatory suppression and of word length suggested that during the articulatory rehearsal process, verbal material is maintained by using a phonological code, comparable to subvocal speech (for an overview, see Refs. 1, 2, and 5). The phonological similarity effect describes participants' inferior performance to recall[24] or recognize[27] phonologically similar verbal material compared to phonologically dissimilar verbal material.

WM for tonal information

The Baddeley and Hitch WM model[1,3] does not specify whether the phonological loop also serves the processing of nonphonological information, or whether different subsystems (a "tonal loop"[13] or a "musical loop"[14]) exist in addition to the phonological loop. As described previously, verbal information can be maintained in verbal WM by internal articulatory rehearsal. But does internal rehearsal also work for pitch information? Studies that investigated this question yielded conflicting results, indicating either a behavioral improvement of WM performance by internal rehearsal[13,16,18] or only a small improvement or no improvement at all.[28–30] However, these studies differed with regard to the degree to which participants could imitate and repeat the experimental stimuli. Experiments that find only a small or no effect of internal rehearsal used tones, whose frequencies did not correspond to the frequencies of the Western chromatic scale;[29,30] tones with a frequency difference smaller than the smallest difference, namely one semitone, used in songs of Western tonal music;[29,30] and/or chords consisting of several simultaneously played sine-wave tones.[28] In comparison, if studies used tones whose frequencies did correspond to the frequencies of the Western chromatic scale,[13,18] or if the frequency differences between the used tones were not smaller than one semitone,[13,16,18] then the observed results support the hypothesis of a rehearsal mechanism underlying WM for tones.

Comparison between verbal and tonal WM

WM for tones is fundamental for music perception and production. However, the majority of research on auditory WM has been carried out using verbal material, namely phonemes, syllables, and words. Research on WM for pitch or the "tonal loop"[13] is rather scant and does not yet provide a consistent picture. Deutsch[31] observed that presenting intervening tones interfered more strongly with a WM task for tones than presenting phonemes, and this was interpreted as evidence for a specialized tonal WM system. Further, Salame and Baddeley[32] showed that instrumental music interfered less with verbal WM compared to vocal music, supporting the theory of two independent WM systems for verbal and tonal stimuli. On the other hand, Semal *et al.*[33] criticized that the frequency relations between the standard tones and the intervening verbal material were not controlled in Deutsch's[31] study, which might explain the missing interference between the standard tones and the intervening verbal material. Pitch similarity of the intervening stimuli (words or tones) had a greater effect on the performance rate than the modality (verbal or tonal) of the intervening stimuli,[33] indicating that pitches for both verbal and tonal stimuli are processed in the same WM system. Along these lines, Chan *et al.*[34] reported that musical training increases verbal WM performance, indicating that rather overlapping mechanisms are underlying verbal and tonal WM. Further support for similarities between auditory verbal and tonal WM comes from an experiment using a suppression paradigm in musically experienced participants.[18] Musical (singing "la") and verbal (producing the words "the") suppression decreased recognition accuracy for both digit and tone sequences, indicating that musical or verbal suppression does not selectively impair verbal or tonal WM. In a recent study, Williamson *et al.*[12] compared WM recall for tones and letters. Their results suggest that well-known characteristics of verbal WM could also be observed for the tonal modality: WM for tonal information showed limited capacity, and nonmusicians, but not musicians, showed a decreased performance if the tone sequences consisted of more proximal pitches compared to more distal pitches, an effect resembling the phonological similarity effect in the verbal WM domain.

Neuroanatomical correlates of WM

WM for verbal information

As seen for the behavioral experiments, most studies investigating the functional neuroarchitecture of auditory WM used verbal material. Neuroimaging studies indicate that mainly Broca's area and premotor areas (as well as pre-SMA and SMA) play a crucial role during the internal rehearsal of verbal material.[35–39] In addition, evidence suggests that both the insular cortex[38,40,41] and the cerebellum[39,42,43] are involved in internal rehearsal of verbal information. Whereas the involvement of Broca's area and the premotor cortex during the internal rehearsal has been supported by numerous studies, the research results regarding the phonological store have been much less conclusive. The phonological store has been suggested to rely on parietal areas, particularly the inferior parietal lobule (IPL; Refs. 35, 37, 38, and 42–46), but also on the superior parietal lobule (SPL, Ref. 35). However, the localization of the phonological store in the parietal lobe is very controversial for several reasons. First, neural activity in this area might also reflect increased engagement of attentional resources.[47,48] Second, the reported coordinates for the phonological store differ greatly between studies,[19] and finally, the IPL is not activated during passive listening,[19,49] which should be the case if this structure is involved in automatically storing incoming auditory information as suggested by the WM model.[1,3,19]

Alternatively, area Spt (Sylvian–parietal–temporal, left posterior planum temporale) has been suggested to be involved in the temporary storage of verbal information during WM tasks.[19] This is because activation in the left Spt has been observed to be enhanced during the delay period of a WM task[17,50] and to be independent of the modality of the presented stimuli (auditory or visual; Ref. 50). On the basis of these findings and because area Spt also supports speech processing, it has been proposed that area Spt acts as an auditory–motor interface for WM.[17,19,50] This proposition fits nicely with the hypothesis of a dual-stream model of speech processing.[51–54] In this model, a ventral stream supports speech comprehension via a lexical access while a left dominant dorsal stream, which comprises also area Spt, enables sensory–motor integration, i.e., the mapping of the perceived speech signals onto articulatory representations.

WM for tonal information

In comparison to the underlying networks of the phonological loop, far fewer neuroimaging studies have investigated WM for tones. In participants who were not selected for musical expertise, Gaab *et al.*[55] showed activation of the supramarginal gyrus (SMG), the SPL, the planum temporale, premotor regions encroaching on Broca's area, and cerebellar regions during a pitch memory task. This network is surprisingly similar to the network subserving the phonological loop described above. A similar network, including the inferior frontal and insular cortex, the planum temporale, and the SMG, had previously been reported to be activated during the active retention of pitch.[56]

Comparison between verbal and tonal WM

To our knowledge, only three neuroimaging studies have directly compared the neural correlates underlying auditory WM for tonal and verbal material.[15–17] Hickok *et al.*[17] compared the neural correlates underlying verbal and tonal WM in nonmusicians using functional magnetic resonance imaging (fMRI). Melodic sequences (tonal condition) and sentences consisting of pseudowords (verbal condition) were presented auditorily, and subsequently participants rehearsed internally the verbal and tonal stimuli. Results showed that internal rehearsal of both verbal and tonal material activated the area Spt, Broca's area, and left premotor regions.[17] Very similar activations were observed in the study by Koelsch *et al.*,[16] in which similarities between the neural components underlying WM for verbal (syllables) and tonal (pitch) material were investigated using a recognition task. During the verbal rehearsal, a neural network comprising the premotor cortex, the anterior insula, the SMG/intraparietal sulcus (IPS), the planum temporale, the inferior frontal gyrus, pre-SMA, and the cerebellum was activated, mainly in the left hemisphere. Importantly, the neural network activated during the tonal rehearsal was virtually identical to that observed during verbal rehearsal. In an fMRI study by Schulze *et al.*,[15] similarities and differences of the functional networks underlying the internal rehearsal of verbal and tonal WM were investigated using a recognition task. Similar to our previous study,[16] both verbal and tonal WM-activated areas typically reported in previous experiments on either verbal[1,35,37,38] or tonal WM[17,55,56] in nonmusicians. The fact that both verbal and tonal WM activated these core structures, namely Broca's area, the left premotor cortex, (pre-)SMA, left insular cortex, and left IPL, corroborates previous results showing considerable overlap of the networks underlying verbal and tonal WM.[16,17] Importantly, only in nonmusicians, all structures involved in tonal WM were also involved in verbal WM; in contrast, verbal but not tonal WM relied on additional structures that have previously been implicated in verbal WM.[1,47,57] This difference in activation of WM resources in nonmusicians is reflected in the behavioral data that showed better performance during verbal compared to tonal WM.

In summary, consistent across studies,[15–17] data obtained from nonmusicians indicate a considerable overlap of neural resources underlying WM for both verbal and tonal information. This common network includes a mainly left-lateralized fronto-parietal network (premotor cortex, Broca's area, and in two of the three studies, the IPL,[15,16] the cerebellum,[15,16] and the planum temporale/area Spt[16,17]).[b]

Comparison between nonmusicians and musicians

Because speech is a fundamental human skill typically acquired during early childhood, nonmusicians can be considered to be trained in processing and producing speech, but they possess less expertise in the music domain. Thus, for a more balanced comparison of verbal and tonal WM, Schulze *et al.*[15] investigated in addition highly trained musicians. Interestingly, many of the structures, namely Broca's area, left premotor cortex, left insular cortex, (pre-)SMA, cingulate gyrus, and left IPL, which were activated more strongly in nonmusicians during verbal compared to tonal WM, were activated more strongly in musicians compared to nonmusicians during tonal WM. That is, the functional network on which nonmusicians relied for verbal WM was also used by musicians for tonal WM. In contrast to nonmusicians, musicians recruited a number of structures exclusively for either verbal or tonal WM. For tonal information these areas were the left cuneus, the right globus pallidus, and the right caudate nucleus, as well as the left cerebellum, and for verbal information the right insular cortex.

In addition, activation differences between verbal and tonal WM were observed in a number of structures in musicians, providing a first indication of the existence of two WM systems, namely a phonological loop maintaining phonological information and a tonal loop dedicated to the maintenance of tonal information. Both systems activated the same core structures of WM and therefore showed considerable overlap, but both systems also differed in that they relied on different neural subcomponents. Importantly, the structural differences between the verbal and tonal loop in musicians could not be explained simply by performance differences between the tonal and the verbal tasks, because several brain structures were recruited selectively for verbal or tonal WM (see discussion in Refs. 15 and 58). One hypothesis, based on the assumption of functional plasticity induced by music, is that musical expertise leads to a network comprising more structures underlying tonal WM, therefore showing a considerable overlap with the functional network

[b]The fMRI studies that detected activation in Spt used continuous scanning,[16,17,50] whereas Schulze *et al.*[15] used a sparse temporal sampling scanning technique that might have not been sensitive enough to capture Spt activation.

subserving verbal WM, but also exhibiting substantial differences.

Sensorimotor codes

To account for the similarities between verbal WM with speech production (and speech perception), the underlying representations of verbal WM have been termed sensorimotor codes.[59] The following results indicate that, indeed, internal verbal rehearsal shares some characteristics with speech production. The word-length effect and the articulatory-suppression effect suggest that verbal WM is comparable to subvocal speech (for an overview see Ref. 1). Furthermore, the phonological loop is conceived as a memory system involving internal articulatory speech actions implemented by motor-related areas such as Broca's area, premotor and insular cortex,[40] (pre-)SMA,[37–39] and the cerebellum.[39,42]

For tonal WM, findings indicate that internal rehearsal mainly improves WM performance for tones if participants are able to imitate and repeat the auditory stimuli,[13,16,18] in contrast to studies in which this might have been more difficult or impossible.[28–30]

Remarkably, the superior performance of nonmusicians during verbal compared to tonal WM, and the better performance of musicians compared to nonmusicians during tonal WM, were primarily associated with activation differences in structures known to be involved in the control, programming, and planning, in addition to execution of actions, such as Broca's area, premotor cortex, (pre-)SMA, left insular cortex, IPS, IPL, and the cerebellum.[15]

The behavioral and neurophysiological differences between WM for verbal and tonal information in nonmusicians were interpreted as a consequence of a more extensive production and rehearsal of verbal information in everyday life. Musicians, on the other hand, might have more elaborate sensorimotor codes underlying the internal rehearsal of tones compared to nonmusicians. This indicates functional plasticity induced by musical training and, more specifically, might be a consequence of musicians' long-term learning of associations between pitch information and motor actions.[60–64]

Previous research has established that Broca's area and the premotor cortex are involved in the planning and controlling of vocal and hand actions[65,66] and in auditory-motor mapping;[67] movement representations for both speech and music are supported by the anterior insula;[51,68,69] and these structures serve voluntary motor control and contribute to the programming, initiation, and execution of movements.[70–73]

Therefore, sensorimotor processes may assist with the representation and manipulation of information, and sensorimotor coding could play an important role for WM processes. This points toward a basic mechanism of auditory WM: to translate the sensory auditory event into a rehearseable sensorimotor code. Action-related sensorimotor codes are assumed to be based on motor knowledge—how to produce the auditory stimulus (e.g., syllable, tone)—and are thought to be involved in the rehearsal and representation of information in auditory, verbal, and tonal WM.[15–17,74]

The dual-stream model of speech perception[51–53] assumes that one of the functions of the dorsal path of the auditory system is sensory–motor integration, i.e., mapping the perceived speech signals onto articulatory representations. The left-dominant dorsal stream for sensory–motor integration involves structures at the parietal–temporal junction and projects to the premotor cortex and Broca's area,[52] structures that were also observed for verbal and tonal WM.[15–17] Speech production requires motor speech representations but also representations of sensory speech targets that are important for comparing between predicted and actual consequences of motor speech acts.[54] Furthermore, in a recent paper it has been suggested that sensorimotor integration also plays a role during singing.[75] In conclusion, we propose that internal rehearsal associated with auditory WM relies on sensorimotor representations, which might also be crucial for singing and speaking.

WM and strategy

The amount of information that can be maintained by the WM system is limited.[1,22,76] However, the use of a strategy, for example chunking the to-be-remembered information,[77,78] can improve WM performance. Chunking refers to a process in which elements of information are organized into one unit or chunk,[76] with stronger associations between items within one chunk than between chunks.[79] This process is assumed to be supported by the episodic buffer enabling features from different sources to be bound into chunks and new information to be integrated into an existing context stored in LTM.[5] Previously, the neural correlates underlying such strategy-based memorization were explored using visual–spatial or verbal material,[80–83] but it was mainly unknown whether a similar network is also involved during strategy-based WM for tones.

By using structured (all tones belonged to one tonality) and unstructured (atonal) five-tone sequences, Schulze *et al.*[84] investigated whether musical structure influences encoding and rehearsal in a nonverbal auditory WM task and how this is reflected in the brain of nonmusicians and musicians. Musicians, but not nonmusicians, showed better performance for structured than for unstructured sequences, indicating that musicians' knowledge about musical regularities[85–88] helped them to keep the structured sequences in WM. The data[84] in musicians showed that a lateral (pre-)frontal–parietal network, including the right inferior precentral sulcus, the premotor cortex, and the left IPS, was more strongly involved during WM for structured compared to unstructured auditory sequences. Previous research reported the involvement of a similar network during strategy-based WM processing for visual and auditory–verbal stimuli,[81–83] therefore pointing toward a modality-independent (pre-)frontal–parietal network subserving strategy-based WM.

In a behavioral study by Schulze *et al.*,[89] participants had to indicate whether two sequences were the same or different, the facilitating effect of tonality (structure) on WM performance for tones could be confirmed, and was also observed for nonmusicians. Tonality, however, only improved WM performance for tones during maintenance (forward task), but not during manipulation (backward task).

Summary and conclusion

This paper reviewed research results indicating differences and similarities between verbal and tonal WM related to the underlying mechanisms and neural correlates. Whereas the core structures, namely Broca's area, premotor cortex, and IPL, show a considerable overlap, these preliminary findings in musicians suggest that there are also different subcomponents activated either during verbal or tonal WM. These results indicate, if confirmed, the existence of both a tonal and a phonological loop in musicians. We further propose a strong link between production and auditory WM. Both verbal and tonal auditory WM appear to be based on the knowledge of how to produce the to-be-remembered sounds, and we therefore suggest that sensorimotor representations are involved in the temporary maintenance of auditory information in WM.

Conflicts of interest

The authors declare no conflicts of interest.

References

1. Baddeley, A.D. 2003. Working memory: looking back and looking forward. *Nat. Rev. Neurosci.* **4:** 829–839.
2. Baddeley, A.D. 1992. Working memory. *Science* **255:** 556–559.
3. Baddeley, A.D. & G.J. Hitch. 1974. Working memory. In *Recent Advances in Learning and Motivation.* Vol VIII. G.A. Bower, Ed.: 47–89. Academic Press. New York.
4. Baddeley, A.D. 2000. The episodic buffer: a new component of working memory? *Trends. Cogn. Sci.* **4:** 417–423.
5. Baddeley, A.D. 2011. Working memory: theories, models, and controversies. *Annu. Rev. Psychol.* doi:10.1146/annurev-psych-120710-100422.
6. Cowan, N. 1988. Evolving conceptions of memory storage, selective attention, and their mutual constraints within the human information-processing system. *Psychol. Bull.* **104:** 163–191.
7. Cowan, N. (1999) An embedded-processes model of working memory. In *Models of Working Memory*. A. Miyake & P. Shah, Eds.: 62–101. University Press. Cambridge.
8. Baddeley, A.D. 2010. Working memory. *Curr. Biol.* **20:** R136–R140.
9. Ericsson, K.A. & W. Kintsch. 1995. Long-term working-memory. *Psychol. Rev.* **102:** 211–245.
10. Jones, D.M. 1993. Objects, streams and threads of auditory attention. In *Attention: Selection, Awareness and Control.* A.D. Baddeley & L. Weiskrantz, Eds.: 87–104. Clarendon. Oxford, U.K.
11. Nairne, J.S. 1990. A feature model of immediate memory. *Mem. Cogn.* **18:** 251–269.
12. Williamson, V.J., A.D. Baddeley & G.J. Hitch. 2010. Musicians' and nonmusicians' short-term memory for verbal and musical sequences: comparing phonological similarity and pitch proximity. *Mem. Cogn.* **38:** 163–175.
13. Pechmann, T. & G. Mohr. 1992. Interference in memory for tonal pitch: implications for a working-memory model. *Mem. Cogn.* **20:** 314–320.
14. Berz, W.L. 1995. Working memory in music: a theoretical model. *Music Percept.* **12:** 353–364.
15. Schulze, K., S. Zysset, K. Mueller, *et al.* 2011. Neuroarchitecture of verbal and tonal working memory in nonmusicians and musicians. *Hum. Brain Mapp.* **32:** 771–783.
16. Koelsch, S. *et al.* 2009. Functional architecture of verbal and tonal working memory: an FMRI study. *Hum. Brain Mapp.* **30:** 859–873.
17. Hickok, G., B. Buchsbaum, C. Humphries & T. Muftuler. 2003. Auditory-motor interaction revealed by fMRI: speech,

music, and working memory in area Spt. *J. Cogn. Neurosci.* **15:** 673–682.
18. Schendel, Z.A. & C. Palmer. 2007. Suppression effects on musical and verbal memory. *Mem. Cogn.* **35:** 640–650.
19. Buchsbaum, B.R. & M. D'Esposito. 2008. The search for the phonological store: from loop to convolution. *J. Cogn. Neurosci.* **20:** 762–778.
20. Cowan, N. 2008. What are the differences between long-term, short-term, and working memory? *Prog. Brain. Res.* **169:** 323–338.
21. Baddeley, A.D., D. Chincotta, L. Stafford & D. Turk. 2002. Is the word length effect in STM entirely attributable to output delay? Evidence from serial recognition. *Q. J. Exp. Psychol. A* **55:** 353–369.
22. Baddeley, A.D., N. Thomson & L. Buchanan. 1975. Word length and the structure of short-term memory. *J. Verbal Learn. Verbal Behav.* **14:** 575–589.
23. Neath, I., A.M. Surprenant & D.C. LeCompte. 1998. Irrelevant speech eliminates the word length effect. *Mem. Cogn.* **26:** 343–354.
24. Surprenant, A.M., I. Neath & D.C. LeCompte. 1999. Irrelevant speech, phonological similarity, and presentation modality. *Memory* **7:** 405–420.
25. Henson, R.N.A., T. Hartley, N. Burgess, *et al.* 2003. Selective interference with verbal short-term memory for serial order information: a new paradigm and tests of a timing-signal hypothesis. *Q. J. Exp. Psychol. A* **56:** 1307–1334.
26. Larsen, J.D. & A.D. Baddeley. 2003. Disruption of verbal STM by irrelevant speech, articulatory suppression, and manual tapping: do they have a common source? *Q. J. Exp. Psychol. A* **56:** 1249–1268.
27. Nimmo, L.M. & S. Roodenrys. 2005. The phonological similarity effect in serial recognition. *Memory* **13:** 773–784.
28. Demany, L., G. Montandon & C. Semal. 2004. Pitch perception and retention: two cumulative benefits of selective attention. *Percept. Psychophys.* **66:** 609–617.
29. Keller, T.A., N. Cowan & J.S. Saults. 1995. Can auditory memory for tone pitch be rehearsed? *J. Exp. Psychol. Learn. Mem. Cogn.* **21:** 635–645.
30. Kaernbach, C. & K. Schlemmer. 2008. The decay of pitch memory during rehearsal. *J. Acoust. Soc. Am.* **123:** 1846–1849.
31. Deutsch, D. 1970. Tones and numbers: specificity of interference in immediate memory. *Science* **168:** 1604–1605.
32. Salame, P. & A.D. Baddeley. 1989. Effects of background music on phonological short-term memory. *Q. J. Exp. Psychol.* **41:** 107–122.
33. Semal, C., L. Demany, K. Ueda & P.A. Halle. 1996. Speech versus nonspeech in pitch memory. *J. Acoust. Soc. Am.* **100:** 1132–1140.
34. Chan, A.S., Y.C. Ho & M.C. Cheung. 1998. Music training improves verbal memory. *Nature* **396:** 128.
35. Awh, E. *et al.* 1996. Dissociation of storage and rehearsal in verbal working memory: evidence from positron emission tomography. *Psychol. Sci.* **7:** 25–31.
36. Fiez, J.A. *et al.* 1996. A positron emission tomography study of the short-term maintenance of verbal information. *J. Neurosci.* **16:** 808–822.
37. Gruber, O. & D.Y. von Cramon. 2003. The functional neuroanatomy of human working memory revisited. Evidence from 3-T fMRI studies using classical domain-specific interference tasks. *NeuroImage* **19:** 797–809.
38. Paulesu, E., C.D. Frith & R.S. Frackowiak. 1993. The neural correlates of the verbal component of working memory. *Nature* **362:** 342–345.
39. Ravizza, S.M., M.R. Delgado, J.M. Chein, *et al.* 2004. Functional dissociations within the inferior parietal cortex in verbal working memory. *NeuroImage* **22:** 562–573.
40. Bamiou, D.E., F.E. Musiek & L.M. Luxon. 2003. The insula (Island of Reil) and its role in auditory processing. Literature review. *Brain. Res. Rev.* **42:** 143–154.
41. Chein, J.M. & J.A. Fiez. 2001. Dissociation of verbal working memory system components using a delayed serial recall task. *Cereb. Cortex* **11:** 1003–1014.
42. Chen, S.H. & J.E. Desmond. 2005. Cerebrocerebellar networks during articulatory rehearsal and verbal working memory tasks. *NeuroImage* **24:** 332–338.
43. Kirschen, M.P., S.H. Chen, P. Schraedley-Desmond & J.E. Desmond. 2005. Load- and practice-dependent increases in cerebro-cerebellar activation in verbal working memory: an fMRI study. *NeuroImage* **24:** 462–472.
44. Crottaz-Herbette, S., R.T. Anagnoson & V. Menon. 2004. Modality effects in verbal working memory: differential prefrontal and parietal responses to auditory and visual stimuli. *NeuroImage* **21:** 340–351.
45. Henson, R.N.A., N. Burgess & C.D. Frith. 2000. Recoding, storage, rehearsal and grouping in verbal short-term memory: an fMRI study. *Neuropsychologia* **38:** 426–440.
46. Jonides, J. *et al.* 1998. The role of parietal cortex in verbal working memory. *J. Neurosci.* **18:** 5026–5034.
47. Cabeza, R. & L. Nyberg. 2000. Imaging cognition II: an empirical review of 275 PET and fMRI studies. *J. Cogn. Neurosci.* **12:** 1–47.
48. Corbetta, M. & G.L. Shulman. 2002. Control of goal-directed and stimulus-driven attention in the brain. *Nat. Rev. Neurosci.* **3:** 201–215.
49. Becker, J.T., D.K. MacAndrew & J.A. Fiez. 1999. A comment on the functional localization of the phonological storage subsystem of working memory. *Brain Cogn.* **41:** 27–38.
50. Buchsbaum, B.R., R.K. Olsen, P. Koch & K.F. Berman. 2005. Human dorsal and ventral auditory streams subserve rehearsal-based and echoic processes during verbal working memory. *Neuron* **48:** 687–697.
51. Hickok, G. & D. Poeppel. 2007. The cortical organization of speech processing. *Nat. Rev. Neurosci.* **8:** 393–402.
52. Hickok, G. 2009. The functional neuroanatomy of language. *Phys. Life Rev.* **6:** 121–143.
53. Rauschecker, J.P. & S.K. Scott. 2009. Maps and streams in the auditory cortex: nonhuman primates illuminate human speech processing. *Nat. Neurosci.* **12:** 718–724.
54. Hickok, G., J. Houde & F. Rong. 2011. Sensorimotor integration in speech processing: computational basis and neural organization. *Neuron* **69:** 407–422.
55. Gaab, N., C. Gaser, T. Zaehle, *et al.* 2003. Functional anatomy of pitch memory–an fMRI study with sparse temporal sampling. *NeuroImage* **19:** 1417–1426.
56. Zatorre, R.J., A.C. Evans & E. Meyer. 1994. Neural mechanisms underlying melodic perception and memory for pitch. *J. Neurosci.* **14:** 1908–1919.

57. Petrides, M., B. Alivisatos, E. Meyer & A.C. Evans. 1993. Functional activation of the human frontal cortex during the performance of verbal working memory tasks. *Proc. Natl. Acad. Sci. USA* **90:** 878–882.
58. Wager, T.D. & E.E. Smith. 2003. Neuroimaging studies of working memory: a meta-analysis. *Cogn. Affect. Behav. Neurosci.* **3:** 255–274.
59. Wilson, M. 2001. The case for sensorimotor coding in working memory. *Psychon. Bull. Rev.* **8:** 44–57.
60. Drost, U.C., M. Rieger, M. Brass, *et al.* 2005. When hearing turns into playing: movement induction by auditory stimuli in pianists. *Q. J. Exp. Psychol. A* **58:** 1376–1389.
61. Drost, U.C., M. Rieger, M. Brass, *et al.* 2005. Action-effect coupling in pianists. *Psychol. Res.* **69:** 233–241.
62. Bangert, M. *et al.* 2006. Shared networks for auditory and motor processing in professional pianists: evidence from fMRI conjunction. *NeuroImage* **30:** 917–926.
63. D'Ausilio, A., E. Altenmüller, M. Olivetti Belardinelli & M. Lotze. 2006. Cross-modal plasticity of the motor cortex while listening to a rehearsed musical piece. *Eur. J. Neurosci.* **24:** 955–958.
64. Haslinger, B. *et al.* 2005. Transmodal sensorimotor networks during action observation in professional pianists. *J. Cogn. Neurosci.* **17:** 282–293.
65. Petrides, M., G. Cadoret & S. Mackey. 2005. Orofacial somatomotor responses in the macaque monkey homologue of Broca's area. *Nature* **435:** 1235–1238.
66. Watkins, K. & T. Paus. 2004. Modulation of motor excitability during speech perception: the role of Broca's area. *J. Cogn. Neurosci.* **16:** 978–987.
67. Lahav, A., E. Saltzman & G. Schlaug. 2007. Action representation of sound: audiomotor recognition network while listening to newly acquired actions. *J. Neurosci.* **27:** 308–314.
68. Mutschler, I. *et al.* 2007. A rapid sound-action association effect in human insular cortex. *PLoS One* **2:** e259.
69. Indefrey, P. & W.J. Levelt. 2004. The spatial and temporal signatures of word production components. *Cognition* **92:** 101–144.
70. Hoover, J.E. & P.L. Strick. 1999. The organization of cerebellar and basal ganglia outputs to primary motor cortex as revealed by retrograde transneuronal transport of herpes simplex virus type 1. *J. Neurosci.* **19:** 1446–1463.
71. Leblois, A., T. Boraud, W. Meissner, *et al.* 2006. Competition between feedback loops underlies normal and pathological dynamics in the basal ganglia. *J. Neurosci.* **26:** 3567–3583.
72. Middleton, F.A. & P.L. Strick. 2000. Basal ganglia output and cognition: evidence from anatomical, behavioral, and clinical studies. *Brain Cogn.* **42:** 183–200.
73. Parent, A. & L.N. Hazrati. 1995. Functional anatomy of the basal ganglia. I. The cortico-basal ganglia-thalamo-cortical loop. *Brain. Res. Rev.* **20:** 91–127.
74. Jacquemot, C. & S.K. Scott. 2006. What is the relationship between phonological short-term memory and speech processing? *Trends Cogn. Sci.* **10:** 480–486.
75. Dalla Bella, S.D., M. Berkowska & J. Sowinski. 2011. Disorders of pitch production in tone deafness. *Front. Psychol.* **2:** 164.
76. Miller, G.A. 1956. The magical number seven plus or minus two: some limits on our capacity for processing information. *Psychol. Rev.* **63:** 81–97.
77. Ericsson, K.A., W.G. Chase & S. Faloon. 1980. Acquisition of a memory skill. *Science* **208:** 1181–1182.
78. Gobet, F. 2005. Chunking models of expertise: implications for education. *Appl. Cogn. Psychol.* **19:** 183–204.
79. Gobet, F. *et al.* 2001. Chunking mechanisms in human learning. *Trends. Cogn. Sci.* **5:** 236–243.
80. Savage, C.R. *et al.* 2001. Prefrontal regions supporting spontaneous and directed application of verbal learning strategies: evidence from PET. *Brain* **124:** 219–231.
81. Bor, D., N. Cumming, C.E. Scott & A.M. Owen. 2004. Prefrontal cortical involvement in verbal encoding strategies. *Eur. J. Neurosci.* **19:** 3365–3370.
82. Bor, D., J. Duncan, R.J. Wiseman & A.M. Owen. 2003. Encoding strategies dissociate prefrontal activity from working memory demand. *Neuron* **37:** 361–367.
83. Bor, D. & A.M. Owen. 2007. A common prefrontal-parietal network for mnemonic and mathematical recoding strategies within working memory. *Cereb. Cortex* **17:** 778–786.
84. Schulze, K., K. Mueller & S. Koelsch. 2011. Neural correlates of strategy use during auditory working memory in musicians and non-musicians. *Eur. J. Neurosci.* **33:** 189–196.
85. Krumhansl, C.L. 1979. The psychological representation of musical pitch in a tonal context. *Cogn. Psychol.* **11:** 346–374.
86. Krumhansl, C.L. & R.N. Shepard. 1979. Quantification of the hierarchy of tonal functions within a diatonic context. (Translated from Eng) *J. Exp. Psychol. Hum. Percept. Perform.* **5:** 579–594.
87. Koelsch, S., E. Schroger & M. Tervaniemi. 1999. Superior pre-attentive auditory processing in musicians. *Neuroreport* **10:** 1309–1313.
88. Koelsch, S., E. Schroger & T.C. Gunter. 2002. Music matters: preattentive musicality of the human brain. *Psychophysiology* **39:** 38–48.
89. Schulze, K., W.J. Dowling & B. Tillmann. 2012. Working memory for tonal and atonal sequences during a forward and a backward recognition task. *Music Percept* **29:** 225–268.

Ann. N.Y. Acad. Sci. ISSN 0077-8923

ANNALS OF THE NEW YORK ACADEMY OF SCIENCES
Issue: *The Neurosciences and Music IV: Learning and Memory*

When right is all that is left: plasticity of right-hemisphere tracts in a young aphasic patient

Lauryn Zipse,[1,2] Andrea Norton,[2] Sarah Marchina,[2] and Gottfried Schlaug[2]
[1]Department of Communication Sciences and Disorders, MGH Institute of Health Professions, Boston, Massachusetts. [2]Department of Neurology, Music, Neuroimaging, and Stroke Recovery Laboratory, Beth Israel Deaconess Medical Center, Boston, Massachusetts

Address for correspondence: Gottfried Schlaug, Department of Neurology, Music, Neuroimaging, and Stroke Recovery Laboratory, Beth Israel Deaconess Medical Center, 330 Brookline Avenue Palmer 127, Boston, MA 02215. gschlaug@bidmc.harvard.edu

Using an adapted version of Melodic Intonation Therapy (MIT), we treated an adolescent girl with a very large left-hemisphere lesion and severe nonfluent aphasia secondary to an ischemic stroke. At the time of her initial assessment 15 months after her stroke, she had reached a plateau in her recovery despite intense and long-term traditional speech-language therapy (approximately five times per week for more than one year). Following an intensive course of treatment with our adapted form of MIT, her performance improved on both trained and untrained phrases, as well as on speech and language tasks. These behavioral improvements were accompanied by functional MRI changes in the right frontal lobe as well as by an increased volume of white matter pathways in the right hemisphere. No increase in white matter volume was seen in her healthy twin sister, who was scanned twice over the same time period. This case study not only provides further evidence for MIT's effectiveness, but also indicates that intensive treatment can induce functional and structural changes in a right-hemisphere fronto-temporal network.

Keywords: aphasia; melodic intonation therapy; brain plasticity; diffusion tensor imaging

Introduction

First described nearly 40 years ago as a treatment for nonfluent aphasia,[1,2] melodic intonation therapy (MIT) was based on the clinical observation that individuals with severe nonfluent aphasia can often sing phrases that they cannot speak.[3,4] MIT capitalizes on this preserved ability by using words or phrases intoned (sung) on two pitches to exaggerate the natural prosody of speech. Over the course of treatment, patients are guided through a hierarchical series of steps designed to increase fluency while decreasing dependence on therapist support.[5,6] The original developers of MIT suggested that the treatment worked by engaging language-capable right-hemisphere regions to compensate for the damaged left hemisphere.[1,2] To further encourage right-hemisphere activity and emphasize the production of each syllable, MIT also includes a motoric element, left hand-tapping, which may help to engage a right-hemispheric sensorimotor network shared by hand and articulatory actions.[7,8] Functional neuroimaging studies have sought to determine whether improvements seen with MIT are in fact associated with increased right-hemisphere activity. The findings have been mixed.[9,10] This inconsistency is most likely due to differences in patient characteristics (e.g., lesion size and location, aphasia diagnosis [Broca's vs. global], time post-onset, etc.) as well as variations in study design (e.g., post-treatment assessment only, cross-sectional studies of singing without actual treatment, longitudinal treatment studies with pre- and postassessments, etc.).

In a recently published case series, we used diffusion tensor imaging (DTI) to demonstrate an association between increases in a right-hemisphere fiber tract and improvements in verbal output following treatment with MIT. The case series included six adult patients with nonfluent aphasia secondary

doi: 10.1111/j.1749-6632.2012.06454.x

to left-hemisphere strokes. These patients showed increases in both volume and number of fibers of the right arcuate fasciculus (AF, a fiber tract that connects the superior temporal lobe to the inferior frontal gyrus and motor/premotor regions in a reciprocal way) that corresponded with speech output improvements after intense, long-term treatment.[11] In the current single case study, we aimed to determine whether functional and structural brain changes could also be observed in a young adolescent girl whose extensive left-hemisphere lesion included the middle and posterior cerebral artery territories and encompassed the classical Broca's and Wernicke's regions. This lesion left the patient, LF, with a nonfluent aphasia, and surprisingly, relatively unimpaired comprehension. In addition, we sought to determine whether plasticity processes would differ in this young patient compared to the previously described group of adult patients because it has been suggested in the literature that recovery processes may be influenced by developmental trajectory.[12] To explore the neural basis of speech and language recovery, we assessed LF's functional communication abilities at several time points before, during, and after treatment and related them to functional and structural imaging changes. We also tracked her improvement on production of trained and untrained phrases during treatment. LF has a healthy twin sister who served as a control case. Her identical twin underwent two DTI studies at time points that corresponded to the interval between LF's baseline and Post80 scans. LF was scanned more frequently and also had a long-term outcome measure.

Our goals were to: (1) test the efficacy of our adaptation of MIT for a patient who did not meet the traditional inclusion criteria for this treatment; and (2) to investigate which structural and functional changes might support MIT-induced improvement when no viable left-hemisphere areas remain to support recovery of speech function.

Methods

Patient

Prior to an ischemic stroke, LF was a healthy, right-handed 11-year-old girl with no significant events in her medical history. Her stroke resulted in a very large left-hemisphere lesion (Fig. 1A). The cause of LF's stroke remains unclear despite an

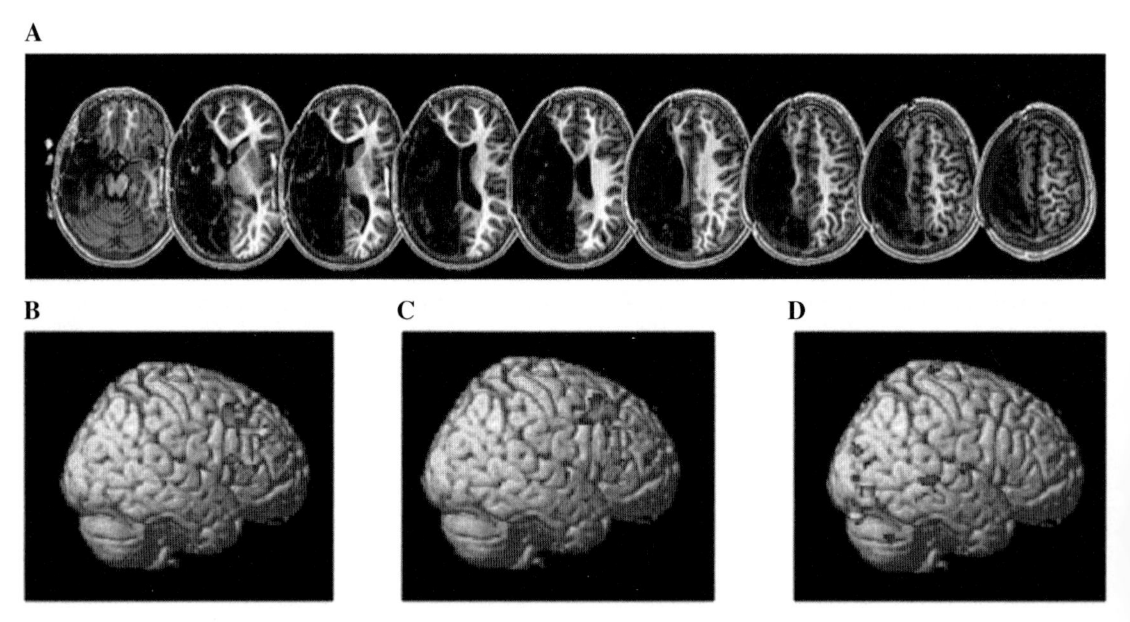

Figure 1. Structural MRI and functional MRI. (A) T1-weighted images of LF's brain, showing the large left-hemisphere lesion. (B) Post40 > Baseline, showing regions of greater activation after 40 treatment sessions with MIT compared to baseline. (C) Post40 > Post80, showing regions of greater activation after 40 treatment sessions compared to after 80 sessions. (D) Baseline > Post80, showing regions of greater activation at baseline compared to after 80 sessions. For functional images, yellow indicates regions of highest activation (thresholded at $P = < 0.001$, uncorrected). Areas of activation are superimposed on a spatially standardized normal brain. All contrasts compare speaking > silence between time points.

extensive work-up done in the acute phase post stroke. No dissection, vasculitis, or hypercoagulable state was detected. Furthermore, her cardiac work-up was negative. Thus, her stroke was classified as one of undetermined etiology. When we first evaluated LF at age 12, 15 months poststroke, she had already received 15 months of intensive traditional speech therapy (at least five times per week). Her nonverbal IQ, as measured with Raven's Coloured Progressive Matrices,[13] was mildly impaired (28/36; norms are only available through age 11). Qualitatively, her speech was nonfluent and agrammatic, with notable word-finding difficulties. Word-finding difficulties were documented in her performance on the Boston Naming Test,[14] in which she scored 17/60. Her repetition was moderately impaired (8/10 on the Word Repetition and 1/2 on the Sentence Repetition subtests of the Boston Diagnostic Aphasia Examination, BDAE;[15] 40th percentile), and her auditory comprehension was relatively preserved (34/37 on the BDAE Word Comprehension, 15/15 on Following Commands, and 9/12 on Complex Ideational Material). On the Apraxia Battery for Adults–2nd edition (*ABA-2*),[16] LF's levels of impairment varied across subtests from "None" to "Severe," but severe deficits were only seen on the repetition of long words and sentences. Somewhat surprisingly considering the size of her lesion, there was no evidence of nonspeech oral apraxia or limb apraxia. Overall, her profile was consistent with a diagnosis of Broca's aphasia.

Experimental design

A multiple baseline design was used to evaluate treatment efficacy in terms of improvement on production of trained versus untrained target phrases. Functional communication abilities were also assessed over time using a picture description task and a semistructured conversational interview. Performance on these functional communication tasks was documented prior to the initiation of treatment (Baselines 1–3), midway through treatment (after 40 treatment sessions; Post40), at the conclusion of treatment (Post80), and one month after the conclusion of treatment (Maintenance). Baselines 1 and 2 were separated by four months, and Baselines 2 and 3 were separated by one month. Functional Magnetic Resonance Imagine (fMRI), used to assess cortical activity during the repetition of spoken phrases, was performed at Baseline, Post40, Post80, and Maintenance. At Baseline, Post40, Post80, and one year after the conclusion of treatment (1year-Post), the size and integrity of fiber tracts associated with speech and language processing were measured using DTI. LF's healthy identical twin sister served as a control for the DTI measures, undergoing scans at two time points that spanned the course of LF's treatment (corresponding to Baseline and Post80).

Treatment

LF underwent an intensive course of adapted MIT, attending five 1.5-h treatment sessions per week for 16 weeks, totaling 80 sessions (120 hours of treatment). Sessions followed a format consistent with the original MIT methodology, with target phrases trained using a hierarchical series of steps that have been described in detail elsewhere.[7] The clinician presented a stimulus picture as a visual cue with each target phrase. Sentences were intoned (sung) on two pitches, while the clinician tapped LF's left hand once per syllable.

Treatment was divided into two phases, with 60 trained phrases per phase. Each phase was subdivided into two levels of difficulty (a and b). During Phase I, LF was introduced to the principles of MIT as it is applied in our clinic.[7] Accordingly, she was instructed to use inner rehearsal (i.e., "*hear* the phrase in your head *before* you sing out loud") and was encouraged to use a slow, steady rate with continuous voicing. Phrases trained during Phase Ia consisted of four, six, and eight syllables, while those trained during Phase Ib contained either eight or ten syllables.

Phase II treatment followed the same general structure as Phase I, but added another element: an emphasis on syntax. Because LF had so readily mastered phrases in Phase I yet still lacked fluency, our rationale for placing special emphasis on structure and sequence in Phase II was to further improve LF's fluency while targeting other areas of difficulty (i.e., gender-specific pronouns, prepositions, etc.) for remediation. By the end of Phase I, LF had succeeded in consistently singing, using a slower rate, and connecting her speech during treatment. Although her conversational speech had improved, it was still notably agrammatic, requiring the listener to interpret LF's telegraphic utterances and/or ask questions for clarification. Phase II was designed to reinforce the MIT principles introduced during Phase I as well as emphasize and train selected syntactic structures. During Phase IIa, the target phrases consisted of 10 intransitive verb sentences (e.g., "She is sneezing"),

10 direct object sentences (e.g., "He is feeding the cat"), and 10 sentences with prepositional phrases (e.g., "They are sledding down the hill"). During Phase IIb, the target phrases consisted of 20 sentences with prepositional phrases and 10 sentences with both direct objects and prepositional phrases (e.g., "She is playing a card game with her mom").

Behavioral data: treatment probes

Probes were conducted after every eight treatment sessions, with six probes during Phase I and four during Phase II. The probe task was repetition: the clinician intoned the target phrase while tapping, and LF was cued to repeat the target. Trained phrases were randomly interspersed with untrained phrases. In Phase I, trained and untrained phrases were matched on phrase length (number of syllables) and number of closed-class words. In Phase II, phrases were matched on phrase length and argument structure of the main verb.

Behavioral data: functional speech and language

Language samples obtained from the picture description and semistructured conversational interview were transcribed and scored. Picture description was presented as a timed task and assessed in terms of communicative efficiency; therefore, performance was measured as the number of correct information units (CIUs) per minute. CIUs are defined as words that are informative, accurate, and relevant.[17] The interview was assessed for LF's ability to produce connected speech and share the conversational burden. These elements were quantified, respectively, as syllables per phrase and CIUs per conversational turn. Twenty-five percent of the language samples were double-scored by a trained research assistant. Interrater agreement was >85% for all measures.

fMRI stimuli and MR data acquisition

For the fMRI task, LF heard two-syllable words or phrases spoken or sung at the rate of one syllable/sec. Following the presentation of each word/phrase, LF repeated what she had heard after an auditory cue. In the control condition, LF waited for the auditory cue, and then took a breath as if she planned to speak or sing, but remained silent. The fMRI paradigm and data acquisition methods were described in detail previously.[18]

fMRI data analysis

fMRI data were analyzed using the SPM5 software package.[19] Preprocessing was done as previously described.[18] A left-hemisphere lesion mask was used to restrict the analysis to brain tissue outside of the large lesion area because any task-related signal changes in the lesioned area would have to have been an artifact due, most likely, to changes in brain pulsation or fluid changes in the pseudocystic area.

A design matrix for each imaging time point was modeled to look at the speaking versus silence (control) condition. In order to look at between-time points effects, we entered contrasts into a fixed-effects design matrix for two imaging time points at a time.

Preprocessing and probabilistic tractography of DTI data

The preprocessing and fiber tracking was done with FSL version 4.1.4.[20] In order to correct for eddy currents and head motion,[21] a 3D affine registration was applied, followed by brain extraction.[22] We then fitted a diffusion tensor model at each voxel using FSL's FMRIB diffusion toolbox to calculate tract volume. Fiber tracking was performed using a probabilistic tractography method based on a multifiber model, and applied using tractography routines implemented in FSL's FDT toolkit.[23,24]

We considered two fiber tracts in LF's right hemisphere that, in the left hemisphere, have been associated with speech and language processing: the arcuate fasciculus (AF) and the uncinate fasciculus (UF). We assessed these same two fiber tracts in the right hemisphere of LF's twin sister. We have previously described details of the anatomical definitions of the AF and UF.[25] In order to account for possible variations in scan quality across scanning sessions, the AF and UF were normalized to the volume of the right-hemisphere corticospinal tract of each twin at each time point. For details on the definition and reconstruction of the corticospinal tract, see our publications by Lindenberg *et al.*[26] and Zhu *et al.*[27] The volumes of the various Regions of Interests (ROIs) used to define the tracts did not differ significantly between time points within either sister.

Results

Treatment probes

LF showed improvement on both the trained and untrained phrases, but with a consistently

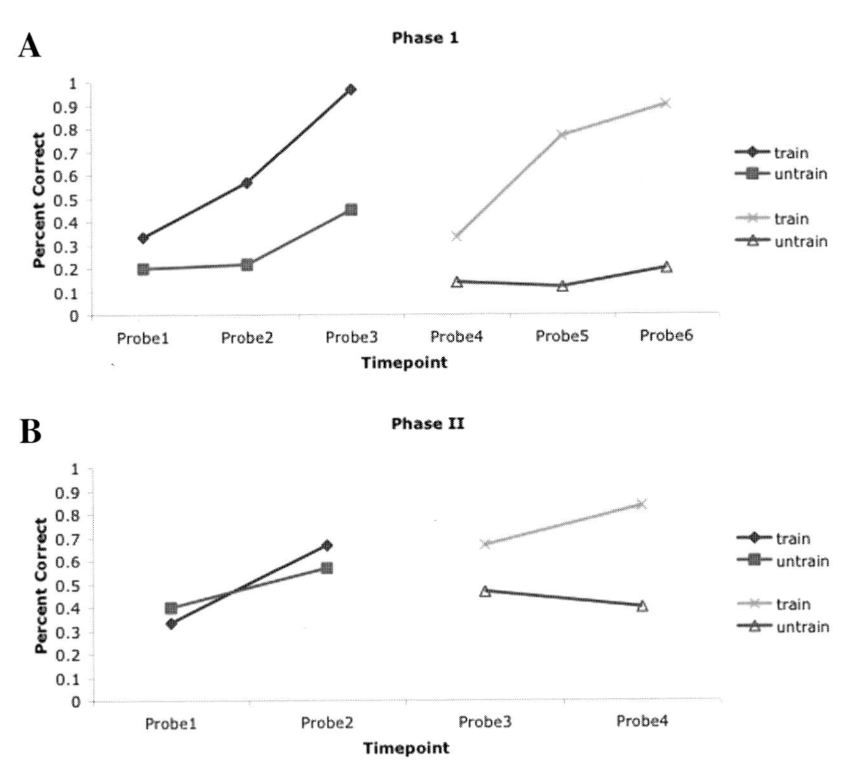

Figure 2. Improvement on trained versus untrained phrases. (A) Treatment in Phase I introduced the principles of MIT as it is applied in our clinic, and emphasized slow, intoned speech with continuous voicing, and the use of inner rehearsal. (B) Treatment in Phase II reinforced the principles introduced in Phase I, and also trained syntactic structures. Across all phases, the rate of improvement was greater for the trained compared to the untrained phrases.

steeper slope of improvement on the trained items (Fig. 2).

Functional speech and language measures

LF's levels of ability to produce fluent speech in conversation (syllables/phrase) and share the conversational burden (CIUs/conversational turn) were stable at baseline, improved notably with treatment, and demonstrated reasonable maintenance of treatment gains (Fig. 3). In contrast, while LF's ability to efficiently describe pictures (CIUs/min) also improved over time, the steepest slope of improvement occurred between the second and third baseline measures, prior to treatment, but the treatment phase still showed pronounced improvements beyond the changes seen in this measure during baseline assessments.

fMRI data

The contrasts revealing activation in right-hemisphere frontotemporal regions were Post40 > Baseline, Post40 > Post80, Post40 > Maintenance, Baseline > Post80, and Baseline > Maintenance. Since regions of activation were essentially identical at Post80 and Maintenance, only the contrasts between Baseline, Post40, and Post80 are shown (Fig. 1B–D). There was an increase in activation in right supplementary motor areas after approximately eight weeks of treatment (Post40), but this additional activation slightly decreased after an additional eight weeks of treatment. The Baseline > Post80 contrast showed higher levels of activation in the right posterior middle temporal gyrus (MTG), occipital cortex, and possibly cerebellum.

DTI data

For LF, both the AF and UF increased in volume between the beginning of treatment and the midpoint of treatment. While AF volume continued to increase between the midpoint and the conclusion of treatment, UF volume appeared to decrease somewhat (Fig. 4) during that period. Both the AF and UF increased in volume between the end of treatment and the scan at one-year posttreatment.

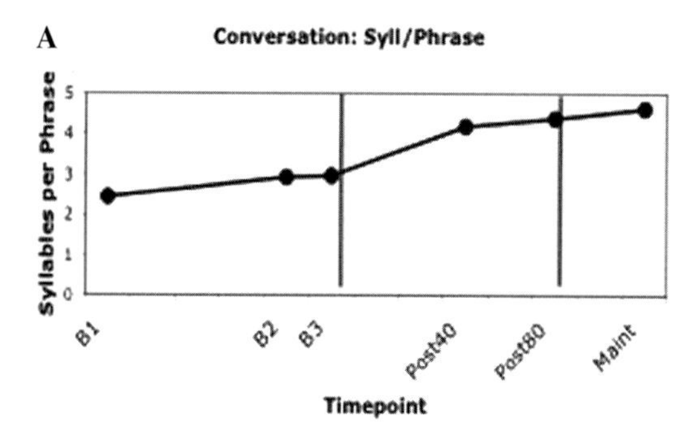

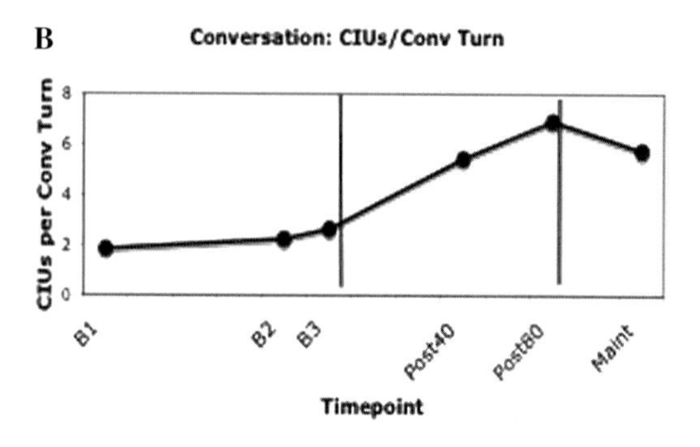

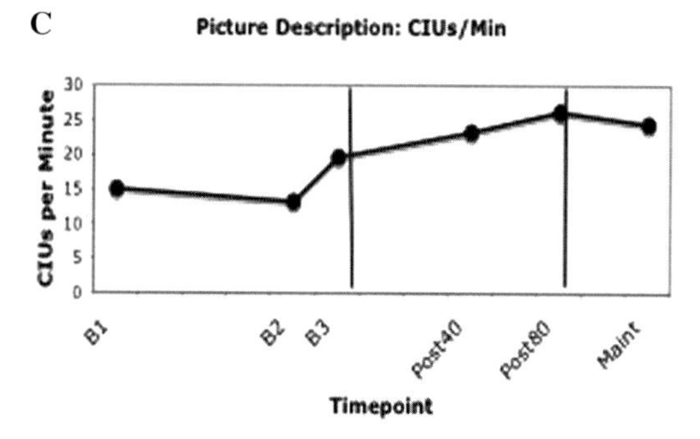

Figure 3. Generalization to functional language tasks. Improvement on a measure of fluency: syllables per phrase (A); ability to share the conversational burden, CIUs per conversational turn, during a semistructured conversation (B); and improvement on a measure of communicative efficiency, CIUs per minute, during a picture description task (C).

LF's healthy twin sister's DTI studies revealed higher volumes of both AF and UF at the baseline imaging study, but then a slight decrease over time.

Discussion

This study provides additional evidence for the efficacy of MIT in an adolescent with a large left-hemisphere stroke, and it supports the adaptation of MIT for use with a broader range of people with aphasia. LF showed greater improvement on trained phrases than untrained phrases across all treatment phases. Although she also showed improvement on untrained phrases, it was more modest, and likely due in part to generalization. Most importantly, LF showed an increased ability to produce connected speech in conversation and share the conversational burden; these gains are both highly functional and clinically significant. As confirmation of this, family members commented positively on the noticeable gains LF made in her ability to communicate during the course of the treatment.

While the measures of LF's abilities in conversation showed stable baselines and steep slopes of improvement during treatment, the measure of communicative efficiency during picture description actually showed the steepest slope of improvement between the second and third baselines. We believe that this is a practice effect and, unfortunately, our study design failed to account for the likelihood of such an effect. In an attempt to keep the measures across time points as controlled as possible, we used the same three pictures at every time point. As a consequence, LF learned to efficiently describe these pictures, with the greatest rate of improvement seen between the two testing sessions that were closest in time, Baselines 2 and 3, which were only a month apart.

As revealed by the fMRI data, LF's pattern of cortical activity during speaking changed over the course of therapy. From baseline to midway through therapy (approximately eight weeks), she showed a strong increase in activation of right posterior middle frontal and inferior frontal areas. This activation then decreased back to baseline levels. While this finding may seem somewhat surprising, it could reflect an increased efficiency of a right-hemisphere frontotemporal system that, as the DTI data showed, changed markedly during that time. Increased right-hemisphere activation following a left-hemisphere lesion has sometimes been interpreted as maladaptive and a consequence of disinhibition.[28] It has also been suggested that increased right-hemisphere activation may persist in people with very large left-hemisphere lesions in order to compensate for the damaged regions.[29] In LF's case, though, the biphasic change in activation occurred in the context of continuous behavioral improvement. We therefore suggest that activation increased initially in response to the demands of treatment, but then decreased as processes became more automatized. In

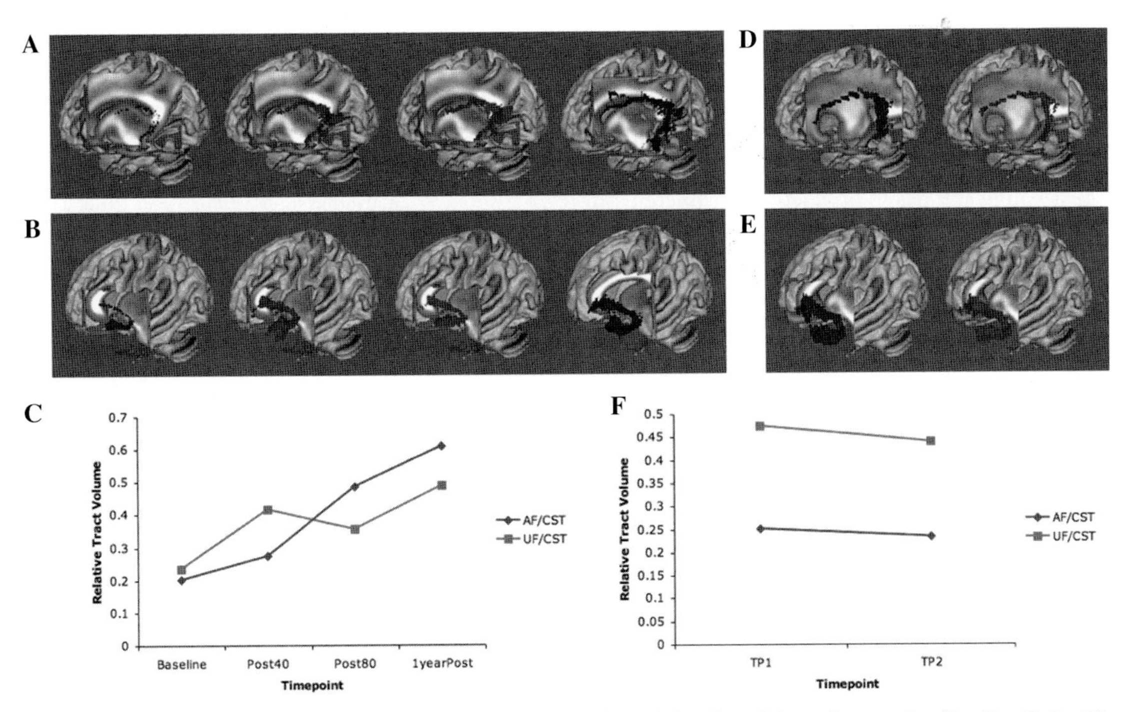

Figure 4. Changes in the AF and UF after treatment with MIT. LF's AF (A) and UF (B) are shown at Baseline, Post40, Post80, and at one-year posttreatment, and (C) volumes for the two tracts are plotted for the same four time points normalized by the corticospinal tract volume of the right hemisphere. For LF's twin sister, the AF (D) and UF (E) are shown at time points 1 and 2, and (F) relative volumes for these tracts are plotted over time. Time points 1 and 2 for LF's twin sister were separated by approximately four months, to match the interval between LF's Baseline and Post80 scans, which is the period during which the maximal structural change occurred.

other words, automatization was supported by the upregulation of the right frontotemporal region, which then became more efficient and required less energy.

The structural white matter increases that we observed in the UF and AF must be interpreted with some caution because these data are from a single patient. However, the observed changes in the AF are consistent with our previous findings in adult subjects undergoing intense treatment with MIT. Furthermore, we had an additional degree of experimental control since LF's healthy twin sister also underwent two DTI scans at time points that corresponded to the interval between LF's Baseline and Post80 scans. LF's twin showed higher AF and UF volumes than LF at baseline, presumably because the twin had typical speech while LF had very limited verbal output in the 15 months between the onset of her stroke and our first evaluation of her. While LF showed increases in her tract volumes from Baseline to Post80, her twin showed decreases in both the AF and the UF. With just two time points, we cannot rule out that these differences could be due to either random variation or differences in scan quality. However, we posit that these volume decreases could be due to synaptic pruning occurring during the prepuberty period.[30]

The increases in this young patient's AF are exceptionally interesting because, in healthy individuals, the left AF links posterior language areas to frontal areas important for motor speech and for the mapping of sounds to motor actions.[8,31] Furthermore, the degree of damage to the left AF has been shown to be correlated with negative speech and language outcomes, at least in patients who sustained strokes as adults.[29] The observed increases in right AF volume may indicate that this tract can play an important role in mapping speech sounds to motor commands when left-hemisphere pathways are no longer available. Since LF no longer has a left perisylvian region, we believe the right hemisphere must have subsumed the processes that support her speech and language function, and that the increases in right AF volume reflect a stronger link between posterior sensory language areas, now presumably located in the right hemisphere, and frontal motor regions.

Although this degree of plasticity might appear surprising, LF's case is rather exceptional in that she is young and has so little left-hemisphere tissue remaining. Thus, it is clear that any observed improvements in speech and language function must be supported by the right hemisphere. It seems likely that LF's age was an important factor in her marked recovery. The impact of age at time of brain insult is a complex issue, since this variable apparently interacts with many others.[33] While some researchers and clinicians have stated that age six is the end of a critical period for language acquisition, at which time language function is firmly instantiated in the left hemisphere, there is evidence that the left-lateralization of language function occurs gradually throughout childhood, at least through age 11 (Ref. 32). The present study provides further evidence that marked plasticity is possible into early adolescence. Some researchers have even suggested that large unilateral lesions may force interhemispheric transfer and lead to better functional outcomes than would be seen with moderate-sized lesions.[33] While LF was still noticeably aphasic after treatment concluded, her improvements in fluency were remarkable.

Acknowledgments

This work was supported by grants from the NIH/NIDCD to G.S. (RO1 DC008796). G.S. also acknowledges support from the Mary Crown and William Ellis Family Fund and the Rosalyn and Richard Slifka Family Fund.

Conflicts of interest

The authors declare no conflicts of interest.

References

1. Albert, M.L., R.W. Sparks & N.A. Helm. 1973. Melodic intonation therapy for aphasia. *Arch. Neurol.* **29:** 130–131.
2. Sparks, R., N. Helm & M. Albert. 1974. Aphasia rehabilitation resulting from melodic intonation therapy. *Cortex* **10:** 303–316.
3. Gerstman, H.L. 1964. A case of aphasia. *J. Speech Hear. Dis.* **29:** 89–91.
4. Geschwind, N. 1971. Current concepts: aphasia. *New Engl. J. Med.* **284:** 654–656.
5. Helm-Estabrooks, N. & M.L. Albert. 2004. *Melodic Intonation Therapy: Manual of Aphasia and Aphasia Therapy.* 2nd ed. Chapt. 16, 221–233. Pro-Ed. Austin, TX.
6. Helm-Estabrooks, N., M. Nicholas & A. Morgan. 1989. *Melodic Intonation Therapy.* Pro-Ed. Austin, TX.
7. Norton, A., L. Zipse, S. Marchina & G. Schlaug. 2009. Melodic intonation therapy: shared insights on how it is done and why it might help. *Ann. N.Y. Acad. Sci.* **1169:** 431–436.
8. Schlaug, G., A. Norton, S. Marchina, *et al.* 2010. From singing to speaking: facilitating recovery from nonfluent aphasia. *Future Neurol.* **5:** 657–665.
9. Belin, P., P. Van Eeckhout, M. Zilbovicius, *et al.* 1996. Recovery from nonfluent aphasia after melodic intonation therapy: a PET study. *Neurology* **47:** 1504–1511.
10. Schlaug, G., S. Marchina & A. Norton. 2008. From singing to speaking: why singing may lead to recovery of expressive language function in patients with Broca's aphasia. *Music Percept.* **25:** 315–323.
11. Schlaug, G., S. Marchina & A. Norton. 2009. Evidence for plasticity in white-matter tracts of patients with chronic Broca's aphasia undergoing intense intonation-based speech therapy. *Ann. N.Y. Acad. Sci.* **1169:** 385–394.
12. Anderson, V., M. Spencer-Smith & A. Wood. 2011. Do children really recover better? Neurobehavioral plasticity after early brain insult. *Brain* **134:** 2197–2221.
13. Raven, J.C. 1995. *Coloured Progressive Matrices.* Oxford Psychologists Press. Oxford, UK.
14. Kaplan, E., H. Goodglass & S. Weintraub. 2001. *Boston Naming Test.* 2nd ed. Lippincott Williams & Wilkins. Baltimore, MD.
15. Goodglass, H., E. Kaplan & B. Barresi. 2001. *The Assessment of Aphasia and Related Disorders.* 3rd ed. Lippincott, Williams and Wilkins. Baltimore, MD.
16. Dabul, B.L. 2000. *Apraxia Battery for Adults.* 2nd ed. Pro-Ed. Austin, TX.
17. Nicholas, L.E. & R.H. Brookshire. 1993. A system for quantifying the informativeness and efficiency of the connected speech of adults with aphasia. *J. Speech Hear. Res.* **36:** 338–350.
18. Ozdemir, E., A. Norton & G. Schlaug. 2006. Shared and distinct neural correlates of singing and speaking. *NeuroImage* **33:** 628–635.
19. http://www.fil.ion.ucl.ac.uk/spm/
20. http://www.fmrib.ox.ac.uk/fsl
21. Jenkinson, M. & S. Smith. 2001. A global optimization method for robust affine registration of brain images. *Med. Image Anal.* **5:** 143–156.
22. Smith, S.M. 2002. Fast robust automated brain extraction. *Hum. Brain Mapp.* **17:** 143–155.
23. Behrens, T.E., H.J. Berg, S. Jbabdi, *et al.* 2007. Probabilistic diffusion tractography with multiple fibre orientations: what can we gain? *NeuroImage* **34:** 144–155.
24. Smith, S.M., M. Jenkinson, M.W. Woolrich, *et al.* 2004. Advances in functional and structural MR image analysis and implementation in FSL. *NeuroImage* **23:** S208–S219.
25. Marchina, S., L.L. Zhu, A. Norton, *et al.* 2011. Impairment of speech production predicted by lesion load of the left arcuate fasciculus. *Stroke* **42:** 2252–2256.
26. Lindenberg, R., V. Renga, L.L. Zhu, *et al.* 2010. Structural integrity of corticospinal motor fibers predicts motor impairment in chronic stroke. *Neurology* **74:** 280–287.

27. Zhu, L.L., R. Lindenberg, M.P. Alexander, & G. Schlaug. 2010. Lesion load of the corticospinal tract predicts motor impairment in chronic stroke. *Stroke* **41:** 910–915.
28. Naeser, M.A., P.I. Martin, M. Nicholas, *et al.* 2005. Improv8ed picture naming in chronic aphasia after TMS to part of right Broca's area: an open-protocol study. *Brain Lang.* **93:** 95–105.
29. Saur, D., R. Lange, A. Baumgaertner, *et al.* 2006. Dynamics of language reorganization after stroke. *Brain* **129:** 1371–1384.
30. Sowell, E.R., P.M. Thompson, C.L. Leonard, *et al.* 2004. Longitudinal mapping of cortical thickness and brain growth in normal children. *J. Neurosci.* **24:** 8223–8231.
31. Lahav, A., E. Saltzman & G. Schlaug. 2007. Action representation of sound: audiomotor recognition network while listening to newly-acquired actions. *J. Neurosci.* **27:** 308–314.
32. Hertz-Pannier, L., C. Chiron, I. Jambaqué, *et al.* 2002. Late plasticity for language in a child's non-dominant hemisphere. *Brain* **125:** 361–372.
33. Anderson, V., M. Spencer-Smith & A. Wood. 2001. Do children really recover better? Neurobehavioral plasticity after early brain insult. *Brain* **134:** 2197–2221.

Ann. N.Y. Acad. Sci. ISSN 0077-8923

ANNALS OF THE NEW YORK ACADEMY OF SCIENCES
Issue: *The Neurosciences and Music IV: Learning and Memory*

The dynamic audio–motor system in pianists

Lutz Jäncke
Division Neuropsychology, Institute of Psychology, University of Zurich, Zurich, Switzerland

Address for correspondence: Prof. Dr. Lutz Jäncke, University of Zurich, Institute of Psychology, Division of Neuropsychology, Binzmühlestrasse 14, 8050 Zurich, Switzerland. l.jaencke@psychologie.uzh.ch

This paper reports a preliminary study based on the theoretical assumption that continuous closed-loop audio–motor control could be disadvantageous for pianists. It is argued that the functional relationship between the intracerebral electrical activations in the auditory and premotor cortex should be rhythmically decreased and increased. To test this hypothesis, intracerebral electrical activations for the auditory and premotor cortex were estimated using scalp EEG and standardized low-resolution electrical tomography (sLORETA). The extracted times series were subjected to a Granger causality analysis, revealing a causal relationship from the auditory cortex to the premotor cortex that was considerably stronger during piano playing and weaker during rest. Importantly, this relationship varied rhythmically during the course of piano playing, with lags (obtained with cross-correlations) between 666 ms and 820 milliseconds. This study thus delivers evidence that the functional coupling between the auditory and premotor cortex varies during piano playing.

Keywords: audio–motor control; pianist; music; expertise; EEG; sLORETA; Granger causality

Introduction

The quality of a musical performance is heavily reliant on the amount of practice. As Ericsson[1] elegantly showed, successful professional musicians practice in excess of 10,000 hours on average by the time they reach the age of 20 years. An increasing number of publications have focused on this extreme level of practice, reporting that practice with a musical instrument results in various neurophysiological and neuroanatomical adaptations.[2–5] In this paper, I will focus on the sensory-motor system in musicians and particularly on the audio–motor system in pianists. During musical practice, pianists (and all musicians) need to establish a strong auditory–motor coupling that enables them to coordinate their motor output with the produced acoustic stimuli and to compare this acoustic output with the intended auditory goals. There should be, therefore, a strong neurophysiological and, possibly, a neuroanatomical link between the auditory and premotor processing modules in the pianist's brain. In fact, some brain-imaging studies lend support to this idea of strong neurophysiological auditory–motor coupling in pianists. For example, Baumann *et al.*[6] demonstrated that the auditory system is strongly activated during piano playing, even when the sound of the electrical piano is turned off. Bangert *et al.*[7] also demonstrated a close link between hemodynamic responses in the auditory cortex and the premotor cortex in pianists. In an earlier paper, the same authors demonstrated that even in nonpianists, a close neurophysiological relationship was found after only 20 min of auditory–motor association training. This association was strongly enhanced after a 5-week training.[8] Thus, there is strong evidence in favor of an extraordinary neurophysiological interplay between the auditory and motor regions in the brain of pianists.

This interplay between the auditory and premotor areas can be conceived of as a kind of correlation between auditory information (auditory afference: a) and the associated motor commands (motor efference: e). These correlations are modulated as a consequence of the amount and quality of practice and are represented in a kind of correlation storage. The output of this correlation storage can be driven by the auditory input. Thus, the efference can be

doi: 10.1111/j.1749-6632.2011.06416.x

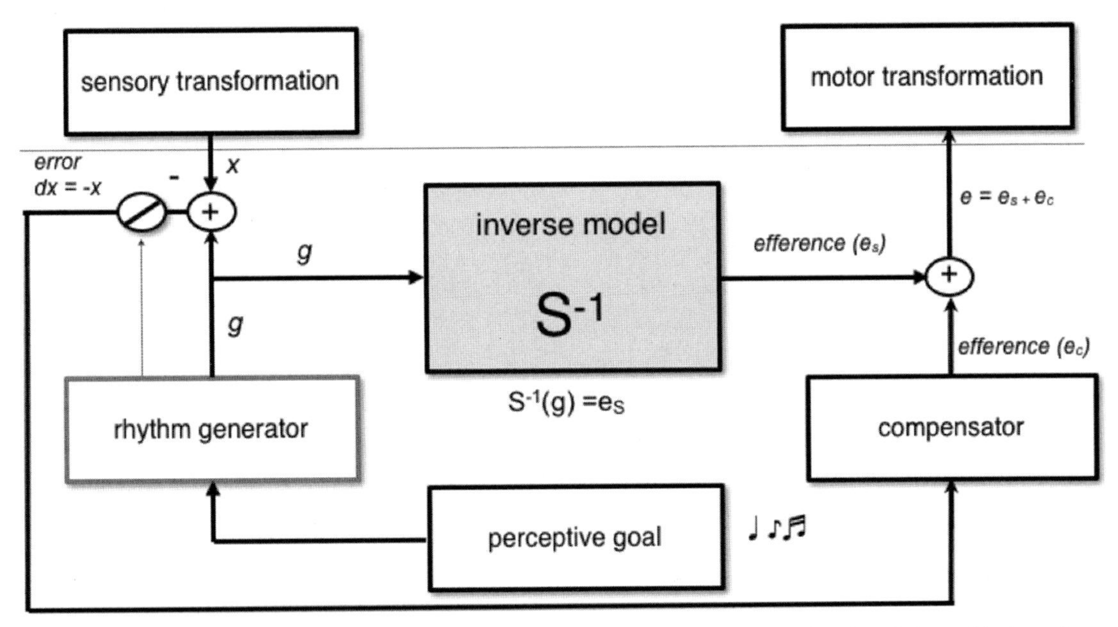

Figure 1. Model for auditory–motor coupling during piano playing. The perceptive goal is thought to be located in the auditory cortex, while the inverse model is thought to be located in the premotor cortex. The rhythm generator is located between the above-mentioned pivotal modules and is thought to modulate the strength of the coupling between the auditory cortex (with the perceptive goal) and the premotor cortex (with the inverse model). Model adapted from Kalveram.[9]

conceived of as a function of the auditory input ($e = f(a)$). The correlation storage is embedded in a more complicated sensorimotor control system and plays a pivotal role in controlling movements and, in this case, in controlling piano playing.[9,10] In accordance with earlier theorists, I use the term "inverse model" to describe the correlation storage. We might consider the following scenario in which a pianist is playing a musical piece (Fig. 1). The pianist (implicitly) imagines what he or she wants to play. The auditory images (let us call them goal afferences: g) are stored in the auditory system and transferred to the inverse model ($S-1$). On the basis of this input, this model generates the appropriate motor commands (efference: $e^s = S-1(g)$) by which the motor areas motor at the end generating the finger and arm movement necessary to manipulate the keyboard and to generate the tones intended to be played. These tones (and melodies) are fed back via sensory transformations and are compared with the intended auditory goals. In the event of a difference (dx) between intended (g) and achieved auditory goals (x), the motor commands are adjusted accordingly (e^c). For the sake of simplicity, let us assume that the perceptual goals are stored in the auditory cortex and the inverse model is located in the dorsal premotor cortex. The dorsal premotor cortex is suggested in this context on the basis of several studies that demonstrate its pivotal role in sensory-motor transformations.[11] This sensory-motor system is a typical closed-loop system. Closed-loop systems are well suited for adapting to changing environments but they are also associated with some problems. First, they are unstable since closed-loop systems oscillate around the reference values and, second, they are prone to be destabilized by extreme input values. In my view, the latter could be disadvantageous when pianists play extremely fast and with high precision. Thus, pianists need to dynamically modulate the strength of the auditory–motor coupling. I assume that they make use of a kind of rhythm generator with which to rhythmically decrease or increase the strength of the audio–motor association. I hypothesize that the functional coupling between the auditory and premotor cortex should be modulated while the pianist is playing the piano, meaning that the auditory–motor coupling should oscillate during the course of piano playing. In order to test whether auditory–motor coupling does oscillate during piano playing, we measured EEG, estimated the time course of intracerebral activation in the auditory and premotor cortex, and calculated the relationship between both areas. To estimate the relationship between the auditory and premotor cortex, we calculated the Granger causality, thus allowing us to infer the directionality of the relationship.

Methods

Six consistently right-handed (assessed with the Annett Handedness Questionnaire) professional pianists from the Zurich Conservatory of Music took part in our study (four men and two women, mean age 25.6 ± 5 years). Pianists were required to play a piano piece by Mozart (*Sonata Facile*) with (Aud+) and without (Aud–) auditory feedback on a Yamaha electrical piano. For the no feedback condition, the sound of the piano was turned off. Before piano playing, the subjects were required to relax and resting EEG was measured for 60 sec (rest). During piano playing, continuous EEG was recorded for 60 seconds. Scalp EEG (32 channels and 2 bipolar eye channels) was recorded with a sampling rate of 500 Hz and a band pass filter (0.1–70 Hz). The electrodes were located at frontal, temporal, parietal, and occipital scalp sites of the international 10–10 system (FP 1/2, F 3/4, F 7/8, FZ, FT 7/8, FC 3/4, T 7/8, C 3/4, Cz, TP 7/8, TP 9/10, CP 3/4, CPZ, P 7/8, P 3/4, PZ, O 1/2, OZ, and FCZ). Recording reference was at FCz, with off-line rereferencing to average reference. The digital sampling rate was 500 Hz, on-line filtering 0.1–100 Hz, off-line filtering 0.5–30 Hz, and impedance was kept below 10 kOhm. Brain Vision Analyzer software (version 1.04; Brainproducts) was used for all steps of digital EEG raw data processing. The data were filtered off-line at 1–30 Hz. Eye movement artifacts (in some cases muscle artifacts) were removed manually where possible or by using an independent component analysis in order to avoid data loss. The processed data were then divided into 2,000 ms segments and subjected to standardized low-resolution electrical tomography (sLORETA) analysis to estimate the inverse solution and the intracerebral sources of brain activity.[12,13] sLORETA computes, from the recorded scalp electric potential differences, the three-dimensional distribution of the electrically active intracerebral neuronal generators as standardized units of current density (A/cm^2) at each voxel by assuming similar activation among neighboring neuronal clusters. In the current implementation of sLORETA, computations were made in a realistic head model,[14] using the MNI152 template,[15] with the three-dimensional solution space restricted to cortical gray matter.[16] The standard electrode positions on the MNI152 scalp were taken from Jurcak *et al.*[17] and Oostenveld and Pramstra.[18] The intracerebral volume is partitioned in 6,239 voxels at 5 mm spatial resolution. Thus, sLORETA images represent the standardized electric activity expressed as the exact magnitude of the estimated current density at each voxel in the MNI space (Montreal Neurological Institute). sLORETA "solves" the inverse problem by taking into account the well-known effects of the head as a volume conductor. Conventional LORETA and the more recent sLORETA analyses have been frequently used in previous experiments to localize brain activations on the basis of EEG or MEG data.[6,19,20] Four regions of interest (ROIs) with a radius of 15 mm each were defined for our analysis: the left- and right-sided premotor cortex (peak points in MNI coordinates: lPMC = –27, –29, 53; rPMC = 26, 35, 53) and the left- and right-sided auditory cortex, including parts of the primary and posterior parts of the secondary auditory regions (lAC = –52, –26, 10; rAC = 53, –26, 10). The locations of these ROIs were chosen on the basis of the Broadmann area defined by sLORETA (for the PMC) and known activation peaks from auditory studies using fMRI measurements.

The current density power time series for each region (lAC, rAC, lPMC, rPMC) and each condition (rest, Aud–, Aud+) was thus estimated (see Fig. 2 for an example of time courses for one subject). For each condition and each subject we obtained four time series with 30 × 1024 time points each (30,720 time points in total). These time series were used for further statistical analysis. Using the R package (http://www.r-project.org/), time series analyses were computed to assess the interrelationship between these time series (R packages MSBVAR, seewave). (1) The average Granger causality was computed across all time series.[21] The Granger causality enables statistical assessment of the causal interrelationship between time series. Here, we calculated the bivariate version, with specific interest in assessing the Granger causality between the AC and the PMC. In case of a significant Granger causality, for example, between lAC to lPMC, we can assume that the time series of lAC predicts (and possibly causes or influences) the time series of lPMC. The Granger causality is advantageous in terms of

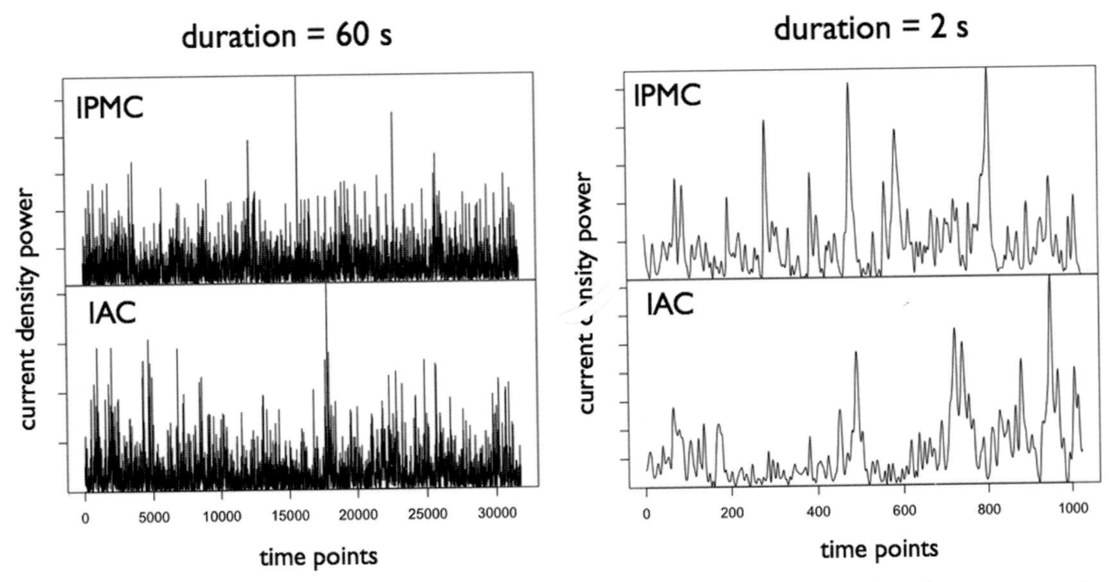

Figure 2. Time courses of intracerebral electrical activations (current density power) in the PMC and the AC for one representative subject. On the left, the time courses are shown for the entire 60-sec epoch. On the right, a segment with a 2-sec duration is shown. The time courses are estimated based on sLORETA.

causality inference compared with classical methods for computing functional connectivities (e.g., coherence measures). (2) The entire time series (of 60 sec duration) of current density power were segmented into 300 segments of 500 ms length. For each segment the Granger causality was computed to delineate the time course of functional interrelationship (see Fig. 3 for an example of one subject).

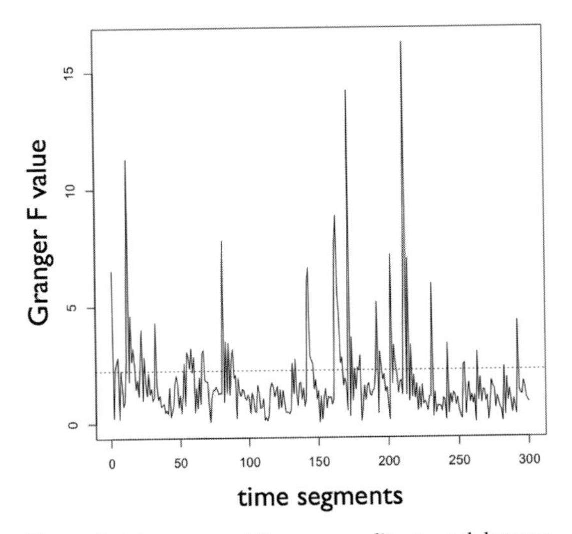

Figure 3. Time course of Granger causality strength between AC and PMC for one subject. The dotted horizontal line demonstrates the strength, which is statistically significant.

(3) The coefficient of variation across the time series of Granger causality measures (SD × 100/mean) was computed in order to statistically describe the variability of Ganger causality. (4) Cross-correlations were also computed to assess the time lag at which maximum cross-correlation occurred. A multivariate MANOVA was also applied to assess differences between hemisphere (right versus left), condition (rest, A–, A+), and direction of causality (AC to PMC, PMC to AC). Because few subjects are examined here, I will focus on the effect size measure Eta^2 and report main effects and interactions, which are associated with a $P \leq 0.1$ and an Eta^2 of at least 0.50. This Eta^2 is known to indicate large effects. When F values are indicated, I will refer to Wilk's lambda.

Results

The time series of the current density power in the left and right AC strongly predicts the time series of the current density power in the left and right PMC during both piano playing conditions. During piano playing, the influence from the AC to the PMC is always larger than the influence from the PMC to the AC (Fig. 4; mean ± S.E.; AC → PMC: 7.4 ± 0.9; PMC → AC 0.56 ± 0.03; main effect direction: $F[1,5] = 50.3$, $P = 0.001$, Eta$^2 = 0.91$). The causal relation between AC → PMC relationship is weaker during the resting state than during piano playing

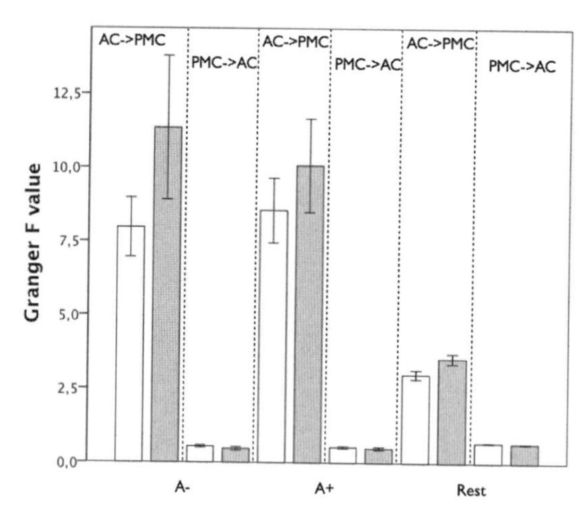

Figure 4. Mean strength (and S.E.) of Granger causality for AC → PMC and PMC → AC. Values for the left hemisphere are shown in white bars, while values representing the Granger causality strength on the right are shown in gray. A–, without auditory feedback; A+, with auditory feedback.

(interaction between direction × condition: $F(2,4) = 11.6$, $P = 0.02$, $\text{Eta}^2 = 0.85$).

The AC–PMC Granger causality variability measures are larger than the variability measures for the PMC → AC relations ($F[1,5] = 20.4$, $P = 0.006$, $\text{Eta}^2 = 0.80$; Fig. 5). The largest AC → PMC variability measures of causality were obtained during rest while the average values are lowest (main effect condition: $F[2,4] = 31.5$, $P = 0.004$, $\text{Eta}^2 = 0.94$). The variability of the AC → PMC causality was a bit larger on the left while the PMC → AC causality variability was a bit larger on the right hemisphere (interaction direction × hemisphere: $F[1,5] = 7.9$, $P = 0.04$, $\text{Eta}^2 = 0.61$). There was also an interaction between direction × condition ($F[2, 4] = 12.8$, $P = 0.018$, $\text{Eta}^2 = 0.865$), which was qualified by larger variabilities for AC–PMC during rest, while the variability for PMC → AC was smallest during rest.

Finally, cross-correlations between the time series were computed, revealing strong maximum cross-correlations between the AC and the PMC during the resting state ranging between $r = 0.64$ and 0.76 at a lag of 2 ms (Table 1). However, these cross-correlations sharply decrease after 2 ms down to cross-correlations between $r = 0.03$ and 0.5 at 500–900 ms. There were significantly strong maximum cross-correlations during piano playing but only at longer lags from 666 ms to 820 ms. However, these maximum cross-correlations were numerically lower than the cross-correlations obtained during the resting state. Inspection of the cross-correlation functions of the time series obtained during piano playing revealed a broader distribution of relatively high cross-correlations around the maximum, while the cross-correlation functions for the resting state time series demonstrate mostly one sharp cross-correlation peak with a sharp drop of correlations after the maximum. Even the correlations after the second maximum for the resting state time series are very low.

Discussion

In this preliminary study, the statistical relationship between the AC and PMC time courses during piano playing and rest have been examined. For this, Granger causality was used, to my knowledge for the first time, in the context of examining the functional relationships between the AC and the PMC. This technique allows causal relationships between brain activations to be inferred, thus providing a fundamental advantage over other methods for estimating functional relationships.[21] With this method, the AC was clearly shown to influence the PMC to a stronger degree during piano playing than during rest. However, even during rest there was a quite strong AC → PMC relationship, which might indicate that this relationship is a firmly established relation. The causal influence of the PMC on the AC

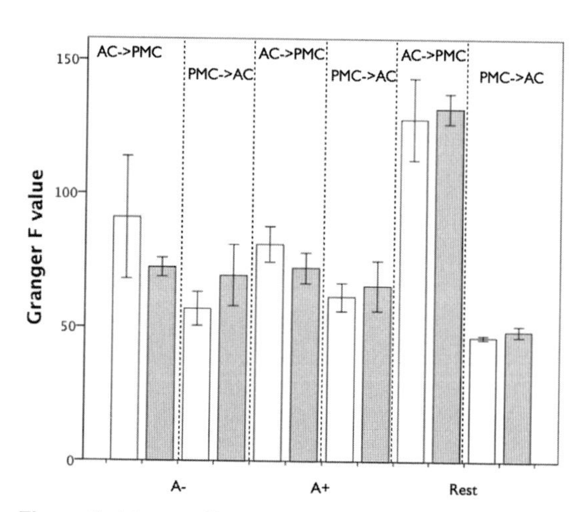

Figure 5. Mean coefficient of variation (and S.E.) of Granger causality for AC → PMC and PMC → AC. Values for the left hemisphere are shown in white bars, while values representing the Granger causality strength on the right are shown in gray. A–, without auditory feedback; A+, with auditory feedback.

Table 1. Maximum cross-correlations and the associated lags between the time series of intracerebral activations of the AC and the PMC

	AC–PMC left	AC–PMC right
Cross-correlation rest	0.76 ± 0.01	0.64 ± 0.02
Cross-correlation A–	0.26 ± 0.02	0.25 ± 0.01
Cross-correlation A+	0.29 ± 0.03	0.25 ± 0.01
Lag cross-correlation rest (ms)	2 ± 0	2 ± 0
Lag cross-correlation A– (ms)	820 ± 40	666 ± 90
Lag cross-correlation A+ (ms)	700 ± 100	750 ± 40

Rest, rest condition; A–, playing piano without auditory feedback; A+, playing piano with auditory feedback. Please note that there was a strong cross-correlation during the rest condition between the AC and the PMC at very short time lags. There was also no variability in this maximum cross-correlation, but after these lags the cross-correlations drop sharply.

during piano playing is much lower even during rest. Thus, this study not only supports recent studies that demonstrated functional relationships between the AC and the PMC,[7,8,22] it also lends weight to the new idea that the AC generates the most important functional input to the motor system.

A major goal of this study was to examine whether the strength of the functional relationship between the AC and the PMC varies during piano playing. This hypothesis is based on theoretical assumptions drawn from theoretical models of motor control.[10] The pivotal idea of these models, which has been transferred to auditory–motor coupling in pianists, is that a closed-loop system needs a kind of rhythmic weakening to allow for open-loop motor control. Besides the advantages of closed-loop control strategies, these strategies suffer from instability in certain circumstances and they operate with time lags. The model I am proposing here assumes a kind of rhythm generator thought to be located between the AC and the PMC. The time courses of the intracerebral electrical activations of the AC and the PMC demonstrate strong variability throughout piano playing and during rest. The PMC → AC connectivity variability was lower than for the AC → PMC relation. Interestingly, the AMC → PMC variability was very strong even during rest, supporting the idea that the AC → PMC connection is variably strong even during rest in pianists.

The cross-correlation analysis also revealed a broad distribution of relatively strong cross-correlations around the maximum, thus indicating the strong statistical relationship between the time series over a wider time range. The maxima of cross-correlations during piano playing occur at lags between 666 ms and 820 ms. It is possible that these lags indicate the periods at which the strength between the AC and PMC is modulated. Further experiments are needed to disentangle more precisely the dynamics of the functional relationships between the AC and the PMC. However, the main finding of this preliminary study is that there is a causal relationship between the AC and the PMC, which dynamically varies across piano playing and which even varies during rest in pianists.

Acknowledgments

This study was supported by a grant of the Swiss National Science Foundation (SNF) to Lutz Jäncke (32003B_121927).

Conflicts of interest

The author declares no conflicts of interest.

References

1. Ericsson, K.A. 2004. Deliberate practice and the acquisition and maintenance of expert performance in medicine and related domains. *Acad. Med.: J. Assoc. Am. Med. Coll.* **79:** S70–S81.
2. Jäncke, L. 2009. The plastic human brain. *Restor. Neurol. Neurosci.* **27:** 521–538.
3. Jäncke, L. 2009. Music drives brain plasticity. *F1000 Biol. Rep.* **1:** 78.
4. Munte, T.F., E. Altenmüller & L. Jancke. 2002. The musician's brain as a model of neuroplasticity. *Nat. Rev. Neurosci.* **3:** 473–478.

5. Schlaug, G. 2001. The brain of musicians. A model for functional and structural adaptation. *Ann. N.Y. Acad. Sci.* **930:** 281–299.
6. Baumann, S., M. Meyer & L. Jancke. 2008. Enhancement of auditory-evoked potentials in musicians reflects an influence of expertise but not selective attention. *J. Cogn. Neurosci.* **20:** 2238–2249.
7. Bangert, M. *et al.* 2006. Shared networks for auditory and motor processing in professional pianists: evidence from fMRI conjunction. *Neuroimage* **30:** 917–926.
8. Bangert, M. & E.O. Altenmüller. 2003. Mapping perception to action in piano practice: a longitudinal DC-EEG study. *BMC Neurosci.* **4:** 26.
9. Kalveram, K.T. 1998. Wie das Individuum mit seiner Umwelt interagiert: Psychologische, biologische und kybernetische Betrachtungen üer die Funktion von Verhalten (Pabst Science).
10. Kalveram, K.T. 2000. Sensorimotor sequential learning by a neural network based on redefined Hebbian Learning. Artificial neural networks in medicine and biology: proceedings of the ANNIMAB-1 Conference, Göteborg, Sweden, 13–16 May 2000, 271.
11. Praeg, E., U. Herwig, K. Lutz & L. Jäncke. 2005. The role of the right dorsal premotor cortex in visuomotor learning: a transcranial magnetic stimulation study. *Neuroreport* **16:** 1715–1718.
12. Pascual-Marqui, R.D. 2002. Standardized low-resolution brain electromagnetic tomography (sLORETA): technical details. *Methods Find Exp. Clin. Pharmacol.* **24**(Suppl D): 5–12.
13. Pascual-Marqui, R.D. 2007. Discrete, 3D distributed, linear imaging methods of electrical neuronal activity. Part 1: exact, zerror error localization WWW document. URL http://arxiv.org/abs/0710.3341.
14. Fuchs, M., J. Kastner, M. Wagner, *et al.* 2002. A standardized boundary element method volume conductor model. *Clin. Neurophysiol.* **113:** 702–712.
15. Mazziotta, J. *et al.* 2001. A probabilistic atlas and reference system for the human brain: International Consortium for Brain Mapping (ICBM). *Philos. Trans. R. Soc. B-Biol. Sci.* **356:** 1293–1322.
16. Lancaster, J.L. *et al.* 2000. Automated Talairach Atlas labels for functional brain mapping. *Human Brain Mapp.* **10:** 120–131.
17. Jurcak, V., D. Tsuzuki & I. Dan. 2007. 10/20, 10/10, and 10/5 systems revisited: Their validity as relative head-surface-based positioning systems. *Neuroimage* **34:** 1600–1611.
18. Oostenveld, R. & P. Praamstra. 2001. The five percent electrode system for high-resolution EEG and ERP measurements. *Clin. Neurophysiol.* **112:** 713–719.
19. Beeli, G., M. Esslen & L. Jäncke. 2008. Time course of neural activity correlated with colored-hearing synesthesia. *Cereb. Cortex* **18:** 379–385.
20. Mulert, C. *et al.* 2004. Integration of fMRI and simultaneous EEG: towards a comprehensive understanding of localization and time-course of brain activity in target detection. *Neuroimage* **22:** 83–94.
21. Bressler, S. & A. Seth. 2011. Wiener-Granger causality: a well established methodology. *Neuroimage* **58:** 323–329.
22. Baumann, S. *et al.* 2007. A network for audio-motor coordination in skilled pianists and non-musicians. *Brain Res.* **1161:** 65–78.

Ann. N.Y. Acad. Sci. ISSN 0077-8923

ANNALS OF THE NEW YORK ACADEMY OF SCIENCES
Issue: *The Neurosciences and Music IV: Learning and Memory*

Tinnitus: the dark side of the auditory cortex plasticity

Christo Pantev, Hidehiko Okamoto, and Henning Teismann

Institute for Biomagnetism and Biosignalanalysis, Münster, Germany

Address for correspondence: Christo Pantev, Institute for Biomagnetism and Biosignalanalysis, Malmedyweg 15, 48149 Münster, Germany. pantev@uni-muenster.de

Music has increasingly been used as a tool for investigation of human cognition and its underlying brain mechanisms. However, music can be used also for neurorehabilitation. Chronic tinnitus is a symptom with high prevalence, especially in industrialized countries. There is evidence that the tinnitus perception is related to unfavorable cortical plastic changes. Maladaptive auditory cortex reorganization may contribute to the generation and maintenance of tinnitus. Because cortical organization can be modified by behavioral training, potentially via reversing maladaptive auditory cortex reorganization, we attempted to reduce tinnitus loudness by exposing chronic tinnitus patients to self-chosen, enjoyable music that was modified ("notched") to contain no energy in the frequency range surrounding the individual tinnitus frequency and thus attracting lateral inhibition to the brain area generating tinnitus. On this basis, we have developed and evaluated a customized music training strategy that appears capable of both reducing cortical tinnitus-related neuronal activity and alleviating subjective tinnitus perception.

Keywords: tinnitus; cortical plasticity; lateral inhibition; human auditory cortex; magnetoencephalography

Introduction

Chronic tinnitus is a prevalent symptom/syndrome that can severely affect a patient's ability to lead a normal life, including the induction of psychiatric distress with even the risk of suicide.[1] Chronic tinnitus is one of the most common auditory disorders, currently affecting 10–15% of the general adult population.[2] Tinnitus can be described as the perception of sound in the absence of external acoustic stimulation. For the patient, it may either be a trivial or a debilitating condition.[3] The quality of the perceived sound can vary enormously, but the majority of cases perceive simple sounds such as whistling or humming. Tinnitus may be perceived in one or in both ears, and the symptom may be continuous or intermittent. Tinnitus is, in most cases, purely subjective, meaning that it cannot be heard by anyone other than the patient. Even though for the patient this auditory perception is real, it can be considered as a phantom perception because there is no corresponding external sound. Tinnitus may be associated with normal standard hearing thresholds or any degree of hearing loss. It can occur at any age; however, there is higher prevalence in the age group between 50 and 70 years.[4] Several treatment protocols have been developed for the management of tinnitus, including masking by sounds, hearing aids, pharmacological regimes, counseling, and acupuncture. To date, however, there is no standard cure.[5] Cognitive behavioral therapy (CBT), for example, includes psychoeducation about tinnitus, applied relaxation and positive imagery, cognitive restructuring of negative beliefs about tinnitus, exposure to sounds, behavioral activation, and mindfulness/attention exercises. The results of a recent meta-analysis indicate that CBT is a moderately effective treatment in reducing annoyance and distress associated with tinnitus.[6]

Recent research[7,8,5] demonstrates that tinnitus is likely the result of maladaptive plasticity in the central auditory pathway, reflecting the "dark side" of cortical plasticity. Originally, the tinnitus perception is probably most often triggered by hearing loss. Due to auditory neural input deprivation, the excitation–inhibition balance in the central auditory pathway is disturbed, most likely the result of the weakening of inhibitory networks. These dysbalances lead to

doi: 10.1111/j.1749-6632.2012.06452.x

reorganizational processes that underlie the subjective tinnitus sensation, most importantly increased spontaneous firing rates of neurons in auditory cortical and subcortical structures, increased synchronization of spontaneous neural activity in the affected frequency regions, and changes in the frequency representation in auditory cortex.

In this study, we suggest that the main reason for initialization, development, and manifestation of tinnitus is the loss of inhibition affecting the functioning of specific populations of auditory neurons. In order to effectively cure tinnitus, these neurons need to be identified and targeted. In patients with tonal tinnitus, usually the tinnitus frequency (i.e., the frequency that sounds most similar to the tinnitus) can be matched, and it has been demonstrated that auditory cortex neurons coding the tinnitus frequency are involved in tinnitus perception.[9–11] Thus, these neurons are a potential treatment target. However, the reliable determination of the tinnitus frequency is not trivial. Pitfalls like octave confusions[12] need to be considered. A previous study[13] demonstrated that listening to spectrally "notched" music is able to reduce cortical activity corresponding to the notch center frequency through lateral inhibition. Motivated by this finding, we developed an innovative tinnitus treatment strategy in which specially filtered music was utilized as a basis for the therapy that was aimed at reducing tinnitus loudness. The treatment regimen consists of regular listening to individual and enjoyable, custom-tailored notched music. We hypothesize that tinnitus is a result of over excitation at the tinnitus frequency in the auditory system and that removal of those frequency components from the musical auditory stimulus will cause the brain to reorganize around that frequency, thereby decreasing the percept of tinnitus through lateral inhibition. Unlike other treatments, this study combined both subjective and objective measures of the tinnitus loudness. A behavioral self-report was performed in addition to two neurophysiological measures of activity of the auditory cortex: the auditory steady state response (ASSR), measuring primary auditory cortex responses, and the N1m, measuring responses in secondary auditory cortical areas.

Methods

Thirty-nine patients matching the criteria of chronic unilateral tonal tinnitus ($\geq$12 months), tinnitus frequency less than 8 kHz, and no severe hearing impairment or neurological or psychiatric complications, between the ages of 18 and 55 years, were treated. The patients were pseudorandomly assigned to one of three groups: a monitoring group (hearing music that was not modified), a target group (hearing music with tinnitus-related modification), and a placebo group (hearing music with placebo modification).

The auditory stimulus was music chosen by the individual participants and was filtered to exclude a one-octave range around the individual's tinnitus

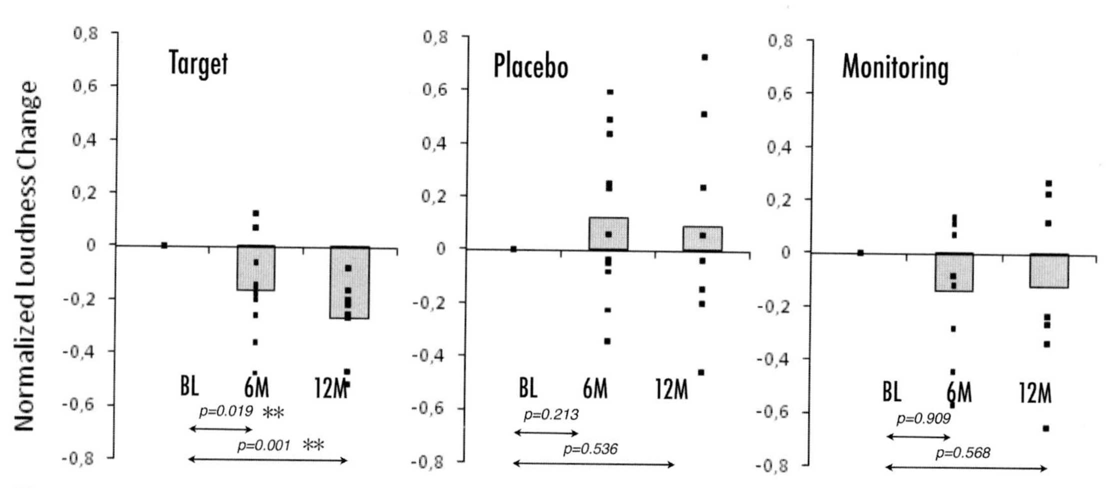

Figure 1. Normalized tinnitus loudness change after 6 and 12 months of treatment (or monitoring) relative to baseline (BL) for the three patient groups (target, placebo, and monitoring). Positive change values reflect impairment; negative change values reflect improvement. The bars indicate group averages; each dot indicates an individual data point.

frequency in cases of tonal tinnitus. The music modification for the tinnitus treatment was performed in the following way. The patients from both treatment groups, target and placebo, provided their favorite music, which was filtered for the target group at fixed notch of one octave corresponding to the tinnitus frequency and for the placebo group by moving notch of one octave not including the tinnitus frequency.[11] The patients of both groups listened to their individually modified treatment music daily via closed headphones with convenient loudness for one to two hours over the course of one year. They were not able to identify to which one of the treatment groups they belonged.

Patients were fully informed about the execution and goals of the study and gave written informed consent in accordance with procedures approved by the Ethics Commission of the Medical Faculty, University of Münster. The tonal tinnitus pitch was multiply-matched to the frequency of a pure tone and the median across pitch matches was considered as the tinnitus frequency. Over the course of the study, additional pitch matches were obtained regularly. Tinnitus loudness was measured weekly on a continuous visual analog scale ranging from 0 (no tinnitus) to 100 (extremely loud tinnitus). To compare treatment effects among subjects, we first normalized the tinnitus loudness means across months 1–6 and months 7–12 relative to the baseline period mean, and then calculated the change of the normalized tinnitus loudness using the formula {[(tinnitus loudness mean˙months 1–6 or 7–12)/(tinnitus loudness mean˙baseline) −1] − 100}. Thus, positive or negative change values indicate tinnitus loudness increment or decrement, respectively. For statistical pre- versus posttreatment comparison, planned contrasts were calculated. In addition, subjective tinnitus annoyance and experienced handicapping by the tinnitus were assessed with the tinnitus questionnaire.[14]

Magnetic fields were measured with a 275-channel MEG system in a magnetically shielded and silent room. The baseline measurement took place before the study; course measurements were performed every six months. Two different sound stimuli (duration of 1-s, 0.3-s pure tone; 0.7-s 40 Hz amplitude-modulated tone) were delivered randomly to either the left or the right ear. The frequency of one stimulus corresponded to a patient's tinnitus frequency; the other stimulus had a frequency of 500 Hz (control stimulus). Both stimuli were matched to 45 dB sensation level. The sound onset asynchrony was randomized between two and three seconds. The tinnitus frequency stimulus evoked activity from a cortical region contributing to the tinnitus perception, the control stimulus from a cortical area not involved in the tinnitus perception. This stimulus design allowed recording simultaneously N1m and ASSR responses from secondary (N1) and primary auditory

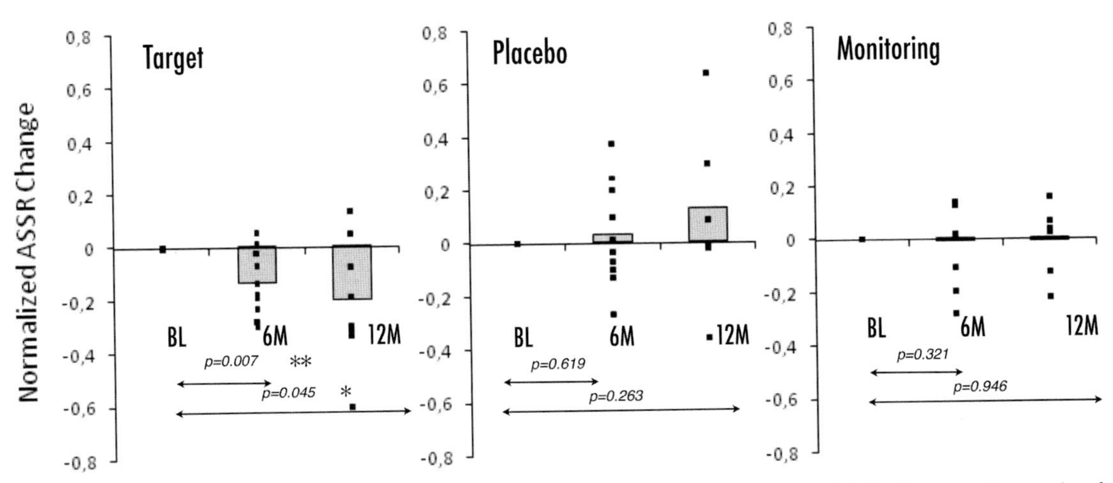

Figure 2. Change of the normalized tinnitus-related auditory-evoked steady-state responses (ASSR) after 6 and 12 months of treatment (or monitoring) relative to baseline (BL) for the three patient groups (target, placebo, and monitoring). Positive change values reflect increment; negative change values reflect decrement. The bars indicate group averages; each dot indicates an individual data point.

cortical structures (ASSR).[15] Because the contour maps of both responses displayed dipolar patterns, a single dipole model was used for source analysis and source space projection technique[16] was performed for calculating the maximal source strength for each condition and hemisphere. These data were then normalized with respect to the baseline data, and then the changes of the normalized ratios {[(source strength elicited by tinnitus frequency at month 6 or 12/source strength elicited by control frequency at month 6 or 12)/(source strength elicited by tinnitus frequency at baseline/source strength elicited by control frequency at baseline) – 1] – 100} were calculated.

Results

After 6 and 12 months of exposure (with an average of 12.4 h per week of listening), those subjects whose tailor-made music excluded their tinnitus frequency showed a significant decrease in both the behavioral and physiological responses compared to controls. Figure 1 clearly demonstrates that the tinnitus loudness of the patients from the target group was significantly reduced ($F(1,7) = 26.1$, $P = 0.001$). In addition, the tinnitus annoyance ($F(1,7) = 13.0$, $P = 0.009$) as well as the experienced handicapping by tinnitus ($F(1,7) = 15.1$, $P = 0.006$) decreased significantly. In contrast, the matched patient groups (placebo and control) did not experience any significant effects in these variables over time (placebo group loudness $F(1,7) = 0.4$, $P = 0.54$; annoyance $F(1,7) = 0.19$, $P = 0.68$; handicapping $F(1,7) = 0.17$, $P = 0.69$); (control group loudness $F(1,6) = 0.9$, $P = 0.38$; annoyance $F(1,6) = 0.14$, $P = 0.72$; handicapping $F(1,6) = 0$, $P = 0.85$). The major advantage of the study was not only that we were able to obtain the subjective data of the three groups of tinnitus patients, but we were also able to evaluate the corresponding auditory-evoked brain waves from the primary and the secondary auditory cortex, reflecting to a high degree the tinnitus perception. This complements and corroborates the subjective tinnitus data, and is demonstrated in Figures 2 and 3. After 6 and 12 months of training, the tinnitus-related neuronal activity of the primary auditory cortex (Fig. 2, ASSR) was significantly reduced in the target group ($F(1,7) = 5.9$, $P = 0.045$), but did not significantly change in the placebo ($F(1,7) = 0.3$, $P = 0.48$) and in the monitoring group ($F(1,6) = 0.0$, $P = 0.95$). The obtained ASSR data from the primary auditory-cortex are in line with the N1m auditory evoked fields, reflecting the activation of secondary auditory structures (Fig. 3). The N1m was significantly reduced in the target group ($F(1,7) = 24.6$, $P = 0.002$). No significant N1m reduction was found in the placebo

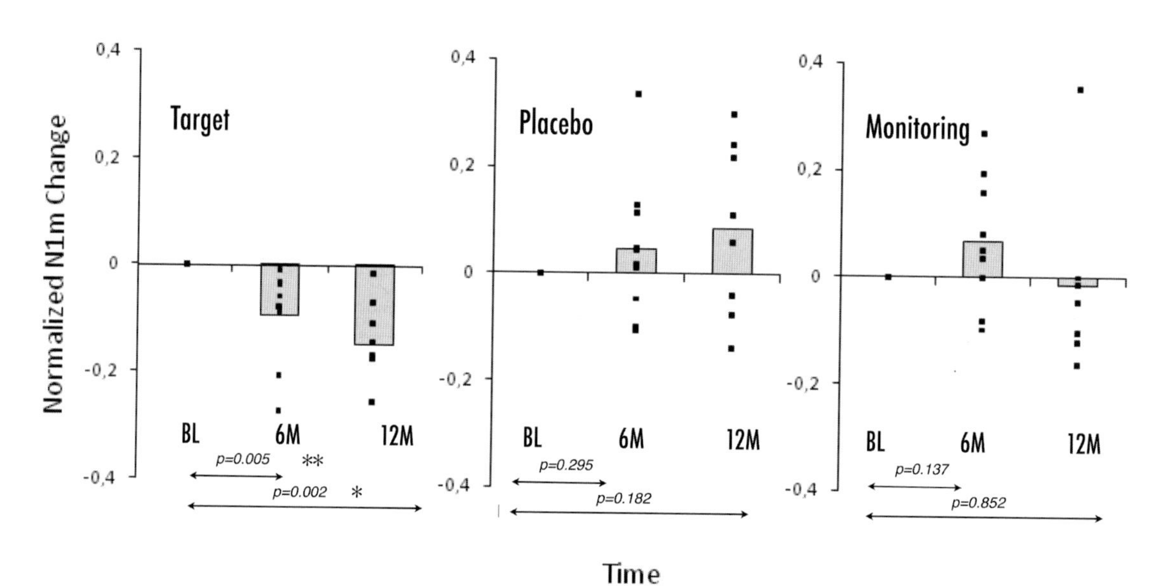

Figure 3. Change of the normalized tinnitus-related late auditory-evoked responses (N1) after 6 and 12 months of treatment (or monitoring) relative to baseline (BL) for the three patient groups (target, placebo, and monitoring). Positive change values reflect increment; negative change values reflect decrement. The bars indicate group averages; each dot indicates an individual data point.

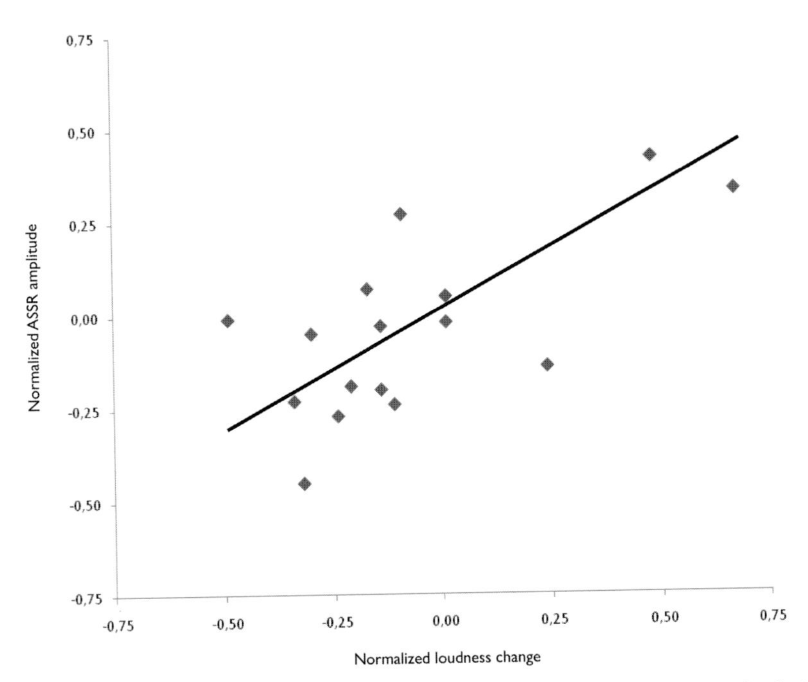

Figure 4. Correlation between the normalized tinnitus loudness change and the normalized ASSR amplitude change.

(N1m $F(1,7) = 2.2$, $P = 0.18$) as well as in the monitoring group ($F(1,6) = 0.0$, $P = 0.85$). Further, an interesting result was the significant linear relationship ($r = 0.69$, $P = 0.003$) between tinnitus loudness change and the evoked neuronal activity change in the primary auditory cortex at individual level (Fig. 4). Patients in whom the tinnitus became less loud exhibited reduced tinnitus-related primary auditory cortex activity, and patients in whom the loudness had not changed or had increased exhibited the corresponding change in tinnitus-related primary auditory cortex activity.

Discussion

We observed in the target group significant reductions in both tinnitus loudness and tinnitus-related auditory cortical activity, a result that we did not observe in the placebo or monitoring groups. Further, the changes in loudness as well as tinnitus-related evoked activity of the auditory cortex were significantly different between the target and placebo groups. Considering these findings, we can conclude that there is no general tinnitus loudness reduction trend over time. Thus, the results imply that the improvement in the target group reflects rather a specific treatment effect of the individual and tinnitus-specific modification of the music. By means of this music modification, we have "reattracted" lateral inhibition into the hyperactive tinnitus neurons, reversing their maladaptive hyperactivity and/or hypersynchrony. The observed reductions in tinnitus loudness, annoyance, and handicapping as well as the reductions in the evoked activity of the auditory cortex appear to be cumulative, which indicates a long-term neuroplastic effect in the corresponding cortical network. Thus, we applied the knowledge about maladaptive cortical reorganization in order to design a training procedure with individually customized music that appears suited to reduce brain activity corresponding to the tinnitus frequency and thus the tinnitus perception. The tailored notched music introduced a functional deafferentation of auditory neurons corresponding to the eliminated frequency band, and because this frequency band overlapped the individual tinnitus frequency, the notched music no longer stimulated the cortical area corresponding to the tinnitus frequency. However, the music still excited the neurons surrounding the tinnitus neurons. Thus, the neurons, which were not stimulated because of the individually set-up notch in the music, were presumably actively suppressed via lateral inhibitory inputs originating from surrounding neurons.[13,17,18] Alternatively, listening to the target notched music could have also induced synaptic and/or cellular plasticity mechanisms.[19,20] For instance, the deprivation from auditory input in the frequency range of the tinnitus frequency could have caused long-term depression of auditory neurons corresponding to the tinnitus frequency.

The customization of the musical frequency spectrum to reverse the maladaptive cortical plasticity related to tinnitus is an important issue. However, of further importance is the evidence that changes in cortical plasticity benefit from focused attention and enjoyment.[21] Thus, it is very advantageous to motivate the brain of the tinnitus patient to process the auditory input information as actively and pleasurably as possible. Therefore, we intentionally use "music" as a broadband acoustic signal containing for each human subject specific meaning and not a broadband "noise," which the brain would probably attempt to "get rid of" or filter out. Furthermore, music is able to absorb attention and elicit positive emotions, which is so important for plastic reorganizational brain processes to occur.[22] Even more, pleasant music can initiate the release of dopamine,[23,24] which evidently promotes cortical plasticity. This is the reason why we are offering the patients the opportunity to choose their favorite music, which enables them to listen with pleasure and for a long period of time. In addition, the tailor-made notched music treatment strategy not only is derived from neuroscientific findings on cortical plasticity and lateral inhibition of the human auditory cortex, targeting the reversion of the maladaptive reorganization of a specific cortical area contributing to the perception of tinnitus, but is also completely noninvasive, enjoyable, and low cost, representing a causal treatment that reduces the tinnitus loudness. A further advantage is that the notched music training can be easily complemented with other indirect psychological treatments.

Conflicts of interest

The authors declare no conflicts of interest.

References

1. Coles, R.R. 1984. Epidemiology of tinnitus: prevalence. *J. Laryngol. Otol.* **9**(Suppl): 7–15.
2. Heller, A.J. 2003. Classification and epidemiology of tinnitus. *Otolaryngol. Clin. North Am.* **36:** 239–248
3. Luxon, L.M. 1993. Tinnitus: its causes, diagnosis, and treatment. *BMJ* **306:** 1490–1491.
4. Davis, A. & A. El Rafaie. 2000. Epidemiology of tinnitus. In *Tinnitus Handbook*. Singular Publishing Group. San Diego.
5. Rauschecker, J.P., A.M. Leaver & M. Mühlau. 2010. Tuning out the noise: limbic-auditory interactions in tinnitus. *Neuron* **66:** 819–826.
6. Hesser, H. 2010. Methodological considerations in treatment evaluations of tinnitus distress: a call for guidelines. *J. Psychos. Res.* **69:** 305–307.
7. Møller, A.R. 2007. The role of neural plasticity in tinnitus. *Prog. Brain Res.* **166:** 37–45.
8. Eggermont, J.J. 2007. Pathophysiology of tinnitus. *Prog. Brain Res.* **166:** 19–35.
9. Muehlnickel, W., T. Elbert, E. Taub & H. Flor. 1998. Reorganization of auditory cortex in tinnitus. *Proc. Natl. Acad. Sci. USA* **95:** 10340–10343.
10. Diesch, E., M. Struve, A. Rupp, *et al.* 2004. Enhancement of steady-state auditory evoked magnetic fields in tinnitus. *Eur. J. Neurosci.* **19:** 1093–1104.
11. Okamoto, H., H. Stracke, W. Stoll & C. Pantev. 2010. Listening to tailor-made notched music reduces tinnitus loudness and tinnitus-related auditory cortex activity. *Proc. Natl. Acad. Sci. USA* **107:** 1207–1210.
12. Moore, B.C., B.R. Glasberg & M.L. Jepsen. 2009. Effects of pulsing of the target tone on the audibility of partials in inharmonic complex tones. *J. Acoust. Soc. Am.* **125:** 3194–3204.
13. Pantev, C., A. Wollbrink, L.E. Roberts, *et al.* 1999. Short-term plasticity of the human auditory cortex. *Brain Res.* **842:** 192–199.
14. Goebel, G. & W. Hiller. 1994. *The tinnitus questionnaire*. A standard instrument for grading the degree of tinnitus. Results of a multicenter study with the tinnitus questionnaire. *HNO* **42:** 166–172.
15. Engelien, A., M. Schulz, B. Ross, *et al.* 2000. A combined functional in vivo measure for primary and secondary auditory cortices. *Hear. Res.* **148:** 153–160.
16. Tesche, C.D., M.A. Uusitalo, R.J. Ilmoniemi, *et al.* 1995. Signal-space projections of MEG data characterize both distributed and well-localized neuronal sources. *Electroencephalogr. Clin. Neurophysiol.* **95:** 189–200.
17. Okamoto, H., R. Kakigi, A. Gunji & C. Pantev. 2007. Asymmetric lateral inhibitory neural activity in the auditory system: a magnetoencephalographic study. *BMC Neurosci.* **8:** 33.
18. Pantev, C., H. Okamoto, B. Ross, *et al.* 2004. Lateral inhibition and habituation of the human auditory cortex. *Eur. J. Neurosci.* **19:** 2337–2344.
19. Feldman, D.E. 2009. Synaptic mechanisms for plasticity in neocortex. *Annu. Rev. Neurosci.* **32:** 33–55
20. Turrigiano, G.G. 2008. The self-tuning neuron: synaptic scaling of excitatory synapses. *Cell* **135:** 422–435.
21. Polley, D.B., E.E. Steinberg & M.M. Merzenich. 2006. Perceptual learning directs auditory cortical map reorganization through top-down influences. *J. Neurosci.* **26:** 4970–4982.
22. Zatorre, R.J. & J. McGill. 2005. Music, the food of neuroscience? *Nature* **434:** 312–315.
23. Blood, A.J. & R.J. Zatorre. 2001. Intensely pleasurable responses to music correlate with activity in brain regions implicated in reward and emotion. *Proc. Natl. Acad. Sci. USA* **98:** 11818–11823
24. Bao, S.W., V.T. Chan & M.M. Merzenich. 2001. Cortical remodelling induced by activity of ventral tegmental dopamine neurons. *Nature* **412:** 79–83.

Ann. N.Y. Acad. Sci. ISSN 0077-8923

ANNALS OF THE NEW YORK ACADEMY OF SCIENCES
Issue: *The Neurosciences and Music IV: Learning and Memory*

Musician's cramp as manifestation of maladaptive brain plasticity: arguments from instrumental differences

Eckart Altenmüller,[1] Volker Baur,[1,2] Aurélie Hofmann,[1] Vanessa K. Lim,[3] and Hans-Christian Jabusch[4]

[1]Institute for Music Physiology and Musicians' Medicine (IMMM), University of Music, Drama and Media, Hannover, Germany. [2]Division Neuropsychology, Institute of Psychology, University of Zurich, Zurich, Switzerland. [3]Department of Psychology, Research Centre for Cognitive Neuroscience, The University of Auckland, Auckland, New Zealand. [4]Institute of Musicians' Medicine, University of Music Carl Maria von Weber, Dresden, Germany

Address for correspondence: Prof. Dr. Eckart Altenmüller, University of Music, Drama and Media Hannover, Institute for Music Physiology and Musicians' Medicine (IMMM), Emmichplatz 1, 30175 Hannover, Germany.
eckart.altenmueller@hmtm-hannover.de

Musician's cramp is a task-specific movement disorder that presents itself as muscular incoordination or loss of voluntary motor control of extensively trained movements while a musician is playing the instrument. It is characterized by task specificity and gender bias, affecting significantly more males than females. The etiology is multifaceted: a combination of a genetic predisposition, termed endophenotype, and behavioral triggering factors being the leading features for the manifestation of the disorder. We present epidemiological data from 591 musician patients from our outpatient clinic demonstrating an influence of fine-motor requirements on the manifestation of dystonia. Brass, guitar, and woodwind players were at greater risk than other instrumentalists. High temporospatial precision of movement patterns, synchronous demands on tonic and phasic muscular activation, in combination with fine-motor burdens of using the dominant hand in daily life activities, constitute as triggering factors for the disorder and may explain why different body parts are affected.

Keywords: musician's cramp; dystonia; maladaptive plasticity; endophenotype; instrument; gender

Introduction

Musician's dystonia (MD), also known as musician's cramp or focal dystonia in musicians, is a task-specific movement disorder that presents itself as muscular incoordination or loss of voluntary motor control of extensively trained movements while a musician is playing the instrument. For those who are affected, focal dystonia is highly disabling and, in many cases, the disorder terminates musical careers.[1]

MD may be classified according to the task specifically involved. For example, embouchure dystonia may affect coordination of lips, tongue, facial, and cervical muscles, and breathing in brass and wind players,[2] whereas pianist's cramp and violinist's cramp may affect the control of finger, hand, or isolated arm movements.[3] Typically, MD occurs without pain, although muscle aching can present after prolonged spasms. The loss of muscular coordination is frequently accompanied by a cocontraction of antagonist muscle groups. For example, in pianist's cramp, the coactivation of wrist flexor and wrist extensor muscles is frequently observed. Other diagnostic criteria are the consistency of dystonic symptoms when playing, although the degree of severity may vary according to general tension and mechanical properties of the instrument. However, as a diagnostic criterion, dystonic symptoms should be apparent shortly after beginning to play the instruments. We therefore do not consider abnormal fatigue or degradation of motor executive functions after prolonged playing as MD. According to recent estimates, one percent of all professional musicians are affected.[1] In contrast, in the general population, prevalence of focal dystonias, including

doi: 10.1111/j.1749-6632.2012.06456.x
Ann. N.Y. Acad. Sci. 1252 (2012) 259–265

writer's cramp, blepharospasm, and cervical dystonia, is estimated as 29.5 per 100,000 in the United States and 6.1 per 100,000 in Japan.[4,5]

With respect to the etiology, it is generally acknowledged that MD may represent a syndrome of central nervous maladaptive plasticity.[6] In healthy musicians, adaptive plastic reorganization of brain circuits has been demonstrated. Sensory–motor cortex receptive fields of single digits are enlarged,[7] and both motor excitability and long-term potentiation/long-term depression–like plasticity are enhanced.[8] It has been speculated that this has behavioral advantages because it may facilitate motor control of adjacent fingers, for example, for the playing of fast passages.

However, in MD patients, the reorganization of sensory–motor networks goes one step further: spatial differentiation of single digits disappears and the topography of receptive fields becomes disorganized.[9] Furthermore, the excitability of projections to all hand muscles is enhanced in an undifferentiated manner.[10] Although this excess reorganization is associated with loss of task-specific motor control, it is not possible to say whether it causes the breakdown in motor control, or whether it is a consequence of persistent abnormal movement patterns. However, the restitution of the topography of receptive fields along with clinical recovery in MD strongly suggests a causal relationship.[11]

Based on neurophysiological, genetic, and epidemiologic findings, we have recently proposed a heuristic model of the manifestation of MD.[12] We assume a genetic predisposition, the endophenotype, which is characterized by a deficit in inhibitory mechanisms in sensory–motor networks on several levels of the central nervous system. This translates into temporal and spatial overshoot in motor activation, leading to less-focused movement patterns and to degradation of sensory discrimination abilities.[13–15]

Whereas in nonmusicians, in most instances, the demands on temporal and spatial precision of movement patterns are not high enough to trigger clinical manifestations of these maladaptive plastic changes; in professional musicians, several conditions seem to contribute to the manifestation of MD: prolonged practice; chronic pain; predefined temporospatial constraints—for example, the need to play "correct notes" in sheet-music;[16] psychological conditions, such as anxiety and perfectionism;[17] gender;[18] and requirements of the musical instrument. All these factors may contribute to trigger the development of a dystonia on the basis of an endophenotype.

It has been found in previous studies that MD is not equally distributed among different instrumentalist groups.[19,20] For example, among string players, focal dystonia seems to be more frequent in high strings than in low strings.[21] Interestingly, in musicians with hand dystonia, an association exists between the instrument group and the localization of focal dystonia. In instruments with different workloads, different complexity of movements, or different temporospatial precision for both hands, focal dystonia appears more often in the more heavily used hand. Keyboard musicians (piano, organ, and harpsichord) and those with plucked instruments (guitar and e-bass) are primarily affected in the right hand. All these instruments are characterized by a higher workload in the right hand. In addition, guitar playing requires higher temporospatial precision in the right hand compared to the left hand. Bowed string players who have a higher workload and complexity of movements in the left hand are predominantly affected in that hand.[22]

In a recent study, we were able to demonstrate that not only the nature of movements required by the respective instruments were related to dystonic movement patterns, but also extra instrumental burdens of daily-life activities: in left-handed musicians, MD occurs more frequently in the left hand as compared to right-handed musicians and vice versa, suggesting that fine-motor activities of the dominant writing hand should be added as an additional triggering factor.[23]

In summary, we suggest that phenotypic occurrence of MD is more heavily influenced by environmental and behavioral factors than previously assumed. To further substantiate this notion, this paper focuses on three aspects. Taking advantage of a very large patient group suffering from MD, we first sought to confirm that the differential workload related to playing different instruments is an important trigger factor. Second, the effect of gender on the prevalence of musicians' cramp was also examined by comparing the patients' records with those of eight music conservatories in Germany. In these respects, the present work is an extension of a previous paper on 183 patients.[18] Third, we included self-declared handedness as a factor in our statistical analysis. The rationale behind this was the idea that handedness might also constitute an additional risk factor, given that sensory–motor workload not only occurs during instrumental playing but also during other skilled movements executed primarily by the dominant hand, such as writing or operating a cell phone. This latter analysis was partly published previously.[23]

Methods

Patients

Patient files of 591 consecutive patients diagnosed with MD between 1994 and 2007 in the outpatient clinic of the Institute of Music Physiology and Musicians' Medicine (IMMM) of the Hanover University of Music, Drama and Media (Hannover, Germany) were evaluated. Patient files comprised diagnosis, description of symptoms, and additional information such as self-declared handedness, which was used as a variable for statistical analysis. All patient files were examined and only those with a clear diagnosis of focal dystonia were included in this study. Exclusion criterion was any other or secondary neurological disorder. Diagnoses were confirmed by a trained movement disorders specialist and neurologist (author E.A.).

Controls

The control group was taken from the previous study of Lim and Altenmüller[18] and consisted of 2,651 students enrolled in studies of all instrumental groups from eight different music conservatories. Mean age differed between the patient and control group because the students were slightly younger than the patients, given that the age of onset of MD is generally in the fourth decade of life.[16]

Data analysis

For statistical analysis, the software package SPSS 15.0 was used. Chi-square (χ^2) tests were employed for comparisons of ratios between groups. In cases of small sample sizes, Fisher's exact test was applied. The level of significance was set at $\alpha = 0.05$ and all P values are reported uncorrected for multiple comparisons because we applied a strongly hypothesis-driven analysis. Instruments were categorized in the following manner: *keyboard* (piano, organ, harpsichord, accordion, and keyboard), *string* (violin, cello, viola, and double bass), *plucked instruments* (guitar, e-bass, and mandoline), *woodwind* (flute, clarinet, saxophone, oboe, bassoon, recorder, and bagpipes), *brass* (trombone, horn, trumpet, tuba, and bariton), and *percussion* (drums and kettledrum).

Results

The majority of the patients were professional musicians (93.9%). From 478 patients, self-declared handedness was noted in the files and was distributed in the following manner: 89.3% right-handed, 9.2% left-handed, and 1.5% ambidextrous. Thirty-two (5.4%) of the patients were simultaneously affected by writer's cramp.

Localization of focal dystonia

Focal dystonia was predominantly located in the upper limb (hand and arm; 83.1%), whereas focal lower-limb dystonia could be seen in four patients and cervical dystonia (e.g., torticollis) could be seen in seven patients. One hundred and twenty (20.1%) patients suffered from embouchure dystonia (cramping or tremor of the lips). There was a higher ratio of patients with embouchure dystonia than upper-limb dystonia when comparing brass with woodwind players ($\chi^2 = 121.76$, $df = 1$, $P < 0.001$). Not surprisingly, percussionists were more likely to have lower-limb dystonia than all other musicians (Fisher's exact test, $P < 0.001$). Among all cases with upper-limb dystonia ($n = 465$), arm dystonia was much less common than hand dystonia (3.64% vs. 96.36%). String players had a higher rate of arm dystonia than all other musicians (16.05% vs. 1.03%; Fisher's exact test, $P < 0.001$). Among patients playing strings, arm dystonia was more often in the right side than left (seven right vs. three left; right is the bowing hand) when compared with hand dystonia (Fisher's exact test, $P = 0.049$).

Laterality of focal upper limb dystonia

Dystonic symptoms among patients with unilateral focal upper-limb dystonia (localized in hand or arm, $n = 452$) occurred on different sides ($\chi^2 = 66.59$, $df = 3$, $P < 0.001$). Because only two brass players and 11 percussionists were affected in the hand or arm, these two instrument groups were excluded from this test. Follow-up χ^2 tests yielded significant differences for keyboard versus string players ($\chi^2 = 49.31$, $df = 1$, $P < 0.001$), keyboard versus woodwind players ($\chi^2 = 31.68$, $df = 1$, $P < 0.001$), string players versus players with plucked instruments ($\chi^2 = 29.87$, $df = 1$, $P < 0.001$), and woodwind players versus players with plucked instruments ($\chi^2 = 17.35$, $df = 1$, $P < 0.001$). This implies that keyboard musicians and those with plucked instruments were predominantly affected in the right hand and string players in the left hand. No general lateralization could be seen in woodwind players (46.2% right, 49.5% left, and 4.4% on both hands). However, within woodwind players, patients playing the flute were mainly affected on the left and those playing oboe or clarinet on the right side ($\chi^2 = 5.87$, $df = 1$, $P = 0.015$). Keyboard musicians suffered more often from bilateral focal upper limb dystonia than musicians of all other instrumental groups (7.3% in keyboards vs. 3% in other instruments, $\chi^2 = 5.30$, $df = 1$, $P = 0.021$).

As already mentioned, an association between handedness and laterality of dystonia could be observed in patients with unilateral hand dystonia ($n = 362$, ambidextrous patients were excluded from this test). That is, the percentage of right-handers was higher among patients with right-sided dystonia compared to left-sided dystonia, and accordingly, the percentage of left-handers was higher among patients with left-sided dystonia compared to right-handed dystonia ($\chi^2 = 4.07$, $df = 1$, $P = 0.044$).[23]

Rare symptoms of focal dystonia

Focal task-specific tremor affected only 3.4% ($n = 20$) of the patients, whereas the majority of the patients showed the typical symptom of cramping. String players had focal tremor more often than all other instrumentalists (Fisher's exact test, $P < 0.001$).[23] This tremor was exclusively (100%) located in the right side, whereas cramping occurred to a lower percentage (30%) on the right side (Fisher's exact test, $P < 0.001$).

Among the patients with cervical dystonia ($n = 7$), three suffered from dystonic tremor in this body region, whereas torticollis, respectively anterocollis, could be seen in the other four patients. Compared to other areas of the body, the ratio of tremor was higher in cervical dystonia than in all other dystonias (42.9% vs. 2.9%, Fisher's exact test, $P < 0.001$; the small sample size of patients with focal tremor should be noted).

Interaction between instrument and population

This comparison involved instrument versus population (dystonia vs. healthy; $\chi^2 = 146.61$, $df = 5$, $P < 0.001$). Gender was not a factor in this analysis. Follow-up two-by-two tests were employed (e.g., keyboard players and nonkeyboard players separated in patients and controls; Fig. 1).

The observed number of patients was higher than the expected number in the following three instrumental groups: plucked instruments ($\chi^2 = 60.48$, $df = 1$, $P < 0.001$), woodwind ($\chi^2 = 5.00$, $df = 1$, $P = 0.025$), and brass ($\chi^2 = 20.79$, $df = 1$, $P < 0.001$). No significant differences between observed and expected numbers could be seen in keyboard players ($\chi^2 = 0.15$, $df = 1$, n.s.) and percussionists ($\chi^2 = 0.39$, $df = 1$, n.s.). Only in the group of string

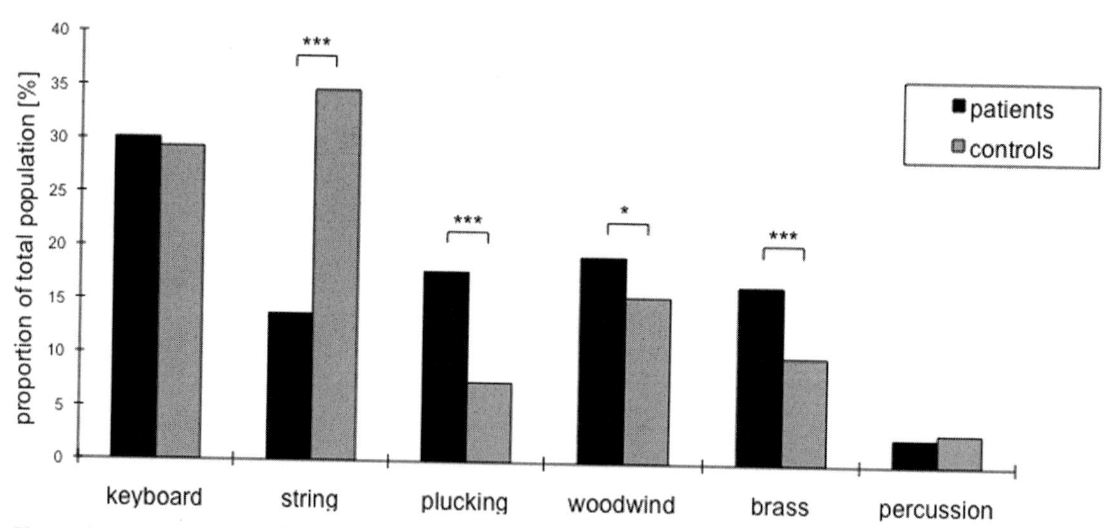

Figure 1. Distribution of instruments in 591 patients with MD as compared to 2,651 healthy musicians displayed as relative ratios. Highly significant differences in both populations ($P < 0.001$) are marked with three asterisks and significant findings ($P < 0.05$) with one asterisk. MD is more frequent than expected in plucking instruments, brass instruments, and woodwinds and less frequent in string players.

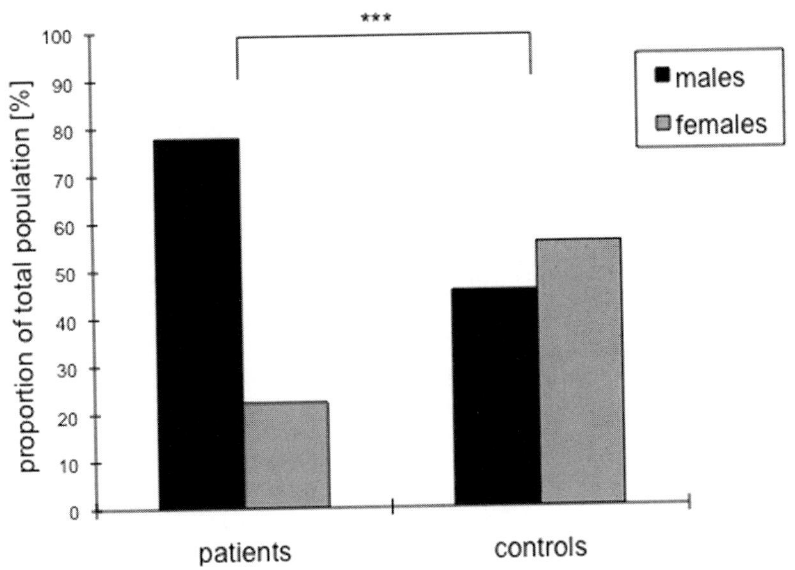

Figure 2. Gender distribution in 591 patients suffering from MD as compared to 2,651 healthy musicians displayed as relative ratios. Significantly more male musicians than females suffer from MD ($P < 0.001$).

players, the observed number of patients was lower than the expected number ($\chi^2 = 99.23$, $df = 1$, $P < 0.001$).

Interaction between gender and population

This interaction involved gender and population, whereas instrumental groups were not split up in this analysis. More males than females were in the dystonia population (77.8% vs. 22.2%, i.e., a ratio of 3.5:1), whereas the proportion of females was higher than that of males in the control group (55.5% vs. 44.5%). This interaction was significant ($\chi^2 = 214.68$, $df = 1$, $P < 0.001$; Fig. 2).

Discussion

The results are consistent with the "trigger hypothesis," which poses that the clinical manifestation of maladaptive brain plasticity in MD is strongly influenced by specific triggering factors, for example, high requirements on manual skills, temporospatial precision of movements, and extra instrumental burdens linked to handedness and skilled use of the dominant hand.[21,23] The cause of the gender prevalence of MD in male musicians still remains unclear. Previous studies report gender ratios between 2:1 and 5:1 (male: female) in focal dystonia patients.[24–26] This study, with its large sample size, reveals a ratio of 3.5:1. Taking into account that there are more females in the control group by a factor of 1.2, it can be concluded that males are at greater risk for developing MD than females by a factor of 4.3. It is an open question why males' susceptibility for MD is higher. There might be hormonal reasons or possibly different psychological attitudes concerning work behavior and coping with stressors in men and women.

Localization of MD

With respect to the higher prevalence of embouchure dystonia in brass and woodwind players, it is obvious that in brass playing, the control of lip movements is more critical for sound production when compared to woodwind players. Brass players are required to fine tune the tension of embouchure muscles to obtain a highly precise control of frequency and amplitude of lip vibrations. In woodwinds, embouchure adjustments do not require lip vibration, and therefore are less time-constrained. Conversely, finger movement patterns are more complex in woodwind than brass players, explaining why hand dystonia is a rare exception in brass players. A similar influence of workload, related to the body part affected, can be seen in arm dystonia: in string players, the right arm is involved in precise bowing movements. Therefore, a relatively high incidence of left-arm dystonia can be observed only in this group.

The results for the laterality of focal upper-limb dystonia indicate a similar influence of workload as a trigger. In keyboard and plucked instruments, movement patterns are more complex in the right upper limb, whereas in string players, there is higher spatial and temporal precision needed in the left hand.[27]

Within woodwind players, flutists have MD mainly on the left side, whereas oboists and clarinetists show their symptoms predominantly in the right hand. This might be due to the combination comprising prolonged support of the instrument with the thumb (left in the flute and right in clarinet and oboe) and fine motor control of the other four fingers. Obviously, this double task in one hand represents a challenge for the motor system since tonic activation and, at the same time, fine-motor phasic control of adjacent fingers in the millisecond range is required.

Although there are different workloads in the hands for keyboard players (76% of them are affected in the right upper limb), this instrumental group has the highest percentage of bilateral dystonia (12%). One explanation is that there are similar movements in the right and left hand.

Interaction between instrument and population

As already shown,[18] the distribution of musicians' cramp across instrumental groups differs and certain instrumental groups are at higher risk for developing MD. In contrast to the previous study, here we show that musicians' cramp occurs more often than expected in brass players as well, whereas percussionists are not at special risk for focal dystonia. The results for the other instrumental groups are in line with the previous study: plucked instruments and woodwind players have higher numbers of observed patients and string players have lower numbers. For keyboard players, observed and expected numbers do not differ significantly. This suggests that workload and complexity of movements may depend on the instrument group, with brass instruments, plucked instruments, and woodwind players showing the highest risk for focal dystonia.

Among the patients who play stringed instruments, high strings (violin and viola) are overrepresented. Compared to the ratio of high and low string players in German orchestras, it can be concluded that high string players are at higher risks for MD than low strings.[28] Again, this suggests that movement patterns requiring higher spatial and temporal precision are a risk factor for the manifestation of MD.

Conclusions

This study confirms in a large sample of musicians suffering from MD that high demands on temporospatial precision and a complex nature of movement patterns are the risk factors for developing MD. Because MD is a disorder based on maladaptive changes in the central nervous sensory-motor networks triggered by a variety of behavioral factors, it seems appropriate to pay specific attention to all young musicians learning to play "instruments at risk" to teach them prevention strategies such as avoidance of overuse and chronic pain.

Conflicts of interest

The authors declare no conflicts of interest.

References

1. Altenmüller, E. 2003. Focal dystonia: advances in brain imaging and understanding of fine motor control in musicians. *Hand Clin.* **19:** 523–538.
2. Frucht, S.J. 2009. Embouchure dystonia: portrait of a task-specific cranial dystonia. *Mov. Disord.* **24:** 1752–1762.
3. Lederman, R.J. 1991. Focal dystonia in instrumentalists: clinical features. *Med. Probl. Perform. Art.* **6:** 132–136.
4. Nutt, J.G., M.D. Muenter, L.J. Melton, *et al.* 1988. Epidemiology of dystonia in Rochester, Minnesota. *Adv. Neurol.* **50:** 361–365.
5. Nakashima, K., M. Kusumi, Y. Inoue & K. Takahashi. 1995. Prevalence of focal dystonias in the western area of Tottori Prefecture in Japan. *Mov. Disord.* **10:** 440–443.
6. Münte, T.F., E. Altenmüller & L. Jäncke. 2002. The musician's brain as a model of neuroplasticity. *Nat. Rev. Neurosci.* **3:** 473–478.
7. Elbert, T., C. Pantev, C. Wienbruch, *et al.* 1995. Increased cortical representation of the fingers of the left hand in string players. *Science* **270:** 305–307.
8. Rosenkranz, K., A. Williamon & J.C. Rothwell. 2007. Motorcortical excitability and synaptic plasticity is enhanced in professional musicians. *J. Neurosci.* **27:** 5200–5206.
9. Elbert, T., V. Candia, E. Altenmüller, *et al.* 1998. Alteration of digital representations in somatosensory cortex in focal hand dystonia. *Neuroreport* **16:** 3571–3575.
10. Rosenkranz, K., A. Williamon, K. Butler, *et al.* 2005. Pathophysiological differences between musician's dystonia and writer's cramp. *Brain* **128:** 918–931.
11. Candia, V., C. Wienbruch, T. Elbert, *et al.* 2003. Effective behavioral treatment of focal hand dystonia in musicians alters somatosensory cortical organization. *Proc. Natl. Acad. Sci. USA* **100:** 7942–7946.
12. Altenmüller, E. & H.C. Jabusch. 2010. Focal dystonia in musicians: phenomenology, pathophysiology and triggering factors. *Eur. J. Neurol.* **17**(Suppl 1): 31–66.

13. Bara-Jimenez, W., P. Shelton, T.D. Sanger & M. Hallett. 2000. Sensory discrimination capabilities in patients with focal hand dystonia. *Ann. Neurol.* **47:** 377–380.
14. Tinazzi, M., T. Rosso & A. Fiaschi. 2003. Role of the somatosensory system in primary dystonia. *Mov. Disord.* **18:** 605–622.
15. Lim, V.K., J.L. Bradshaw, M. Nicholls & E. Altenmüller. 2003. Perceptual differences in sequential stimuli across patients with musicians and writer's cramp. *Mov. Disord.* **11**: 1286–1293.
16. Altenmüller, E. & H.C. Jabusch. 2009. Focal hand dystonia in musicians: phenomenology, etiology, and psychological trigger factors. *J. Hand Ther.* **22:** 144–154.
17. Enders, L., J. Spector, E. Altenmüller, *et al.* 2011. Musician's dystonia and comorbid anxiety: two sides of one coin? *Mov. Disord.* **15:** 539–542.
18. Lim, V. & E. Altenmüller. 2003. Musicians cramp: instrumental and gender differences. *Med. Probl. Perform. Art.* **18:** 21–27.
19. Brandfonbrener, A.G. & C.A. Robson. 2002. A review of 111 musicians with focal dystonia seen at a performing artist's clinic 1985–2002. *Mov. Disord.* **17:** 1135–1138.
20. Tubiana, R. & P. Chamagne. 2000. Prolonged rehabilitation treatment of musicians' focal dystonia. In *Medical Problems of the Instrumentalist Musician.* R. Tubiana, P.C. Amadio, Eds.: 244–269. Martin Dunitz. London.
21. Jabusch, H.C. & E. Altenmüller. 2006. Epidemiology, phenomenology and therapy of musician's cramp. In *Music, Motor Control and the Brain.* E. Altenmüller, J. Kesselring, M. Wiesendanger, Eds.: 265–282. Oxford University Press. Oxford.
22. Conti, A.M., S. Pullman & S.J. Frucht. 2008. The hand that has forgotten its cunning—lessons from musicians' hand dystonia. *Mov. Disord.* **23:** 1398–1406.
23. Baur, V., H.C. Jabusch & E. Altenmüller. 2011. Behavioral factors influence the phenotype of musician's dystonia. *Mov. Disord.* **26:** 1780–1781.
24. Fahn, S., C.D. Marsden & D.B. Calne. 1987. Classification and investigation of dystonia. In *Movement Disorders 2.* C.D. Marsden, S. Fahn, Eds.: 332–358. Butterworths. London.
25. Lim, V.K., E. Altenmüller & J.L. Bradshaw. 2001. Focal dystonia: current theories. *Hum. Mov. Sci.* **20:** 875–914.
26. Jankovic, J. & A. Ashoori. 2008. Movement disorders in musicians. *Mov. Disord.* **23**: 1957–1965.
27. Jabusch, H.C., D. Zschucke, A. Schmidt, *et al.* 2005. Focal dystonia in musicians: treatment strategies and long-term outcome in 144 patients. *Mov. Disord.* **20:** 1623–1626.
28. Jabusch, H.C. & E. Altenmüller. 2006. Focal Dystonia in musicians: From phenomenology to therapy. *Adv. Cogn. Psychol.* **2:** 207–220.

Ann. N.Y. Acad. Sci. ISSN 0077-8923

ANNALS OF THE NEW YORK ACADEMY OF SCIENCES
Issue: *The Neurosciences and Music IV: Learning and Memory*

Music listening after stroke: beneficial effects and potential neural mechanisms

Teppo Särkämö[1,2] and David Soto[3]

[1]Cognitive Brain Research Unit, Institute of Behavioural Sciences, University of Helsinki, Helsinki, Finland. [2]Finnish Centre of Excellence in Interdisciplinary Music Research, University of Jyväskylä, Jyväskylä, Finland. [3]Department of Medicine, Centre for Neuroscience, Imperial College London, London, United Kingdom

Address for correspondence: Teppo Särkämö, Cognitive Brain Research Unit, Institute of Behavioural Sciences, Siltavuorenpenger 1 B, P.O. Box 9, FI-00014, University of Helsinki, Finland. teppo.sarkamo@helsinki.fi

Music is an enjoyable leisure activity that also engages many emotional, cognitive, and motor processes in the brain. Here, we will first review previous literature on the emotional and cognitive effects of music listening in healthy persons and various clinical groups. Then we will present findings about the short- and long-term effects of music listening on the recovery of cognitive function in stroke patients and the underlying neural mechanisms of these music effects. First, our results indicate that listening to pleasant music can have a short-term facilitating effect on visual awareness in patients with visual neglect, which is associated with functional coupling between emotional and attentional brain regions. Second, daily music listening can improve auditory and verbal memory, focused attention, and mood as well as induce structural gray matter changes in the early poststroke stage. The psychological and neural mechanisms potentially underlying the rehabilitating effect of music after stroke are discussed.

Keywords: music listening; stroke; emotion; cognition; rehabilitation; neuroplasticity

Besides being an enjoyable leisure activity, music conveys one of the most powerful and multifaceted sensory, motor, cognitive, and emotional experiences for the human brain.[1–3] Music can be considered a universal human trait that reaches deep into our species' past, develops early in life, and spans across all known human cultures.[4–6] Through the rapid development of portable music media, such as iPods, and increased music distribution channels, music has, however, never been more available and easily accessible as today. Today, most people interact with music on a daily basis, either by listening, singing, dancing, or playing, and music is valued by many, especially for its capacity to evoke and regulate emotions, provide enjoyment and comfort, and relieve stress.[7–9] Music has an important role in life during adolescence, especially in mood regulation,[10] but also in adults and aging people, who regard music as an important way to maintain self-esteem, feel competent and independent, and avoid feelings of loneliness and isolation,[11,12] suggesting that the emotional impact and importance of music applies to all ages.

In its daily use, music listening often accompanies normal ongoing activities, and even at work it may be used to enhance performance by improving our psychological state. Depending on the need, music can be used to energize or relax, to help maintain attentional focus or distract us, to help us remember or forget, and to isolate us from the environment or unite us with others—or simply just to make us feel good. During the past years, the subjective experience of emotional and cognitive change by music listening has increasingly been coupled with evidence that music listening can evoke emotions and influence our autonomic nervous system and neuroendocrine systems[7,13–18] and that music listening can also stimulate and enhance our mental performance in many various cognitive tasks.[19–31] In this paper, we will first summarize previous studies on the emotional and cognitive effects of music listening as well as on the neural basis of music processing in both healthy persons and in neurological patients. Then, we will review our recent studies on the short- and long-term effects of music listening on the recovery of function in stroke survivors and discuss

doi: 10.1111/j.1749-6632.2011.06405.x

the psychological and neural mechanisms potentially underlying the rehabilitating effect of music on the recovering brain.

Emotional and cognitive effects of music listening

Music listening can evoke strong emotional experiences, such as happiness, joy, peacefulness, and nostalgia.[7,13] These emotions are often accompanied by physiological reactions such as changes in heart rate (HR), respiration, skin temperature and conductance, and hormone (e.g., cortisol and testosterone) secretion.[14–18] By inducing positive affect and heightened arousal,[19] exposure to pleasant and enjoyable music can also temporarily enhance performance in many cognitive domains, including psychomotor or information processing speed,[20–23] reasoning,[19,24–26] attention and memory,[27–31] and creativity[22] in healthy subjects. Evidence from learning studies in healthy subjects also indicates that verbal material presented in a musical context (such as song lyrics) is learned and recalled better than spoken verbal material[32,33] (however, see Ref. 31 for conflicting results). It should, however, be noted that not all studies have reported any beneficial effects of music exposure on cognition, especially on verbal learning and memory performance.[35–37] Interestingly, some animal studies have also reported that frequent exposure to an enriched sound environment, which contains complex sounds or music, can improve auditory functions[38–41] as well as nonauditory learning and memory.[42–44] In many studies, music has directly been coined with positive affect and positive arousal states (or, in fact, used experimentally to induce it), although this may not be the case with animal studies because it is still unclear whether or not animals actually have affective responses to music.[45,46] Positive affect has been suggested to have an important influence on cognition and has been observed to systematically influence performance on many cognitive tasks, including episodic memory, working memory, and creative problem solving tasks, possibly due to increased brain dopamine (DA) levels associated with positive affect.[47–49]

In addition to transient stimulating effects in healthy subjects, music listening can also have short-term beneficial effects on cognition in many neurological disorders. For example, music listening has been shown to temporarily improve arithmetic performance in children with attention-deficit hyperactivity disorder[50] as well as enhance autobiographical recall in Alzheimer's disease patients.[51,52] Interestingly, nonverbal auditory stimulation and phasic alerting tones have also been observed to temporarily alleviate neglect,[53,54] suggesting that improved alertness has a role in enhancing spatial attention. As in healthy subjects, the learning and recall of sung verbal material is better than that of spoken verbal material also in patients with multiple sclerosis[55] or Alzheimer's disease.[56] There are also case reports of patients with aphasia that show improved speech when the patients sing familiar music lyrics relative to when the patients merely speak excerpts of familiar lyrics.[57] Similarly, aphasic patients have also been observed to repeat and recall more words when singing than when speaking along an auditory model.[58] In summary, music, both when played in the background and when used as a mnemonic device, seems to improve cognitive performance both in healthy subjects and in various clinical populations.

Neural basis of music listening

The emotional and cognitive effects of music can also be understood by assessing the neural mechanisms involved in music listening. These have been investigated by means of functional neuroimaging methods, such as electroencephalography (EEG), magnetoencephalography (MEG), positron emission tomography (PET), and functional magnetic resonance imaging (fMRI), as well as studies of persons suffering from amusia, a deficit of music perception and memory caused by a lesion (acquired amusia) or abnormal brain development (congenital amusia). Regarding the perception of basic acoustical features in music, evidence from fMRI and PET studies suggests that the auditory cortex (AC) and other temporal lobe areas, especially in the right hemisphere, are active during passive listening to melodies[59–61] and respond to small pitch changes[62–64] as well as changes in timbre.[65,66] Similarly, studies on amusia indicate that the right AC, the planum temporale, the anterior superior temporal gyrus (STG), the temporoparietal junction, and the insula as well as right frontal areas, such as the inferior frontal gyrus (IFG), are crucial for perceiving melodies, pitch, and timbre.[67–74] By contrast, the perception of temporal acoustical cues (e.g., duration, rhythm, and tempo) seems to activate the AC in a more left-lateralized or bilateral fashion[63,64] and, according to lesion studies, also appears to involve many temporal, frontal, and temporal-parietal areas in both the left and the right hemispheres.[71,74–77] Corroborating evidence for the involvement of the AC in encoding basic acoustical features (e.g., pitch and duration) in music also comes from EEG and MEG studies using the mismatch negativity (MMN) response.[78]

Music listening triggers also many cognitive, emotional, and motor responses, involving neural activity that extends well beyond the AC and adjacent temporal lobe areas. First of all, the processing of more complex musical attributes (e.g., chords, harmony, and tonality), which requires a rule-based analysis of simultaneous and sequential pitch structures, engages the IFG, the medial prefrontal cortex, the premotor cortex, and the anterior STG.[79–84] Second, a similar frontoparietal network, including inferior frontal, dorsolateral prefrontal, precentral, anterior cingulate, and intraparietal areas, is also activated when we follow a melody and keep it in mind for a short period of time, activities requiring attention and working memory.[60,61,85–87] Third, hearing familiar music activates a rather widespread frontotempoparietal network, especially involving medial, orbital, and inferior prefrontal areas; supplementary motor areas; superior and middle temporal areas; the planum temporale; and the angular gyrus.[88–93] In addition, lesion studies suggest that deficits in recognizing familiar tunes tend to occur especially after damage to the anterior STG and the insula.[74,94] Fourth, music that evokes emotions engages virtually the entire limbic/paralimbic system, including the amygdala, the hippocampus, the parahippocampal gyrus, the nucleus accumbens (NAc), the ventral tegmental area (VTA), the anterior cingulate, and the orbitofrontal cortex.[95–102] Medial temporal lobe structures, such as the insula, the amygdala, and the parahippocampal cortex, as well as the posterior temporal lobe, have also been implicated as typical lesion areas resulting in loss of emotional reactivity to music.[74,103,104] Finally, the perception of rhythm involves areas in the cerebellum, the basal ganglia, and the motor cortex.[105–107] Taken together, this converging evidence from studies of both healthy subjects and amusic persons suggests that listening to music involves distributed cortical systems extending well beyond the AC and includes a vast network of temporal, frontal, parietal, subcortical, and cerebellar areas.

Use of music in stroke rehabilitation

Given that music widely engages brain regions involved in emotion, cognition, and motor functions, and that music therapy is used in many other clinical fields, it is surprising that the potential rehabilitating effects of musical activities in patients with neural damage have received relatively little scientific attention. Rhythmic auditory stimulation (RAS) and other interventions, which utilize the rhythm embedded in music to entrain motor behavior, have been observed to improve the recovery of gait and arm movements in hemiparetic stroke patients.[108–112] Similarly, training with musical instruments can improve the speed, precision, and smoothness of arm movements after a stroke.[113–115] Another application of music in rehabilitation is melodic intonation therapy (MIT),[116,117] which uses the musical elements (melody and rhythm) to train speech production in nonfluent aphasic patients. Case studies suggest that MIT may improve spontaneous speech output, articulation, and naming in aphasic patients.[118–121] Finally, different forms of active music therapy have also been used to improve mood and to increase social interaction and participation in rehabilitation after stroke.[122,123] A common theme across these intervention studies is that they all have used music as a specific therapeutic tool in a controlled, therapist-led clinical setting. Very little is actually known about the potential rehabilitative effects of normal music listening as an everyday leisure activity. This topic is explored in the following two studies.

Study I: effect of pleasant music on visual awareness after stroke

Visual neglect is a common neuropsychological disease that affects a high proportion of patients after stroke. Brain lesions usually in the right hemisphere, around the inferior parietal lobe, the temporoparietal junction, and the IFG, can lead to visual "neglect" characterized by a deficit in awareness for information presented on the side of space that is contralateral to the site of the brain lesion (e.g., impaired awareness of the left visual field following a right hemisphere lesions). Awareness deficits are more pronounced in visually crowded conditions, particularly when there are stimuli in the "good" visual field of the patients competing for attentional resources against stimuli in the contralesional "bad" side of space, leading to visual "extinction" of contralesional items.[124] There have been many studies examining the role of stimulus and task factors on the expression of visual neglect.[125] Attempts to rehabilitate visual neglect have focused on retraining of cognitive strategies to improve attention and facilitate selection of items in the contralesional visual field, the use of brain stimulation, and also pharmacological interventions by dopaminergic and noradrenergic agonists.[126] The motivation for the study presented here[127] was the absence of research on the role of emotional state in modulating visual processing in patients with neglect. But as reviewed above, positive emotional states can enhance a wide range of cognitive skills in the healthy, including, but not limited to, visual attention. Because there are no established effective treatments for visual neglect, the potential use of pleasant music to aid recovery of neglect may be timely and should inform the development of therapeutic approaches.

We used music to modulate the emotional state of three patients with visual neglect and/or extinction as they performed different visual tasks. Patients were aged between 60 and 74 years (mean age 67 years). Patient MP had damage to the right inferior frontal, superior temporal, and posterior parietal regions. Patient RH had a lesion in the left temporoparietal junction. Patient AS had damage to the right inferior and middle frontal regions and medial temporal cortex. Patients MP and AS showed left neglect, while patient RH showed right neglect.

The pleasant music was selected by the patient based on preference to maximize the induction of positive emotional responses. The unpreferred music was selected by the experimenter following a prior debriefing stage where patients listened to different music pieces of different genres. We obtained subjective measures of the effect of the different music conditions on the emotional state. The patients were asked to complete visual analogue scale (VAS) for ratings of enjoyment, mood, and arousal as they listened to the music. The ratings given by the patients in the VAS scales showed that, relative to the unpreferred music, the preferred music listening was consistently associated with heightened positive affect state in our three patients.[127] The preferred and the unpreferred music did not consistently differ in the ratings of subjective alertness. MP showed higher arousal ratings in the unpreferred music case, and the opposite was the case for RH; AS showed no differences between preferred and unpreferred music in arousal ratings.

The first study assessed the role of music exposure on visual neglect. The patients performed a computerized visual task as they listened to either preferred or unpreferred music or in a silence condition. The task required perceptual report of the color and shape of different geometric shapes presented in the left and/or right visual field (i.e., "blue circle," "green triangle"). The patients showed visual neglect. Perceptual report of contralesional visual targets was impaired relative to targets in the "good" ipsilateral field ($F = 579.4$, $P < 0.0001$).[127] Interestingly, there were significant effects of music condition ($F = 15.54$, $P < 0.0001$), with preferred music leading to improved perceptual identification of contralesional items relative the unpreferred music or silence conditions.[127] Enhanced awareness for contralesional items, through pleasant music listening, occurred both when a single item was presented in the contralesional field and when a competing ipsilesional item appeared.

In a separate study, we assessed the effects of music on physiological arousal by recording galvanic skin responses (GSR) and HR as the patients listened to preferred or unpreferred music or in silence. Figure 1 depicts the average of the GSRs across time for each of the music conditions and for each patient. In line with the VAS ratings, there were no consistent differences in GSR and HR across the conditions.[127] We note that, if anything, the highest level of arousal appeared with unpreferred music relative to the other conditions, though this pattern was not observed across the three patients, across the different time periods (i.e., see the graph for RH) or across the GSR and HR data. This result suggests that arousal alone was unlikely to account for the effect of pleasant music on visual neglect.

The enhancement of visual awareness through music was replicated in an additional paradigm where patients had to detect a red visual signal presented either in the contra- or ipsilateral visual field. The same detection paradigm was used in a subsequent single-case functional MRI study (see later). Figure 2A depicts the proportion of visual target detection as a function of the target visual field and the music conditions. Detection performance was

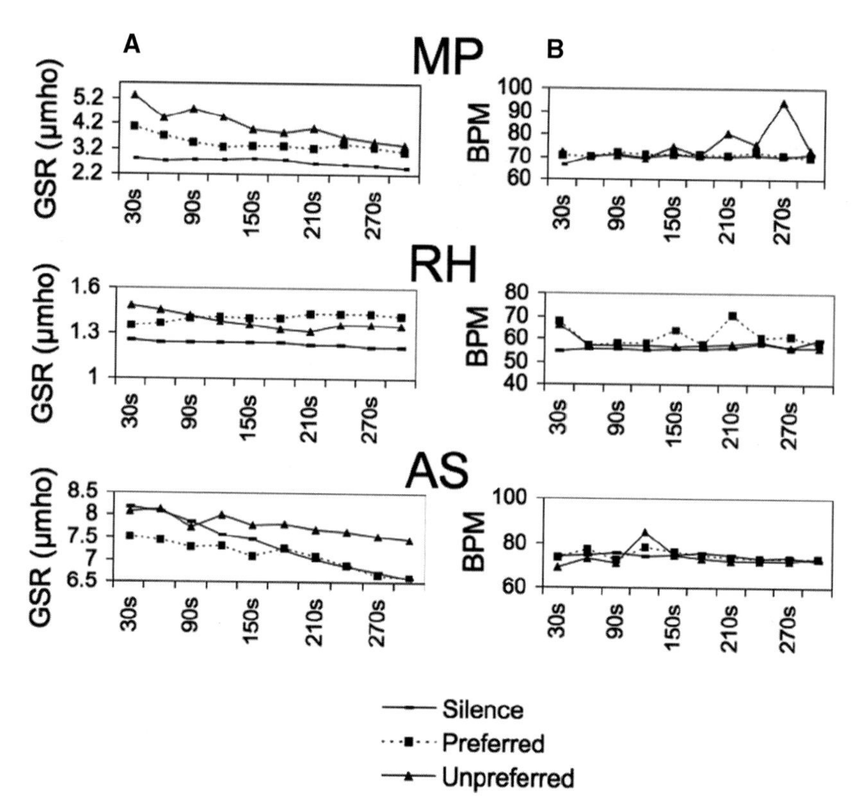

Figure 1. Temporal courses of the GSR (A) and HR (B) in beats per minute (BPM). Copyright (2009) National Academy of Sciences (www.nasonline.org).

impaired for targets in the contralesional field ($F = 579.4$, $P < 0.0001$), but this effect was critically modulated by music condition ($F = 131.3$, $P < 0.0001$) with the degree of visual neglect of contralesional targets attenuated by preferred music listening.[127] Similar results were obtained with patient MP in a separate study where the music was not played during the task but positive affect was induced prior to the task by exposing MP to a musical video of the patient's favorite artist. Following the induction of positive affect, there was a dramatic reduction in visual neglect.[127]

The neural correlates of the music effect were delineated by means of functional MRI in one of the patients (MP). The visual task used in the scanner was identical to the detection paradigm described above and for which behavioral performance is given in Figure 2A. Pleasant music listening led to increased activation in the left IFG (including Broca's area), the left dorsolateral prefrontal cortex, and the cingulate gyrus ($P < 0.001$, uncorrected, Fig. 2B).[127] Notably, enhanced awareness of contralesional targets, through pleasant music listening, correlated with activity in the left orbitofrontal cortex and early visual areas ($P < 0.001$, uncorrected, Fig. 2C).[127] Furthermore, the results of functional connectivity analysis indicated that the left orbitofrontal cortex, a region involved in positive affect responses by music,[96,97] was functionally coupled with areas of the right posterior parietal cortex and early visual cortex in the preferred relative to the unpreferred music condition (Fig. 2D, $P < 0.001$, FDR corrected for multiple comparison).[127] This result is consonant with the idea that positive affect, through music, can increase attention resources available for visual processing.

Study II: effect of daily music listening on long-term stroke recovery

In this project,[128–132] we recruited 60 stroke patients from the Department of Neurology of the Helsinki University Central Hospital. The patients' ages were between 35 years and 75 years (mean 58.9 years, SD 8.8 years) and had an acute ischemic stroke in the

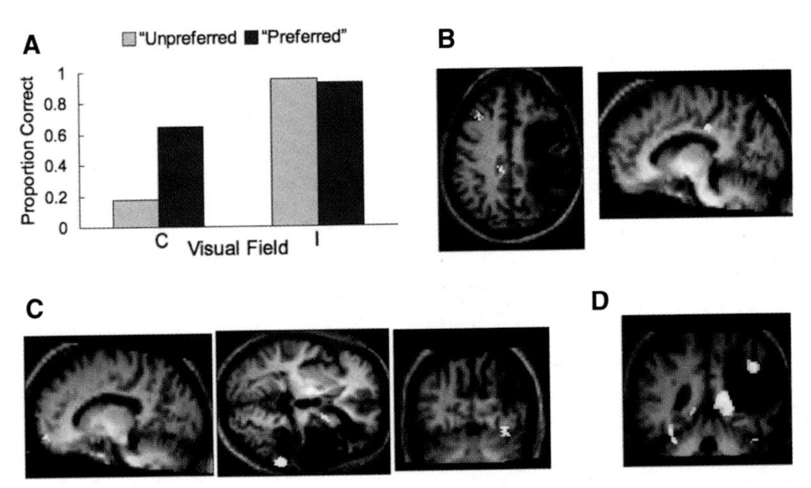

Figure 2. (A) Behavioral performance of patient MP in the visual detection task used in the functional MRI study. The proportions of correct responses as a function of the target visual field and the music listening condition are depicted. (B) Increases in neural activity in the preferred music relative to the unpreferred music condition. (C) Activity increases elicited by contralesional visual targets in the preferred music condition. (D) Functional coupling between the orbitofrontal cortex and perilesional areas of the right posterior parietal cortex and also intact early visual regions on correct detection trials in the preferred music compared with the unpreferred music condition. Copyright (2009) National Academy of Sciences (www.nasonline.org).

left ($n = 29$) or right ($n = 31$) middle cerebral artery (MCA) territory. As soon as possible after their hospitalization, the patients were randomly assigned to one of three groups: a music group, an audio book group, or a control group ($n = 20$ in each). In the music group, a music therapist provided the patients with portable CD players and CDs of their own favorite music in any musical genre (mostly popular music with lyrics but also jazz, folk, or classical music). Similarly, the therapist provided the audio book group with portable players and self-selected narrated audio books. The control group was not given any listening material. All groups received standard treatment for stroke in terms of medical care and rehabilitation, and there were no demographic or clinical differences between the groups.[128,129] Patients in the music and audio book groups were trained in using the players and were instructed to listen to the material by themselves daily (for a minimum of one hour per day) for the following two months. After this intervention period, they were encouraged to continue listening to the material on their own. In order to ensure that the patients were able to engage in the listening protocol, the therapist kept close weekly contact with the patients, the nurses, and relatives of the patients were asked to help. The patients also kept a listening diary. In order to determine the potential effect of the music listening on recovery, all patients underwent a neuropsychological assessment and an auditory MEG measurement one week, three months, and six months poststroke, as well as a structural MRI within two weeks of stroke onset and six months poststroke. All measures were performed and analyzed blind to the group allocation of the patients. In addition, the music and audio group patients were interviewed about their subjective experiences related to the listening experience and their recovery. All in all, 54 patients completed the study.

The neuropsychological assessment consisted of cognitive tests measuring short-term and working memory, verbal memory, language skills, visuospatial cognition, music perception, executive functions, and attention, as well as questionnaires related to mood and quality of life. As shown in Figure 3A, the groups differed on the recovery of verbal memory ($F = 4.7$, $P < 0.005$) and focused attention ($F = 3.9$, $P < 0.05$), with better improvement in the music group than in both the audio book and control groups.[128] In addition, there were group effects on the depression ($F = 3.7$, $P < 0.05$) and confusion ($F = 3.3$, $P < 0.05$) scores on the Profile of Mood States (POMS) after the intervention period, with the music group experiencing less depression and confusion than the control group (Fig. 3B).[128] Thus, the results suggest that daily music

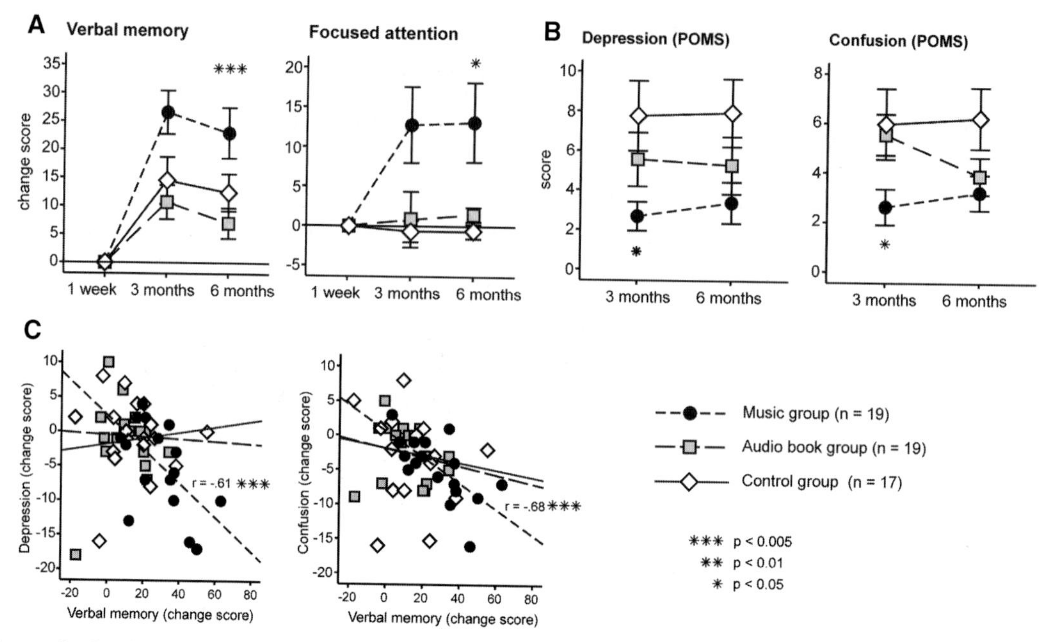

Figure 3. Cognitive performance and mood of patients at different stages of stroke recovery. (A) Group differences (mean ± SEM) in verbal memory and focused attention scores from the one-week to the six-month poststroke stage. (B) Group differences (mean ± SEM) in the POMS Depression and Confusion scores three months and six months poststroke. (C) Scatter plots depicting the correlation between the changes in verbal memory and mood. Panels A and B adapted from Ref. 128, with permission of the copyright holder (Oxford University Press).

listening can enhance cognitive recovery and prevent negative mood in the early poststroke stage. Interestingly, there were significant correlations between improved verbal memory and reduced depression ($r = -0.61$, $P = 0.005$) and confusion ($r = -0.68$, $P = 0.001$) within the music group,[130] suggesting that the improvements in memory and mood were related (Fig. 3C).

In order to gain more insight about the psychological factors underlying the therapeutic effects of music listening after stroke, the patients in the music and audio book groups were also interviewed on how they felt the listening had contributed to their recovery during the first three months after the stroke. Based on a qualitative content analysis, the responses of the patients were thematically grouped into six categories and the proportion of patients in each group whose responses fell into each of these categories was compared. Results showed that patients in the music group reported more often than the patients in the audio book group that the listening experience had helped them to relax ($\chi^2 = 25.8$, $P < 0.0001$), increase their motor activity (e.g., helping to do household chores, dancing; $\chi^2 = 31.8$, $P < 0.0001$), and improve their mood ($\chi^2 = 31.4$, $P < 0.0001$).[131] In contrast, both the music group and the audio book group experienced the listening experience as providing refreshing stimulation and evoking thoughts and memories about the past to the same extent.[131] Another small interview study of professional nurses ($n = 5$) who were working with stroke patients using music also suggested that music listening was beneficial for the mood, arousal, and concentration of the patient as well as for the communication and interaction between the patient and the nurse.[132]

In addition to the neuropsychological and interview data, both MEG and MRI were used to determine if there were any neuroplastic changes evoked by the listening intervention. In the MEG part of the study, a simple auditory oddball task was used to expose the patients with a sequence of harmonically rich tones, which consisted of frequently occurring standard tones and infrequently occurring deviant tones with either higher frequency or shorter duration than in the standard tone. The patients were instructed to ignore the sound stimuli and focus on watching a silent film. Contrasting the responses

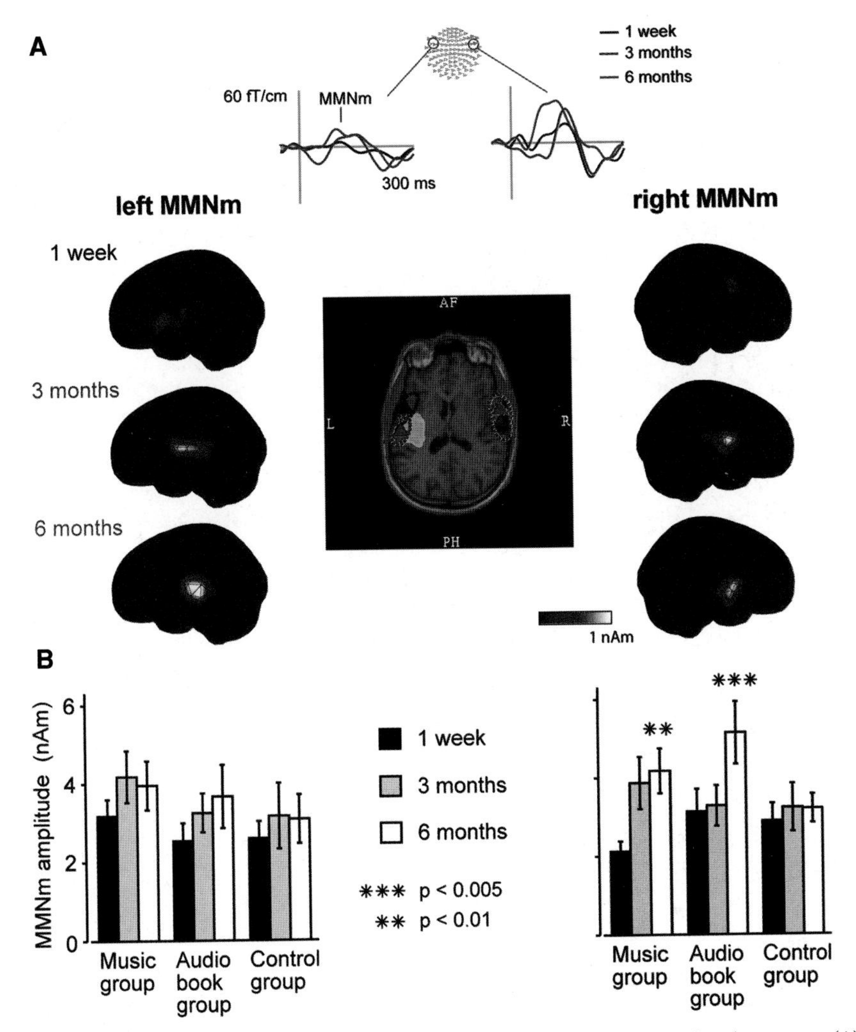

Figure 4. The mismatch negativity (MMNm) elicited by a frequency change at different stages of stroke recovery. (A) Case example of a stroke patient with changes in MMNm shown from individual MEG channels and with minimum current estimation (MCE) source models. The MRI image depicts the location of the lesion (white) as well as the region of interest (ellipsoids) used in the MCE model. (B) Group differences (mean ± SEM) in the left and right hemisphere MMNm amplitude from the one-week to the six-month poststroke stage. Adapted from Ref. 129 with permission of the copyright holder (MIT press).

to the deviant tones in this sequence relative to control sequences where the same tones were presented alone yielded a MMN (or its magnetic counterpart, MMNm) response,[133] which is an index of auditory discrimination or memory (Fig. 4A). We found that the groups differed on the recovery of the MMNm to both frequency ($F = 2.7$, $P < 0.05$) and duration ($F = 4.2$, $P < 0.005$) changes in the right hemisphere. Post-hoc testing indicated that the frequency–MMNm amplitude increased more in both music and audio book groups than in the control group (Fig. 4B).[129] In contrast, the duration–MMNm amplitude increased more in the audio book group than in the control and music groups.[129] As with the mood scores, the MMNm amplitudes were also found to correlate with improvement in tests of verbal memory ($r = 0.46$, $P < 0.05$) and focused attention ($r = 0.61$, $P < 0.05$) in the music group,[129] suggesting that cognitive recovery may also be linked to the enhancement of auditory sensory memory.

In the MRI part of the study, a 3D set of T1-weighted anatomical images were acquired from the patients within two weeks of stroke onset and six months poststroke using a 1.5 tesla scanner. Currently, we are performing voxel-based morphometry (VBM) analyses[134] with cost function masking[135] of the data to determine if there are any group differences in gray matter volume from the two-week to the six-month stage. The analyses are still ongoing, but preliminary results indicate that there were gray matter volume changes in the music group compared to the control group in the right insula (BA 13), the right IFG (BA 44), and the right precentral gyrus (BA 6). The changes in the right insula were also observed when comparing the music group to the audio book group. Thus, based on these tentative results, it seems that daily music listening after stroke may also induce long-term neuroplastic changes in the brain, especially in brain areas that have been implicated in the processing of music-related emotion,[97–102,136] musical structure,[80,137,138] and musical memory.[91]

Potential neurobiological mechanisms underlying the therapeutic effect of music listening after stroke

The two studies presented here provide the first evidence that music listening can influence stroke recovery by temporarily enhancing visual awareness neglect in patients as well as by improving the recovery of memory and attention, preventing negative mood, providing both relaxation and motor stimulation, as well as inducing functional and structural plastic changes in the right hemisphere. These findings suggest that music listening may be a beneficial leisure activity that could potentially provide an individually targeted, easy-to-conduct, and inexpensive way to help stroke patients cope with the adverse emotional and psychological impacts of stroke as well as to support their cognitive recovery, especially in the early poststroke stage when active rehabilitation or other hobbies are not yet possible. The question, however, remains as to what psychological and neural mechanisms could underlie the positive effect of music listening on cognition after stroke. Next, we will discuss four such potential mechanisms.

The first explanation for the positive effect of music on cognition after stroke is that it improves mood and arousal through modulation of the dopaminergic mesolimbic system, also known as the pleasure or reward system of the brain. This system is crucial for different emotional and cognitive processes, such as arousal, motivation, memory, attention, and executive functioning.[47] Activity of dopaminergic mesolimbic areas, such as the VTA, the NAc, the amygdala, and the hippocampus, has been associated with music-evoked emotions.[95–102] Recently, using a combination of psychophysical measures, PET, and fMRI, Salimpoor *et al.* found that endogenous DA release in striatal areas, such as the NAc, the caudate, and the putamen, was associated with emotional arousal during music listening,[139] thus providing the first direct evidence of a linkage between DA, music, and emotions. Interestingly, animal studies have also shown that music listening can increase dopaminergic neurotransmission and neostriatal DA concentrations[140] as well as enhance the effect of MDMA (ecstasy) on DA levels in the NAc.[141] In humans, DA has been found to mediate many cognitive functions, including working memory, attention, and executive functioning,[142] as well as regulate motor activity, motivation, and reward behavior.[143] Pharmacological studies have shown that DA agonists and stimulants that increase DA levels in the brain can improve working memory and executive functions in healthy subjects[144] and in patients with brain damage.[145] Thus, it is possible that listening to pleasurable music acts as a natural stimulant that increases DA secretion and thereby temporarily enhances cognitive performance. Another neurotransmitter that may possibly be linked to the alertness- and attention-stimulating effects of music is noradrenaline.[146,147] Noradrenergic drugs, such as guanfacine, can enhance memory and attention in both monkeys[148] and healthy humans.[149] There is also some evidence with a small number of patients that guanfacine can improve sustained attention in patients with visual neglect.[150] However, there is yet little evidence to support the possible link between music exposure and noradrenergic transmission and how this may vary as a function of music preference and the degree of enjoyment.

The DA hypothesis can explain the short-term, moment-to-moment effect of music on cognition, but the improved cognitive recovery after stroke may be the result of a more long-term positive effect of music on mood. Generally, the emotional and cognitive outcomes of stroke are highly interrelated, with one often affecting the other in the long run.[151,152] In study II above, music listening was found to reduce depression and confusion during the first three poststroke months,[128] and this reduction also correlated with the improvement in memory in the music listeners.[130] Similarly, study

I showed that heightened positive mood through music—and not merely arousal—accounted for the short-term remediation of visual neglect.[127] Study II also indicates that music listening, relative to the audio book listening, was experienced by the patients as the most mood-enhancing and relaxing context,[131] suggesting that it could alleviate the anxiety and psychological stress experienced by the patients. Stress is a common experience after stroke, and in the first months following the stroke, patients typically have chronically elevated cortisol levels (hypercortisolism), which is associated with poor cognitive function and depression.[153–155] Prolonged stress can also have maladaptive effects on neural plasticity, causing, for example, dendritic atrophy, excitatory synapse loss, and decreased neurogenesis in the hippocampus.[156] Listening to calming, pleasant music has been shown to transiently reduce cortisol levels, both after experimentally induced stress[15,18,157] and after stress induced by a medical procedure.[158,159] Thus, it is possible that, in the long run, daily listening to music could promote cognitive recovery after stroke by preventing depression and reducing stress.

A third mechanism that could account for the positive effect of music listening on both cognition and the MMNm response in study II is enhanced glutamatergic neurotransmission. Glutamate is the primary excitatory amino transmitter in the cortex and plays a critical role in learning and memory through its action at NMDA receptors.[160] Pharmacological studies have demonstrated that NMDA function is important for MMN generation in animals[161,162] and humans.[163,164] Furthermore, the MMN has also been shown to correlate with many cognitive skills, such as working memory, learning, executive functioning, and verbal skills, in both children and adults.[165–170] Animal studies have demonstrated that an enriched sound environment can enhance the expression and receptor function of glutamate in the AC and the anterior cingulate,[40,171] whereas auditory deprivation can decrease NMDA receptor expression levels in the AC.[172] Because changes in glutamate transmission also parallel the recovery from stroke,[173] glutamate may therefore be one crucial mechanism underlying the positive effect of music on the recovery of memory and the MMNm.

Finally, the fourth potential mechanism is increased neural plasticity evoked by environmental enrichment. This refers to any molecular or structural changes in the brain induced by the music environment. During the first weeks after stroke, the sensory and motor environment has an important role on recovery, and an enriched environment can lead to many forms of plastic changes in the recovering brain, including increased dendritic spine density, neurotrophic factor levels, and neurogenesis.[174–176] Moreover, developmental animal studies suggest that an enriched sound environment can shape the structure and function of the AC[38–41,171,177] as well as enhance neurogenesis and neurotrophin production in the brain.[42–44,178,179] In humans, these findings are paralleled by studies of children and adult musicians, which indicate that musical training can increase gray and white matter volumes in many cortical and subcortical areas controlling auditory, motor, and cognitive skills, which are needed for playing one's instrument.[180–185] Because music listening activates a widespread, mostly bilateral network of temporal, prefrontal, motor, parietal, cerebellar, and subcortical areas and has been observed to increase cerebral blood flow in the MCA,[186–188] after stroke it may perhaps work by stimulating the brain areas surrounding the ischemic lesion and thereby enhancing the adaptive plastic changes that typically occur in these areas after stroke.[189] This mechanism may also underlie the music-induced gray matter volume changes in the right insula and surrounding regions that were tentatively observed in study II.

In summary, the positive effect of music listening on cognition after stroke may be related to improved mood and arousal mediated by the dopaminergic mesolimbic system (and possibly the noradrenaline system), reduced depression and stress, increased glutamate-based neurotransmission, and enhanced molecular and structural brain plasticity. These suggested four mechanisms are by no means mutually exclusive and most likely work in concert to bring about the music-induced enhancement of recovery observed in the present studies. Owing to the relative novelty of the research field, however, there is currently no direct experimental evidence to support these hypothesized mechanisms. Thus, more studies utilizing physiological measures (e.g., HR, blood pressure), hormonal markers (e.g., cortisol), and structural and functional neuroimaging methods (e.g., fMRI, PET) in a joint manner are clearly needed in order to elucidate the mechanisms supporting the effect of music listening on stroke recovery and to delineate the specific role of different types of music and their optimal timing and intensity, with the aim of understanding how the interplay between music, emotion, and cognition is implemented in the brain in both health and disease and the ultimate goal of improving rehabilitation approaches through music in the injured brain.

Conflicts of interest

The authors declare no conflicts of interest.

References

1. Sacks, O. 2006. The power of music. *Brain* **129:** 2528–2532.
2. Trainor, L. 2008. Science & music: the neural roots of music. *Nature* **453:** 598–599.
3. Zatorre, R.J. 2005. Music, the food of neuroscience? *Nature* **434:** 312–315.
4. Merriam, A.P. 1964. *The Anthropology of Music*. Northwestern University Press. Evanston.
5. Nettl, B. 2000. An ethnomusicologist contemplates universals in musical sound and musical culture. In *The Origins of Music*. N.L. Wallin, B. Merker & S. Brown, Eds.: 463–472. MIT Press. Cambridge.
6. Trehub, S.E. 2003. The developmental origins of musicality. *Nat. Neurosci.* **6:** 669–673.
7. Juslin, P.N. & P. Laukka. 2004. Expression, perception, and induction of musical emotions: a review and a questionnaire study of everyday listening. *J. New Music Res.* **33:** 217–228.
8. Saarikallio, S. 2011. Music as emotional self-regulation throughout adulthood. *Psychol. Music* **39:** 307–327.
9. Sloboda, J. & S. O'Neill. 2001. Emotions in everyday listening to music. In *Music and Emotion: Theory and Research*. P. Juslin & J. Sloboda, Eds.: 415–429. Oxford University Press. Oxford.
10. Saarikallio, S. & J. Erkkilä. 2007. The role of music in adolescents mood regulation. *Psychol. Music* **35:** 88–109.
11. Cohen, A., B. Bailey & T. Nilsson. 2002. The importance of music to seniors. *Psychomusicology* **18:** 89–102.
12. Hays, T. & V. Minichiello. 2005. The meaning of music in the lives of older people: a qualitative study. *Psychol. Music* **33:** 437–451.
13. Zentner, M., D. Grandjean & K.R. Scherer. 2008. Emotions evoked by the sound of music: characterization, classification, and measurement. *Emotion* **8:** 494–521.
14. Fukui, H. & M. Yamashita. 2003. The effects of music and visual stress on testosterone and cortisol in men and women. *Neuroendocrinol. Lett.* **24:** 173–180.
15. Khalfa, S., S. Dalla Bella, M. Roy, *et al.* 2003. Effects of relaxing music on salivary cortisol level after psychological stress. *Ann. N.Y. Acad. Sci.* **999:** 374–376.
16. Krumhansl, C.L. 1997. An exploratory study of musical emotions and psychophysiology. *Can. J. Exp. Psychol.* **51:** 336–352.
17. Lundqvist, L.O., F. Carlsson, P. Hilmersson, *et al.* 2009. Emotional responses to music: experience, expression, and physiology. *Psychol. Music* **37:** 61–90.
18. Suda, M., K. Morimoto, A. Obata, *et al.* 2008. Emotional responses to music: towards scientific perspectives on music therapy. *NeuroReport* **19:** 75–78.
19. Thompson, W.F., E.G. Schellenberg & G. Husain. 2001. Arousal, mood, and the Mozart effect. *Psychol. Sci.* **12:** 248–251.
20. Clark, D.M. & J.D. Teasdale. 1985. Constraints on the effect of mood on memory. *J. Pers. Soc. Psychol.* **48:** 1595–1608.
21. Escoffier, N., D.Y. Sheng & A. Schirmer. 2010. Unattended musical beats enhance visual processing. *Acta. Psychol.* **135:** 12–16.
22. Schellenberg, E.G., T. Nakata, P.G. Hunter, *et al.* 2007. Exposure to music and cognitive performance: tests of children and adults. *Psychol. Music* **35:** 5–19.
23. Wood, J.V., J.A. Saltzberg & L.A. Goldsamt. 1990. Does affect induce self-focused attention? *J. Pers. Soc. Psychol.* **58:** 899–908.
24. Chabris, C.F. 1999. Prelude or requiem for the "Mozart effect"? *Nature* **400:** 826–827.
25. Rauscher, F.H., G.L. Shaw & K.N. Ky. 1993. Music and spatial task performance. *Nature* **365:** 611.
26. Rowe, G., J.B. Hirsh & A.K. Anderson. 2007. Positive affect increases the breadth of attentional selection. *P. Natl. Acad. Sci. USA* **104:** 383–388.
27. Beh, H.C. & R. Hirst. 1999. Performance on driving-related tasks during music. *Ergonomics* **42:** 1087–1098.
28. Greene, C.M., P. Bahri & D. Soto. 2010. Interplay between affect and arousal in recognition memory. *PLoS One* **5:** e11739.
29. Hallam, S., J. Price & G. Katsarou. 2002. The effects of background music on primary school pupils' task performance. *Educ. Stud.* **28:** 111–122.
30. Mammarella, N., B. Fairfield & C. Cornoldi. 2007. Does music enhance cognitive performance in healthy older adults? The Vivaldi effect. *Aging Clin. Exp. Res.* **19:** 394–399.
31. Thompson, R.G., C.J. Moulin, S. Hayre, *et al.* 2005. Music enhances category fluency in healthy older adults and Alzheimer's disease patients. *Exp. Aging Res.* **31:** 91–99.
32. Schön, D., M. Boyer, S. Moreno, *et al.* 2008. Songs as an aid for language acquisition. *Cognition* **106:** 975–983.
33. Wallace, W.T. 1994. Memory for music: effect of melody on recall of text. *J. Exp. Psychol. Learn.* **20:** 1471–1485.
34. Racette, A. & I. Peretz. 2007. Learning lyrics: to sing or not to sing? *Mem. Cognition* **35:** 242–253.
35. Boyle, R. & V. Coltheart. 1996. Effects of irrelevant sounds on phonological coding in reading comprehension and short-term memory. *Q. J. Exp. Psychol. A.* **49:** 398–416.
36. Jäncke, L. & P. Sandmann. 2009. Music listening while you learn: no influence of background music on verbal learning. *Behav. Brain Func.* **6:** 1–14.
37. Salamé, P. & A. Baddeley. 1989. Effects of background music on phonological short-term memory. *Q. J. Exp. Psychol-A.* **41:** 107–122.
38. Engineer, N.D., C.R. Percaccio, P.K. Pandya, *et al.* 2004. Environmental enrichment improves response strength, threshold, selectivity, and latency of auditory cortex neurons. *J. Neurophysiol.* **92:** 73–82.

39. Percaccio, C.R., N.D. Engineer, A.L. Pruette, *et al.* 2005. Environmental enrichment increases paired-pulse depression in rat auditory cortex. *J. Neurophysiol.* **94:** 3590–3600.
40. Xu, F., R. Cai, J. Xu, *et al.* 2007. Early music exposure modifies GluR2 protein expression in rat auditory cortex and anterior cingulate cortex. *Neurosci. Lett.* **420:** 171–183.
41. Xu, J., L. Yu, R. Cai, *et al.* 2009. Early auditory enrichment with music enhances auditory discrimination learning and alters NR2B protein expression in rat auditory cortex. *Behav. Brain Res.* **196:** 49–54.
42. Angelucci, F., M. Fiore, E. Ricci, *et al.* 2007. Investigating the neurobiology of music: brain-derived neurotrophic factor modulation in the hippocampus of young adult mice. *Behav. Pharmacol.* **18:** 491–496.
43. Chikahisa, S., H. Sei, M. Morishima, *et al.* 2006. Exposure to music in the perinatal period enhances learning performance and alters BDNF/TrkB signaling in mice as adults. *Behav. Brain Res.* **169:** 312–319.
44. Kim, H., M.H. Lee, H.K. Chang, *et al.* 2006. Influence of prenatal noise and music on the spatial memory and neurogenesis in the hippocampus of developing rats. *Brain Dev.* **28:** 109–114.
45. McDermott, J. & M. Hauser. 2005. The origins of music: innateness, uniqueness, and evolution. *Music Percept.* **23:** 29–59.
46. Rickard, N.S., S.R. Toukhsati & S.E. Field. 2005. The effect of music on cognitive performance: insight from neurobiological and animal studies. *Behav. Cogn. Neurosci. Rev.* **4:** 235–261.
47. Ashby, F.G., A.M. Isen & A.U. Turken. 1999. A neuropsychological theory of positive affect and its influence on cognition. *Psychol. Rev.* **106:** 529–550.
48. Isen, A.M. 1985. Asymmetry of happiness and sadness in effects on memory in normal college students: comment on Hasher, Rose, Zacks, Sanft, and Doren. *J. Exp. Psychol. Gen.* **114:** 388–391.
49. Isen, A.M., K.A. Daubman & G.P. Nowicki. 1987. Positive affect facilitates creative problem solving. *J. Pers. Soc. Psychol.* **52:** 1122–1131.
50. Abikoff, H., M.E. Courtney, P.J. Szeibel, *et al.* 1996. The effects of auditory stimulation on the arithmetic performance of children with ADHD and nondisabled children. *J. Learn. Disabil.* **29:** 238–246.
51. Foster, N.A. & E.R. Valentine. 2001. The effect of auditory stimulation on autobiographical recall in dementia. *Exp. Aging Res.* **27:** 215–228.
52. Irish, M., C.J. Cunningham, J.B. Walsh, *et al.* 2006. Investigating the enhancing effect of music on autobiographical memory in mild Alzheimer's disease. *Dement. Geriatr. Cogn.* **22:** 108–120.
53. Hommel, M., B. Peres, P. Pollak, *et al.* 1990. Effects of passive tactile and auditory stimuli on left visual neglect. *Arch. Neurol.* **47:** 573–576.
54. Robertson, I.H., J.B. Mattingley, C. Rorden, *et al.* 1998. Phasic alerting of neglect patients overcomes their spatial deficit in visual awareness. *Nature* **395:** 169–172.
55. Thaut, M.H., D.A. Peterson & G.C. McIntosh. 2005. Temporal entrainment of cognitive functions: musical mnemonics induce brain plasticity and oscillatory synchrony in neural networks underlying memory. *Ann. N.Y. Acad. Sci.* **1060:** 243–254.
56. Simmons-Stern, N.R., A.E. Budson & B.A. Ally. 2010. Music as a memory enhancer in patients with Alzheimer's disease. *Neuropsychologia* **48:** 3164–3167.
57. Straube, T., A. Schulz, K. Geipel, *et al.* 2008. Dissociation between singing and speaking in expressive aphasia: the role of song familiarity. *Neuropsychologia* **46:** 1505–1512.
58. Racette, A., C. Bard & I. Peretz. 2006. Making non-fluent aphasics speak: sing along! *Brain* **129:** 2571–2584.
59. Patterson, R.D., S. Uppenkamp, I. Johnsrude, *et al.* 2002. The processing of temporal pitch and melody information in auditory cortex. *Neuron* **36:** 767–776.
60. Brown, S. & M.J. Martinez. 2007. Activation of premotor vocal areas during musical discrimination. *Brain Cogn.* **63:** 59–69.
61. Zatorre, R.J., A.C. Evans & E. Meyer. 1994. Neural mechanisms underlying melodic perception and memory for pitch. *J. Neurosci.* **14:** 1908–1919.
62. Hyde, K.L., I. Peretz & R.J. Zatorre. 2008. Evidence for the role of the right auditory cortex in fine pitch resolution. *Neuropsychologia* **46:** 632–639.
63. Schönwiesner, M., R. Rübsamen & D.Y. von Cramon. 2005. Hemispheric asymmetry for spectral and temporal processing in the human antero-lateral auditory belt cortex. *Eur. J. Neurosci.* **22:** 1521–1528.
64. Zatorre, R.J. & P. Belin. 2001. Spectral and temporal processing in human auditory cortex. *Cereb. Cortex* **11:** 946–953.
65. Belin, P., R.J. Zatorre, P. Lafaille, *et al.* 2000. Voice-selective areas in human auditory cortex. *Nature* **403:** 309–312.
66. Warren, J.D., A.R. Jennings & T.D. Griffiths. 2005. Analysis of the spectral envelope of sounds by the human brain. *Neuroimage* **24:** 1052–1057.
67. Hyde, K.L., R.J. Zatorre, T.D. Griffiths, *et al.* 2006. Morphometry of the amusic brain: a two-site study. *Brain* **129:** 2562–2570.
68. Hyde, K.L., J.P. Lerch, R.J. Zatorre, *et al.* 2007. Cortical thickness in congenital amusia: when less is better than more. *J. Neurosci.* **27:** 13028–13032.
69. Johnsrude, I.S., V.B. Penhune & R.J. Zatorre. 2000. Functional specificity in the right human auditory cortex for perceiving pitch direction. *Brain* **123:** 155–163.
70. Liégeois-Chauvel, C., I. Peretz, M. Babaï, *et al.* 1998. Contribution of different cortical areas in the temporal lobes to music processing. *Brain* **121:** 1853–1867.
71. Milner, B.A. 1962. Laterality effects in audition. In *Interhemispheric Relations and Cerebral Dominance.* V.B. Mountcastle, Ed.: 177–195. Johns Hopkins Press. Baltimore.
72. Peretz, I. 1990. Processing of local and global musical information by unilateral brain-damaged patients. *Brain* **113:** 1185–1205.
73. Samson, S. & R.J. Zatorre. 1988. Melodic and harmonic discrimination following unilateral cerebral excision. *Brain Cogn.* **7:** 348–360.

74. Stewart, L., K. von Kriegstein, J.D. Warren, *et al.* 2006. Music and the brain: disorders of musical listening. *Brain* **129:** 2533–2553.
75. Robin, D.A., D. Tranel & H. Damasio. 1990. Auditory perception of temporal and spectral events in patients with focal left and right cerebral lesions. *Brain Lang.* **39:** 539–555.
76. Schuppert, M., T.F. Münte, B.M. Wieringa, *et al.* 2000. Receptive amusia: evidence for cross-hemispheric neural networks underlying music processing strategies. *Brain* **123:** 546–559.
77. Shapiro, B.E., M. Grossman & H. Gardner. 1981. Selective musical processing deficits in brain damaged populations. *Neuropsychologia* **19:** 161–169.
78. Tervaniemi, M. 2001. Musical sound processing in the human brain. Evidence from electric and magnetic recordings. *Ann. N.Y. Acad. Sci.* **930:** 259–272.
79. Janata, P., J.L. Birk, J.D. Van Horn, *et al.* 2002. The cortical topography of tonal structures underlying Western music. *Science* **298:** 2167–2170.
80. Koelsch, S., T. Fritz, K. Schulze, *et al.* 2005. Adults and children processing music: an fMRI study. *Neuroimage* **25:** 1068–1076.
81. Levitin, D.J. & V. Menon. 2003. Musical structure is processed in "language" areas of the brain: a possible role for Brodmann Area 47 in temporal coherence. *Neuroimage* **20:** 2142–2152.
82. Maess, B., S. Koelsch, T.C. Gunter, *et al.* 2001. Musical syntax is processed in Broca's area: an MEG study. *Nat. Neurosci.* **4:** 540–545.
83. Tillmann, B., P. Janata & J.J. Bharucha. 2003. Activation of the inferior frontal cortex in musical priming. *Cogn. Brain Res.* **16:** 145–161.
84. Koelsch, S. & W.A. Siebel. 2005. Towards a neural basis of music perception. *Trends Cogn. Sci.* **9:** 578–584.
85. Gaab, N., C. Gaser, T. Zaehle, *et al.* 2003. Functional anatomy of pitch memory: an fMRI study with sparse temporal sampling. *Neuroimage* **19:** 1417–1426.
86. Griffiths, T.D., I. Johnsrude, J.L. Dean, *et al.* 1999. A common neural substrate for the analysis of pitch and duration pattern in segmented sound? *Neuroreport* **10:** 3825–3830.
87. Janata, P., B. Tillmann & J.J. Bharucha. 2002. Listening to polyphonic music recruits domain-general attention and working memory circuits. *Cogn. Affect. Behav. Neurosci.* **2:** 121–140.
88. Groussard, M., F. Viader, V. Hubert, *et al.* 2010. Musical and verbal semantic memory: two distinct neural networks? *NeuroImage* **49:** 2764–2773.
89. Janata, P. 2009. The neural architecture of music-evoked autobiographical memories. *Cereb. Cortex* **19:** 2579–2594.
90. Peretz, I., N. Gosselin, P. Belin, *et al.* 2009. Music lexical networks. The cortical organization of music recognition. *Ann. N.Y. Acad. Sci.* **1169:** 256–265.
91. Plailly, J., B. Tillmann & J.P. Royet. 2007. The feeling of familiarity of music and odors: the same neural signature? *Cereb. Cortex* **17:** 2650–2658.
92. Platel, H., J.C. Baron, B. Desgranges, *et al.* 2003. Semantic and episodic memory of music are subserved by distinct neural networks. *Neuroimage* **20:** 244–256.
93. Satoh, M., K. Takeda, K. Nagata, *et al.* 2006. Positron-emission tomography of brain regions activated by recognition of familiar music. *Am. J. Neuroradiol.* **27:** 1101–1106.
94. Ayotte, J., I. Peretz, I. Rousseau, *et al.* 2000. Patterns of music agnosia associated with middle cerebral artery infarcts. *Brain* **123:** 1926–1938.
95. Baumgartner, T., M. Esslen & L. Jäncke. 2006. From emotion perception to emotion experience: emotions evoked by pictures and classical music. *Int. J. Psychophysiol.* **60:** 34–43.
96. Blood, A.J., R.J. Zatorre, P. Bermudez, *et al.* 1999. Emotional responses to pleasant and unpleasant music correlate with activity in paralimbic brain regions. *Nat. Neurosci.* **2:** 382–387.
97. Blood, A.J. & R.J. Zatorre 2001. Intensely pleasurable responses to music correlate with activity in brain regions implicated in reward and emotion. *P. Natl. Acad. Sci. USA* **98:** 11818–11823.
98. Brown, S., M.J. Martinez & L.M. Parsons. 2004. Passive music listening spontaneously engages limbic and paralimbic systems. *Neuroreport* **15:** 2033–2037.
99. Koelsch, S., T. Fritz, D.Y. von Cramon, *et al.* 2006. Investigating emotion with music: an fMRI study. *Hum. Brain Mapp.* **27:** 239–250.
100. Menon, V. & D.J. Levitin. 2005. The rewards of music listening: response and physiological connectivity of the mesolimbic system. *Neuroimage* **28:** 175–184.
101. Mitterschiffthaler, M.T., C.H. Fu, J.A. Dalton, *et al.* 2007. A functional MRI study of happy and sad affective states evoked by classical music. *Hum. Brain Mapp.* **28:** 1150–1162.
102. Koelsch, S. 2010. Towards a neural basis of music-evoked emotions. *Trends Cogn. Sci.* **14:** 131–137.
103. Gosselin, N., S. Samson, R. Adolphs, *et al.* 2006. Emotional responses to unpleasant music correlates with damage to the parahippocampal cortex. *Brain* **129:** 2585–2592.
104. Griffiths, T.D., J.D. Warren, J.L. Dean, *et al.* 2004. "When the feeling's gone": a selective loss of musical emotion. *J. Neurol. Neurosur. Ps.* **75:** 344–345.
105. Grahn, J.A. & M. Brett. 2007. Rhythm and beat perception in motor areas of the brain. *J. Cogn. Neurosci.* **19:** 893–906.
106. Popescu, M., A. Otsuka & A.A. Ioannides. 2004. Dynamics of brain activity in motor and frontal cortical areas during music listening: a magnetoencephalographic study. *Neuroimage* **21:** 1622–1638.
107. Rao, S.M., A.R. Mayer & D.L. Harrington. 2001. The evolution of brain activation during temporal processing. *Nat. Neurosci.* **4:** 317–323.
108. Thaut, M.H., G.C. McIntosh & R.R. Rice. 1997. Rhythmic facilitation of gait training in hemiparetic stroke rehabilitation. *J. Neurol. Sci.* **151:** 207–212.
109. Thaut, M.H., G.P. Kenyon, C.P. Hurt, *et al.* 2002. Kinematic optimization of spatiotemporal patterns in paretic arm training with stroke patients. *Neuropsychologia* **40:** 1073–1081.
110. Thaut, M.H., A.K. Leins, R.R. Rice, *et al.* 2007. Rhythmic auditory stimulation improves gait more than NDT/Bobath training in near-ambulatory patients early poststroke: a single-blind, randomized trial. *Neurorehab. Neural Re.* **21:** 455–459.

111. Jeong, S. & M.T. Kim. 2007. Effects of a theory-driven music and movement program for stroke survivors in a community setting. *Appl. Nurs. Res.* **20:** 125–131.
112. Schauer, M. & K.H. Mauritz. 2003. Musical motor feedback (MMF) in walking hemiparetic stroke patients: randomized trials of gait improvement. *Clin. Rehabil.* **17:** 713–722.
113. Altenmüller, E., J. Marco-Pallares, T.F. Münte, *et al.* 2009. Neural reorganization underlies improvement in stroke-induced motor dysfunction by music-supported therapy. *Ann. N.Y. Acad. Sci.* **1169:** 395–405.
114. Schneider, S., P.W. Schönle, E. Altenmüller, *et al.* 2007. Using musical instruments to improve motor skill recovery following a stroke. *J. Neurol.* **254:** 1339–1346.
115. Schneider, S., T. Münte, A. Rodriguez-Fornells, *et al.* 2010. Music-supported training is more efficient than functional motor training for recovery of fine motor skills in stroke patients. *Music Percept.* **27:** 271–280.
116. Albert, M.L., R.W. Sparks & N.A. Helm. 1973. Melodic intonation therapy for aphasia. *Arch. Neurol.* **29:** 130–131.
117. Norton, A., L. Zipse, S. Marchina, *et al.* 2009. Melodic intonation therapy: shared insights on how it is done and why it might help. *Ann. N.Y. Acad. Sci.* **1169:** 431–436.
118. Sparks, R., N. Helm & M. Albert. 1974. Aphasia rehabilitation resulting from melodic intonation therapy. *Cortex* **10:** 303–316.
119. Wilson, S.J., K. Parsons & D.C. Reutens. 2006. Preserved singing in aphasia: a case study of the efficacy of the Melodic Intonation Therapy. *Music Percept.* **24:** 23–36.
120. Schlaug, G., S. Marchina & A. Norton. 2008. From singing to speaking: why singing may lead to recovery of expressive language function in patients with Broca's aphasia. *Music Percept.* **25:** 315–323.
121. Schlaug, G., S. Marchina & A. Norton. 2009. Evidence for plasticity in white-matter tracts of patients with chronic Broca's aphasia undergoing intense intonation-based speech therapy. *Ann. N.Y. Acad. Sci.* **1169:** 385–394.
122. Magee, W.L. & J.W. Davidson. 2002. The effect of music therapy on mood states in neurological patients: a pilot study. *J. Music Ther.* **39:** 20–29.
123. Nayak, S., B.L. Wheeler, S.C. Shiflett, *et al.* 2000. Effect of music therapy on mood and social interaction among individuals with acute traumatic brain injury and stroke. *Rehabil. Psychol.* **45:** 274–283.
124. Karnath, H.O. 1988. Deficits of attention in acute and recovered visual hemi-neglect. *Neuropsychologia* **26:** 27–43.
125. Driver, J. & P. Vuilleumier. 2001. Perceptual awareness and its loss in unilateral neglect and extinction. *Cognition* **79:** 39–88
126. Singh-Curry, V. & M. Husain. 2010. Rehabilitation in practice: hemispatial neglect: approaches to rehabilitation. *Clin. Rehabil.* **24:** 675–684.
127. Soto, D., M.J. Funes, A. Guzmán-García, *et al.* 2009. Pleasant music overcomes the loss of awareness in patients with visual neglect. *P. Natl. Acad. Sci. USA* **106:** 6011–6016.
128. Särkämö, T., M. Tervaniemi, S. Laitinen, *et al.* 2008. Music listening enhances cognitive recovery and mood after middle cerebral artery stroke. *Brain* **131:** 866–876.
129. Särkämö, T., E. Pihko, S. Laitinen, *et al.* 2010. Music and speech listening enhance the recovery of early sensory processing after stroke. *J. Cogn. Neurosci.* **22:** 2716–2727.
130. Särkämö, T. 2011. Music in The Recovering Brain. Doctoral dissertation. University of Helsinki, Finland. URL https://helda.helsinki.fi/handle/10138/24940 [Accessed on 9 January 2012].
131. Forsblom, A., T. Särkämö, S. Laitinen, *et al.* 2010. The effect of music and audiobook listening on people recovering from stroke: the patient's point of view. *Music Med.* **2:** 229–234.
132. Forsblom, A., S. Laitinen, T. Särkämö, *et al.* 2009. Therapeutic role of music listening in stroke rehabilitation. *Ann. N.Y. Acad. Sci.* **1169:** 426–430.
133. Näätänen, R., P. Paavilainen, T. Rinne, *et al.* 2007. The mismatch negativity (MMN) in basic research of central auditory processing: a review. *Clin. Neurophysiol.* **118:** 2544–2590.
134. Ashburner, J. 2009. Computational anatomy with the SPM software. *Magn. Reson. Imaging* **27:** 1163–1174.
135. Brett, M., A.P. Leff, C. Rorden, *et al.* 2001. Spatial normalization of brain images with focal lesions using cost function masking. *NeuroImage* **14:** 486–500.
136. Omar, R., S.M. Henley, J.W. Bartlett, *et al.* 2011. The structural neuroanatomy of music emotion recognition: evidence from frontotemporal lobar degeneration. *Neuroimage* **56:** 1814–1821.
137. James, C.E., J. Britz, P. Vuilleumier, *et al.* 2008. Early neuronal responses in right limbic structures mediate harmony incongruity processing in musical experts. *Neuroimage* **42:** 1597–1608.
138. Mutschler, I., B. Wieckhorst, S. Kowalevski, *et al.* 2009. Functional organization of the human anterior insular cortex. *Neurosci. Lett.* **457:** 66–70.
139. Salimpoor, V.N., M. Benovoy, K. Larcher, *et al.* 2011. Anatomically distinct dopamine release during anticipation and experience of peak emotion to music. *Nat. Neurosci.* **14:** 257–262.
140. Sutoo, D. & K. Akiyama. 2004. Music improves dopaminergic neurotransmission: demonstration based on the effect of music on blood pressure regulation. *Brain Res.* **1016:** 255–262.
141. Feduccia, A.A. & C.L. Duvauchelle. 2008. Auditory stimuli enhance MDMA-conditioned reward and MDMA-induced nucleus accumbens dopamine, serotonin and locomotor responses. *Brain Res. Bull.* **77:** 189–196.
142. Nieoullon, A. 2002. Dopamine and the regulation of cognition and attention. *Prog. Neurobiol.* **67:** 53–83.
143. Knab, A.M. & J.T. Lightfoot. 2010. Does the difference between physically active and couch potato lie in the dopamine system? *Int. J. Biol. Sci.* **6:** 133–150.
144. Mehta, M.A. & W.J. Riedel. 2006. Dopaminergic enhancement of cognitive function. *Curr. Pharm. Design* **12:** 2487–2500.
145. Bales, J.W., A.K. Wagner, A.E. Kline, *et al.* 2009. Persistent cognitive dysfunction after traumatic brain injury: a dopamine hypothesis. *Neurosci. Biobehav. R.* **33:** 981–1003.
146. Panksepp, J. & G. Bernatzky. 2002. Emotional sounds and the brain: the neuro-affective foundations of musical appreciation. *Behav. Process.* **60:** 133–155.
147. Smith, A. & D. Nutt. 1996. Noradrenaline and attention lapses. *Nature* **380:** 291.
148. Franowicz, J.S. & A.F. Arnsten, 1998. The alpha-2a noradrenergic agonist, guanfacine, improves delayed response

performance in young adult rhesus monkeys. *Psychopharmacology* **136**: 8–14.
149. Jäkälä, P., M. Riekkinen, J. Sirviö, *et al.* 1999. Guanfacine, but not clonidine, improves planning and working memory performance in humans. *Neuropsychopharmacology* **20:** 460–470.
150. Malhotra, P.A., A.D. Parton, R. Greenwood, *et al.* 2006. Noradrenergic modulation of space exploration in visual neglect. *Ann. Neurol.* **59**: 186–190.
151. Nys, G.M., M.J. van Zandvoort, H.B. van der Worp, *et al.* 2006. Early cognitive impairment predicts long-term depressive symptoms and quality of life after stroke. *J. Neurol. Sci.* **247:** 149–156.
152. Rasquin, S., J. Lodder & F. Verhey. 2005. The association between psychiatric and cognitive symptoms after stroke: a prospective study. *Cerebrovasc. Dis.* **19:** 309–316.
153. Åström, M., T. Olsson & K. Asplund. 1993. Different linkage of depression to hypercortisolism early versus late after stroke. A 3-year longitudinal study. *Stroke* **24:** 52–57.
154. Franceschini, R., G.L. Tenconi, F. Zoppoli, *et al.* 2001. Endocrine abnormalities and outcome of ischaemic stroke. *Biomed. Pharmacother.* **55:** 458–465.
155. Lee, B.K., T.A. Glass, M.J. McAtee, *et al.* 2007. Associations of salivary cortisol with cognitive function in the Baltimore Memory Study. *Arch. Gen. Psychiat.* **64:** 810–818.
156. Radley, J.J. & J.H. Morrison. 2005. Repeated stress and structural plasticity in the brain. *Ageing Res. Rev.* **4:** 271–287.
157. Kreutz, G., S. Bongard, S. Rohrmann, *et al.* 2004. Effects of choir singing or listening on secretory immunoglobulin A, cortisol, and emotional state. *J. Behav. Med.* **27:** 623–635.
158. Nilsson, U. 2009. The effect of music intervention in stress response to cardiac surgery in a randomized clinical trial. *Heart Lung* **38:** 201–207.
159. Schneider, N., M. Schedlowski, T.H. Schürmeyer, *et al.* 2001. Stress reduction through music in patients undergoing cerebral angiography. *Neuroradiology* **43:** 472–476.
160. Cotman, C.W., D.T. Monaghan & A.H. Ganong. 1988. Excitatory amino acid neurotransmission: NMDA receptors and Hebb-type synaptic plasticity. *Annu. Rev. Neurosci.* **11:** 61–80.
161. Javitt, D.C., M. Steinschneider, C.E. Schroeder, *et al.* 1996. Role of cortical N-methyl-D-aspartate receptors in auditory sensory memory and mismatch negativity generation: implications for schizophrenia. *P. Natl. Acad. Sci. USA* **93:** 11962–11967.
162. Tikhonravov, D., T. Neuvonen, A. Pertovaara, *et al.* 2008. Effects of an NMDA-receptor antagonist MK-801 on an MMN-like response recorded in anesthetized rats. *Brain Res.* **1203:** 97–102.
163. Korostenskaja, M., V.V. Nikulin, D. Kičić, *et al.* 2007. Effects of NMDA receptor antagonist memantine on mismatch negativity. *Brain Res. Bull.* **72:** 275–283.
164. Umbricht, D., R. Koller, F.X. Vollenweider, *et al.* 2002. Mismatch negativity predicts psychotic experiences induced by NMDA receptor antagonist in healthy volunteers. *Biol. Psychiat.* **51:** 400–406.
165. Ahveninen, J., I.P. Jääskeläinen, E. Pekkonen, *et al.* 1999. Suppression of mismatch negativity by backward masking predicts impaired working-memory performance in alcoholics. *Alcohol. Clin. Exp. Res.* **23:** 1507–1514.
166. Baldeweg, T., A. Klugman, J. Gruzelier, *et al.* 2004. Mismatch negativity potentials and cognitive impairment in schizophrenia. *Schizophr. Res.* **69:** 203–217.
167. Ilvonen, T.M., T. Kujala, A. Kiesiläinen, *et al.* 2003. Auditory discrimination after left-hemisphere stroke: a mismatch negativity follow-up study. *Stroke* **34:** 1746–1751.
168. Kujala, T., K. Karma, R. Ceponiene, *et al.* 2001. Plastic neural changes and reading improvement caused by audiovisual training in reading-impaired children. *P. Natl. Acad. Sci. USA* **98:** 10509–10514.
169. Mikkola, K., E. Kushnerenko, E. Partanen, *et al.* 2007. Auditory event-related potentials and cognitive function of preterm children at five years of age. *Clin. Neurophysiol.* **118:** 1494–1502.
170. Toyomaki, A., I. Kusumi, T. Matsuyama *et al.* 2008. Tone duration mismatch negativity deficits predict impairment of executive function in schizophrenia. *Prog. Neuropsychopharmacol. Biol. Psychiatry* **32:** 95–99.
171. Nichols, J.A., V.P. Jakkamsetti, H. Salgado, *et al.* 2007. Environmental enrichment selectively increases glutamatergic responses in layer II/III of the auditory cortex of the rat. *Neuroscience* **145:** 832–840.
172. Bi, C., Y. Cui, Y. Mao, *et al.* 2006. The effect of early auditory deprivation on the age-dependent expression pattern of NR2B mRNA in rat auditory cortex. *Brain Res.* **1110:** 30–38.
173. Keyvani, K. & T. Schallert. 2002. Plasticity-associated molecular and structural events in the injured brain. *J. Neuropath. Exp. Neur.* **61:** 831–840.
174. Nithianantharajah, J. & A.J. Hannan. 2006. Enriched environments, experience-dependent plasticity and disorders of the nervous system. *Nat. Rev. Neurosci.* **7:** 697–709.
175. Maegele, M., M. Lippert-Gruener, T. Ester-Bode, *et al.* 2005. Multimodal early onset stimulation combined with enriched environment is associated with reduced CNS lesion volume and enhanced reversal of neuromotor dysfunction after traumatic brain injury in rats. *Eur. J. Neurosci.* **21:** 2406–2418.
176. Maegele, M., M. Lippert-Gruener, T. Ester-Bode, *et al.* 2005. Reversal of neuromotor and cognitive dysfunction in an enriched environment combined with multimodal early onset stimulation after traumatic brain injury in rats. *J. Neurotraum.* **22:** 772–782.
177. Bose, M., P. Muñoz-Llancao, S. Roychowdhury, *et al.* 2010. Effect of the environment on the dendritic morphology of the rat auditory cortex. *Synapse* **64:** 97–110.
178. Angelucci, F., E. Ricci, L. Padua, *et al.* 2007. Music exposure differentially alters the levels of brain-derived neurotrophic factor and nerve growth factor in the mouse hypothalamus. *Neurosci. Lett.* **429:** 152–155.
179. Chaudhury, S. & S. Wadhwa. 2009. Prenatal auditory stimulation alters the levels of CREB mRNA, p-CREB and BDNF expression in chick hippocampus. *Int. J. Dev. Neurosci.* **27:** 583–590.
180. Amunts, K., G. Schlaug, L. Jäncke, *et al.* 1997. Motor cortex and hand motor skills: structural compliance in the human brain. *Hum. Brain Mapp.* **5:** 206–215.

181. Bengtsson, S.L., Z. Nagy, S. Skare, *et al.* 2005. Extensive piano practicing has regionally specific effects on white matter development. *Nat. Neurosci.* **8:** 1148–1150.
182. Gaser, C. & G. Schlaug. 2003. Brain structures differ between musicians and non-musicians. *J. Neurosci.* **23:** 9240–9245.
183. Hyde, K.L., J.P. Lerch, A. Norton, *et al.* 2009. Musical training shapes structural brain development. *J. Neurosci.* **29:** 3019–3025.
184. Schlaug, G., L. Jäncke, Y. Huang, *et al.* 1995. Increased corpus callosum size in musicians. *Neuropsychologia* **33:** 1047–1055.
185. Sluming, V., T. Barrick, M. Howard, *et al.* 2002. Voxel-based morphometry reveals increased gray matter density in Broca's area in male symphony orchestra musicians. *Neuroimage* **17:** 1613–1622.
186. Antić, S., I. Galinović, A. Lovrečić-Huzjan, *et al.* 2008. Music as an auditory stimulus in stroke patients. *Coll. Antropol.* **32:** 19–23.
187. Matteis, M., M. Silvestrini, E. Troisi, *et al.* 1997. Transcranial doppler assessment of cerebral flow velocity during perception and recognition of melodies. *J. Neurol. Sci.* **149:** 57–61.
188. Vollmer-Haase, J., K. Finke, W. Hartje, *et al.* 1998. Hemispheric dominance in the processing of J.S. Bach fugues: a transcranial Doppler sonography (TCD) study with musicians. *Neuropsychologia* **36:** 857–867.
189. Kreisel, S.H., H. Bäzner & M.G. Hennerici. 2006. Pathophysiology of stroke rehabilitation: temporal aspects of neuro-functional recovery. *Cerebrovasc. Dis.* **21:** 6–17.

Ann. N.Y. Acad. Sci. ISSN 0077-8923

ANNALS OF THE NEW YORK ACADEMY OF SCIENCES
Issue: *The Neurosciences and Music IV: Learning and Memory*

The involvement of audio–motor coupling in the music-supported therapy applied to stroke patients

Antoni Rodriguez-Fornells,[1,2,3] Nuria Rojo,[1,2] Julià L. Amengual,[4] Pablo Ripollés,[1,2] Eckart Altenmüller,[5] and Thomas F. Münte[6]

[1]Cognition and Brain Plasticity Group, Bellvitge Biomedical Research Institute, L'Hospitalet de Llobregat, Barcelona, Spain. [2]Department of Basic Psychology, Campus Bellvitge, University of Barcelona, L'Hospitalet de Llobregat, Barcelona, Spain. [3]Catalan Institution for Research and Advanced Studies, Barcelona, Spain. [4]Neurodynamic Laboratory, Department of Psychiatry and Clinical Psychobiology, University of Barcelona, Barcelona, Spain. [5]Institute of Music Physiology and Musicians' Medicine, University of Music and Drama Hannover, Hannover, Germany. [6]Department of Neurology, University of Lübeck, Lübeck, Germany

Address for correspondence: Antoni Rodriguez-Fornells, Department of Basic Psychology, Campus Bellvitge, University of Barcelona, 08097, L'Hospitalet de Llobregat, Barcelona, Spain. antoni.rodriguez@icrea.es

Music-supported therapy (MST) has been developed recently to improve the use of the affected upper extremity after stroke. MST uses musical instruments, an electronic piano and an electronic drum set emitting piano sounds, to retrain fine and gross movements of the paretic upper extremity. In this paper, we first describe the rationale underlying MST, and we review the previous studies conducted on acute and chronic stroke patients using this new neurorehabilitation approach. Second, we address the neural mechanisms involved in the motor movement improvements observed in acute and chronic stroke patients. Third, we provide some recent studies on the involvement of auditory–motor coupling in the MST in chronic stroke patients using functional neuroimaging. Finally, these ideas are discussed and focused on understanding the dynamics involved in the neural circuit underlying audio–motor coupling and how functional connectivity could help to explain the neuroplastic changes observed after therapy in stroke patients.

Keywords: music-supported therapy; auditory–motor coupling; plasticity; stroke; functional connectivity

Music and neurorehabilitation in stroke patients

Research on brain plasticity during the last decades has provided evidence of the capacity to induce plastic changes and repair in the adult damaged brain from discoveries concerning neurogenesis and learning,[1,2] neuroimaging,[3–5] neuroscience, and epigenetics.[6–8] Clear evidence already exists in neuroscientific literature that after brain damage, new neuronal connections and pathways can be formed in the brain, be reshaped, or be rewired.[9–11] This new research provides a more optimistic view regarding the adult learning brain[12–14] and the importance of designing new strategic interventions that target residual learning abilities in patients. For example, motor disabilities after stroke have been the target of several recently developed therapies that have proven to be more effective than standard rehabilitation approaches.[15,16] Using this approach, the constraint-induced therapy (CIT) induces the use of the paretic limb over extended periods of time, leading to marked clinical improvements that are accompanied by neuroplastic changes.[17]

Music learning could be conceived as one of these potential intervention strategies,[18–22] mostly because of the extensive brain network engaged in music listening and performance, which is indeed necessary for processing multimodal information conveyed by music (coordinating information from auditory, visual, and sensorimotor information). Moreover, music training shapes the development of the brain by producing long-lasting changes in children and adults (see Refs. 23–26) and several neuroscience studies have shown that music training produces rapid changes in motor-related brain areas.[27–32]

doi: 10.1111/j.1749-6632.2011.06425.x

Against this background, a new motor rehabilitation therapy has been developed recently (music-supported therapy (MST)) for the rehabilitation of motor deficits in neurological patients.[33] Musical instruments (an electronic piano and an electronic drum set designed to produce piano tones) are used to train fine (piano) and gross (drums) motor functions in patients suffering from mild-to-moderate paresis after stroke. In two large samples of acute stroke patients, this therapy showed highly significant and clinically relevant improvements (see details in Refs. 33 and 34).

The MST rehabilitation program was designed on the following principles, considering previous studies on motor rehabilitation and brain plasticity:[33,35] (1) *Massive repetition* and exercising of simple finger and arm movements;[36] (2) *audio–motor coupling and integration*: reinforcement of movement effects due to immediate auditory feedback supporting the precise timing and control of movements and coupling of movements to auditory events permits the development of multimodal auditory–sensorimotor corepresentations of movements;[27,29,37] (3) *shaping*: adapting the complexity of the required movements according to the individual's progress; and (4) *emotion–motivation effects*: increased motivation of the patients due to the playfulness and emotional impact of making music and acquiring a new skill.

This last aspect might be very important because according to animal studies, cortical plasticity is increased by the behavioral relevance of the stimulation and its motivational impact.[12] Emotional effects induced by music listening, learning, and performance could engage reward–learning networks and corresponding neurotransmitter systems in the brain,[38,39] helping to consolidate new information, increasing the amount of reward experienced during the rehabilitation program, and increasing the probability of voluntary practicing the new movement exercises. In agreement with these ideas, a recent study of Särkämo and colleagues[40] showed that music listening significantly enhances cognitive functioning in the domains of verbal memory and focused attention in a music group compared to a control group. The music group also experienced less depression and confusion than the control groups. These results have been replicated recently in our study conducted on 20 chronic stroke patients (middle cerebral artery stroke; mean age, 59 ± 9 years; mean number of months after the first stroke, 30 months) with slight–moderate upper-extremity hemiparesis receiving MST (one-month intensive intervention program, 30-min daily music training sessions) (N. Rojo, J. Amengual, P. Ripolles, *et al.*, unpublished results). As shown in Figure 1A, we see a clear and significant improvement of positive mood (pleasure) in our patient sample using daily evaluations of affective valence (using the Self-Assessment Manikin—SAM, a nonverbal pictorial assessment technique[41]). When this patient group was contrasted with a well-matched healthy sample (14 control participants; mean age, 56 ± 9 years; matched for age, sex, and education) and evaluated two times with the same interval between both assessments (approximately one month and a half), the patient group showed a clear reduction of depressive symptoms (assessed using the Beck Depression Inventory Scale) and significant improvement of positive affect (using the Positive and Negative Affect Schedule (PANAS);[42] see Fig. 1B). These results converge with the previous findings from Särkämo *et al.*[40] and support the positive emotional effects induced by MST in a chronic stroke group. More importantly, the chronic group also showed motor improvements in the Action Research Arm Test (ARAT[43]) after the MST program, this test being one of the most widely used in the evaluation of motor function in the upper extremity.[44] Thus, even considering the limits for amelioration in the motor domain in these patients, MST clearly helped to improve their fine and gross motor skills. Indeed, the comparison of the three studies in which MST has been used (see Fig. 1C) shows that the amount of improvement of the chronic group (N. Rojo, J. Amengual, P. Ripolles, *et al.*, unpublished results) is about half of the effect observed in the acute patients for the ARAT test.[33,34] Notice, however, that the chronic patient group has a better initial score in the ARAT test due to residual and moderate physical deficits, but they have a smaller range of improvement compared with the acute group.

In sum, until now, different studies[33–35,45] have used MST in acute and chronic stroke patients and suggest a potential for exploiting residual learning abilities in these patients through indirect, intact brain pathways engaged by music performance. We will show in this study how neuroimaging techniques and, specifically, functional

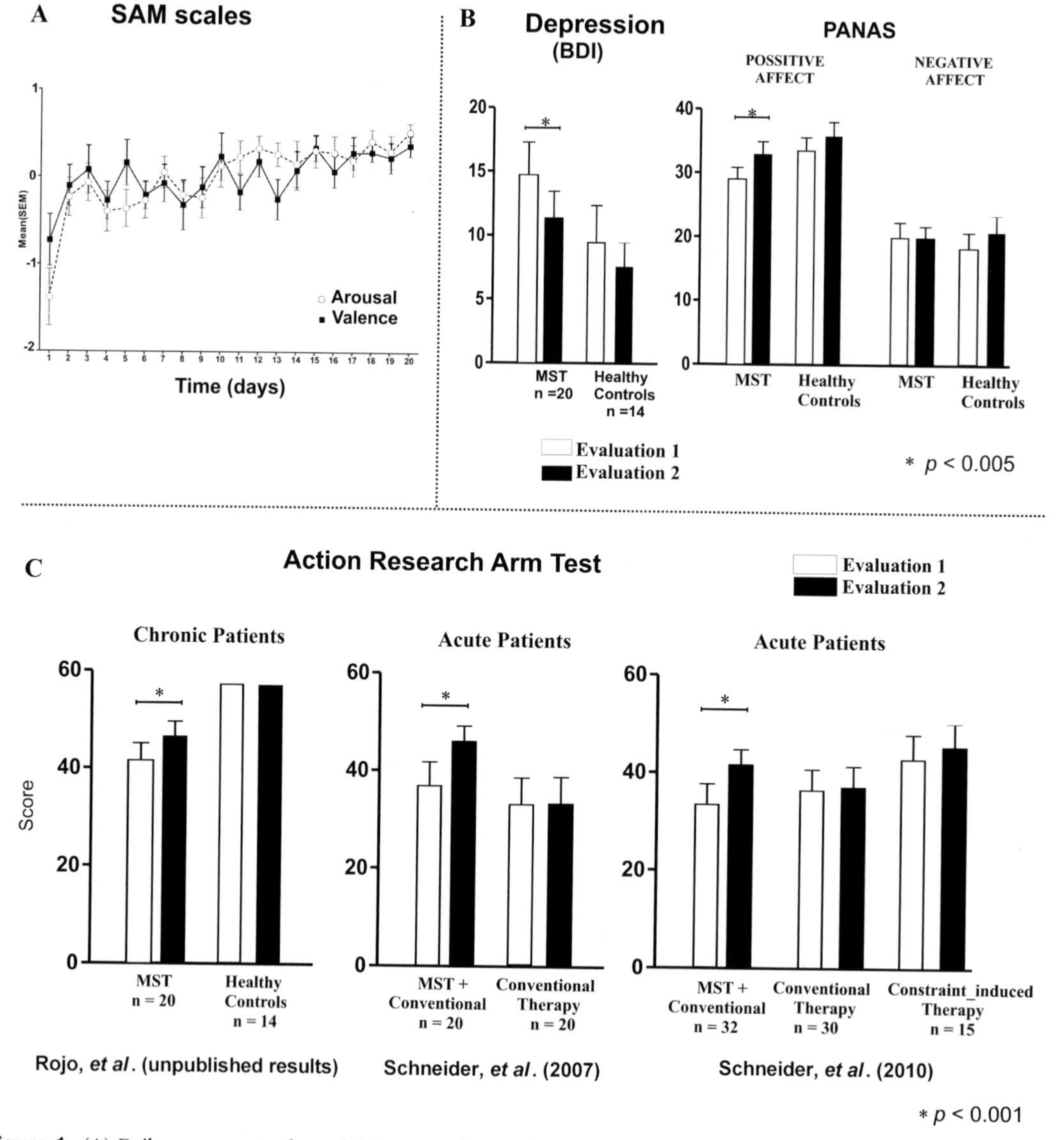

Figure 1. (A) Daily assessments of mood (pleasure and arousal) during each therapy sessions (intraindividual normalized Z mean values $\pm$ SEM) for the chronic stroke patients (Rojo *et al.*, unpublished results; $n = 20$). A significant improvement was observed during the course of the therapy. (B) Significant reduction of the depressive symptoms posttherapy (second evaluation) when compared to pretherapy evaluations (first evaluation). Significant improvement was also observed for the positive affect but not the negative affect scales (PANAS). Notice that for the control group (healthy group that did not receive MST), no differences were observed for the BDI and the PANAS. (C) Comparison of the motor improvement effects evaluated with the ARAT of the three cohorts of stroke patients who received MST and the comparison with the respective control groups.

connectivity approaches, may help us to understand the neuroplastic changes observed after this new neurorehabilitation strategy. Finally, we will highlight the importance of audio–motor plasticity as a plausible mechanism to explain the success of MST for the neurorehabilitation of stroke patients.

Audio–motor coupling hypothesis in MST

Music performance is an extremely complex process that requires, in some cases, the integration between the auditory system, proprioceptive feedback, visual information, and motor control. The hypothesis of audio–motor coupling and an integration mechanism is based on the idea that music performance requires the creation of fast feedforward and feedback loops to precisely coordinate auditory and motor information.[22] Playing an instrument requires fine-grained mapping between a musical note (or a sound) and the motor movement that will be executed to produce that note. Forward information is important as it has been proposed that internal representations in motor control (or "internal models";[35,46,47] see Ref. 48) can be created to predict the outcome of a particular action using "efference copy". Feedforward (bottom up) information could be transferred from the auditory system to the premotor cortices (PMC) using an internal model of the desired "sound" that will influence and modulate the motor output. Thus, well-trained motor responses associated with a specific sound will be primed or facilitated. However, the reverse is also possible; on-line motor actions might create internal representations of their actions ("efference copy") and send them back to the auditory regions to evaluate the appropriateness of these actions for the goal of producing a specific note. In this last alternative, a learned action triggers top-down auditory expectations that will facilitate and refine auditory processing. This top-down influence is plausible considering data showing that auditory neurons are suppressed during vocalizations in monkeys and human speech production.[49,50]

These feedforward and feedback connections between motor and auditory systems after learning to play an instrument could be used ultimately as fast nonconscious error-monitoring and correction systems that would allow the execution and correction of very fast movements in music performance (see Refs. 51, 52). These mechanisms might be especially relevant for string instrument players who have a continuum of note pitches without clear visual and/or kinaesthetic cues. In this particular case, fast corrections of movements could be necessary and be implemented via external auditory feedback mechanisms: the production of a partial erroneous note in the middle of a musical passage would be heard and correction would be implemented by the auditory cortex modulating the motor output. Indeed, this seems to be the case. For example, skilled cellists seem to require some constant motor recalibration using acoustic feedback guidance.[53]

A growing number of studies have investigated the relevance of these auditory–motor coupling and integration mechanisms (for a review, see Refs. 22, 27, 29, 37). On the one hand, motor and PMC activation (including the supplementary motor area (SMA)) can be elicited in musicians after passive listening of known melodies using functional magnetic resonance imaging or when playing familiar music pieces without auditory feedback (functional MRI (fMRI)[27,28] and similar transcranial magnetic stimulation (TMS) evidence[54]). In the study by Bangert *et al.*,[27] activation was also observed in the opercular part of the inferior frontal gyrus (IFG, BA 44), planum temporale, and supramarginal gyrus (SMG). More importantly, in a longitudinal study using nonmusicians, it was possible to follow the creation of these common auditory–motor representations during piano training. In this study, and after 20 min of training, the first signs of increased neuronal coupling between auditory and motor brain regions were observed when nonmusicians replayed the trained melodies. After five weeks, listening to piano tunes produced additional activity in the central and left sensorimotor regions. Finally, when trained participants played on a mute (soundless) keyboard, they produced additional activity in the auditory regions of both temporal lobes.[37] This experiment nicely demonstrates how fast and dynamic this auditory–motor plasticity mechanism can be.

Furthermore, in a similar and interesting fMRI study, Lahav *et al.*[29] showed that posterior middle ventral (PMC), but not the primary motor region (M1), was activated specifically when passive listening of trained simple but unfamiliar melodies in a group of nonmusicians (five-day training) was compared to a different untrained combination of the same notes. An important aspect of this study is the robust activation observed bilaterally in the left opercular part of the IFG as well as bilateral supramarginal activation when participants were listening to the trained melodies. The authors interpret these results in the left IFG as a support for the involvement of the mirror system, considering that this region is the homolog of area F5 (ventral PMC cortex) in monkeys where the mirror neurons have been located.[55] Thus, this region might be activated due its involvement in creating cross-modal repetitions and, specifically, for auditory–motor integration; these findings are also in agreement with previous ideas about the role of Broca's area in sensorimotor integration.[28,56]

However, the right IFG was constantly activated in all the conditions presented (trained and untrained), suggesting a different role of this region in music listening, most probably reflecting cognitive operations involved in music perception. Strong activations in this IFG region have also been observed in auditory object discrimination tasks.[57]

Previous studies clearly suggest the possible role of this auditory–motor coupling mechanism in MST rehabilitation. Recently, in a single-case study, we provided preliminary evidence for a possible benefit of the MST in a chronic stroke patient who showed clinical improvement and an increase in the quality of rapidly alternating movements.[45] This result and new ones for the whole cohort evaluated (N. Rojo, J. Amengual, P. Ripolles, *et al.*, unpublished results)[41] suggest that MST, like other recently developed therapies such as constrained induced therapy,[15] is capable of improving motor functions in patients with chronic stroke. Clinical improvements were accompanied by profound neural changes evidenced by both fMRI and TMS (J. Amengual, N. Rojo, M. Veciana, et al., unpublished results), suggesting plastic changes in the contralateral sensorimotor cortex after therapy (see Fig. 2A). fMRI of hand movements showed a significant decrease of activation in the contra- and ipsilateral sensorimotor areas and PMC regions after therapy.[58,59]

In addition, using a similar design inspired in previous studies,[27,29] we showed the first evidence in favor of the idea that the mechanism that contributes to the efficacy of MST (besides massive practice of the paretic arm) is audio–motor coupling. A music listening task was carried out in the MRI scanner (pre- and posttherapy) in which the patient had to passively listen to short and well-known familiar monophonic piano songs (songs that were going to be trained during the rehabilitation therapy) as well as well-known familiar songs (that were not going to be trained during the therapy). Alternating blocks of familiar-trained, familiar-untrained pieces, and rest blocks were presented inside the scanner.[45] As observed in Fig. 2B, the patient did not show activation in the PMC and IFG in the pretherapy session, but these regions were activated after therapy (notice also the activation observed in the SMA). The activation observed in this patient for passive music listening was replicated in the final protocol that consisted of 20 chronic stroke patients (14 patients could complete both scanning sessions) (N. Rojo, J. Amengual, P. Ripolles, *et al.*, unpublished results). Interestingly, increased activation was observed after therapy on the caudal and ventral part of the PMC, posterior IFG, and SMA regions after music training, this being an effect significantly larger in the trained song condition. Thus, these results provided the first evidence of the involvement of audio–motor coupling and integration mechanisms in MST.

Functional connectivity evidence of audio–motor coupling in music processing

One question that arises from the previous experiments in MST and passive listening to familiar songs is to what degree the lack of activation in the chronic stroke patients in the pretherapy sessions could be

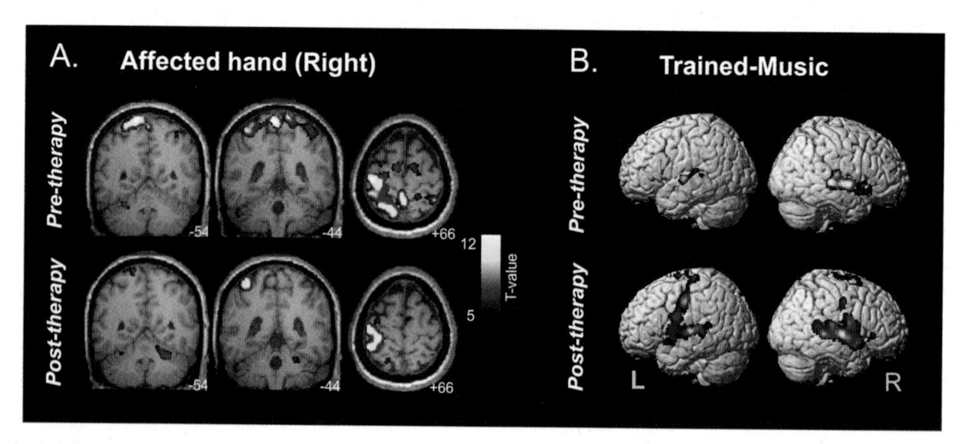

Figure 2. (A) fMRI activations in the motor task (superimposed on the patient's T1 image in standard stereotactic space ($P < 0.05$, family-wise error (FWE) corrected); (B) fMRI activation in the music listening task showing bilateral activation of motor-related brain regions when the patient was listening to trained music posttherapy but not pretherapy ($P < 0.05$, FWE corrected). Adapted from Rojo *et al.* (2011).

due to a decrease in connectivity in the auditory–motor circuit for music listening after the stroke instead of a clear effect of MST in increasing the activation of these feedforward–feedback loops for trained songs? In the previous experiments we observed that the neural network involved in these auditory–motor mechanism comprises the superior temporal lobe, inferior parietal lobe (probably SMG), ventral-caudal PMC, sensorimotor, and IFG regions. However, this strong coupling between auditory–motor regions observed in musicians and trained nonmusicians could also be observed in untrained people (see Fig. 2A depicted by Bangert *et al.* (2006) for nonmusicians[27]). For example, ventral PMC activation has been observed during melodic discrimination[60] and while listening to consonant excerpts.[61] Interestingly, in this last cited study, the activation in the PMC cortex was not observed when listening to unpleasant music. Besides, as we commented earlier, strong activation was also observed in the vPMC and IFG at the right hemisphere in the study by Lahav *et al.*—even for untrained songs. Thus, the automatic activation of the PMC/IFG cortex in passive listening in untrained musicians could be related to other cognitive functions in which this region is also involved; for example, working memory and rehearsal of tonal information,[62,63] automatic transformation of auditory information in motor–premotor representations ((as has also been proposed in language learning)[64–66]), the mirror neuron sensorimotor integrator,[56] or unconscious internal simulation and prediction of sequential auditory information.[67,68]

In the rehabilitation study on chronic stroke and MST mentioned earlier (N. Rojo, J. Amengual, P. Ripolles, *et al.*, unpublished results), we evaluated, using neuroimaging, a control healthy group well matched for age, sex, and education. This group was also evaluated two times (separated by approximately one and a half months) using the same fMRI passive listening task of well-known, familiar songs reported in Ref. 45. In Figure 3A, we can see the preliminary results of the activations of seven representative control subjects.[a] The most important aspect here is that in both sessions the control healthy participants showed strong robust activations when listening to popular songs in the STG, IFG, and ventral PMC. This activation pretherapy is not observed in the patient reported in Rojo *et al.* (Fig. 2B) or a subsample of the three chronic stroke patients reported (N. Rojo, J. Amengual, P. Ripolles, *et al.*, unpublished results) (see Fig. 4A and B for the lesions in these patients). Instead, patients showed restored activation after therapy on this circuit (see Fig. 4A), including activation in the precentral gyrus, inferior frontal locations, and SMA.

The question that arises when observing these results is to what degree the connectivity of this audio–motor circuit is not reduced or dampened down because of the lesion? To answer this question, we evaluated the functional connectivity in the group of seven control healthy participants and three patients (N. Rojo, J. Amengual, P. Ripolles, *et al.*, unpublished results). This idea is plausible considering recent findings of large incidences of acquired amusia in patients with middle cerebral artery stroke,[35,69,71] being at about 35% of the patients in chronic cases.[71]

To study the dynamics of the auditory–motor mechanism, we investigated the functional connectivity between the regions involved in this circuit, which is defined as the covariation between spatially remote neurophysiological processes.[72] The underlying idea is that areas that are involved in the same brain network should show consistent correlations between their respective time courses.[38] We investigated functional connectivity in the control group shown in Figure 3 using first an *a priori* seed-based approach (selecting a specific region of interest in the network and correlating its time-course with the other target brain regions)[b] and a multivariate approach (independent component analysis, ICA).[c]

---

[a]fMRI data was analyzed using SPM8 (Wellcome Trust Centre for Neuroimaging, University College London, London). Images were realigned between them and coregistered to their respective T1 before being normalized to the standard MNI452 template. Finally, a smoothing kernel of 8 mm was applied. For each participant a statistical model was computed by applying a canonical hemodynamic response, and the conditions of interest were modeled in a GLM (see Ref. 45 for details).

[b]Functional connectivity analysis was conducted first defining four anatomical regions of interest (ROIs) in the auditory–motor circuit, and using picktalas software.[73,74] The regions included the primary auditory cortex (PAC, BA 41, 42), IFG, SMA, and the precentral gyrus (including primary motor cortex, M1) in the affected hemisphere, according to the Talairach Daemon database atlases[75,76] and the AAL atlas.[77] Once these areas were defined, and

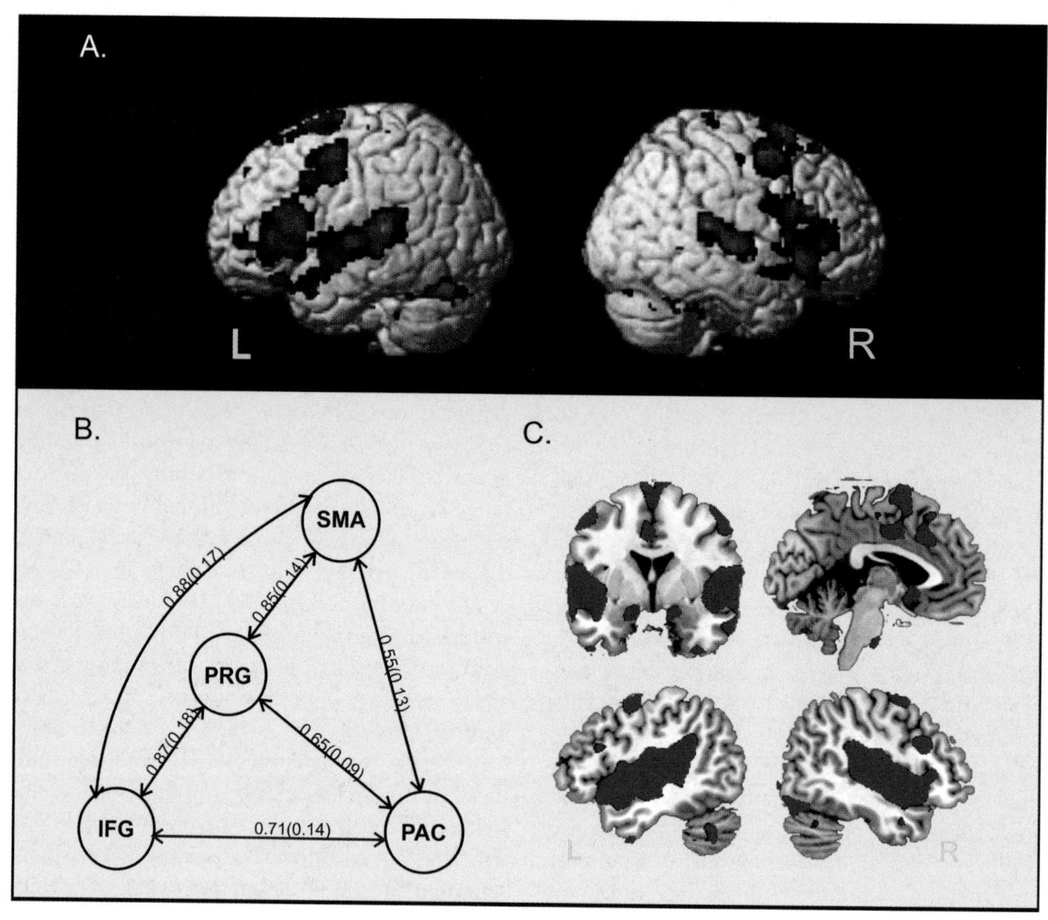

Figure 3. (A) Functional activation patterns in the passive music listening task in a sample of seven healthy participants and in the "Listening to Music" versus "Rest" contrast. Images were superimposed on a group-averaged structural MRI image in standard stereotactic space ($P < 0.01$, $n = 50$; uncorrected). (B) Mean functional connectivity (Fisher's z-transformed) within the auditory–motor network (selected regions of interest: SMA, supplementary motor area; PRG, precentral gyrus (BA6/4); IFG, inferior frontal gyrus; PAC, primary motor cortex). (C) Audio–motor network component reconstructed using independent component analysis (ICA) on seven healthy participants.

for seven control participants (two evaluations) and three patients (patients were evaluated pre- and posttherapy), a set of voxels was selected inside an ROI using the fMRI results from the condition *Listening to Trained Music* in the patients or *Listening to Music* in controls. For each anatomical ROI, only the time courses from the voxels, which were activated in these contrasts, were extracted. Before averaging, all the time courses had the linear trend removed and were low-pass filtered using a MATLAB toolbox for functional connectivity.[78] Functional connectivity was calculated separately for each participant as a correlation between the activation time courses averaged over all the activated voxels in a pair of ROIs (as detailed in Ref. 79). Thus, six different correlations (PAC-SMA, PAC-IFG, PAC-M1, M1-SMA, M1-IFG, and IFG-SMA) between ROIs were computed for each subject and Fisher's r-to-z transformation was applied to each of them.

[c]Independent component analysis (ICA). Functional images from the *Listening* experiment from 7 controls (previously preprocessed) were imported into the Group ICAfMRI Toolbox (GIFT v. 1.3i), and an ICA analysis was performed (see details in Refs. 80 and 81). To select a component related to the audio-frontal-motor network detected with parametrical maps, a functional template was created from the fMRI activations of the control group for the *Listening to Music* condition. This template was spatially correlated with all the ICA components, and the

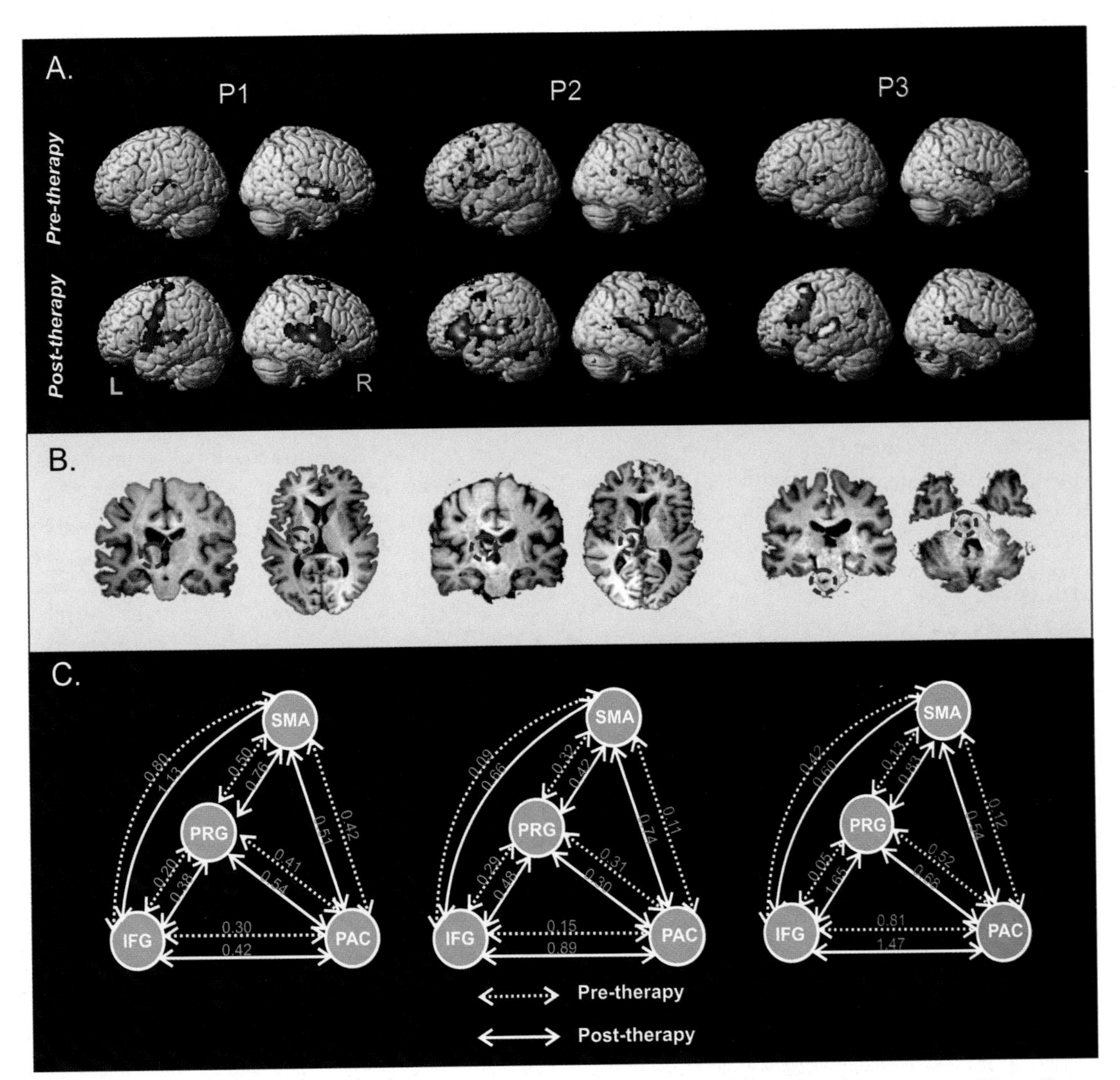

Figure 4. (A) fMRI activation in the music listening task in three chronic patients showing bilateral activation of the auditory–motor circuit when listening to trained music after but not before therapy (patient 1 (P1), $P < 0.05$, FWE corrected; P2, $P < 0.05$, uncorrected; and P3, $P < 0.001$, uncorrected). P1 is the same patient shown in Figure 2 (from Rojo *et al.*, 2011). (B) T1-weighted images showing the lesions of the three chronic patients (P1 includes thalamus, internal capsule, and posterior putamen; P2 thalamus; and P3 pons). (C) Individual functional connectivity (Fisher's z-transformed) within the audio–motor network reconstructed for each patient and for pre- and posttherapy (SMA, supplementary motor area; PRG, precentral gyrus; IFG, inferior frontal gyrus; PAC, primary auditory cortex). Notice the increase of functional connectivity in the different auditory–motor pathways involved in music listening.

In the standard *a priori* covariate approach, an initial ROI was selected in the primary auditory cortex, and we correlated its time course with the time course of the other regions involved in the pathway (precentral gryus, including premotor and primary motor regions, the inferior frontal region, and SMA) in the passive music listening task. As can be seen in Figure 3B, strong connectivity was observed between the premotor and IFG regions (SMA-precentral, SMA-IFG, precentral-IFG) as well

component with the highest correlation ($r = 0.48$) was selected.

as between the primary auditory cortex (PAC) and IFG (the same pattern was observed in both scanning sessions). The connectivity between the superior temporal cortex and premotor regions is somehow reduced, likely because this pathway is mediated by the dorsal connections between the posterior superior temporal cortex, inferior parietal lobe (probably through the SMG), and then through the connection to the prefrontal cortex via the superior longitudinal fasciculus.[22,35,65,82–84] It is important also to reiterate that no direct connections exist between the STG and the primary motor region;[85] this influence is likely mediated via the dorsal route. This dorsal route might constitute the anatomical basis of the feedforward sounds-to-action loop. Importantly, the SMA is also highlighted in this pathway as a possible hub of strong connectivity in this auditory–motor network (Fig. 3B).

Especially relevant here is the anterior–ventral processing stream because of the strong connectivity observed between the STG and the IFG. This route is important as it has been proposed to be involved in auditory pattern recognition of complex sounds and auditory object identification.[57,65,86] Important evidence also exists for the structural connectivity between the anterior temporal regions and the inferior frontal cortex from anatomical studies,[87–89] also demonstrated by recent diffusion tensor imaging studies in humans.[90–94] For example, a ventral connection through the extreme capsule (EmC) has been proposed connecting the middle section of the STG with the frontal operculum (FOP) close to the insular cortex.[95] White-matter individual differences in this pathway have been recently associated with success in a language learning task.[96] Thus, this pathway might be very relevant in the functional network related to music perception and learning and the implementation of audio–motor coupling loops (see Refs. 22, 65).

Finally, in Figure 3C we report the functional connectivity results but using the multivariate data-driven approach, ICA. This method allows for the decomposition of neuroimaging data into a set of spatial modes that capture the greatest amount of variance expressed over time and, therefore, identifying functional networks.[3] As can be observed, ICA identified a very similar network when compared to the standard connectivity seed-based approach. This identified component comprised the STG (from more anterior to posterior regions), PMC, IFG, SMA, and anterior cingulate regions (as well as the amygdala). Thus, both approaches showed strong convergence about the involvement of this network in music perception.

Finally, in Figure 4C we applied the same standard *a priori* connectivity approach to identify the weights of the connections between the different pathways in the three patients selected from N. Rojo, J. Amengual, P. Ripolles, *et al.* (unpublished results). The connectivity was assessed pre- and posttherapy. As demonstrated by the three connectivity pathways, an increase in connectivity between these regions was observed in practically all connections. Although this data is preliminary, the largest increases were observed between SMA and precentral and inferior frontal regions, as well as between the STG and the IFG (anteroventral stream). Thus, after MST, an increase in functional connectivity is exhibited in the three patients, suggesting a restoration of the inherent dynamics of the auditory–motor loops involved in music processing.

Conclusions

This study is an initial and preliminary report suggesting that MST could be effective in chronic stroke and acute patients and that it has a direct impact in the functional connectivity of the audio–motor networks that support musical perception and learning. One important result of the present research is that a clear increase in connectivity is evident in the auditory–motor circuit after MST. This finding suggests that MST affects the re-establishment of the default dynamics of this circuit in patients. This idea is convergent with the strong activation and connectivity observed in healthy untrained musicians in dorsal and anterior–ventral routes in the established auditory–motor circuit for music processing. Further studies are needed to understand how this auditory–motor plasticity mechanism helps in the amelioration of motor problems and positive affect and to clarify the impact of the different processes involved in MST (e.g., the role of auditory or propioceptive feedback). In this regard, functional and structural connectivity neuroimaging information will be crucial to better understand the impact of certain training programs and to allow for the design of future specific neurorehabilitation programs based on preserved and undamaged brain connectivity (see Refs. 21, 97, 98).

Acknowledgments

This project has been supported by la Fundacio La Marato TV3 (Spain) and the DZNE (German Center for Neurodegenerative Diseases).

Conflicts of interest

The authors declare no conflicts of interest.

References

1. Deng, W., J.B. Aimone & F.H. Gage. 2010. New neurons and new memories: how does adult hippocampal neurogenesis affect learning and memory? *Nat. Rev. Neurosci.* **11:** 339–350.
2. Shors, T.J. 2008. From stem cells to grandmother cells: how neurogenesis relates to learning and memory. *Cell Stem Cell.* **3:** 253–258.
3. Boyke, J., J. Driemeyer, C. Gaser, *et al.* 2008. Training-induced brain structure changes in the elderly. *J. Neurosci.* **28:** 7031–7035.
4. Draganski, B., C. Gaser, V. Busch, *et al.* 2004. Neuroplasticity: changes in grey matter induced by training. *Nature* **427:** 311–312.
5. Scholz, J., M.C. Klein, T.E. Behrens & H. Johansen-Berg. 2009. Training induces changes in white-matter architecture. *Nat. Neurosci.* **12:** 1370–1371.
6. Franklin, T.B., H. Russig, I.C. Weiss, *et al.* 2010. Epigenetic transmission of the impact of early stress across generations. *Biol. Psychiatry.* **68:** 408–415.
7. Jiang, Y., B. Langley, F.D. Lubin, *et al.* 2008. Epigenetics in the nervous system. *J. Neurosci.* **28:** 11753–11759.
8. McGowan, P.O., A. Sasaki, A.C. D'Alessio, *et al.* 2009. Epigenetic regulation of the glucocorticoid receptor in human brain associates with childhood abuse. *Nat. Neurosci.* **12:** 342–348.
9. Dancause, N., S. Barbay, S.B. Frost, *et al.* 2005. Extensive cortical rewiring after brain injury. *J. Neurosci.* **25:** 10167–10179.
10. Hihara, S., T. Notoya, M. Tanaka, *et al.* 2006. Extension of corticocortical afferents into the anterior bank of the intraparietal sulcus by tool-use training in adult monkeys. *Neuropsychologia* **44:** 2636–2646.
11. Yoshida, M., Y. Naya & Y. Miyashita. 2003. Anatomical organization of forward fiber projections from area TE to perirhinal neurons representing visual long-term memory in monkeys. *Proc. Natl. Acad. Sci. USA* **100:** 4257–4262.
12. Buonomano, D.V. & M.M. Merzenich. 1998. Cortical plasticity: from synapses to maps. *Annu. Rev. Neurosci.* **21:** 149–186.
13. Defelipe, J. 2006. Brain plasticity and mental processes: cajal again. *Nat. Rev. Neurosci.* **7:** 811–817.
14. Stiles, J. 2000. Neural plasticity and cognitive development. *Dev. Neuropsychol.* **18:** 237–272.
15. Taub, E., G. Uswatte & T. Elbert. 2002. New treatments in neurorehabilitation founded on basic research. *Nat. Rev. Neurosci.* **3:** 228–236.
16. Woldag, H. & H. Hummelsheim. 2002. Evidence-based physiotherapeutic concepts for improving arm and hand function in stroke patients: a review. *J. Neurol.* **249:** 518–528.
17. Liepert, J., W.H. Miltner, H. Bauder, *et al.* 1998. Motor cortex plasticity during constraint-induced movement therapy in stroke patients. *Neurosci. Lett.* **250:** 5–8.
18. Chen, J.L., V.B. Penhune & R.J. Zatorre. 2009. The role of auditory and premotor cortex in sensorimotor transformations. *Ann. N.Y. Acad. Sci.* **1169:** 15–34.
19. Munte, T.F., E. Altenmüller & L. Jancke. 2002. The musician's brain as a model of neuroplasticity. *Nat. Rev. Neurosci.* **3:** 473–478.
20. Pantev, C. & S.C. Herholz. 2011. Plasticity of the human auditory cortex related to musical training. *Neurosci. Biobehav. Rev.*
21. Wan, C.Y. & G. Schlaug. 2010. Music making as a tool for promoting brain plasticity across the life span. *Neuroscientist* **16:** 566–577.
22. Zatorre, R.J., J.L. Chen & V.B. Penhune. 2007. When the brain plays music: auditory-motor interactions in music perception and production. *Nat. Rev. Neurosci.* **8:** 547–558.
23. Bengtsson, S.L., Z. Nagy, S. Skare, *et al.* 2005. Extensive piano practicing has regionally specific effects on white matter development. *Nat. Neurosci.* **8:** 1148–1150.
24. Gaser, C. & G. Schlaug. 2003. Brain structures differ between musicians and non-musicians. *J. Neurosci.* **23:** 9240–9245.
25. Hyde, K.L., J. Lerch, A. Norton, *et al.* 2009. Musical training shapes structural brain development. *J. Neurosci.* **29:** 3019–3025.
26. Schlaug, G., A. Norton, K. Overy & E. Winner. 2005. Effects of music training on brain and cognitive development. *Ann. N.Y. Acad. Sci.* **1060:** 219–230.
27. Bangert, M., T. Peschel, G. Schlaug, *et al.* 2006. Shared networks for auditory and motor processing in professional pianists: evidence from fMRI conjunction. *Neuroimage* **30:** 917–926.
28. Baumann, S., S. Koeneke, C.F. Schmidt, *et al.* 2007. A network for audio-motor coordination in skilled pianists and non-musicians. *Brain Res.* **1161:** 65–78.
29. Lahav, A., E. Saltzman & G. Schlaug. 2007. Action representation of sound: audiomotor recognition network while listening to newly acquired actions. *J. Neurosci.* **27:** 308–314.
30. Meyer, M., S. Elmer, S. Baumann & L. Jancke. 2007. Short-term plasticity in the auditory system: differential neural responses to perception and imagery of speech and music. *Restor. Neurol. Neurosci.* **25:** 411–431.
31. Pascual-Leone, A., D. Nguyet, L.G. Cohen, *et al.* 1995. Modulation of muscle responses evoked by transcranial magnetic stimulation during the acquisition of new fine motor skills. *J. Neurophysiol.* **74:** 1037–1045.
32. Rosenkranz, K., A. Williamon & J.C. Rothwell. 2007. Motorcortical excitability and synaptic plasticity is enhanced in professional musicians. *J. Neurosci.* **27:** 5200–5206.
33. Schneider, S., P.W. Schonle, E. Altenmüller & T.F. Munte. 2007. Using musical instruments to improve motor skill recovery following a stroke. *J. Neurol.* **254:** 1339–1346.
34. Schneider, S., T.F. Munte, A. Rodriguez-Fornells, *et al.* 2010. Music-supported training is more efficient than functional motor training for recovery of fine motor skills in stroke patients. *Music Perception* **27:** 271–280.

35. Altenmüller, E., J. Marco-Pallares, T.F. Munte & S. Schneider. 2009. Neural reorganization underlies improvement in stroke-induced motor dysfunction by music-supported therapy. *Ann. N.Y. Acad. Sci.* **1169:** 395–405.
36. Sterr, A., T. Elbert, I. Berthold, *et al.* 2002. Longer versus shorter daily constraint-induced movement therapy of chronic hemiparesis: an exploratory study. *Arch. Phys. Med. Rehabil.* **83:** 1374–1377.
37. Bangert, M. & E.O. Altenmüller. 2003. Mapping perception to action in piano practice: a longitudinal DC-EEG study BMC. *Neurosci.* **4:** 26.
38. Camara, E., A. Rodriguez-Fornells, Z. Ye & T.F. Munte. 2009. Reward networks in the brain as captured by connectivity measures. *Front Neurosci.* **3:** 350–362.
39. Salimpoor, V.N., M. Benovoy, K. Larcher, *et al.* 2011. Anatomically distinct dopamine release during anticipation and experience of peak emotion to music. *Nat. Neurosci.* **14:** 257–262.
40. Särkämo, T., M. Tervaniemi, S. Laitinen, *et al.* 2008. Music listening enhances cognitive recovery and mood after middle cerebral artery stroke. *Brain* **131:** 866–876.
41. Bradley, M.M. & P.J. Lang. 1994. Measuring emotion: the Self-Assessment Manikin and the Semantic Differential. *J. Behav. Ther. Exp. Psychiatry* **25:** 49–59.
42. Watson, D., L.A. Clark & A. Tellegen. 1988. Development and validation of brief measures of positive and negative affect: the PANAS scales. *J. Pers. Soc. Psychol.* **54:** 1063–1070.
43. Lyle, R.C. 1981. A performance test for assessment of upper limb function in physical rehabilitation treatment and research. *Int. J. Rehabil. Res.* **4:** 483–492.
44. Sathian, K., L.J. Buxbaum, L.G. Cohen, *et al.* 2011. Neurological principles and rehabilitation of action disorders: common clinical deficits. *Neurorehabil. Neural Repair* **25:** S21–S32.
45. Rojo, N., J. Amengual, M. Juncadella, *et al.* 2011. Music-supported therapy induces plasticity in the sensorimotor cortex in chronic stroke: a single-case study using multimodal imaging (fMRI-TMS). *Brain Inj.* **25:** 787–793.
46. Grush, R. 2004. The emulation theory of representation: motor control, imagery, and perception. *Behav. Brain Sci.* **27:** 377–396.
47. Wolpert, D.M., K. Doya & M. Kawato. 2003. A unifying computational framework for motor control and social interaction. *Philos. Trans. R. Soc. Lond. B Biol. Sci.* **358:** 593–602.
48. Buxbaum, L.J., S.H. Johnson-Frey & M. Bartlett-Williams. 2005. Deficient internal models for planning hand-object interactions in apraxia. *Neuropsychologia* **43:** 917–929.
49. Eliades, S.J. & X. Wang. 2008. Neural substrates of vocalization feedback monitoring in primate auditory cortex. *Nature* **453:** 1102–1106.
50. Numminen, J., R. Salmelin & R. Hari. 1999. Subject's own speech reduces reactivity of the human auditory cortex. *Neurosci. Lett.* **265:** 119–122.
51. Rodriguez-Fornells, A., A.R. Kurzbuch & T.F. Munte. 2002. Time course of error detection and correction in humans: neurophysiological evidence. *J. Neurosci.* **22:** 9990–9996.
52. Ruiz, M.H., H.C. Jabusch & E. Altenmüller. 2009. Detecting wrong notes in advance: neuronal correlates of error monitoring in pianists. *Cereb. Cortex* **19:** 2625–2639.
53. Chen, J., M.H. Woollacott, S. Pologe & G.P. Moore. 2008. Pitch and space maps of skilled cellists: accuracy, variability, and error correction. *Exp. Brain Res.* **188:** 493–503.
54. D'Ausilio, A., E. Altenmüller, B.M. Olivetti & M. Lotze. 2006. Cross-modal plasticity of the motor cortex while listening to a rehearsed musical piece. *Eur. J. Neurosci.* **24:** 955–958.
55. Rizzolatti, G. & M.A. Arbib. 1998. Language within our grasp. *Trends Neurosci.* **21:** 188–194.
56. Binkofski, F. & G. Buccino. 2006. The role of ventral premotor cortex in action execution and action understanding. *J. Physiol. Paris* **99:** 396–405.
57. Zatorre, R.J., M. Bouffard & P. Belin. 2004. Sensitivity to auditory object features in human temporal neocortex. *J. Neurosci.* **24:** 3637–3642.
58. Pineiro, R., S. Pendlebury, H. Johansen-Berg & P.M. Matthews. 2001. FMRI detects posterior shifts in primary sensorimotor cortex after stroke: evidence for adaptive reorganization? *Stroke* **32:** 1134–1139.
59. Rossini, P.M., C. Altamura, F. Ferreri, *et al.* 2007. Neuroimaging experimental studies on brain plasticity in recovery from stroke. *Eura. Medicophys.* **43:** 241–254.
60. Brown, S. & M.J. Martinez. 2007. Activation of premotor vocal areas during musical discrimination. *Brain Cogn.* **63:** 59–69.
61. Koelsch, S. 2006. Significance of Broca's area and ventral premotor cortex for music-syntactic processing. *Cortex* **42:** 518–520.
62. Hickok, G., B. Buchsbaum, C. Humphries & T. Muftuler. 2003. Auditory-motor interaction revealed by fMRI: speech, music, and working memory in area. *Spt. J. Cogn. Neurosci.* **15:** 673–682.
63. Koelsch, S. 2009. A neuroscientific perspective on music therapy. *Ann. N.Y. Acad. Sci.* **1169:** 374–384.
64. Cunillera, T., E. Camara, J.M. Toro, *et al.* 2009. Time course and functional neuroanatomy of speech segmentation in adults. *Neuroimage* **48:** 541–553.
65. Rauschecker, J.P. & S.K. Scott. 2009. Maps and streams in the auditory cortex: nonhuman primates illuminate human speech processing. *Nat. Neurosci.* **12:** 718–724.
66. Warren, J.E., R.J. Wise & J.D. Warren. 2005. Sounds do-able: auditory-motor transformations and the posterior temporal plane. *Trends Neurosci.* **28:** 636–643.
67. Huettel, S.A., P.B. Mack & G. Mccarthy. 2002. Perceiving patterns in random series: dynamic processing of sequence in prefrontal cortex. *Nat. Neurosci.* **5:** 485–490.
68. Schubotz, R.I. & D.Y. Von Cramon. 2002. Predicting perceptual events activates corresponding motor schemes in lateral premotor cortex: an fMRI study. *Neuroimage* **15:** 787–796.
69. Särkämo, T., M. Tervaniemi, S. Soinila, *et al.* 2009. Cognitive deficits associated with acquired amusia after stroke: a neuropsychological follow-up study. *Neuropsychologia* **47:** 2642–2651.
70. Schuppert, M., T.F. Munte, B.M. Wieringa & E. Altenmüller. 2000. Receptive amusia: evidence for cross-hemispheric neural networks underlying music processing strategies. *Brain* **123**(Pt 3): 546–559.
71. Ayotte, J., I. Peretz & K. Hyde. 2002. Congenital amusia: a group study of adults afflicted with a music-specific disorder. *Brain* **125:** 238–251.

72. Grefkes, C. & G.R. Fink. 2011. Reorganization of cerebral networks after stroke: new insights from neuroimaging with connectivity approaches. *Brain* **134:** 1264–1276.
73. Maldjian, J.A., P.J. Laurienti & J.H. Burdette. 2004. Precentral gyrus discrepancy in electronic versions of the Talairach atlas. *Neuroimage* **21:** 450–455.
74. Maldjian, J.A., P.J. Laurienti, R.A. Kraft & J.H. Burdette. 2003. An automated method for neuroanatomic and cytoarchitectonic atlas-based interrogation of fMRI data sets. *Neuroimage* **19:** 1233–1239.
75. Lancaster, J.L., M.G. Woldorff, L.M. Parsons, *et al.* 2000. Automated Talairach atlas labels for functional brain mapping. *Hum. Brain Mapp.* **10:** 120–131.
76. Lancaster, J.L., L.H. Rainey, J.L. Summerlin, *et al.* 1997. Automated labeling of the human brain: a preliminary report on the development and evaluation of a forward-transform method. *Hum. Brain Mapp.* **5:** 238–242.
77. Tzourio-Mazoyer, N., B. Landeau, D. Papathanassiou, *et al.* 2002. Automated anatomical labeling of activations in SPM using a macroscopic anatomical parcellation of the MNI MRI single-subject brain. *Neuroimage* **15:** 273–289.
78. Zhou, D., W.K. Thompson & G. Siegle. 2009. MATLAB toolbox for functional connectivity. *Neuroimage* **47:** 1590–1607.
79. Prat, C.S., T.A. Keller & M.A. Just. 2007. Individual differences in sentence comprehension: a functional magnetic resonance imaging investigation of syntactic and lexical processing demands. *J. Cogn. Neurosci.* **19:** 1950–1963.
80. Calhoun, V.D., T. Adali, V.B. Mcginty, *et al.* 2001. fMRI activation in a visual-perception task: network of areas detected using the general linear model and independent components analysis. *Neuroimage* **14:** 1080–1088.
81. Calhoun, V.D., T. Adali, G.D. Pearlson & J.J. Pekar. 2001. A method for making group inferences from functional MRI data using independent component analysis. *Hum. Brain Mapp.* **14:** 140–151.
82. Catani, M., D.K. Jones & D.H. Ffytche. 2005. Perisylvian language networks of the human brain. *Ann. Neurol.* **57:** 8–16.
83. Glasser, M.F. & J.K. Rilling. 2008. DTI tractography of the human brain's language pathways. *Cereb. Cortex* **18:** 2471–2482.
84. Poremba, A., R.C. Saunders, A.M. Crane, *et al.* 2003. Functional mapping of the primate auditory system. *Science* **299:** 568–572.
85. Nieuwenhuys, R., J. Voogd & C. Van Huijzen. 2007. *The Human Central Nervous System: A Synopsis and Atlas*, Steinkopff, Amsterdam.
86. Patterson, R.D., S. Uppenkamp, I.S. Johnsrude & T.D. Griffiths. 2002. The processing of temporal pitch and melody information in auditory cortex. *Neuron* **36:** 767–776.
87. Hackett, T.A., I. Stepniewska & J.H. Kaas. 1999. Prefrontal connections of the parabelt auditory cortex in macaque monkeys. *Brain Res.* **817:** 45–58.
88. Romanski, L.M., B. Tian, J. Fritz, *et al.* 1999. Dual streams of auditory afferents target multiple domains in the primate prefrontal cortex. *Nat. Neurosci.* **2:** 1131–1136.
89. Seltzer, B. & D.N. Pandya. 1989. Frontal lobe connections of the superior temporal sulcus in the rhesus monkey. *J. Comp. Neurol.* **281:** 97–113.
90. Anwander, A., M. Tittgemeyer, D.Y. Von Cramon, *et al.* 2007. Connectivity-based parcellation of Broca's area cereb. *Cortex* **17:** 816–825.
91. Croxson, P.L., H. Johansen-Berg, T.E. Behrens, *et al.* 2005. Quantitative investigation of connections of the prefrontal cortex in the human and macaque using probabilistic diffusion tractography. *J. Neurosci.* **25:** 8854–8866.
92. Frey, S., J.S. Campbell, G.B. Pike & M. Petrides. 2008. Dissociating the human language pathways with high angular resolution diffusion fiber tractography. *J. Neurosci.* **28:** 11435–11444.
93. Parker, G.J., S. Luzzi, D.C. Alexander, *et al.* 2005. Lateralization of ventral and dorsal auditory-language pathways in the human brain. *Neuroimage* **24:** 656–666.
94. Saur, D., B.W. Kreher, S. Schnell, *et al.* 2008. Ventral and dorsal pathways for language. *Proc. Natl. Acad. Sci. U.S.A* **105:** 18035–18040.
95. Makris, N. & D.N. Pandya. 2009. The extreme capsule in humans and rethinking of the language circuitry. *Brain Struct. Funct.* **213:** 343–358.
96. Lopez-Barroso, D., R. de Diego-Balaguer, T. Cunillera, *et al.* 2011. Language learning under working memory constraints correlates with microstructural differences in the ventral language pathway. *Cereb. Cortex* **21:** 2742–2750.
97. Schlaug, G., S. Marchina & A. Norton. 2009. Evidence for plasticity in white-matter tracts of patients with chronic Broca's aphasia undergoing intense intonation-based speech therapy. *Ann. N.Y. Acad. Sci.* **1169:** 385–394.
98. Wan, C.Y., T. Ruber, A. Hohmann & G. Schlaug. 2010. The therapeutic effects of singing in neurological disorders music. *Percept.* **27:** 287–295.

Ann. N.Y. Acad. Sci. ISSN 0077-8923

ANNALS OF THE NEW YORK ACADEMY OF SCIENCES
Issue: *The Neurosciences and Music IV: Learning and Memory*

Changes in neuromagnetic beta-band oscillation after music-supported stroke rehabilitation

Takako Fujioka,[1] Jon Erik Ween,[2,3] Shahab Jamali,[4] Donald T. Stuss,[1] and Bernhard Ross[1,4]

[1]Rotman Research Institute, Baycrest Centre, Toronto, Canada. [2]Stroke and Cognition Clinic, Baycrest Centre, Toronto, Canada. [3]Stroke Program, Rockwood/Deaconess Medical Center, Spokane, Washington. [4]Department of Medical Biophysics, University of Toronto, Toronto, Canada

Address for correspondence: Takako Fujioka, Ph.D., Rotman Research Institute, 3560 Bathurst Street, Toronto, Ontario, M6A 2E1, Canada. tfujioka@rotman-baycrest.on.ca

Precise timing of sound is crucial in music for both performing and listening. Indeed, listening to rhythmic sound sequences activates not only the auditory system but also the sensorimotor system. Previously, we showed the significance of neural beta-band oscillations (15–30 Hz) for the timing processing that involves such auditory–motor coordination. Thus, we hypothesized that motor rehabilitation training incorporating music playing will stimulate and enhance auditory–motor interaction in stroke patients. We examined three chronic patients who received Music-Supported Therapy following the protocols practiced by Schneider. Neuromagnetic beta-band activity was remarkably alike during passive listening to a metronome and during finger tapping, with or without the metronome, for either the paretic or nonparetic hand, suggesting a shared mechanism of the beta modulation. In the listening task, the magnitude of the beta decrease after the tone onset was more pronounced at the posttraining time point and was accompanied by improved arm and hand skills. The present case data give insight into the neural underpinnings of rehabilitation with music making and rhythmic auditory stimulation.

Keywords: event-related desynchronization (ERD); stroke rehabilitation; magnetoencephalography (MEG); rhythmic auditory stimulation; timing processing

Introduction

Recent research has demonstrated that musical training leads to reorganization of brain function and structure related to sensorimotor, auditory, and visual information processing, as well as integration of the processing with executive control and emotion.[1–5] A critical question is how this knowledge can be translated into remediation and rehabilitation using music.[6] Current studies in rehabilitation science explore effective therapeutic approaches for motor and cognitive functions to overcome the substantial and long-term disadvantages in quality of life in stroke survivors.[7] A novel short-term physical training program, termed Music-Supported Therapy (MST), uses the goals of musical activities with percussion and keyboard instruments. MST seems to be more effective in improving gross and fine motor skills in the paretic hand in subacute patients than conventional physical therapy or the constraint-induced movement therapy.[8–10] Substantial motor improvement was also reported in a chronic case[11] and in a group of chronic patients.[12] In MST, playing melodic sequences requires coordination of complex bimanual movements in a time-sensitive manner. The instructor demonstrates an exercise and the patient replicates it. This means that the patient first has to memorize the spatial and acoustical aspects of the sequence, including timing information, then use this association for planning actions, and finally compare the auditory feedback to the memorized sequence. The remarkable motor improvement after MST has been attributed to the enhanced multimodal feedbacks, "shaping" (e.g., a gradual increase of challenges), enjoyable and goal-oriented tasks, and the motivational value built into the exercises. Indeed, personally meaningful goal setting plays an important in successful motor learning[13] and physical therapy.[14–16]

However, the neural mechanisms underlying the effects of MST are less well understood. A current hypothesis is that music making in MST enhances

doi: 10.1111/j.1749-6632.2011.06436.x

the functional interactions within the sensorimotor system during hand movements, as shown by cortico-cortical coherence,[9] and between the motor system and the auditory system as indicated by activation in the premotor areas during listening to the trained melodic materials.[11,12] Such auditory–motor association has been demonstrated in musicians and healthy, naive subjects after short-term keyboard training.[17,18]

More insight into the unique way of engaging the sensorimotor system in MST may come from findings about neural representation of temporal information. Rhythmic events in music are often organized within a hierarchical structure based on a basic isochronous unit called beat, or tactus, where the typical beat frequency ranges around 1–3 Hz.[19] Neural structures important for the processing of temporal information in such an interval range are known to involve both the basal ganglia and the cerebellum, which play crucial roles in the motor system.[20,21] Encoding the auditory rhythms and internalizing the subjective metric structure also activated motor-related cortical and subcortical areas[22–25] even when the subjects were not familiar with the motor and/or auditory sequences. Thus, auditory processing of timing information may involve obligatory auditory–motor interactions. Our recent magnetoencephalography (MEG) studies have found that beta-band oscillations (15–30 Hz), which are crucially related to sensorimotor functions, also index the brain processes of automatic and predictive timing, even when no actual movement is performed or intended.[26,27] In these studies, we examined MEG in healthy, musically untrained adults who passively listened to a metronome beat. We found that the amplitude of beta activity in bilateral auditory and motor-related brain areas (including the sensorimotor cortex, supplementary motor area [SMA], and inferior frontal gyrus [IFG]) was synchronously modulated with the rate of the metronome. This suggests a distributed network for the processing of metronome-dependent timing information. The time course of beta modulation was similar to the beta activity in SM1 and SMA during repetitive finger motion in healthy subjects[28,29] and well-recovered stroke survivors.[30] Thus, in the current study, we investigated the dynamics of neuromagnetic beta activity in chronic stroke patients to examine functional reorganization of auditory–motor communication. Specifically, we examined whether periodic modulation of beta activities was present in the patients during passive listening to a metronome and repetitive finger tapping with or without the metronome, and whether the cortical networks involved in the beta activities during listening and tapping was modified after the MST training.

Materials and methods

Patients

Three male patients (all right-handed) referred from our outpatient stroke clinic participated in the study (see demographics in Table 1, with the National Institute of Health Stroke Scale [NIHSS] and the Chedoke McMaster Impairment Inventory arm and hand subsections).[31,32] The patients reported no history of psychological, hearing, or language problems before or after their stroke for verbal communication. Pure tone audiometry (250–2,000 Hz) confirmed age-related mild to moderate hearing loss (patient 1: right ear 35–55 dB, left ear 50–70 dB; patient 2: right ear 20–30 dB, left ear 25–30 dB; patient 3: right ear 5–15 dB, left ear 10–15 dB). Although none had formal musical training, patients 2 and 3 would occasionally sing in a church choir. The Research Ethics Board at Baycrest Center approved the study and participants signed consent forms after written information about the study had been provided.

MST intervention

MST was administered in a one-to-one setting in a total of 15 sessions, each for 30 min per day, two to five times per week, over a five-week period. The MST apparatus consisted of an eight-pad electronic drum set and an electronic piano, programmed with MIDI (Musical Instrument Digital Interface) software (Logic Express v.8, Apple Inc., Cupertino, CA) to produce the piano sounds G, A, B, C, D, E, F, and G′, respectively. The eight drum pads were arranged from left to right in a half circle so that the patient could reach all from the center position with their paretic hand/arm. Patients 2 and 3 played the piano during the session. Exercises for MST were arranged with stepwise increases of complexity, so that the number of tones, velocity, order, and limb used for playing were increasingly demanding. Also, the hand posture for hitting the drum was varied by using the open palm, the fist, or a drum stick with pronation or supination. To accommodate the

Table 1. Patient demographics

Patient	Age (years)	Poststroke onset	Lesion	Sensorimotor symptoms	NIHSS motor score	Chedoke McMaster stroke assessment	MMSE	Total years of education
1	87	4y7m	Right posterior limb of the internal capsule lacune with scattered microvascular infarcts	Left upper-limb weakness and no sensory loss	3	Arm/hand 3/4 (Left)	28	13
2	69	1y10m	Right inferolateral thalamic hemorrhage and mild microvascular changes in the hemispheres	Pain, numbness, and clumsiness on the left side but no weakness	2	Arm/hand 4/5 (Left)	30	15
3	39	3y5m	Left dorsolateral cord infarct at C1 level. No brain lesions	Left upper- and lower–limb sensory ataxia but no weakness	4	Arm/hand 6/7 (Right) 5/5 (Left)	30	19

y, year; m, month.

variety of the nature of disability in the paretic limb, the training sessions were tailored to each patient such that patient 2 received equal amounts of drum and keyboard exercises, whereas patient 1 practiced exclusively on drums and patient 3 practiced on keyboard only. The week before the first intervention session and the week after the last intervention session were dedicated to pre- and posttraining assessments. Behavioral assessments included motor skills in the paretic hand examined by Action Research Arm Test (ARAT),[33] Finger Tapping Test,[34] Grooved Pegboard Test,[35] and Purdue Pegboard Test.[36]

Stimuli and task for MEG recording

Auditory stimuli were 262-Hz tones with 10 ms rise and fall times and 40 ms steady-state duration, which were presented in an isochronous sequence as a metronome beat. During MEG recording, five experimental conditions were tested. First, the patients passively listened to the metronome beat without any tasks. Thereafter, the patients were asked to tap in synchrony with the metronome with the right index finger (e.g., less-affected side) onto a response switch. Third, the patients were asked to continue finger tapping while maintaining the same (onset-to-onset) interval after the metronome was turned off. The tapping procedure was repeated in the same manner for the left index finger (e.g., affected side). For patient 1, we used a metronome beat with onset-to-onset intervals of 1,000 ms. For patients 2 and 3, three different metronome tempi were used in separate blocks: 400 ms ("fast"), 800 ms ("medium"), and 1,200 ms ("slow"). For patients 2 and 3, the blocks with listening to the three metronome tempi were recorded first. Then the finger tapping tasks with medium tempo were performed using the right and left index finger both with and without the metronome beat. Finally, the tapping tasks using fast and slow tempi were performed in the same manner. About 100–130 tones were presented in each recording block. The stimulus intensity was set to 40 dB above individual sensation threshold, which was measured immediately before each MEG recording. All stimulus sequences were presented under control of STIM software (Neuroscan Inc., El Paso, TX). The sound was delivered binaurally through Etymotic ER3A transducers connected with 1.5 m of matched plastic tubing and foam insert earphones to the participant's ears.

Neuromagnetic activity was recorded with a 151-channel whole-head axial gradiometer type MEG system (OMEGA, VSM MedTech, Coquitlam, Canada) in a quiet, magnetically shielded room. The MEG was recorded continuously with a sampling rate of 2,083 Hz after 600-Hz low-pass filtering. Patients 1 and 2 used a seated position with their head resting in the helmet-shaped MEG device whereas patient 3 used a supine position. The electromyogram (EMG) of the first dorsal interosseous muscle (FDI) in both hands were recorded using Ag/AgCl electrodes to ensure that there were no hand movements during the auditory listening conditions. EMG data were acquired simultaneously with the MEG and the timing of the button presses, and stored continuously. All trials contaminated with EMG activity during the listening condition were rejected (<2% of trials). In the finger-tapping tasks, we confirmed through video monitoring that no explicit mirror movements occurred and monitored the EMG activities in the nontask hand. Again, the number of rejected trials was small. The EMG signals from the task hand were used to mark the timing of muscle activity for each finger tap in the tapping conditions so that it coincided at the timing of the button press. This was necessary for some trials when finger movements were not strong enough to depress the button.

Data analysis

We examined functional coordination between auditory and sensorimotor systems indexed by beta oscillations. We conducted source analysis using a spatial filter called MEG beamformer to extract source activity in each volume element (8-mm cube) inside the brain volume. For each patient, the volume was calculated using the surface shape determined by the individual MRI. Thereafter, we calculated the beta-band power at each source location and finally used comodulation of the beta magnitudes as a measure of functional communication across the whole brain. This MEG activity map was overlaid on each individual MRI. The MRI of patient 1 was obtained as a clinical scan with the FLAIR protocol one year nine months before the study participation, whereas for patients 2 and 3 the standard T1 structural MRI was obtained (the first week of participation, and four months before participation, respectively).

First, we segmented the MEG data into 2,000 ms stimulus-related epochs beginning 400 ms before both the auditory stimulus for the listening conditions and the button press for the tapping conditions. Eye-blink artifacts were eliminated using principal component analysis (PCA)[37] in which components exceeding the threshold of 1.5 pT at any MEG sensor were subtracted using a house-made program developed in MATLAB (MathWorks, Natick, MA). In patient 3, excessive eye movement-related activities were further removed with independent component analysis (ICA) based on visual inspection of the topography[38] using a function provided in the EEGLAB package.[39] We applied the beamformer algorithm called synthetic aperture magnetometry (SAM) to the magnetic field data to calculate the time series of local source activity at 8 mm × 8 mm × 8 mm volume elements covering the whole brain based on the individual MRI first for single trials. The SAM approach uses a linearly constrained minimum variance beamformer algorithm,[40,41] normalizes source power across the whole cortical volume,[42] and is capable of revealing deep brain sources.[43,44] Then we calculated the mean signal power across trials in the beta band (15–30 Hz) at each volume element by applying a band-pass filter followed by Hilbert transformation. Our analysis was focused on non–phase-locked "induced" oscillatory activity. Therefore, we obtained separately the phase-locked "evoked" activity by averaging single trials before filtering and subtracted its signal power from the mean signal power described earlier. We applied a spatiotemporal PCA to describe the main effect of temporally correlated beta amplitude modulation by the single time course of the largest principal component (PC) and a corresponding volumetric brain map as temporal and spatial components, respectively. The normalized intensity at each location in the map indicated how strongly the activity at that location contributed to the identified PC. The variance of the data accounted for by the first PC was about 40% or more in each individual and each condition. For demonstrating consistency across different task blocks, we applied the PCA to individual data and compared the time courses as shown in Figures 1 and 2 (top panels for each). Because the time courses were similar under the five conditions (auditory listening, auditory-guided right finger tapping, self-guided right finger tapping, auditory-guided left finger tapping,

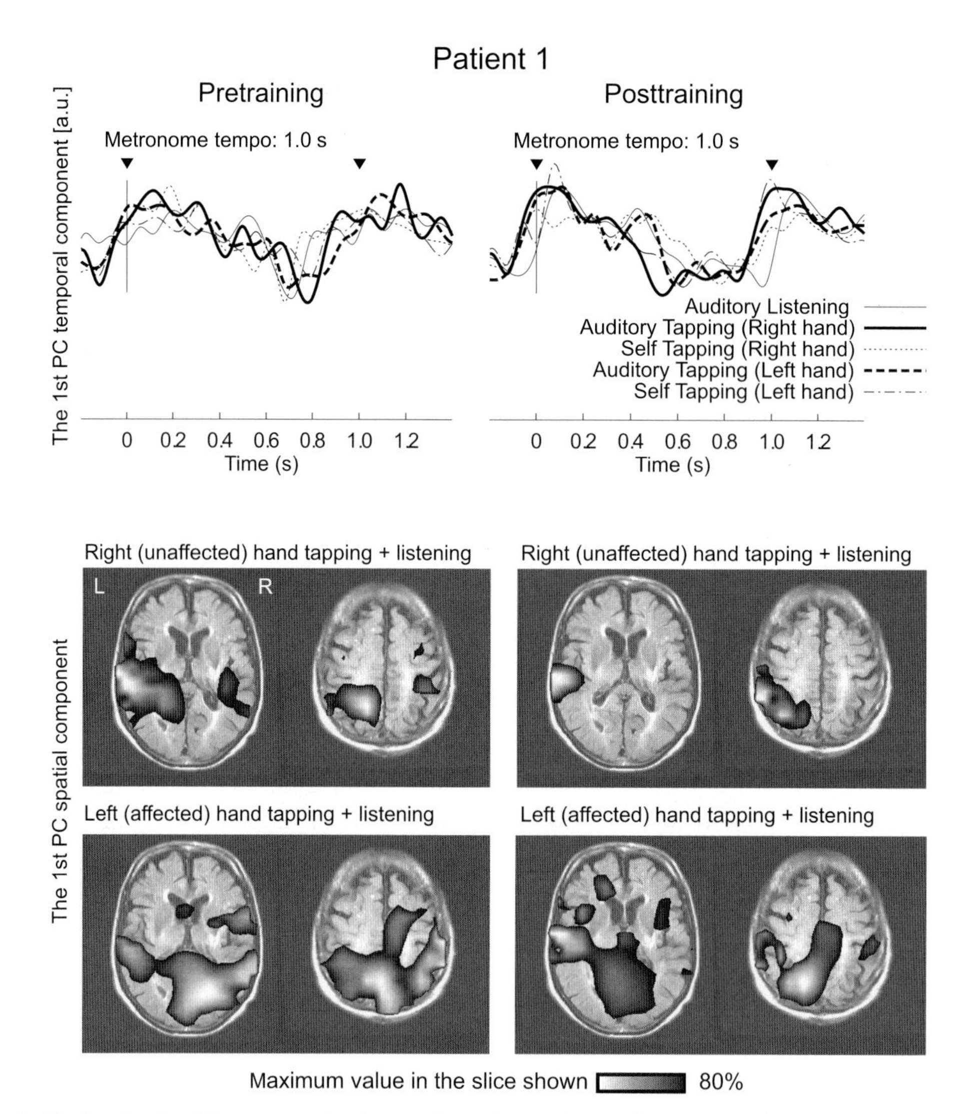

Figure 1. The beta-band activity as event-related power change from patient 1. The time series of the first PC (i.e., the largest) in all five listening and tapping conditions are shown for pretraining (top, left) and posttraining (top, right). The filled triangle marks the timing of each cycle of auditory stimulus presentation or designated finger tapping. Brain maps corresponding to the time series are shown in overlay of MRI axial slices for the auditory and sensorimotor areas for pretraining (bottom, left) and posttraining (bottom, right) for the right- and left-hand tasks separately.

and self-guided left finger tapping), we constructed two shared maps for right- and left-hand tapping, respectively.

We further examined the beta power change in the auditory listening condition in regions of interest (ROI) in bilateral auditory cortices and sensorimotor cortices. The ROIs were selected as volumes with 15-mm radius around auditory and sensory source locations as identified in previous studies on normal subjects. The center of auditory sources was chosen as the mean Talairach coordinates of Heschl's gyrus and superior temporal gyrus ([R–L, A–P, I–S]: right [−57, 27, 12], left [58, 13, 11]).[45] The center of the sensorimotor ROI was defined by the mean source coordinates of somatosensory-evoked and movement-related beta-band response for the index finger, observed from beta-band event-related synchronization and desynchronization (right [36, 24, 41], left [−36, 24, 41]).[46] The difference in source activities between pre- and posttraining in the ROIs

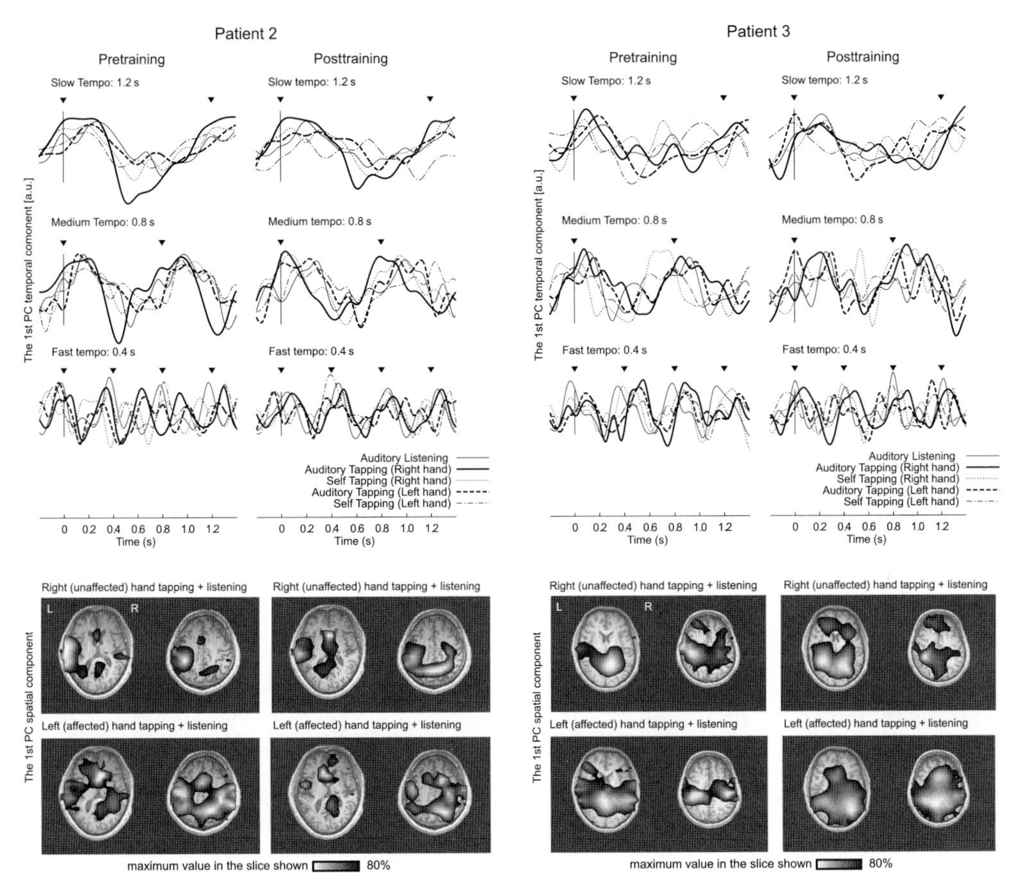

Figure 2. The beta-band activity as event-related power change from patient 2 (left panel) and 3 (right panel). For all three different metronome tempi (slow, medium, and fast) the time course of the first PC is shown for pretraining (top panel, left) and posttraining (top, right) for all five listening and tapping conditions. The filled triangle marks the timing of each cycle of auditory stimulus presentation or designated finger tapping. Brain maps associated with the time series are shown in overlay of MRI axial slices for the auditory and sensorimotor areas for pretraining and posttraining (bottom) for the right- and left-hand tasks separately.

were statistically compared to pre–post source activity differences in randomly selected locations across the entire brain volume within individual patients. We performed a student's t-test to examine whether the mean of the event-related beta power change differed in amplitude between pre- and posttraining within each individual. For this analysis, we used the data from the "medium" metronome tempo (0.8 sec) condition in patients 2 and 3, whereas the data from patient 1 was recorded with the 1.0-sec metronome tempo.

Results

Behavioral motor skills

Table 2 shows the score of the ARAT and Finger Tapping Test (an average number of taps for five trials within five points of each other) for each patient. The ARAT score showed a small improvement in each patient, reflecting the everyday life skills in the paretic hand such as grasping, reaching, and pinching. For the finger tapping, although patients 1 and 2 improved the score for the right hand after the intervention, none showed substantial change in the left finger. For the fine motor-skill assessment, we have administered the Grooved Pegboard Test and Purdue Pegboard Test, but patients 1 and 2 were not able to perform them in the affected left hand.

Functional network of event-related power change in beta oscillation

We used spatiotemporal PCA for each stimulus condition to identify the first PC with its temporal and spatial components; the former reflects common temporal patterns of beta power and the latter

Table 2. Behavioral data

Patient	ARAT score		Finger tapping (a number of taps in 10 sec, average of five blocks)			
			Right		Left	
	Pre	Post	Pre	Post	Pre	Post
1	50	54	32.4	42.6	12.4	14.8
2	37	39	43.2	51.6	15.8	15.4
3	46	52	45.8	44.8	36.2	37.8

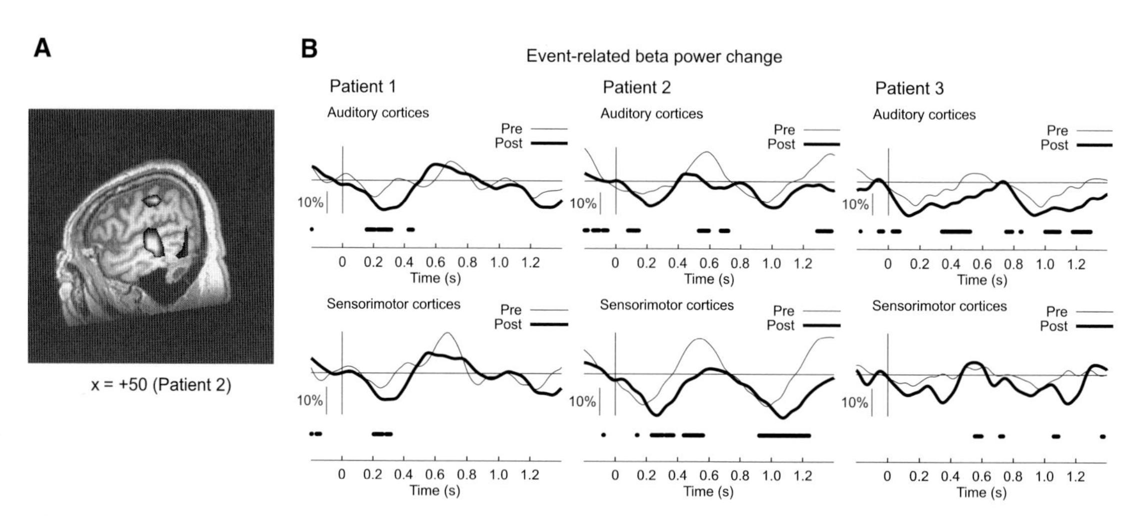

Figure 3. (A) The first principal component spatial map in patient 2 in the auditory listening condition with the medium tempo stimulus. The map shows a sagittal plane (Talairach coordinate $x = 50$ mm) in the left hemisphere. Note that the sensorimotor and auditory cortices have spatially well-separated local maxima showing that the source activities in the two areas in the beta band are independent from each other, and that the similarity of the beta band modulation is not caused by one common source spatially smeared into adjacent areas as an artifact of the beamformer source analysis. (B) The mean signal of the event-related power change in bilateral auditory and sensorimotor cortices in the auditory listening condition (patient 1: metronome tempo 1.0 sec; patients 2 and 3: metronome tempo 0.8 sec). For all three patients, the beta dynamics show deeper power decrease after the MST training (thick line) compared to the data in the same condition at pretraining (thin line) for both areas. The dots at the bottom of each plot illustrate the time point at which the comparison was statistically significant ($P < 0.0001$).

indicates the strength of the temporal pattern at each brain location. Figures 1 and 2 illustrate each patient's data, respectively, describing the time series of the first PC for each metronome tempo, and the corresponding spatial component. The first PC explained the majority of the variance in beta-power changes for each individual and all conditions (patient 1: 85–99%; patient 2: 40–99%; patient 3: 40–94%). Furthermore, the time courses of the first PC were similar across different conditions including auditory listening and finger tapping in the paretic and nonparetic hand. Most notably, the characteristic patterns of beta modulation, which synchronized to the metronome and/or tapping tempo (Figs. 1 and 2, top panel of each), were shared across conditions. The data from patient 1 showed a marked similarity in the time course of the first PC (Fig. 1). Also, patients 2 and 3, who were tested with three different metronome tempi, exhibited the periodic synchronized temporal pattern for all the tempi.

For each patient, the bottom panels in Figures 1 and 2 illustrate the PCA spatial map of brain areas contributing to the beta time course mainly in the bilateral auditory cortices and sensorimotor cortices close to the somatotopic hand area, respectively. The maps at pre- and posttraining time points are clearly different from each other for both right- and left-hand tasks. Importantly, the contribution of the auditory areas seems to be changed after the training.

Furthermore, we examined whether the MST intervention had significantly influenced the dynamics of the beta power modulation in the ROIs in bilateral auditory and sensorimotor cortices in the auditory listening condition. Figure 3A shows an example of the PCA map of the beta activity during the auditory listening in which the auditory and sensorimotor sources are clearly separated in space, indicating that the source activities in the two areas are independent from each other. This means that the similarity in the beta time course between these areas is not caused by one common source activity nearby smeared into the two locations as an artifact of the beamformer source analysis. Figure 3B shows that in all patients, the intervention had a significant effect on the amplitude of the beta power change. Specifically, the beta decrease after the sound onset in the posttraining became deeper and the subsequent beta rebound became smaller compared to the pretraining, consistently across patients.

Discussion

We presented three main findings regarding beta oscillations in stroke patients who participated in MST intervention. First, as expected from our previous study with healthy subjects, during passive listening beta oscillations showed a periodic modulation, which consisted of an amplitude decrease immediately after the beat onset and a subsequent rebound in both pre- and posttraining MEG recordings. Moreover, this time course was reproduced during finger tapping tasks with and without isochronous sound stimuli, in either the paretic or nonparetic hand. This suggests a robust existence of the shared beta modulation mechanism across the different task conditions. Second, temporally correlated beta modulation involved auditory and motor-related cortical areas. The weighting of contribution across the brain indexed by the spatial map of the PC changed between MEG recordings before and after the MST, suggesting functional cortical reorganization in the beta-band networks. Finally, in the auditory listening condition, the event-related beta decrease was enhanced after the MST training.

The time courses of the event-related beta power changes were remarkably consistent across patients, task conditions, and pre- and postintervention times. Even though the number of patients is small, it is still striking that their data shared the same characteristics despite having quite different lesions. This observation extends our previous finding in healthy young adults during auditory listening.[26,27] We note that the beta dynamics for the actual tapping movements agree well with typical findings in the literature in healthy subjects[28,29,47] and stroke patients.[9,30] The beta-band modulation is one of the most basic neurophysiological phenomena associated with a change in the status of the sensorimotor functions.[48,49] Because the present case data are to our knowledge the first to compare the beta activity between listening and tapping within individuals, future research is required to confirm this in a larger normal and patient cohort. Nevertheless, the finding offers an insight into a possible novel explanation of how an auditory pacing sound or musical beat could stimulate auditory–motor connections and thus provide positive effects in motor rehabilitation, such as in MST described earlier or rhythmic auditory stimulation (RAS).[50–52] Importantly, "auditory–motor coupling" in the context of music cognition has been discussed in two slightly different ways. The first is that the learning experience while making music helps establish an action–sound association perhaps through a "mirror-neuron" system in the premotor cortex.[17,18] Another is that perception of rhythms and extracting musical beat information from a sound sequence directly engages basal-ganglia-thalamo-cortical motor circuits involving the dopamine system and the related motor networks, but does not actually require preexisting knowledge about the sound materials in doing so.[22–24,53] Here, we propose that these two are not mutually exclusive. Rather, we think that the latter of auditory–motor obligatory interaction likely facilitates building a more explicit association between actions and sounds in the former. In fact, beta-band activities have also been associated with mirror-neuron system because beta oscillatory activities are modulated by visual observation of purposeful actions and listening to the sounds produced by the actions.[54,55] Our data are in line with the latter, more basic mechanism and support that neural processing of timing information associated with auditory listening and music making likely offers a direct gateway to access to the auditory–motor functional network. The auditory metronome listening paradigm may provide another useful way to examine the integrity of sensorimotor beta-band oscillatory network functions in passive stimulation without confounds of actual movements, in addition to existing experimental techniques such as transcranial magnetic stimulation–elicited motor evoked beta-band response,[56,57] somatosensory-related beta-band activity,[46,58,59] and spontaneous (e.g., resting-state) beta-band activity.[60,61]

Our patient cases have also demonstrated the substantial functional reorganization in the spatial configuration of temporally correlated beta-band activities in each patient, although individual patterns varied quite drastically from case to case. In neuroimaging studies, stroke patients in general show a wide variety of activation patterns for simple motor tasks or somatosensory stimulations for the tasks for both paretic and nonparetic hands.[62,63] Researchers have commonly observed bilateral or ipsilateral sensorimotor cortex activation to the paretic hand task, and interpreted it as a compensatory recruitment of the ipsilateral sensorimotor pathways.[62–65] Previous MEG studies also confirmed that there are clear correlations between functional recovery and interhemispheric asymmetry between the activities in the sensorimotor cortices.[66–68] According to the converging evidence from different imaging techniques and cross-sectional/longitudinal studies, current consensus is that a shift in laterality of the sensorimotor cortical activities in general is a robust indicator of functional reorganization and successful recovery of behavioral functions.[69,70] Our data based on the beta oscillatory activities are in line with this by demonstrating that the spatial maps of the first PC mostly show a shift from a bilateral pattern at the pretraining time point to the more lateralized pattern at the posttraining time points (except the data in patient 2 for the right [unaffected] hand tapping tasks). The PCA is clearly useful in assessing "spatial" reorganization of the beta-band network.

Here, we make the interesting observation that the beta decrease after the sound onset by auditory stimulation was enhanced after the MST intervention. This resonates with the previous findings where larger amplitude of beta modulation likely characterizes more efficient sensorimotor functional communication. For example, the enhanced beta desynchronization related to the paretic hand movements is found in patients who substantially improved their motor skills by receiving the MST during the intervention.[9] Also, the patients who had already made a good recovery still show a reduced range of the event-related beta power change during finger tapping compared to normal controls.[30] In the aging literature, older subjects show a substantial reduction of the beta rebound after the completion of movement tasks compared to young subjects.[71] However, overall research examining the beta-band oscillation in stroke patients has been sparse,[9,60,61,72] and because of the variety of paradigms it is still premature to reach an agreement in how it is related to functional recovery.

In conclusion, our case-based study has examined three chronic stroke patients regarding the effect of music-supported rehabilitation on the beta-band neuromagnetic oscillatory activities. The data showed functional reorganization in both spatial and temporal patterns of the beta activity, which were shared by auditory and sensorimotor modalities, despite an only subtle improvement in motor skills after the short five-week period of intervention in each patient. It is noteworthy that for chronic-stage patients, the MST intervention schedule may require a longer time frame to observe substantial improvements than we did here, and that our experimental procedure and analysis methods may potentially help future research to identify what could be unique about MST compared to other motor therapies for stroke survivors.

Acknowledgments

This work was supported by the Canadian Institutes of Health Research, the Centre for Stroke Rehabilitation of the Heart and Stroke Foundation of Ontario, the James S. McDonnell Foundation, and the Canadian Foundation for Innovation. The authors sincerely thank Eckert Altenmüller and Sabine Schneider for their guidance in MST protocols, Deirdre Dawson and Sara McEwen for their insightful discussions about the study design, Panteha Razavi and Susan Gillingham for their assistance in organizing testing materials and data collection, Michelle Nurwandi for her assistance in behavioral data analysis, Ellen Cohen for her assistance in clinical assessment for the motor skills, and Patricia Van Roon and Brian Fidali for their editorial assistance in the earlier version of the manuscript.

Conflicts of interest

The authors declare no conflicts of interest.

References

1. Schlaug, G. 2001. The brain of musicians. A model for functional and structural adaptation. *Ann. N.Y. Acad. Sci.* **930:** 281–299.
2. Münte, T.F., E. Altenmüller & L. Jäncke. 2002. The musician's brain as a model of neuroplasticity. *Nat. Rev. Neurosci.* **3:** 473–478.
3. Peretz, I. & R.J. Zatorre. 2005. Brain organization for music processing. *Annu. Rev. Psychol.* **56:** 89–114.
4. Jäncke, L. 2009. The plastic human brain. *Restor. Neurol. Neurosci.* **27:** 521–538.
5. Pantev, C. *et al.* 2009. Auditory-somatosensory integration and cortical plasticity in musical training. *Ann. N.Y. Acad. Sci.* **1169:** 143–150.
6. Koelsch, S. 2009. A neuroscientific perspective on music therapy. *Ann. N.Y. Acad. Sci.* **1169:** 374–384.
7. Johansson, B.B. 2011. Current trends in stroke rehabilitation. A review with focus on brain plasticity. *Acta Neurol. Scand.* **123:** 147–159.
8. Schneider, S. *et al.* 2007. Using musical instruments to improve motor skill recovery following a stroke. *J. Neurol.* **254:** 1339–1346.
9. Altenmüller, E. *et al.* 2009. Neural reorganization underlies improvement in stroke-induced motor dysfunction by music-supported therapy. *Ann. N.Y. Acad. Sci.* **1169:** 395–405.
10. Schneider, S. *et al.* 2010. Music-supported training is more efficient than functional motor training for recovery of fine motor skills in stroke patients. *Music Percept.* **27:** 271–280.
11. Rojo, N. *et al.* 2011. Music-supported therapy induces plasticity in the sensorimotor cortex in chronic stroke: a single-case study using multimodal imaging (fMRI-TMS). *Brain Inj.* **25:** 787–793.
12. Rodriguez-Fornells, A. *et al.* 2012. The involvement of audio-motor coupling in the music-supported therapy applied to stroke patients. *Ann. N.Y. Acad. Sci.* **1252:** 282–293 (this volume).
13. Krakauer, J. W. 2006. Motor learning: its relevance to stroke recovery and neurorehabilitation. *Curr. Opin. Neurol.* **19:** 84–90.
14. Dawson, D.R. *et al.* 2009. Using the cognitive orientation to occupational performance (CO-OP) with adults with executive dysfunction following traumatic brain injury. *Can. J. Occup. Ther.* **76:** 115–127.
15. McEwen, S.E. *et al.* 2010. 'There's a real plan here, and I am responsible for that plan': participant experiences with a novel cognitive-based treatment approach for adults living with chronic stroke. *Disabil, Rehabil.* **32:** 540–550.
16. McEwen, S.E. *et al.* 2009. Exploring a cognitive-based treatment approach to improve motor-based skill performance in chronic stroke: results of three single case experiments. *Brain Inj.* **23:** 1041–1053.
17. Bangert, M. *et al.* 2006. Shared networks for auditory and motor processing in professional pianists: evidence from fMRI conjunction. *Neuroimage* **30:** 917–926.
18. Lahav, A., E. Saltzman & G. Schlaug. 2007. Action representation of sound: audiomotor recognition network while listening to newly acquired actions. *J. Neurosci.* **27:** 308–314.
19. Large, E.W. 2008. Resonating to musical rhythm: theory and experiment. In *The Psychology of Time*. S. Grondin, Ed.: 189–231. Emerald. Cambridge.
20. Ivry, R.B. & R.M. Spencer. 2004. The neural representation of time. *Curr. Opin. Neurobiol.* **14:** 225–232.
21. Mauk, M.D. & D.V. Buonomano. 2004. The neural basis of temporal processing. *Annu. Rev. Neurosci.* **27:** 307–340.
22. Grahn, J.A. & M. Brett. 2007. Rhythm and beat perception in motor areas of the brain. *J. Cogn. Neurosci.* **19:** 893–906.
23. Chen, J.L., V.B. Penhune & R.J. Zatorre. 2008. Listening to musical rhythms recruits motor regions of the brain. *Cereb. Cortex* **18:** 2844–2854.
24. Bengtsson, S.L. *et al.* 2009. Listening to rhythms activates motor and premotor cortices. *Cortex* **45:** 62–71.
25. Fujioka, T., B.R. Zendel & B. Ross. 2010. Endogenous neuromagnetic activity for mental hierarchy of timing. *J. Neurosci.* **30:** 3458–3466.
26. Fujioka, T. *et al.* 2009. Beta and gamma rhythms in human auditory cortex during musical beat processing. *Ann. N.Y. Acad. Sci.* **1169:** 89–92.
27. Fujioka, T. , L.J. Trainor, E.W. Large, and B. Ross. 2012 Internalized timing of isochronous sounds is represented in neuromagnetic beta oscillations. *J. Neurosci.* **32:** 1791–1802.
28. Toma, K. *et al.* 2002. Movement rate effect on activation and functional coupling of motor cortical areas. *J. Neurophysiol.* **88:** 3377–3385.
29. Gerloff, C. *et al.* 1998. Functional coupling and regional activation of human cortical motor areas during simple, internally paced and externally paced finger movements. *Brain* **121**(Pt 8): 1513–1531.
30. Gerloff, C. *et al.* 2006. Multimodal imaging of brain reorganization in motor areas of the contralesional hemisphere of well recovered patients after capsular stroke. *Brain* **129:** 791–808.
31. Barreca, S.R. *et al.* 2005. Test-retest reliability, validity, and sensitivity of the Chedoke arm and hand activity inventory: a new measure of upper-limb function for survivors of stroke. *Arch. Phys. Med. Rehabil.* **86:** 1616–1622.
32. Barreca, S. *et al.* 2004. Development of the Chedoke arm and hand activity inventory: theoretical constructs, item generation, and selection. *Top Stroke Rehabil.* **11:** 31–42.
33. Yozbatiran, N., L. Der-Yeghiaian & S.C. Cramer. 2008. A standardized approach to performing the action research arm test. *Neurorehabil. Neural Repair* **22:** 78–90.
34. Reitan, R.M. & D. Wolfson. 1993. *The Halstead-Reitan Neuropsychological Test Battery: Theory and Clinical Interpretation.* Neuropsychology Press. Tucson, AZ.
35. Bornstein, R.A. 1986. Normative data on intermanual differences on three tests of motor performance. *J. Clin. Exp. Neuropsychol.* **8:** 12–20.
36. Tiffin, J. & E.J. Asher. 1948. The Purdue Pegboard: norms and studies of reliability and validity. *J. Appl. Psyc.* **32:** 234–247.
37. Kobayashi, T. & S. Kuriki. 1999. Principal component elimination method for the improvement of S/N in evoked neuromagnetic field measurements. *IEEE Trans. Biomed. Eng.* **46:** 951–958.

38. Jung, T.P. *et al.* 2000. Removing electroencephalographic artifacts by blind source separation. *Psychophysiology* **37:** 163–178.
39. Delorme, A. & S. Makeig. 2004. EEGLAB: an open source toolbox for analysis of single-trial EEG dynamics including independent component analysis. *J. Neurosci. Methods* **134:** 9–21.
40. Robinson, S.E. & J. Vrba. 1999. Functional neuroimaging by synthetic aperture magnetometry. In *Recent Advances in Biomagnetism.* T. Yoshimoto *et al.*, Eds.: 302–305. Tohoku University Press. Sendai.
41. Van Veen, B.D. *et al.* 1997. Localization of brain electrical activity via linearly constrained minimum variance spatial filtering. *IEEE Trans. Biomed. Eng.* **44:** 867–880.
42. Robinson, S.E. 2004. Localization of event-related activity by SAM(erf). *Neurol. Clin. Neurophysiol.* **2004:** 109.
43. Vrba, J. 2002. Magnetoencephalography: the art of finding a needle in a haystack. *Physica C: Superconductivity Appl.* **368:** 1–9.
44. Vrba, J. & S.E. Robinson. 2001. Signal processing in magnetoencephalography. *Methods* **25:** 249–271.
45. Ross, B., S.A. Hillyard & T.W. Picton. 2010. Temporal dynamics of selective attention during dichotic listening. *Cereb. Cortex* **20:** 1360–1371.
46. Bardouille, T., T.W. Picton & B. Ross. 2010. Attention modulates beta oscillations during prolonged tactile stimulation. *Eur. J. Neurosci.* **31:** 761–769.
47. Stancak, A., Jr. & G. Pfurtscheller. 1996. Event-related desynchronisation of central beta-rhythms during brisk and slow self-paced finger movements of dominant and nondominant hand. *Brain Res. Cogn. Brain Res.* **4:** 171–183.
48. Pfurtscheller, G. & F.H. Lopes da Silva. 1999. Event-related EEG/MEG synchronization and desynchronization: basic principles. *Clin. Neurophysiol.* **110:** 1842–1857.
49. Hari, R. & R. Salmelin. 1997. Human cortical oscillations: a neuromagnetic view through the skull. *Trends Neurosci.* **20:** 44–49.
50. Thaut, M.H. *et al.* 2002. Kinematic optimization of spatiotemporal patterns in paretic arm training with stroke patients. *Neuropsychologia* **40:** 1073–1081.
51. Thaut, M.H. *et al.* 2007. Rhythmic auditory stimulation improves gait more than NDT/Bobath training in near-ambulatory patients early poststroke: a single-blind, randomized trial. *Neurorehabil. Neural Repair* **21:** 455–459.
52. Malcolm, M.P., C. Massie & M. Thaut. 2009. Rhythmic auditory-motor entrainment improves hemiparetic arm kinematics during reaching movements: a pilot study. *Top Stroke Rehabil.* **16:** 69–79.
53. Geiser, E. & A. Kaelin-Lang. 2011. The function of dopaminergic neural signal transmission in auditory pulse perception: evidence from dopaminergic treatment in Parkinson's patients. *Behav. Brain Res.* **225:** 270–275.
54. Hari, R. 2006. Action-perception connection and the cortical mu rhythm. *Prog. Brain Res.* **159:** 253–260.
55. Caetano, G., V. Jousmaki & R. Hari. 2007. Actor's and observer's primary motor cortices stabilize similarly after seen or heard motor actions. *Proc. Natl. Acad. Sci. USA* **104:** 9058–9062.
56. Van Der Werf, Y.D. & T. Paus. 2006. The neural response to transcranial magnetic stimulation of the human motor cortex. I. Intracortical and cortico-cortical contributions. *Exp. Brain Res.* **175:** 231–245.
57. Maki, H. & R.J. Ilmoniemi. 2010. EEG oscillations and magnetically evoked motor potentials reflect motor system excitability in overlapping neuronal populations. *Clin. Neurophysiol.* **121:** 492–501.
58. Pfurtscheller, G. 1981. Central beta rhythm during sensorimotor activities in man. *Electroencephalogr. Clin. Neurophysiol.* **51:** 253–264.
59. Cheyne, D. *et al.* 2003. Neuromagnetic imaging of cortical oscillations accompanying tactile stimulation. *Brain Res. Cogn. Brain Res.* **17:** 599–611.
60. Tecchio, F. *et al.* 2005. Rhythmic brain activity at rest from rolandic areas in acute mono-hemispheric stroke: a magnetoencephalographic study. *Neuroimage* **28:** 72–83.
61. Tecchio, F. *et al.* 2006. Long-term effects of stroke on neuronal rest activity in rolandic cortical areas. *J. Neurosci. Res.* **83:** 1077–1087.
62. Cao, Y. *et al.* 1998. Pilot study of functional MRI to assess cerebral activation of motor function after poststroke hemiparesis. *Stroke* **29:** 112–122.
63. Ward, N.S. *et al.* 2003. Neural correlates of motor recovery after stroke: a longitudinal fMRI study. *Brain* **126:** 2476–2496.
64. Marshall, R.S. *et al.* 2000. Evolution of cortical activation during recovery from corticospinal tract infarction. *Stroke* **31:** 656–661.
65. Tombari, D. *et al.* 2004. A longitudinal fMRI study: in recovering and then in clinically stable sub-cortical stroke patients. *Neuroimage* **23:** 827–839.
66. Rossini, P.M. *et al.* 2001. Interhemispheric differences of sensory hand areas after monohemispheric stroke: MEG/MRI integrative study. *Neuroimage* **14:** 474–485.
67. Huang, M. *et al.* 2004. MEG response to median nerve stimulation correlates with recovery of sensory and motor function after stroke. *Clin. Neurophysiol.* **115:** 820–833.
68. Tecchio, F. *et al.* 2007. Interhemispheric asymmetry of primary hand representation and recovery after stroke: a MEG study. *Neuroimage* **36:** 1057–1064.
69. Calautti, C. & J.C. Baron. 2003. Functional neuroimaging studies of motor recovery after stroke in adults: a review. *Stroke* **34:** 1553–1566.
70. Rossini, P.M. *et al.* 2007. Neuroimaging experimental studies on brain plasticity in recovery from stroke. *Eura. Medicophys.* **43:** 241–254.
71. Labyt, E. *et al.* 2006. Oscillatory cortical activity related to voluntary muscle relaxation: influence of normal aging. *Clin. Neurophysiol.* **117:** 1922–1930.
72. Wilson, T.W., A. Fleischer & D. Archer. 2011. Oscillatory MEG motor activity reflects therapy-related plasticity in stroke patients. *Neurorehabil. Neural Repair* **25:** 188–193.

Ann. N.Y. Acad. Sci. ISSN 0077-8923

ANNALS OF THE NEW YORK ACADEMY OF SCIENCES
Issue: *The Neurosciences and Music IV: Learning and Memory*

Making music after stroke: using musical activities to enhance arm function

Frederike van Wijck,[1] Don Knox,[2] Colin Dodds,[2] Gianna Cassidy,[2] Gillian Alexander,[3] and Raymond MacDonald[1]

[1]Institute for Applied Health Research and School of Health and Life Sciences, Glasgow Caledonian University, Glasgow, United Kingdom. [2]School of Engineering and Built Environment, Glasgow Caledonian University, Glasgow, United Kingdom. [3]NHS Greater Glasgow and Clyde, Glasgow, United Kingdom

Address for correspondence: Frederike van Wijck, Institute for Applied Health Research and School of Health and Life Sciences, Glasgow Caledonian University, Cowcaddens Road, Glasgow G0 4BA, UK. Frederike.vanWijck@gcu.ac.uk

A common long-term consequence of stroke is impaired arm function, which affects independence and quality of life in a considerable proportion of stroke survivors. There is a growing need for self-management strategies that enable stroke survivors to continue their recovery after rehabilitation has ceased. Interventions with high-intensity, repetitive task training and feedback are most likely to improve function. Achieving the required amount of self-practice is challenging, however. Innovative approaches are required to translate therapies into rewarding activities that can be undertaken independently. This paper describes the key principles and development of a novel intervention that integrates individuals' preferred music with game technology in upper limb rehabilitation. The "tap tempo" paradigm, which uses rhythmic auditory cueing, provides repetitive upper limb task training, which can be tailored to individual goals and progress (e.g., in terms of movement range and complexity), while providing sensitive quantitative feedback to promote skill acquisition and enhance self-management.

Keywords: stroke; rehabilitation; arm function; preferred music; game technology

Introduction

The purpose of this paper is to describe the key principles and the development of a prototype intervention, designed to enable stroke survivors to enhance functional recovery of their affected upper limb through a game activity integrating evidence from music psychology, clinical rehabilitation, neuroscience, movement science, and audio technology.

Stroke is defined by the World Health Organization as "an acute neurologic dysfunction of vascular origin with sudden (within seconds) or at least rapid (within hours) occurrence of symptoms and signs corresponding to the involvement of focal areas in the brain."[1] Across the world, 15 million people suffer a stroke each year, leaving five million with permanent disabilities.[2] Up to 80% of stroke survivors experience impaired arm function at three months after stroke,[3] which is often persistent, disabling,[3–5] and affects quality of life.[6] Common upper limb impairments include weakness or paresis,[7] abnormal force production,[8] problems with muscle coactivation[9] and coordination,[10] often compounded by impaired sensation[3] and proprioception[11] as well as pain. Weakness, impaired coordination, and spasticity together with soft tissue stiffness (hypertonia) may reduce range of movement and result in permanent deformities[12] that affect the ability to undertake activities of daily living.

A number of different treatments have been shown to be effective in improving arm function after stroke.[13] These comprise Constraint-Induced Movement Therapy (CIMT), mental practice, robotics, and electromyographic biofeedback. Repetitive task training and electrostimulation have also demonstrated an effect, although marginal.[13] Acknowledging the need for further research, the broad conclusion of this comprehensive systematic review was that interventions characterized by high-intensity, repetitive task-specific practice that

doi: 10.1111/j.1749-6632.2011.06403.x

provide feedback are most likely to be effective. Studies using neuroimaging techniques have also shown that task-specific therapeutic activities have the ability to forge neuroplastic changes in the brain.[14] The findings from studies that have explored the impact of CIMT on patterns of neural activity are too diverse for any clear conclusions, however.[15] Furthermore, the principles of intensive practice, task-specific training, and feedback have long been recognized as cornerstones of skill acquisition in healthy people.[16]

Taken together, converging evidence from stroke rehabilitation, neuroscience, and motor learning research indicates that activities that are functional, practiced repeatedly, and include feedback have the potential to drive long-term neuroplastic changes in the brain that in turn can enhance functional recovery after stroke. A key challenge for clinicians working with stroke survivors is therefore to provide the required amount of practice as well as feedback.

Health care resources are, however, under threat in the current global recession. Contemporary United Kingdom clinical initiatives and health policies are aimed at facilitating early supported discharge after stroke and shifting the balance of care from hospital to self-management in the community.[17] There is therefore an urgent need to improve self-management strategies for people with stroke, who continue to experience impaired arm function.

Exciting new opportunities for stroke rehabilitation are emerging from the field of neurologic music therapy (NMT), which aims to improve cognitive, sensory, and motor function in people with neurological conditions through the therapeutic application of music. Active ingredients of NMT include auditory motor synchronization, an entrainment function whereby the auditory rhythm is synchronized with movement execution as well as motivational aspects.[18] Rhythmic auditory cueing can be effective in improving motor recovery, particularly gait after stroke, as shown in a Cochrane systematic review by Bradt *et al.*[19] This review, however, only comprised two studies examining the effects of music or rhythm on arm function after stroke, including a total of 41 participants. Although both studies showed short-term improvements, it is clear that there is a need for further robust clinical trials.

With regards to the utility of music listening for therapeutic purposes, a crucial issue is the nature of the music.[20] Specifically, a significant body of work highlights the effects of preselected music upon a variety of psychological variables.[21] For example, music with particular structural features (moderate tempo and complexity) may be effective in reducing pain and anxiety in clinical contexts.[22] More recently, a number of studies have highlighted the importance of preference in determining therapeutic potential, demonstrating that the liking of a particular piece of music can be a key element in predicting positive clinical outcomes. For example, Mitchell *et al.* reported that participants listening to favorite music were able to keep their hands in cold water for longer and reported less pain than participants listening to experimenter-selected "anxiolytic music."[23] These authors argue that listening to preferred music renders music with varying structural features "functionally equivalent," that is, music with different features, such as tempo, complexity, and key signature, can have a similar therapeutic effect. Current work is investigating the interplay between music structure and preference,[24] which indicates a sensitive interplay between variables that relate to preference and key structural features of the music. This relationship suggests that it is not sufficient to only consider musical structure (e.g., tempo) but, additionally, that musical preferences of listeners must also be taken into consideration when evaluating the effects of music. Indeed, it is vital to consider if participants enjoy the music that is being listened to during experiments that examine the effects of music.

The prototype intervention described here addresses these issues in a number of ways. In the first instance, the intervention is designed to allow each participant to select their preferred music to enable them to listen to music that is familiar and enjoyed. The hypothesis to be tested is that participants may enjoy participating in the intervention more than with preselected music, which may have a positive effect on motivation and may in turn increase the amount of time spent engaging with the intervention. Furthermore, it will also be possible to examine the participants' musical choices using innovative software specifically designed to analyze the relationship between key structural features and the emotional impact of the music that may be central when listening to music for therapeutic purposes.[24]

Another emerging field of interest to neurological rehabilitation is game play, which is receiving increasing attention as a potential platform for health promotion and self-management and addressing psychological and behavioral barriers to optimal health care and participation.[25,26] Saposnik[27] suggested that the parameters of effective treatment, that is, repetition, task-specific training, and challenge, are those that games scaffold intrinsically.[28] From commercial "serious games" to teach patients and clinicians,[29] to tailor-made games built as an adjunct to therapy,[30] the potential of games to innovate and enrich existing practice is emerging. Game play can help increase patient adherence and self-management, aid physical and psychological recovery, and enhance patient and clinician knowledge in a range of contexts.[26] In general, the literature emphasizes the potential of games to increase patient motivation, learning through repetition and an enriched environment, confidence through reinforcement and immediate feedback, and positivity through achievement and social interaction.[31–38] Games may be particularly valuable in supporting stroke rehabilitation where effective rehabilitation must be early, intensive, and repetitive.[39] In a study by Merians *et al.*,[40] 12 patients who had experienced a stroke at least six months previously played an off-the-shelf game aimed at improving eye–hand coordination, grip, and individual finger movement, for two to three hours a day over eight days. Patients also wore special robotic gloves (CyberGlove) as the game controller and to assist and measure hand movement. After eight days, patients showed an average improvement in their standard clinical scores of 20–22%, displaying better stability, greater smoothness of movement, and improved control over their fingers, in comparison to a control group. Saposnik *et al.*[27] reviewed 12 studies investigating the effects of electronic games on upper arm strength and mobility, concluding that patients who played video games including Wii and Playstation were up to five times more likely to show improved arm motor function in comparison to those receiving standard therapy. Further, a recent Cochrane systematic review[41] on virtual reality for stroke rehabilitation, comprising 19 studies with 565 stroke patients, found that playing interactive games may improve arm function and the management of activities of daily living. Within this review, the analysis of seven trials comparing game play with traditional therapy indicated that virtual reality resulted in better arm function.

Although these findings are promising, the relative dearth of rigorous evaluation in this body of research, the bias toward younger and less severely affected patients, and the limited inclusion of a range of commercial and tailored game activities raises a need for further robust and ecologically valid investigation. Moreover, this should comprise the entire trajectory, from the design of game activities to the evaluation of their impact on patient experiences and clinical outcomes. Importantly, one pervasive feature of game activity omitted in the current literature is music, often concurrent to game play. In general, the power of music to benefit psychological and physical health is widely acknowledged throughout health care literature—as explained earlier. Specifically, evidence indicates that concurrent music can have ergogenic and psychophysical benefits in a range of activity contexts, including game play.[42–45]

In summary, there is an urgent need to develop innovative approaches to self-management of upper limb rehabilitation after stroke. Converging evidence in stroke rehabilitation, neuroscience, and motor learning indicates that interventions incorporating repetitive practice of functionally relevant tasks are most likely to improve functional recovery. However, for these interventions to be clinically effective, they need to stimulate intrinsic motivation to encourage stroke survivors to undertake and maintain the required amount of self-practice. By using preferred music, rhythmic auditory cueing, and game play, our aim is to design a prototype intervention that enables stroke survivors to improve the function of their affected arm through self-directed practice. Its specifications are that the intervention should be safe and enjoyable, relevant to the individual, clinically effective, and provide options for individual tailoring and progression. Measures of motor performance used for feedback purposes should be valid, reliable, and easy for the stroke survivor to understand, the interface should be user friendly, and the prototype should be acceptable to stroke survivors and therapists.

Methods and results

The prototype intervention, written in C#, is a simple graphical interface consisting of a set of concentric targets, one of which is highlighted to coincide with each bar/beat of the music currently playing. The task for the stroke survivor is to position the

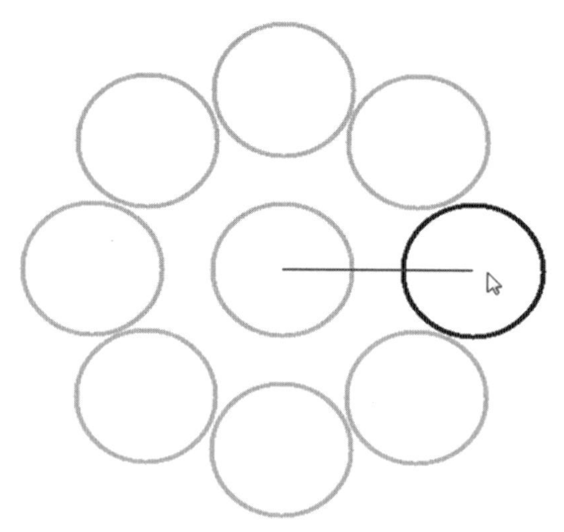

Figure 1. Screenshot of the tap tempo game. Target circles are highlighted in synchronization with music tempo.

on-screen cursor (through movement of their affected arm) above the currently highlighted circle. Thus, the game uses a simple "tap tempo" paradigm (Fig. 1).

Technology

The system uses inexpensive, off-the-shelf Nintendo Wii game technology. This is a flexible and popular system, and has been suggested as a suitable technological means for facilitating game-based stroke rehabilitation in the home for participants with limited range of movement.[46] In this implementation, the Wii remote controls are used as sensors (using the built-in IR cameras) and thus the accelerometers are not required. This setup means that the stroke survivor does not need to hold the Wii remote in their hand. It also allows for freedom of placement of sensors at multiple points on the arm, hand, or wrist. Upper limb movements, which translate to actions in the game, are tracked using small, lightweight, wireless infrared (IR) sources (LED and battery). These are placed on the participant's hand or arm, depending on the therapeutic goal. Signals from the IR sources are picked up by two IR sensors with a 1024×768 pixel resolution and an approximate 40° field of view. The two sensors are placed at right angles to the participant to enable the IR sources to be tracked in three dimensions.[47] The best response from the sensors is achieved when placed at a distance of between 0.5 m and 3 m from the participant.[48] Up to four IR sources can be tracked at one time, and the spatial accuracy of source tracking has been shown to be acceptable.[49] Sensor data are read and translated into game input data by scripts running on the GlovePie input emulator program (http://glovepie.org). These scripts can be used to tailor movement range, direction, and speed to best match the requirements of the participant.

Integrating therapeutic goals

The game comprises a number of parameters that enable it to be tailored to individual goals and capabilities. Importantly, the activity is based on the individual's musical preference. Speed of movement can be altered by setting the distance between targets on the screen, the gain between the participant's movements and movements displayed on screen, and the beat pulse to which participants must synchronize their movement (e.g., bar, quarter). In addition, the direction and range of movement can be tailored to individual needs; for example, the activity can be set to emphasize shoulder flexion and external rotation together with elbow extension. Further, accuracy can be tailored by altering the size of targets (Fig. 2), while the complexity of the activity as a whole can be adjusted by changing speed and accuracy as well as the order (e.g., blocked, random) in which targets need to be touched. The parameters of speed and accuracy, range, and direction of movement, as well as task complexity, can all be selected by the therapist and stored in an individual patient file to facilitate set-up for follow-up sessions and independent practice.

Music selection

Music selected by the participant is preprocessed in order to extract tempo and rhythm information. This is achieved through use of onset detection algorithms implemented in the Sonic Visualizer audio analysis and visualization program.[50] The output of this analysis is a vector of time values, labeled with bar/beat locations in the digital music file. The vector is stored in an output comma separated values file that is used by the game software, along with the digital music file. As the music plays during the game, the game software reads the vector of time values and synchronizes the game graphics with the music (Fig. 3).

Feedback

Data on several aspects of the participant's performance are recorded during game play. These include the music chosen by the participant, the time spent

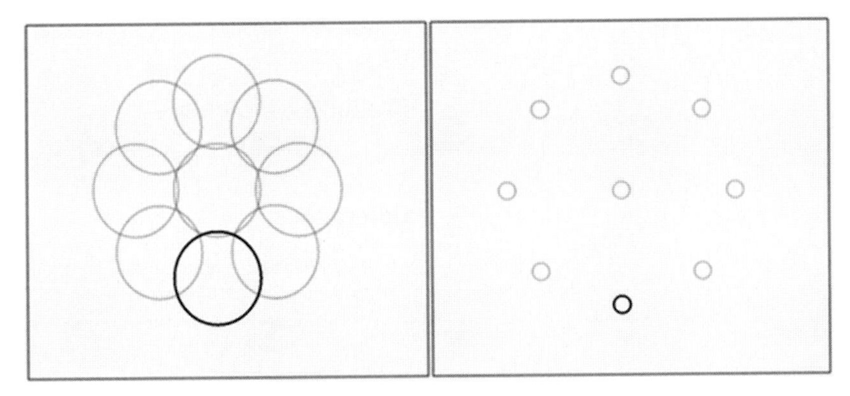

Figure 2. Changing the demand for movement accuracy by adjusting the target size.

playing the game, the number of times it was played, and the game "score" (the number of times the correct target was hit).

In addition, the input emulator software records the movement of the IR sources in three dimensions for post-hoc analysis. These data can be compared to the music beat/pulse data vector to examine the participant's movements in comparison to the music and provide knowledge of results (i.e., information about the degree to which the environmental goal was achieved[16]) in terms of spatial and temporal accuracy. Cursor position is logged at each point in time coinciding with the bar/beat of the music, allowing production of scatter plots showing the accuracy with which the participant moves the cursor in time to the music (Fig. 4).

Next steps

The next step in the development of this prototype is seeking the opinions of stroke survivors and therapists in the evaluation of safety and appropriateness of the intervention, as well as its impact on patient motivation. Our aim is to make the interface simple to use and accessible for stroke survivors with sensory, cognitive, and/or communication impairments, since a key aim of this technology is to support its use in the home by a wide range of stroke survivors. In addition, we will explore a range of music game models, including creating music as well as playing along existing music. Further work will examine the accuracy and reliability of upper limb movement, captured with this IR technology, to explore its use in knowledge of performance (i.e.,

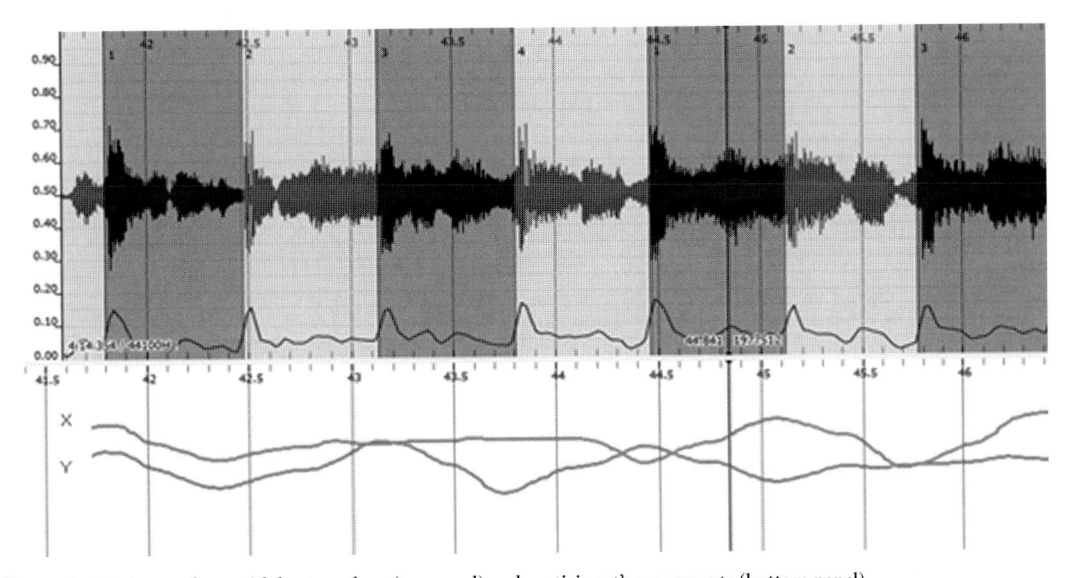

Figure 3. Music waveform with beat markers (top panel) and participant's movements (bottom panel).

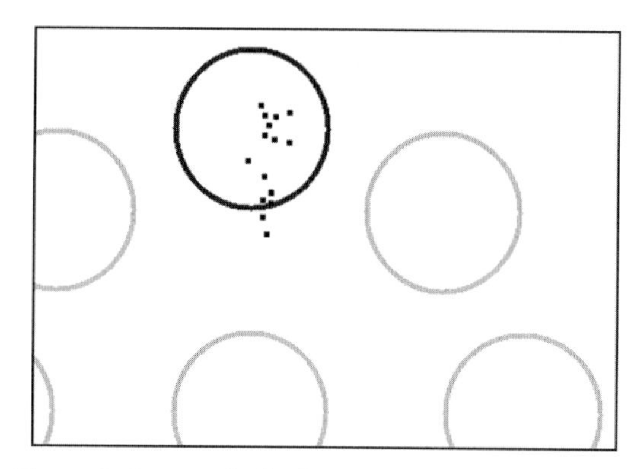

Figure 4. Summary feedback on movement accuracy from several attempts.

feedback on joint kinematics) to further facilitate motor learning. Next, clinical trials will be conducted to compare the effects and experiences of the game activity played together with individuals' preferred music, played with preselected music with the same rhythm, with rhythm only, and without any auditory stimulus. This also includes an RCT to compare the impact of this music game intervention with current conventional rehabilitation interventions in terms of clinical outcomes, motivation, and self-efficacy. Finally, our aim is to add a social dimension by enabling people to play together.

Conclusions

A prototype music game intervention has been designed for stroke survivors with impaired arm function, based on principles of music psychology, game technology, rehabilitation, neuroscience, movement science, and audiotechnology. The intervention can be tailored to individuals with a range of upper limb impairments and activity limitations, as well as musical preferences. One of the strengths of this technology is that the therapeutic input can be quantified, while outcomes can also be measured in detail. Several iterations of user involvement and feedback are now underway to test and further develop the prototype. This intervention is anticipated to open up novel opportunities to tailor an off-the-shelf technology to the specific preferences and rehabilitation needs of individuals with stroke, for use in the home environment as a strategy for self-managed upper limb rehabilitation.

Acknowledgments

This study was supported by a pump-priming research fund from the Institute for Applied Health Research, Glasgow Caledonian University, Glasgow, UK.

Conflicts of interest

The authors declare no conflicts of interest.

References

1. World Health Organisation Task Force on Stroke and Other Cardiovascular Disorders. 1989. Special report from the World Health Organization. Stroke-1989. Recommendations on stroke prevention, diagnosis, and therapy. *Stroke* **20:** 1407–1431.
2. Mackay, J. & G. Mensah. 2004. *Atlas of Heart Disease and Stroke*. Geneva, Switzerland: World Health Organisation Press.
3. Parker, V.M., D.T. Wade & R. Langton Hewer. 1986. Loss of arm function after stroke: measurement, frequency, and recovery. *Int. Rehabil. Med.* **8:** 69–73.
4. Broeks, J.G., G.J. Lankhorst, K. Rumping & A.J. Prevo. 1999. The long-term outcome of arm function after stroke: results of a follow-up study. *Disabil. Rehabil.* **21:** 357–364.
5. Lai, S.M., S. Studenski, P.W. Duncan & S. Perera. 2002. Persisting consequences of stroke measured by the Stroke Impact Scale. *Stroke* **33:** 1840–1844.
6. Wyller, T.B., U. Sveen, K.M. Sodring, A.M. Pettersen, *et al.* 1997. Subjective well-being one year after stroke. *Clin. Rehabil.* **11:** 139–145.
7. Boissy, P., D. Bourbonnais, M.M. Carlotti, D. Gravel, *et al.* 1999. Maximal grip force in chronic stroke subjects and its relationship to global upper extremity function. *Clin. Rehabil.* **13:** 354–362.
8. Dewald, J.P.A. & R.F. Beer. 2001. Abnormal joint torque patterns in the paretic upper limb of subjects with hemiparesis. *Muscle Nerve* **24:** 273–283.
9. Gowland, C., H. DeBruin, J.V. Basmajian, N. Plews, *et al.* 1992. Agonist and antagonist activity during voluntary upper-limb movement in patients with stroke. *Phys. Ther.* **72:** 624–633.
10. Archambault, P., P. Pigeon, A.G. Feldman & M.F. Levin. 1999. Recruitment and sequencing of different degrees of freedom during pointing movements involving the trunk in healthy and hemiparetic subjects. *Exp. Brain Res.* **126:** 55–67.
11. Carey, L.M., L.E. Oke & T.A. Matyas. 1996. Impaired limb position sense after stroke: a quantitative test for clinical use. *Arch. Phys. Med. Rehabil.* **77:** 1271–1278.
12. Edwards, S. 2002. Abnormal tone and movement as a result of neurological impairment: considerations for treatment. In *Neurological Physiotherapy: A Problem-Solving Approach.* S. Edwards, Ed.: 89–120. Churchill Livingstone. Edinburgh.
13. Langhorne, P., F. Coupar & A. Pollock. 2009. Motor recovery after stroke: a systematic review. *Lancet Neurol.* **8:** 741–754.
14. Hubbard, I.J., M.W. Parsons, C. Neilson & L.M. Carey. 2009. Task-specific training: evidence for and translation to clinical practice. *Occup. Ther. Int.* **16:** 175–189.
15. Huang, W.C., Y.J. Chen, C.L. Chien, H. Kashima & K.C. Li. 2011. Constraint-induced movement therapy as a paradigm of translational research in neurorehabilitation: reviews and prospects. *Am. J. Transl. Res.* **3:** 48–60.

16. Schmidt, R.A. & C.A. Wrisberg. 2008. *Motor Learning and Performance: A Situation-Based Learning Approach*. 4th ed. Human Kinetics. Champaign, IL.
17. Scottish Government. 2009. Improving the health and wellbeing of people with long term conditions in Scotland: A national action plan [online]. Available at: http://www.scotland.gov.uk/Publications/2009/12/03112054/11 (Accessed 24 September, 2011).
18. Thaut, M.H. 2008. *Rhythm, Music, and the Brain: Scientific Foundations and Clinical Applications*. Routledge. New York.
19. Bradt, J., W.L. Magee, C. Dileo, *et al.* 2010. Music therapy for acquired brain injury. *Cochrane Database Syst. Rev.*: CD006787, doi:10.1002/14651858.CD006787.pub2.
20. Pothoulaki, M., R.A.R. MacDonald & P. Flowers. 2008. An investigation of the effects of music on anxiety and pain perception in patients undergoing haemodialysis treatment. *J. Health Psych.* **13:** 912–920.
21. Bernatzky, G., S. Strickner, M. Presch, *et al.* (in press). Music as non-pharmacological pain management in clinics. In *Music Health and Being*. R.A.R. MacDonald, G. Kreutz & L. Mitchell, Eds. Oxford University Press. Oxford.
22. Spintge, R. (in press). Clinical use of music in operating theatres. In *Music Health and Being*. R.A.R. MacDonald, G. Kreutz & L. Mitchell, Eds. Oxford University Press. Oxford.
23. Mitchell, L.A., R.A.R. MacDonald & C. Knussen. 2008. An investigation of the effects of music and art on pain perception. *Psych. Aesth. Creat. Arts.* **2:** 162–170.
24. Knox, D., S. Beveridge, L.A. Mitchell & R.A. MacDonald. 2011. Acoustic analysis and mood classification of pain-relieving music. *J. Acoust. Soc. Am.* **130:** 1673–1682.
25. Kato, P. 2010. Video games in health care: closing the gap. *Am. Psych. Ass.* **14:** 113–121.
26. Kato, P.M., S.W. Cole, A.S. Bradlyn & B. Pollock. 2008. A video game improves behavioral outcomes in adolescents and young adults with cancer: a randomized trial. *Pediatrics* **122:** 305–317.
27. Saposnik, G., R. Teasell, M. Mamdani, *et al.* 2010. Effectiveness of virtual reality using Wii gaming technology in stroke rehabilitation: a pilot randomized clinical trial and proof of principle. *Stroke* **41:** 1477–1484.
28. Gee, J.P. 2003. *What Video Games have to Teach Us About Learning and Literacy*. Palgrave Macmillan. New York.
29. Annetta, L. 2010. The "I's" have it: A framework for serious educational game design. *Rev. Gen. Psych.* **14:** 105–112.
30. Hoffman, H.G., D.R. Patterson, E. Seibel, *et al.* 2008. Virtual reality pain control during burn wound debridement in the hydrotank. *Clin. J. Pain* **24:** 299–304.
31. Fitzgerald, S. & R. Cooper. 2004. The GAME(Cycle) exercise system: comparison with standard ergometry. *J. Spin. Cord Med.* **27:** 453–459.
32. Graves, L., G. Stratton, N. Ridgers & N. Cable. 2008. Energy expenditure in adolescents playing new generation computer games. *Br. J. Sports Med.* **42:** 592–594.
33. Kato, P.M. & I.L. Beale. 2006. Factors affecting acceptability to young cancer patients of a psychoeducational video game about cancer. *J. Pediatr. Onc. Nurs.* **23:** 269–275.
34. Krichevets, A., E. Sirotkina, I. Yevsevecheva & L. Zeldin. 1994. Computer games as a means of movement rehabilitation. *Disabil. Rehabil.* **17:** 100–105.
35. Lanningham-Foster, L., T. Jensen, R. Foster, *et al.* 2006. Energy expenditure of sedentary screen time compared with active screen time for children. *Pediatrics* **118:** 1831–1835.
36. Lieberman, D.A. 1997. Interactive video games for health promotion: effects on knowledge, self-efficacy, social support, and health. In *Health Promotion and Inter-active Technology: Theoretical Applications and Future Directions*. R.L. Street, W.R. Gold & T. Manning, Eds.: 103–120. Erlbaum. Mahwah, N.J.
37. Rosas, R., M. Nussbaum, P. Cumsille, *et al.* 2003. Beyond Nintendo: design and assessment of educational video games for first and second grade students. *Comput. Educ.* **40:** 71–94.
38. Sharar, S., G. Carrougher, D. Nakamura, *et al.* 2007. Factors influencing the efficacy of virtual reality distraction analgesia during postburn physical therapy: preliminary results from 3 ongoing studies. *Arch. Phys. Med. Rehabil.* **88:** 43–49.
39. Burke, J.W., M.D. McNeill, D.K. Charles, *et al.* 2009. Optimising engagement for stroke rehabilitation using serious games. *Vis. Comput.* **25:** 1085–1099.
40. Merians, A.S., G.G. Fluet, Q. Qiu, *et al.* 2011. Robotically facilitated virtual rehabilitation of arm transport integrated with finger movement in persons with hemiparesis. *J. Neuroeng. Rehabil.* **8:** 27.
41. Laver, K.E., S. George, S. Thomas, *et al.* 2011. Virtual reality for stroke rehabilitation. *Cochrane Database Syst. Rev.*: CD008349, doi:10.1002/14651858.CD008349.pub2.
42. Cassidy, G. & R.A. MacDonald. 2008. The role of music in videogames: the effects of self-selected and experimenter-selected music on driving game performance and experience. In *Proceedings of the 10th International Conference on Music Perception and Cognition*, 25–29 August, Sapporo, Japan.
43. Cassidy, G.G. & R.A.R. MacDonald. 2009. The effects of music choice on task performance: A study of the impact of self-selected and experimenter-selected music on driving game performance and experience (ESCOM Young Researcher of the Year Award).
44. Cassidy, G.G. & R.A. MacDonald. 2010. The effects of music on time perception and performance of a driving game. *Scand. J. Psyc.* **51:** 455–464.
45. Knox, D., G. Cassidy, S. Beveridge & R. MacDonald. 2008. Music emotion classification by audio signal analysis: analysis of self-selected music during game play. In *Proceedings of the 10th International Conference on Music Perception and Cognition*, 25–29 August 2008, Sapporo, Japan.
46. Alankus, G., A. Lazar, M. May & C. Kelleher. 2010. Towards customizable games for stroke rehabilitation. In *CHI 2010: Therapy and Rehabilitation*. Atlanta, GA, USA.
47. Murgia, A., R. Wolff, P.M. Sharkey & B. Clark. 2008. Low-cost optical tracking for immersive collaboration in the CAVE using the Wii Remote. In *Proceedings of the 7th ICDVRAT with ArtAbilitation*, Maia, Portugal.
48. Schou, T. & H.J. Gardner. 2007. A Wii remote, a game engine, five sensor bars and a virtual reality theatre. In *Proceedings of OzCHI* 2007, Adelaide, Australia.
49. Vannoni, M. & S. Straulino. 2007. Low-cost accelerometers for physics experiments. *Eur. J. Phys.* **28:** 781–787.
50. Cannam, C., C. Landone & M. Sandler. 2010. Sonic Visualiser: an open source application for viewing, analysing, and annotating music audio files. In *Proceedings of the ACM Multimedia 2010 International Conference, MM'10*, Firenze, Italy.

Ann. N.Y. Acad. Sci. ISSN 0077-8923

ANNALS OF THE NEW YORK ACADEMY OF SCIENCES
Issue: *The Neurosciences and Music IV: Learning and Memory*

Effective music therapy techniques in the treatment of nonfluent aphasia

Concetta M. Tomaino

Institute for Music and Neurologic Function, Beth Abraham Family of Health Services, Bronx, New York

Address for correspondence: Concetta M. Tomaino, Institute for Music and Neurologic Function, Beth Abraham Family of Health Services, 612 Allerton Avenue, Bronx, NY 10467. ctomaino@bethabe.org

In music therapy for nonfluent aphasia patients who have difficulty producing meaningful words, phrases, and sentences, various benefits of singing have been identified: strengthened breathing and vocal ability, improved articulation and prosody of speech, and increased verbal and nonverbal communicative behaviors. This paper will introduce these various techniques used in clinical music therapy, and summarize findings based on our recent study to illustrate the strength of different techniques emphasizing rhythm, pitch, memory, and vocal/oral motor components dealing with different symptoms. The efficacy of each component is enhanced or diminished by the choice of music and the way it is interactively delivered. This indicates that neural mechanisms underlying speech improvement vary greatly with available acoustic and social cues in aphasic brain.

Keywords: singing; stroke; oral motor rehabilitation; prosody; rhythmic cues

Introduction

It is well documented that many people with nonfluent aphasia, also known as Broca's aphasia, can sing even though they cannot speak, although some patients who have lost musical abilities would exhibit little or no language-related problems.[1] Historically, this dissociation has been seen as evidence of the differentiation between speech and music, i.e., that these are distinct abilities with little overlap in neurological function. However, recently, as the view of brain mechanisms supporting language processing has moved away from focusing on Broca/Wernicke's areas to a network perspective that emphasizes the contribution from several brain regions in various aspects of language,[2,3] different musical features have also been uncovered that may also be processed in widely distributed neural networks.[4,5] Accordingly, neuroimaging research has shown evidence for the shared neural pathways of speech and music.[6–8] In addition, various neurological disorders may benefit differently from singing exercises,[9] including nonfluent aphasia.[10] This corresponds well with clinical observations and trial findings in music therapy that in aphasia treatments, the preserved speech abilities are enhanced through various types of music-based speech therapy exercises.[11–14] Although the best known of the music-based speech therapy techniques is melodic intonation therapy (MIT),[15] which has been studied extensively with neuroimaging,[16–18] other techniques have not been examined in terms of specific effects that link behavioral and neurological recoveries. More importantly, however, it has yet to be systematically examined as to which techniques can be most beneficial in improving what types of speech deficits in aphasia and what is the "dosage" that could lead to substantial and sustainable recovery. The aim of this paper is to give an overview of various techniques used in music therapy and summarize our findings of the outcomes and benefits of these techniques in our previous research in aphasic patients.

In music therapy, therapists consider how to address not only speech deficits but also other sensory, motor, and cognitive issues by using a variety of therapy techniques to individually optimize the outcomes. For example, facial gestures or exaggerated tonal/rhythmic inflection are used to stimulate existing networks involved in verbal communication by engaging nonverbal communication

doi: 10.1111/j.1749-6632.2012.06451.x

pathways, likely related to the role of mirror neurons in communication in general,[19,20] as speculated previously by Condon.[21] Moreover, the patient's ability to participate fully in and benefit from traditional speech therapy can be limited because of problems with poor attention, depression, poor motivation, deficits in motor timing (inability to tap to rhythm), and cognitive slowing. Music therapists are often concerned with prioritizing how these barriers to treatment are addressed within the therapy session to enhance the patient's capacity for active involvement in their recovery of communication skills. It is important to note that the knowledge from these clinical observations are actually in line with the view of multifaceted and distributed networks involved in speech and music processing as described above.

Detailed clinical observations that document the abilities and deficits in aphasic individuals in the context of music therapy can inform research more generally because it can identify which music-based techniques (temporal structure/rhythm, melodic contour, prelearned song lyrics) are beneficial to which clinical conditions. For example, even though people with aphasia may be able to sing only fragments of song lyrics immediately after their stroke, many are able to improve in the number of correct lyrics with consistent repetition. It has been observed that as patients improve in the number of correct lyrics sung, their ability to speak the lyrics without singing also improves. Does this improvement in correct lyric singing generalize to other word-finding abilities? Is this skill correlated to the patients' general recovery? Another observation is the increased ability for naming tasks and word retrieval after singing a familiar song. Does singing "prime" word retrieval mechanisms for patients with nonfluent aphasia? If so, what are the underlying mechanisms that allow for this "release" in initiation? Some patients need multiple cues such as exaggerated facial cues along with singing and finger tapping to improve outcomes, although others need only simple cues such as finger tapping alone. Such observations can lead to better assessment tools to identify which patients will benefit from a specific music-based speech technique or combination thereof.

To better understand the impact of various music-based speech techniques in individuals with nonfluent aphasia, we have undertaken several clinical studies to gain further insight into the efficacy of these techniques. In the following sections we will first review these techniques used in clinical music therapy and our finding in the outcome performance in seven patients in our recent study. Thereafter, we will introduce our subsequent study which enrolled a larger cohort of patients in a comparison between music-based speech therapy and nonverbal visually-based communication therapy to enhance communication skills.

Study I: music-based speech therapy techniques for nonfluent aphasia

In this study we investigated the effectiveness of various music-based speech techniques by using an in-depth analysis of video of music therapy sessions.[22] The analysis particularly evaluated the effectiveness of the music-based technique on articulation, fluency, prosody, and breath support. Seven patients participated in the study (2 males and 5 females) with nonfluent aphasia (9 months to over 20 years poststroke). Each participated in 8–12 individual, 30-min music therapy sessions three times a week for four weeks.

The therapist sat facing the patient to provide proper facial cues when needed. A descriptive analysis of 66 video sessions was completed independently by two research associates, a music therapist, and a neuropsychology fellow, and all findings were cross-examined and finally synthesized into a protocol containing the seven most effective techniques along with guidelines for use. Following are highlights of these techniques. For more detailed descriptions of the techniques, as well as analysis of tested outcomes, see the original article.[22]

Singing familiar songs

The patient is cued to sing a familiar song, with the music therapist initiating the melody and singing along with the lyrics. The lyrics of the song that are easiest for the patient to produce are repeated. The tempo is adjusted to maximize success. The dynamics of tempo, intensity, and modulation are used to draw the patient into an interaction with the therapist. It was frequently observed that the rhythmic flow of the patients' singing was relatively intact despite severely impaired speech rhythm. There was also a positive correlation between the temporal entrainment and the fluency rate in the patients' singing.

Breathing into single-syllable sounds

The patient is asked to follow breathing into single-syllable sounds. The vocal sounds move from vowels to consonants or bilabial (/m/, /b/), to tip of the tongue (/d/, /t/), to tongue-based sounds ("g" and /k/). Focus on breathing patterns helped the patients relax, with the therapist often suggesting naturally occurring vocal responses, such as sighing, yawning, or clearing voices. This was especially obvious when the therapist mirrored and matched the patients' naturally occurring breathing patterns rather than imposing a breathing pattern. By drawing on the preserved skills of the patient, the therapist can extend this skill to a closely related task. For example, single-syllable sounds were elicited most successfully when there were gradual steps moving from simple breathing to natural vocal responses, to sighing vowels, and finally to single syllables. The patients' vocal quality was enhanced when each sound was superimposed on a slow and long exhale. In addition, visual cueing, such as hand movement, helped the patients initiate and sustain syllable sounds in synchrony with exhaling.

Musically assisted speech

Commonly used conversational phrases are associated with musical melodies that are familiar to the patient (e.g., "Hello, how are you today?" with the tune "Swing Low, Sweet Chariot"). The patients show signs of better recognition of the melody when an original lyric phrase of a tune is presented first before it was associated with a daily speech phrase. Once there is more familiarity with the tune, increased motivation and successful outcomes can be observed.

Dynamically cued singing

The intentional use of musical dynamics is introduced in familiar song singing. This exercise creates a strong sense of anticipation for a certain part of the song to facilitate interpersonal interaction and emotional expression. The dynamic cues include dynamic variations of the song flow and pausing at the last words of each phrase of the song. This "gap" allows for the patient to complete the phrase with the correct word, although the music dynamics provide additional cues to alert the patient that they will be completing the lyric. This type of back and forth engagement between the therapist and patient improves the patient's motivation as well as provides an effective "conversation-like" context that could stimulate similar skills outside of the music.

Rhythmic speech cueing

The patient is guided to clap or tap the speech rhythm to the phrase that is being practiced. The therapist may use song lyric phrases, daily conversational phrases, or phrases that are related to the immediate context. Although this is similar to the finger tapping in MIT, the rhythmic self-cueing is done with either the right or left hand—depending on which is more functional (usually the left hand/finger as there may be a right-hand paresis in someone with Broca's aphasia). The rhythmic cues include slow steady beats that are gauged to the patient's speech tempo, prosodic rhythms of the speech phrases, and musical phrase rhythms for the song lyrics.

Oral motor exercises

The therapist presents a short portion of a familiar song in an exaggerated manner of mouthing and tongue movements. The patient is asked to closely watch and follow the therapist's facial and oral movements. For the patient to successfully mimic these expressions, it is crucial for the therapist to allow time for that response. It has also been found that the patients perform better if the therapist coordinated the exaggerated expression to the temporal order/rhythm of the familiar song lyric.

Vocal intonation

The use of intoned phrases helped the patients improve the melodic aspect of speech prosody. Through the repeated exercise of intoned phrases over time, the patients can regain some of the ability to differentiate inflection, pitch, and intensity in their speech that approximated those of normal conversational prosody. Intentional variations of the intonation of daily speech phrases are introduced to help the patient better convey different meanings in speech. The intonations are exaggerated to provide additional contextual cues. Visual cues, such as head movement or hand conducting, may be added to assist the patient's perception of intonation contours and to enhance contextual meanings. Visual cueing, such as up-and-down hand movement, reinforced the auditory cues of the vocal intonation. This was shown to enhance a more natural pattern of word production.

Table 1. Summary of each patient's performance during each therapy technique

	Techniques						
Subject	Singing familiar songs	Breathing and single-syllable sounds	Musically assisted speech	Dynamic-cued singing	Rhythmic-cued speech	Oral motor exercises	Vocal intonation
1	Able to sing with and without cueing	Mimicry leads to better success	Prosody and intonation are intact	Follows dynamics and fills in correct word on cue	More success if slow and includes motor tapping	Follows with 80% success	Follows with motor cues
2	Rhythm and intonation intact; misses many words	Mimicry leads to better success over time	Prosody and intonation are slow and deliberate	Accuracy of fill-ins over time	More success if slow and includes motor tapping	Mimicry leads to better success	Follows with motor cues intonation/ expression
3	Sings when prompted; dynamics developed over time	Mimicry needed for success	Initially monotone, but more on target words over time	Monotone	More success if slow and includes motor tapping	Follows with correct rhythms	Follows with correct rhythm
4	Melody and intonation intact; substitutes words	Mimicry leads to better success over time	Prosody and intonation present if slow and deliberate	Relies on cues	Better with rhythmic cues but misses words	–	Intonation and rhythm intact but words are not exact
5	Mouths some words but is in audible, although rhythm is intact	Hums with no articulation	Rhythm and intonation intact	Relies on cues with minimum success; no words	Does not follow cues	Mimics with exaggeration	Mimics with exaggeration
6	Mouths some words although rhythm intact	Mimics better success with rhythmic cue	Relies on cues although intonation and rhythm intact	with slow tempo; incorrect words	Prosody and rhythm intact	–	Intonation intact
7	Eye contact needed for response	Mimics with minimal motion	Relies on cues but intonation and rhythm intact	Successful with slow tempo and eye contact; words are incorrect but the vocalization still contains same sounds and syllables	Minimal response	–	Intonation intact

Results

Table 1 summarizes the clinical observations of each patient's performance during each therapy technique over time. Patients who were successful in motor cueing were observed to be also successful in singing the correct rhythm of the song.

Study II: music-based speech therapy techniques for nonfluent aphasia

Recently we conducted another study that investigated the impact of using familiar songs and rhythmic speech-motor entrainment for patients with nonfluent aphasia who have been discharged from speech therapy. Specifically, we chose outcome measures in these patients based on standardized measures of expressive and receptive speech (C.M. Tomaino & M. Kim, unpublished results). This study randomly assigned 40 subjects to either a one-on-one music therapy intervention or a one-on-one picture-based conversation intervention. All participants had completed at least one course of speech therapy, were no longer receiving speech therapy, and all demonstrated good comprehension skills at the time of enrollment in the study. Each participated in three 30-min sessions per week for 12 weeks. Of the initial subjects, 18 participants in the music therapy session completed the study, but only eight in the picture-based intervention completed all sessions. The difference in the rate of attrition between the two intervention groups suggests that music therapy might be easier or more encouraging for the patients at this level of speech recovery. The protocol for the music therapy session included singing familiar songs and speech-motor entrainment through exercising finger tapping to the rhythms of the familiar songs, although the comparison session included picture-based conversational speech exercises using visual cues and the participants' verbal responses to these cues. Two evaluation tools were used: the Western Aphasia Battery and the Test of Adult and Adolescent Word Finding. Subcategories of these tests included the following tasks: following command, repetition, sentence completion, and naming nouns. Each participant was given a total of three evaluations: one before the start of intervention, another at six weeks into the intervention, and a postevaluation in close proximity to the last session.

Paired sample *t*-tests revealed an increase in the expressive language scores (repetition, sentence completion, naming nouns) for the music therapy participants, which proved to be significant (preintervention score of 60.6 and postintervention score of 67.2). Although the scores for the picture-based intervention also improved (preintervention score of 46.8 and postintervention score of 53.6), the small sample size of this comparison group diminished the potential for any statistical comparison between groups. Nevertheless, participants' adherence to music therapy might be supported by meaningful communication that likely emerged throughout the intervention.

Conclusions

These and other studies indicate that music therapy and music-based speech protocols provide useful tools for the rehabilitation of patients with nonfluent aphasia. The temporal and rhythmic components are particularly important because of the close proximity to normal speech rhythms. For both speech and singing, the careful manipulation of rhythm to match and enhance the patients' expectancy is an important factor in achieving improved word retrieval, prosody, and articulation. The use of prelearned lyrics also aid in stimulating word retrieval. Predictable lyrics aid in word retrieval for speech. The interplay between the music therapist and patient plays a role in enhancing clinical outcomes as well because it can approximate a conversational exchange by allowing for pauses, or exaggerating dynamics, or melodic contour to convey meaning and facilitate expectation. From these therapeutic techniques and clinical studies, we can derive future research questions as to what aspects of music are interrelated to distinct language functions in the brain. In particular, further studies should address the neurological underpinnings of the possible priming effect of singing familiar song lyrics on word retrieval, and the effect of rhythmic speech-motor entrainment on disinhibition of speech production in patients with nonfluent aphasia. At the same time, examining these questions from a cognitive neuroscience perspective will certainly inform clinical music therapy. As more research from both music therapy and cognitive neuroscience advances our understanding of music and the brain, new targeted protocols can be developed to improve clinical outcomes for those with nonfluent aphasia and other specific language-related disorders.

Conflicts of interest

The author declares no conflicts of interest.

References

1. Marin, O.S.M. & D.W. Perry. 1999. Neurological aspects of music perception and performance. In *The Psychology of Music*. D. Deutch, Ed.: 653–724. Academic Press. New York.
2. Mesulam, M.M. 1990. Large-scale neurocognitive networks and distributed processing for attention, language, and memory. *Ann. Neurol.* **28:** 597–613.
3. Damasio, A.R. 1992. Aphasia. *N. Engl. J. Med.* **326:** 531–539.
4. Peretz, I. & M. Coltheart. 2003. Modularity of music processing. *Nat. Neurosci.* **6:** 688–691.
5. Peretz, I. & R.J. Zatorre. 2005. Brain organization for music processing. *Annu. Rev. Psychol.* **56:** 89–114.
6. Ozdemir, E., A. Norton & G. Schlaug. 2006. Shared and distinct neural correlates of singing and speaking. *NeuroImage* **33:** 628–635.
7. Maess, B. *et al.* 2001. Musical syntax is processed in Broca's area: an MEG study. *Nat. Neurosci.* **4:** 540–545.
8. Patel, A.D. 2003. Language, music, syntax and the brain. *Nat. Neurosci.* **6:** 674–681.
9. Wan, C.Y. *et al.* 2010. The therapeutic effects of singing in neurological disorders. *Music Percept.* **27:** 287–295.
10. Racette, A., C. Bard & I. Peretz. 2006. Making non-fluent aphasics speak: sing along! *Brain* **129:** 2571–2584.
11. Michel, D.E. & N.H. May. 1974. The development of music therapy procedures with speech and language disorders. *J. Music Ther.* **11:** 74–80.
12. Rogers, A. & P.L. Fleming. 1981. Rhythm and melody in speech therapy for the neurologically impaired. *Music Ther.* **1:** 33–38.

13. Cohen, N.S. & R.E. Masse. 1993. The application of singing and rhythmic instruction as a therapeutic intervention for persons with neurogenic communication disorders. *J. Music Ther.* **30:** 81–99.
14. Magee, W. 1999. Music therapy within brain injury rehabilitation: to what extent is our clinical practice influenced by the search of outcome? *Music Ther. Persp.* **17:** 20–26.
15. Albert, M.L., R.W. Sparks & N.A. Helm. 1973. Melodic intonation therapy for aphasia. *Arch. Neurol.* **29:** 130–131.
16. Norton, A. *et al.* 2009. Melodic intonation therapy: shared insights on how it is done and why it might help. *Ann. N.Y. Acad. Sci.* **1169:** 431–436.
17. Belin, P. *et al.* 1996. Recovery from nonfluent aphasia after melodic intonation therapy: a PET study. *Neurology* **47:** 1504–1511.
18. Schlaug, G., S. Marchina & A. Norton. 2009. Evidence for plasticity in white-matter tracts of patients with chronic Broca's aphasia undergoing intense intonation-based speech therapy. *Ann. N.Y. Acad. Sci.* **1169:** 385–394.
19. Fadiga, L. *et al.* 1995. Motor facilitation during action observation: a magnetic stimulation study. *J. Neurophysiol.* **73:** 2608–2611.
20. Fadiga, L., L. Craighero & A. Roy. 2006. Broca's region: a speech area? In *Broca's Region.* Y. Grodzinsky & K. Amunts, Eds.: 137–152. Oxford University Press. New York.
21. Condon, W.S. 1982. Cultural microrhythms. In *Interaction Rhythms: Periodicity in Communicative Behavior*. M. Davis, Ed.: 53–76. Human Sciences Press. New York.
22. Kim, M. & C.M. Tomaino. 2010. Protocol evaluation for effective music therapy for persons with nonfluent aphasia. *Top. Stroke Rehabil.* **15:** 555.

Ann. N.Y. Acad. Sci. ISSN 0077-8923

ANNALS OF THE NEW YORK ACADEMY OF SCIENCES
Issue: *The Neurosciences and Music IV: Learning and Memory*

Music: a unique window into the world of autism

Istvan Molnar-Szakacs[1] and Pamela Heaton[2]

[1]Department of Psychiatry and Biobehavioral Sciences, Semel Institute for Neuroscience and Human Behavior, University of California, Los Angeles, California. [2]Goldsmiths College, Department of Psychology, University of London, London, United Kingdom

Address for correspondence: Istvan Molnar-Szakacs, UCLA—Psychiatry and Behavioral Sciences, 760 Westwood Blvd, Los Angeles, CA 90095. imolnar@ucla.edu

Understanding emotions is fundamental to our ability to navigate the complex world of human social interaction. Individuals with autism spectrum disorders (ASD) experience difficulties with the communication and understanding of emotions within the social domain. Their ability to interpret other people's nonverbal, facial, and bodily expressions of emotion is strongly curtailed. However, there is evidence to suggest that many individuals with ASD show a strong and early preference for music and are able to understand simple and complex musical emotions in childhood and adulthood. The dissociation between emotion recognition abilities in musical and social domains in individuals with ASD provides us with the opportunity to consider the nature of emotion processing difficulties characterizing this disorder. There has recently been a surge of interest in musical abilities in individuals with ASD, and this has motivated new behavioral and neuroimaging studies. Here, we review this new work. We conclude by providing some questions for future directions.

Keywords: ASD; alexithymia; insula; mirror neurons; emotion

The ability to enjoy music is a universal human trait and we engage with it spontaneously and effortlessly. Music is unique in the extent that it triggers memories, awakens emotions, and intensifies our social experiences. Through music we are exposed to the thoughts, emotions, and ideas of others, and the experience of music listening enables us to learn how to combine elements of sound in a coherent communicative stream that may be as important as learning how to "talk scientifically" in science lessons.[1] Music allows us to develop self-knowledge, self-identity, and group identity, enabling us to share thoughts, emotions, and feelings, and understand those subtle and unique human experiences that cannot easily be put into words.[2]

Although music psychologists have long emphasized the intensely social nature of musical activities, considerably less thought has been given to the intrapersonal consequences of musical engagement. Unfortunately, social isolation is not uncommon, and for individuals with psychiatric or psychological disorders, this may be a constant experience. In these cases, music may become increasingly important as a means of alleviating loneliness and suffering, as well as improving communicative and cognitive abilities.[3]

Autism spectrum disorder (ASD) is an umbrella term used to describe a continuum of diagnoses that include autism, Asperger's disorder, and pervasive developmental disorder-not otherwise specified (PDD-NOS). Collectively, ASD is characterized by deficits in communication, impairments in social interactions, and restricted and repetitive patterns of behavior.[4] Theoretical accounts of the social and communicative difficulties in ASD have tended to differ in the degree of importance allocated to cognitive or affective factors. The theory of mind (ToM) account of autism, first described by Baron-Cohen *et al.*,[5] proposes that ASD is characterized by a fundamental difficulty with representing the mental states of others. In contrast, Hobson has stressed the importance of intersubjectivity, and the developmental sequelae of an early disturbance in the ability to form emotional bonds with others.[6] To help explain the

doi: 10.1111/j.1749-6632.2012.06465.x

wide range of impairments in ASD, Dawson *et al.* have proposed the social motivation hypothesis, which holds that the social impairments in ASD are only secondary to a primary deficit in social motivation.[7,8] More recently Ramachandran and Oberman have proposed a "broken mirrors" hypothesis of autism.[9] According to this account, the ability to understand the intentions and actions of others depends upon an intact mirror neuron system (MNS), which is compromised in ASD.[10,11]

Improvements in the early identification of autism have enabled researchers to study social and communicative behaviors in infants and young children at risk for autism and to look for early precursors of later emerging social and communication abnormalities. These studies have revealed a number of interesting findings. Numerous studies have reported imitation deficits in autism.[10,12,13] Rogers and Pennington suggested that motor imitation may be one of the primary deficits, and they highlighted two subcomponents of imitation that might underlie the imitative deficit in autism—self–other correspondence and the sequencing of intentional movements.[12] In support of this theory, Vanvuchelen *et al.* found that individuals with autism showed impaired performance in both gestural imitation and general motor skills, suggesting that their imitation deficits may be part of a broader perceptual–motor problem.[14] Furthermore, Perra *et al.* found that children with autism performed worse than typically developing children and children with general developmental delay on imitation and ToM tasks.[15] Deficits in imitation can have extensive ripple effects, as imitation scaffolds the development of language, cultural transmission, and social communication.

Studies have also shown that the characteristic interest in faces shown by typically developing infants is significantly reduced in infants and children with autism.[16–19] Social referencing, a tendency to look to significant others in ambiguous situations, is also atypical in children diagnosed with autism. Whereas typically developing infants use social referencing to make sense of confusing or threatening situations,[20] children with autism are less likely to reference another person in these same instances.[21] Young children with autism are also unlikely to voluntarily share their experience with others, and often avoid initiating interactions with multiple social partners, even when guided to do so.[22] Studies that have specifically focused on the role of emotion processing deficits in social understanding and reciprocity have shown that individuals with ASD display less positive emotion to their social partners[23,24] when compared to typically developing peers and experience more negative affective exchanges with others.[25] Reduced attention to other people's emotional cues may result in negative peer interactions and increased difficulties in resolving conflict with others. Although some experimental studies directly testing emotion recognition have failed to observe deficits,[26–29] the findings from the majority of studies are consistent with clinical reports of poor emotion understanding.

An important recent avenue of research into emotion processing deficits has investigated the co-occurrence of ASD and alexithymia. Alexithymia is a disorder characterized by reduced or absent affective responses (type I alexithymia) or difficulties in understanding and ascribing affective labels to one's own physiological states of arousal, even when affective arousal is present (type II alexithymia).[30] Whereas compromised emotional awareness, or type II alexithymia, is estimated to affect around 10% of the general population,[30] a much higher prevalence rate of 85% has been observed in a sample of intellectually high-functioning adults with ASD.[31–34]

Using functional neuroimaging, Silani *et al.* showed that high levels of alexithymia, measured using the Toronto Alexithymia Scale,[35] were associated with hypoactivation in the anterior insula in individuals with high-functioning autism.[36] Furthermore, there was a significant correlation between activity in the insular cortex not only with alexithymia scores, but also with scores on empathic concern and perspective-taking scales. In a more recent study, the same authors measured empathic brain responses in participants with ASD and neurotypical controls while they witnessed another person experiencing pain.[37] The results were consistent with those of the original study, showing that the levels of alexithymia, but not a diagnosis of autism, were associated with the degree of empathic brain activation in anterior insula. These results are important in showing that the empathy deficit widely attributed to ASD can be explained by the extent of alexithymic traits and does not constitute a universal social impairment in autism.

One interesting question that arises from this recent work is how alexithymia would affect musical appreciation. Behavioral studies that have compared children with autism and age- and intelligence-matched typical controls on their ability to match musical extracts with visual representations of different emotions have failed to reveal group differences.[38,39] This work suggests that difficulties in recognizing emotions in social stimuli (voices and faces) do not generalize to the domain of music. In a study that specifically investigated the impact of alexithymic traits on music perception, Allen *et al.* asked groups of intellectually able adults, with and without ASD, to select words that described their personal responses to music. Although the high-functioning adults with autism did select fewer emotion words than controls, the data showed that the severity of the participants' alexithymic traits, rather than their diagnostic status, explained this result. In the study, galvanic skin response measures were taken while the participants listened to musical excerpts. The analyses of these data failed to observe a between-group difference in autonomic responses to music, showing that a physiological-level response is also preserved in the population with ASD.[40]

This work, providing evidence for unimpaired appreciation of music's affective qualities in ASD, raises interesting questions about how such understanding is acquired and what the consequences of it may be. Many infants and children with ASD fail to attend to faces and voices (e.g., Ref. 19), and this reduces their opportunities to learn about other people's emotions. Studies have shown that mothers mimic their infant's facial expressions of emotion, and infants who are underresponsive to such maternal feedback may fail to associate their own bodily states of arousal with the emotion categories mirrored by their mothers.[41,42] Atypical attention at early stages of development may therefore impoverish both inter- and intrapersonal emotional understanding. But what about music? In Kanner's first case account of autism he described an unusual preoccupation with music that co-occurred alongside the social and communicative disabilities that are still core criteria for autism.[43]

In a recent observational pilot study, we tested patterns of attention in response to musical and other auditory stimuli in 20 autistic and typically developing children engaged in play activity at school. The stimuli were either short sentences (e.g., "those look like nice toys"), environmental noises (door opening and closing), or short excerpts of classical music. The children's responses to the sounds were recorded using a video camera and coded for positive affect, vocalization, orientation–coordination, and anticipation–increased focus. The results showed that the children with autism were more responsive to all stimuli than the typical children and this may reflect sensory and/or attentional difficulties. However, there was also a significant between-group difference in the pattern of performance across the three different conditions. Whereas the typical children showed a similar pattern of response to music, speech, and environmental noise, the children with ASD showed a statistically significant increase in their responses to music compared with speech and environmental noise. These results strongly suggest that music elicits special attention for children with autism and may help explain the feats of musical memory initially described by Kanner.[43]

These findings also increase our understanding of how children with ASD come to understand musical emotions when emotions expressed on faces and in voices are so difficult to interpret. It appears that music is spared the characteristic early "neglect" accorded to social stimuli. Given the potential therapeutic implications of this, questions about the nature of spared musical emotion recognition are important. To address the question of whether children with ASD, who recognize emotions in music, will recognize those same emotions in verbal information, we carried out an experimental study of domain-general auditory emotion recognition. In the study, 15 intellectually able children with autism and age- and intelligence-matched typical controls were asked to identify happy, sad, and fearful emotions in short musical excerpts, nonverbal vocalizations (affective vocal gestures), and vocalized three-digit numbers (e.g., 523) spoken with affective intonation. Consistent with other recent studies (e.g., Refs. 28 and 29), the results failed to reveal emotion recognition deficits in the ASD group. However, an interesting finding was revealed by the correlations carried out on the data from the different experimental conditions. For typical children, good categorization of nonverbal emotional vocalizations was strongly associated with good categorization of music. It appeared that for them, affective cues in music and vocal gesture were similarly accessible but that this was not the case for verbal information. For the children with ASD, all correlations were significant, so the ability to extract expressive cues from music and vocal gesture also generalized to speech. This finding could have implications for therapies aimed at remediating impoverished perception of affective prosody in ASD.

Although most influential theoretical accounts of ASD focus on explaining specific aspects of the disorder (e.g., ToM[5]), others have been more ambitious in scope. For example, the broken mirrors hypothesis has attempted to account for wide ranging difficulties in joint attention, social orienting, imitation, emotional responsiveness, and face processing under the umbrella of a compromised MNS.[9] The MNS functions as a sort of "neural Wi-Fi," which allows an observer to understand the actions of another by activating the same regions of their brain as if they were performing the actions themselves through a process called simulation (for reviews, see Refs. 44 and 45). As a result, the MNS has been conceptualized as a neural hub that plays a seminal role in action perception, imitation, language, intention understanding, perspective taking, and self–other discrimination.

Based on the simulation mechanism implemented by the human MNS, Molnar-Szakacs and Overy have developed the Shared Affective Motion Experience (SAME) model of emotional music perception.[3,46,47] The SAME model suggests that musical sound is perceived not only in terms of the auditory signal, but also in terms of the intentional, hierarchically organized sequences of expressive motor acts behind the signal. Expression *and* perception of these "musical motor acts" (clapping, singing, and plucking strings) lead to recruitment of the same neural systems in the musician making music and the listener perceiving it—allowing for a shared affective motion experience.

At the neural level, SAME relies on a network whose key nodes are the human MNS, the anterior insula, and the limbic system. The auditory features of the musical signal are processed primarily in the superior temporal gyrus (STG) and are combined with structural features of the expressive motion information within the MNS. The anterior insula forms a "communicative channel" between the MNS and the limbic system, allowing incoming information to be evaluated in relation to the perceiver's autonomic and emotional state. This then combines with top–down cognitive information and can lead to a complex affective or emotional response to the music.[3,46,47]

A recent study was designed to test the neural model proposed by SAME.[48] High-functioning children with ASD and age-matched controls listened to happy, sad, and peaceful music[49] and made button-press responses to indicate whether they thought the musical excerpt sounded happy, sad, or peaceful while their brain activity was recorded using functional magnetic resonance imaging (fMRI). In both groups of children, preliminary results revealed a strong activation in the STG bilaterally and within a network, including right posterior inferior frontal gyrus (IFG) and premotor cortex—areas composing part of the human MNS—while they listened to emotional music compared to rest. The thalamus—known to play an important role in preprocessing auditory signals and projecting the information into the auditory cortex—was also recruited. Activations were also seen in the amygdala, medial orbitofrontal cortex, and anterior cingulate cortex—structures implicated in the processing and regulation of emotion.[50,51]

Behavioral data collected during the study showed that children with ASD identified the emotional musical excerpts as well as the neurotypical control participants. Consistent with results from other studies (e.g., Ref. 49), both groups were significantly faster at identifying happy music than sad or peaceful music.

Although these results are preliminary, they correspond well to similar studies in neurotypical individuals, which have found that affective responses to music recruit a network of paralimbic regions (cingulate cortex) and temporal lobe regions (superior temporal sulcus/superior temporal gyrus),[52–54] and also the posterior inferior frontal gyrus and premotor cortices.[53,54] These data provide support for the SAME model of affective musical experience, whereby recruitment of the MNS allows children with ASD to experience and understand emotional music.[48] Furthermore, these results challenge the idea that mirror regions in the autistic brain are "broken," as has previously been proposed.[9,10]

In another recent fMRI study, Caria *et al.* investigated emotion processing in individuals with ASD during the processing of happy and sad music excerpts.[55] Overall, fMRI results indicated that while listening to both happy and sad music, individuals with ASD activated cortical and subcortical brain regions known to be involved in emotion processing and reward. Interestingly, a comparison of participants with ASD and neurotypical individuals demonstrated decreased brain activity in the premotor area and in the left anterior insula, especially in response to happy music excerpts. The authors conclude that individuals with ASD are able to perceive simple emotions in the musical domain, and impairments of emotion processing in ASD appear to be stimulus specific (i.e., faces[56–58]).

In a recent comprehensive meta-analysis of functional neuroimaging studies of social processing in ASD, Di Martino *et al.* demonstrated that across a group of 24 studies examining various aspects of social processing ranging from face processing to ToM, one of the regions consistently showing significant hypoactivity in ASD was the right anterior insula.[59] The right anterior insula has been associated with subjective perception of emotional states[60,61] and awareness of emotionally salient stimuli.[62,63] It has been posited that the anterior insula, along with the anterior cingulate cortex, serve as a key substrate for the conscious experience of emotion and for the central representation of autonomic arousal, as they integrate visceral, attentional, and emotional information.[64,65] Evidence from brain network analyses suggests that the anterior insula can be considered as part of a "salience network," which serves to integrate sensory data with visceral, autonomic, and hedonic information,[66] making it a region that is critically involved in social–emotional processing.

In the study by Caria *et al.*, hypoactivation of the left anterior insula in response to music in the ASD group compared with the neurotypical group provides further confirmation of the importance of this region as the site of differences in sensitivity to emotion-inducing stimuli in autism.[55] These results are echoed in the neuroimaging studies on alexithymia.[36,37] Both of these studies showed that high levels of alexithymia were associated with hypoactivation in the anterior insula in individuals with high-functioning autism. Furthermore, several studies investigating the perception of faces[59,67] and eye gaze[68] were also associated with hypoactivity of the anterior insula in ASD. These data show that the role of this neural region in emotion perception and understanding is not specific to music, but is rather domain general. Considerable evidence exists about the crucial role of the anterior insular cortex in the representation of internal bodily states of arousal as well as emotional awareness or second-order (interoceptive) awareness.

Taken together, these data suggest that the anterior insula plays a seminal role in the emotion–perception related deficits seen in autism. As discussed earlier, the SAME model assigns a central role to the anterior insula in emotional music perception as a communication hub and center of integration of information coming from the MNS and the limbic system. Future neuroimaging work should address the functional and structural connectivity of the anterior insula. In a complex disorder such as ASD, it is likely that disruptions in interactions within and between large-scale brain networks, rather than focal deficits, underlie the symptoms.[66] Critically, the anterior insula serves a fundamental function with respect to representing and evaluating salient stimuli, and is uniquely positioned as a hub, mediating interactions between large-scale brain networks involved in attentional and self-directed processes.

Based on this review of recent behavioral and neuroimaging work in autism, we conclude that impaired emotion recognition, characteristically observed within socioemotional and interpersonal domains, does not generalize to music. Results showing that musical understanding and appreciation is intact in individuals with ASD may help to explain the efficacy of music therapies carried out with this group.[3] Music, as a form of nonverbal communication, constitutes a domain of preserved skills and interest and is a powerful and accessible affective stimulus that captures and emotionally rewards individuals with ASD.

Acknowledgments

The authors would like to acknowledge the Fondazione Mariani and the University of Edinburgh for their support of The Neurosciences and Music conference, which made this collaboration possible. P.H. would like to acknowledge Laura Bott and Kathy Filer for data collection and analysis. I.M.-Sz. would like to acknowledge the GRAMMY Foundation for their support of the work described here.

Conflicts of interest

The authors declare no conflicts of interest.

References

1. MacDonald, R.A.R. & D. Miell. 2000. Creativity and music education: the impact of social variables. *Int. J. Music Educ.* **36:** 58–68.
2. Hodges, D.A. 2005. Why study music? *Int. J. Music Educ.* **23:** 111–115.
3. Molnar-Szakacs, I., V. Green Assuied & K. Overy. 2012. Shared affective motion experience (SAME) and creative, interactive music therapy. In *Musical Imaginations: Multidisciplinary Perspectives on Creativity, Performance and Perception.* D.J. Hargreaves, D.E. Miell & R.A.R. MacDonald, Eds.: 313–331 Oxford University Press. Oxford, UK.
4. APA. 2000. *Diagnostic and Statistical Manual of Mental Disorders*, 4th ed., Text Revision (DSM-IV-TR). American Psychiatric Association. Washington, DC.
5. Baron-Cohen, S., A.M. Leslie & U. Frith. 1985. Does the autistic child have a "theory of mind"? *Cognition* **21:** 37–46.
6. Hobson, R.P. 1986. The autistic child's appraisal of expressions of emotion. *J. Child Psychol. Psychiatry* **27:** 321–342.
7. Dawson, G., S.J. Webb & J. McPartland. 2005. Understanding the nature of face processing impairment in autism: insights from behavioral and electrophysiological studies. *Dev. Neuropsychol.* **27:** 403–424.
8. Dawson, G. *et al.* 2002. Neurocognitive function and joint attention ability in young children with autism spectrum disorder versus developmental delay. *Child Dev.* **73:** 345–358.
9. Ramachandran, V.S. & L.M. Oberman. 2006. Broken mirrors: a theory of autism. *Sci. Am.* **295:** 62–69.
10. Dapretto, M. *et al.* 2006. Understanding emotions in others: mirror neuron dysfunction in children with autism spectrum disorders. *Nat. Neurosci.* **9:** 28–30.
11. Iacoboni, M. & M. Dapretto. 2006. The mirror neuron system and the consequences of its dysfunction. *Nat. Rev. Neurosci.* **7:** 942–951.
12. Rogers, S.J. & B.F. Pennington. 1991. A theoretical approach to the deficits in infantile autism. *Dev. Psychopathol.* **3:** 137–162.
13. Williams, J.H. *et al.* 2001. Imitation, mirror neurons and autism. *Neurosci. Biobehav. Rev.* **25:** 287–295.
14. Vanvuchelen, M., H. Roeyers & W. De Weerdt. 2007. Nature of motor imitation problems in school-aged boys with autism: a motor or a cognitive problem? *Autism* **11:** 225–240.
15. Perra, O. *et al.* 2008. Imitation and 'theory of mind' competencies in discrimination of autism from other neurodevelopmental disorders. *Res. Autism Spectr. Disord.* **2:** 456–468.
16. Baird, G. *et al.* 2000. A screening instrument for autism at 18 months of age: a 6-year follow-up study. *J. Am. Acad. Child Adolesc. Psychiatry* **39:** 694–702.
17. Baron-Cohen, S. *et al.* 1996. Psychological markers in the detection of autism in infancy in a large population. *Br. J. Psychiatry* **168:** 158–163.
18. Cohen, D.J. & F.R. Volkmar. 1997. *Handbook of Autism and Pervasive Developmental Disorders.* John Wiley & Sons. New York.
19. Osterling, J. & G. Dawson. 1994. Early recognition of children with autism: a study of first birthday home videotapes. *J. Autism Dev. Disord.* **24:** 247–257.
20. Hornik, R., N. Risenhoover & M. Gunnar. 1987. The effects of maternal positive, neutral, and negative affective communications on infant responses to new toys. *Child Dev.* **58:** 937–944.
21. Kasari, C. & M. Sigman. 1996. Expression and understanding of emotion in atypical development: autism and Down syndrome. In *Emotional Development in Atypical Children.* M. Lewis & M.W. Sullivan, Eds.: 109–130. Erlbaum Associates. Mahwah, NJ.
22. Baron-Cohen, S. 1989. Perceptual role taking and protodeclarative pointing in autism. *Br. J. Dev. Psychol.* **7:** 113–127.
23. Dawson, G. *et al.* 1990. Affective exchanges between young autistic children and their mothers. *J. Abnorm. Child Psychol.* **18:** 335–345.
24. Snow, M.E., M.E. Hertzig & T. Shapiro. 1987. Expression of emotion in young autistic children. *J. Am. Acad. Child Adolesc. Psychiatry* **26:** 836–838.
25. Yirmiya, N. *et al.* 1989. Facial expressions of affect in autistic, mentally retarded and normal children. *J. Child Psychol. Psychiatry* **30:** 725–735.
26. Castelli, F. 2005. Understanding emotions from standardized facial expressions in autism and normal development. *Autism* **9:** 428–449.
27. Grossman, J.B. *et al.* 2000. Verbal bias in recognition of facial emotions in children with Asperger syndrome. *J. Child Psychol. Psychiatry* **41:** 369–379.
28. Williams, D. & F. Happe. 2010. Recognising 'social' and 'non-social' emotions in self and others: a study of autism. *Autism* **14:** 285–304.
29. Jones, C.R. *et al.* 2011. A multimodal approach to emotion recognition ability in autism spectrum disorders. *J. Child Psychol. Psychiatry* **52:** 275–285.
30. Bermond, B. 1997. Brain and alexithymia. In *The (Non)Expression of Emotions in Health and Disease.* A. Vingerhoets, F. Bussel & J. Boelhouwer, Eds.: 115–130. Tillburg University Press. Tillburg, Netherlands.
31. Hill, E., S. Berthoz & U. Frith. 2004. Brief report: cognitive processing of own emotions in individuals with autistic spectrum disorder and in their relatives. *J. Autism Dev. Disord.* **34:** 229–235.
32. Berthoz, S. & E.L. Hill. 2005. The validity of using self-reports to assess emotion regulation abilities in adults with autism spectrum disorder. *Eur. Psychiatry* **20:** 291–298.
33. Ben Shalom, D. *et al.* 2006. Normal physiological emotions but differences in expression of conscious feelings in children with high-functioning autism. *J. Autism Dev. Disord.* **36:** 395–400.
34. Rieffe, C., M. Meerum Terwogt & K. Kotronopoulou. 2006. Awareness of single and multiple emotions in high-functioning children with autism. *J. Autism Dev. Disord.* **37:** 455–465.
35. Bagby, R.M., J.D. Parker & G.J. Taylor. 1994. The twenty-item Toronto alexithymia scale–I. Item selection and cross-validation of the factor structure. *J. Psychosom. Res.* **38:** 23–32.
36. Silani, G. *et al.* 2008. Levels of emotional awareness and autism: an fMRI study. *Soc. Neurosci.* **3:** 97–112.
37. Bird, G. *et al.* 2010. Empathic brain responses in insula are modulated by levels of alexithymia but not autism. *Brain* **133:** 1515–1525.

38. Heaton, P., B. Hermelin & L. Pring. 1999. Can children with autistic spectrum disorders perceive affect in music? An experimental investigation. *Psychol. Med.* **29:** 1405–1410.
39. Heaton, P. *et al.* 2008. Do social and cognitive deficits curtail musical understanding? Evidence from autism and Down syndrome. *Br. J. Dev. Psychol.* **26:** 171–182.
40. Allen, R. 2010. A Comparative Study of the Effects of Music on Emotional State in Normal Adults and those with High functioning Autism. Ph.D. Thesis, Goldsmiths, University of London, London.
41. Marcelli, D. *et al.* 2000. A depressed mother observes her child. A study of the search for behavioral signs which differentiate between depressed mothers and a control group of mothers in the neonatal period. *Psychiatrie de l'Enfant* **43:** 541–586.
42. Trevarthen, C. 1979. Communication and cooperation in early infancy: a description of primary intersubjectivity. In *Before Speech: The Beginning of Interpersonal Communication.* M. Bullowa, Ed.: 321–348. Cambridge University Press. Cambridge, UK.
43. Kanner, L. 1943. Autistic disturbances of affective contact. *Nerv. Child* **2:** 217–250.
44. Rizzolatti, G. & C. Sinigaglia. 2010. The functional role of the parieto-frontal mirror circuit: interpretations and misinterpretations. *Nat. Rev. Neurosci.* **11:** 264–274.
45. Rizzolatti, G. & L. Craighero. 2004. The mirror-neuron system. *Annu. Rev. Neurosci.* **27:** 169–192.
46. Molnar-Szakacs, I. & K. Overy. 2006. Music and mirror neurons: from motion to 'e'motion. *Soc. Cogn. Affect Neurosci.* **1:** 235–241.
47. Overy, K. & I. Molnar-Szakacs. 2009. Being together in time: musical experience and the mirror neuron system. *Music Percept.* **26:** 489–504.
48. Molnar-Szakacs, I. *et al.* 2011. The neural correlates of emotional music perception: an fMRI study of the shared affective motion experience (SAME) model of musical experience. June 9–12, *Presented at The Neurosciences and Music IV: Learning and Memory*, Edinburgh, Scotland, UK.
49. Vieillard, S. *et al.* 2008. Happy, sad, scary and peaceful musical excerpts for research on emotions. *Cogn. Emot.* **22:** 720–752.
50. Koelsch, S. & W.A. Siebel. 2005. Towards a neural basis of music perception. *Trends Cogn. Sci.* **9:** 578–584.
51. Devinsky, O., M.J. Morrell & B.A. Vogt. 1995. Contributions of anterior cingulate cortex to behavior. *Brain* **118**(Pt 1): 279–306.
52. Blood, A.J. & R.J. Zatorre. 2001. Intensely pleasurable responses to music correlate with activity in brain regions implicated in reward and emotion. *Proc. Natl. Acad. Sci.* USA **98:** 11818–11823.
53. Koelsch, S. *et al.* 2005. Adults and children processing music: an fMRI study. *NeuroImage* **25:** 1068–1076.
54. Menon, V. & D.J. Levitin. 2005. The rewards of music listening: response and physiological connectivity of the mesolimbic system. *NeuroImage* **28:** 175–184.
55. Caria, A., P. Venuti & S. de Falco. 2011. Functional and dysfunctional brain circuits underlying emotional processing of music in autism spectrum disorders. *Cereb. Cortex* **12:** 2838–2849.
56. Schultz, R.T. *et al.* 2000. Abnormal ventral temporal cortical activity during face discrimination among individuals with autism and Asperger syndrome. *Arch. Gen. Psychiatry* **57:** 331–340.
57. Pelphrey, K.A. *et al.* 2002. Visual scanning of faces in autism. *J. Autism Dev. Disord.* **32:** 249–261.
58. Spezio, M.L., *et al.* 2007. Analysis of face gaze in autism using "Bubbles". *Neuropsychologia* **45:** 144–151.
59. Di Martino, A. *et al.* 2009. Functional brain correlates of social and nonsocial processes in autism spectrum disorders: an activation likelihood estimation meta-analysis. *Biol. Psychiatry* **65:** 63–74.
60. Craig, A.D. 2002. How do you feel? Interoception: the sense of the physiological condition of the body. *Nat. Rev. Neurosci.* **3:** 655–666.
61. Craig, A.D. 2003. Interoception: the sense of the physiological condition of the body. *Curr. Opin. Neurobiol.* **13:** 500–505.
62. Critchley, H.D. *et al.* 2004. Neural systems supporting interoceptive awareness. *Nat. Neurosci.* **7:** 189–195.
63. Craig, A.D. 2008. Interoception and emotion: a neuroanatomical perspective. In *Handbook of Emotions.* M. Lewis & L. Feldman-Barrett, Eds.: 272–288. Guilford. New York.
64. Dalgleish, T. 2004. The emotional brain. *Nat. Rev. Neurosci.* **5:** 583–589.
65. Pessoa, L. 2008. On the relationship between emotion and cognition. *Nat. Rev. Neurosci.* **9:** 148–158.
66. Uddin, L.Q. & V. Menon. 2009. The anterior insula in autism: under-connected and under-examined. *Neurosci. Biobehav. Rev.* **33:** 1198–1203.
67. Hubl, D. *et al.* 2003. Functional imbalance of visual pathways indicates alternative face processing strategies in autism. *Neurology* **61:** 1232–1237.
68. Dichter, G.S. & A. Belger. 2007. Social stimuli interfere with cognitive control in autism. *NeuroImage* **35:** 1219–1230.

Ann. N.Y. Acad. Sci. ISSN 0077-8923

ANNALS OF THE NEW YORK ACADEMY OF SCIENCES
Issue: *The Neurosciences and Music IV: Learning and Memory*

Auditory-musical processing in autism spectrum disorders: a review of behavioral and brain imaging studies

Tia Ouimet,[1,3] Nicholas E.V. Foster,[1,2,3] Ana Tryfon,[1,2,3] and Krista L. Hyde[1,2,3]

[1]Montreal Children's Hospital Research Institute and McGill University, Psychiatry, Montreal, Quebec, Canada. [2]Montreal Neurological Institute, McGill University, Neurology and Neurosurgery, Montreal, Quebec, Canada. [3]International Laboratory for Brain, Music, and Sound Research, McGill and University of Montreal, Neurology and Neurosurgery, Montreal, Quebec, Canada

Address for correspondence: Krista L. Hyde, Montreal Children's Hospital Research Institute and McGill University, Psychiatry, 4060 Sainte-Catherine Street West, Room 322, Montreal, Quebec, Canada H3Z 2Z3. krista.hyde@mcgill.ca

Autism spectrum disorder (ASD) is a complex neurodevelopmental condition characterized by atypical social and communication skills, repetitive behaviors, and atypical visual and auditory perception. Studies in vision have reported enhanced detailed ("local") processing but diminished holistic ("global") processing of visual features in ASD. Individuals with ASD also show enhanced processing of simple visual stimuli but diminished processing of complex visual stimuli. Relative to the visual domain, auditory global–local distinctions, and the effects of stimulus complexity on auditory processing in ASD, are less clear. However, one remarkable finding is that many individuals with ASD have enhanced musical abilities, such as superior pitch processing. This review provides a critical evaluation of behavioral and brain imaging studies of auditory processing with respect to current theories in ASD. We have focused on auditory-musical processing in terms of global versus local processing and simple versus complex sound processing. This review contributes to a better understanding of auditory processing differences in ASD. A deeper comprehension of sensory perception in ASD is key to better defining ASD phenotypes and, in turn, may lead to better interventions.

Keywords: auditory; autism; music; pitch; brain

Introduction

Autism spectrum disorder (ASD) is a lifelong and heritable neurodevelopmental condition that affects about 1 in 100 individuals.[1] The core features of ASD are atypical social cognition, communication, and repetitive behaviors. In addition, many individuals with ASD have atypical sensory processing, particularly in terms of a hypersensitivity to intense visual (e.g., bright lights) or auditory stimuli (e.g., crowd noises). Studies on visual processing have generally found enhanced processing of "local" (detailed) visual features, with either diminished or intact processing of "global" (holistic) features in individuals with ASD relative to typically developing (TD) controls (for a review, see Ref. 2). The processing of simple visual features (e.g., static visual information) is enhanced, but the processing of complex visual features (e.g., dynamic visual information) is diminished in ASD versus TD (for a review, see Ref. 3). However, relative to the visual domain, less is known about auditory processing differences in ASD. Moreover, there are inconsistent findings in terms of global–local processing distinctions and the relationship between stimulus complexity and performance on auditory tasks in ASD. One finding that stands out in auditory research in ASD is that many individuals have exceptional musical abilities (e.g., enhanced pitch discrimination and memory).[4] In contrast to these strengths, individuals with ASD generally show diminished processing of social and more complex sounds such as speech. Since music is nonverbal, it provides a unique way to study individuals with ASD who have atypical language profiles. Given the special engagement that many individuals with ASD show for music, the study of

doi: 10.1111/j.1749-6632.2012.06453.x

auditory-musical processing is promising in the development of new auditory-based interventions in ASD.

This review provides an updated report of behavioral and brain imaging studies of auditory processing in ASD. (Note that unless otherwise specified, the data reviewed here are from individuals with ASD who have typical intelligence quotients.)[5] A special focus was made to review auditory-music-like processing in ASD in terms of auditory global–local processing, as well as the effect of stimulus complexity on auditory processing in ASD. We also present preliminary data from our laboratory on these topics. Findings are discussed with respect to influential perceptual and neurobiological theories of ASD. We conclude by providing some key future research directions on the study of auditory processing in ASD.

Overview of perceptual and neurobiological theories in ASD: from vision to audition

Various theories have been proposed to explain atypical sensory perception in ASD. These theories have generally been elaborated based on findings from the visual domain. In turn, studies in the visual domain can guide us on what to expect in terms of auditory processing differences in ASD. Here, we review current, influential, perceptual, and neurobiological theories that are relevant for a better understanding of both visual and auditory perception in ASD.

TD individuals tend to process global (holistic) features of a stimulus faster and more accurately than local (detailed) features. For example, in vision, we typically see the image of a forest (global feature) before a single tree (local feature). This perceptual phenomenon is termed the *global precedence effect*, and has been consistently replicated in the visual domain in TD individuals (for a review, see Ref. 6). In contrast, individuals with ASD tend to process local visual features better than global features (for a review, see Ref. 2). On the basis of such findings from the visual domain, Frith and colleagues proposed the *weak central coherence theory*.[7–11] This theory proposes that perceptual differences between ASD and TD result from poor integration of local features into larger ensembles at the global level, and therefore predicts enhanced local but diminished global processing in ASD. However, evidence for the weak central coherence theory in the auditory domain is not clear. Moreover, this theory is limited in the sense that it does not address the effect of stimulus complexity on perceptual processing in ASD.

A more recent theory, the *enhanced perceptual functioning model*,[12,13] also predicts locally oriented perception in ASD, but not impaired global processing. In addition, the enhanced perceptual functioning model proposes that low-level perception is enhanced in ASD and that performance is inversely related to the degree of neural complexity required to process the stimulus features (*neural complexity hypothesis*[3]). The neural complexity hypothesis was elaborated based on findings from the visual domain of enhanced simple feature analyses (e.g., static visual information that is processed in primary visual cortex), but diminished complex visual perception (e.g., dynamic visual information that is processed in secondary visual cortex) in ASD (for a review, see Ref. 3). This hypothesis was recently extended to the auditory domain[14] and predicts enhanced simple auditory processing (e.g., pure-tone discrimination), but diminished complex auditory processing (e.g., speech in noise with large spectral–temporal variation) in ASD. Bertone *et al.*[3] proposed that atypical brain connectivity between primary and nonprimary perceptual brain regions might account for differential processing between simple and complex features in ASD. Consistent with this idea, the *under-connectivity hypothesis*[15] proposes that complex information processing in ASD is impaired due to impoverished long-range connections between brain regions. However, evidence for the neural complexity hypothesis of ASD is not yet clear in the auditory domain, particularly with regard to complex auditory processing (as will be discussed below).

Most recently, the *intense world theory*[16] provides a unifying framework that encompasses the previously discussed theories in ASD. In concordance with earlier theories, it predicts enhanced local and simple feature processing, with diminished global and more complex information processing in ASD. However, the intense world theory explains atypical perception in ASD differently than previous theories, proposing that it arises from hyperfunctioning of local neural microcircuits that become "autonomous and memory trapped." In turn, this "hyperfunctionality" leads to the core cognitive consequences of hyperperception and other core features of ASD.[16]

In sum, each of the above theories has strengths and limitations, and no one theory can account for all aspects of atypical auditory processing in ASD. In the next section, we critically evaluate these theories above with respect to research on auditory processing in ASD. In particular, we focus on auditory-musical global–local processing and the role of stimulus complexity on auditory processing in ASD.

Global versus local auditory-musical processing in ASD

In the auditory domain, global and local processing distinctions have typically been investigated in the context of music. In the standard musical global–local task called an "interval-contour" task,[17] listeners judge whether two melodies are the same or different based on a local (interval) pitch change of one note, or global (contour) change in the overall pattern of ups and downs of the melody. Studies using such interval-contour tasks have shown that TD individuals detect global (contour) changes faster and more accurately relative to local (interval) changes.[18,19] In contrast, individuals with ASD demonstrate enhanced local processing, but typical global processing on interval-contour tasks[20–22] (but see Foxton *et al.*[23] who used a modified interval-contour task and found evidence of atypical global processing in ASD).

However, the use of interval-contour stimuli for studying auditory global–local processing has been challenged. Justus and List[24] pointed out that it is impossible to test the extent of independence between the global and local levels using these types of stimuli, since changing the global contour of the melody will necessarily result in a local interval change. Thus, Justus and List[24] proposed a new set of melodic stimuli that allow the independent manipulation of global and local levels. These new stimuli carry the additional advantage of being structurally analogous to previously used visual global–local stimuli,[25] thus facilitating cross-domain comparison that is not possible with interval-contour stimuli.

We recently adapted the Justus and List[24] stimuli to investigate auditory global–local processing in TD adults,[26] as well as in ASD versus TD children.[27] Participants were presented with 9-tone melodies, each comprised of three triplet (3-tone) sequences. The global pattern was defined as the first tone of each triplet pattern, and the local pattern was defined as a single triplet. Participants were asked to discriminate between ascending and descending pitch direction on either the local or global level. As expected, we found that TD adults have the typical global advantage wherein global patterns were processed faster and more accurately relative to local patterns.[26] Preliminary findings from our lab indicate that TD children have a similar global advantage.[27] However, in comparison, ASD children show a trend for a less pronounced global advantage, which appears to be driven by enhanced processing of local structure.[27] These findings are generally consistent with previous results of enhanced local processing on interval-contour tasks in individuals with ASD.[22] However, because these stimuli allow the independent manipulation of global and local structure, we can conclude with greater confidence that global–local auditory processing differs in ASD versus TD. Our stimuli also permit a more direct comparison between auditory and visual global–local processing. Our preliminary findings favor the enhanced perceptual functioning model[12,13] of ASD (which predicts enhanced local processing in ASD but intact global processing) over the weak central coherence theory[7–11] (which also predicts enhanced local processing, but with diminished global processing). However, additional studies in larger samples are required before stronger conclusions on auditory global–local processing differences in ASD can be made.

Studies of brain function and anatomy are also important to provide insight on global–local processing differences in ASD, particularly in terms of brain lateralization and connectivity. It is first necessary to consider how the TD brain processes global–local information. Brain imaging studies of visual global–local processing in TD[28] have shown that global and local processing are lateralized such that the left hemisphere is critical for processing local features and the right hemisphere is critical for processing global features.[28–31] Relative to vision, fewer studies have investigated the brain lateralization of auditory global–local processing, but some studies indicate a similar pattern of brain lateralization. For example, findings from electrophysiological[19] and brain-lesion work[17] on interval-contour processing in TD adults have also shown global versus local auditory processing to be right versus left lateralized. In contrast, findings from a recent brain imaging study on visual processing indicate that this pattern is reversed in ASD, with local visual processing occurring in the right hemisphere and global visual processing in the left.[32] No studies have yet examined the neural basis of auditory global–local processing in ASD. Based on findings from the visual domain,[32] we can predict that a similar atypical brain lateralization will be found for auditory global–local processing in ASD.

Global–local processing differences in ASD may also be explained by atypical brain connectivity between distal brain regions (as proposed by the under-connectivity hypothesis[15]) and/or hyperfunctioning of local circuits (as proposed by the intense world theory[16]). Specifically, a global processing deficit in ASD may be related to impaired connectivity between specialized networks located in different brain regions (e.g., frontal and auditory cortices), and enhanced local processing may be related to enhanced connections within local networks (e.g., regions within the auditory cortex). Moreover, atypical connectivity in ASD is thought to be associated with larger brain size. For example, recent findings from the visual domain indicate that individuals with ASD who show a local processing bias in vision also have larger brains.[33] An important issue that remains to be investigated is whether brain functional differences in auditory global–local processing in ASD are related to brain structural–differences. To this aim, we are currently conducting brain structural–functional investigations of auditory global–local processing in ASD versus TD.

Simple versus complex auditory processing in ASD

In the auditory domain, the enhanced perceptual functioning model[12,13] and the neural complexity hypothesis[3] predict that performance on auditory perceptual tasks will be inversely related to the level of stimulus complexity involved. In support of these theories, enhanced auditory perception in ASD has been reported for very simple auditory material, such as pure tones, and also for musical tones that contain minimal spectrotemporal variation. For example, individuals with ASD have shown superior processing of such simple auditory material on pitch identification and discrimination tasks,[34–37] identification of local pitch changes in a melody,[22,23,36] identification of individual pitches in a chord,[38] and in terms of pitch memory.[36,38] In addition, members of our group have found that individuals with ASD maintain superior pitch perception of pure tones even when pitch changes are presented at fast temporal rates.[39,40] At a neural level, individuals with ASD have shown enhanced electrophysiological brain activity in response to pitch changes in simple auditory sound sequences[41–43]). We have also found increased cortical thickness in the auditory cortex of ASD adults,[44] as well as a positive correlation between auditory cortical volumes and enhanced pitch direction perception.[45]

While there is consistent evidence for enhanced pitch processing of simple auditory material in ASD, the evidence for diminished complex auditory processing in ASD is less clear, and appears to depend on the degree and nature of stimuli and task complexity. Individuals with ASD have shown diminished processing of complex auditory material with large spectrotemporal variations, such as speech in noise that contains temporal dips.[46,47] At a neural level, individuals with ASD have shown atypical brain responses to complex speech-like stimuli,[48,49] human vocal sounds,[50] and speech prosody,[51] as well as to nonsocial complex auditory material.[52] However, a recent behavioral study[35] found (unexpectedly) that adults with ASD could indeed discriminate both spectrotemporally complex vocal-like and nonvocal sounds. The authors suggested that their auditory stimuli may not have been sufficiently complex to reveal predicted deficits in complex auditory processing in ASD. They proposed that more complex stimuli (e.g., speech in noise with temporal dips) in combination with more complex cognitive (e.g., auditory disembedding tasks) and/or attentionally demanding tasks may be required to uncover auditory processing difficulties in ASD.

Taken together, the findings reviewed above on simple versus complex auditory processing in ASD only partially support the enhanced perceptual functioning model.[12,13] Specifically, these findings provide strong support for the prediction of enhanced processing of simple auditory material in ASD, but the findings on complex auditory processing in ASD are mixed. Further studies on complex auditory processing in ASD, specifically using more complex stimuli in the context of higher-level tasks, are required. At a neural level, findings of enhanced simple auditory material in ASD are consistent with the neural complexity hypothesis[3] and the intense world theory[16] of ASD, which propose that hyperconnectivity within local auditory neurocircuits may explain enhanced processing of physically simple sound information. However, future brain imaging studies are required to better understand the neural correlates of more complex auditory processing in ASD.

Future research directions

Developmental changes in auditory processing in ASD

The TD brain continues to develop well into adolescence,[53] and these maturational changes coincide with the development of auditory processing abilities.[54] Brain development in ASD is also dynamic from childhood through adolescence, but differs from the TD developmental trajectory (for a review, see Ref. 55). Importantly, brain imaging studies have found age-related brain structural and functional differences between ASD and TD in auditory and language-based areas.[56–58] However, no study has yet examined the brain-behavioral correlates of auditory global–local processing or the effect of stimulus complexity on auditory processing in ASD as a function of development. We are currently conducting such studies in our laboratory.

The role of genes on auditory processing in ASD

ASD has a strong genetic component (with a heritability factor as high as 90% in twins).[59] However, very little is known about how genes influence brain development and behavior (and specifically sensory processing) in ASD. Many individuals with ASD present exceptional musical abilities such as absolute pitch (a rare ability to identify or produce the pitch of a tone without reference to an external standard), which has been associated to genetic origins.[60,61] From this perspective, the study of auditory-musical pitch processing in ASD provides a unique opportunity to investigate the genetic contributions of behavioral phenotypes in ASD. In our laboratory, we have the rare chance to study a large ($n = 200$) longitudinal sample of children with ASD; we are currently examining the link between genes, brain structure and function, and auditory-musical abilities.[62]

Auditory-based interventions in ASD

There has been growing interest in the use of sensory-based training in various developmental disorders—for example in the use of auditory spectral temporal-based training as a potential remediation tool in dyslexia.[63] In the case of ASD, auditory-based training paradigms appear to have a positive effect on the brain[64] and on speech production in nonverbal children with ASD.[65] However, no one has yet examined the effects of auditory training paradigms in the context of global versus local or simple versus complex sound processing in ASD. We are currently undertaking such auditory-based training studies in our laboratory.

Conclusion

Here, we have reviewed current behavioral and brain imaging studies on auditory processing in ASD with a special focus on musical pitch perception. Consistent with findings from the visual domain, individuals with ASD show enhanced processing of local auditory features, but global processing appears to be intact. Also consistent with results from vision, individuals with ASD show enhanced processing of simple auditory material, but findings on complex auditory processing in ASD are unclear. No present theory fully explains auditory processing differences in ASD; however, the results reviewed previously are most consistent with the enhanced perceptual functioning model[12,13] of ASD. At a neural level, global versus local and simple versus complex auditory processing distinctions in ASD may be explained by atypical brain connectivity and hyperfunctioning of local neural microcircuits as proposed in the intense world theory.[16] However, future brain imaging studies are required to specifically investigate these aspects of auditory perception in ASD with respect to the intense world theory. Overall, this review enriches our knowledge of the behavioral and brain basis of auditory-musical perception and cognitive styles in ASD. Moreover, this review contributes to a better comprehension of basic sensory processing differences in ASD. This is key for understanding the complex social and cognitive impairments observed in ASD, and may contribute to better diagnoses and interventions.

Acknowledgments

We would like to thank Laurent Mottron, Fabienne Samson, and Andree-Anne Simard-Meilleur for their contribution on some of the research described in this review. This work was funded by grants from the National Centers of Excellence ("NeuroDevNet") and by the Canadian Institutes of Health Research to K. Hyde.

Conflicts of interest

The authors declare no conflicts of interest.

References

1. Baron-Cohen, S., *et al.* 2009. Prevalence of autism-spectrum conditions: UK school-based population study. *Br. J. Psychiatry* **194:** 500–509.
2. Simmons, D.R., *et al.* 2009. Vision in autism spectrum disorders. *Vision Res.* **49:** 2705–2739.
3. Bertone, A., L. Mottron, P. Jelenic & J. Faubert. 2005. Enhanced and diminished visuo-spatial information processing in autism depends on stimulus complexity. *Brain* **128:** 2430–2441.
4. Heaton, P. 2009. Assessing musical skills in autistic children who are not savants. *Philos. Trans. R. Soc. Lond. Ser. B Biol. Sci.* **364:** 1443–1447.
5. Tsai, L.Y. 1992. Diagnostic issues in high-functioning autism. In *High-Functioning Individuals with Autism.* Eric Schopler & Gary B. Mesibov, Eds.: 11–40. Plenum Press. New York.
6. Kimchi, R. 1992. Primacy of wholistic processing and global/local paradigm: a critical review. *Psychol. Bull.* **112:** 24–38.
7. Frith, U. 1989. *Autism: Explaining the Enigma.* Basil Blackwell. Oxford, UK; Cambridge, MA.
8. Frith, U. & F. Happe. 1994. Autism: beyond "theory of mind." *Cognition* **50:** 115–132.
9. Happe, F. 1999. Autism: cognitive deficit or cognitive style? *Trends Cogn. Sci.* **3:** 216–222.
10. Frith, U. 2003. *Autism: Explaining the Enigma.* Blackwell Publishing. Malden, MA; Oxford, UK.
11. Happe, F. & U. Frith. 2006. The weak coherence account: detail-focused cognitive style in autism spectrum disorders. *J. Autism Dev. Disord.* **36:** 5–25.
12. Mottron, L. & J. Burack. 2001. Enhanced perceptual functioning in the development of autism. In *The Development of Autism: Perspectives from Theory and Research.* T. Charman, J.A. Burack, N. Yirmiya & P.R. Zelazo, Eds. Erlbaum. Mahwah, NJ.
13. Mottron, L., M. Dawson, I. Soulieres, *et al.* 2006. Enhanced perceptual functioning in autism: an update, and eight principles of autistic perception. *J. Autism Dev. Disord.* **36:** 27–43.
14. Samson, F., L. Mottron, B. Jemel, *et al.* 2006. Can spectro-temporal complexity explain the autistic pattern of performance on auditory tasks? *J. Autism Dev. Disord.* **36:** 65–76.
15. Just, M.A., V.L. Cherkassky, T.A. Keller & N.J. Minshew. 2004. Cortical activation and synchronization during sentence comprehension in high-functioning autism: evidence of underconnectivity. *Brain* **127:** 1811–1821.
16. Markram, K. & H. Markram. 2010. The intense world theory—a unifying theory of the neurobiology of autism. *Front Hum. Neurosci.* **4:** 224.
17. Peretz, I. 1990. Processing of local and global musical information by unilateral brain-damaged patients. *Brain* **113:** 1185–1205.
18. Fujioka, T., L.J. Trainor, B. Ross, *et al.* 2004. Musical training enhances automatic encoding of melodic contour and interval structure. *J. Cogn. Neurosci.* **16:** 1010–1021.
19. Schiavetto, A., F. Cortese & C. Alain. 1999. Global and local processing of musical sequences: an event-related brain potential study. *NeuroReport* **10:** 2467–2472.
20. Heaton, P., L. Pring & B. Hermelin. 1999. A pseudo-savant: a case of exceptional musical splinter skills. *Neurocase* **5:** 503–509.
21. Heaton, P. 2005. Interval and contour processing in autism. *J. Autism Dev. Disord.* **35:** 787–793.
22. Mottron, L., I. Peretz & E. Menard. 2000. Local and global processing of music in high-functioning persons with autism: beyond central coherence? *J. Child Psychol. Psychiatry* **41:** 1057–1065.
23. Foxton, J.M., *et al.* 2003. Absence of auditory 'global interference' in autism. *Brain* **126:** 2703–2709.
24. Justus, T. & A. List. 2005. Auditory attention to frequency and time: an analogy to visual local-global stimuli. *Cognition* **98:** 31–51.
25. Navon, D. 1977. Forest before trees: the precedence of global features in visual perception. *Cogn. Psychol.* **9:** 353–383.
26. Ouimet, T., N.E. Foster & K.L. Hyde. 2011. The effect of attention and musical training on auditory local-global processing. In *Conference Proceedings from 'The Neurosciences and Music—IV Learning and Memory'.* June 9–12. Edinburgh, Scotland.
27. Ouimet, T., N.E. Foster, A. Tryfon & K.L. Hyde. 2012. Auditory global–local processing in children with autism spectrum disorders. In *Conference Proceedings from the 'International Meeting for Autism Research'.* May 17–19. Toronto, Canada.
28. Fink, G.R. *et al.* 1996. Where in the brain does visual attention select the forest and the trees? *Nature* **382:** 626–628.
29. Kimchi, R. & I. Merhav. 1991. Hemispheric processing of global form, local form, and texture. *Acta Psychologica* **76:** 133–147.
30. Martinez, A. *et al.* 1997. Hemispheric asymmetries in global and local processing: evidence from fMRI. *Neuroreport* **8:** 1685–1689.
31. Weissman, D.H. & M.G. Woldorff. 2005. Hemispheric asymmetries for different components of global/local attention occur in distinct temporo-parietal loci. *Cereb. Cortex* **15:** 870–876.
32. Manjaly, Z.M. *et al.* 2007. Neurophysiological correlates of relatively enhanced local visual search in autistic adolescents. *NeuroImage* **35:** 283–291.
33. White, S., H. O'Reilly & U. Frith. 2009. Big heads, small details and autism. *Neuropsychologia* **47:** 1274–1281.
34. Bonnel, A. *et al.* 2003. Enhanced pitch sensitivity in individuals with autism: a signal detection analysis. *J. Cogn. Neurosci.* **15:** 226–235.
35. Bonnel, A. *et al.* 2010. Enhanced pure-tone pitch discrimination among persons with autism but not Asperger syndrome. *Neuropsychologia* **48:** 2465–2475.

36. Heaton, P., B. Hermelin & L. Pring. 1998. Autism and pitch processing: a precursor for savant musical ability. *Music Perception* **15**: 291–305.
37. O'Riordan, M. & F. Passetti. 2006. Discrimination in autism within different sensory modalities. *J. Autism Dev. Disord.* **36:** 665–675.
38. Heaton, P. 2003. Pitch memory, labelling and disembedding in autism. *J. Child Psychol. Psychiatry* **44:** 543–551.
39. Hyde, K.L., N.E. Foster, A.A. Simard-Meilleur & L. Mottron. 2011. Enhanced perception of pitch direction In young adults with autism spectrum disorder. In *Conference Proceedings from the '10th International Meeting for Autism Research'*. May 12–14. San Diego, CA.
40. Foster, N.E., T. Ouimet, A. Tryfon & K.L. Hyde. 2011. Enhanced pitch direction perception in children with autism spectrum disorder. In *Conference Proceedings from the '11th International Meeting for Autism Research'*. May 12–14, 2011.
41. Ferri, R. *et al.* 2003. The mismatch negativity and the P3a components of the auditory event-related potentials in autistic low-functioning subjects. *Clin. Neurophysiol.* **114:** 1671–1680.
42. Gomot, M., M.H. Giard, J.L. Adrien, *et al.* 2002. Hypersensitivity to acoustic change in children with autism: electrophysiological evidence of left frontal cortex dysfunctioning. *Psychophysiology* **39:** 577–584.
43. Lepistö, T. *et al.* 2005. The discrimination of and orienting to speech and non-speech sounds in children with autism. *Brain Res.* **1066:** 147–157.
44. Hyde, K.L., F. Samson, A.C. Evans & L. Mottron. 2010. Neuroanatomical differences in brain areas implicated in perceptual and other core features of autism revealed by cortical thickness analysis and voxel-based morphometry. *Hum. Brain Mapp.* **31:** 556–566.
45. Foster, N.E., L. Mottron, F. Samson & K.L. Hyde. 2011. Auditory cortical structure is related to enhanced pitch processing in autism spectrum disorder. In *Conference Proceedings from '17th Annual Meeting of the Organization for Human Brain Mapping'*. June 26–30. Quebec, Canada.
46. Alcántara, J.I., E.J. Weisblatt, B.C. Moore & P.F. Bolton. 2004. Speech-in-noise perception in high-functioning individuals with autism or Asperger's syndrome. *J. Child Psychol. Psychiatry* **45:** 1107–1114.
47. Groen, W.B. *et al.* 2009. Intact spectral but abnormal temporal processing of auditory stimuli in autism. *J. Autism Dev. Disord.* **39:** 742–750.
48. Boddaert, N. *et al.* 2003. Perception of complex sounds: abnormal pattern of cortical activation in autism. *Am. J. Psychiatry* **160:** 2057–2060.
49. Boddaert, N. *et al.* 2004. Perception of complex sounds in autism: abnormal auditory cortical processing in children. *Am. J. Psychiatry* **161:** 2117–2120.
50. Gervais, H. *et al.* 2004. Abnormal cortical voice processing in autism. *Nat. Neurosci.* **7:** 801–802.
51. Kujala, T., T. Lepistö, T. Nieminen-von Wendt, *et al.* 2005. Neurophysiological evidence for cortical discrimination impairment of prosody in Asperger syndrome. *Neurosci. Lett.* **383:** 260–265.
52. Samson, F. *et al.* 2011. Atypical processing of auditory temporal complexity in autistics. *Neuropsychologia* **49:** 546–555.
53. Poulsen, C., T.W. Picton & T. Paus. 2009. Age-related changes in transient and oscillatory brain responses to auditory stimulation during early adolescence. *Dev. Sci.* **12:** 220–235.
54. Moore, J.K. & F.H. Linthicum, Jr. 2007. The human auditory system: a timeline of development. *Int. J. Audiol.* **46:** 460–478.
55. Courchesne, E., E. Redcay & D.P. Kennedy. 2004. The autistic brain: birth through adulthood. *Curr. Opin. Neurol.* **17:** 489–496.
56. Groen, W.B., J.K. Buitelaar, R.J. van der Gaag & M.P. Zwiers. 2011. Pervasive microstructural abnormalities in autism: a DTI study. *J. Psychiatry Neurosci.* **36:** 32–40.
57. Keller, T. A., R.K. Kana & M.A. Just. 2007. A developmental study of the structural integrity of white matter in autism. *NeuroReport* **18:** 23–27.
58. Lee, J.E. *et al.* 2007. Diffusion tensor imaging of white matter in the superior temporal gyrus and temporal stem in autism. *Neurosci. Lett.* **424:** 127–132.
59. Rutter, M., J. Silberg, T. O'Connor & E. Simonoff. 1999. Genetics and child psychiatry: I advances in quantitative and molecular genetics. *J. Child Psychol. Psychiatry* **40:** 3–18.
60. Zatorre, R.J. 2003. Absolute pitch: a model for understanding the influence of genes and development on neural and cognitive function. *Nat. Neurosci.* **6**: 692–695.
61. Theusch, E. & J. Gitschier. 2011. Absolute pitch twin study and segregation analysis. *Twin Res. Hum. Genet.* **14:** 173–178.
62. Zwaigenbaum, L. *et al.* 2011. The NeuroDevNet Autism Spectrum Disorders Demonstration Project. *Semin. Pediatr. Neurol.* **18:** 40–48.
63. Gaab, N., J.D. Gabrieli, G.K. Deutsch, *et al.* 2007. Neural correlates of rapid auditory processing are disrupted in children with developmental dyslexia and ameliorated with training: an fMRI study. *Restor. Neurol. Neurosci.* **25:** 295–310.
64. Russo, N.M., J. Hornickel, T. Nicol, *et al.* 2010. Biological changes in auditory function following training in children with autism spectrum disorders. *Behav. Brain Funct.* **6:** 60.
65. Wan, C.Y. *et al.* 2011. Auditory-motor mapping training as an intervention to facilitate speech output in nonverbal children with autism: a proof of concept study. *PLoS One* **6:** e25505.

Ann. N.Y. Acad. Sci. ISSN 0077-8923

ANNALS OF THE NEW YORK ACADEMY OF SCIENCES
Issue: *The Neurosciences and Music IV: Learning and Memory*

Atypical hemispheric asymmetry in the arcuate fasciculus of completely nonverbal children with autism

Catherine Y. Wan, Sarah Marchina, Andrea Norton, and Gottfried Schlaug
Department of Neurology, Music, Neuroimaging, and Stroke Recovery Laboratories, Beth Israel Deaconess Medical Center and Harvard Medical School, Boston, Massachusetts

Address for correspondence: Catherine Wan, Ph.D., Department of Neurology, Beth Israel Deaconess Medical Center and Harvard Medical School, 330 Brookline Avenue, Boston, MA 02215. cwan@bidmc.harvard.edu

Despite the fact that as many as 25% of the children diagnosed with autism spectrum disorders are nonverbal, surprisingly little research has been conducted on this population. In particular, the mechanisms that underlie their absence of speech remain unknown. Using diffusion tensor imaging, we compared the structure of a language-related white matter tract (the arcuate fasciculus, AF) in five completely nonverbal children with autism to that of typically developing children. We found that, as a group, the nonverbal children did not show the expected left–right AF asymmetry—rather, four of the five nonverbal children actually showed the reversed pattern. It is possible that this unusual pattern of asymmetry may underlie some of the severe language deficits commonly found in autism, particularly in children whose speech fails to develop. Furthermore, novel interventions (such as auditory-motor mapping training) designed to engage brain regions that are connected via the AF may have important clinical potential for facilitating expressive language in nonverbal children with autism.

Keywords: autism; nonverbal; language; arcuate fasciculus; asymmetry; auditory-motor mapping training

Introduction

Autism spectrum disorder (ASD) is a developmental condition that affects one in 110 children. One of the core diagnostic features of ASD relates to impairments in language and communication. It has been estimated that up to 25% of the individuals with ASD lack the ability to communicate with others using speech sounds, and many of them have limited vocabulary in any modality, including sign language.[1,2] Severe deficits in communication not only diminish the quality of life for affected individuals, but also present a lifelong challenge for their families.

The ability to communicate verbally is considered to be a positive prognostic indicator for children with ASD.[3] Unfortunately, few techniques are available that can reliably produce improvements in speech output in nonverbal children with ASD.[4] Recently, our laboratory has developed a novel intonation-based intervention, auditory-motor mapping training (AMMT), which aims to facilitate speech output and vocal production in nonverbal children with ASD.[5,6] One of the unique features of this intervention is that it promotes speech production directly. The acquisition of basic vocal output will allow these children to eventually participate in speech therapy that focuses on verbal expression as a primary means of communication.

In addition to the behavioral data collected as part of the ongoing treatment study, we have also acquired neuroimaging data in a subset of our nonverbal participants to identify potentially atypical language-related characteristics in this understudied population. Previous neuroimaging studies have shown that high-functioning verbal individuals with autism typically have larger brains, more gray matter, and possibly more local connections, but fewer long-range connections than typically developing controls.[7,8] Interestingly, a reversal of the usual left–right asymmetry (found in typically developing individuals) is present in the inferior frontal gyrus, with larger volumes in the right

doi: 10.1111/j.1749-6632.2012.06446.x

hemisphere of individuals with autism.[9,10] In contrast, a smaller right volume in autism has also been reported.[11] Some studies have shown smaller volumes of the left planum temporale,[12,13] whereas other studies have reported a reduction in volume in both hemispheres.[14] A recent report suggested a general loss of typical lateralization in tracts that interact with the fusiform gyrus.[15] Furthermore, functional imaging studies of high-functioning verbal individuals with autism have shown relatively normal temporal lobe activation, but reduced inferior frontal activation during semantic language tasks.[16] However, no studies to date have examined the brains of nonverbal children with ASD.

Here, we report preliminary data from our neuroimaging study on completely nonverbal children with autism. We investigated possible differences in a language-related white matter tract, the arcuate fasciculus (AF), using diffusion tensor imaging (DTI) and comparing their results to those of age-matched, typically developing children. Specifically, the AF is a language-related tract that connects brain regions involved in feedforward and feedback control of vocal output, and the mapping of sounds to articulatory motor actions.[17–21] Because of its role in speech production, we investigated structural differences in the AF of nonverbal children with autism with those of typically developing children.

Method

Participants

A total of 10 children underwent MRI scanning. There were five completely nonverbal children with ASD and five typically developing children (see Table 1 for participant characteristics). For the nonverbal children, the diagnosis of autism was made by pediatric neurologists and neuropsychologists prior to enrollment. We confirmed the participants' diagnoses using the childhood autism rating scale (CARS). "Nonverbal" was defined as having the complete absence of intelligible words. All participants had previously received speech therapy for at least 18 months and demonstrated minimal progress in speech acquisition (i.e., no intelligible words) based on speech-language pathologist and parent reports. Their minimal speech production was also confirmed by our intake assessment, based on clinical observations, and results from a phonetic inventory and the expressive vocabulary test.[22] The typically developing children were recruited as part of the laboratory's longitudinal study on children,[23] and all of them were classified as nonmusicians at the time of the scan. The study was approved by the Institutional Review Board of Beth Israel Deaconess Medical Center. The parents of all children gave written informed consent prior to their participation, and all procedures were conducted according to the approved protocol.

Image acquisition

All participants underwent scanning using a 3-tesla General Electric (Fairfield, CT) scanner, which included T1-weighted, gradient-echo anatomical images (resolution: 0.93 × 0.93 × 1.5 mm3), and diffusion tensor images. Diffusion-weighted images were acquired using single-shot, spin-echo, echo-planar imaging sequence (TE = 86.9 ms, TR = 10,000 ms, FOV = 240 mm, matrix size = 128 × 128 voxels, slice thickness = 5 mm (resolution: 1.87 × 1.87 × 5.0 mm^3), no skip, NEX = 1, axial acquisition, 25 noncollinear directions with b-value = 1000 s/mm^2, 1 image with b-value = 0 s/mm^2).

Preprocessing of DTI data

The diffusion data were preprocessed using FSL version 4.1.4 (www.fmrib.ox.ac.uk/fsl). Using FSL's FMRIB diffusion toolbox (FDT), we first corrected the diffusion data for eddy current and head motion artifacts by affine multiscale two-dimensional registration. We then fitted a diffusion tensor model at each voxel, which yielded lambda values for each principal eigenvector and fractional anisotropy.

Table 1. Participant characteristics

	Nonverbal ASD ($n = 5$)			Typically developing ($n = 5$)			Comparison
	Mean	SD	Range	Mean	SD	Range	P-value
Age (years)	6.7	1.2	5.8–8.8	7.0	0.9	6.2–8.5	0.175
Gender	3 males, 2 females			3 males, 2 females			

Fiber tracking parameters were estimated using a probabilistic tractography method based on a multifiber model, and applied using tractography routines implemented in FSL's FDT toolkit (5,000 streamline samples, 0.5-mm step lengths, curvature threshold of 0.2, and modeling 2 fibers per voxel to take into account crossing fibers).[24,25]

Tractography of the arcuate fasciculus.
A single rater drew the regions of interest (ROI) of the arcuate fasciculus—a curved fiber bundle that connects the posterior portion of the temporal lobe and the temporo-parietal junction with the inferior frontal lobe[26]—on both hemispheres in diffusion space. This rater was blind to whether participants were typically developing or had autism. The seed ROI was drawn in the white matter underlying the pars opercularis of the posterior inferior frontal gyrus on a sagittal slice of the FA map. Two waypoint ROIs were drawn: one on a coronal slice in the sensory-motor region covering the superior longitudinal fasciculus and another on a sagittal slice in the white matter underlying the posterior middle temporal gyrus. Exclusion masks were drawn axially in the external capsule, coronally in the region posterior to the temporal gyrus, and sagittally in the region medial to the fiber bundle in order to exclude fiber projections that were not part of the AF. The volume of each anatomical ROI was constrained such that a similar size ROI was used across the two hemispheres to minimize potential bias ($P > 0.9$).

Results

Fiber tracking reliably identified the arcuate fasciculus on both the left and right hemispheres of all 10 children. Figure 1 shows the average volumes of the left and right AF for the two groups of children. For each child, we calculated a laterality index (LI) = (left AF volume – right AF volume)/(left AF volume + right AF volume), where indices greater than zero indicate leftward asymmetry, and indices less than zero indicate rightward asymmetry. Consistent with the literature on the AF,[27] the typically developing children in our sample exhibited the usual leftward asymmetry (median LI = 0.166, range = 0.013 to 0.586). The nonverbal children, however, did not show the usual leftward pattern of asymmetry (median LI = –0.168, range = –0.940 to 0.120). To confirm that the two groups showed different patterns of laterality, a nonparameteric Mann–Whitney test was conducted, which revealed significant distributions of the two groups ($P = 0.016$). As illustrated in Figure 2, all five typically developing children showed greater AF tract volume in the left hemisphere compared to the right hemisphere, whereas four of the five nonverbal children with autism showed the reversed pattern of asymmetry (larger right than left AF volumes).

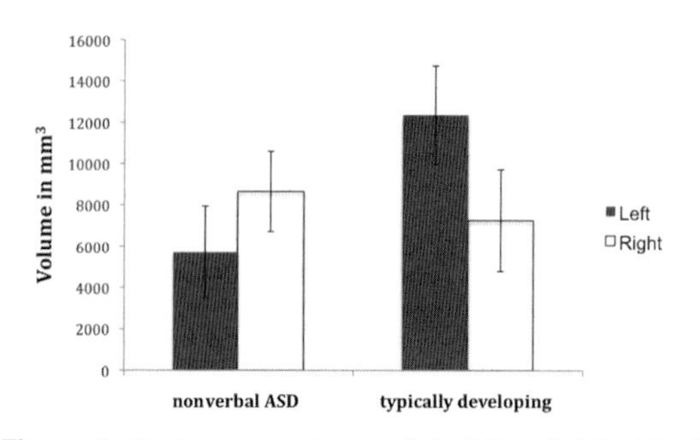

Figure 1. Total average volumes of the left and right AF of the nonverbal children with ASD and the typically developing children.

Discussion

In this study, we sought to examine potential structural brain abnormalities in completely nonverbal children with autism. Using DTI, we found that the arcuate fasciculus, a major language-related white-matter pathway in the brain, showed an overall hemispheric asymmetry reversal in a group of completely nonverbal children with autism compared to typically developing children. This abnormal structure of the AF may underlie some of the severe language deficits in autism, particularly in children who never develop speech.

Results from the present study converge with some of the previous imaging studies on high-functioning verbal individuals with autism. Using structural MRI, a reversal of the usual left–right asymmetry has been observed in the right inferior frontal gyrus of high-functioning individuals with autism, although smaller right frontal volumes in autism have also been reported. Similarly, smaller volumes of the left planum temporale have been observed, but other research has reported a reduction in both hemispheres.[13,28] The inconsistent findings reported by these structural imaging studies may be attributable, in part, to the heterogeneity in linguistic abilities among individuals included in these studies. Indeed, individuals with Asperger's

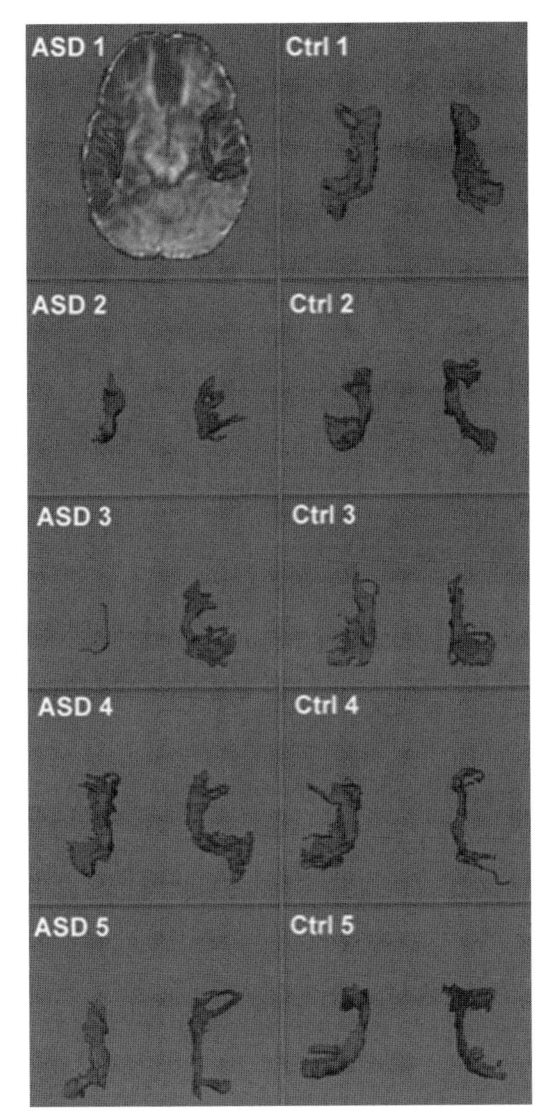

Figure 2. Images showing the left (yellow) and right (green) AF of all 10 children (left panel = nonverbal ASD; right panel = typically developing controls). For display purposes, only the tracts of one child are superimposed onto their own FA image.

syndrome (who have no language delay) have been found to have less gray matter than individuals with autism (who have atypical language development).[29] Thus, this finding highlights the importance of assessing language skills as a differentiating variable.

Two recent studies have also investigated the role of the arcuate fasciculus in autism. However, one study scanned a more heterogeneous group of children, including those who were not only verbal but also carried the diagnosis of Asperger's syndrome,[30] while another study scanned only verbal high-functioning adolescents.[31] Relative to controls, individuals on the autism spectrum were found to have longer fibers in the right arcuate fasciculus[30] and less lateralized fractional anisotropy,[31] but no differences in volume were observed. Our study extends these previous findings in two important ways. First, we recruited a relatively homogenous group of individuals who were completely nonverbal. Second, we used a probabilistic (rather than a deterministic) algorithm, which is less susceptible to regional measurement artifacts and problems concerning crossing fibers.[24] Probabilistic tractography approaches have been shown to produce better results in areas of crossing fibers, with superior reconstruction of fibers at borders of anatomical structures, and significantly more sensitive than deterministic approaches.[32]

The AF is a major white-matter tract involved in language and speech processing, and thus, may also be associated with the integration of auditory and motor functions. Because it runs through the premotor and motor cortex, it has been implicated in the mapping of sounds to articulatory actions, the coordination and planning of motor actions for speech production, as well as the monitoring of speech production and language learning.[19,33] Based on observations made in our laboratory, many nonverbal children with autism have speech–motor planning difficulties, and their deficits could be explained, in part, by the abnormal asymmetry of the AF found in the present study. More importantly, because AMMT is an intervention that trains the association between sounds and articulatory actions through rhythmic bilateral motor activities and repetitions of intoned words,[5] its potential in facilitating speech output in nonverbal children with autism may lie in its ability to engage a network of brain regions (e.g., the AF) that may be dysfunctional in autism.[34,35]

In the present study, a relatively short DTI sequence (less than 5 min) was used to minimize movement artifacts in our group of nonverbal children with autism. The resulting voxels from this sequence were nonisotropic, which means that partial volume effects and angular resolution could vary along different axes, thus limiting the degree to which comparisons can be made between tracts that are oriented along these different axes. Despite this limitation, however, our data revealed significant volume differences across the two hemispheres between nonverbal children with autism and typically developing children, while any systematic errors associated with our DTI sequence should be evident across all individuals.

This is a preliminary study that examined structural abnormalities in the arcuate fasciculus specifically in completely nonverbal children with autism. Future studies on this understudied population could test a larger sample of children; use higher resolution and isotropic scanning parameters if possible, considering that these children may not be sedated during scanning sessions; and examine the structure of other language-related tracts, such as the uncinate fasciculus and the extreme capsule fiber tract. Finally, it would be interesting to examine whether the atypical asymmetry observed in the nonverbal children is also present in other family members. Identifying brain abnormalities in these children will help with the development and refinement of effective treatment programs.

Acknowledgments

This research was supported by The Nancy Lurie Marks Family Foundation, Autism Speaks, and the Deborah Munroe Noonan Memorial Research Fund. We sincerely thank all the families who participated.

Conflicts of interest

The authors declare no conflicts of interest.

References

1. Koegel, L.K. 2000. Interventions to facilitate communication in autism. *J. Autism. Dev. Disord.* **30:** 383–391.
2. Turner, L.M., W.L. Stone, S.L. Pozdol & E.E. Coonrod. 2006. Follow-up of children with autism spectrum disorders from age 2 to age 9. *Autism* **10:** 243–265.
3. Luyster, R., S. Qiu, K. Lopez & C. Lord. 2007. Predicting outcomes of children referred for autism using the macarthur-bates communicative development inventory. *J. Speech Lang. Hear. Res.* **50:** 667–681.
4. Francis, K. 2005. Autism interventions: a critical update. *Develop. Med. Child Neurol.* **47:** 493–499.
5. Wan, C.Y., K. Demaine, L. Zipse, *et al.* 2010. From music making to speaking: engaging the mirror neuron system in autism. *Brain Res. Bull.* **82:** 161–168.
6. Wan, C.Y., T. Rüber, A. Hohmann & G. Schlaug. 2010. The therapeutic effects of singing in neurological disorders. *Mus. Percept.* **27:** 287–295.
7. Herbert, M.R., D.A. Ziegler, C.K. Deutsch, *et al.* 2005. Brain asymmetries in autism and developmental language disorder: a nested whole-brain analysis. *Brain* **128:** 213–226.
8. Just, M.A., V.L. Cherkassky, T.A. Keller & N.J. Minshew. 2004. Cortical activation and synchronization during sentence comprehension in high-functioning autism: evidence of underconnectivity. *Brain* **127:** 1811–1821.
9. De Fosse, L., S.M. Hodge, N. Makris, *et al.* 2004. Language-association cortex asymmetry in autism and specific language impairment. *Ann. Neurol.* **56:** 757–766.
10. Herbert, M.R., G.J. Harris, K.T. Adrien, *et al.* 2002. Abnormal asymmetry in language association cortex in autism. *Ann. Neurol.* **52:** 588–596.
11. McAlonan, G.M., V. Cheung, C. Cheung, 2005. *et al.* Mapping the brain in autism. A voxel-based MRI study of volumetric differences and intercorrelations in autism. *Brain* **128:** 268–276.
12. Rojas, D.C., S.L. Camou, M.L. Reite & S.J. Rogers. 2005. Planum temporale volume in children and adolescents with autism. *J. Autism. Dev. Disord.* **35:** 479–486.
13. Rojas, D.C., S.D. Bawn, T.L. Benkers, *et al.* 2002. Smaller left hemisphere planum temporale in adults with autistic disorder. *Neurosci. Lett.* **328:** 237–240.
14. Boddaert, N., N. Chabane, H. Gervais, *et al.* 2004. Superior temporal sulcus anatomical abnormalities in childhood autism: a voxel-based morphometry MRI study. *Neuroimage* **23:** 364–369.
15. Conturo, T.E., D.L. Williams, C.D. Smith, *et al.* 2008. Neuronal fiber pathway abnormalities in autism: an initial MRI diffusion tensor tracking study of hippocampo-fusiform and amygdalo-fusiform pathways. *J. Int. Neuropsychol. Soc.* **14:** 933–946.
16. Harris, G.J., C.F. Chabris, J. Clark, *et al.* 2006. Brain activation during semantic processing in autism spectrum disorders via functional magnetic resonance imaging. *Brain Cogn.* **61:** 54–68.
17. Bohland, J.W. & F.H. Guenther. 2006. An FMRI investigation of syllable sequence production. *Neuroimage* **32:** 821–841.
18. Duffau, H. 2008. The anatomo-functional connectivity of language revisited. New insights provided by electrostimulation and tractography. *Neuropsychologia* **46:** 927–934.
19. Glasser, M.F. & J.K. Rilling. 2008. DTI tractography of the human brain's language pathways. *Cereb. Cortex* **18:** 2471–2482.
20. Ozdemir, E., A. Norton & G. Schlaug. 2006. Shared and distinct neural correlates of singing and speaking. *Neuroimage* **33:** 628–635.
21. Saur, D., B.W. Kreher, S. Schnell, *et al.* 2008. Ventral and dorsal pathways for language. *Proc. Natl. Acad. Sci. USA* **105:** 18035–18040.
22. Williams, K.T. 1997. *Expressive Vocabulary Test*. American Guidance Service. Circle Pines, MN.
23. Hyde, K.L., J. Lerch, A. Norton, *et al.* 2009. The effects of musical training on structural brain development a longitudinal study. *Neurosci. Mus.* 182–186.
24. Behrens, T.E., H.J. Berg, S. Jbabdi, *et al.* 2007. Probabilistic diffusion tractography with multiple fibre orientations: what can we gain? *Neuroimage* **34:** 144–155.
25. Behrens, T.E., M.W. Woolrich, M. Jenkinson, *et al.* 2003. Characterization and propagation of uncertainty in

diffusion-weighted MR imaging. *Magn. Reson. Med.* **50:** 1077–1088.
26. Catani, M., D.K. Jones & D.H. Ffytche. 2005. Perisylvian language networks of the human brain. *Ann. Neurol.* **57:** 8–16.
27. Vernooij, M.W., M. Smits, P.A. Wielopolski, *et al.* 2007. Fiber density asymmetry of the arcuate fasciculus in relation to functional hemispheric language lateralization in both right- and left-handed healthy subjects: a combined FMRI and DTI study. *Neuroimage* **35:** 1064–1076.
28. McAlonan, G.M., V. Cheung, C. Cheung, *et al.* 2005. Mapping the brain in autism. A voxel-based MRI study of volumetric differences and intercorrelations in autism. *Brain* 128:268–276.
29. McAlonan, G.M., J. Suckling, N. Wong, *et al.* 2008. Distinct patterns of grey matter abnormality in high-functioning autism and Asperger's syndrome. *J. Child Psychol. Psychiatr.* **49:** 1287–1295.
30. Kumar, A., S.K. Sundaram, L. Sivaswamy, *et al.* 2010. Alterations in frontal lobe tracts and corpus callosum in young children with autism spectrum disorder. *Cereb. Cortex* **20:** 2103–2113.
31. Fletcher, P.T., R.T. Whitaker, R. Tao, *et al.* 2010. Microstructural connectivity of the arcuate fasciculus in adolescents with high-functioning autism. *Neuroimage* **51:** 1117–1125.
32. Klein, J., A. Grötsch, D. Betz, *et al.* 2010. Qualitative and quantitative analysis of probabilistic and deterministic fiber tracking. *Proc. SPIE Med. Imag.* **7623:** 76232A-76231–76232A-76238.
33. Hickok, G. & D. Poeppel. 2004. Dorsal and ventral streams: a framework for understanding aspects of the functional anatomy of language. *Cognition* **92:** 67–99.
34. Wan, C.Y. & G. Schlaug. 2010. Neural pathways for language in autism: the potential for music-based treatments. *Future Neurol.* **5:** 797–805.
35. Wan, C.Y., L. Bazen, R. Baars, *et al.* 2011. Auditory-motor mapping training as an intervention to facilitate speech output in nonverbal children with autism: a proof of concept study. *PlosOne* **6:** e25505.

Ann. N.Y. Acad. Sci. ISSN 0077-8923

ANNALS OF THE NEW YORK ACADEMY OF SCIENCES
Issue: *The Neurosciences and Music IV: Learning and Memory*

Memory disorders and vocal performance

Simone Dalla Bella,[1,2,3] Alexandra Tremblay-Champoux,[3] Magdalena Berkowska,[2] and Isabelle Peretz[3]

[1]EuroMov, Movement to Health Laboratory, University of Montpellier-1, Montpellier, France. [2]Department of Cognitive Psychology, University of Finance and Management, Warsaw, Poland. [3]International Laboratory for Brain, Music, and Sound Research (BRAMS), Montreal, Canada

Address for correspondence: Simone Dalla Bella, EuroMov, Movement to Health Laboratory (M2H), University of Montpellier-1, 700 Avenue du Pic Saint Loup, 34090 Montpellier, France. simone.dalla-bella@univ-montp1.fr

The ability to carry a tune, natural for the majority, is underpinned by a complex functional system (i.e., the vocal sensorimotor loop, VSL). The VSL involves various components, including perceptual mechanisms, auditory-motor mapping, motor control, and memory. The malfunction of one of these components can bring about poor-pitch singing. So far, disturbed perception and deficient sensorimotor mapping have been treated as important causes of poor singing. Yet, memory has been paid relatively little attention. Here, we review results obtained from both occasional singers and individuals suffering from congenital amusia, who were asked to produce from memory or imitate a well-known melody under conditions with different memory loads. The findings point to memory as a relevant source of impairment in poor-pitch singing and to imitation as a useful aid for poor singers.

Keywords: amusia; vocal performance; memory disorders; music performance

Introduction

The majority of people can carry a tune. Most of us can sing from memory well-known songs (e.g., "Happy Birthday") and imitate short novel melodies quite accurately,[1–3] although not always precisely.[4] This observation is not totally surprising. Singing is as natural as speaking for humans. This ability emerges early during development without needing vocal training or musical tutoring.[5–7] Moreover, singing, due to its participatory character, is thought to foster group bonding, an advantage that may have had some adaptive value in the course of evolution.[8–10] In sum, singing is a fundamental function that is likely to be deeply rooted in human biology.

Some individuals (i.e., tone deaf), representing approximately 10–15% of the general population, are particularly inaccurate in producing pitch. When asked to produce familiar melodies or imitate single pitches, intervals, or short melodies, they typically sing quite far from the target.[1–3] This estimate rises to about 55% of the population when the consistency of repeated attempts to produce pitches (i.e., precision) is taken into account.[4] A few recent studies have aimed to characterize poor-pitch singing and elucidate its causes.[2,11,12] Different sources of impairments may underline inaccurate singing including motor control, perceptual processes, auditory-motor mapping, and memory. In this paper, we will particularly focus on the latter by providing evidence in favor of the idea that memory deficits often accompany poor-pitch singing. To this aim, the mechanisms underlying normal singing will be presented first, and the different explanations of poor-pitch singing briefly addressed.

The vocal sensorimotor loop

People generally tend to underestimate their ability to carry a tune. Almost 60% of 1,000 university students report that they cannot accurately imitate melodies.[2] Moreover, self-declared tone-deaf individuals (around 17% of the student population) believe that they cannot sing proficiently.[13] This picture is clearly too defeatist though. When singing is assessed systematically, we find that about 85–90% of the general population can sing accurately.[1–3,11]

doi: 10.1111/j.1749-6632.2011.06424.x

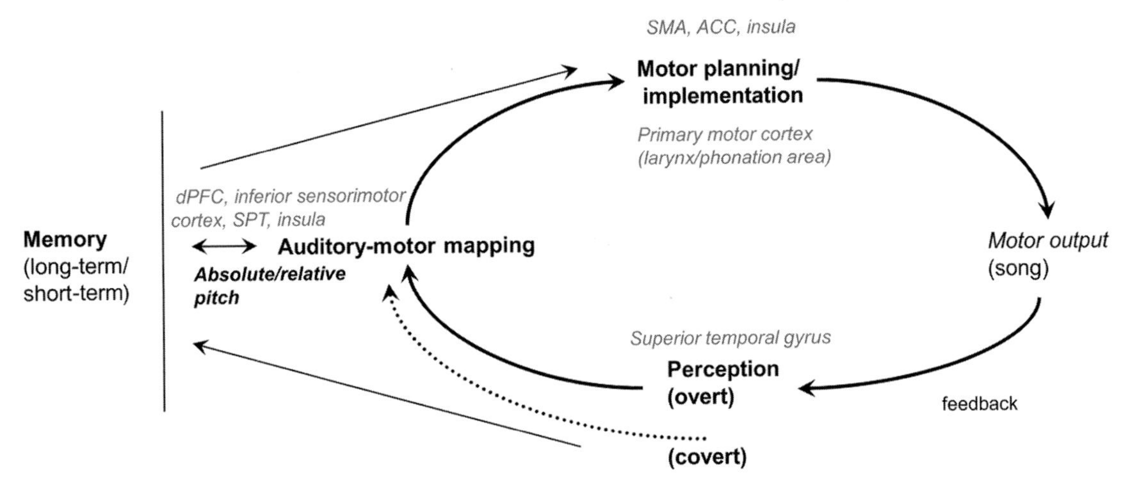

Figure 1. The vocal sensorimotor loop (VSL).[11,14] SMA, supplementary motor area; ACC, anterior cingulate cortex; dPFC, dorsal prefrontal cortex; SPT, cortex of the dorsal Sylvian fissure at the parietal–temporal junction.

Proficient singing in the general population has been shown across a variety of tasks focusing on vocal performance, spanning from singing from memory of a familiar melody, imitation (e.g., single-pitch matching or melody matching), and imitation with reduced or augmented feedback.

Which functional mechanisms and brain circuitry support the widespread ability to sing in tune? The functional components of the song system underlying accurate pitch production have been summarized using the VSL,[13–15] as illustrated in Figure 1. The VSL comprises four main components: perceptual mechanisms, motor planning/implementation, auditory-motor mapping, and memory (short term and long term). Pitch production in the aforementioned tasks can be accounted for by the VSL. Singing of familiar melodies requires the retrieval from long-term memory of pitch and temporal information, directed to fine motor planning/implementation mechanisms (e.g., articulation and eventually phonation). The output of the system (i.e., vocal production) is then fed back to perception. Matching the percept of the produced melody with the memory representation of the intended melody (auditory-motor mapping) warrants error correction if there are discrepancies, thus ultimately affecting planning of upcoming events. Imitation of novel pitch sequences engages similar mechanisms; however, it relies more on short-term memory, without tapping retrieval from long-term memory. The target pitches to be imitated are perceptually analyzed, stored in short-term memory, and eventually mapped into motor gestures. Accurate imitation is made possible by self-monitoring of vocal performance through feedback analysis, and, in some cases, error correction. Note that overt and covert pathways for pitch perception are included in the VSL. The overt pathway is engaged when performing explicit judgments of pitch differences (e.g., pitch discrimination). The covert pathway is supposed to account for cases in which participants are very inaccurate in judging pitch differences, but still exhibit proficient singing (i.e., revealing an implicit ability to make the difference between pitches).[16–18]

The components forming the VSL are underpinned by a complex neuronal network, as revealed by neuroimaging studies (Fig. 1).[11] Singing recruits motor areas (e.g., primary motor cortex), in particular, the mouth region and the larynx/phonation area,[19,20] the anterior cingulate cortex (ACC), the insula, and the supplementary motor area (SMA). The ACC and the insula are linked with initiation of vocalization and articulatory processes, respectively.[19,21–23] The SMA, rather, is involved in high-level motor control, which guarantees efficient motor planning.[24] Sensory areas (e.g., the superior temporal gyrus) are also engaged by vocal performance.[19,22,25] Finally, auditory-motor mapping involves various areas such as the dorsal prefrontal cortex, inferior sensorimotor cortex, the superior temporal gyrus and sulcus,[23,26,27] and area SPT (i.e., cortex of the dorsal Sylvian fissure at the parietal–temporal junction).[28,29] To sum up, brain imaging contributes to uncovering the major components of the song system by providing the neural substrates of the mechanisms identified at a functional level in the VSL.

Our understanding of the song system paves the way for explaining poor-pitch singing. Indeed, poor-pitch singing is not a monolithic deficit and can be brought about by the malfunctioning of any of the components in the VSL. Here, the most relevant accounts will be summarized.[2,11,12,14] Poor-pitch singing is often treated as the outcome of perceptual deficits (e.g., in congenital amusics), resulting from the impaired extraction of pitch information from the auditory input (herein, the "perceptual account"). Impaired perception makes monitoring of the ongoing performance more difficult, thus hindering error correction and leading to reduced accuracy. However, perceptual deficits are not a *sine qua non* for poor-pitch singing. The fact that inaccurate singing can occur despite spared perception points to postperceptual mechanisms as a cause of poor-pitch singing.[1,2] An alternative explanation is that auditory-motor mapping is not carried out properly (the "sensorimotor account").[2,30] A correct auditory representation of the vocal performance can be inaccurately mapped to motor representations for phonation. This mismap concerns local musical features (absolute pitch and secondary pitch intervals) without affecting global features (e.g., melodic contour).[2] Note that mapping relative and absolute musical features (i.e., absolute pitches and intervals) to phonation may engage at least partly independent mechanisms. This possibility is supported by the observation that imitation of absolute and relative pitch can be selectively disrupted in poor-pitch singers,[3] and by differential effects of feedback on pitch accuracy (i.e., choral singing enhances pitch accuracy in producing intervals and contour, but is detrimental for producing absolute pitch).[2] Finally, the sensorimotor account gained some neurobiological plausibility after the discovery of abnormally reduced connectivity through the fasciculus arcuatus (i.e., a pathway connecting temporal and frontal brain regions) in tone deafness.[31]

Another possibility, although little examined in the literature, is that poor-pitch singing results from (or is linked to) a faulty memory system. Despite spared perception and normal motor control, difficulties with encoding, storage, or retrieval of pitch information during reproduction (e.g., in imitation tasks) may bring about inaccurate singing. The possibility that memory disorders are associated with poor-pitch singing is plausible. Congenital amusics, who are typically poor-pitch singers,[16,32] also exhibit impaired memory.[32–35] Pitch representations in amusic individuals are typically less stable and decay more rapidly. It is worth noting that amusics do not suffer from general memory difficulties, as observed, for example, in amnesic patients or in patients with dementia. Amusics' memory disorder is highly selective and is likely to be confined to music.

One way to examine the role of memory in poor-pitch singing is to observe the effect of manipulating memory demands. If poor-pitch singing stems from weak memory traces, reducing memory demands (e.g., via additional feedback or by reducing stimulus length) should alleviate deficits in imitating and producing pitch. For example, Wise and Sloboda[36] asked self-reporting tone-deaf individuals to imitate short-pitch sequences (i.e., formed by one up to five notes). In one condition, the target was presented before participants' imitation; in another condition, the participants were asked to sing along (i.e., synchronized condition), a situation that reduces the memory load. All participants were less accurate with longer sequences. However, increasing the length of the sequence had a more disruptive effect on the performance of tone-deaf individuals as compared with non-tone-deaf participants. Moreover, tone-deaf individuals were aided more than controls by singing along. These findings suggest that tone-deaf individuals cannot maintain short-term memory traces of the presented sequences as compared with non-tone-deaf individuals. Similar stimulus manipulations were adopted by Pfordresher and Brown[2] to test a group of occasional singers. In addition, normal feedback, augmented feedback (i.e., singing along), and masked feedback (i.e., pink noise presented during sung performance) were provided. However, neither an effect of stimulus length nor an advantage due to augmented feedback was observed for poor-pitch singers. In sum, previous studies manipulating memory demands yield a conflicting picture.

In this paper, we briefly review recent studies in which pitch accuracy was examined when the memory load was manipulated, by asking congenital amusics and occasional singers to sing a familiar song from memory either with lyrics or on a syllable, and by presenting a model to be imitated.

In the first case, it is assumed that recalling a familiar melody without lyrics (i.e., when singing on a syllable) will put less demand on long-term memory, thereby allowing the singers to focus on retrieval of the correct melodic information and enhancing auditory-motor mapping. In the second, it is expected that presenting a model to be imitated will enrich the memory trace of the melody to be reproduced (i.e., facilitating its retrieval from memory), thus eventually leading to improved singing accuracy (via auditory-motor mapping).

Study 1: Singing with lyrics versus on a syllable

In a previous experiment we asked 11 congenital amusics (as assessed with the Montreal Battery of Evaluation of Amusia, MBEA[37]) and 11 matched controls to sing the chorus of a familiar melody from memory (i.e., "Gens du Pays") with lyrics, and on a syllable (i.e., /la/ or /ta/).[16] In another experiment, 50 occasional singers (35 females and 15 males, mean age = 25.1 years), mostly university students without musical training, were asked to perform a similar task with three familiar melodies ("Brother John," "Jingle Bells," and "Sto lat"; see also[3,38] for preliminary results). Accuracy in pitch production was computed using acoustical analyses.[1,3] This method, based on acoustical segmentation of sung recordings, provides reliable estimates of pitch heights and note onset times; this data is used to compute various measures of singing accuracy. For the sake of simplicity, we report here only two measures of singing accuracy in terms of relative pitch, which have proven to be very sensitive to individual differences in the general population.[1,3] The first measure is the *number of pitch interval errors*, namely the number of produced intervals that depart from the intended intervals (i.e., based on the musical notation) by at least one semitone. The second measure is *pitch interval deviation*, a continuous measure of pitch deviation obtained by averaging the absolute difference in semitones between the produced intervals and the intervals prescribed by musical notation. Small deviation indicates high accuracy in relative pitch.

All participants in the two experiments were able to produce complete performances with lyrics (i.e., including the full set of notes). Amusics were typically more impaired than controls, though, exhibiting on average 13 pitch-interval errors (vs. 3.6 errors for controls), and pitch interval deviation of 1.3 semitones (vs. 0.5 semitones). The difference between amusics and controls was even more striking, however, when amusics sang on a syllable. Six out of 11 amusics could not produce complete renditions of the melody; their performance was limited to just a few pitches. Note that they were the most impaired amusics, and that they had impaired incidental memory, as revealed by the MBEA. Pitch accuracy for the remaining five amusics, as compared to controls and to 50 occasional singers is reported in Figure 2. As can be seen, these five amusics were still on average less accurate than controls ($P_s < 0.05$, by the Mann–Whitney test) and occasional singers ($P_s < 0.01$). It is worth noting that the opposite dissociation (i.e., higher accuracy when singing on

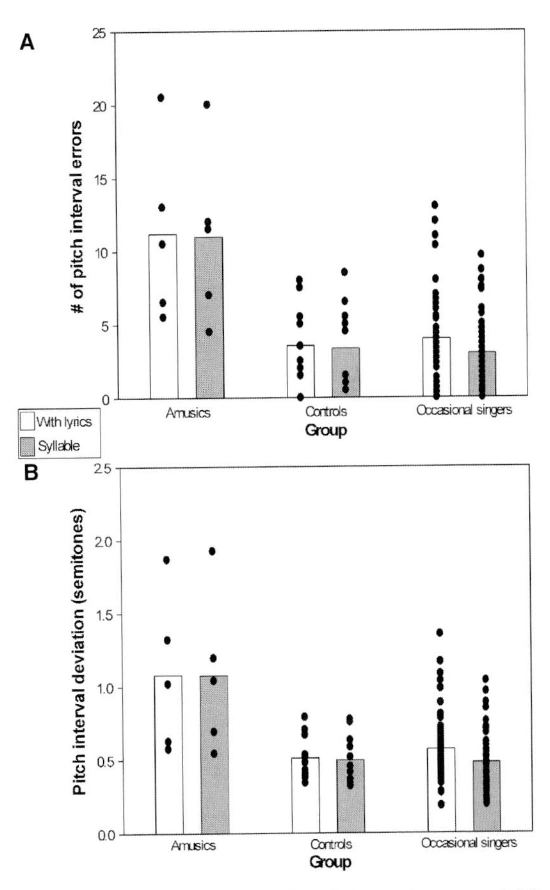

Figure 2. Mean (A) number of pitch interval errors and (B) pitch interval deviations for congenital amusics ($n = 5$), controls ($n = 11$), and occasional singers ($n = 50$) when singing familiar songs from memory with lyrics and on a syllable. For amusics and controls, the maximum possible number of errors was 31; for occasional singers, 25.

a syllable than when singing with lyrics) was observed in the group of 50 occasional singers (for the number of pitch interval errors, $t(49) = 5.0$, $P < 0.001$; for pitch interval deviation, $t(49) = 5.1$, $P < 0.001$).

In sum, these findings indicate that memory disorders and impoverished perception coexist in congenital amusia. The fact that congenital amusics have major difficulties when asked to sing on a syllable can result from weak memory traces of the musical component of songs. Severe amusics can still produce complete performances with lyrics, although inaccurately, probably owing to the strong association between melody and text. However, when an association of a well-known melody to a new speech segment is required (like when singing on a syllable), the melody information may become more difficult to retrieve.

Study 2: Effects of imitation on singing proficiency in congenital amusia

In another study we tested congenital amusics with the goal of examining whether providing an aid to memory (i.e., asking amusics to imitate a model) enhances singing accuracy.[39] Eleven congenital amusics and 11 matched controls were asked to sing a well-known melody from memory ("Gens du pays") with lyrics and on the syllable /la/. In one condition ("memory"), the melody was sung from memory without any model. In two other conditions, a model to be imitated was presented (i.e., an accurate performance of the melody by a nonprofessional singer). The melody was sung after the model was presented ("after model") and, in another condition, along with it (i.e., "in unison").

As before, the number of pitch interval errors and pitch interval deviation were computed based on acoustical analyses of the sung renditions. The results showed that pitch accuracy when singing with lyrics was not affected by the fact that imitation occurred after the model or in unison. Nevertheless, the amusic individuals benefited from the imitation of a model as compared to singing from memory. This effect is visible in particular when considering the number of pitch interval errors ($F(1,9) = 13.5$, $P < 0.01$); a tendency is found with pitch interval deviation (see Fig. 3). Remarkably, for 6 of 11 amusics, imitation led to pitch productions as stable as those observed in controls.[39] When singing on a syllable, only five amusics were able to produce complete performances (i.e., the full set of 32 notes), confirming what we observed in the previous study.[16] The six amusics who could not sing complete performances were the most impaired in terms of incidental memory, as indicated by the MBEA. Interestingly, these individuals sang more notes when they imitated a model, in particular when singing in unison. In sum, these findings confirm that poor-pitch singing is often associated with impaired memory. Providing a memory support (i.e., a model to be imitated) aids congenital amusics to sing more accurately, in particular for individuals exhibiting a faulty memory system.

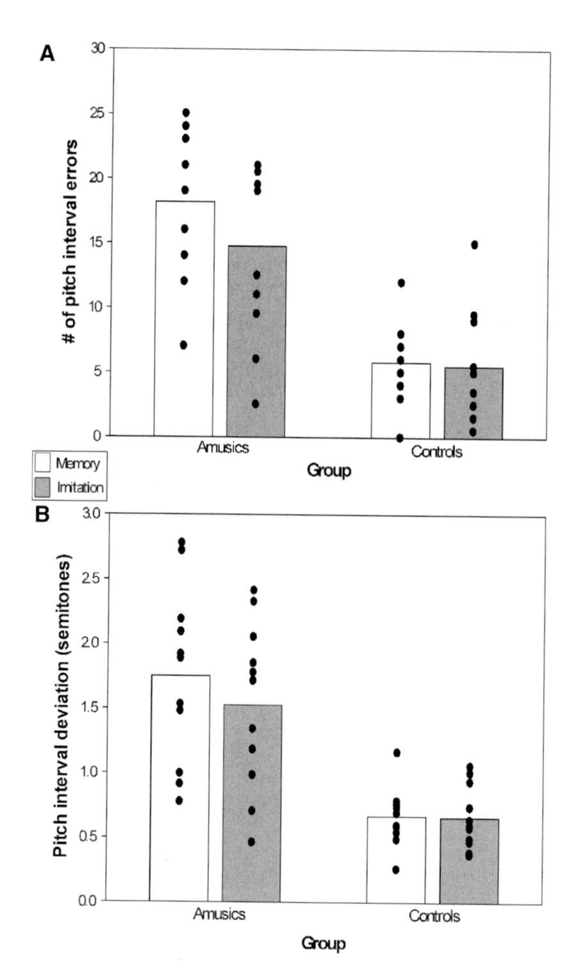

Figure 3. Mean (A) number of pitch interval errors and (B) pitch interval deviations for congenital amusics ($n = 11$) and controls ($n = 11$) when singing from memory (*Memory* condition) and when imitating a model (*imitation*, average of the *After Model* and *in Unison* conditions). The maximum possible number of errors was 31.

Conclusions

In this article, we sought to briefly illustrate the mechanisms underlying accurate singing using the VSL, and to review the existing explanations of poor-pitch singing. The most popular causes of poor-pitch singing rely on impoverished perception (the perceptual account) or disrupted auditory-motor mapping (the sensorimotor account). Less attention has been paid to memory disorders. Here, we reviewed recent findings showing that poor-pitch singing is often associated with memory disorders by studying pitch accuracy in individuals with congenital amusia. Note that this does not indicate that a faulty memory system is the primary cause of poor-pitch singing. Some congenital amusics (i.e., with dramatically impaired pitch perception), albeit they do not exhibit major memory disorders, are still poor-pitch singers.

Because poor-pitch singers often have weaker memory traces, providing a memory support (e.g., asking them to imitate a model) is expected to alleviate the retrieval of melodic information. We provided compelling evidence that this is true, by showing that singing after a model is presented, or in unison, enhances pitch accuracy in poor-pitch singers. Thus, a memory aid can act as a mediator for improving pitch accuracy in poor-pitch singers, a possibility that should be thoroughly examined in future studies.

Acknowledgments

This research was supported by a grant from the European Commission to S.D.B., and by grants from the Natural Sciences and Engineering Research Council of Canada, the Canadian Institutes of Health Research, and a Canada Research Chair to I.P.

Conflicts of interest

The authors declare no conflicts of interest.

References

1. Dalla Bella, S., J-F. Giguère & I. Peretz. 2007. Singing proficiency in the general population. *J. Acoust. Soc. Am.* **121:** 1182–1189.
2. Pfordresher, P.Q. & S. Brown. 2007. Poor-pitch singing in the absence of 'tone-deafness'. *Music Percept.* **25:** 95–115.
3. Dalla Bella, S. & M. Berkowska. 2009. Singing proficiency in the majority: normality and "phenotypes" of poor singing. *Ann. N.Y. Acad. Sci.* **1169:** 99–107.
4. Pfordresher, P.Q., S. Brown, K.M. Meier, *et al.* 2010. Imprecise singing is widespread. *J. Acoust. Soc. Am.* **128:** 2182–2190.
5. Ostwald, P.F. 1973. Musical behavior in early childhood. *Dev. Med. Child Neurol.* **15:** 367–375.
6. Papoušek, H. 1996. Musicality in infancy research: biological and cultural origins of early musicality. In *Musical Beginnings*. I. Deliège & J. Sloboda, Eds.: 37–55. Oxford University Press, Oxford.
7. Welch, G.F. 2006. Singing and vocal development. In *The Child as Musician: A Handbook of Musical Development*. G. McPherson, Ed.: 311–329. Oxford University Press, New York.
8. Huron, D. 2001. Is music an evolutionary adaptation? *Ann. N.Y. Acad. Sci.* **930:** 43–61.
9. Mithen, S. 2006. *The Singing Neanderthals*. Harvard University Press, Cambridge, MA.
10. Wallin, N.L., B. Merker & S. Brown. 2000. *The Origins of Music*. MIT Press, Cambridge, MA.
11. Dalla Bella, S., M. Berkowska & J. Sowiński. 2011. Disorders of pitch production in tone deafness. *Front. Psychol.* **2:** 164.
12. Hutchins, S. & I. Peretz. 2011. A frog in your throat or in your ear? Searching for the causes of poor singing. *J. Exp. Psychol. Gen.*, in press. doi: 10.1037/a0025064
13. Cuddy, L.L., L-L. Balkwill, I. Peretz, *et al.* 2005. Musical difficulties are rare. A study of "tone deafness" among university students. *Ann. N.Y. Acad. Sci.* **1060:** 311–324.
14. Berkowska, M. & S. Dalla Bella. 2009. Acquired and congenital disorders of sung performance: a review. *Adv. Cogn. Psychol.* **5:** 69–83.
15. Dalla Bella, S. & M. Berkowska. 2009. Singing and its neuronal substrates: evidence from the general population. *Contemp. Music Rev.* **28:** 1–13.
16. Dalla Bella, S., J-F. Giguère & I. Peretz. 2009. Singing in congenital amusia. *J. Acoust. Soc. Am.* **126:** 414–424.
17. Loui, P., F. Guenther, C. Mathys, *et al.* 2008. Action-perception mismatch in tone-deafness. *Curr. Biol.* **18:** R331–R332.
18. Griffiths, T.D. 2008. Sensory systems: auditory action streams? *Curr. Biol.* **18:** R387–R388.
19. Brown, S., M.J. Martinez, D.A. Hodges, *et al.* 2004. The song system of the human brain. *Cogn. Brain Res.* **20:** 363–375.
20. Brown, S., E. Ngan & M. Liotti. 2008. A larynx area in the human motor cortex. *Cereb. Cortex* **18:** 837–845.
21. Dronkers, N.F. 1996. A new brain region for coordinating speech articulation. *Nature* **384:** 159–161.
22. Kleber, B., N. Birbaumer, R. Veit, *et al.* 2007. Overt and imagined singing of an Italian aria. *Neuroimage* **36:** 889–900.
23. Zarate, J.M. & R.J. Zatorre. 2008. Experience-related neural substrates involved in vocal pitch regulation during singing. *Neuroimage* **40:** 1871–1887.
24. Turkeltaub, P.E., G.F. Eden, K.M. Jones, *et al.* 2002. Meta-analysis of the functional neuroanatomy of single-word reading: method and validation. *Neuroimage* **16:** 765–780.
25. Perry, D.W., R.J. Zatorre, M. Petrides, *et al.* 1999. Localization of cerebral activity during simple singing. *Neuroreport* **10:** 3979–3984.

26. Gunji, A., R. Ishii, W. Chau, *et al.* 2007. Rhythmic brain activities related to singing in humans. *Neuroimage* **34:** 426–434.
27. Özdemir, E., A. Norton & G. Schlaug. 2006. Shared and distinct neural correlates of singing and speaking. *Neuroimage* **33:** 628–635.
28. Hickok, G., B. Buchsbaum, C. Humpries, *et al.* 2003. Auditory-motor interaction revealed by fMRI: speech, music, and working memory in area Spt. *J. Cogn. Neurosci.* **15:** 673–682.
29. Pa, J. & G. Hickok. 2008. A parietal-temporal sensory-motor integration area for the human vocal tract: evidence from an fMRI study of skilled musicians. *Neuropsychologia* **46:** 362–368.
30. Mandell, J., K. Schulze & G. Schlaug. 2007. Congenital amusia: an auditory-motor feedback disorder. *Restor. Neurol. Neurosci.* **25:** 323–334.
31. Loui, P., D. Alsop & G. Schlaug. 2009. Tone deafness: a new disconnection syndrome? *J. Neurosci.* **29:** 10215–10220.
32. Ayotte, J., I. Peretz & K. Hyde. 2002. Congenital amusia: a group study of adults afflicted with a music-specific disorder. *Brain* **125:** 238–251.
33. Tillmann B., K. Schulze & J.M. Foxton. 2009. Congenital amusia: a short-term memory deficit for nonverbal, but not verbal sounds. *Brain Cogn.* **71:** 259–264.
34. Gosselin, N., P. Jolicoeur & I. Peretz. 2009. Impaired memory for pitch in congenital amusia. *Ann. N.Y. Acad. Sci.* **1169:** 270–272.
35. Williamson, V.J., C. McDonald, D. Deutsch, *et al.* 2010. Faster decline of pitch memory over time in congenital amusia. *Adv. Cogn. Psychol.* **6:** 15–22.
36. Wise, K.J. & J.A. Sloboda. 2008. Establishing an empirical profile of self-defined 'tone deafness': perception, singing performance and self-assessment. *Musicae Scientiae* **12:** 3–23.
37. Peretz, I., S. Champod & K. Hyde. 2003. Varieties of musical disorders: the Montreal Battery of Evaluation of Amusia. *Ann. N.Y. Acad. Sci.* **999:** 58–75.
38. Berkowska, M. & S. Dalla Bella. 2009. Reducing linguistic information enhances singing proficiency in occasional singers. *Ann. N.Y. Acad. Sci.* **1169:** 108–111.
39. Tremblay-Champoux, A., S. Dalla Bella, J. Phillips-Silver, *et al.* 2010. Singing proficiency in congenital amusia: imitation helps. *Cogn. Neuropsychol.* **26:** 463–476.

Ann. N.Y. Acad. Sci. ISSN 0077-8923

ANNALS OF THE NEW YORK ACADEMY OF SCIENCES
Issue: *The Neurosciences and Music IV: Learning and Memory*

Is there potential for learning in amusia? A study of the effect of singing intervention in congenital amusia

Susan Anderson,[1] Evangelos Himonides,[2] Karen Wise,[3] Graham Welch,[2] and Lauren Stewart[1]

[1]Department of Psychology, Goldsmiths, University of London, London, United Kingdom. [2]Institute of Education, University of London, London, United Kingdom. [3]Faculty of Music, University of Cambridge, Cambridge, United Kingdom

Address for correspondence: Lauren Stewart, Goldsmiths, University of London-Psychology, Whitehead Building, New Cross, SE14 6NW, United Kingdom. l.stewart@gold.ac.uk

Congenital amusia is a neurodevelopmental disorder of musical perception and production. Much research has focused on characterizing the deficits within this special population; however, it is also important from both a psychological and educational perspective to determine which aspects of the disorder may be subject to change because this will also constrain theorizing about the nature of the disorder, as well as facilitating possible future remediation programs. In this small-scale study, a professional singing teacher used a broad-brush intervention approach with five individuals diagnosed with congenital amusia. The compensatory elements were designed to enhance vocal efficiency and health, singing technique, musical understanding, pitch perception, and production. Improvements were observed in most individuals in perception, indexed via the Montreal Battery for the Evaluation of Amusia scale subtest and in the vocal performance of familiar songs. The workshop setting gave a unique opportunity for observation and discussion to inform further investigations of this disorder.

Keywords: congenital amusia; poor-pitch singing; intervention

Introduction

Congenital amusia is a developmental disorder that impacts negatively on the perception of music, which has been estimated to occur in about 4% of otherwise normal individuals.[1] In everyday life, these individuals have difficulties in detecting anomalous pitches in melodies, judging dissonance in musical excerpts, and recognizing melodies without lyrics.[2] These deficits are diagnosed and defined in perceptual terms using the Montreal Battery of Evaluation of Amusia (MBEA).[1] Additional features reported in these individuals are higher thresholds than controls for the discrimination of pitch direction,[3] and pitch memory deficits.[3,4]

The pitch production of those diagnosed with congenital amusia is less commonly assessed, but there is a general picture of impairment: nearly all amusics are unable to sing in tune.[2] However, clear differentiation is made in the literature between congenital amusia and "tone-deafness," a self-reported problem that almost always carries the meaning of "can't sing."[5] Poor-pitch singing is relatively common; about 15% of typical populations consider themselves "tone-deaf," whereas musical perceptual difficulties as diagnosed by the MBEA occur less frequently.[6] The phenomenon of poor-pitch singing without accompanying perceptual deficits has been referred to as "purely vocal tone-deafness."[7] Poor-pitch singing has been studied in educational terms since the 19th century from both perceptual and production perspectives.[8] Production deficits may be attributable to a variety of causes, both physical and cognitive, in addition to perceptual causes, such as poor vocal-motor planning and adjustment, lack of tonal schema, reduced vocal flexibility, misunderstanding of speech/singing differentiation,[9] and is most likely to arise from a stagnation of vocal development arising from childhood self-perception as a "nonsinger."[10]

Counterintuitively, proficient singing despite impaired perception has been noted in several studies of amusia.[2] A mismatch between implicit and explicit pitch awareness has been observed in behavioral and neuroimaging studies,[12] leading to the possibility that there may be residual implicit pitch abilities in the amusic population. These factors, coupled with the confirmation that neural plasticity

doi: 10.1111/j.1749-6632.2011.06404.x

is now known to be lifelong,[13] and that any potential improvements in production may also benefit perception according to premotor theories of perception,[14] motivated a singing intervention study in which both perception and production were monitored. From a pedagogical perspective, many singing teachers have anecdotal evidence of major improvements in singing ability with very challenged singers, and are often of the opinion that anyone can be taught to sing.

In this study, a wide range of activities were used with five individuals with congenital amusia, in an attempt to improve both pitch production and potentially also perception. The program was devised and delivered in seven weekly singing workshops by a professional singing teacher with many years of experience working with poor-pitch singers. It focused on vocal flexibility, musical awareness, development of tonal schema, and used "Sing and See,"[15] a real-time visual feedback program,[16] that allowed participants to visualize the pitches they produced. The five amusic participants were matched with five nonamusic controls on gender, age, years of education, and years of musical training; all participants were asked to complete a "musical experience" chart, highlighting positive and negative musical experiences during their lives. Although the controls were assessed only once, the amusics were assessed pre- and postintervention.

Hypothesis

It was hypothesized that an intervention program, in the form of singing workshops, would improve the pitch production abilities and potentially also perception abilities of the amusic participants.

Method

Design

The design was a classic intervention study, with assessments made before and after a group of five amusic participants had taken part in seven weekly singing workshops. It was both a within-subjects design, comparing assessment results pre- and postintervention, and a between-subjects design, as the amusic group results, preintervention, were also compared with those of matched controls.[1]

Participants

Five amusic female adults (mean age 54.8) and five female matched controls (mean age 58.0) participated in the study. Participants were drawn from a pool of amusics and controls recruited via an online version of the scale subtest from the MBEA, and subsequently tested on four subtests of the battery under controlled laboratory conditions. Participants had also undergone audiometry tests, as well as completing National Adult Reading Test and digit span tests. Status as amusic or nonamusic was based on a composite score across the three pitch-based subtests of MBEA.[1] A summary of relevant participant characteristics is given in Table 1. Informed consent was obtained from all participants.

Materials and procedure

Assessment sessions

All testing took place in a sound-attenuated booth. The amusic participants were tested individually across two assessment sessions, once before the intervention workshops began, and once after the last workshop session. The control group was tested in a single assessment session. All assessment sessions followed a protocol devised previously by an experienced singing teacher.[17] Each session began with basic vocal tasks such as exaggerated speech patterns, glides, and stepwise exploration of vocal range, which served as a warm-up and familiarized participants with the researcher and the testing situation.

The actual assessment sessions included both perception and production tasks. In terms of perception, performance was measured on the MBEA scale subtest (online) as well as on a computer-based pitch-matching task (details below). In terms of production, singing accuracy was measured using a standard, well-known song ("Happy Birthday") as well as a self-chosen song, and the ability to imitate and match sequences of tones, which ranged in length from one to five notes.

Computer pitch-matching task

The computer pitch-matching task (CPM) was designed to allow participants to engage in active pitch matching, but without the requirement to use the voice. They were instructed to use two keys on a computer keyboard to play first a "target" tone and then a "comparison" tone, presented through headphones. By turning a dial clockwise to raise pitch and anticlockwise to lower it, the pitch of the comparison note could be brought into alignment with that of the target tone. Participants were able to

Table 1. Matching of five amusics with five controls on all preintervention measures

Group	N		Age (years)	Education (years)	Musical training (years)	MBEA pitch composite (out of 90)	CPM (cents deviation)	Song rating average (1–8)	VPM average (cents deviation)
Amusic	5	μ	53.80	16.60	2.80	55.20	25.66	3.04	256
		σ	9.70	0.89	4.70	6.98	9.45	1.50	136
Control	5	μ	57.00	15.80	0.80	81.20	33.83	6.37	43
		σ	3.80	1.64	1.30	6.98	27.40	0.58	7
		U	10.50	10.00	9.50	<0.001	10.00	0.50	<0.001
		Z	−0.42	−0.57	−0.67	−2.61	−0.52	−2.53	−2.61
		P	0.69	0.69	0.55	0.01	0.69	0.01	0.01
		r	−0.39	−0.18	−0.19	−0.83	−0.17	−0.80	−0.83

play only one tone at any one time, but they could switch between target and comparison notes at will. Target notes spanned the chromatic scale from G^3 to G^4, and comparison tones were a semitone (100 cents), major 3rd (400 cents), or a perfect 5th (700 cents) above or below the target note. The pitch adjustments made by turning the dial were sampled approximately every 93 milliseconds. The difference (in cents) between the final pitch of the comparison tone and the target tone was calculated as a measure of perceptual pitch-matching accuracy.

Singing tasks

Songs. Participants were asked in advance of the first assessment to think of a song that they knew well, which they would be willing to sing to the researcher. No starting note was given for their own choice song, allowing each participant to choose a comfortable singing range. This was followed by two performances of "Happy Birthday," first without, then with, accompaniment. The starting pitch chosen for the accompanied version was dependent on the vocal range established in the warm-up, and the accompaniment added simple harmony as well as doubling the vocal line. Sung performances were recorded using Audacity, extracted from the Audacity files, and copied in random order onto CD. Evaluation was carried out by two independent experts, using an accuracy rating scale developed for use with adults by Wise and Sloboda.[17]

Vocal pitch matching (echo and synchronize conditions). The stimuli used for this task were taken from a previous study,[17] using single pitches and meaningful melodic fragments. The stimuli, recorded by an accurate but untrained female singer, were heard in two separate blocks, each consisting of single notes, pairs of notes, groups of three notes, and finishing with groups of five-note patterns. In the first block (echo condition), the participant heard the stimulus and was asked to sing it back in their own time; in the second block (synchronize condition), a click track cued the participant's synchronized response. All notes and patterns were in D major, in the range B^3 to B^4, using simple tuneful relationships and avoiding any single interval larger than a perfect 5th. They were analyzed for pitch accuracy using Praat, with the frequency of the steadiest portion of each note being recorded.

Intervention phase

The five participants met with the researcher (SA) for seven weekly group-singing workshops, each lasting one and a half hours, with worksheets and CDs providing guided practice between sessions. The structure and content of the workshops was designed to remediate deficits identified from the literature and from pedagogical experience as possibly underlying the phenomenon of amusia. They were intended to be relaxed, fun, nonjudgmental, vocally and intellectually stimulating, and aimed to improve vocal confidence and efficient voicing—regardless of the research outcome.

All workshops began with vocal warm-ups that were not pitch specific, such as lip-flutters (motorbike sounds), glides on /u/, and a "fun" physical and vocal warm-up that incorporates the basic children's chant featuring the falling minor 3rd (scale pitches 53 653). Elements that contribute to strong, efficient voicing were reinforced, such as use of the transverse abdominis muscles to support the vocal sound, derived from Accent Method techniques,[18] and an open, unconstricted laryngeal posture.[19] Simple vocal anatomy was introduced, with anatomical and physiological explanations given whenever appropriate. An additional intervention to simulate pitch perception was the playing "by ear" of simple well-known melodies on pianos/keyboards in the department, and free composition.

Between sessions, participants were asked to practice three or four times a week, for about 15 min per session, using specially prepared CDs and progressive exercises with Sing and See. The final feedback forms indicated compliance with these requirements. The exercises progressed from glides throughout the vocal range, to steady sounds on specified "target pitches" around middle C (C^4), a comfortable note for all participants. The descending minor 3rd (5–3) was introduced in the children's chant mentioned above, before descending major triads, 5 3 1. The use of numbers was used consistently to encourage awareness of pitch relationships.

Results

Owing to the small number of participants involved in this study (five participants in each group), nonparametric statistics were appropriate; the Mann-Whitney test for two unrelated samples was used to compare amusic participants' preintervention with controls, and the Wilcoxon signed-rank test was used for pre- and postintervention comparison. In both cases, effect sizes can be more meaningfully interpreted than P values.[20] Following group comparisons, first amusic preintervention (amusic1) versus controls, and then amusics pre- versus postintervention (amusic1 and amusic2), we focus on some of the individual profiles within the amusic group data.

Preintervention amusics and controls

Perception

By definition, the MBEA pitch composite score was significantly different between the amusic and control groups before intervention (amusic1 [Mdn = 56] and controls [Mdn = 83] giving $U = 0.000$, $P = 0.008$, $r = 0.83$ [large effect size]). In contrast, there was a lack of group differences on the CPM task (amusics Mdn = 31.4, controls Mdn = 27.1, $U = 10$, $P = 0.65$, $r = -0.17$ [small effect size]; Table 1).

Production

In all singing tasks, controls scored significantly higher than the amusic group before intervention ($U = 0.5$, $z = -2.530$, $P = 0.008$, $r = 0.8$ [large effect size]), although one amusic (HR) showed a level of accuracy that was almost comparable to that seen in the control group. In vocal pitch matching, amusics were significantly less accurate than the control group in both conditions, displaying a wider spread of scores and, in the echo condition, there was a greater decrement in performance as the sequences increased in length (Fig. 1).

Amusic results pre- and postintervention

Perception

In comparing the online scale subtest of the MBEA before and after the intervention period, we observed no significant improvement at a group level, but important individual changes that contributed to a large effect size (amusic1 Mdn = 17, amusic2 Mdn = 21, $z = 1.5$, $P = 0.19$, $r = -0.51$ [large effect size]). Improvement on the scale subtest was seen in all but one amusic participant. There were no significant changes in the computerized pitch-matching scores, which, even preintervention, were similar to the control scores (Table 2).

Production

The intervention resulted in the greatest improvements for the song ratings, seen in all songs (own choice, "Happy Birthday" solo, and "Happy Birthday" accompanied). In terms of changes in performance at a group level, performance of one's own choice song yielded the least improvement, whereas "Happy Birthday" (accompanied) showed the most positive change (Fig. 2). The vocal pitch-matching task showed little significant change, although there was a large effect size nearing significance for the synchronized condition when scores were averaged across all sequence lengths ($z = -1.826$, $P = 0.068$, $r = -0.61$).

Individual profiles

With such a small and heterogeneous group, noting individual differences in results is very relevant for the integrity of the study (Table 2). One of the

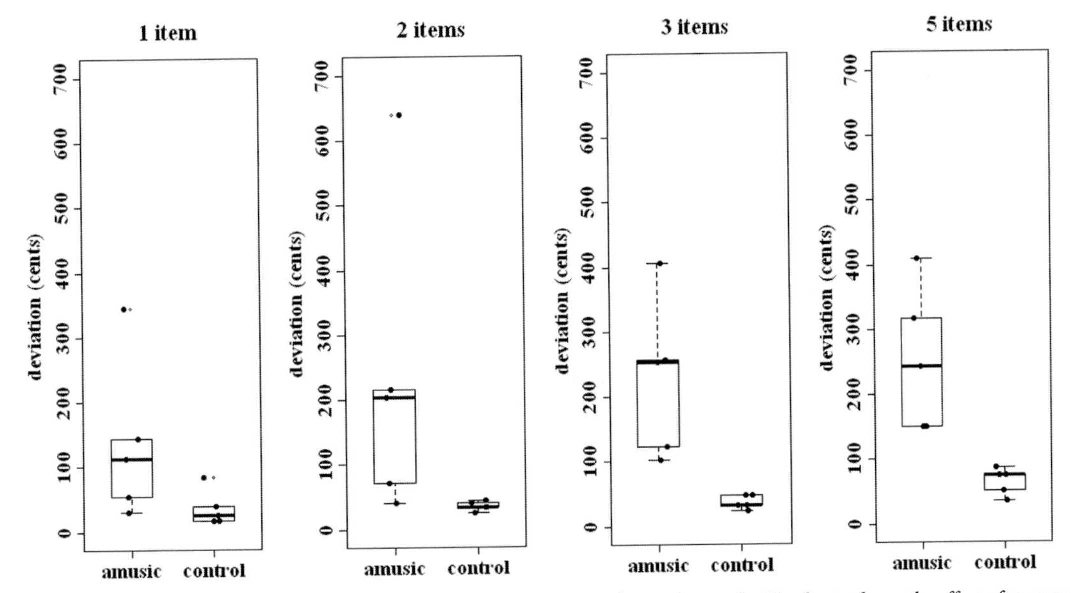

Figure 1. Vocal pitch matching (echo condition) in amusics preintervention and controls. The figure shows the effect of sequence length as deviation in cents from the target pitch.

participants who showed the weakest performance in all baseline conditions (ABL) reported hearing very little music in her preschool years, as did JM and JP (amusics) and one control. After intervention, ABL's online MBEA score improved from 15 to 21, and her song rating for "Happy Birthday" (accompanied) saw the highest single change, from 2/8 to 6/8. JM began from a low MBEA score and the lowest singing rating. After the intervention, modest improvements were made in both conditions of the vocal pitch-matching task, with minimal positive change in song ratings. Participant JP also made several positive changes from a low baseline score; her MBEA online score improved by four points, her song rating average increasing by 0.66, and vocal pitch-matching performance was also more accurate, postintervention, in both conditions.

At the preintervention assessment, HR's production profile bore more resemblance to the control group than the amusic group. Her scores improved in all tasks postintervention, taking her online MBEA score from 23 to 28, increasing her average song ratings from 5.67 to 6.17, and reducing the deviation in VPM from 78 to 42 cents. Her computerized pitch-matching score remained consistent with an average small deviation of 10 cents (1/10 of a semitone). ABK made a small improvement in song ratings, but her performance worsened for both MBEA score and vocal pitch matching. In particular, this participant was unable to complete the vocal pitch-matching task (synchronized condition), remarking that she was distressed by it going "too fast."

Discussion

This study represents an exploratory investigation into the potential for change in musical behavior in a small group of individuals with congenital amusia. Although genetic findings[21,22] indicate that there may be innate factors contributing to the disorder, recent research regarding the potential for learning and plasticity to occur throughout the lifespan[13] suggests that behavioral changes may indeed be possible. Despite the limitations of group size and short intervention period, some positive changes were seen in both perception and/or production abilities in three of the five individuals, implying that congenital amusia may not be entirely immune to environmental modification. Below, we consider the main findings of the study and discuss their relevance to theories of the disorder.

Profiles of performance in congenital amusia

The five participants exhibited different profiles of perception and production at the outset across the measures we assessed. For example, JM showed consistently accurate pitch matching in the

Table 2. Amusics, individual performances on all measures pre- and postintervention

	MBEA online (/30)	CPM (cents deviation)	Song ratings average (1–8)	Own choice Solo (1–8)	Happy Birthday Acc (1–8)	Happy Birthday (1–8)	VPM average (cents deviation)	VPM echo (cents deviation)	VPM sync (cents deviation)
Preintervention									
ABL	15	27	2.67	3.5	2.5	2.0	312	234	390
ABK	21	34	2.67	3.5	2.0	2.5	213	254	173
HR	23	10	5.67	5.0	6.0	6.0	78	103	54
JM	15	26	2.00	2.0	2.0	2.0	447	332	562
JP	17	33	2.17	2.5	2.5	1.5	231	225	238
μ	18.20	25.66	3.04	3.30	3.00	2.80	256.20	229.60	283.40
σ	3.63	9.45	1.50	1.15	1.70	1.82	135.81	82.39	137.39
Postintervention									
ABL	21	32	4.17	3.5	3.0	6.0	313	265	351
ABK	19	41	3.50	4.0	3.0	3.5	357	357	N/A
HR	28	10	6.17	6.0	5.0	6.5	42	47	38
JM	17	26	2.17	2.0	2.0	2.5	315	274	357
JP	21	59	2.83	2.0	4.0	2.5	127	116	138
μ	21.2	28.21	3.77	3.50	3.40	4.20	230.80	211.80	221.00
σ	4.15	11.97	1.54	1.66	1.14	1.92	138.02	126.61	158.93
z	−1.63	−1.10	−2.02	−0.82	−0.92	−2.04	−0.41	−0.67	−1.83
P	0.10	0.27	0.04	0.41	0.36	0.04	0.69	0.50	0.07
r	−0.51	−0.35	−0.64	−0.26	−0.29	−0.65	−0.13	−0.21	−0.61

computerized perception task, but her sung imitation of novel pitch patterns and familiar songs was poor, whereas ABK had poor pitch perception, coupled with poor sung performance. In contrast, participant HR performed equally as well as controls in all measures. Such variance in profile is typical across developmental disorders in general and suggests that different individuals may benefit from different aspects of a broad-brush intervention approach. Overall, there was less change in perception measures pre- and postintervention, whereas production scores improved for all participants in the song ratings, and for some in the vocal pitch-matching task as well.

MBEA and CPM tasks

The finding of improved scores on the scale subtest of the MBEA in all but one participant is noteworthy. Although HR scored 23 (above the cutoff for that particular subtest preintervention),[1] she was considered amusic on the basis of a composite score across all three pitch subtests. The average change in score was three points, with participant ABL making a substantial positive change of six points; HR, five points; JP, four points; and ABK making a negative change of two. Although a small increase in score may be attributed to a test–retest factor, such effects are typically modest and would be unable to account for the magnitude of improvement seen in three participants. As the test is specifically found to be suitable to "measure evolution of performance after recovery or therapy,"[1] these positive changes can be viewed as individual improvements in musical perception.

The computerized pitch-matching program was designed to test the ability of participants to detect, compare, and match small differences in pitch, while minimizing the potential impact of memory. Accurate performance in individuals with amusia is consistent with findings of typical pitch detection thresholds (approximately 0.2 semitones[4]) in the context of psychophysical tasks requiring participants to indicate which of three tones is the "odd one out."[3] This study showed that individuals with

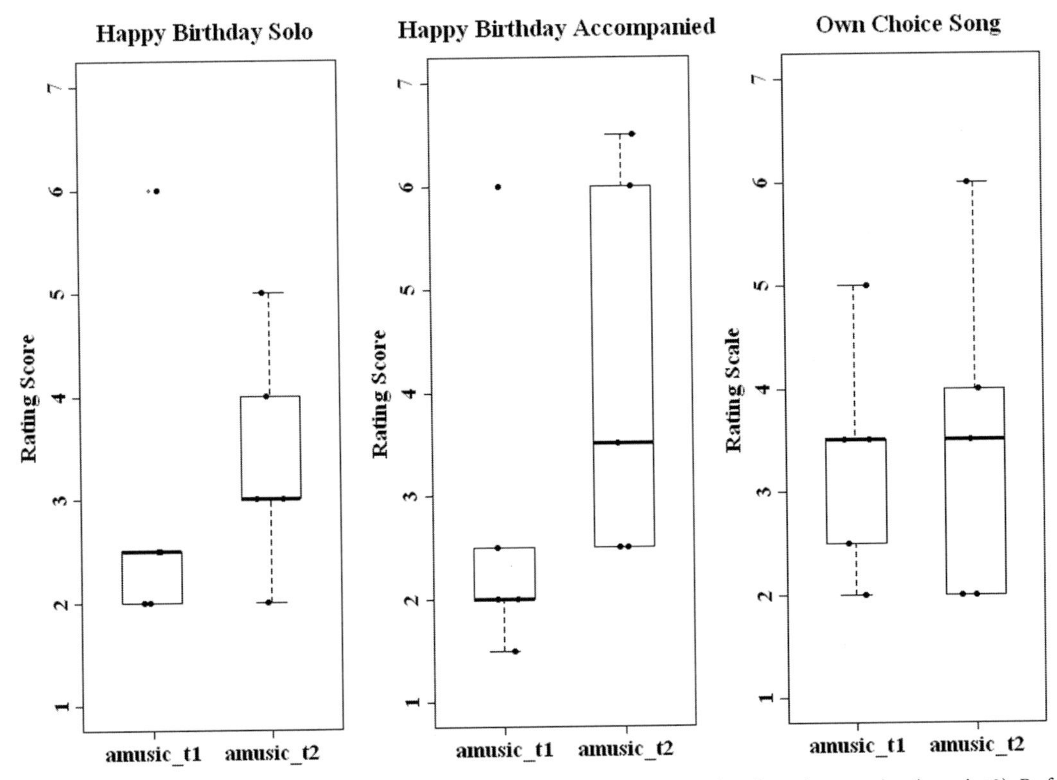

Figure 2. Ratings for the singing of amusic participants, preintervention (amusic t1) and postintervention (amusic t2). Performances were rated on a scale of 1–8, 2 indicating that words and rhythm were mainly correct, but pitches were random and there were errors in contour; and 6 indicating that key is maintained throughout and that the melody is fairly accurately represented.

amusia could not only detect small pitch differences, but could actively match two differing tones, relying on an iterative process of feedback and comparison. Performance on this task did not change significantly following the intervention, being already equal to that of controls.

Singing and vocal imitation tasks

The most notable improvements in production accuracy occurred in line three of "Happy Birthday," a phrase consisting of an octave leap followed by a descending major triad; postintervention, this was sung accurately by ABL in the accompanied condition and JP in the solo condition. These shapes, but not in this context, were featured in the intervention as basic building tools of musical patterns, essential for the development of tonal schema. The improvements reported suggest that this element of musical understanding had been absent before the intervention. The pedagogical expectation that support (either accompaniment in "Happy Birthday" or synchronized sounds in vocal pitch-matching tasks) would improve intonation is true of self-defined "tone-deaf" individuals,[17] although Hutchins *et al.* found that the performance of amusic participants in a pitch-matching trial was not affected by feedback condition.[23] There were mixed responses regarding the role of support, a finding echoing that of Wise,[9] with the two participants above (ABL and JP) consistently having more pitch deviation in vocal pitch matching (synchronized condition) compared with vocal pitch matching (echo condition). Timing constraints in the synchronized condition may have had a negative impact on these individuals in terms of vocal production as they had very little experience with singing control and coordination. This coordination typically develops in childhood, with roughly 30% of children aged seven unable to sing accurately, reducing to about 4% four years later.[7] Development is often curtailed when children are criticized during these formative years, and they then stop taking part in singing activities. Many adults attending groups for nonsingers recall an abiding sense of "deep emotional upset" associated with their childhood singing experiences, subsequently avoiding social situations involving singing.[13] In the feedback document completed after the second assessment session, all amusic participants indicated that Sing and See had been one of the most important elements of the intervention. They also commented informally that they now listened to music instead of ignoring it.

It is notable that three amusic participants, ABL, JM, and JP, reported no preschool musical experience. Although it has been suggested that pitch processing hinges upon the need to experience spectrally and temporally rich sound to wire brain circuits for pitch processing,[24] one cannot assume that this lack of musical input was responsible for a failure to develop musical awareness in the normal way because one of the control groups was also similar in this regard. However, it is certainly possible that musical difficulties arising because of biological factors were exacerbated by the lack of musical stimulation in early childhood. Lack of musical stimulus in childhood may happen for many reasons, including those of parental choice and lack of electricity cited by three of the amusic participants, but seems less likely to occur now than 40 to 50 years ago, when many of the participants were growing up. Indeed, the decreasing presence of congenital amusia in younger generations may be because of the greatly increased "musical stimulation and experience" in contemporary life.[25]

Conclusions

A workshop environment allowed close observation of individuals and their relationship with music and singing over a sustained period of time. The small numbers and heterogeneity of the group made statistical analysis difficult, highlighting individual variations rather than illuminating overall "amusic" group characteristics. However, changes of varying degrees were found in the song production of all amusic participants, and all except one participant improved in their online MBEA score. The research project has suggested both productive compensatory tactics for the singing teacher working with poor-pitch singers, and fruitful lines for further laboratory-based research.

Conflicts of interest

The authors declare no conflicts of interest.

References

1. Peretz, I., A.S. Champod & K. Hyde. 2003. Varieties of musical disorders–The Montreal Battery of Evaluation of Amusia. *Ann. N.Y. Acad Sci.* **999:** 58–75.
2. Dalla Bella, S., J.-F. Giguère & I. Peretz. 2009. Singing in congenital amusia. *J. Acoust. Soc. Am.* **126:** 414–424.
3. Foxton, J.M., J.L. Dean, R. Gee, *et al.* 2004. Characterisation of deficits in pitch perception underlying 'tone deafness'. *Brain* **127:** 801–810.
4. Willliamson, V.J. & L. Stewart. 2010. Memory for Pitch in congenital amusia: beyond a fine-grained pitch discrimination problem. *Memory* **18:** 657–669.
5. Sloboda, J., K.J. Wise & I. Peretz. 2005. Quantifying tone deafness in the general population. *Ann. N.Y. Acad Sci.* **1060:** 255–261.
6. Cuddy, L.L., L.-L. Balkwill, I. Peretz & R.R. Holden. 2005. Musical difficulties are rare: a study of 'tone-deafness' among university students. *Ann. N.Y. Acad Sci.* **1060:** 311–324.
7. Dalla Bella, S., J.-F. Giguère & I. Peretz. 2007. Singing proficiency in the general population. *J. Acoust. Soc. Am.* **121:** 1182–1189.
8. Welch, G.F. 1979. Poor-Pitch singing: a review of the literature. *Psychol. Music* **7:** 50–58.
9. Wise, K.J. 2009. Understanding 'Tone-Deafness': A Multi-Compositional Analysis of Perception, Cognition, Singing and Self-Perception in Adults Reporting Musical Difficulties. Unpublished PhD thesis, Keele University.
10. Welch, G.F. 2006. Singing and vocal development. In *The Child as Musician.* G.E. McPherson, Ed.: 311–329. Oxford University Press. Oxford.
11. Loui, P., F.H. Guentehr, C. Mathys & G. Schlaug. 2008. Action-perception mismatch in tone-deafness. *Current Biology* **18:** R331–R332.
12. Peretz, I., E. Brattico, M. Järvenpää & M. Tervaniemi. 2009. The amusic brain: in tune, out of key, and unaware. *Brain* **132:** 1277–1286.
13. Pellicciari, M.C., C. Miniussi, P.M. Rossini & L. De Gennaro. 2009. Increased cortical plasticity in the elderly: changes in the somato-sensory cortex after paired associative stimulation. *Neuroscience* **163**: 266–276.
14. Brown, S. & M.J. Martinez. 2007. Activation of premotor vocal areas during musical discrimination. *Brain Cogn.* **67:** 59–69.
15. Wilson, P.H., K. Lee, J. Callaghan & C.W. Thorpe. 2008. Learning to sing in tune: does real-time visual feedback help? *J. Interdiscipl. Music Stud.* **2:** 157–172.
16. Welch, G.F., D.M. Howard & C. Rush. 1989. Real-time visual feedback in the development of vocal pitch accuracy in singing. *Psychol. Music* **17:** 146–157.
17. Wise, K.J. & J.A. Sloboda. 2008. Establishing an empirical profile of self-defined 'tone-deafness': perception, singing performance and self-assessment. *Musicae Scientae* **12:** 3–26.
18. Chapman, J.L. 2006. *Singing and Teaching Singing: A Holistic Approach to Classical Voice.* Plural Publishing Inc. San Diego, Oxford, Brisbane.
19. Shewell, C. 2009. *Voice Work: Art and Science in Changing Voices.* Wiley-Blackwell. Chichester.

20. Field, A. 2005. *Discovering Statistics Using SPSS*. Sage Publications. London, Thousand Oaks, New Delhi.
21. Drayna, D., A. Manichaikul, M. de Lange, *et al.* 2001. Genetic correlates of musical pitch recognition in humans. *Science* **291:** 1969–1972.
22. Peretz, I., S. Cummings & M. Dubé. 2007. The genetics of congenital amusia (tone deafness): a family aggregation study. *Am. J. Hum. Genet.* **81:** 582–588.
23. Hutchins, S., J.M. Zarate, R.J. Zatorre & I. Peretz. 2010. An acoustical study of vocal pitch matching in congenital amusia. *J. Acoust. Soc. Am.* **127:** 504–512.
24. Trainor, L.J. 2005. Are there critical periods for musical development? *Dev. Psychobiol.* **46:** 262–278.
25. Peretz, I. 2008. Musical disorders: from behaviour to genes. *Curr. Direct. Psychol. Sci.* **17:** 329–333.

Ann. N.Y. Acad. Sci. ISSN 0077-8923

ANNALS OF THE NEW YORK ACADEMY OF SCIENCES
Issue: *The Neurosciences and Music IV: Learning and Memory*

Impaired learning of event frequencies in tone deafness

Psyche Loui and Gottfried Schlaug
Beth Israel Deaconess Medical Center and Harvard Medical School, Boston, Massachusetts

Address for correspondence: Psyche Loui, Beth Israel Deaconess Medical Center and Harvard Medical School, 330 Brookline Ave, Palmer 127, Boston, MA 02215. ploui@bidmc.harvard.edu

Musical knowledge is ubiquitous, effortless, and implicitly acquired all over the world via exposure to musical materials in one's culture. In contrast, one group of individuals who show insensitivity to music, specifically the inability to discriminate pitches and melodies, is the tone-deaf. In this study, we asked whether difficulties in pitch and melody discrimination among the tone-deaf could be related to learning difficulties, and, if so, what processes of learning might be affected in the tone-deaf. We investigated the learning of frequency information in a new musical system in tone-deaf individuals and matched controls. Results showed significantly impaired learning abilities in frequency matching in the tone-deaf. This impairment was positively correlated with the severity of tone deafness as assessed by the Montreal Battery for Evaluation of Amusia. Taken together, the results suggest that tone deafness is characterized by an impaired ability to acquire frequency information from pitched materials in the sound environment.

Keywords: statistical learning; frequency; probability; tone deafness; amusia; music; pitch

Introduction

Music is celebrated across all cultures and across all ages. Sensitivity to the basic principles of music is present regardless of formal musical training. The ubiquity of music begs the question of where our knowledge in music might originate at multiple levels of analysis:[1] At a computational level, what types of information are represented in music? At an algorithmic level, how is music represented? And at an implementation level, what are the brain structures that give rise to musical knowledge?

While these questions get at the core of music cognition and neuroscience, more generally, answers have been offered at many levels as well. Most would agree that part of musical competence is the ability to perceive pitch,[2] which is at least partially disrupted in tone-deaf (TD) individuals.[3,4] But we also know that pitches do not exist in isolation, but are strung together with different frequencies and probabilities that give rise to structural aspects of music such as melody and harmony. Thus, the frequencies and probabilities that govern the cooccurrences of pitches are extremely important toward our understanding of the source of musical knowledge, and may offer a unified view between the constructs of pitch, melody, and harmony.

In the attempt to understand how knowledge of various musical constructs is acquired, researchers have pursued developmental approaches[5,6] as well as cross-cultural approaches.[7,8] However, both of these approaches are difficult to interpret: development unfolds with huge variability due to maturational constraints as well as differences in the complexities of environmental exposure, whereas cross-cultural differences may arise from multiple historical, cultural, and genetic and environmental causes. Even within the controlled environment of the laboratory, most test subjects have already had so much exposure to Western music that even people without musical training show implicit knowledge of the frequencies and probabilities of Western musical sounds as evidenced by electrophysiological studies.[9–11] To study the source of musical knowledge as it emerges *de novo*, we need a new system of pitches with frequencies and probabilities that are different from Western music, but that are not confounded by long-term memory and other factors. A

doi: 10.1111/j.1749-6632.2011.06401.x

new musical system would give us a high degree of experimental control, so that we can systematically manipulate the frequencies and probabilities within subjects' environment, and then compare TD and control subjects to see how TD individuals might differ in the way they learn the statistics of music.

To that end, in the past few years we have developed a new musical system based on an alternative musical scale known as the Bohlen–Pierce scale.[4,12,13] The Bohlen–Pierce scale is different from the existing musical systems in important ways. Musical systems around the world are based on the octave, which is a 2:1 frequency ratio; thus, an octave above a tone that is 220 Hz is 2×220 Hz $= 440$ Hz. The equal-tempered Western musical system divides the 2:1 frequency ratio into 12 logarithmically even steps, resulting in the formula $F = 220 \times 2^{(n/12)}$, where n is the number of steps away from the starting point of the scale (220 Hz) and F is the resultant frequency of each tone. In contrast, the Bohlen–Pierce scale is based on the 3:1 frequency ratio, so that one "tritave" (instead of one octave) above 220 Hz is 3×220 Hz $= 660$ Hz, and within that 3:1 ratio there are 13 logarithmically even divisions, resulting in the formula $F = 220 \times 3^{(n/13)}$. The 13 divisions were chosen such that dividing the tritave into 13 logarithmically even divisions results in certain tones, such as 0, 6, and 10 (each of which plugs into n in the Bohlen–Pierce scale equation to form a tone with a single frequency), which form approximate low-integer ratios relative to each other, so that when played together these three tones sound psychoacoustically consonant, i.e., "smooth" like a chord. Together, the numbers 0, 6, and 10 form the first chord. Three other chords can follow this first chord to form a four-chord progression. We can use this chord progression to compose melodies by applying rules of a finite-state grammar:[13] given the 12 chord tones that represent nodes of a grammar, each tone can either repeat itself, move up or down within the same chord, or move forward toward any note in the next chord. This can be applied to all of the 12 tones in the chord progression, resulting in thousands of possible melodies that can be generated in this new musical environment. We can also conceive of an opposite but parallel musical environment where we also have four chords, but they are in retrograde (reversed order). These two musical environments are equal in event frequency—that is, each note happens the same number of times—but are different in conditional probability, in that the probability of each note given the one before it is completely different.

Having defined a robust system with which we can compose many possible melodies, we can now ask a number of questions with regard to music learning in TD individuals and controls. The first question we ask is: Can TD subjects learn from event frequencies of pitches in their sound environment? Frequency, also known as zero-order probability, represents the number of events in a sequence of stimuli. As an example in language, the word "the" has the highest frequency in the English language. In contrast to first-order and higher-order probabilities, which refer to the probabilities of events given other events that occur before them, zero-order probability is simply a count of the occurrence of each event given a corpus of input.

Frequency information is an important source of musical knowledge[14] as well as linguistic knowledge and competence.[15,16] Sensitivity to frequency and probability in music has been demonstrated using behavioral[17] and electrophysiological techniques.[12,18] In particular, the probe tone paradigm is a reliable behavioral indicator of sensitivity or implicit knowledge of event frequencies in music, so much so that it has been described as a functional listening test for musicians.[19] In this study, we compared frequency learning performance between TD subjects and controls by using the probe tone test[17] adapted for the new musical system.[13]

Methods

Subjects

Sixteen healthy volunteers, eight TD individuals and eight non-tone-deaf (NTD) controls, were recruited from schools and online advertisements from the Boston area for this study. The two groups were matched for age (mean $\pm$ SE: TD $= 27 \pm 1.0$, NTD $= 26 \pm 2.1$), sex (TD = five females, NTD = four females), amount of musical training (TD $= 1.8 \pm 1.3$ years, NTD $= 1.3 \pm 0.8$ years), and scaled IQ as assessed by the Shipley scale of intellectual functioning[20] (TD $= 117 \pm 2.6$, NTD $= 118 \pm 2.0$). However, pitch discrimination, as assessed by a three-up one-down staircase procedure around the center frequency of 500 Hz, showed significantly higher thresholds in the TD group (TD $= 24 \pm 10.7$ Hz, NTD $= 11 \pm 3.4$ Hz), and performance on the Montreal Battery for Evaluation of Amusia (MBEA) was

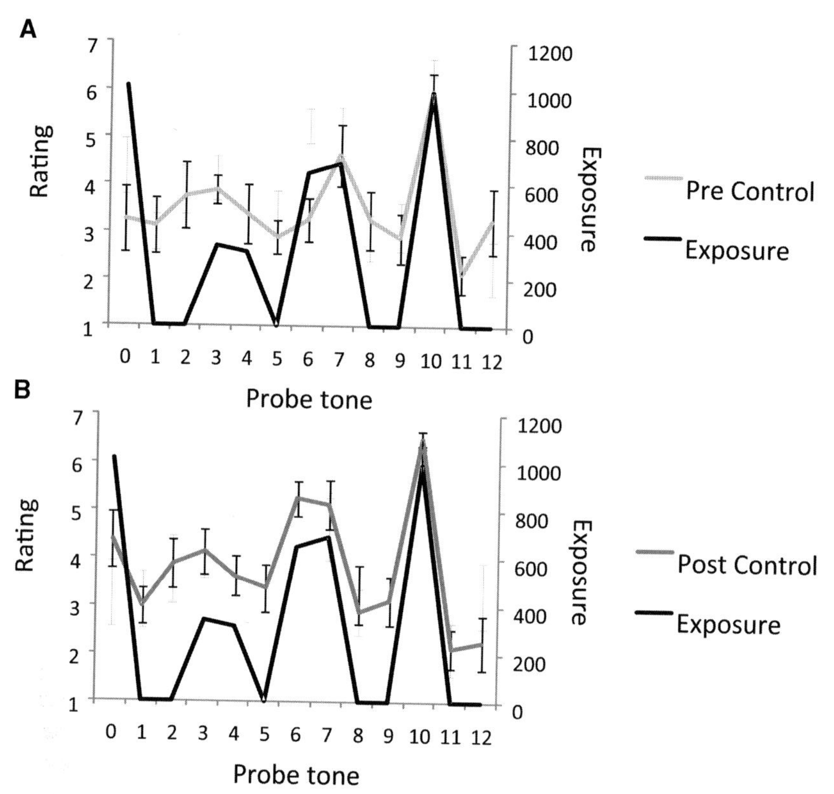

Figure 1. Probe tone ratings and exposure profiles in control subjects. (A) Pre-exposure. (B) Postexposure. Ratings are plotted as a function of probe tone, where each of the tones corresponds to a single note in the Bohlen–Pierce scale. In both panels, the black line is a histogram of what subjects heard during the exposure phase (i.e., the "ground truth"). The gray line is subjects' averaged ratings. Error bars indicate ±1 between-subject standard error. The postexposure ratings are highly correlated with the ground truth, suggesting that within half an hour of exposure, subjects acquired sensitivity to the frequency structures of the new musical system.

significantly impaired in the TD group (average performance on the three melody subtests (scale, interval, and contour): TD = 63 ± 3%, NTD = 85 ± 2%), thus confirming that the TD group was impaired in pitch and melody discrimination.

Stimuli

All auditory stimuli were tones in the new musical system, which were based on the Bohlen–Pierce scale as outlined in the introduction (see also Ref. 13). Each tone was 500 ms, with rise and fall times of 5 ms each. Melodies consisted of eight tones each, and a 500 ms silent pause was presented between successive melodies during the exposure phase. All test and exposure melodies were generated and presented using Max.[21]

Procedure

The experiment was conducted in three phases: pretest, exposure, and posttest. The pre-exposure test was conducted to obtain a baseline level of performance prior to exposure to the new musical system. This was followed by a half-hour exposure phase, during which subjects heard 400 melodies in the new musical system. After the exposure phase, a posttest was conducted in the identical manner as the pretest. The pre- and posttests consisted of 13 trials each in the probe tone paradigm.[17] In each trial, subjects heard a melody followed by a "probe" tone, and their task was to rate how well the probe tone fit the preceding melody. Probe tone ratings were done on a scale of 1–7, 7 being the best fit and 1 being the worst. Previous results had shown that the profiles of subjects' ratings reflect the frequencies of musical composition.[17]

Data analysis

Ratings from both pre- and posttests were regressed on the exposure corpus (i.e., the "ground truth") to

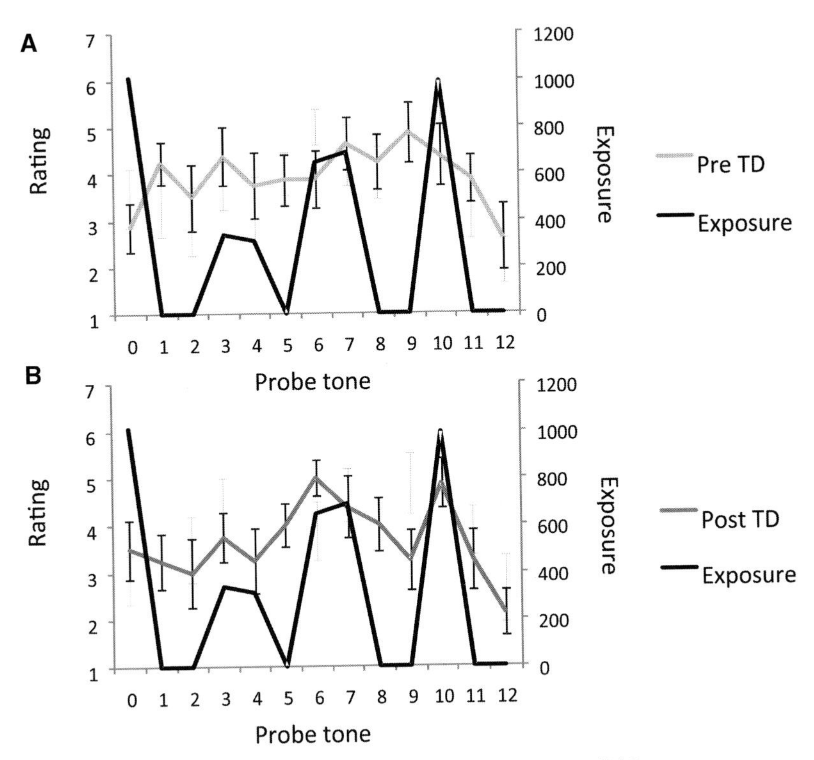

Figure 2. Probe tone ratings and exposure profiles in TD subjects. (A) Pre-exposure. (B) Postexposure.

obtain a single r-value score as a measure of performance in frequency matching. Postexposure test scores (r-values) were then compared against the pre-exposure scores to assess learning due to exposure. Finally, pre-exposure and postexposure test scores were compared between groups to determine whether TD subjects were indeed impaired in learning frequency information. In an additional analysis, the melodic context used to obtain the probe tone ratings was partialed out from the correlation between pre- or postexposure ratings and the exposure corpus, such that the contribution of the melody presented before the probe tone ratings was removed.[13] The resultant partial-correlation scores were again compared before and after exposure, and between TD and NTD groups.

Results

Figure 1A shows the results of NTD control subjects' ratings done before exposure, whereas Figure 1B shows the same subjects' ratings after exposure. The exposure profile ("ground truth") is plotted in black for comparison purposes in both cases. Compared to pretest ratings, posttest ratings are more highly correlated with exposure, suggesting that within half an hour of exposure, subjects acquired sensitivity to the frequency structures of the new musical system.

In contrast, ratings by the TD individuals were not so highly correlated with exposure (Fig. 2A shows pre-exposure ratings; Fig. 2B shows postexposure ratings). When the two groups of subjects are compared in their frequency-matching scores (calculated as an r-value expressing the correlation between ratings and exposure profile; Fig. 3), it becomes clear that the controls acquired sensitivity to tone frequencies after exposure as evidenced by the improvement in correlation between probe tone ratings and exposure, whereas TD individuals showed no such improvement, suggesting that TD individuals are impaired in frequency learning.

A t-test comparing pre- and postexposure frequency-matching scores was significant ($t(15) = 2.55$, $P = 0.022$) for NTD controls, confirming successful learning in the controls (Fig. 1). In contrast, TD subjects were unable to learn the frequency structure of the new musical system, as shown by ratings that were uncorrelated with exposure frequencies both before and after exposure (Fig. 2), as well as statistically indistinguishable performance between

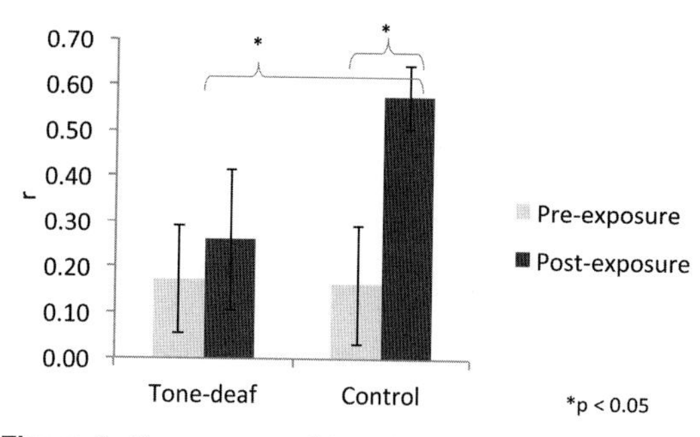

Figure 3. Frequency-matching scores for pre- and postexposure tests comparing tone-deaf and control groups.

pre- and postexposure frequency-matching scores ($t(15) = 1.41$, $P = 0.18$). A direct comparison of postexposure frequency-matching scores between TD subjects and controls was significant ($t(15) = 2.25$, $P = 0.038$), confirming that TD individuals performed worse than controls after exposure (Fig. 3). With small sample sizes with nonparametric distributions, the Friedman test is appropriate as an alternative to the two-way analysis of variance (ANOVA). The Friedman test is marginally significant for the correlation scores ($Q(28) = 3.45$, $P = 0.06$); however, when the effects of the melodic context are partialed out, then the Friedman's is highly significant ($Q(28) = 22.6$, $P < 0.001$), suggesting that NTD controls learned the frequency structure of the new musical system, whereas the TD subjects did not.

If tone deafness is truly linked to a disrupted ability to learn from the event frequencies of different pitches, then individuals who are more TD should be worse learners. To test the hypothesis that individuals who are more severely TD might be more severely impaired in frequency learning, we correlated the performance score (r-values) in the postexposure test with the scale subtest of the MBEA.[22] Results showed a significant inverse correlation between MBEA score and learning performance score (Fig. 4). This confirms that the more severely TD subjects are less able to learn the event frequencies of pitches from exposure, consistent with the hypothesis that tone deafness is characterized by the inability to acquire the frequency structure of music.

Discussion

While NTD subjects learned the frequency structure of the new musical system within half-an-hour of exposure, TD individuals failed to learn the same frequency structure. Furthermore, learning scores as obtained from the present frequency-matching probe tone task is correlated with the MBEA, confirming that the impairment of the statistical learning mechanism is related to the severity of tone deafness.

Taken together, the results suggest that insensitivity to musical pitch in the TD could arise from learning difficulties, specifically in the learning of event frequency information. To answer our original question of what gives rise to the lack of musical knowledge in the TD population, we have shown that TD individuals, who possess known disabilities in pitch perception, are also impaired in frequency learning (but not necessarily in probability learning). This frequency learning ability is crucial in music acquisition as frequency information governs how pitches are combined to form melodies and harmonies, structural aspects of music that are fundamental to the perception and cognition of music.

One important design aspect of the current study is that it distinguishes frequency learning from probability learning, a distinction that has been made in the previous literature on language acquisition.[23] While conditional probability (also known as first-order probability, i.e., the probability that one event follows another) is important for speech segmentation[16] and is most commonly tested in statistical learning studies, event frequency (also known as zero-order probability, i.e., the number of occurrences within a corpus) is important in language for word learning[24,25] as well as in music for forming a sense of key.[26] Sensitivity to frequency information can be assessed using the probe tone ratings task.[14,17,19] The present results show that sensitivity to event frequency is impaired in tone deafness; in this regard, our results dovetail with results from Peretz, Saffran, Schön, and Gosselin (in this volume), who show an inability to learn from conditional probabilities among congenital amusics. At a computational level, results relate to language-learning abilities and provide further support for the relationship between tone deafness and language-learning difficulties such as dyslexia.[27] At an implementation level, results on the learning of the new musical system are compatible with diffusion tensor imaging results that relate pitch-related learning abilities to the temporal-parietal junction in

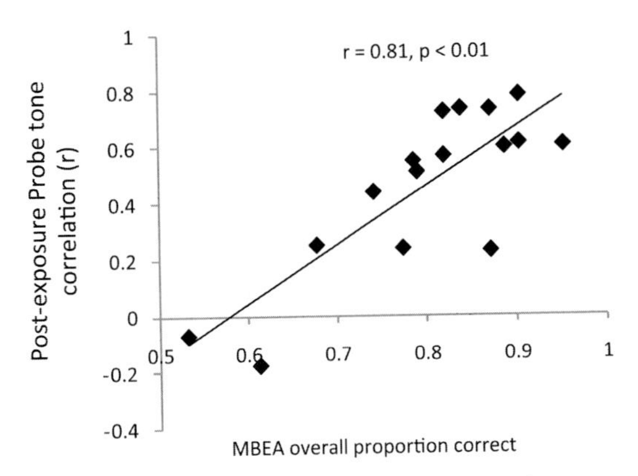

Figure 4. Relationship between postexposure frequency-matching scores and performance on the scale subtest of the MBEA.

the right arcuate fasciculus,[28] thus offering a possible neural substrate that enables the music-learning ability.

Taken together, the current results contribute to a growing body of literature in linking music learning to language learning, and more specifically in linking tone deafness—a type of musical disability—to learning disorders that are frequently described in language development. By investigating the successes and failures in which our biological hardware enables us to absorb the frequencies and probabilities of a novel musical system, we hope to capture both the universality and the immense individual variability of the human capacity to learn from events in our sound environment.

Acknowledgments

This research was supported by funding from NIH (R01 DC 009823), the Grammy Foundation, and the John Templeton Foundation.

Conflicts of interest

The authors declare no conflicts of interest.

References

1. Marr, D. 1982. *Vision: A Computational Investigation into the Human Representation and Processing of Visual Information.* W.H. Freeman and Company. New York, NY.
2. Krumhansl, C.L. & F.C. Keil. 1982. Acquisition of the hierarchy of tonal functions in music. *Mem. Cognit.* **10:** 243–251.
3. Foxton, J.M., J.L. Dean, R. Gee, *et al.* 2004. Characterization of deficits in pitch perception underlying 'tone deafness'. *Brain* **127:** 801–810.
4. Loui, P. & D.L. Wessel. 2008. Learning and liking an artificial musical system: effects of set size and repeated exposure. *Music. Sci.* **12:** 207–230.
5. Trainor, L. & S.E. Trehub. 1994. Key membership and implied harmony in Western tonal music: developmental perspectives. *Percept. Psychophys.* **56:** 125–132.
6. Hannon, E.E. & L.J. Trainor. 2007. Music acquisition: effects of enculturation and formal training on development. *Trends Cogn. Sci.* **11:** 466–472.
7. Krumhansl, C.L. *et al.* 2000. Cross-cultural music cognition: cognitive methodology applied to North Sami yoiks. *Cognition* **76:** 13–58.
8. Castellano, M.A., J.J. Bharucha & C.L. Krumhansl. 1984. Tonal hierarchies in the music of north India. *J. Exp. Psychol. Gen.* **113:** 394–412.
9. Koelsch, S., T. Gunter, A.D. Friederici & E. Schroger. 2000. Brain indices of music processing: "non-musicians" are musical. *J. Cogn. Neurosci.* **12:** 520–541.
10. Loui, P., T. Grent-'t-Jong, D. Torpey & M. Woldorff. 2005. Effects of attention on the neural processing of harmonic syntax in Western music. *Cogn. Brain Res.* **25:** 678–687.
11. Winkler, I., G.P. Haden, O. Ladinig, *et al.* Newborn infants detect the beat in music. *Proc. Natl. Acad. Sci. USA* **106:** 2468–2471.
12. Loui, P., E.H. Wu, D.L. Wessel & R.T. Knight. 2009. A generalized mechanism for perception of pitch patterns. *J. Neurosci.* **29:** 454–459, doi:10.1523/jneurosci.4503-08.2009.
13. Loui, P., D.L. Wessel & C.L. Hudson Kam. 2010. Humans rapidly learn grammatical structure in a new musical scale. *Music Percept.* **27:** 377–388.
14. Huron, D. 2006. *Sweet Anticipation: Music and the Psychology of Expectation.* 1st ed, Vol. 1. MIT Press. Cambridge, MA.
15. Hudson Kam, C.L. 2009. More than words: adults learn probabilities over categories and relationships between them. *Lang. Learn. Dev.* **5:** 115–145, doi: 10.1080/15475440902739962.
16. Saffran, J.R., R.N. Aslin & E. Newport. 1996. Statistical learning by 8-month-old infants. *Science* **274:** 1926–1928.
17. Krumhansl, C. 1990. *Cognitive Foundations of Musical Pitch.* Oxford University Press. New York, NY.
18. Kim, S.G., J.S. Kim & C.K. Chung. 2011. The effect of conditional probability of chord progression on brain response: an MEG study. *PLoS ONE* **6:** e17337, doi: 10.1371/journal.pone.0017337.
19. Russo, F.A. 2009. In Towards a functional hearing test for musicians: the probe tone method. *Hearing Loss in Musicians.* M. Chasin, Ed.: 145–152. Plural Publishing. San Diego, CA.
20. Shipley, W.C. 1940. A self-administering scale for measuring intellectual impairment and deterioration. *J. Psychol.* **9:** 371–377.
21. Zicarelli, D. 1998. In *Proceedings of the International Computer Music Conference.* 463–466. University of Michigan. Ann Arbor, MI, USA.
22. Peretz, I., A.S. Champod & K. Hyde. 2003. Varieties of musical disorders. The Montreal Battery of Evaluation of Amusia. *Ann. N.Y. Acad. Sci.* **999:** 58–75.
23. Aslin, R., J.R. Saffran & E. Newport. 1998. Computation of conditional probability statistics by 8-month old infants. *Psychol. Sci.* **9:** 321–324.
24. Hall, J.F. 1954. Learning as a function of word-frequency. *Am. J. Psychol.* **67:** 138–149.

25. Hochmann, J.-R., A.D. Endress & J. Mehler. 2010. Word frequency as a cue for identifying function words in infancy. *Cognition* **115:** 444–457.
26. Temperley, D. & E.W. Marvin. 2008. Pitch-class distribution and the identification of key. *Music Percept.* **25:** 193–212, doi:10.1525/mp.2008.25.3.193.
27. Loui, P., K. Kroog, J. Zuk, *et al.* 2011. Relating pitch awareness to phonemic awareness in children: implications for tone-deafness and dyslexia. *Front. Psychol.* **2:** doi:10.3389/fpsyg.2011.00111.
28. Loui, P., H.C. Li & G. Schlaug. 2011. White matter integrity in right hemisphere predicts pitch-related grammar learning. *NeuroImage* **55:** 500–507.

Ann. N.Y. Acad. Sci. ISSN 0077-8923

ANNALS OF THE NEW YORK ACADEMY OF SCIENCES
Issue: *The Neurosciences and Music IV: Learning and Memory*

Statistical learning of speech, not music, in congenital amusia

Isabelle Peretz,[1] Jenny Saffran,[1,2] Daniele Schön,[3] and Nathalie Gosselin[1]

[1]International Laboratory of Brain, Music and Sound Research (BRAMS), Department of Psychology, University of Montreal, Montreal, Canada. [2]University of Wisconsin, Madison, Wisconsin. [3]Institut de Neurosciences Cognitives de la Méditerranée, Marseille, France

Address for correspondence: Isabelle Peretz, BRAMS-Pavillon 1420 Mont-Royal, Université de Montréal, Montreal, Quebec, Canada H2V 4P3. Isabelle.Peretz@umontreal.ca

The acquisition of both speech and music uses general principles: learners extract statistical regularities present in the environment. Yet, individuals who suffer from congenital amusia (commonly called tone-deafness) have experienced lifelong difficulties in acquiring basic musical skills, while their language abilities appear essentially intact. One possible account for this dissociation between music and speech is that amusics lack normal experience with music. If given appropriate exposure, amusics might be able to acquire basic musical abilities. To test this possibility, a group of 11 adults with congenital amusia, and their matched controls, were exposed to a continuous stream of syllables or tones for 21-minute. Their task was to try to identify three-syllable nonsense words or three-tone motifs having an identical statistical structure. The results of five experiments show that amusics can learn novel words as easily as controls, whereas they systematically fail on musical materials. Thus, inappropriate musical exposure cannot fully account for the musical disorder. Implications of the results for the domain specificity of statistical learning are discussed.

Keywords: statistical learning; speech; music; pitch disorder; amusia

Introduction

Statistical learning is the process by which learners rapidly acquire structured information from variable environmental inputs in the absence of explicit feedback.[1] This mechanism operates in numerous domains, such as speech, music, and visual geometric figures. This powerful mechanism is considered to be fundamental to the acquisition of language because statistical learning plays a critical role in speech segmentation.[2,3] Several studies demonstrate comparable learning with tones,[4] musical timbres,[5] and sung syllables.[6] Because the type of computation appears to operate equivalently across syllables and tones, statistical learning is conceived as a domain-general mechanism that encompasses the acquisition of both music and language.

This raises the question of why such a domain-general mechanism does not function properly in the case of congenital amusia. Amusic individuals typically display normal language abilities and poor musical skills. For example, they can normally recognize the lyrics of songs, although they fail to recognize the melody component of the same song.[7] One possibility is that amusics have had impoverished learning experiences with music because their brain was not "ready" for or tuned to music at the appropriate time during development. This delay may have been exacerbated by a lifelong history of musical failures and lack of interest for music. Indeed, Conway *et al.*[8] show that a period of auditory deprivation may impair general sequencing abilities, which in turn may explain why some deaf children still struggle with language following cochlear implantation. The goal of the present study was to test whether amusics might be able to acquire basic musical abilities by way of statistical learning if given appropriate exposure.

To address this aim, a group of 11 adults with congenital amusia and their matched controls were

doi: 10.1111/j.1749-6632.2011.06429.x

exposed to a continuous stream of syllables or tones that were organized according to a set of simple statistical regularities (e.g., the syllable /pa/ tends to be followed by the syllable /bi/, or the tone A4 tends to be followed by F#4). After this familiarization phase, the participants were asked to judge the familiarity of three-syllable nonsense words or three-tone motifs, defined by the same transitional probabilities. As mentioned earlier, normal adults rapidly detect the regularities that link together the syllables by discriminating familiar groups of syllables (or tones) better than syllables (tones) that span the group boundary. This effect was first discovered by Saffran *et al.*[2,4] Thus, as a starting point, we used Saffran's material for adults in Experiment 1 (syllables) and in Experiment 2 (pure tones). Because amusics failed to learn the pure tone structure, we tested them further with a reduced set of pitches (Experiment 3) and of targets (Experiment 4). Finally, we tested the learning of pitch structure indirectly by presenting sung syllables in Experiment 5.

General method

Participants

Eleven amusic adults and 13 matched controls (who had no musical education and no musical impairment) were tested (see Table 1 for their background information). They were considered amusic (or not) on the basis of their scores on the Montreal Battery of Evaluation of Amusia (MBEA).[9] The battery involves six tests (180 trials) that assess various music processing components. Three tests assess the ability to discriminate pitch changes in melodies (melodic tests), and three tests assess the ability to discriminate rhythmic changes, meter, and memory, respectively. Each amusic participant obtained a global score and a melodic score that were both two standard deviations below the control participants (Table 1).

Procedure

All subjects were francophone and tested individually. They were instructed to listen to a nonsense language. They were further told that the language contained words (or tone motifs), but no meaning or grammar. Their task was to figure out where the words (motifs) began and ended. Note that we could not deceive the subjects as we used a within-subjects design and, thus, after having been tested with syllables (Experiment 1), they were also later tested with tones, although in different sessions. After the familiarization phase, subjects were presented with a forced-choice test. They were instructed to indicate which of the two sequences sounded more like a word (motif) from the familiarized language. To do so, they pressed either "1" or "2" on the keyboard of the computer, corresponding to whether the familiar sequence was played first or second in that trial. The two test sequences were separated by 750 ms, with an intertrial of five seconds. Stimuli were presented over speakers at a comfortable level.

Experiment 1: syllables

The first step was to determine whether amusics can learn novel words determined by adjacent probabilities. To this aim, we used the material and procedure of Ref. 3. It consists of four consonants (p, t, b, d) and three vowels (a, i, u) combined into 12 CV syllables making up six words (babupu, bupada,

Table 1. Mean age, education, and gender for amusic and control participants; mean percentages of correct response (standard deviations) obtained on the MBEA (global score) for the 11 amusics and the 13 matched controls. Chance level is 50% in all tests. Group differences were assessed by way of bilateral *t* tests.

Participants	Amusics	Controls	*t* tests
Gender	4M, 7F	5M, 8F	
Age (SD)	59.1 (9.5)	58.9 (6.9)	$t(22) = 0.05$, *n.s.*
Years of education (SD)	16.9 (1.6)	17.2 (2.5)	$t(22) = -0.28$, *n.s.*
Musical experience level[a] (SD)	1.9 (1.0)	1.9 (0.8)	$t(21) = -0.02$, *n.s.*
MBEA% global score (SD)	64.5 (7.6)	88.0 (3.7)	$t(22) = -9.85$, $P < 0.001$

[a]level 1 = less than 1 year; 2 = between 1 and 3 years; 3 = between 4 and 6 years; 4 = between 7 and 10 years; 5 = more than 10 years.

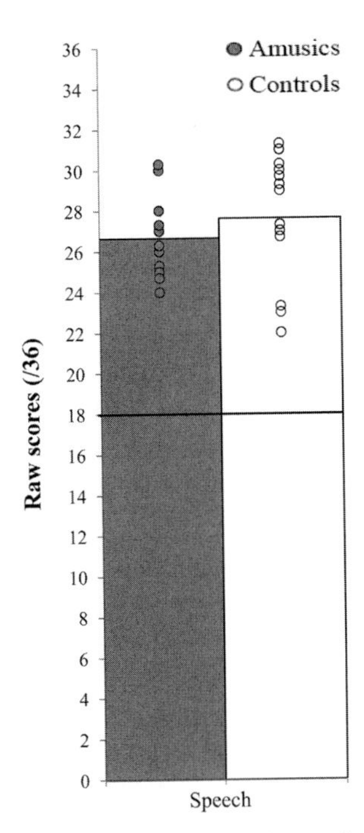

Figure 1. Performance by amusic and control participants in Experiment 1. Circles represent the number correct (out of a possible 36) for individual subjects; columns represent the mean; chance = 18.

dutaba, patubi pidabu, and tutibu). Note that some syllables occurred in more words than others; for example, "bu" occurred in four words, while "ta" occurred in only one. Thus, the transitional probabilities between syllables within words varied (range: 0.3–1.0) and were higher than the transitional probabilities between syllables spanning a word boundary (range: 0.1–0.2). The words were randomized and combined into a 21-minute stream of continuous speech with no pauses between words. The test consisted of 36 items. Each item paired a word from the stream with a part-word sequence, which consisted of a sequence that crossed a word boundary (e.g., dutaba–bapatu).

Results are presented in Figure 1. The individual scores obtained by both amusics and controls were significantly different from chance, where chance equals 18 (50%; $t\,[10] = 14.23$ and $t\,[12] = 11.01$ for amusics and controls, respectively, both $P < 0.001$ by two-tailed tests). Moreover, amusics (74.0% correct) performed as well as controls (76.7%; $t[22] = 0.887$). The results are very similar to those (76%) obtained in university students in Ref. 3. Thus, amusics can learn novel words on the basis of transitional probabilities as well as controls.

Experiment 2: pure tones

Here we tested whether the same amusic participants can track transitional probabilities in a tone sequence analogous in structure to the speech materials from Experiment 1. To do this, we used the material and procedure of Saffran *et al.*,[4] who substituted a distinct tone for each of the 12 syllables from which the novel words were created in Experiment 1 (e.g., BU became the musical note D). The 12 tones were pure tones in the same octave, starting at middle C within a chromatic set, with a duration of 330 milliseconds. Each of the six trisyllabic nonsense words (e.g., BUPADA) from the novel speech stream was thereby translated into a sequence of three musical tones (e.g., DFE). Six three-tone motifs (ADB, DFE, GG#A, FCF#, D#ED, and CC#D) were then concatenated together, in a random order and with no silence between motifs, to generate a continuous stream identical in statistical structure to the speech stream tested in Experiment 1. The tones stream was presented for 21 min and followed by 36 test trials. Each trial consisted of a pair of three-tone motifs, one of which was a tone motif from the familiarization stream, and the other one straddled a motif boundary.

The pure tone motifs were impossible to learn for 6 out of 7 controls as well as for 7 out of 10 amusics. None of the groups performed above chance level (60%; $t\,[6] = 0.00$; and 48%; $t\,[9] = 0.25$, for controls and amusics, respectively, both $P_s > 0.05$). This was unexpected given that Saffran *et al.*[4] found successful learning with similar tone stimuli with college-aged participants. However, Evans, Saffran and Robe-Torres[10] also found that these tone sequences were particularly difficult for children with specific language impairment to learn. Indeed, the children performed above chance after 40 min of exposure to a similar speech stream, while failing to acquire the tone stream after the same amount of exposure. Thus, to make the tones easier to learn, we changed the material in Experiments 3 and 4 while keeping the same procedure.

Experiment 3: diatonic piano tones

In order to simplify the task at least for the musically unimpaired controls, we reduced the 11 chromatic pitches to 8 diatonic tones. All pitches were

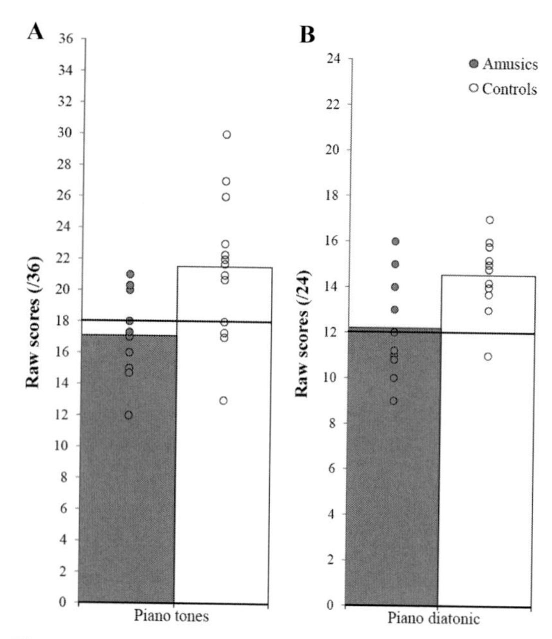

Figure 2. (A) Performance by amusic and control participants in Experiment 3. Circles represent the number correct (out of a possible 36) for individual subjects; columns represent the mean; chance is 18. (B) Performance by amusic and control participants in Experiment 4. Circles represent the number correct (out of a possible 24) for individual subjects; columns represent the mean; chance = 12.

diatonic and taken from the octave above middle C. The six motifs were EDC, C*FG, CEF, GC*D, DGC, and FED (* = one octave above middle C). Furthermore, to facilitate the forced choice between the motifs and the foils, the transitional probabilities between the individual pitches in the test foils were low (CGF, ECC*, FDF, DFC*, CDE, and C*GD). That is, adjacent pitches in a foil did not occur in the tone motifs. For example, in the test foil CGF, learners did not hear CG or GF in the target motifs (as done in Ref. 11, to be discussed later). As previously, the foil pitches spanned motif boundaries. Finally, instead of using pure tones, the pitches were presented with the timbre of a piano. Otherwise, the task was identical. Participants were presented with the six motifs for 21 min after which they had to select the familiar motif in 36 test trials.

The results are presented in Figure 2A. This time, controls succeeded to perform above chance as a group with 59.8% ($t[12] = 2.78$, $P < 0.05$). None of the 11 amusics performed above chance (with 47.5%; $t[9] = 1.03$) and performed significantly below controls ($t[21] = 2.694$, $P < 0.02$). Thus, amusics failed to learn the novel motifs whereas most of their matched controls learned them. However, four out of the 12 controls performed at chance despite the fact that the diatonic set allowed them to use tonal encoding of pitch as a possible memory code.

Experiment 4: four motifs

In order to boost performance so that all musically unimpaired controls could achieve recognition of the novel tone motifs above chance, we reduced the number of targets to recognize in the test trials. Familiarization consisted of the same material as in Experiment 3 and lasted 21 minutes. Only the test pairs were changed. The number of motifs to recognize in the test was decreased from six to four. These four motifs (EDC, C*FG, DGC, and CEF) were selected so as to be as distinct as possible in terms of pitch contour. The foils were the reverse orderings of the motifs so as to keep the same intervals; the transitional probabilities between tones in the foils were always zero. The test phase consisted of 24 trials instead of 36.

As can be seen in Figure 2B, the changes were successful in bringing the scores of all but one matched control above chance ($t[12] = 2.78$, $P < 0.05$). Amusics' scores remained at chance ($t[9] = 1.03$) and below controls ($t[21] = 2.69$, $P < 0.05$). Among the four amusics who obtained scores above chance here, only two obtained similar scores in Experiment 3. Note that we also tried 40 min of exposure with a few amusics, with no more success.

Experiment 5: syllables and sung tones

Amusics repeatedly failed to learn novel tone motifs, although successfully learned novel words organized along the same statistical principles. This result indicates that lack of adequate musical exposure is probably not the critical factor, at least within the limited time window of exposure used here. Rather, the results suggest that statistical learning of speech does not involve the same codes or processes as statistical learning of music.

However, it is generally easier to extract statistical rules from speech than music (see Ref. 12 for a review). Because tones are more difficult to learn than syllables, the same statistical mechanism may operate at different levels of efficiency in both controls and amusics; for one reason or another, amusics

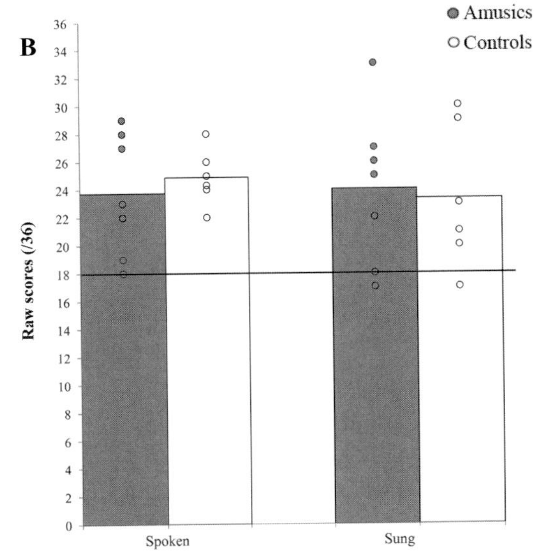

Figure 3. (A) Representation of words and motifs used as targets in Experiment 5; (B) Performance by amusic and control participants after familiarization with spoken or sung materials in Experiment 5. Circles represent the number correct (out of a possible 36) for individual subjects; columns represent the mean; chance = 18.

would not fully engage in the tone motifs processing.

One solution to avoid this efficiency difference is to test novel word recognition and examine whether the addition of melodic structure improves learning. This has been previously studied by Schön *et al.*,[6] who showed that novel words were better learned when sung, provided that the motif structure coincided with the word boundaries, both defined in terms of transitional probabilities.

Similarly, here amusic and control participants were presented with 21 min of either spoken syllables (as in Experiment 1) or sung tones. The grouping of the spoken syllables and sung syllables were determined by transitional probabilities. Although the familiarization phase consisted of either spoken words or sung words, it was followed by the same test phase. In the latter, both words and foils were presented in a spoken format (not sung; Fig. 3A).

The advantage of this procedure is that the learning of the motif structure is assessed indirectly. This may provide a better test of amusics' ability to extract pitch structure as they can show evidence of statistical learning without awareness.[13]

Method

The structure of the continuous stream of spoken syllables was identical to that used in Experiment 1. The stream was generated using a male French voice setting on the Mbrola speech synthesizer (http://tcts.fpms.ac.be/synthesis/mbrola.html). The result was a monotone and continuous stream of syllables. The foils consisted of either the last syllable of a word plus the first syllable pair of another word, or the last syllable pair of a word plus the first syllable of another word. For instance, the last syllable of the word "dutaba" was combined with the first two syllables of "patubi" to create the partial word foil "bapatu."

In the sung version, each of the 11 syllables was associated with a distinct tone (Fig. 3A). The eleven tones were C5, D5, F5, G5, A5, B5, C6, Db6, D6, E6, and F6. Thus, the statistical structure of the spoken and sung sequence was identical and superimposed in the sung version, with constant syllable–pitch mapping. The sung sequence was also synthesized using Mbrola.

Seven amusics and seven controls were tested with the spoken and sung version. The order of presentation of the spoken material and sung material was counterbalanced across participants.

Results and comments

Like Experiment 1, amusics performed above chance in the spoken version of the material (66%; $t[6] = 3.45$ $P < 0.05$). They also performed above chance with the sung version (67%; $t[6] = 2.87$, $P < 0.05$; Fig. 3B). This provides converging evidence that amusics can learn novel words on the basis of statistical learning. Their scores did not differ from the scores obtained by their matched controls who also performed above chance in both the spoken condition (68%; $t[6] = 3.45$, $P < 0.005$) and in the sung condition (65%; $t[5] = 2.53$, $P = 0.053$).

An ANOVA including conditions (spoken, sung) as within-subjects factor and groups (amusics, controls) as between-subjects factor revealed no significant effects for condition ($F\ [1, 11] = 0.463$, $\eta^2 = 0.04$) or group ($F < 1$), nor an interaction between these two factors ($F < 1$). Thus, we failed to replicate the advantage of the sung over the spoken words obtained by Schön *et al.*[6]

This failure might be because of two differences in the experimental design. The first is that the duration of the familiarization phase was three times longer in the current study than in Schön *et al.*'s study. Thus, participants may reach a plateau with both materials. The second difference is that while Schön *et al.*[6] used a between-subjects design; we used a within-subjects design. Some transfer of learning may have taken place. Marcus *et al.*[14] have shown that infants are better able to extract rules from sequences of musical tones if they first hear those rules instantiated in sequences of speech. We did not find support for such an effect here. The performance on the sung words of the subjects who started with the spoken version was 65.9% and did not differ from the 68.9% obtained by the subjects who started with the sung version ($t[10] = 0.35$, $P = 0.73$). Note that unlike previous studies conducted with students, we tested small samples of subjects. We computed that we would need to test 36 amusics to show an advantage for the sung material over the spoken one. This is simply not realistic.

Like Schön *et al.*,[6] we observed that words with higher transitional probabilities tended to be better recognized than words with lower transitional probabilities (70.5% and 62.8%, respectively), across conditions and groups ($F\ [1, 12] = 3.74$, $P = 0.077$). Amusics did not differ from controls ($F\ [1, 11] = 0.01$, $P = 0.918$) and the mode of familiarization (sung versus spoken) did not influence learning ($F\ [1, 11] = 0.21$, $P = 0.653$).

General discussion and conclusions

Individuals with congenital amusia can learn novel words but fail to learn novel pitch motifs organized according to the same statistical properties. These results suggest that statistical learning is not mediated by a single processing system but by two systems specialized for processing syllables and tones, respectively. It is likely that the speech and musical input codes adjust statistical learning to their processing needs. In other words, the input and output of the statistical computation may be domain-specific while the learning mechanism is not.[15,16]

The present findings also suggest that amusics will be unable to develop a normal capacity for music, although they can acquire a novel language. This dissociation was clearly present in the data despite the fact that amusics had equal opportunities to learn the two materials. However, amusics may show evidence of statistical learning for pitch intervals[11] without awareness.[13] Thus, the experimental setting used here may not be adequate to reveal implicit learning in small samples of amusic adults. Alternatively, the musical representations acquired implicitly may not be robust enough to sustain the interference created by the different test trials.

Acknowledgments

Preparation of this paper was supported by grants from the Natural Sciences and Engineering Research Council of Canada and from a Canada Research Chair to the first author.

Conflicts of interest

The authors declare no conflicts of interest.

References

1. Aslin, R.N. & E.L. Newport. 2009. What statistical learning can and can't tell us about language acquisition. In *Infant Pathways to Language: Methods, Models, and Research Disorders*. J. Colombo, P. McCardle & L. Freund, Eds.: 15–29. Erlbaum. New York.
2. Saffran, J.R., R.N. Aslin & E.L. Newport. 1996. Statistical learning by 8-month-old infants. *Science* **274:** 1926–1928.
3. Saffran, J.R., E.L. Newport & R.N. Aslin. 1996. Word segmentation: the role of distributional cues. *J. Mem. Lang.* **35:** 606–621.
4. Saffran, J.R. *et al.* 1999. Statistical learning of tone sequences by human infants and adults. *Cognition* **70:** 27–52.
5. Tillmann, B. & S. McAdams. 2004. Implicit learning of musical timbre sequences: statistical regularities confronted with acoustical (dis)similarities. *J. Exp. Psychol. Learn.* **30:** 1131–1142.
6. Schon, D. *et al.* 2008. Songs as an aid for language acquisition. *Cognition* **106:** 975–983.
7. Ayotte, J., I. Peretz & K.L. Hyde. 2002. Congenital amusia: a group study of adults afflicted with a music-specific disorder. *Brain* **125:** 238–251.
8. Conway, C.M. *et al.* 2011. Implicit sequence learning in deaf children with cochlear implants. *Dev. Sci.* **14:** 69–82.
9. Peretz, I., A.S. Champod & K.L. Hyde. 2003. Varieties of musical disorders. The Montreal Battery of Evaluation of Amusia. *Ann. N.Y. Acad. Sci.* **999:** 58–75.
10. Evans, J., J.R. Saffran & K. Robe-Torres. 2009. Statistical learning in children with Specific Language Impairments. *J. Speech Lang. Hear. Res.* **52:** 321–335.
11. Omigie, D. & L. Stewart. 2011. Preserved statistical learning of tonal and linguistic material in congenital amusia. *Front. Psychol.* **2:** 109.
12. Francois, C. & D. Schon. 2011. Musical expertise boosts implicit learning of both musical and linguistic structures. *Cereb. Cortex.* **21:** 2357–2365.

13. Peretz, I. *et al.* 2009. The amusic brain: in tune, out of key, and unaware. *Brain* **132:** 1277–1286.
14. Marcus, G.F., K.J. Fernandes & S.P. Johnson. 2007. Infant rule learning facilitated by speech. *Psychol. Sci.* **18:** 387–391.
15. Peretz, I. 2006. The nature of music from a biological perspective. *Cognition* **100:** 1–32.
16. Saffran, J.R. & E.D. Thiessen. 2006. Domain-general learning capacities. In *Handbook of Language Development*. E. Hoff & M. Shatz, Eds.: 68–86. Blackwell. Cambridge.